The Phylogenetic System of Ephemeroptera

The Phylogenetic System of Ephemeroptera

by

Nikita Kluge

Biological Faculty,
St-Petersburg State University, St-Petersburg, Russia

KLUWER ACADEMIC PUBLISHERS
DORDRECHT / BOSTON / LONDON

A C.I.P. Catalogue record for this book is available from the Library of Congress.

ISBN 1-4020-1974-2

Published by Kluwer Academic Publishers,
P.O. Box 17, 3300 AA Dordrecht, The Netherlands.

Sold and distributed in North, Central and South America
by Kluwer Academic Publishers,
101 Philip Drive, Norwell, MA 02061, U.S.A.

In all other countries, sold and distributed
by Kluwer Academic Publishers,
P.O. Box 322, 3300 AH Dordrecht, The Netherlands.

Printed on acid-free paper

Printed in the Netherlands.

TABLE OF CONTENTS

GENERAL

SYSTEMATIC ACCOUNT OF TAXA

FOREWORD

The first draft version of this book, written in English and named "Draft revision of supraspecies taxa of Ephemeroptera (without Atalophlebiinae)", was prepared as an Internet publication in 1998; thanks to Dr. Michael Hubbard (Florida A & M University), it has been available from the web site "Ephemeroptera Galactica" since 1998. The second version, written in Russian and named "Revision of supra-species taxa of Ephemeroptera (except for Atalophlebia/fg1)", was prepared as an Internet publication in 2000; from 2001 till now it has been available from the web site of Biological Faculty of S-Petersburg University, http://www.bio.pu.ru/win/entomol/KLUGE/EPHEMER_/content_.htm. Both versions contain the systematics of all mayflies except for Atalophlebia/fg1. The present Volume 1 does not contain the systematics of Turbanoculata (i. e. Baetidae auct.) nor of all Leptophlebia/fg1 (i. e. Leptophlebiidae auct.), the largest part of which belongs to Atalophlebia/fg1. The reasons for such a restriction are objective and subjective difficulties in baetid and leptophlebiid systematics. Both groups are really difficult for taxonomic revision, being geologically young, abundant and richly represented in poorly investigated tropical areas. Last years, after revisions of Holarctic baetid taxa (Novikova & Kluge 1987, Waltz & McCafferty 1987, et al.) a lot of new baetid "genera" have been described from non-Holarctic areas; many of these descriptions are insufficient and based on a few specimens only. While there is some progress in supraspecies baetid taxonomy (Kluge 1991b, 1992a, 1992b, 1997c; Novikova & Kluge 1987, 1994; Kluge & Novikova 1992), for a comprehensive revision examination of more material is necessary. That's why the systematics of Turbanoculata is excluded from this volume. As for leptophlebiids, their supraspecies revision is only started (Kluge 1994b, 1994c) and is far from finished; here also a lot of material should be examined. In future, the author hopes to prepare for publication two other volumes of this book – "Volume 2. Turbanoculata" and "Volume 3. Leptophlebia/fg1".

Parallel with the mayfly revision, a work on the book on general insect systematics is continued. The first volume of this book (which does not include systematics of Neoptera) had two different editions in Russian – in 1999 and in 2000 (Kluge 2000). English version of this first volume is also prepared, but not published yet.

NEW TERMS, TAXA NAMES AND SYNONYMS

New terms: plesiomorphon (instead of wrongly used "plesion", p.16); basitornal margin, tornoapical margin, amphitornal margin of wing or wing bud (p.31); tergalius (instead of formerly used "tergalia" or "tergalium", p.35); caudalius, tricaudale, basitricaudale, cercotractor (p.38).

For some morphological terms type taxa are designated. Till now this was not practised, but in some cases the type taxon can help to dissolve disagreements concerning usage of the term. As well as the type taxon of a taxon's name, the **type taxon of a morphological term** is an arbitrarily chosen taxon, for which the original usage of this term should be regarded to be correct independently of subsequent opinions concerning homology of this part of the body.

New circumscriptional names: Discoglossata (p. 136), Geminovenata (p.139), Fimbriatotergaliae (p. 218), Pantricorythi (p.316), Tricoryptera (p.326), Afrotricorythi (p.327), Tricorygnatha (p.327).

New family-group name: Epeorus/f (p.201).

New genus-group names: *Ecdyogymnurus* (p.181), *Thamnodontus* (p.182), *Himalogena* (p.195), *Sibirigena* (p.196), *Proepeorus* (p.204), *Albertiron* (p.205), *Sinephemera* (p.235), *Tillyardocaenis* (p. 287), *Notacanthella* (p.306).

New synonymy in species names: *alexandrae* Brodsky 1930 [*Ameletus*] = *asiacentralis* Soldán 1978 [*Ameletus*] SYN. NOV.; *ignota* Walker 1853 [*Baetis*] = *hainanensis* She et You 1988 [*Isonychia*] SYN. NOV.; *pallida* Hagen 1855 [*Oligoneuria*] = *mongolica* Soldán et Landa 1977 [*Oligoneuriella*] SYN. NOV.; *tskhomelidzei* Sowa et Zosidze 1973 [*Oligoneuriella*] = *baskale* Soldán et Landa 1977

[*Oligoneuriella*] SYN. NOV. = *zanga* Soldán et Landa 1977 [*Oligoneuriella*] SYN. NOV.; *kurenzovi* Bajkova 1962 [*Cinygma*] = *kaszabi* Landa et Soldán 1983 [*Cinygmula*] SYN. NOV.; *stackelbergi* Sinitshenkova 1973 [*Rhithrogena*] = *angulata* Braasch 1980 [*Rhithrogena*] SYN. NOV.; *pellucida* Brodsky 1930 [*Cinygma*] = *tshernovae* Braasch 1979 [*Epeorus*] SYN. NOV.; *pseudorivulorum* Keffermuller 1960 [*Caenis*] = *beskidensis* Sowa 1973 [*Caenis*] SYN. NOV.; *lepnevae* Tshernova 1949 [*Ephemerella*] = *fusongensis* Su et Gui 1995 [*Drunella*] SYN. NOV.; *submontana* Brodsky 1930 [*Ephemerella*] = *traverae* Allen et Edmunds 1963 [*Ephemerella (Drunella)*] SYN. NOV. = *nasiri* Ali 1971 [*Ephemerella*] SYN. NOV.; *ignita* Poda 1761 [*Ephemera*] = *antuensis* Su et You 1989 [*Ephemerella (Ephemerella)*] SYN. NOV.

COLLECTION DEPOSITION

Most of the mayflies mentioned in "Material examined" are deposited in Saint-Petersburg (Russia). The place of permanent deposition of this collection is the Laboratory of Insect Systematics of the Zoological Institute of Russian Academy of Sciences (RAS), but recently most part of this collection is located in the Department of Entomology of Saint-Petersburg State University, where the author works. Material deposited in other places is specially mentioned in the text as following: **Albany Mus.** – Albany Museum, Grahamstown, South Africa; **Entomol. Inst.** – Entomological Institute, Czech Academy of Sciences, České Budějovice, Czech Republic; **Florida A&M Univ.** – Florida Agricultural and Mechanical University, Tallahassee, USA.; **Kazan. Univ.** – Geological Faculty of Kazan' State University, Tatarstan, Russian Federation; **Mus. Nat. Hist. of Inst. Syst. Evol. Anim. in Krakow** – Museum of Natural History of Institute of Systematics and Evolution of Animals, Polish Academy of Sciences, Krakow; **Paleontol. Inst.** – Palaeontological Institute, Russian Academy of Sciences, Moscow; **Purdue Univ.** – Purdue University, West Lafayette, USA.

ACKNOWLEDGEMENTS

The work was done in the Department of Entomology of S.-Petersburg State University, in close collaboration with the Laboratory of Insect Systematics of Zoological Institute RAS (Saint-Petersburg) and the Laboratory of Arthropods of Palaeontological Institute RAS (Moscow).

Mayflies used in this study partly were collected by the author during expeditions of S.-Petersburg State University and Zoological Institute RAS to various parts of the USSR, Mongolia and Cuba. However, this work would be impossible without examination of mayflies from various parts of the World, which the author was unable to collect himself; sincerest thanks are expressed to everybody, who gave necessary material – Y.J. Bae (South Korea), H. Barber-James (South Africa), D.E. Baumgardner (USA), C. Belfiore (Italy), S.K. Burian (USA), J.W. Early (New Zealand), S. Elliott (Chile), T. Fujitani (Japan), P. Getwongsa (Thailand), M.T. Gillies (Great Britain), T. Gonser (Switzerland), A.V. Gorokhov (Russia), V.I. Gusarov (Russia), C. Hofmann (Switzerland), S.-I. Ishiwata (Japan), V.D. Ivanov (Russia), L.M. Jacobus (USA), A.E. Korolev (Russia), P. Landolt (Switzerland), Yu.M. Marusik (Russia), W.P. McCafferty (USA), S. Melnitsky (Russia), M. Mercado (Chile), J.C. Naranjo (Cuba), J.G. Peters (USA), W.L. Peters (USA), A.V. Przhiboro (Russia), N.D. Sinitshenkova (Russia), T. Soldán (Czechia), A.H. Staniczek (Germany), D. Studemann (Switzerland), L. Sun (USA), I.V. Tatarenko (Russia), T.M. Tiunova (Russia), I. Tomka (Switzerland), X. Tong (Hong Kong), V.V. Zherichin (Russia), L.A. Zhiltzova (Russia), J. Zloty (USA).

Special thanks to M.D. Hubbard for editorial suggestions, to J.G. Peters for comments on the manuscript, and everybody who helped the author by critical comments and useful information.

Great significance had discussions during the ephemeropterological conferences; the author expresses special thanks to W.L. Peters, J. Alba-Tercedor, T. Soldán, K.E. Gibbs, P. Landolt, D. Studemann, M. Sartori, E. Gaino, C. Belfiore and other members of the Permanent Committee of the International Conferences on Ephemeroptera and the committees of 4th–10th International Conferences on Ephemeroptera.

The work was partly supported by Russian Federal Program for Support of Leading Scientific Schools, grant 2232.2003.4, and by federal program "Universities of Russia", grant 07.01.056.

WARNINGS

In contrast to other taxonomic papers, this one is done in such an improved form, which makes wordily taxa diagnoses meaning. The traditional layout does not allow to do this, so readers of taxonomic papers use to study illustrations, keys, legends for phylogenetic dendrograms and other supplementary materials rather than the main text. In contrast to such literature, here all useful information is concentrated in the main text (Chapters I-VIII). The book is supplied with figures; these are illustrations for the text, but not an atlas, and the reader should not try to learn mayfly system watching the figures only.

The "Index of characters" represents an important component of the layout used here. But the reader should be prevented from possible mistakes: 1) This is an index of characters, but not an account of them; it only sends the reader to certain places of the main text, but does not give final information about taxa and their characters. For example, if there is written "Larva ... flattened dorsoventrally: Acanthametropus/fg1 (6), *Ametropus* (1), *Chiloporter* (2), Heptagennota (1)", it does not mean that larvae of these taxa are especially flattened, and others are not; here are only indicated places in the main text, where something is written about flatness of larva (the character No.6 in the characteristics of Acanthametropus/fg1, etc.). It is the same as a well-known alphabetic index, but with subjects accounted not alphabetically, but according to parts of the body. 2) While this index somewhat resembles a matrix of characters used in numeric cladistics, do not try to count the characters in this index! Only those who do not understand Darwin's theory, Hennig's method, and elementary mathematics, think that counting characters can be useful (for detailed explanation see Kluge 2000 and Internet publication http://www.bio.pu.ru/win/entomol/ KLUGE /sys-ins. htm).

Instead of traditional rank-based nomenclature, here is used a system of two non-ranking nomenclatures: the hierarchical nomenclature based on the ICZN, and the circumscriptional nomenclature; the first one is applied for each taxon, the second – for a few taxa only. All former attempts to write a serious comprehensive taxonomic analysis of large groups using the rank-based nomenclature only were not successful; this made me use a more complicated but more powerful nomenclatural system. In contrast to some other nomenclatural systems proposed in the last years, this one does not contradict the International Code of Zoological Nomenclature and allows the more effective use of all names accumulated in zoological literature since 1758. Principles of nomenclature are explained here in the Introduction.

INTRODUCTION

A BRIEF HISTORY OF MAYFLY CLASSIFICATION

In the early classifications (Linnaeus 1758, et al.) all mayflies, constituting a single holophyletic genus *Ephemera* Linnaeus 1758 (placed to artificial order Neuroptera), were divided into two groups according to the number of imaginal caudalii – 3 or 2. Each of these groups was actually polyphyletic. The imaginal paracercus is developed in the majority of European Furcatergaliae and vestigial in the majority of European Tridentiseta and Branchitergaliae; thus if one studies superficially the European species only, an impression could appear that this character allows one to divide mayflies into natural groups. However, more detailed examination of mayflies reveals that representatives with 3 and 2 caudalii occur in many evidently holophyletic taxa (see Index of characters [2.3.20]). After Latreille (1802) introduces a rank of family to zoological systematics, it became possible to raise the rank of mayflies from genus to family and to attribute generic ranks to subordinated groups. In the beginning of the XIX century there were attempts to divide mayflies into subordinate groups based of presence or absence of hind wings (Leach 1815, et al.); all mayflies were divided into 4 genera: *Ephemera* (3 caudalii and 4 wings), *Brachycercus* Curtis 1834 (3 caudalii and 2 wings), *Baetis* Leach 1815 (2 caudalii and 4 wings) and *Cloeon* Leach 1815 (2 caudalii and 2 wings). However, the type species of the generic names *Baetis* and *Cloeon* appear to be related (recently both are placed to Turbanoculata), and the genus *Baetis* in such sense appears to be very heterogenous. Later, the number of genera was increased (Pictet 1843–1845, et al.), but the classification remained artificial.

Eaton (1883–1888 et al.) made a comprehensive revision of mayfly species and suggested a new classification. His taxa diagnoses are based on adult characters only and are rather formal; larval structures are excellently illustrated but insufficiently described; for many taxa larvae were unknown or associated wrongly. It is even difficult to understand how such detailed and absolutely correct drawings of larvae could be made by the investigator, who did not know the taxonomic significance of many characters shown on them. Many supraspecies taxa established by Eaton were natural, although they did not have sufficient diagnoses.

Later (Lestage 1917, et al.) ephemeropterologists paid more and more attention to larval characters rather than to imaginal ones, and established classifications based mainly or solely on larval characters.

Since the artificial Linnaean order Neuroptera was completely divided into smaller natural orders (the process started by Burmeister 1829, and finished by Packard 1886 and Handlirsch 1903), mayflies got ordinal rank and were divided into a number of families and superfamilies, which in large degree corresponded to sections, series and groups proposed by Eaton (1883–1888) to the former family Ephemeridae. Basing mainly on larval characters, authors of new classifications changed many of these taxa to make the classification more natural and suggested different phylogenetic schemes (Ulmer 1920b, Edmunds & Traver 1954, Demoulin 1958, Tshernova 1970, Landa 1969, Riek 1973, et al.). Recently it is usual to accept several superfamilies, approximately from 10 to 40 families and several hundred genera.

Attempts to divide mayflies into highest taxa has undergone the following evolution.

1) McCafferty and Edmunds (1979) divided all mayflies into Pannota and Schistonota, regarding Pannota to be holophyletic, and Schistonota to be paraphyletic. Even if one agrees with the phylogenetic hypothesis of these authors, this classification is not good, because here the paraphyletic taxon is larger than the holophyletic one.

2) Because of this, Kluge (1989), based on the same phylogenetic theory, suggested dividing mayflies into Furcatergalia (which included Pannota and a part of Schistonota) and Costatergalia (which included a part of Schistonota), regarding Furcatergalia to be holophyletic, and Costatergalia to be paraphyletic. In this classification, the two taxa regarded to be holophyletic and paraphyletic have subequal species numbers, which is also not good, but better than the previous classification.

3) The next step was made by McCafferty (1991), who divided the paraphyletic taxon Costatergalia into

(1991), who divided the paraphyletic taxon Costatergalia into two, thus dividing mayflies into three taxa of equal rank – Pisciforma (paraphyletic), Setisura (assumed to be holophyletic) and Rectracheata (subequal to Furcatergalia and assumed to be holophyletic).

4) The next step in this direction was made by Kluge (oral presentation 1992 and publication 1998), who united the taxon Eusetisura (subequal to Setisura) with Furcatergaliae (subequal to Furcatergalia and Rectracheata) to form a large holophyletic taxon Bidentiseta.

Such subsequent change of classification agrees with the general rule, according to which in course of investigation, classification always approximates to the cladistic ideal never reaching it (Kluge 2000).

If one ignores disagreements concerning the systematic position of some small taxa (*Baetisca*, *Prosopistoma*, *Pseudiron* and Vetulata) which made us create new circumscriptional names for higher taxa, the change of classification can be shown as following (TABLE 1).

However, the phylogenetic hypotheses on which the classifications by McCafferty & Edmunds (1979), Kluge (1989) and McCafferty (1991) were based, are now regarded to be not quite correct: attributing *Baetisca* and *Prosopistoma* to taxa Pannota, Furcatergalia and Rectracheata now is regarded to be wrong (Kluge 1992–1998, McCafferty 1997); attributing *Pseudiron* to Setisura now is also regarded to be wrong (Kluge 1992–1998, Wang & McCafferty 1995); the taxa Ractracheata (uniting Vetulata with Furcatergalia) and Pannota are polyphyletic.

Taking into account these corrections, a new classification was suggested, where mayflies are divided into Posteritorna and Anteritorna, and the later – to Tridentiseta and Bidentiseta (Kluge 1992–1998). In the present monograph, this classification is accepted.

The general classification of mayflies can be correctly understood only by taking into account particular classifications of subordinate taxa. A lot of papers on systematics of selected mayfly taxa were published; especially useful ones appeared in the last decades. Review of this literature would take too much space; the reader can find references to the most important papers in the Special Part of this book, where references for each taxon under consideration are given. Several taxa that are not recognized in the presented classification, are here also characterized and supplied with references (see divisions "Classifications of ..." and "Systematic position of ..." in the Special Part).

Table 1. Simplified version of development of higher mayfly classification from 1979 to 1998. Taxa shown in the same line, do not exactly mach in circumscription, thus they have different circumscriptional names. Names of wittingly paraphyletic taxa are shown by bold. For other explanations see text and alphabetic Index of supraspecies taxa names.

<table>
<tr><th rowspan="2">Phylogeny</th><th rowspan="2">Ranking names of 6 taxa which include most part of Ephemeroptera</th><th colspan="4">General classification of Ephemeroptera by:</th></tr>
<tr><th>McCafferty & Edmunds 1979</th><th>Kluge 1989</th><th>McCafferty 1991</th><th>Kluge 1992–1998</th></tr>
<tr><td rowspan="6">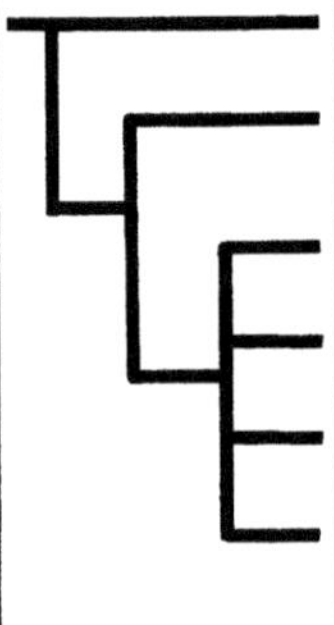</td><td>Baetis/fg</td><td rowspan="4">Schistonota</td><td rowspan="2">Costatergalia</td><td>Pisciforma</td><td>Tridentiseta</td></tr>
<tr><td>Heptagenia/fl= Oligoneuria/gl</td><td>Setisura</td><td rowspan="5">Bidentiseta</td></tr>
<tr><td>Leptophlebia/fgl</td><td rowspan="4">Furcatergalia</td><td rowspan="4">Rectracheata</td></tr>
<tr><td>Ephemera/fg</td></tr>
<tr><td>Caenis/fl= Brachycercus/gl</td><td rowspan="2">Pannota</td></tr>
<tr><td>Ephemerella/fgl</td></tr>
</table>

METHODS OF ASSOCIATING LARVAE AND ADULTS

Mayfly systematics is based on a combination of larval, subimaginal and imaginal characters; however, larvae and winged stages (subimago and imago) are so different, that their association represents a special problem. In literature one can find many mistakes, when such association was made wrongly, so that larvae and imagoes of different species were regarded as belonging to the same species, and vice versa, that of the same species were described as different species. Such confusions were made not only on the species level, but on the level of supra-species taxa (genera and families) as well.

Some features of winged stages can be found in the larval stage; this helps to associate them. As in other Pterygota, wing venation is the same in larval wing buds and adult wing (FIGS 37:A; 75:A). Some authors confuse venation with tracheation; tracheation strongly changes during larval development, while venation remains constant. In the larval wing bud venation represents a system of internal hypodermal channels, which can be seen on translucent slide as light lines. For this purpose it is necessary to take such larva, which is not preparing to moult to next instar or to subimago, in other case wing hypodermis with its channels is crumpled under the larval cuticle.

Mature larvae, which are preparing to moult to subimago, can be used to extract and study subimaginal features – thorax sclerotization (FIG.102:A–B), caudalii, sometimes genitals (FIG.84:B–C) and others. Subimaginal wings extracted from mature larva, can be spread after treating by alkali. Structure of subimaginal tarsus with claws can be studied on translucent slide of total larval leg in Canadian balsam; in shortly-moulting mayflies the same slide allows to study structure of imaginal tarsus with claws as well (FIG.98:A–B).

Examination of mature larva allows one to study only selected adult characters, but not all of which are necessary; particularly, in many mayfly species crumpled subimaginal genital buds inside mature larvae differ from imaginal genitals so strongly, that their comparison does not allow the association of larvae and adults correctly.

In most cases exact association of larvae and imagoes can be made by rearing only. The aim of rearing is to get such specimens, each of which has: (1) exuviae of mature larva of last instar, (2) exuviae of subimago, (3) imago. In order to do this, it is necessary to collect mature larvae, put them in a cage with water, wait until larva moults to subimago, then take its larval exuviae to alcohol and move subimago to another cage without water, than wait until subimago moults to imago and put imago and its subimaginal exuviae in alcohol together with the larval exuviae. Here are given suggestions how to do it successfully.

Some larval mayflies, especially those, which normally inhabit stagnant waters, can be kept in a simple aquarium or a can, until moulting to subimago. But larvae normally inhabiting running waters, especially mountain streams, can not live in stagnant water for a long time. For all species it is best to use special water-cages made of net and put into natural running water (not obligatory to the same stream where these larvae inhabit).

The author uses water-cages of original construction, which can be folded and packed compactly when travelling, and can be used in all kinds of running waters, including mountain streams and greatest rivers, independently of weather and water level. Here are shown two variants of such water-cage: The 1st variant (FIG.1:A–B) has 2 frames made of aluminium attached from outside by threads to an integral cube of polyamide (kapron) net, which has 5 equal walls – 4 side walls + bottom. Thc 2nd variant (FIG.1:C–D) has 2 walls made of translucent plastic and 3 walls (2 side walls + bottom) made of an integral band of polyamide net.

In the both cases the net cube (either made of the polyamide net or the polyamide net and translucent plastic) continues above by a tube made of textile. Framework (made of aluminium or translucent plastic) is supplied with 4 floats made of foamy plastic, two of which are attached to the framework, and other two are removable and serve to make the whole construction rigid when ready for work. Removable bracket with a string on its top allows closing the textile part and at the same time supporting it, in order not to allow emerged subimagoes to fly away. A plummet under the water-cage prevents overturning by wind (that is especially likely in rain when the textile above is wet and heavy). A long polyamide cord serves to tie the water-cage to something on the bank.

If there is no necessity to pack water-cages compactly for travelling, the water-cage can have simpler construction. In all cases it must have following details. The water-cage is supplied with floats in such a manner that when floating, a half of the water-cage is located in the water, and a half above the water, to allow subimagoes escape from the water. Water-cage must be not large, about 12 cm high (without the textile part) and 12 cm with: in larger water-cage it is difficult to find larval exuviae of small species, and smaller water-cage does not allow looking into it by both eyes when searching for larval exuviae and emerged adults. Walls (at least partly) are made of fine polyamide (kapron) net (cell 0.4 mm). The bottom is made of the same net, as the walls: the dust brought by water current through the walls must fall down thorough the bottom not accumulating on it. Inside the water-cage, there must be no places where larvae or their exuviae can be hidden; the framework and sutures must be outside, but not inside. No objects (stones, sand, leaves, sticks, etc.) should be putted into the water-cage; the polyamide net is a comfortable substrate for mayfly larvae, and water current brings enough food though the net for that larvae, which are not mature enough to stop feeding. The polyamide part of the water-cage is opened from above and continued by a textile tube, which has the same width as the polyamide part and subequal length. This allows to open and close the water-cage wider or narrower depending on behaviour of the emerged subimagoes, not allowing them to fly away. Cotton textile is a comfortable substrate for subimagoes. When closed, the textile part should have the form of a high roof to give more space for subimagoes, and to protect them from rain. The water-cage floating in running water should be fixed by a cord to something on the bank (tree branch, stone or something other).

Many (up to several dozens) larvae can be placed at once into the same water-cage, but it is better to use several water-cages.

It is preferable to check water-cages often, not less than three times a day, in extreme cases not less that once a day: if larval exuviae stay in water longer than one day, they are destroyed. Emerged subimagoes should be removed from the water-cage to an air-cage.

Such air-cage can be a glass tube about 10 cm length and about 2.5-3 cm in diameter, and is closed by cotton-wool (FIG.1:E). Inside the tube it is necessary to put a piece of paper, which is somewhat shorter than the tube and somewhat wider than its diameter; this paper must be immovably pressed to a wall of tube. It has three functions: (1) a comfortable substrate for subimago, that makes it to sit quietly and not to spend energy for flying; (2) water absorption; (3) label, on which a specimen number is written by pencil (the same number has the tube with alcohol containing larval exuviae of this specimen). It is necessary to move subimagoes from the water-cage to the air-cage, not touching them by fingers or pincer. Subimago sitting on the wall of the water-cage, can be covered by the air-cage; subimago sitting on the water surface can be moved with help of thin stick, to which it crawls if place the stick in front of subimago. It is very important to note, that glass air-cage with subimago never should be kept at direct sun light, even for a moment; it must be always in shade or in a box. Most subimagoes (apart of shortly-moulting ones) develop in about 24 hours – some times less or some times more, quicker in the warm and longer in the cold.

A usual error made when rearing, is confusing of larval exuviae and adults of externally similar species when several larvae are kept together in the same water-cage. In order to avoid this, it is necessary either to use individual water-cages for each specimen (that is rather difficult), or to take out carefully all emerged subimagoes and larval exuviae each time when check the water-cage. For this purpose, it is important not to confuse exuviae of mature larvae from which subimagoes emerged, with exuviae of younger larvae, from which larvae of next instar emerged.

All stages of mayflies can be preserved in alcohol of high concentration (75% and higher). As these insects are delicate, it is better to put them into tubes full of alcohol and closed by cotton wool, and to put these tubes into a hermetically closed glass with alcohol: in this case tubes should not contain air bulbs, which could destroy mayflies if shaken.

For examination, it is useful to make slides of all parts of mayflies and their exuviae. Separated parts can be mounted in Canadian balsam; this allows to study not only cuticle, but muscles as well. Delicate translucent cuticular parts – colourless wings, tergalii, subimaginal exuviae and cuticular parts treated by alkali – can be mounted in glycerine to see better their details (in this case cover glass can be glued to the mount by Canadian balsam by sides).

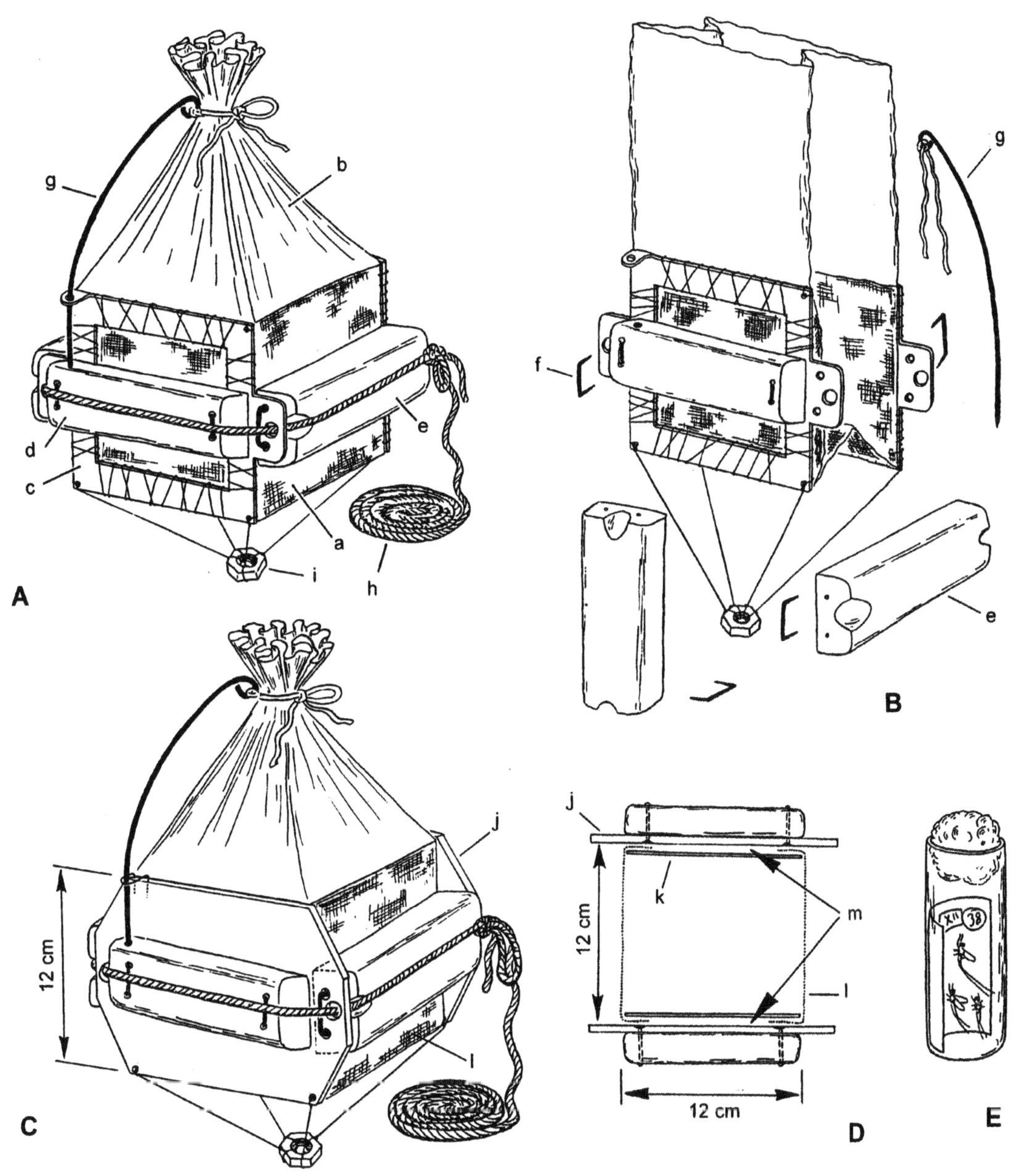

Figure 1. Cages for rearing mayflies.

A–D – Two different variant of water-cages for rearing subimagoes from larvae: **A–B** – 1st variant of water-cage, made of polyamide net with aluminium framework: **A** – in assembled condition ready to use; **B** – in disassembled condition; **C–D** – 2nd variant of water-cage, made of polyamide net and translucent plastic: **C** – in assembled condition ready to use; **D** – scheme of horizontal section, to show method of fastening of plastic walls, net walls and non-removable floats. **E** – air-cage for rearing imagoes from subimagoes.

a – cube (without upper wall) made of polyamide net; **b** – textile tube; **c** – 2 aluminium frameworks attached to polyamide net by threads; **d** – 2 foamy plastic floats attached to framework; **e** – 2 removable foamy plastic floats; **f** – 4 metallic staples to fix removable foamy plastic floats; **g** – removable metallic bracket with string; **h** – polyamide cord; **i** – plummet; **j** – 2 walls of thick translucent plastic; **k** – 2 square plates of thin translucent plastic; **l** – band of polyamide net forming two walls and bottom; **m** – schematically shown two spaces, where thick and thin plastic plates are glued together by the same translucent plastic dissolved in chloroform.

PRINCIPLES OF NON-RANKING ZOOLOGICAL NOMENCLATURE

Classification of living organisms always reflects ideas about phylogeny (or "natural system" of older authors), independently which explanation to principles of systematics was given by these or that authors. The main problem in phylogenetic investigation, is a necessity to use a great number of facts concerning characters and their distribution among organisms. There are no direct methods for reconstructing phylogeny; thus we have to use a single known indirect method – cladistic analysis (it can be called also scientific cladistic analysis, in order to avoid confusion with the numerical cladistic analysis). The scientific cladistic analysis, or analysis of apomorphies, never gives a correct final result, but allows an approximation of it; the process of such approximation is endless when more and more characters are taken into account. In this situation strict principles of text layout and nomenclature of taxa become important components of scientific work. C. Linnaeus elaborated such principles for the XVIII century scientific level, and they allowed reaching great success during the subsequent quarter of the millennium. Thanks to this, a great number of facts was accumulated, and phylogenetic theory was elaborated, which, in its turn, made the Linnaean principles out of date.

Particularly, the ranking zoological nomenclature elaborated by Linnaeus (1758) and Latreille (1802) and adopted by the modern International Code of Zoological Nomenclature (2000) in our days is not enough to supply with names all taxa in a serious scientific classification. Working on the present monograph and the book "Modern systematics of insects", the author had to restudy general principles of biological nomenclatures and to elaborate a new system of non-ranking nomenclatures for zoology (Kluge 1999a, 1999b, 1999c).

Problems of reconstructing phylogeny, principles of systematics and principles of all zoological nomenclatures are discussed in detail in the book "Modern systematics of insects. Part I" (Kluge 2000); the English translation of the chapter on systematics and nomenclature from this book is available from the Internet, http://www.bio.pu.ru/win/entomol/KLUGE/sys-ins.htm. Here are repeated only the most important explanations for the non-ranking nomenclature used in the present book.

Why do different types of nomenclature coexist?

Classification cannot be permanent; instead, it is subject to incessant change, because it is based on phylogeny, and there is no direct way to reconstruct phylogeny; as all methods of reconstructing phylogeny are indirect and rely on the entire body of biological knowledge, and the latter is continuously growing, the process of adjusting our idea of phylogeny, and hence changing classification, will be endless as well. So there is no hope that a perfect and final classification of living organisms would ever be built. Should a constant classification appear, rules of nomenclature would become redundant, as the names of all taxa in such classification will only need to be validated once and for all. It is the inability to create such a classification that forces us to set universal rules of naming taxa.

All principles of nomenclature pursue a single aim: to have names firmly stick to taxa. Yet any taxon has many different attributes, which include its circumscription, diagnosis, rank, position in the classification, etc. It is impossible to make a name refer to all such attributes at once, because any change of the classification entails changes in these attributes. For example, in different classifications taxa of the same circumscription may have different ranks, different diagnoses or be assigned to different higher taxa; and vice versa, taxa of the same rank can have different circumscriptions, and so on. Nomenclature must support ever-changing classification, which implies that a name can only be associated with just one attribute of a taxon.

Based on the attribute with which a name is associated, several fundamentally different types of nomenclatures can be recognised, viz. rank-based, circumscription-based, description-based, phylogeny-based, hierarchy-based, etc. Among them, only rank-based, hierarchy-based, and circumscription-based ones are meaningful. Their difference is shown in FIG. 2.

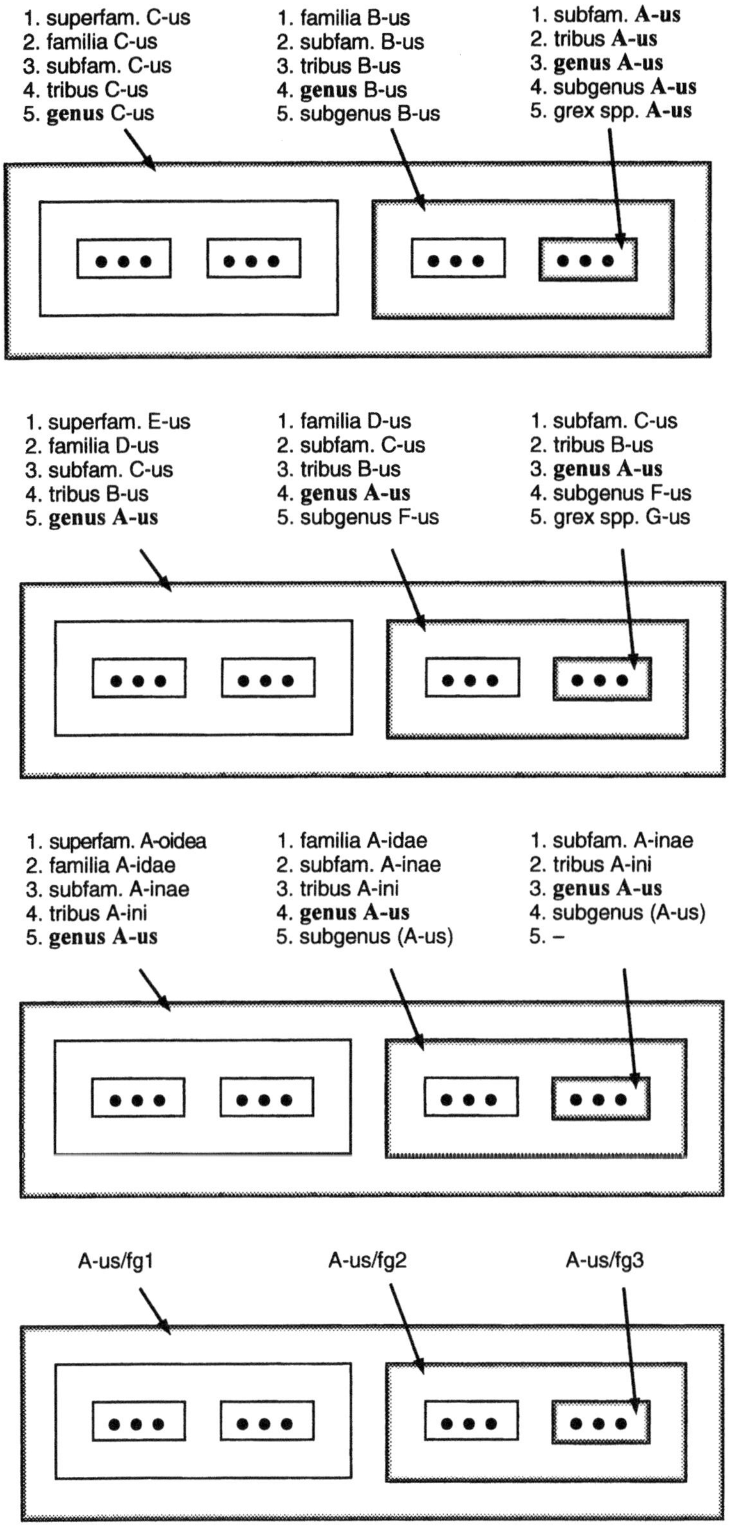

Circumscriptional nomenclature

Five variants of ranks arrangement (1–5) are shown; one rank (**genus**) and one name (**A-us**) are everywhere marked by bold to demonstrate that certain name (A-us) always belongs to taxa of the same circumscription (the darkest rectangle), independently of their ranks.

Ranking non-typified nomenclature

Five variants of ranks arrangement (1–5) are shown; one rank (**genus**) and one name (**A-us**) are everywhere marked by bold to demonstrate that certain name (A-us) always belongs to taxa of the same rank (genus), while these taxa can have different circumscriptions.

Ranking typified nomenclature

Five variants of ranks arrangement (1–5) are shown; one rank (**genus**) and one name (**A-us**) are everywhere marked by bold (see above). A-oidea, A-idae, A–inae and A-ini are typified names formed from the generic name A-us; rank-depending endings are given according to ICZN (in different ranking typified nomenclatures different endings are used).

Hierarchical nomenclature

Figure 2. Difference between circumscriptional, ranking and hierarchical nomenclatures. Different nomenclatures are applied here for the same classification of 12 species (black dots) united in 7 hierarchically subordinated supra-species taxa (rectangles) of 3 ranks; names are applied to 3 subordinated supra-species taxa (rectangles with shaded borders), which have 3 different ranks.

Rank-based (ranking) nomenclatures

In these nomenclatures a name is associated with a certain rank of taxon (such as genus, family, etc.) and is subject to change whenever the rank changes, but remains the same when other attributes (such as circumscription or position) change (FIG.2). The ranking nomenclature still plays a major role in taxonomy, because all international codes, including the International Code of Zoological Nomenclature, are based on this principle. A significant shortcoming of ranking nomenclature is that names are associated with a purely conventional taxon's attribute, i. e. its rank. In different classifications the same ranking name can be assigned to taxa of different circumscription while taxa consisting of the same members (i. e. having identical circumscriptions) should be given different names within the same ranking nomenclature if such taxa have different ranks. As a result, ranking nomenclature may cause confusion (Kluge 1996c, 1999a–c, 2000).

Hierarchy-based (hierarchical) nomenclatures

In hierarchy-based nomenclature a name is associated with the taxon's placement within hierarchical classification and does not depend on rank (FIG.2). This nomenclature is based on the recently enacted International Code of Zoological Nomenclature (ICZN), yet overcomes some important flaws of the ICZN's ranking principle.

Generic names, that fall under rules of ICZN and, thus, are rank-based by definition, can be used not only in rank-based nomenclature but to derive hierarchy-based names as well, in which case type species of generic names, authorship, priority and co-ordination all work just as provided by the ICZN. Unlike in the ICZN's nomenclature, in the hierarchy-based nomenclature no name is assigned absolute rank (such as genus, family, etc.), but refers rather to a relative rank indicated by the number of higher hierarchically subordinated taxa. That's why the hierarchy-based nomenclature can be used in non-ranked classification.

The **hierarchy-based**, or **hierarchical name** (NOMEN HIERARCHICUM) consists of an available generic name (or, better, of an available name of genus-group in terms of the ICZN) to which, after a slash, are attached the letter(s) "f" and/or "g" and a number ("1" or higher). Here is the procedure to create a hierarchy-based name: first we pick the oldest generic name within the taxon in question. Since in our classification we use no ranks and just disregard them, the priority of the generic name can be established based on authorship of either genus-group or family-group names (according to ICZN, they may not be the same). After the slash we insert either a "**g**" (as in "genus") if the priority gets established based on genus-group names, or an "**f**" (as in "familia" – family) if we establish it based on family-group names. Number "**1**" is attached to the taxon, which in our hierarchical classification is the largest (highest) among taxa, for which this generic name is the oldest within the chosen group of names (i. e. within the genus-group for names followed by "g" or within the family-group for names followed by "f"). Subordinated taxa with the same generic name are numbered according to their order of subordination in such a way that the smaller (lower) is the taxon, the higher is the number.

For example, the taxon uniting *Behningia ulmeri*, *Dolania americana* and *Protobehningia asiatica*, in hierarchical nomenclature can be named Behningia/g1. The generic name *Behningia* Lestage 1930 is the oldest among genus-group names whose type species are included into this taxon (i. e. *Dolania* Edmunds & Traver 1959 and *Protobehningia* Tshernova & Bajkova 1960). The number "1" is attributed to this taxon, because this is the highest taxon for which the generic name *Behningia* is the oldest. If we take a higher taxon (in our classification it is Fossoriae), it will include the type species of the older name *Ephemera*, thus its hierarchical name should be formed from "*Ephemera*", but not from "*Behningia*". The taxon subordinated to Behningia/g1 can be named Behningia/g2, and the taxon subordinated to it – Behningia/g3:

```
                 Behningia/g1
                  /        \
  Protobehningia          Behningia/g2
                           /        \
                     Dolania       Behningia/g3
```

The taxa with the same oldest generic name are numbered from the highest to the lowest, not vice versa, because the highest one can be easily identified based on priority, while taxa splitting can be unlimited.

Based on genus-group priority rules alone, one can assign unique hierarchy-based names to all taxa within a classification. However, as the current Code provides for a separate priority for family-group names, the hierarchy-based names generated under the rules for the genus-group give no idea what should taxa names be in a rank-based nomenclature if assigned family-group ranks (i. e. if we treat these taxa as tribes, subfamilies, families, or superfamilies). In this book hierarchy-based names are presented in such a format that allows converting them into familiar-looking rank-based names without recurring to any additional information. All we need to make a hierarchy-based name convertible to rank-based one is adding to it, with an "=" in between (no spaces), another hierarchy-based name, this time the one obtained based on family-group priority. In this case the name base is spelled out in its original form (i. e. as a generic name, without family-group endings) followed by a slash and an "f" instead of "g". Both the generic name and the number established based on genus- and family-group rules often match; in such cases we just write down one generic name with the letters "**fg**" and the number.

For example, the generic name *Behningia* is a base for family name Behningiidae Motas & Bacesco 1937 and other available family-group names with the same authorship; thus, in the example given above, taxa names can be written as following:

```
            Behningia/fg1
            /           \
Protobehningia       Behningia/fg2
                      /          \
                  Dolania     Behningia/fg3
```

If the generic names are identical but the numbers are not, we insert into the hierarchy-based name both numbers with their respective letters, separated by an "=" without spaces. If the generic names are not identical we write down both generic names (separated by an "=" without spaces) with their respective letters and numbers.

For example, there is a taxon that includes type species of two generic names: *Caenis* Stephens 1835 and *Brachycercus* Curtis 1834, of which the latter is older; however, the oldest family-group name derived from the name *Caenis* – Caenidae Newman 1853 – is older than the oldest family-group name derived from *Brachycercus* – Brachycercidae Lestage 1924. The taxon including both type species will have the hierarchy-based name Caenis/f1=Brachycercus/g1. Such spelling means that under ICZN rules, if this taxon is assigned genus-group rank its name will be *Brachycercus*, while if assigned family-group rank its name should be derived from the generic name *Caenis*. One of subordinated taxa within Caenis/f1=Brachycercus/g1 also includes both type species, and its hierarchy-based name will be Caenis/f2=Brachycercus/g2. In rank-based nomenclature this taxon also can be either named *Brachycercus* or get a typified name derived from *Caenis*, depending on whether we consider it a genus-group or a family-group taxon. This taxon, in turn, contains two taxa: one including the type species of the generic name *Caenis*, and another – the type species of the generic name *Brachycercus*. The hierarchy-based names of these taxa will be Caenis/f3=g1 and Brachycercus/f1=g3, respectively. Hierarchy of these taxa looks as follows:

```
            Caenis/f1=Brachycercus/g1
               /              \
Neoephemera/fg1      Caenis/f2=Brachycercus/g2
                          /              \
                   Caenis/f3=g1     Brachycercus/f1=g3
```

As classification changes, numbering in hierarchy-based names also shifts, so depending on classification taxa of the same circumscription may have different names, while taxa of different circumscriptions may be named identically. Hierarchy-based and rank-based nomenclature share such disadvantage, only circumscription-based nomenclature (see below) is free of it. The important benefit of hierarchy-based nomenclature is that the names shift only if there are changes in the classification, i. e. if the subordination of taxa is modified, while in rank-based nomenclature names change with any rank shift as well. Unlike rank changes, always purely discretional, classification changes are always based on evidence and can be discussed.

If a hierarchy-based name is used, it may be helpful to provide details on how taxa are arranged in this classification, as a comment on the name's number. This can be done when the name is first mentioned, listing (in parentheses) generic names of closest excluded taxa (using "**sine**", Latin for "without") and those of directly subordinated taxa (using "**incl.**" – incluso, including).

For example:
Behningia/fg2 (sine *Protobehningia*; incl. *Dolania*).

If the name mentioned for the first time has number much larger than "1", it can be helpful to list also excluded generic names of higher taxa.

For example, in this book classification starts with the taxon Gnathopoda Lankester 1881, which hierarchical name is written as "Araneus/fg6 ([2]sine *Spongia*; [3]sine *Fasciola*, *Medusa* et al.; [4]sine *Homo*; [5]sine *Limax*; [6]sine *Lumbricus*; incl. *Peripatus, Macrobiotus, Linguatula*)". This means that the highest taxon, to which the oldest generic name *Araneus* Clerck 1958 can be attributed, is

Araneus/fg1 (incl. *Spongia*) (circumscriptional name – Metazoa Haeckel 1874); its subordinated taxon is Araneus/fg2 (sine *Spongia*; incl. *Fasciola, Medusa* et al.) (circumscriptional name – Eumetazoa Bütschli 1910); its subordinated taxon is Araneus/fg3 (sine *Fasciola, Medusa* et al.; incl. *Homo*) (in circumscription matches Coelomaria sensu Haeckel 1898); its subordinated taxon is Araneus/fg4 (sine *Homo*; incl. *Limax*) (circumscriptional name – Trochozoa Beklemischev 1944); its subordinated taxon is Araneus/fg5 (sine *Limax*; incl. *Lumbricus*) (circumscriptional name – Polymeria Hadži 1953; widely used name – Articulata auct.); its subordinated taxon is Araneus/fg6 (sine *Lumbricus*; incl. *Peripatus, Macrobiotus, Linguatula*) (circumscriptional name – Gnathopoda Lankester 1881).

Circumscription-based (circumscriptional) nomenclatures

Under this approach a name is associated with a certain circumscription of a taxon without regard of its rank or position (FIG.2). Special circumscription-based names have a wider circulation. Such a name can be called either **circumscription-based**, or **circumscriptional** (NOMEN CIRCUMSCRIBENS); in the previous papers this term was translated from Russian as **"volumetric name"** (Kluge 1999a, 1999b). Each circumscription-based name is attached to a taxon of a given circumscription (i. e. a certain set of members included, but not the number of species). Most generally accepted circumscription-based names are those of major taxa, but there were no rules governing their usage until recently, when such rules were suggested (Kluge 1999) based on long established and well-proven custom.

To make a decision on using a circumscription-based name, we assess the name based on three criteria: 1) availability, 2) circumscription match, and 3) validity (rank-based names are assessed by availability and validity only). The key concept of circumscription-based nomenclature is the "admissible membership", based on which one can decide whether several taxa are identical in terms of circumscription. To determine whether taxa match in circumscription, the following parameters have been introduced (Kluge 1996):

Listed membership: the set of species, explicitly or implicitly referred to in the publication as members of the taxon.

Net membership: listed membership less (1) species whose position is considered uncertain in the publication, and (2) species erroneously listed as members of the taxon contrary to the diagnosis provided for that taxon in the same publication.

Excluded membership: a set of species explicitly or implicitly (see above) referred to in the publication as non-members of the taxon or as members of taxa other than those subordinated to it.

Net excluded membership: excluded membership less (1) species whose position is considered uncertain in the publication, and (2) species erroneously listed as members of another taxon contrary to the diagnoses provided in the same publication.

Admissible membership, or **admissible circumscription**: any set containing all species of net membership and no species of net excluded membership.

Original listed membership, original net membership, original excluded membership, original net excluded membership, and **original admissible membership** – all refer to their respective parameters in the context of the publication where the name in question was first published.

Criteria of availability for circumscription-based names. Since rank- and circumscription-based nomenclatures are inherently incompatible, it would be helpful to effectively separate rank-based names from circumscription-based ones. We propose to do so using such criteria of availability that would make names available as rank-related unavailable for purposes of circumscription-based nomenclature, and vice versa.

It would be appropriate to consider available for purposes of circumscription-based nomenclature all scientific names published since 1758 other than species-, genus-, and family-group names subject to ICZN, and other than typified names (those derived from type-genera names using only suffixes, endings and/or the stems "-form-" and "-morph-"). In that case all species-, genus-, and family-group names and all typified names (including both family-group names and typified names of higher taxa) would be available only as rank-based ones.

> Examples: **Odonata, Odonatoidea, Odonatoptera** and **Odonatopterata** are available circumscription-based names, as they are derived from "Odonata", which is not a generic name (though some of these names have been first introduced as rank-based). **Ephemeroptera, Ephemeropteria** and **Ephemeropteroidea** are available circumscription-based names, as they are derived by adding the stem"-pter-" to the generic name

scription-based names, as they are derived by adding the stem"-pter-" to the generic name "*Ephemera*". **Oniscomorpha** may be either an available circumscription-based name if assigned to a taxon within Diplopoda (where there is no genus named *Oniscus*), or a typified rank-based name if assigned to an eucrustacean group containing the genus *Oniscus*.

Circumscription match. Depending on how well names match taxa in terms of circumscription, names may be mismatching, non-univocally matching, or exactly matching.

Mismatching name is such an available circumscription-based name whose admissible circumscription is inconsistent with the taxon's circumscription. In circumscription-based nomenclature such a name cannot be valid for this taxon by definition.

Non-univocally matching name (NOMEN CIRCUMSCRIBENS NON UNIVOCUM) is such an available circumscription-based name whose admissible circumscription is not inconsistent with the circumscriptions of both the taxon in question and another taxon (taxa) within the same classification.

Exactly matching name (NOMEN CIRCUMSCRIBENS UNIVOCUM) is such an available circumscription-based name whose admissible circumscription is not inconsistent with the circumscription of the taxon while being inconsistent with the circumscription of any other taxon within the classification.

Example: for the name Plectoptera Packard 1886, the admissible circumscription would be any set including its original listed membership (the family Ephemeridae as it was generally accepted at that time) and not including the original excluded membership (all other insect taxa, which Packard mentioned in the same paper as not belonging to his Plectoptera). Adding new mayfly taxa described thereafter into Plectoptera is not inconsistent with the original admissible circumscription of the name Plectoptera, because Packard, being unaware of them, didn't refer to them as non-members of Plectoptera. Likewise, Packard said nothing on whether *Triplosoba*, *Protereisma* and other extinct Palaeozoic groups remotely related to extant mayflies are part of Plectoptera. This means that if we use a classification where a taxon includes all mayflies and Palaeozoic groups, while its subordinate taxon includes all extant mayflies but no Palaeozoic groups, the name Plectoptera should be considered non-univocally matching either taxon.

The name Agnatha Cuvier 1798, which was used by some authors (Börner 1904, Martynov 1924–1938 and others) for the same taxon as Plectoptera, actually has original admissible circumscription mismatching circumscription of the taxon under consideration. Originally this name (as "Agnathes" in French spelling) was attributed to a taxon uniting the genus *Phryganea* (i. e. caddisflies – Trichoptera) and the genus *Ephemera* (i. e. mayflies). Because of this, its subsequent attributing to mayflies only was regarded to be wrong (Handlirsch 1904); this decision agrees with recently proposed principle of circumscriptional nomenclature (that can not be said about other Handlirsch's nomenclatural suggestions).

Circumscription-based synonymy. Different names matching the same taxon are **circumscription-based synonyms**. Such synonyms may be **exact** (SYNONYMA CIRCUMSCRIBENTIA UNIVOCA) or **non-univocal** (SYNONYMA CIRCUMSCRIBENTIA NON UNIVOCA). Circumscription-based synonymy fundamentally differs from the rank-based synonymy, including synonymy as defined by ICZN; unlike rank-based synonyms, circumscription-based synonyms apply to taxa identical in circumscription, but not necessarily in rank. Therefore, whenever we mean circumscription-based synonyms we should always make it clear. The valid circumscription-based name should be chosen among circumscription-based synonyms, if any.

Validity of circumscription-based names. The valid circumscription-based name is the oldest one among exactly or non-univocally matching available names of the taxon. If an exactly matching name is also the oldest, it becomes the only valid name of the taxon; if the oldest name is only non-univocally matching, then the oldest exactly matching name becomes the second valid name of the taxon.

Thus, unlike the rules for rank-based names, those for circumscription-based names allow for more than one valid name for a taxon.

For example, the taxon uniting extant mayflies excluding Palaeozoic groups, has the oldest univocally matching name Euplectoptera Tillyard 1932, and more old non-univocally matching name Plectoptera Packard 1886.

In the circumscription-based nomenclature, exceptions can be made in usage of the principle of priority, but never in usage of the principle of circumscriptional matching. In exceptional cases a younger name can be used instead of the oldest one; reason for this can be homonymy, hemihomonymy, or other inconvenience of the oldest name, or general acceptance and stability of the younger name.

The name Plectoptera Packard 1932, being not preoccupied, nevertheless has an inconvenient similarity with the name Plecoptera Burmeister 1839 (stoneflies); it was substituted by a younger name Ephemeroptera Hyatt & Arms 1891, which has the same admissible membership and in our days is generally accepted.

Some authors (Demoulin 1956b, Tshernova 1970) adopted the name Plectoptera for mayflies including Permian representatives, but excluding Carboniferous ones, and the name Ephemeroptera – for mayflies in the widest sense, including Permian and Carboniferous representatives. Other authors (Tillyard 1932) used the name Plectoptera for mayflies in the widest sense. According to the principle of circumscriptional nomenclature, the names Plectoptera and Ephemeroptera having the same original admissible membership, can not be arbitrarily used for different taxa of the same classification.

The taxon uniting recent mayflies can be called either Euplectoptera (if it is important to indicate that it does not include the Palaeozoic groups), or Ephemeroptera (= Plectoptera) (in the cases when this is not important, for example in faunistic lists of recent species).

Combining circumscription-based and hierarchy-based nomenclatures

Use of different nomenclatures. Among supraspecific taxa, the hierarchy-based nomenclature based on the natural hierarchy of the phylogenetic tree works better than the rank-based one relying on purely artificial absolute ranks, while on the species level the rank-based nomenclature is adequate because it relies on evidence-based definition of species.

Therefore it would be only appropriate to switch to hierarchy-based nomenclature dealing with supraspecific taxa while retaining the rank-based nomenclature for species-group names. However, circumscription-based nomenclature is better equipped to satisfy taxonomic needs than hierarchy-based nomenclature, so if there is a choice between hierarchy-based and circumscription-based names, the latter should be preferred. Hierarchy-based nomenclature has only one, yet very important advantage over the circumscription-based one: to codify the circumscription-based nomenclature new rules are to be introduced (including homonymy rules – see Kluge 1999), published names catalogued and many new names created, while to adopt the hierarchy-based nomenclature all we need is current ICZN rules and already available pool of genus-group and family-group names, so we may go ahead converting these rank-based names into hierarchy-based ones using a simple procedure.

Format of species name in non-rank-based nomenclature. Rejecting absolute ranks in supraspecific taxa would mean getting rid of the generic rank as well. But under ICZN any species name may only exist as a binomen, i. e. combined with a name of a genus (but not of a taxon of another rank). Specific epithet can not be used unless in such a combination since many epithets are used more than once throughout the nomenclature; it is also impossible to replace generic names in binomina with names of supraspecific taxa of another rank(s) because this would upset homonymy.

In a non-rank-based nomenclature, a species name might have the following format: first goes the specific epithet, then authorship and year (parenthesis is never used: secondary binomina do not exist since there is no binomina whatsoever), then the original generic name in square brackets (whether it reflects currently recognized position of the species or not); if the original combination included a subgeneric name, such name (in parentheses) may follow the generic name in square brackets. This or similar format is widely used in catalogues. The generic name works here as a surname: initially given based on kinship, it is not subject to change and is used for identification purposes no matter how accurately it describes actual relationship.

To indicate current position of the species we may add, before the epithet, a hierarchy-based name of any higher taxon where this species belongs; such names will not be mistaken for generic part of a binomen because of their distinct hierarchy-based format. If the species is a type for the hierarchical name, number in the hierarchical name before such species name can be substituted by asterisk – "*".

The same species name can be written either as
Behningia/fg3 *ulmeri* [*Behningia*],
or:
Behningia/fg* *ulmeri* [*Behningia*].

Unlike the original generic name (the one in square brackets) which can never be altered, the hierarchy-based name in front is subject to change to reflect progress in taxonomy.

Such format has obvious advantages over the one currently in use. The generic name within a binomen is charged with two conflicting functions: (1) ensuring the uniqueness of the species name, and (2) reflecting the taxonomist's opinion on the species position. Each of these tasks is difficult in itself, and trying to accomplish both at once it hardly practical. As species position changes, homonyms may emerge or vanish, thus creating the need to replace epithets; this may lead to extremely puzzling situations where the very identity of the species under discussion is not clear and downright confusing. On the other hand, there are cases requiring either more or less details about species position, so generic name alone may not be enough.

For example, in two different papers the names of two obviously different species look like this:
species 1:
Epeorus znojkoi (Tshernova 1938),
species 2:
Epeorus znojkoi (Tshernova 1938).
Such name format sends confusing message on which species is referred to in each case. The following format carries much more information:
species 1:
nojkoi Tshemova 1938 [*Ecdyonurus*],
species 2:
znojkoi Tshernova 1938 [*Iron*].
If our knowledge about these two species would be limited to original descriptions and type specimens (adults only), their respective positions could be presented as follows:
species 1:
Rhithrogena/fg2 *znojkoi* Tshernova 1938 [*Ecdyonurus*],
species 2:
Epeorus/g2 *znojkoi* Tshernova 1938 [*Iron*],
since the taxa Rhithrogena/fg2 and Epeorus/g2 are well-defined based on adult stage. In current classification, Rhithrogena/fg2 is divided into Cinygmula/g1 and Rhithrogena/fg3; Cinygmula/g1 and Rhithrogena/g3 are defined based on larvae only. Epeorus/g2 is divided into Ironodes/g(1) and Epeorus/g3, and the latter into Epeorus/g4, Caucasiron/g(1) and other subordinated taxa; *Ironodes*, Epeorus/f4 and *Caucasiron* are defined based on larvae only. Now that the larvae of both species are known, we may elaborate:
species 1:
Rhithrogena/fg3 *znojkoi* [*Ecdyonurus*],
species 2:
Caucasiron/g(1) *znojkoi* [*Iron*];
or in a more detailed way:
Epeorus/g3 Caucasiron/g(1) *znojkoi* [*I.*].

All these alternative names showing the position of species 2 are correct and differ only in the amount of details they provide. In contrast, traditional name format makes all but one binomen incorrect, while the only "correct" binomen turns to be different depending on publication.

The traditional binary format is very concise and convenient whenever species taxonomy is not an issue, e. g., in papers on ecology, physiology and other non-taxonomic texts. In papers dealing with species position and status it is advisable to use more elaborate non-rank-based name format.

New names. In order to confer availability of a new genus or family group name, this name should be once published satisfying all provisions of availability proposed by the ICZN. Theoretically, it would be not important to indicate rank of the taxon in such publication, but enough to indicate if the new name belongs to the genus group or to the family group, because in all ICZN's rules the original rank is ignored. However, it is not clearly said in the ICZN, if a taxon name originally introduced without a rank, becomes available, or not. In order to avoid confusion, in this book new names are supplied with arbitrary ranks. For example, a new family group name Epeorini is here arbitrarily and temporarily supplied with a tribal rank, in spite of the fact that its highest taxon Epeorus/fg1 is a subordinate taxon inside Rhithrogena/fg1, which in its turn was earlier proposed to be a tribe. In the same manner, new genus group names are here arbitrarily and temporarily supplied with subgeneric ranks, while they can belong to taxa subordinate to traditional subgenera or/and divided into traditional subgenera.

Sliding binomina and polynomina. To show the hierarchical subordination concisely, names may be presented as binomina or polynomina. To do so, names of two or more taxa are to be arranged consecutively starting from the highest one; names of circumscription-based, hierarchy-based and/or rank-based nomenclatures may be used. There can be either no punctuation marks between names, just like between elements of binomina in the current nomenclature, or hyphens can be inserted.

For example, names of selected taxa from this book can be written as following:
Ephemeroptera Anteritorna Bidentiseta Branchitergaliae;
Branchitergaliae Heptagennota *Pseudiron*;
Heptagennota Pentamerotarsata;
Pentamerotarsata Radulapalpata Rhithrogena/fg2;

Rhithrogena/fg2 Cinygmula/g1 *cavum* Ulmer 1927 [*Cinygma*];
or:
Ephemeroptera-Anteritorna-Bidentiseta-Branchitergaliae;
Branchitergaliae-Heptagennota-*Pseudiron*;
Heptagennota-Pentamerotarsata;
Pentamerotarsata-Radulapalpata-Rhithrogena/fg2;
Rhithrogena/fg2-Cinygmula/g1-*cavum* [*Cinygma*].

The layout of a taxonomic paper. Modern taxonomic papers of Linnaean tradition tend to give each taxon a separate diagnosis, a description, and a differential diagnosis (also called comparison or discussion); the diagnosis is supposed to contain a summary of diagnostic characters, the description to provide a detailed characteristics of the taxon, and the differential diagnosis to pinpoint what distinguishes it from other individual taxa. Such reiteration of the same characters is impractical, since description, diagnosis and differential diagnosis are but vaguely special. As a matter of fact, the description is just an elaborate diagnosis, while the latter is nothing but a concise description: both reflect only a part of the taxon's characters (the number of characters of any taxon being infinitely large), and only those of taxonomic importance. Differential diagnosis is believed to be special in that the characters are described in comparison with those of other taxa. However, any character is meaningful only in a comparative context, so the only thing which sets the differential diagnosis aside is that it presents the comparison expressly, while in both the description and diagnosis the comparison is implicit. Yet a scientific paper is no place for implicit statements: its very aim is to expose the author's findings and opinions. That's why it is essential for any taxonomic text (be it a description, a diagnosis, differential diagnosis or whatever you choose to call it) to provide comparison in an explicit form.

Within the Linnaean tradition taxonomic diagnoses usually follow a certain plan whose standard vary depending on rank. For example, if a diagnosis of an order says that fore wings are transformed into elytra, the diagnoses of other orders within the same class will state that the fore-wings are not so transformed, while the diagnoses of superorders may not mention this character. Such an approach makes it easier for the reader to find a character of interest while avoiding redundancy in diagnoses of consecutively subordinated taxa. But in papers following such a plan the diagnoses of taxa would depend on purely artificial absolute ranks, thus inconsistent with the goal of building a natural system where taxa are supposed to be natural entities.

In this book we use another approach to meet taxonomic requirements. Each supraspecific taxon is characterized following a universal rank-independent scheme allowing to do without assigning absolute ranks to taxa, i. e. to switch to non-rank-based post-Linnaean classification. First the autapomorphies are listed, then the characters of unclear phylogenetic status, then the plesiomorphies; finally, or perhaps among plesiomorphies, variable characters of the taxon may be mentioned (usually referring to lower taxa not discussed here). There is no special paragraph for synapomorphies: all the apomorphies shared with any other taxa are listed under "Autapomorphies" of a higher taxon rather than among characteristics of this one. All doubtful synapomorphies are listed as "Characters of unclear phylogenetic status"; this paragraph may be subdivided as necessary. Each character is not only described but also compared, whenever possible, to other taxa; its unique or otherwise status is indicated. If necessary, references are given [in square brackets] to the description of a more general character in a higher taxon.

All characters are numbered with the only purpose to facilitate looking for similar items in descriptions, not for counting characters. Character counting used in all recent cladistic computer programs deprives such programs of any scientific meaning. One cannot assign numbers to characters unless for the sake of convenience, as we do when we arrange a text into sentences, paragraphs, chapters, etc.; the numbers may not be attributed any biological meaning.

A general "Index of Characters" is attached to the main text (taxa characteristics). In the index, the characters are arranged by structural parts to which they belong; the arrangement of the structural pats follows the usual pattern. Under each character entry, taxon names are listed, each followed by a reference number under which the character is described. To facilitate the search, some characters are mentioned in the index more than once. From the outside, the index may look like the character list and matrix of a cladistic paper, but it is just an index (like a table of contents or an alphabetic index), not a base for conclusions. The index lists all hierarchically subordinated taxa discussed in the text, not only "operational units" to be included into a matrix (the cladistic "operational unit" is rank-related, which makes it artificial and hardly meaningful). The numbers of structural parts used in the index are by no means a universal character numbering; each character is given a number to refer to the taxon description.

Authorship of each supraspecies taxon name is given just below the corresponding title. In order not to repeat several times the same authorship for hierarchical names based on the same generic name, authorship is everywhere given only for the hierarchical name with number "1". Together with authorship, there are given objective synonyms and those subjective synonyms, which are based on type species regarded to be conspecific.

Other subjective supraspecies synonyms are absent in hierarchical nomenclature. In ranking nomenclature, synonyms are taxa of the same rank. In circumscriptional nomenclature, synonyms are taxa of the same circumscription. In hierarchical nomenclature, name has no fixed rank or circumscription; thus it has no subjective synonyms. Taxa can have one or several genus-group names included (in this book, all these names are listed in the end of the taxon characteristics); these are not synonyms, but names for subordinate taxa which can be established.

Fonts. In agreement with recommendation of the ICZN (Supplement B6), all names of genus-group and species-group are given in *Italics*, and all other names, including hierarchical names, are given in normal font; it is not necessary to use Italics for hierarchical names, because the symbol "/g" indicates availability of the genus-group name.

Citation of authors and dates. Citation of authors of taxa names used in this book, may seem to be non-traditional. Actually there are no good rules or traditions for such citation. Even in modern literature, we can often see citation of author's name without date, such as "*Ephemera* L.", or "*Ephemera danica* Müller". Such kind of citation comes from tradition existed in the XVIII and beginning of XIX century, when it was well justified: At that time rules of nomenclature did not exist, and there were no concepts of availability, validity, priority, and starting point of nomenclature. Many names came to zoology from ancient Latin and Greek languages, and no one zoologist could be regarded as the author of such name. If different zoologists applied different names for the same animal, or applied the same name for different animals, no one of these zoologists could be regarded to be wrong, and each zoologist continued to use the same nomenclature in all his publications. So, in order to clarify which animal is understood under the certain name, one had to cite the author after the animal's name. Such author's name was often separated from the animal's name by a comma, and it was referred to the author in whose sense the name us used, but not to the author responsible for the name's availability. In this case date was not important, because the same author often used the same name in all his publications, and it was not important who used this name earlier.

Recently, according to the Code (51.2), when the author is cited after the taxon's name without punctuation mark, this is the author responsible for the name's availability, but not for its current usage. The availability is conferred by a certain publication, which has a certain date. In accordance with principle of priority, validity of an available name is provided by the date of publication, but not by personal services of the author (as it was proposed by Linnaeus). Thus, in fact, the author's name is cited in order to cite the date, and its citation without date is meaningless.

Nevertheless, there are no strict rules for the date citation; it can be written without punctuation mark, but often is separated from author's name by a comma. When a taxon name is used as a part of the sentence (for example, as a subject), such comma inside it can lead to grammatical confusion. The same happens when there are more than two authors of the publication conferring the name availability, and their names are separated by commas.

In order to avoid commas, and at the same time clearly distinguish taxon's name from author's names, and to distinguish names of several authors (which can consist of more than one word each), here date is used without comma, and between all author's names ampersand is inserted: *rubromaculata* You & Wu & Gui & Hsu 1981 [*Cinygmina*], or Coryphoridae Molineri & Peters & Zuñiga de Cardoso 2001.

Phylogenetic status of taxon

It is generally accepted to divide all taxa into **holophyletic** [the term introduced by Ashlock (1971) instead of "**monophyletic**" sensu Hennig], **paraphyletic** [the term introduced by Hennig] and **polyphyletic** [the term introduced by Haeckel] ones (for detailed explanation – see Kluge 2000 and English translation of this chapter in Internet, http://www.bio.pu.ru/win/entomol/KLUGE/syst_1_2.htm). Phylogenetic (cladistic) principle of systematics declares that all taxa should be holophyletic only, but never polyphyletic or paraphyletic. However, besides evidently holophyletic, paraphyletic and polyphyletic taxa, there are many those which phylogenetic status is not clarified yet, and they exist in all classifications together with holophyletic ones. Some authors call them "plesions", to distinguish from holophyletic taxa; in the preliminary text of this revision, the term "plesion" was also used in this meaning. Dr J.G. Peters paid my attention to the fact that in modern literature the term "plesion" is used in various meanings, and helped to find its original definition.

Indeed, Patterson and Rosen (1977) who introduced the term "**plesion**" (in plural "**plesions**"), applied it for taxa of any phylogenetic status, including holophyletic ones, and used it to represent a phylogenetic tree as a sequence of taxa rather than a hierarchy of taxa: in their classification, several plesions written one after another, mean that this is a pectinate portion of phylogenetic tree, in which each plesion is a sister group of all those that succeed it. In this sense, the term "plesion" indicates systematic position of the taxon and substitutes traditional Linnaean ranks, but does not characterize the taxon itself. In the ancient Greek, the word "**plesion**" (plural "**plesia**") means neighbour in direct and indirect meanings, that is close to the meaning of the term "plesion" proposed by Patterson and Rosen. In the same paper (Patterson & Rosen 1977:163) it was suggested to put into quotation marks name of a taxon which is "nonmonophyletic, or ... no longer contains its type genus, or both". In order to avoid confusion, it would be better not to use the term "plesion" and quotation marks to signify paraphyletic taxa.

Instead, here is suggested a new term "**plesiomorphon**" (in English plural "**plesiomorphons**", but not Greek "plesiomorpha"): plesiomorphon is a taxon characterized by plesiomorphies only; thus, its holophyly is not proven, and possibly (or probably) this taxon is paraphyletic, but its paraphyly is also not proven yet. Plesiomorphon can be a taxon of any size and systematic position in ranking or non-ranking classification, living or fossil, with any name (ranking, hierarchical, circumscriptional or other).

GENERAL PART

Chapter I

Systematic position of Ephemeroptera

Comprehensive discussion on various opinions concerning the systematic position of mayflies among other insects, and the position of insects within arthropods is given in the book "Modern Systematics of Insects" (Kluge 2000).

Here is a brief account of selected subordinated taxa mentioned in this book:

1. **Gnathopoda** Lankester 1881, or **Arthropoda s. l.** Circumscriptional synonyms: **Lobopoda** Snodgrass 1938, **Podophora** Waggoner 1996, **Panarthropoda** Zrzavy & Stys 1997, **Antennopoda** De-Haro 1998. Hierarchical name: **Araneus/fg6** (^{2}sine *Spongia*; ^{3}sine *Fasciola, Medusa* et al.; ^{4}sine *Homo*; ^{5}sine *Limax*; ^{6}sine *Lumbricus*; incl. *Peripatus, Macrobiotus, Linguatula*). Holophyly of this taxon is proven by autapomorphies – reduction of coelome, opened blood system, presence of chitin-containing cuticle, and others. An opposite idea about polyphyletic origin of different arthropod groups from different annelids in course of independent "atrhropodization" is not grounded. Particularly, absence of coelomic sacs in all arthropods and their presence in all annelids, vertebrates, et al., can be explained neither by their function, nor by animal's mode of life, as each this group contains extremely diverse animals.

Gnathopoda are divided into **Onychophora** Grube 1853 (or **Peripatus/fg1**), **Linguatula/fg1**, **Tardigrada** Spallanzani 1776 (or **Macrobiotus/fg1** [nomen oblitum: **Arctiscon/fg1**]), and **Euarthropoda**.

1.1. **Euarthropoda** Lankester 1904, or **Arthropoda s. str.** Hierarchical name: **Araneus/fg7** (sine *Peripatus* et al.; incl. *Calymene, Scarabaeus*). Holophyly of this taxon is proven by autapomorphies – presence of sclerites and articulated legs, unique complicate visual system consisting of ocelli and facetted oculi, unique arthropod setae and others. An opposite idea about polyphyly of this taxon (repeatedly erected in the second half of the XIX century and in 60–70th years of the XX century) is not grounded by any apomorphies.

Euarthropoda are divided into a plesion **Trilobitomorpha** Størmer 1944 (or **Calymene/fg1**) and two holophyletic taxa – **Pseudognatha** Latreille 1821 (or **Araneus/fg8**) and **Mandibulata**.

1.1.1. **Mandibulata** Snodgrass 1935 (non Mandibulata Clairville 1798, nec Mandibulata Latreille 1825). Circumscriptional synonyms: **Polygnatha** Latreille 1821 (non Polygnatha Cuvier 1805), **Cormogradi** Crampton 1928, **Entomomorpha** Lameere 1936. Hierarchical name: **Scarabaeus/fg1** (incl. *Cancer*). Assumed autapomorphies of this taxon are: peculiar structure of ommatidium of facetted eye; mandibles bearing incisor, kinetodontium and mola (see Chapter II); presence of superlinguae (see ibid.) and some others. However, these characters are well expressed in Eucrustacea and Hexapoda, but not in Myriapoda; their absence in Myriapoda can be explained either as a secondary reduction, or as an independent origin of Myriapoda and **Pancrustacea** Zrzavy & Stys 1997 (the taxon uniting Eucrustacea and Hexapoda). The last idea contradicts the assumption about holophyly of Atelocerata.

With some doubt, we accept division of Mandibulata into **Eucrustacea** Kingsley 1894 (or **Cancer/fg1**) and **Atelocerata**.

1.1.1.1. **Atelocerata** Heymons 1901. In circumscription matches: **Insecta** SENSU Cuvier 1817 (non Linnaeus 1758), **Tracheata** SENSU Pocock 1893 (non Haeckel 1866). Hierarchical name: **Scarabaeus/fg2** (sine *Cancer*; incl. *Scolopendra*). In favour of holophyly of this taxon testify reduction of first postoral (tritocerebrum-innervated) limbs, reduction of pretarsal extensor and some other apomorphies.

Atelocerata are divided into holophyletic taxa **Myriapoda** Latreille 1802 (non Myriapodes Latreille 1796) (or **Scolopendra/fg1**) and **Hexapoda.**

1.1.1.1.1. **Hexapoda** Blainville 1816. In circumscription matches: **Insecta** SENSU Leach 1815 (non Linnaeus 1758). Hierarchical name: **Scarabaeus/ /fg3** (sine *Scolopendra*; incl. *Podura*). Holophyly of this taxon is well-proven by unique autapomorphies, and no one fact is known that would testify against; in spite of this, some authors regard Hexapoda to be polyphyletic, that is a result of confusion provoked by formerly used nomenclature (Kluge 1996c, 1999c, 2000).

Hexapoda are divided into holophyletic taxa **Entognatha** Stummer-Traunfels 1891 (or **Podura/ /fg1**) and **Amyocerata.**

1.1.1.1.1.1. **Amyocerata** Remington 1954. Circumscriptional synonym: **Ectognathata** Boudreaux 1979; in circumscription also matches: **Ectognatha** SENSU Hennig 1953 (non Stummer-Traunfels 1891), **Insecta** SENSU Handschin 1958 (non Linnaeus 1758). Hierarchical name: **Scarabaeus/fg4** (sine *Podura*, incl. *Lepisma*). Holophyly is proven by peculiar structure of antennae and some other apomorphies.

Amyocerata are divided into Triplura and Pterygota.

1.1.1.1.1.1-1. **Triplura** Ewing 1942. Circumscriptional synonyms: **Ectotropha = Ectotrophi** Grassi 1888, **Ectognatha** Stummer-Traunfels 1891, **Euthysanura** Schepotieff 1909, **Thysanuradelphia** Crampton 1916, **Phanerognatha** Krausse & Wolff 1919, **Panthysanura** Crampton 1928, **Apterentoma** Chen 1958, **Thysanurata** Bey-Bienko 1962, **Zygoentomata** Brusca & Brusca 1990. In circumscription also matches: **Thysanura** SENSU Lameere 1895 (non Latreille 1796) and **Apterygota** SENSU Scharov 1966 (non Lang 1888). Hierarchical name: **Lepisma/ fg1.**

Triplura unite **Zygentoma** Börner 1904 (or **Lepisma/fg2**) and **Microcoryphia** Verhoeff 1904 (or **Machilis/fg1**). Possibly, Triplura is a holophyletic taxon; an opposite idea about holophyly of **Dicondylia** Hennig 1953 (uniting Pterygota and Zygentoma) is not grounded (Kluge 2000).

1.1.1.1.1.1-2. **Pterygota** Lang 1888. Older circumscriptional synonym: **Pterygogenea** Brauer 1885; younger circumscriptional synonyms: **Eupleomerentomata** Krausse & Wolff 1919, **Pterentoma** Chen 1958. Hierarchical name: **Scarabaeus/fg5** (sine *Lepisma*; incl. *Ephemera*). Besides wings, which have quite mysterious origin) and undoubtedly evolved once (Kluge 1989a, holophyly of Pterygota is proven by some other autapomorphies – integral tentorium, presence of lateral cervical sclerites, characteristic thoracic apodemes and sutures, and others.

Pterygota are divided into Ephemeroptera (see Chapters II–III) and Metapterygota.

The huge taxon **Metapterygota** Börner 1909 units all winged insects except for mayflies and is regarded to be holophyletic on the basis of the following apomorphies (Kluge 2000).

(1) Only one winged stage is present – imago; i. e. the moult, at which acting wings appear, is the last moult in cycle of development. It is assumed, that the imago of Metapterygota corresponds to the subimago of Ephemeroptera, which lost ability to moult and got ability to reproduce (Edmunds & McCafferty 1988).

Some authors regard the presence of single winged stage in all Metapterygota to be not a synapomorphy, but a convergence; according to their opinion, in different phylogenetic branches of Pterygota the primitive representatives had stages of subimago and imago, and during further evolution the stage of subimago was lost. This opinion is based on the fact that for some extinct Palaeozoic representatives of Palaeodictyoptera and Polyneoptera, forms regarded to be "subimagoes" were described(Sharov 1957, Sinitshenkova 1979, et al.). These forms have smaller sizes of wings and smaller number of branches of some veins, than in "imago" attributed to the same species. In this respect, they in essence differ from the subimago of Ephemeroptera, which wings always have the same size and venation as in imago. If it is assumed that the Palaeozoic long-winged "imagoes" were developed from the short-winged "subimagoes", we would have to admit that during ontogenesis there was disproportionate growth of selected fields of the wing, where new vein branches appeared. However, in ontogenesis of recent insects such does not occur, wing buds grow proportionally, retaining constant arrangement of veins. Most probably, the Palaeozoic short-winged insects are not "subimagoes", but definitive forms. Similar definitive short-winged non-flying forms are present also in many recent groups of insects (including Ephemeroptera, among which there is one undescribed short-winged species with variable size of wings – FIG.8:E–F). For some Palaeodictyoptera "subimagoes" of another kind were described: Their wings have size and venation as in imago, but with tops curved backward, like wing buds of the larvae of the same species (Kukalova-Peck & Peck 1976). In recent insects the similar form of wing is observed in the imago when a wing spreads only partially after moult. Thus, there are no insects, besides Ephemeroptera, where the stage of subimago would be discovered; it permits one to think that the assumption about holophyly of Metapterygota is correct.

(2) Mandible base is diminished; because of this the posterior (lateral) condyle is approximated to the anterior (medial) condyle and located behind it; due to this, mandibles close nearly exactly towards one another. Mandibular-tentorial muscles attached inside the mandibles (i. e. the ventral mandibular muscles) are partly lost (Staniczek 2000), sometimes disappear completely. In this character Metapterygota differ from Ephemeroptera, primary wingless insects, eucrustaceans and myriapods, where the mandible retains well-developed ventral muscles and long basis with the posterior condyle located far from the anterior one.

(3) Superlinguae (initially present in Mandibulata) are lost. Non-unique apomorphy: the superlinguae had also reduced several times in some extremely specialized carnivorous mayfly larvae; they are absent in Zygentoma and Protura. In literature statements are found that some Metapterygota have superlinguae. In these cases processes of hypopharynx for were taken the superlinguae; in contrast to true superlinguae, these processes arise from the hypopharynx itself, but not from ventral wall of the head. Similar processes are present also in some larvae of Ephemeroptera, particularly in majority of Atalophlebia/fg1, where they exist together with the true superlinguae.

(4) Structure of caudalii is simplified: in contrast to Triplura and Ephemeroptera, the paracercus and basi-basal muscles are lost (only muscles going from cerci bases to tergite X, are retained). Besides Metapterygota, reduction of paracercus took place in various groups of Euplectoptera also.

A statement was expressed that the paracercus is retained in some Metapterygota, particularly in larvae of Amphinotic stoneflies Austroperla/fg1; in connection with this, a theory was suggested about common origin of Ephemeroptera and Plecoptera from Palaeozoic Protephemeroidea (Illies 1960). In this case for the paracercus was taken a secondary gill process. Some other representatives of Metapterygota also have an unpaired caudal process, which has no relation to the paracercus.

(5) Into each wing, branches from two different leg tracheae come, one of which arises from the intersegmental spiracle lying ahead of the wing, and another from the intersegmental spiracle lying behind the wing (in contrast to Ephemeroptera where all wing tracheae arise from the ahead lying spiracle). We can not exclude that this character is not an apomorphy of Metapterygota, but vice versa, an apomorphy of Ephemeroptera (or of Euplectoptera only).

The alternative opinion is to unite Ephemeroptera and **Odonatoptera** Lameere 1900 (or **Odonata** Fabricius 1793 **s. l.**, or **Libellula/fg1**) in a taxon **Subulicornes** Latreille 1807 (circumscriptional synonyms: **Raphiacera** Billberg 1820, **Subulicornia** Burmeister 1839 and **Hydropalaeoptera** Rohdendorf 1968), which in its turn is united with an extinct Palaeozoic taxon **Protorrhynchota** Rohdendorf 1968 (or **Dictyoneura/fg1**) to a taxon **Palaeoptera** Martynov 1923.

The following characters were named as possible apomorphies of Subulicornes or Palaeoptera.

(1) In all Palaeoptera wings are unable to fold on back, thus in rest are either spread by sides or raised upwards. Many authors follow the opinion by Martynov, who regarded this character to be a plesiomorphy. However, some investigators accept another hypothesis about wing evolution, according to which the wings were initially able to fold on the back, but in Ephemeroptera, Odonata and Protorrhynchota this ability was lost in connection with specialization (Rasnitsyn 2002). This opinion can be supported by the fact that Ephemeroptera retain a suture separating the posterior articulatory process of basal plate (FIG.6:APP) from the remaining part of the wing basal plate; this suture is invariably present in all mayflies, but does not enable any mobility or flexibility. Possibly, the posterior articulatory process of Ephemeroptera is homologous to the 3rd axillary sclerite of Neoptera (Brodsky 1974). In Neoptera, the 3rd axillary sclerite is movably articulated with the wing base, and when turning over, makes the wing to fold on back.

(2) In all Palaeoptera longitudinal veins arising from wing base, alternate as concave and convex (while in Neoptera such regular alternation is characteristic only for anteriormost veins – marginal

C, concave Sc, and convex RA). In Subulicornes, besides this, triad principle of vein branching is present, which serves regular alternating of all longitudinal veins, including branches and intercalaries (Fig.7:B); in Ephemeroptera triad principle is expressed in branching of the majority of veins, while in Odonata in branching of RS only. Both in Ephemeroptera and Odonata, the vein RS (which in all Subulicornes except for Protephemeroidea arises not from RA, but from MA) has the same manner of branching: it divides into two concave branches and a convex intercalary between them (RSa, iRS, RSp), and then the anterior branch divides into two concave branches and a convex intercalary between them (RSa_1, iRSa, RSa_2) (FIG.7:C).

(3) It is assumed that the maxillary galea is fused with lacinia: Ephemeroptera have an integral biting lobe and a segmented palp; Odonata have an integral biting lobe and a non-segmented appendage, which can be regarded either as a palp or as a galea. In the both cases, the biting lobe bears on its inner side two parallel longitudinal rows of denticles and stout setae (FIG.3:E) (while in many other Hexapoda denticles and setae form a single row). Maxillae of Protorrhynchota are highly modified, and their structure is poorly known.

(4) In Subulicornes (but not in Protorrhynchota) imaginal (but not larval) antennae are diminished, with flagellum vestigial. Non-unique apomorphy. Most probably antennae were reduced independently in Ephemeroptera and in Odonata. In Ephemeroptera the antennae retain primitively indeterminate number of segments (FIG.76:B), while in Odonata they are constantly 7-segmented.

(5) Larvae are aquatic and sharply differ from imagoes. Most probably, this feature appeared independently in Ephemeroptera and Odonata, as larvae of these two taxa have no any common morphological characters connected with their aquatic inhabitancy.

Chapter II

Morphology of mayflies

Table 2. Size range in main mayfly taxa

	Fore wing length (mm) 111112222233 24680246802468 02>
† Protephemerida	□□□□□□□□□■□□□□□□□
† Permoplectoptera	□□■■■■■■■■■■■■■■□
Prosopistoma/f1=g2	■■■□□□□□□□□□□□□□□
Baetisca/f2=g1	□□■■■■■■□□□□□□□□□
Siphlonurus/fg1	□□□□■■■■■■■■□□□□□
Dipteromimus	□□□□□□■■□□□□□□□□□
Ameletus/fg1	□□□■■■■■□□□□□□□□□
Metretopus/fg1	□□□□■■■■□□□□□□□□□
Acanthametropus/fg1	□□□□□■■■■■□□□□□□□
Ametropus	□□□□■■■■■■■□□□□□□
Tetramerotarsata	■■■■■■□□□□□□□□□□□
Nesameletus/f1=Metamonius/g1	□□□□□■■■□□□□□□□□□
Oniscigaster/fg2	□□□□□□□■■■□□□□□□□
Tasmanophlebia/g1	□□□□■■■□□□□□□□□□□
Ameletopsis/fg1	□□□□□□□□■■□□□□□□□
Rallidens	□□□□■■■□□□□□□□□□□
Coloburiscus/fg1	□□□□□■■■■■□□□□□□□
Isonychia/fg1	□□□■■■■■□□□□□□□□□
Chromarcys	□□□□□□□■■■■■■□□□□
Geminovenata	□□□■■■■■■■■□□□□□□
Heptagennota	□□■■■■■■■■□□□□□□□
Potamanthus/fg1	□□□■■■■■■□□□□□□□□
Euthyplocia/fg1	□□■■■■■■■■■■■■■■■
Ichthybotus	□□□□□□□■■■■□□□□□□
Ephemera/fg9	□□□■■■■■■■■■■■■□□
Protobehningia	□□■■□□□□□□□□□□□□□
Behningia/fg2	□□□□□□■■■■■■□□□□□
Palingenia/f2=g1	□□□□□■■■■■■■■■■□□
Polymitarcys/f1=Ephoron/g2	□□■■■■■■■■□□□□□□□
Neoephemera/fg1	□□■■■■■■□□□□□□□□□
Caenoptera	■■■■□□□□□□□□□□□□□
Ephemerella/fg2	□□■■■■■■■□□□□□□□□
Vietnamella	□□□□□□□■□□□□□□□□□
Austremerella	□□□□□■□□□□□□□□□□□
Melanemerella/fg1	□□■■■■□□□□□□□□□□□
Teloganodes/fg1	□■■■■■□□□□□□□□□□□
Tricoryptera	■■■■■□□□□□□□□□□□□
Leptophlebia/fg1	■■■■■■■□□□□□□□□□□

Morphological characteristics given below are based mainly on recent representatives, i. e. on Euplectoptera only (see Chapter III); for many derived characters it is unclear if they are autapomorphies of Euplectoptera, or that of Euephemeroptera, or that of Panephemeroptera.

Mayflies are insects of medium size, body length ranges from 2 mm to more than 40 mm (TABLE 2); size is species-specific, being either equal in both sexes, or larger in females; sometimes specimens developed in warmer water are smaller that those developed in colder water.

STAGES OF DEVELOPMENT

The characteristic feature of Pterygota is splitting of postembryogenesis (which, as in all Gnathopoda, consists of several instars separated by moults) to sharply different wingless and winged stages: the wingless stage(s) never have movable wings, while the winged stage(s) have acting wings, modified pterothorax and other features connected with ability to fly. Some authors believe that ancient insects had stages intermediate between wingless and winged ones, with movable but not fully grown wings; however such instars (erroneously called "subimago") have not been found in any living or fossil insects, and probably could not exist (Kluge 2000). Ephemeroptera differ from all other insects by having not one, but two winged stages separated by moult – **subimago** and **imago** (FIG.4). Both subimago and imago have completely developed wings with equal size and venation.

LARVA
(Fig. 3)

In contrast to subimagoes and imagoes, wingless stages of mayflies are always aquatic and obtain oxygen dissolved in the water. Their tracheostia (i. e. mouths of tracheal system) are closed, and other adaptations for aquatic respiration are present: abdominal tergalii are often used as tracheal gills, and other tracheal gills can appear independently on various parts of the body (see Index of characters [1]). This aquatic wingless stage is called either **larva**, **nymph**, or **naiad**; younger instars (lacking wing buds) can be called larva, and older ones (with wing buds) – nymph or naiad. The youngest instar(s) without tergalii and wing buds can be called **larvula**. Here we use the term **larva** for all wingless instars. Larva has large and often indeterminate number of instars; first instar never has tergalii and wing buds (as in all other insects), and during subsequent moults tergalii and wing buds appear and increase gradually, so that it is difficult to mark boundaries between larvula, larva in strict sense, and nymph.

Initially mayfly larvae have a peculiar swimming **siphlonuroid specialization** (well-expressed in Siphlonurus/fg1): the body is slender, with long abdomen and relatively small thorax; legs are able to be pressed to the body, being stretched posteriorly; abdomen is elongate and able to make undulate dorsoventral swimming movements; caudalii are not long, much shorter than in imago, with primary swimming setae forming a horizontal caudal flipper (see below). Pressing its legs to the body and making undulate movements up and down by its abdomen, the larva can swim rapidly for a short time (FIG.9:A–B). This specialization is initially present at least in Euephemeroptera (larvae of Protephemeroidea are unknown): it is present in Permoplectoptera (FIG.14:C–D), in majority of Mesozoic representatives of Euplectoptera and in many recent mayflies (see Index of characters [1]). In some cases this swimming specialization is secondarily lost: larvae of many mayflies are adapted not for swimming, but for fixing on stones in rapid stream, or for burrowing, or for other modes of life in aquatic environment. But the primary swimming setae are retained in many non-related groups of mayflies (see Index of characters [1.3.66]); sometimes the swimming specialization disappears, but vestigial primary swimming setae are retained. A very constant character of Ephemeroptera is a manner of swimming: mayfly larvae move by their abdomen up-and-down, in contrast to aquatic larvae of Odonata and Plecoptera, which when swimming, always move by their abdomen from side to side. The abdominal movement up-and-down is very effective in the case when the larva has the siphlonuroid specialization, but it appears to be useless if there is another shape of body and caudalii. In spite of this, such kind of movement is retained even in some mayfly larvae that have completely lost siphlonuroid specialization (FIG.9: C–D, F–H). Only in rare cases have larvae lost ability of the dorsoventral swimming movements (FIG.9:E) (Kluge & al. 1984).

SUBIMAGO AND IMAGO
(Fig. 4)

Both winged stages – subimago and imago, in contrast to larva, are non-feeding, able to fly and inhabit air environment; they sharply differ from larva in structure of head, thorax, legs, abdomen and caudalii, have functional wings and lack tergalii (FIG.4). Transformation from larva to subimago is supplied by great changes comparable with complete metamorphosis, but in contrast to it, is not supplied by immobility: during the whole development before ecdysis the subimaginal leg anlage is located inside the larval leg cuticle in such a manner that subimaginal and larval knee articulations always coincide, and allows active mobility of the leg (FIG.3:A).

Subimago

Subimago has the same shape and size as imago, but differs in cuticle structure and setation; male subimagoes, besides this, have less expressed sexual characters in structure of eyes, fore legs, genitals and caudalii (see below).

Subimaginal cuticle in most part is covered with microtrichia – densely and evenly situated small (about 0.01 mm) immobile crescent-shaped cuticular processes resembling setae; each microtrichion arises from the centre of cuticular area produced by one hypodermal cell. Subimaginal wings are always entirely covered with microtrichia (because of this they look dull); in imago the microtrichia are always absent, and at least the wing membrane is always bare. Possibly, the microtrichia play a positive role, as they keep a layer of air, which protects the wing against getting wet when the insect moults from larva to subimago on water surface or under the water. Besides mayfly subimagoes, similar microtrichia on wings are present in imagoes of many other insects, and allows one to conclude that subimago of mayflies corresponds to imago of Metapterygota.

Subimaginal and imaginal pterothoracic sclerites

usually have different outlines (about differences in mesonotum and mesopleuron sclerites – see below).

Subimaginal cuticle of wing is uniform, thin, elastic, non-sclerotized, equal on veins and on membrane (in contrast to the imaginal one, which is thickened and sclerotized on veins); this allows the insect to take off imaginal wing from the subimaginal cuticle when it moults from subimago to imago. As subimaginal wing veins lack any sclerotization, rigidity of wing necessary for flight is served only by goffered wing form (due to alternating of convex and concave veins – see below) and possibly by hemolymph pressure in veins. Subimago is able to spread wings and fly immediately after escaping from larval exuviae.

In subimago posterior margin of wing always has a row of setae (which are longer than microtrichiae and have different structure), while in imago these setae are nearly always absent, being present only in some specialized groups (see Index of characters [2.2.27]).

When the subimago transforms to imago, its exuviae are taken away as an integral cover (as at other moults), together with covers of wings. Only in some specialized forms that have short-living imago with non-functional legs, the moult to imago is lost, and the subimago becomes a reproductive stage (see Index of characters [2]); in some of these short-living forms males moult in air and throw their exuviae only partly.

Imago

Imaginal organization is adopted to the peculiar mating flight of male: it flies vertically upward, and then passively parachutes down, keeping its wings semi-spread in a V-shape, abdomen turned somewhat upward, and cerci in a V-shape turned to sides and somewhat upward. The male repeats such flying up and down above the same place, sometimes in a swarm with other males of the same species, attracting females. Upon seeing a female coming to the swarm, the male flies to it from beneath, orienting with help of its dorsal eyes (see below). Mating can take place at flight or on ground (FIG.10:G); the male is located under the female by its dorsal side directed upward and holds wing bases of the female by its fore tarsi, arching them dorsally-posteriorly (see below); abdominal apex of the male is curved dorsally-anteriorly, thus gonostyli appear to be directed upward and fix the female's abdominal apex, overlapping it from sides, and the penis also appears to be directed upward and is inserted into the female genital opening, located between sterna VII and VIII (about genital structure – see below). Such mating flight and manner of copulation are peculiar at least for Euplectoptera, and only in some mayfly taxa are secondarily changed (Brodsky 1973). Possibly, some of peculiarities in structure of pterothorax and wings of mayflies (see below) evolved in connection with the mating behaviour of males, but are present in both sexes.

HEAD

EYES

Facetted **eyes (oculi)** and all 3 **ocelli** are always developed in mayfly larvae, imagoes and subimagoes (FIG.3–4) (presence of the facetted eyes and ocelli is an autapomorphy of Euarthropoda, and reduction of ocelli number to three is a peculiarity of Amyocerata). Sexual dimorphism in structure of facetted eyes is initial for Ephemeroptera (at least for Euplectoptera), being expressed in majority of mayflies: eyes of male are enlarged and divided into two portions – dorsal and ventral ones; the ventral portion is more or less similar to the eye of females (FIG.4:B–C). The dorsal portion of the male eye is the largest in imago, can be smaller in subimago, smaller in mature larva and absent in young larva. In the most primitive case (which is characteristic for the majority of mayflies) the division of male eye into two portions is only slightly expressed (FIG.4:B). Evolutionary changing of male eyes took place in two opposite directions, independently several times in each direction: (1) in some cases the dorsal portion of male eye is strongly enlarged, separated from the ventral portion and transformed to **turban eye** (in Turbanoculata and some Leptophlebia/fg1); (2) in other cases the dorsal portion is diminished or disappears at all, thus male eye becomes similar to that of female (see Index of characters [2.1.3]).

ANTENNAE

Antenna of Amyocerata consists of **scapus** (first muscle-bearing segment), **pedicellus** (second, muscle-less, sensory segment) and **flagellum** (third, muscle-less, secondarily segmented part). Structure of mayfly antennae is primitive for Amyocerata: flagellum is bristle-like and consists of indeterminate number of segments, which become narrower toward apex of flagellum and multiply by division of the proximalmost segment. Larval antennae are well-developed, multisegmented (FIGS 3:A; 76:B), while in imago and subimago flagellum is vestigial, segmentation of flagellum often being indistinct or absent (FIG.4). Reduction of antennae in winged stages is probably an autapomorphy of Euplectoptera: long multisegmented antennae are reported for winged *Protereisma* (Carpenter 1933, 1979).

MOUTH APPARATUS
(Fig. 3)

Mouth apparatus is developed in larva, and all features of mouthparts described below relate to larva only. In imago and subimago of Euplectoptera mouth apparatus is always absent; anterior margin of the frons forms a projected lamella – **face fold** (FIG.4:B); this lamella is usually directed ventrally and limits anteriorly a concavity, which corresponds to the area of clypeus and mouth apparatus. Sometimes more or less developed soft non-functional processes arise from this concavity (FIG.50:C); these processes are not vestigial mouthparts of the imago (as some investigators assumed), but are remainders of larval mouthparts that had not disappeared during metamorphosis: their structure repeats that of specialized larval mouthparts of the same specimen. Thus, in phylogenesis these processes originated independently many times, and their presence is not a plesiomorphy. In larva of the last instar, subimaginal mouthpart tissues located under larval cuticle diminish gradually, thus, at moult, only empty cuticle sheds from them (in contrast to tergalii and sometimes paracercus, which cuticle sheds together with remainders of tissues). Absence of mouth apparatus in winged stages is probably an autapomorphy of Euplectoptera, as developed sclerotized mouthparts are reported for *Protereisma* (Carpenter 1979).

Mandibles

The majority of Ephemeroptera have a mandibular structure (FIG.3:F–G) which is probably initial for Mandibulata and occurs in various groups of eucrustaceans. Similarly to Eucrustacea and wingless insects (Entognatha and Triplura), but differing from Metapterygota, the mandibular basis is long, so the **posterior condylus** is far from the biting edge. In Ephemeroptera, besides the posterior one, two more condyli are present, being situated in one line: a **middle condylus** has a form of concavity on mandible, into which a projection of margin of the head capsule enters (it corresponds to the anterior condylus of Metapterygota); an **anterior condylus** has a form of sclerotized projection of mandible, which enters into concavity on the head capsule margin. Here we shall use the term "**mandible flatness**" for the flatness, in which lie all three mandible condyli and incisor.

In many mayflies the mandible has well-expressed incisor, kinetodontium, prostheca and mola.

Incisor (usually called "apical canine") represents an apical-median process of mandible, usually pointed and dentate; in contrast to kinetodontium, it is never separated from the mandible corpus by a suture.

Kinetodontium (usually called "subapical canine") is the same as lacinia mobilis of eucrustaceans. The term "kinetodontium" proposed by Kukalova-Peck (1991:151) is more convenient for transliteration from Latin to other languages, than the old term "lacinia mobilis". The kinetodontium represents a median process of mandible proximad of the incisor; base of the kinetodontium closely adjoins the incisor base. Like the incisor, the kinetodontium is usually pointed and dentate. In contrast to the incisor, the kinetodontium is often separated from mandible corpus by a suture; sometimes it has mobile articulation with the mandible corpus (FIG.26:C); but in some mayflies, as well as in majority of insects, the kinetodontium is completely fused with the mandible corpus, sometimes it is fused also with the incisor (FIG.29:B). Besides insects, the kinetodontium is present in some Eucrustacea – Peracarida, Thermosbaena/fg1 and Remipedia; probably, it is initial for Mandibulata (Kluge 1999d, 2000). In all cases the kinetodontium never has muscle and is unable to make active movements.

Prostheca is located on median margin of mandible proximad of the kinetodontium; this is an appendage separated by a suture from the mandibular corpus. Usually the prostheca is short and distally divided into a bunch of setiform processes (FIGS 3:F–G; 26:C); sometimes (in many Turbanoculata) prostheca has a form of integral stick with dentate apex (FIG.29:B); sometimes the prostheca is vestigial or lost (FIG.54:E–F) (see Index of characters [1.1.24] and [1.1.25]). Many authors (Snodgrass 1935, and others) erroneously took the prostheca for lacinia mobilis (i. e. kinetodontium); actually it is probably a result of fusion of group of setae. Because of this error in identification of the kinetodontium in mayflies (and insects in general), there were stated doubts concerning the possibility of comparing mandibles of Hexapoda with mandibles of Eucrustacea, that led to doubts concerning common origin of mandibles and monophyly of Mandibulata.

Mola represents a proximal-median projection of the mandible; distal surface of the mola, faced toward the mola of opposed mandible, has a form of grater with dense dentate ridges stretching perpendicular to the mandible flatness. Probably such mola, as well as the incisor and the kinetodontium, is initial for Mandibulata: it is present in many Eucrustacea and Hexapoda. The overwhelming majority of mayflies have mola well-developed, only in some specialized carnivorous mayflies has the mola lost its grater, become dentate, or is completely lost (see Index of characters [1.1.26]).

Mandibles are asymmetrical. As well as in other Hexapoda (and probably in other Mandibulata in general), in Ephemeroptera the mandible with mola most projected in its distal part, is the left mandible, and the mandible with mola most projected in proximal part, is the right one (FIG.3:F–G). An exception is made only by selected taxa (supra-species taxa, species and infra-species taxa) among Pentamerotarsata, which mandibles look as mirror reflection of the normal ones. In rare cases the asymmetry of mandibles is lost (see Index of characters [1.1.17]).

Superlinguae

A pair of well-developed superlinguae is present in nearly all mayflies, except for a few highly specialized carnivorous groups where superlinguae are reduced (see Index of characters [1.1.27]). Superlinguae are a pair of non-segmented appendages belonging to mandibular segment and situated between mandibles and maxillae (FIG.3:A, C) (they are often erroneously regarded as lateral parts of hypopharynx, while hypopharynx belongs to maxillary segment). Probably superlinguae are characteristic for Mandibulata, being known in Eucrustacea under names "paragnatha" or "labium"; among Hexapoda superlinguae are developed, besides Ephemeroptera, only in Entognatha and Microcoryphia (in other Hexapoda they are not described or described erroneously, when lateral parts of hypopharynx are taken for superlinguae). Thus, the presence of superlinguae in Ephemeroptera is a unique plesiomorphy among Pterygota.

Maxillae

Maxilla of Ephemeroptera always has only one biting lobe (FIG.3:B), which is regarded by many authors to be a result of complete fusion of galea and lacinia (which are initially peculiar for Hexapoda); but another assumption is possible, that this lobe is lacinia without galea, while galea is completely lost.

Apex of maxilla bears **maxillary canines** (the term introduced by Kluge 1994c: 35, type of the term is *Habrophlebiodes americana*, designated here) – tooth-like processes, which are not separated from corpus of maxilla. Usually there are 3 maxillary canines (FIG.3:E), and probably this number is initial for Ephemeroptera; sometimes the number of maxillary canines is less than three, or they are lost (see Index of characters [1.1.33]); sometimes maxillary canines have additional denticles.

In various non-related groups of Ephemeroptera the distal margin of the maxilla (laterad of the maxillary canines) bears a regular row of more or less pectinate setae directed distally or ventrally; here this row is called an **apical-ventral row** (see Index of characters [1.1.31]).

The inner (median, or biting) margin of maxilla proximad of the canines, nearly in all Ephemeroptera bears 2 longitudinal rows of setae – the **inner-dorsal** and the **inner-ventral** rows. Setae of these rows can be modified in variable manner; some setae can be thickened, immovable and tooth-like. Modified setae in distal part of the inner-dorsal (but not inner-ventral!) row are named **dentisetae** (the term introduced by Kluge 1994c: 35, type of the term is *Habrophlebiodes americana*, designated here). Number and structure of the dentisetae is constant for large taxa (see Index of characters [1.1.37]–[1.1.40]).

Maxillary palps are 3-segmented (FIG.3:B). The 1st segment contains two muscles – adductor and abductor of 2nd segment; the 2nd segment lacks muscles. Sometimes the 2nd and the 3rd segments are fused together; in this case maxillary palp is 2-segmented, with distal segment representing 2nd+3rd segment (see Index of characters [1.1.42]). Only in one taxon – Ameletopsis/fg1 – maxillary palp is secondarily multisegmented (but has no muscles) (FIG.34:D). In some taxa maxillary palp lacks muscles and can be vestigial up to complete disappearance (see Index of characters [1.1.41]).

The 3-segmented maxillary palp is an apomorphic character (being autapomorphy of Euplectoptera, or Euephemeroptera, or probably Panephemeroptera in general), but not a unique apomorphy of this taxon. Initial for Amyocerata, and probably for Hexapoda in general, is 5-segmented maxillary palp: it is characteristic for Zygentoma, Polyneoptera, Zoraptera and many Oligoneoptera. In many other Amyocerata, as well as in all Entognatha, the number of maxillary palp segments is diminished, more rarely increased.

Labium

Majority of Ephemeroptera have typical for Hexapoda labium structure with division of unpaired portion to **submentum** (sometimes named **postmentum**) and **mentum** (sometimes named **prementum**), bearing paired **glossae**, **paraglossae** and **palps**; in some taxa glossae and/or paraglossae are fused (see Index of characters [1.1.50], [1.1.52]).

Labial palps are 3-segmented (FIG.3:D). The 1st segment contains two muscles – adductor and abductor of 2nd segment; the 2nd segment contains a single muscle – adductor of 3rd segment (sometimes this muscle is absent). Sometimes the 2nd and the 3rd segments are fused together, in this case maxillary palp is 2-segmented, with distal segment representing 2nd+3rd segment (see Index of characters [1.1.55]); only in one taxon – Ameletopsis/fg1 – labial palp is secondarily multisegmented (FIG.35:A).

In contrast to maxillary palp (see above), 3-segmented labial palp is a plesiomorphy within Amyocerata and probably within Hexapoda in general: it is characteristic for Microcoryphia, Polyneoptera, Zoraptera, and many Oligoneoptera. In other taxa of Amyocerata, as well as in all Entognatha, number of labial palp segments is diminished, more rarely increased.

THORAX

GENERAL STRUCTURE

Thorax of Ephemeroptera is integral, mobility between three thoracic segments and first abdominal segment is limited or lost because of following modifications:

In larva posterior-lateral angles of pronotum and anterior-lateral angles of mesonotum are brought together (FIG.3P:A) or even fused (FIG.37:A), and articulatory membrane is well-developed only in median part of pronotum-mesonotum joint; because of this, prothorax can make only limited dorsoventral movements relatively to mesothorax. Imaginal and subimaginal mesonotum strongly differs from larval one, has no anterior-lateral angles and no direct connection with pronotum (FIGS 4–6); however, a nearly immobile connection of prothorax and mesothorax is served by means of prealar bridge (see below).

Both in larva and winged stages sterno-pleural areas of mesothorax and metathorax are connected immobile. Furcasternum of metathorax is completely fused with first abdominal sternite, without any trace of suture between them (while suture between metanotum and first abdominal tergite is retained) (FIGS 4–5; 35:A).

Probably, initially for Pterygota, thoracic segments have following apodemes: paired **furca** (or **sternal apodemes**) in each segment; paired **pleural apodemes** in each segment; **unpaired spina** behind furca in prothorax and mesothorax only. In Ephemeroptera spinae are completely lost. In connection with this, most sternal thoracic muscles are lost, and only muscles inserted on furcae are retained; muscles connecting mesothoracic and metathoracic furcae are also lost. Pleural apodemes and muscles connected with them are also lost on all segments (in larvae of Furcatergaliae propleura are transformed to secondary apodemes, which are not homologous to the pleural apodemes of other Pterygota).

PTEROTHORAX OF IMAGO AND SUBIMAGO
(Figs 5–6)

Structure of mayfly pterothorax is discussed in the separate paper (Kluge 1994a); for all terms that are introduced in that paper and marked there as "new term", the type taxon should be *Siphlonurus aestivalis* (designated here).

Prealar bridge

Prealar bridge of mesothorax (PAB) represents a sclerotized ring, which firmly connects anterior end of mesonotum with anterior end of mesosternum; the stenothoracic spiracle (anterior most spiracle of Hexapoda, initially located on the boundary between prothorax and mesothorax) is located behind the prealar bridge – i. e. in limits of the mesothorax. In this respect the prealar bridge of mayflies differs from the prealar bridge of some other Pterygota, in which it passes behind the spiracle (the prealar bridge is not present in all Pterygota and probably independently evolved in various groups). The prealar bridge of mayflies consists of the dorsal, lateral, and ventral arcs. The **dorsal arc** (PAB:DA) (term by Kluge 1994a) may contain the anterior and posterior costae separated by a groove – **anterior phragma** (PhA); so the anterior costa of the dorsal arc belongs to acrotergite, and its posterior costa – to notum. The **lateral arc** of prealar bridge (PAB:LA) (term by Kluge 1994a) may also consist of two or three costae separated by grooves. The **ventral arc** (PAB:VA) (term by Kluge 1994a) is known also as **presternite**. From the lateral and dorsal arcs begins a pair of **posterior arms** of prealar bridge (PAB:PA) (term by Kluge 1994a); each of these arms goes posteriorly toward the wing base and joint with a small distinct emargination on the lateral margin of prelateroscutum (PLS – see below). Anteriad of this joint, between the prealar bridge, the posterior arm, and the prelateroscutum, is located a narrow membranous area. Only in Branchitergaliae the posterior arms of prealar bridge are strongly shortened, do not reach the emarginations of prelateroscutum margins, while these emarginations are retained (FIG.45:A, C).

Mesonotum

In anterior part of mesonotum just behind the dorsal arc of prealar bridge there is an unpaired **anteronotal protuberance** (ANp) (term by Kluge 1994a) separated from the remainder part of notum by the **anteronotal transverse impression** (ANi) (see Index of characters [2.2.5] and [2.2.6]).

Along the median line of notum goes the **median** (or **median longitudinal**) **suture** (MLs). It is distinctly developed along the largest part of notum, but disappears in its anterior part (usually near the anteronotal impression) and in its posterior part (usually near the scuto-scutellar impression). Usually the median suture is concave, but in selected Tetramerotarsata it is convex.

Laterad of the median suture there is a pair of **medioparapsidal sutures** (MPs) (term by Kluge 1994a). These narrow concave sutures separate unpaired convex **medioscutum** (MS) (term by Kluge 1994a) (which contains the anterior bases of the pair of largest **median tergal muscles** – MTm) from paired convex **submedioscutum** (SMS) (term by Kluge 1994a) (which contains the dorsal bases of the pair of large **scuto-episternal muscle** – S.ESm).

Laterad of the medioparapsidal sutures there is a pair of **lateroparapsidal sutures** (LPs) (term by Kluge 1994a) (see Index of characters [2.2.9]). These deep wide strongly sclerotized concave sutures, or furrows, bear mechanical function and at the same time separate the submedioscutum from the paired convex **sublateroscutum** (SLS) (term by Kluge 1994a), which contains the dorsal bases of the **anterior** and **posterior scuto-coxal muscles** (S.CmA and S.CmP) (see Index of characters [2.2.10]). Lateroparapsidal sutures can go exactly between muscle bases or somewhat touch them (FIGS 56:L–M; 63:D–E).

Anteriorly the lateroparapsidal suture turns to **antelateroparapsidal suture** (ALPs) (term by Kluge 1994a), which sets off anteriorly the submedioscutum, separating it from the **anterolateral scutal costa** (ALSC) (term by Kluge 1994a). The anterolateral scutal costa is well developed in all mayflies, separating the dorsal side of notum from the narrow **prelateroscutum** (PLS) (term by Kluge 1994a), which is usually not visible from above. As said earlier, prelateroscutum usually has articulation with the hind end of the posterior arm of prealar bridge. Posteriorly prelateroscutum is connected with suralare (SrA), sublateroscutum (SLS) and lateroscutum (LS), which can be separated by more or less developed sutures or ridges of various forms.

Suralare (SrA) is a portion of scutum which bears the **anterior notal wing process**; it can be separated from the remainder scutum by the **anteronotal scutal suture** (ALSs) (Matsuda 1970: Fig.4).

Lateroscutum (LS) (term by Kluge 1994a) is separated from the sublateroscutum by the **lateroscutal suture** (LSs) (term by Kluge 1994a) (see Index of characters [2.2.12]) and contains in its anterior portion the dorsal base of the **scuto-trochanteral muscle** (S.Trm).

Posteriad of the sublateroscutum is usually present a pair of **posterior scutal protuberances** (PSp) (term by Kluge 1994a) – large convex areas, usually indistinctly outlined, which contain the dorsal bases of large scuto-lateropostnotal muscles (S.LPNm) (see Index of characters [2.2.11]).

Behind the posterior scutal protuberances, is situated the prominent **scutellum** (SL) (term adopted by Audouin 1824), which is separated from the posterior scutal protuberances by a shallow **scuto-scutellar impression** (SSLi). Laterally scutellum is separated from parascutellum (PSL) by an indistinct invagination, which is called **recurrent scuto-scutellar suture** (RSSLs) (Matsuda 1970).

Parascutellum (PSL) (term used by Crampton 1914) is a large area laterad of scutellum, which bears the **posterior notal wing process** and contains a single small base of the **parascutellar-coxal muscle** (PSL.Cm). Parascutellum is separated from sublateroscutum and lateroscutum by the **scuto-parascutellar suture** (SPSLs). This suture allows to bend the notum when the median tergal muscles contract, that leads to wing depression; it is well developed on mesothorax of all mayflies. Lateral margin of parascutellum bears a sclerotized costa – **parascutellar lateral convexity** (PSLcvx), which is separated from the remainder part of parascutellum by a groove – **parascutellar lateral concavity** (PSLccv).

Behind the notum, winged stages have a sclerotized **postnotum**, which corresponds to an intersegmental articulatory membrane of larva. Postnotum consists of an unpaired infrascutellum, unpaired mediopostnotum and a pair of lateropostnota. The **infrascutellum** (ISL) (term by Kluge 1994a) represents a transverse shelf-like sclerotized convexity located on the deeply concave hind wall of notum under scutellum (FIG.89:A); usually it is separated from mediopostnotum by a transverse membranous suture. Laterally infrascutellum can be produced as a pair of **infrascutellar-postsubalar arms**, which unite it with posterior-dorsal angles of postsubalar sclerites belonging to lateropostnota. Sometimes infrascutellum is reduced (see Index of characters [2.2.13]). The **mediopostnotum** (MPN) lies behind infrascutellum, and continues posteriorly-ventrally as an anterior wall of **middle phragma** – i. e. phragma between mesonotum and metanotum (thus it is also called **phragmanotum**); laterally mediopostnotum is continued as a pair of **lateropostnota** (LPN), uniting there with the infrascutellar-postsubalar arms. About the structure of lateropostnotum see below, in characteristic of mesopleuron.

Besides the sutures whose position is fixed by their mechanical role or by position of muscle bases, there is a suture whose position is not determined by any internal causes – it is the **mesonotal suture** (MNs) (term by Kluge 1994a). In the primitive case the mesonotal suture goes across scutum in its anterior part, behind the anteronotal transverse impression, and laterally connects with the anterior ends of medioparapsidal sutures (FIG.61:A–B). Sometimes the mesonotal suture is stretched backward medially in its point of crossing with the median suture (FIG.6). In other cases lateral parts of the mesonotal suture are strongly curved and stretched backward (FIG.7:E). Sometimes these lateral portions of mesonotal suture are so strongly shifted backward, that nearly reach the posterior scutal protuberances; in this case it seems that there is not a single suture, but two pairs of longitudinal sutures, the median of which goes parallel to the median suture, and the lateral ones go parallel to the lateroparapsidal sutures close to them (FIG.83:F). Sometimes the mesonotal suture, being strongly curved and stretched backward, is indistinct in imago, and can be seen only in subimago because in front of it is located a pigmented field with microtrichia, and behind it – a light field without microtrichia. Sometimes such mesonotal suture is non-expressed both in imago and subimago. In other cases the mesonotal suture disappears without curvation and stretching backward (see Index of characters [2.2.8]).

In literature when structure of the insect thorax is described, the terms "praescutum", "praescutal suture" and "parapsidal suture" are often used, whose meanings are initially indeterminate (Kluge 1994a). The term "**praescutum**" was introduced by Audouin (1824), and as its type should be regarded the beetle *Dytiscus circumflexus*, because only its structure is illustrated. Originally, on the beetle mesothorax the term "praescutum" was attributed to the anterior phragma, while on the beetle metathorax the same term was attributed to the medioscutum. The term "**parapsides**" (in plural) was introduced by MacLeay (1830) for a pair of lateral lobes of mesonotum in the vesp *Polistes billardieri*; later pair of sutures separating these lobes were called "**parapsidal sutures**". Among Hymenoptera some species have the parapsidal sutures, some species have notaulici – another pair of sutures, which correspond to the medioparapsidal sutures of mayflies, and some species have both pairs of sutures – the parapsidal sutures and the notaulici (Tulloch 1929). Authorship of the term "**notaulix**" (plural "**notaulici**") or "**notaulus**" (plural "**notauli**") is unclear.

Sclerotization of mesonotum is markedly different in subimago and imago. Imaginal mesonotum is nearly evenly sclerotized (if there are colour patterns, they have hypodermal origin); subimaginal mesonotum has distinctly outlined sclerotized pigmented areas and light areas between them. Intensity of pigmentation of these sclerotized areas can strongly vary individually, but their shape allows to characterize supra-species taxa (see Index of characters [2.2.14]–[2.2.15]). Usually there is an unpaired **anterior pigmented area,** limited from behind by the mesonotal suture, and a paired **lateral pigmented area.** Probably, initially the lateral pigmented area bifurcates backward, forming a **lateroparapsidal stripe,** which stretches along the lateroparapsidal suture, and a **lateral portion,** which occupies antero-lateral part of sublateroscutum and whole lateroscutum (FIG.18:E). Such shape of the lateral pigmented area is peculiar for selected groups both among Tridentiseta (Siphlonurus/fg1, Vetulata, *Siphluriscus*) and Branchitergaliae (*Coloburiscoides*, Heptagennota). In other taxa lateral pigmented area is larger, occupying sublateroscutum and sometimes other areas.

Wing base

Wing base (FIG.6) is connected with lateral margin of notum by two movable sclerites: the **anterior axillary sclerite** (AxA) (term used by Becker 1954: aAx) and **posterior axillary sclerite** (AxP) (term by Kluge 1994a). Both of them are movably connected with the wing base and with notal wing processes: AxA with the anterior notal wing process of suralare, and AxP with the posterior notal wing process of parascutellum. In the FIG.6 wing base is shown stretched with axillary membrane torn, so the both movable sclerites AxA and AxP are visible; on intact wing base one of these axillary sclerites is turned over and appears under the corresponding wing process, while another one is stretched. It allows the wing to move forward (when AxA is turned over) and backward (when AxP is turned over). AxA is flat, not so strongly sclerotized as AxP, its form differs among mayfly taxa. The **proximal axillary sclerite** (AxPr) (term by Kluge 1994a) may be either well-developed (FIG.6), or vestigial, or absent. It is connected with lateroscutum and can not make such movements as AxA and AxP. The **middle axillary sclerite** (AxM) (term by Kluge 1994a) is movably connected with AxA and with the **middle articulatory process** of basal plate (APM) (term by Kluge 1994a). Form of AxM is similar in all mayflies; it has in its posterior part a distinct projection directed medially. The **basal plate of wing** represents a large roundish sclerite convex dorsally and concave ventrally; it consists of immovably fused together **basisubcostale** (BSc), **basiradiale** (BR), the **middle articulatory process** (APM) and the **posterior articulatory process** (APP).

Mesopleuron

Lateral surface of mesothorax has following structure. The most developed suture is a suture composed of the dorsal part of the pleural suture (i.e. **superior pleural suture** – PLsS) and the anterior part of the paracoxal suture (i.e. **anterior paracoxal suture** – PCxA); this combined suture represents a deep wide sclerotized groove running from the **pleural wing process** (PWP) to the episternum; it prevents the pleuron from deformation during contracting of the scuto-episternal muscle (S.ESm). Judging by the form of the anterior paracoxal suture in various mayflies, we can assume that the plesiomorphy is the condition when it is complete, i. e. crosses the whole episternum, completely dividing it to **anepisternum** (AES) and **katepisternum** (KES), turns to its ventral side and reaches the sternite. In some taxa the anterior paracoxal suture is incomplete, i. e. does not turn to the ventral side of episternum and does not divide it completely (see Index of characters [2.2.19]). The remaining parts of the pleural and the paracoxal sutures, i.e. the **inferior pleural suture** (PLsI) and the **posterior paracoxal suture** (PCxsP) are weak and sometimes disappear. In contrast to mayflies, in majority of other Pterygota the mostly developed suture of the pleurite is the whole pleural suture, running from the pleural wing process to the dorsal coxal articulation, and dividing the pleurite into **episternum** and **epimeron** (see Index of characters [2.2.20]). **Subalar sclerite** (SA) is usually large, with its lower portion containing the dorsal base of large **subalar-sternal muscle** (SA.Sm) (see below). The portion of **lateropostnotum** (LPN) situated exactly under the wing base is named **postsubalar sclerite** (PSA) (= posterior subalare: Crampton 1914). Posterior-dorsal angle of the postsubalar sclerite can continue dorsally as an infrascutellar-postsubalar arm (see above). Ventrad of the postsubalar sclerite, along the lateropostnotum in dorsoventral direction usually runs a **lateropostnotal crest** (LPNC) (term by Kluge 1994a). Often outlines of the postsubalar sclerite and the lateropostnotal crest are most distinctly expressed in subimago (see Index of characters [2.2.16]).

Mesosternum

Sternite of mesothorax in Ephemeroptera has an especially strongly developed **furcasternum** (area behind furcal pits); this is connected with the fact that in contrast to Neoptera, it includes bases of large **subalar-sternal muscles** (SA.Sm). The portions of furcasternum, which contain the bases of SA.Sm are strongly convex and are named **furcasternal protuberances** (PSp). The furcasternal protuberances may be brought together (FIG.5:C) or separated by means of **furcasternal longitudinal impression** (FSi). Form of this impression depends upon the position of bases of the subalar-sternal muscles (SA.Sm), while their position depends upon structure of the nerve system. In primitive cases the metathoracic nerve ganglion is located in metathorax, being connected with the mesothoracic ganglion (located in mesothoracic basisternum) by a pair of long slender connectives, which lie at some distance of body wall, thus allowing the bases of SA.Sm to connect medially (FIG.8:B); in this case furcasternal impression is absent (FIG.39:B) or represented by a slender line (FIG.23:E). In some mayfly taxa the metathoracic nerve ganglion is transferred into furcasternum of mesothorax, nearer to the mesothoracic ganglion, and lies between the bases of SA.Sm separating them (Fig.8:D); in this case between the furcasternal protuberances appears a more or less wide furcasternal impression (FIGS 32:D; 34:C). If the metathoracic ganglion is located in the hind part of mesothoracic furcasternum, the furcasternal impression is narrow in its fore part and widened posteriorly (FIG.56:C–D); if the ganglion is transferred into the middle or anterior part of furcasternum, the furcasternal impression becomes wide all over its length (FIG.57:A–B) (see Index of characters [2.2.23]–[2.2.24]). Among Ephemeroptera only in Caenoptera are the subalar-sternal muscles completely lost, but even in this case the furcasternal protuberances are retained, being diminished and widely separated (FIG.87:F).

Metathorax

Metanotum of all Euplectoptera is diminished in connection with anteromotority and diminishing or disappearance of hind wings. Relatively complete development of metathoracic structures is shown in FIG.5. In some mayfly taxa the metathorax is more strongly reduced: pleural wing process and subalar sclerite may disappear; alinotum (scutum + scutellum) becomes shorter while mediopostnotum may become longer, or the whole metathorax becomes shorter. The wing indirect musculature of metathorax may be nearly completely developed (FIG.5:B) or more or less reduced. In metathorax of all Euplectoptera the direct wing depressor – subalar-sternal muscle (SA.Sm) – is lost. In different taxa reduction of hind wings, metathoracic exoskeleton, and metathoracic wing musculature has unequal rate. For example, many Turbanoculata have no vestiges of hind wings or hind wing buds at any stages of development, however their metathorax is rather large and contains very strong wing musculature, which can not function (FIG.8:D); in some other mayflies the hind wings are relatively large, but the metathorax is strongly shortened and its wing musculature is very weak (for example in Posteritorna – FIG.16:H). Most constant metathoracic wing muscles are the median tergal muscle (MTmIII) and scuto-epistenal muscle (S.ESmIII); they undergo reduction only in Caenoptera and Tricorythodes/fg1 (see Index of characters [2.2.26]).

WINGS
(Figs 7–8)

Nearly in all mayflies fore wings are well developed, and length of fore wing is subequal to trunk length (FIG.8:A, C) (because of this, in taxa characteristics fore wing length should be given rather than body length). In contrast to many other insects, in mayflies wing length never exceeds markedly trunk length. The reason is that mayflies have to moult from subimago to imago and shed subimaginal exuviae by abdominal movements. If during the moult imaginal abdominal tip becomes free from subimaginal cuticle earlier than wing tips, the wing tips remain in the subimaginal cuticle forever; such insect can not fly and dies.

Only a few mayfly species are flightless and can have fore wings shorter than trunk (Fig.8:F) (see Index of characters [2]).

In Euplectoptera the hind wings are reduced, their length never exceeds 1/2 of fore wing lengths; in flight they are coupled with fore wings, because the basitornal (hind-proximal) margin of fore wing is bent ventrally, and the costal (fore) margin of hind wing is bent dorsally; in some mayflies the hind wing bears a special **costal process**. In many groups of Euplectoptera independent reduction of hind wings takes place up to their complete disappearance (see Index of characters [2.2.59]).

Fore wing usually has characteristic triangular form with more or less prominent obtuse hind angle – **tornus**; this angle separates the hind-proximal portion of wing margin, which couples with hind wing, from the rest forewing margin (the same in many other non-related anteromotoric insects which fore wing is able to couple with hind wing).

In entomological literature wing margins are usually called "anterior" (or costal), "outer", and "posterior" ones. Such terminology is not convenient when used for wing buds of Ephemeroptera larva: in this case the margin of wing bud corresponding to outer margin of wing is directed inward, and the margin of wing bud corresponding to posterior margin of wing is directed anteriorly. Below, the following terms are used: **Costal margin** – anterior margin of wing and lateral (or ventral) margin of wing bud, from base to apex. **Basitornal margin** (new term) – hind-proximal margin of fore wing and anterior margin of fore wing bud, from base to tornus. **Tornoapical margin** (new term) – outer (hind-distal) margin of fore wing and median (or dorsal) margin of fore wing bud, from tornus to apex. **Amphitornal margin** (new term) – basitornal and tornoapical margins combined, independently if the tornus is expressed or not.

In winged stages (imago and subimago) at rest the wings are never folded; usually they are raised upwards but some mayflies keep their wings spread laterally.

Larval wing buds

In all Pterygota larval wing buds represent immobile outgrowths of notum margin (i. e. paranota), appear in certain larval instar (but never in the first instar) and subsequently transform to adult wings. In recent mayflies larval wing buds arise from the posterior margin of notum and are directed by their apices posteriorly, by costal margin laterally-ventrally, and by dorsal surface dorsally-laterally (FIG.3:A) – thus, they have the same pose as folded wings of Neoptera, while adult mayfly wings never can strike such an attitude.

Based on wing buds position of recent mayflies and on Handlirsch's reconstruction of Permian *Phtharthus* (in which posteriorly directed wing buds were shown), some authors believed that this was the initial position of wing buds, and even assumed that insect wings evolved from outgrowths of posterior margin of the notum.

Actually the most primitive insects, including Permian mayflies – *Protereisma* – have wing buds arising not from posterior, but from lateral margins of the notum (FIG.14:D). All three specimens, on which the description of *Phtharthus* was based, have no wing buds preserved (FIG.14:C) (that is rather strange, because usually wing buds are well-preserved on fossils, and all three specimens of *Phtharthus* have well-preserved meso- and metanotal relief typical for Pterygota).

Among recent mayflies, in the primitive case wing buds are attached to the body only by their bases (Fig.25:A). Hind wing buds always retain this condition, but fore wing buds can be more strongly fused with mesonotum: in many taxa basitornal margins of fore wing buds are fused with posterior margin of mesonotum; in some taxa tornoapical margins are also partly or completely fused with notum or one with another (see Index of characters [1.2.5]). Even being strongly integrated with notum, the wing bud retains its outline as a relief line on the surface of the notum, and when the subimaginal wing develops, it is crumpled inside this outline; only in Posteritorna are wing buds completely integrated with notal shield (FIG.15).

Wing venation

In the larva wing venation appears at the earliest stages of development of wing buds as a net of lacunas (canals) inside the wing bud. Sometimes certain or all veins are visible as convexities on surface of the wing bud (Fig.3:A). Larval wing bud venation matches imaginal wing venation (FIGS 37:A; 75:A–B); in exceptional cases larval venation can be even more complete than imaginal one [see Geminovenata (3)]. Some authors mix veins and tracheae, which penetrate into some of the veins, which leads to wrong conclusions on vein homology.

Homology and nomenclature of insect wing veins is a subject of long-term discussion. Comstock and Needham (1898–1899 and later publications) proposed a universal usage of insect vein abbreviations **C** (costa), **Sc** (subcosta), **R** (radius), **M** (media), **Cu** (cubitus), **1stA** (first analis), **2ndA** and **3rdA**; their R divides into $\mathbf{R_1}$ and **Rs** (radius sector); these names were taken from older literature, where they were differently used for different groups of insects. Recently Comstock's interpretation is most widely accepted for wing venation of many insect groups, but not Ephemeroptera. For Ephemeroptera, the most generally accepted vein abbreviations are C, Sc, R, MA, MP, CuA, CuP, 1A. The names **MA** (media anterior), **MP** (media posterior), **CuA** (cubitus anterior) and **CuP** (cubitus posterior) were introduced by Martynov (1924), but their recently used interpretation for Ephemeroptera was suggested by Tillyard (1932).

Possibly most of the veins in wings of mayflies and other Palaeoptera are not homologous to any vein in Neoptera: in Palaeoptera each longitudinal vein is either convex or concave, and can not change this feature in course of evolution; in Neoptera a homologous vein in various representatives can be convex, concave or neutral. In order to avoid confusion, probably it would be expedient to use Comstock's vein nomenclature for Neoptera only, designating a stonefly *Nemoura* sp. (Comstock & Needham 1898, p.238, Fig.8) as a type taxon for the vein names C, Sc, R, Rs, M, Cu, 1A, 2A and 3A.

For Ephemeroptera, here are used following names: **C** and **Sc** – both homologous to that of Neoptera; **RA** (term by Kukalova-Peck 1983) – a separate vein homologous to Comstock's R-R_1 of Neoptera; **RS** – a separate vein, which homology with Rs of Neoptera is unclear; **MA, MP, CuA, CuP** – four veins, possibly not homologous to branches of M and Cu of Neoptera; **AA** and **AP** (terms by Kukalova-Peck 1983) – two veins corresponding to Tillyard's 1A and 2A, possibly not homologous to 1A and 2A of Neoptera. Tillyard regarded RA (=R_1) and RS to be secondarily separated branches of the same vein R, and because of this supplied branches of RS with numbers 2, 3 and 4+5. Here branches of RS are supplied with letters "**a**" (anterior) and "**p**" (posterior) and numbers (see below), to avoid confusion with the Tillyard's numbers (Kluge 2000).

In Ephemeroptera convex and concave veins are alternating forming triads. The **triad** is such a form of branching, when a convex vein is branched to two convex branches with a concave intercalary between them, and a concave vein is branched to two concave branches with a convex intercalary between them (such triads are characteristic for Subulicornes, i. e. Odonata + Ephemeroptera). Veins Sc (concave, as in other Pterygota) and RA (convex, as in other Pterygota) are non-branched, at least on fore wings go parallel to the costal margin (which is armed by the costal vein), reaching the wing apex. On fore wing distal part of the field between C and Sc has membrane slightly thickened and, thus, represents a **pterostigma**. Veins Sc and RA are firmly fused with a sclerotized plate in wing base; near wing base C, Sc and RA are connected together by a **costal brace** (see Index of characters [2.2.29]). Other veins have soft bases or are secondarily firmly fused with the base of RA. In Euephemeroptera RS and MA are fused in proximal part. Vein RS is concave and is branched forming subordinate triads: its first triad contains concave branches **RSa** and **RSp** and a convex intercalary **iRS**; RSa forms a second triad, which contains concave branches **RSa$_1$** and **RSa$_2$** and a convex intercalary **iRSa**; RSa$_2$ forms a third triad, which contains concave branches **RSa$_2$'** and **RSa$_2$"** and a convex intercalary **iRSa$_2$** [on hind wing only the first of these triads is present – see Euplectoptera (1) below]. Vein MA is convex and forms a single triad with convex branches **MA$_1$** and **MA$_2$** and a concave intercalary **iMA**. Vein MP is concave and forms a triad with concave branches **MP$_1$** and **MP$_2$** and a convex intercalary **iMP**. Vein CuA is convex; in Euplectoptera it is either non-branched or has one or several secondary branches arising posteriorly [see below, Anteritorna (1)]. Vein CuP is concave. Vein AA is convex, vein AP is concave; behind them two or more alternating convex and concave veins can be present. In some triads the intercalary vein incorporates basally with one of branches, thus looking not like intercalary, but like a branch; sometimes, vice versa, a branch becomes free in its basis and looks like intercalary; in rare cases some branches and intercalaries are lost (see Index of characters [2.2.32]–[2.2.54]). Sometimes between the longitudinal veins, their branches and intercalaries, there are present additional intercalary veins (see Index of characters [2.2.55]–[2.2.56]). Usually longitudinal veins are connected by large indeterminate number of cross veins (except for a few extremely specialized groups – see Index of characters [2.2.57]).

LEGS
(Figs 3–4; 10)

While dorsally the coxa is always articulated with katapleurite (as in other Hexapoda), ventral coxal articulation is variable among mayflies: Mesothorax and metathorax always lack trochantines, and coxae are articulated either directly to sternite (mesothorax in FIG.52:B), or to movable paired sclerites articulated with sternite (FIG.35 and metathorax in FIG.52: B); non-functional vestiges of these sclerites can be present on prothorax as well (FIG.35). Prothorax can have a pair of trochantines, which serve ventral coxal articulations (FIG.35), or trochantines are lost, and coxa have direct articulation with sternite (FIG.70:A–B).

As well as in all other Hexapoda, the leg of Ephemeroptera consists of **coxa**, **trochanter**, **femur**, **tibia** (sometimes called metatibia – see below), **tarsus** (see below) and **pretarsus** (i. e. claw or claws – see below).

Tibia

Probably the **tibia** (or **metatibia**) of Hexapoda is formed as a result of fusion of **patella** and **telotibia** (or tibia itself). In the majority of Hexapoda, including all known primary wingless insects (Entognatha and Triplura), fusion of patella and telotibia is complete, without trace of suture between them. But in Ephemeroptera and Odonata vestigial **patella-tibial suture** is retained. This suture is non-functional, patella and telotibia are connected immobile. Patella-tibial suture is strongly oblique, so patella is very short on its outer side, being several times longer on its inner side. On outer side of leg, the patella-tibial suture always has a form of distinct wide transverse concavity; it can be continued on anterior (dorsal) side and sometimes on other sides – in larva in a form of distinct narrow oblique grove, in subimago and imago in a form of indistinct longitudinal-oblique concavity. Sometimes such oblique grove or concavity is absent (everywhere below, the sentence "patella-tibial suture is absent" means that only the concavity on outer side is present).

Most Euplectoptera have patella-tibial suture on middle and hind legs only, while on fore legs it is absent (FIGS 3–4). In selected taxa patella-tibial suture disappears also on middle and/or hind legs (see Index of characters [1.2.18] and [2.2.82]). Only in two non-related taxa (Tridentiseta-Turbanoculata-Anteropatellata and Bidentiseta-Rhithrogena/fg3) the patella-tibial suture has secondarily restored on larval fore legs. Even in the cases when larval fore tibiae have the same structure as middle and hind tibiae, adults often (but not always) retain distinct vestiges of patella-tibial suture on middle and hind legs only.

Such difference of fore leg from middle and hind leg occurs in all principal phylogenetic branches of Ephemeroptera, being present in majority of species, independently of their leg specialization. In contrast to Ephemeroptera, in Odonata patella-tibial suture is equally developed on all legs. This allows one to conclude that reduction of the patella-tibial suture on fore legs only is an autapomorphy of Ephemeroptera (either Euplectoptera, or Euephemeroptera, or Panephemeroptera, as structure of extinct Protephemeroidea and Permoplectoptera is unknown).

Tarsus

Tarsi of Ephemeroptera have peculiar structure. Tarsus is immovable or slightly movable: usually tarsi of middle and hind legs lack adductors and abductors (being moved only by adductor of claw) and tarsus of fore leg has a single adductor; sometimes this muscle is also absent. Tarsus has different structure in larva and winged stages. In winged stages (i. e. imago and subimago) the first tarsal segment is usually immobile fused with tibia, while other tarsal segments are joined mobile (FIG.4).

In contrast to winged stages, larval tarsus (including its first segment) is mobile joined with tibia, but all tarsal segments are immobile fused together. Often larval tarsus is non-segmented, without any traces of segmentation; in Siphlonurus/fg1 and some others, slightly visible traces of tarsal segmentation are retained (FIG.3:A); only in Ameletopsis/fg1 are several (but not all) tarsal segments separated by more or less developed articulations (FIG.35:A). Probably non-segmented larval tarsus is an autapomorphy of Euplectoptera, as for the known larva of Permoplectoptera (*americana* [*Kukalova*]) segmented tarsi are described. This apomorphy is not unique, as non-segmented tarsus occurs also in some other Hexapoda.

In Pentamerotarsata and some other mayflies, imaginal and subimaginal tarsus has 1st segment mobile articulated with tibia (see Index of characters [2.2.84]) and externally looks like primitive insect tarsus (probably movable 5-segmented tarsus is initial for Amyocerata). Because of this, one can think that among Ephemeroptera such a completely segmented tarsus should be a plesiomorphy, and fusion of 1st tarsal segment with tibia – an apomorphy; but in this case we would have to assume, that in different phylogenetic branches of Ephemeroptera the same fusion of 1st tarsal segment with tibia took place independently, while in other insects such tendency is not expressed. It is much more probable that the common ancestor of Ephemeroptera had 1st tarsal segment fused with tibia, while in some taxa it became secondarily separated; in all cases tibio-tarsal muscles remain to be reduced. The restoration of the adult tibia-tarsal joining in some mayflies does not contradict to the principle of irreversibility of evolution, as all mayflies retain mobility of tibia-tarsal joining in larval stage.

Usually winged stages of Ephemeroptera have 5 tarsal segments (including the first segment fused with tibia), but sometimes number of tarsal segments is less than five (see Index of characters [2.2.78] and [2.2.83]). 5-segmented tarsus is probably plesiomorphic within Amyocerata (and possibly within Hexapoda in general), as 5-segmented tarsi occur in many groups of Amyocerata, and number of tarsal segments never increases five.

Pretarsus

Pretarsus of Ephemeroptera has peculiar structure and differs in different stages. As well as in majority of Pterygota and in some other insects, in winged stages of Ephemeroptera pretarsus consists of two claws articulated with a single unguitractor. In majority of mayflies one of these claws (the **anterior** one, if the leg is directed laterally with its knee articulation directed dorsally) is blunt, while another claw (the **posterior** one) has form typical for a claw – pointed, curved and sclerotized. Everywhere below this claw structure is called **ephemeropteroid claws**. Such structure is probably an autapomorphy of Euplectoptera, or Euephemeroptera, or Panephemeroptera; this apomorphy is unique, being never found in other insects. In selected taxa of Euplectoptera both claws are similar – pointed, curved and sclerotized (see Index of characters [2.2.85]). Some authors regarded this structure of pretarsus to be plesiomorphic, because it is the same as in the outer-group – many non-ephemeropterous Hexapoda; but this assumption requires that ephemeropteroid claws appeared independently many times among Ephemeroptera, but never appeared in other insect groups. Much more reliable is the assumption that ephemeropteroid claws appeared once, being an autapomorphy of Ephemeroptera, but all Ephemeroptera retain genetic potentiality to form ancestral pointed claws, and this potentiality is realised independently in some taxa of Ephemeroptera.

In contrast to the winged stages, in larvae of all Euplectoptera the pretarsus consists of a single claw; only on the fore leg of Metretopus/fg1 is the claw bifurcate (Fig.22:C), but this bifurcation probably is not connected with double claws of adults. Probably the single claw is an autapomorphy of Euplectoptera, as for the known larva of Permoplectoptera (*americana* [*Kukalova*]) double claws are described. This apomorphy is not unique, as a single claw occurs in some other insect groups. Some authors believe that the single claw of Ephemeroptera larvae is a plesiomorphic condition, because in many arthropods only a single unpaired claw is present – in Eucrustacea, Diplopoda, Chilopoda, Ellipura and marine Pseudognatha. At the same time, paired claws are secondarily substituted by unpaired claw in larvae of many Oligoneoptera and some other insects. In Ephemeroptera this character also can be a secondary one.

Fore leg of male

Fore leg of the male imago is specialized for grasping female at copulation. It is elongate, usually tibia and tarsal segments are especially long. Articulation of tibia and tarsus has such a construction, which allows to turn the tarsus around at 180^0 (FIG.10:A–F); thanks to this, the tarsus can be arched upward to hold the female wing base at copulation (FIG.10:G). Claws of male imaginal fore legs can have the same structure as claws of other legs; but in some mayflies they have another structure, being blunt (this character appears independently several times – see Index of characters [2.2.77]).

ABDOMEN

GENERAL STRUCTURE

In all stages the abdomen consists of ten segments – condition initial for Hexapoda. Many authors assume that the abdomen of Hexapoda, and particularly that of Ephemeroptera, consists of 11 or 12 segments, regarding some structures at the end of abdomen to be vestiges of segments XI and XII; however, such assumptions are not proved (see below).

In the winged staged each of segments I–IX has tergite and sternite distinctly separated by soft pleura. In the larva the sutures between tergite, pleura and sternite are lost, so borders of these parts of segment can be found only by tracing how inside them the corresponding parts of the subimago develop. Posterolateral angles of abdominal segments are usually stretched forming paired flat denticles or spines; in the primitive case (characteristic for majority of mayflies) such posterolateral spines are larger in larva and smaller in adults, and are the largest on segment IX, being progressively smaller on previous segments; sometimes they are modified or lost.

Abdominal tergites and sternites of the imago and subimago are weakly sclerotized and lack setation (in subimago they are covered by microtrichia – see above). In the larva the abdominal cuticle has the same degree of sclerotization as that on its head and thorax, varying from moderate in the majority of mayflies, to rather hard in some taxa, and often bears peculiar setation. The posterior margin of the larval tergite (and sometimes sternite) is often armed with a regular row of small flat denticles (possibly modified setae), which project posteriorly as a continuation of the tergite surface and overlap the intersegmental membrane; in many cases these denticles are vestigial or absent (for some of them – see Index of Characters [1.3.5]).

TERGALII
(Fig. 13)

In the larva abdominal segments bear paired movable joined appendages – **tergalii** (singular – **tergalius**). In previous publications this term was used either as feminine – "tergaliae" in plural and "tergalia" in singular (Kluge 1989a: 49; 1996: 73), or as neuter – "tergalia" in plural (Kluge 1989a: 77). In order to avoid confusion between plural and singular, gender of the Latin term is now changed to masculine (Tiunova & Kluge & Ishiwata 2003), while in Russian it remains to be feminine ("тергалия" in singular, "тергалии" in plural). Type of the term is *Siphlonurus lacustris* (FIG.13:A, reproduced from Kluge 1989a: Fig.4; designated here). Tergalii are often called "tracheal gills"; the term "tergalii" is attributed to a set of homologous organs, while the "tracheal gills" are analogous organs of various origin (Kluge 1989a, 1996a, 2000). A tergalius may or may not serve as a gill, and a gill may or may not be a tergalius; sometimes the tergalius bears a special gill (FIG.36:B), sometimes gills are present on other body parts (see Index of characters [1], [1.3.25] and [1.3.30-32]).

In winged stages tergalii are absent, so here all characters connected with tergalii structure are attributed to larvae only (see Index of characters [1.3.19]–[1.3.59]). In the larva of 1st instar tergalii are never present, they appear after one of next moults. Tergalii of young larva can strongly differ in their structure and number from tergalii of mature larva; so here in descriptions of taxa all characters connected with tergalii are attributed only to mature larva (several last instars) and would be wrong if apply them to young larvae.

In Euplectoptera seven pairs of tergalii can be present on abdominal segments I–VII. In some euplectopteran taxa number of tergalii pairs is less, as the tergalii are retained only on some of these segments (see Index of characters [1.3.19]–[1.3.20]); only in abnormal specimens tergalii can be present on abdominal segment VIII. In extinct Permoplectoptera nine pairs of tergalii were present on abdominal segments I–IX. Here certain pairs of tergalii are indicated by Roman numerals corresponding to abdominal segments; for example, "tergalius III" means tergalius of third abdominal segment, independently of the presence or absence of tergalii on the two first abdominal segments.

Tergalii are joined at the sides of the posterior margin of the tergite, nearly always on the dorsal side of the body; only in rare cases are their bases translocated together with the lateral margin of tergite to the ventral side; in some specialized mayflies the bases of some tergalii are shifted to the anterior part of the tergite (see Index of characters [1.3.22]).

Tergalius always has mobile articulation with the body, being articulated to it by narrow base and moved by special tergalial muscles located inside the segment (inside the tergalius itself muscles are absent). Tergalial muscles are the most lateral group of muscles of the segment; they are more lateral than dorsoventral muscles and run from the basis of tergalius obliquely anteriorly-ventrally, to the ventral wall of the segment – sternopleuron. In some cases each tergalius has only one tergalial muscle (FIG.13:C), in other cases a bunch of 2-4 parallel muscles which can work as antagonists arises from the basis of each tergalius. In larvae of many mayflies tergalii are able to make fast rhythmic fluctuations and are used by the larva to create a water current around its body. Such an ability to create a water current is very important for respiration of larvae inhabiting stagnant waters, but has no practical significance for rheophilous (lentic) larvae inhabiting fast streams. In some rheophilous mayfly larvae tergalii have the same mobility, in others they are able only to slow movements, and can not create a water current. Mobility of tergalii is an important systematic character of some taxa (see Index of characters [1.3.30]).

As a whole, the character of musculature and the

places of attachment of tergalii on the abdomen in mayfly larvae correspond to musculature and places of attachment of wings on mesothorax and metathorax of adult Pterygota (Kluge 1989a), so the tergalii are most probably the serial homologues of wings. The bunch of tergalial muscles probably corresponds to a complicatedly differentiated complex of wing muscles of direct action running from the basalar, subalar and axillary sclerites to the pleurite, sternite and furca (sternal apodeme). Some researchers stated the alternate point of view – that these appendages on the abdomen of mayfly larvae are homologous not to the wings, but to the limbs, thus the places of their attachment were considered to be located not on the tergite, but on the sternopleurite. It is quite difficult to compare a segment of abdomen with a segment of thorax, because of the great difference in their structure, but it is possible to compare structure of different abdominal segments and to trace development of a segment from larva to imago (FIG.13:A–B). Such comparison shows that styliger and gonostyli of male (being homologous to coxites and styli, that is, limb derivatives) have another, more ventral, position on a segment than tergalii do; that is, tergalii can not be homologous to limbs. With the idea about homology of tergalii and wings some theories about origin of wings and phylogeny of Pterygota are connected. Some authors, naming tergalii "tracheal gills", compare them with tracheal gills of other insects, in particular, with paired abdominal gills of some Odonata, Plecoptera, Megaloptera and some other insects. However tergalii essentially differ from these gills, as they are articulated to the tergite, and the muscles, which move them, run not dorsally, but ventrally to the sternum. Though it is supposed that tergalii have a very ancient origin, their homologues in other groups of insects are not found. Ideas about the phylogeny of Ephemeroptera strongly depend upon point of view on tergalii origin (if they are homologous to wings or to legs), as in these cases the initial plan of tergalii structure is assumed differently.

Tergalial form and structure are diverse. Usually the tergalius is lamellate, its dorsal surface is directed dorsally or anteriorly, and its ventral surface is directed ventrally or posteriorly. One of margins (from tergalius base to its apex) is named here **costal margin**; this is the margin which can be directed anteriorly or ventrally (if tergalius apex is directed laterally) or laterally (if tergalius apex is directed posteriorly). The opposite of it is an **anal margin** – this is the margin, which can be directed posteriorly or dorsally (if tergalius apex is directed laterally) or medially (if tergalius apex is directed posteriorly). As a rule, the tergalius has two sclerotized ribs – a **costal rib** (running from tergalius basis by its costal margin or at some distance from it) and an **anal rib** (running from tergalius basis by the anal margin, or at some distance from it) (FIG.13:C–G); sometimes these ribs are vestigial or lost (see Index of characters [1.3.27] and [1.3.28]). Inside tergalius, more or less advanced tracheae pass. In difference from wings, in which tracheae pass inside sclerotized veins, in tergalius tracheae always pass irrespectively of sclerotized ribs, so tergalii have no true veins.

Basing on a wrong reconstruction of the Permian *Phtharthus*, where ventral stylus-like abdominal appendages were shown (Handlirsch 1904a, 1906–1908, 1925), some authors believed that ventral attachment (occurring in recent Behningia/fg2 as well) and slender shape (characteristic for recent Pinnatitergaliae in general) were initial features of the mayfly abdominal appendages, which they regarded to be limb derivatives. This led to the assumption of a very ancient origin of the Pinnatitergaliae. Actually abdominal appendages of *Phtharthus* have posterior-lateral-dorsal attachment typical for mayfly tergalii, and probably lamellate shape (FIG.14:C), as well as tergalii of another Permian mayfly – *Protereisma* (FIG.14:D), that is most probably the initial tergalial structure.

Functions of tergalii are various. In some mayflies they create a water current necessary for respiration. In many cases tergalii are used as tracheal gills (as far as they increase the body surface and this facilitates respiration). Tergalii can execute a role of organs of attachment (overlapping one another by their edges and forming one large sucker in larvae of some Holarctic Radulapalpata and Australian Atalophlebia/fg1 – see Index of characters [1.3.31]). Sometimes tergalii are transformed into protective gill opercula (see Index of characters [1.3.32]). In Coloburiscus/fg1 tergalii, being sclerotized and covered by large spine-like setae, probably, execute a protective role. In a many cases tergalii lack any function, but nevertheless are retained together with the tergalial musculature.

GENITALS
(Figs 11, 18:A–D)

In the female imago, the sexual aperture opens between abdominal sternites VII and VIII. Usually on this place no external morphological structures are present (FIG.18:A–B); the sternite VII can be produced posteriorly, forming a **pregenital plate**; in rare cases (in some Leptophlebia/fg1) the pregenital plate forms a tubular process – an unpaired secondary ovipositor. If pregenital plate is present, it is expressed only in imago and subimago, but not in larva. Mayflies have no any vestiges of the primary ovipositor (inherent in many other Amyocerata).

Abdominal sternum IX is produced posteriorly in the form of a plate. In the female this is a simple plate called **postgenital**, or **preanal** plate (see Index of characters [2.3.6]). In the male this plate, named **styliger** (see Index of characters [2.3.7]), bears a pair of mobile appendages – **gonostyli**, or **forceps** (see Index of characters [2.3.8]–[2.3.14]). Gonostyli are used by the male imago at copulation for holding female abdomen (FIG.10:G–H). Gonostylus is a derivative of the abdominal stylus – such styli are developed on abdominal segments I–IX (or at least some of them) in Triplura, on abdominal segments I–VII (or some of them) in Diplura, and only on abdominal segment IX of males and/or females in some Pterygota. In all insects which have abdominal styli or their derivatives the stylus is non-segmented; in some Diplura and Triplura the stylus bears an apical pointed appendage – tarsellus (only in some palaeontological publications segmented styli are described for extinct insects, but these descriptions are quite doubtful, not being supported by fossils). Gonostylus of Ephemeroptera looks segmented, but its segments are secondary ones, they have no active mobility and no muscles or apodemes inside. Gonostylus is moved only by a muscle located in the styliger and attached to its first segment. A lateral paired portion of styliger, which contains the muscle of gonostylus, is named here a **pedestal of gonostylus**; in some mayflies the styliger is strongly reduced, but its gonostyli pedestals are prominent, segment-like, thus sometimes they are erroneously taken for proximal segments of gonostyli. Gonostylus usually consists of the following 4 secondary segments: a short thick 1st (proximal) segment is immobile connected with a long 2nd segment, further follow two passively-mobile articulated **distal** segments – 3rd and 4th ones. In some mayfly taxa the number of gonostylus segments is reduced, in more rare cases it is increased (see Index of characters [2.3.10]–[2.3.14]). Inner surface of the gonostylus often bears numerous mechanoreceptorial globular papillae representing modified setae (Gaino & Rebora 2002).

The projection of abdominal sternum IX (the subanal plate of female and the styliger with gonostyli of male) is better developed in the imago, and usually is present not only in imago and subimago, but in the larva as well (in contrast to the subanal plate of female). Larval gonostyli are small and have no more than one distal segment, from which the both subimaginal distal segments are developed; often larval gonostyli are non-segmented, sometimes reduced or fused with styliger. In the majority of mayflies structure of abdominal sternum IX allows to distinguish male and female larvae; only in Turbanoculata larval gonostyli are reduced, and in Caenoptera larval gonostyli are completely fused with styliger, thus in these two taxa sexual dimorphism in larval abdominal sternum IX is not expressed (see Index of characters [1.3.60]).

In the male imago, from a membrane between styliger base and paraproct bases (i. e. from the boundary of segments IX and X), a **penis** arises. Cuticle laterad of penis base is sclerotized in such a manner that forms a pair of curved sclerotized **penial arms.** Each penial arm has a lateral-ventral angle articulated with a peculiar small proximal-dorsal projection of styliger, and a lateral-dorsal end articulated with posterior margin of tergite IX somewhat mediad of its lateral-posterior corner. Styliger is able to bent ventrally by contraction of longitudinal sternal muscles. At rest, the articulation of lateral styliger margin with immobile lateral margin of sternite is located somewhat distad of the articulation of styliger with penial arm; thanks to this, when styliger bents ventrally, penis is protracted posteriorly and dorsally (FIG.11). The penial arms are well developed in the overwhelming majority of mayflies, with exception for a few taxa (see Index of characters [2.3.17]).

Penis is usually paired (in contrast to majority of other insects); its left and right lobes can be either completely separated, or more or less fused together. Paired seminal ducts usually open on penis by a pair of gonopores, rarely by an unpaired gonopore (particularly, in *fragilis* [*Ametropus*]); seminal ducts can be paired all aver their length (FIG.23:G) or are fused in penial base (FIG.93:C) (see below).

Form and structure of the penis are extremely diverse, it can have complex musculature and movable spines – **titillators** (see Index of characters [2.3.15]–[2.3.17]).

Subimaginal and larval penis (FIG.18:D) never has sclerotized arms and can have other differences if compared with the imaginal one – its structure can be more simple, rarely more complex than in imago; in a few taxa larval penis is lost (see Index of characters [1.3.60]).

ABDOMINAL SEGMENT X AND CAUDALII (FIG.12)

The last, tenth abdominal segment has a well-developed tergite, whose lateral-anterior angles are produced ventrally more strongly than that of preceding tergites, but do not meet on the ventral side (in contrast to Microcoryphia, Odonata, some Plecoptera and some other insects). Tenth tergite is well-outlined both in adults and larvae (in contrast to preceding tergites, which are not laterally outlined in larva); its posterior margin is produced posteriorly as a flap above bases of caudalii.

Ventral wall of the tenth segment is formed by a pair of **paraprocts**. In the larva paraprocts have a form of distinct sclerites (see Index of characters [1.3.62]), while in the imago they are usually soft and indistinct. As the tenth tergite is interrupted ventrally, paraprocts are directly articulated to sternum IX.

Posterior wall of the tenth segment is formed by a **tricaudale** (new term) – integral sclerotized formation consisting of a **basitricaudale** (new term) – transverse sclerite of body wall, and **caudalii** (new term) – three processes arising from the basitricaudale in caudal direction. Formerly caudalii of mayflies were called "caudal filaments", as they often have a thread-like form, especially in adults. Lateral paired caudalii are **cerci**, and median unpaired caudalius is **paracercus**. Between lateral margin of basitricaudale, lateral-posterior margin of paraproct and lateral-posterior margin of tergite, body wall is formed by a paired sclerite – **cercotractor** (new term). Type taxon of the new terms tricaudale, caudalius, basicaudale and cercotractor is *aestivalis* [*Siphlurus*] in FIG.12:A–E. Usually the cercotractor has triangular shape, is movably connected with tergite, movably articulated with lateral base of cercus and fused with paraproct (FIG.12:A–F); but in some taxa the cercotractor has another shape, can be separated from paraproct (FIG. 12:G) and/or fused with cercus (see Index of characters [1.3.62] and [2.3.18]). Basitricaudale has a pair of deep dorsoventral grooves, which serve its flexibility and divide it into three portions each bearing one caudalius; **direct caudalial muscles** stretch from tergite to these grooves and serve as adductors of cerci. Probably no primary direct abductors of cerci are present in insects. Abduction is served by **tergo-cercotractoral muscles** (Figs.12: A–E); in the cases when cercotractors are fused with cercal bases, the tergo-cercotractoral muscles look as direct cercal abductors (FIG.12:G). Each caudalius has a **basi-basal muscle,** which connects dorsal and ventral edges of its base. Such basi-basal muscles are well developed in all mayflies, being retained even in vestigial paracercus of that mayflies, which look as two-tailed. Besides Ephemeroptera, basi-basal muscles are developed in Triplura (both in Zygentoma and Microcoryphia), but lost in Metapterygota.

Some authors (Snodgrass 1935, et al.) erroneously regard paraprocts to be coxites of abdominal segment XI, cerci to be leg derivatives of abdominal segment XI (i. e. appendages of paraprocts), and the paracercus to be a dorsal appendage of another origin. This assumption is based on examination of Microcoryphia and some other insects with specialized abdomen, where abdominal tergite X forms an integral ring, separating paraprocts from sternum IX (so the ring formed by tergite X is taken for a fusion of tergite X and sternite X). Such homologization contradicts to muscles arrangement, as no special sternal muscles are attached to the "sternite X" (which is actually a ventral part of the tergite X). Abdominal structure of Ephemeroptera (as well as that of Zygentoma and some other insects) is more primitive, so the sternite X (pair of paraprocts) is situated here not behind, but ventrad of the tergite X and just behind the sternum IX. In all Amyocerata the cerci and paracercus are in the equal manner articulated with abdominal tergite X and all muscles going from their bases are attached to the tergite X only. Most probably, cerci and paracercus are organs of the same origin, being dorso-posterior appendages of tergite X (Kluge 1999d, 2000). In contrast to leg derivatives, the caudalii never have primary segmentation, never have muscles or apodemes inside.

In Ephemeroptera caudalii have such a kind of secondary segmentation, which is most primitive among Amyocerata, being the same as in all Triplura: the number of segments is large and indeterminate; at each moult it increases thanks to division of proximal segments half-and-half; each caudalius is thickest in its base, and becomes thinner toward apex (i. e. has a bristle-like shape); proximal-most segments are shortest and indistinctly divided one from another, and in distal part segments become longer and distinctly separated. This structure and kind of growth resembles that of the antennal flagellum of primitive representatives of Amyocerata, including Triplura and Ephemeroptera (but not of antenna as a whole, which have besides the flagellum, also scapus and pedicellus).

Besides Ephemeroptera, cerci are retained in many other Pterygota, but the paracercus is lost in all Matapterygota (in some Plecoptera and some other Metapterygota presence of paracercus was erroneously stated in literature). In Ephemeroptera cerci are always developed, and the paracercus can be as long as cerci or even somewhat longer, or it is more or less reduced, up to a non-segmented vestige (see Index of characters [1.3.64] and [2.3.20]). In many mayflies the paracercus is reduced only in winged stages, being developed in the larva; in this case, when subimaginal tissues are developed under larval cuticle, hypodermal paracercus narrows and breaks near base, thus subimaginal vestige of paracercus develops only from the basal part of larval paracercus, and at moult larval cuticle shads together with remainder of hypoderm of most part of paracercus (in contrast to mouthparts – see above). Sometimes the paracercus is reduced in larva of first instar, being developed in mature larva; sometimes it is developed in larva of first instar, being reduced in mature larva and winged stages; sometimes it is reduced in all stages.

In male imagoes nearly of all Ephemeroptera caudalii are longer than the body (see Index of characters [2.3.18]) and are used in the mating flight: most mayflies have in their mating flight a stage of parachuting, when the insect passively moves down with its wings are V-like elevated upwards, and its cerci are widely divergent. In male subimagoes caudalii are not so long as in the imago. In female imagoes caudalii are less long, little longer than the body or shorter than it.

Larval caudalii often have denticles on posterior margins of segments, similar to denticles on posterior margins of abdominal tergites (see above).

In larvae of the primitive siphlonuroid type (see above), caudalii have a peculiar structure (FIG.28: A): they are not long (much shorter than in imago, shorter than the body); paracercus is subequal to cerci; cerci have oblique margins of segments, so each segment on lateral (outer) side is situated more distally, than on median (inner) side; **primary swimming setae** are present – these are setae arranged in four regular rows – one row on median (inner) side of each cercus and a pair of rows on lateral sides of paracercus. Such structure of caudalii allows larva to swim, moving by its abdomen up-and-down (FIG.9:A–B). In various mayfly taxa this primary siphlonuroid specialization is secondarily lost or changed to other specialization. Sometimes on lateral (outer) sides of cerci **secondary swimming setae** can be developed, they differ in structure from the primary swimming setae (see Index of characters [1.3.67]). Sometimes primary swimming setae are reduced (see Index of characters [1.3.66]) or substituted by secondary swimming setae, which have the same structure on both lateral and median sides of cerci and lateral sides of paracercus. Margins of segments of cerci can be not oblique, paracercus can be more or less shortened, and cerci elongate, being more similar to cerci of winged stages; such modification is especially usual for rheophilous larvae, which lost ability to active swimming.

Based on a wrong reconstruction of Permian *Phtharthus*, where cerci were shown as fringed by setae on both sides (Handlirsch 1904a, 1906–1908, 1925), some authors believed that such setation (occurring in recent Pinnatitergaliae as well) was initial for mayflies. Actually caudalii of *Phtharthus* have typical siphlonuroid setation with cerci bearing setae on median sides only (FIG.14:C), as well as that of another Permian mayfly – *Protereisma* (FIG. 14:D). The same siphlonuroid setation is most common for Mesozoic and Recent mayflies (see Index of characters [1.3.66]), which leads to the assumption of its primary nature.

INTERNAL ANATOMY

ALIMENTARY CANAL AND MALPIGHIAN TUBES

The alimentary canal is functional in larvae and non-functional in subimagoes and imagoes; it is straight and simple; the stomodaeum is slightly separated or non-separated from the mesenteron, thin-walled and lacking sclerotized formations (characteristic for ectodermal proventriculus of many other insects); the proctodaeum is more differentiated (Needham et al. 1935: Pl.6), varying among mayfly taxa (Landa & Soldán 1985: Figs 44-59).

Malpighian tubes are numerous (from several dozens to several thousand) and have unique structure: each tube consists of a distal portion usually coiled spirally or S-like, and of a very thin duct arising from the inner end of the spiral (Needham & al. 1935:Pl.7:5–10). Ducts of Malpighian tubes fall either directly into the intestine, or into special projections of the intestine – **trunks of Malpighian**

tubes (Landa 1969b:Fig.12; Kluge 1993a:Figs 1–19; 1998:Figs 32–34). The trunks of Malpighian tubes occur in many (but not all) mayfly taxa and have various number, length and arrangement, can be simple or branched; most of the trunks are directed anteriorly. Number, arrangement and branching of the trunks were regarded to be characters of high level taxa (Landa 1969b, Landa & Soldán 1985), but actually the number of trunks and their branches is under great individual variability; it can differ in specimens of the same species and in left and right halves of the same specimen (Kluge 1993a:Figs 1–19). Most constant are longest trunks, while short trunks and short branches can easily appear and disappear, varying individually (Kluge 1993a, 1998). In many taxa examined (Posteritorna, Isonychia/fg2, Fimbriatotergaliae) there are 2 longest lateral trunks directed anteriorly; sometimes anterior end of each trunk bears a peculiar straight Malpighian tube partly fused with its duct (Kluge 1998: Figs 32–34); an identical pair of peculiar Malpighian tubes directed anteriorly occur also in some mayflies, which have no trunks – Turbanoculata (Landa 1968:Fig.12BR) and some Ephemerella/fg1. So, the lateral paired position of Malpighian tubes is usual for many non-related groups of Ephemeroptera. Some other mayflies, particularly Radulapalpata, instead of one pair, have 2 equal pairs of longest trunks directed anteriorly (Kluge 1993a: Figs 1–19).

TRACHEAL SYSTEM

Mayflies have all 10 pairs of **tracheostia** (mouths of tracheal system) that are initial for Amyocerata: 2 intersegmental thoracic pairs – **stenothoracic** (between prothorax and mesothorax) and **cryptothoracic** (between mesothorax and metathorax), and 8 segmental abdominal pairs – one pair on each abdominal segment I–VIII (Kluge 2000). All tracheostia are lateral, each abdominal tracheostium is located at the anterior part of its segment (FIG.13). In subimago and imago the both pairs of thoracic tracheostia have a form of widely opened spiracles lacking closing apparatus, and the abdominal tracheostia are either closed or have a form of small spiracles (FIG.4). In the larva all tracheostia are closed, but at each moult serve for escaping of old tracheal intima thorough them.

The tracheal system of mayflies is described and figured in detail by Landa (1948). Tracheae originating from different tracheostia are connected by a single pair of thick lateral **longitudinal trunks** (the same in many other insects). Left and right trunks are connected one with another only by **transverse anastomoses,** which have no passage for the air: each transverse anastomose is formed by a pair of tracheal branches meeting medially and fused by their apical cuticular thickenings. One of such transverse anastomoses, named **Palmen's body**, is located in the head dorsad of oesophagus and is formed by fusion of apices of two pairs of tracheae meeting at one point; some other transverse anastomoses can be present in the head, thorax and abdomen. Abdominal anastomoses, if present, are located ventrad of the intestine close to the nerve cord, no more than one anastomose per a segment. They can be present in abdominal segments VIII and/or IX only, or in other abdominal segments as well. Arrangement of transverse abdominal anastomoses was regarded as an important character of high rank taxa (Landa 1968b; Landa & Soldán 1985; McCafferty 1991a); however, the number of anastomoses varies individually. During ontogenesis, new anastomoses are added, thus their number is less in the young larva and more in the mature one.

Arrangement of visceral tracheae was regarded as another character of high rank taxa (Landa 1968b; Landa & Soldán 1985). Sometimes tracheae penetrating into the same internal organs or muscles originate from different tracheostia; this can vary individually or in the left and right sides of the same individual. Taking into account that examination of thin tracheae is rather difficult and needs special methods, characters connected with the tracheal system are hardly usable in taxonomy.

Thoracic and abdominal tracheae arising from different pairs of tracheostia are connected by the single pair of lateral trunks only, and have no other longitudinal anastomoses; particularly, in contrast to Metapterygota, there are no loops connecting stenothoracic and cryptothoracic tracheostia and giving rise to mesothoracic leg and wing tracheae, and no loops connecting cryptothoracic and first abdominal tracheostia and giving rise to metathoracic leg and wing tracheae. Instead of this, each leg and each wing is supplied by a single trachea; mesothoracic leg and wing get trachea from stenothoracic tracheostium, and metathoracic leg and wing get trachea from cryptothoracic tracheostium only.

The single trachea coming into the wing divides into several branches, which penetrate through the wing base either passing as a single bunch anteriad of the basal wing plate, or a branch going to MP and CuA passes separately from others posteriad of the basal plate, and then unites with others just before the place where tracheae diverge penetrating into RA, RS+MA and MP. Among the taxa examined, only Campsurus/fg1 have unusual separate entering of the trachea into CuA (see Index of characters [2.2.50] and FIGS 79–80).

NERVE SYSTEM

As in other Hexapoda, the central nerve system of Ephemeroptera initially consists of a supraoesophageal synganglion (fused preoral brain and tritocerebrum), suboesophageal ganglion (fused ganglia of mandibular, maxillary and labial segments), 3 thoracic ganglia and 8 abdominal ganglia (last of which is probably a synganglion of abdominal segments VIII–X). This or that thoracic or abdominal ganglion can be shifted anteriorly, and is sometimes fused with ganglion of the preceding segment. Thus, the 1st abdominal ganglion is often fused with the metathoracic ganglion, and the last two abdominal ganglia can be fused together; nerve connectives can be fused together partly or completely (Landa & Soldán 1985:Figs 1–4). Position of the metathoracic ganglion in adults is well-indicated externally, thanks to the structure of the mesothoracic furcasternal protuberances (see paragraph "Mesosternum" and Index of characters [2.2.23]). Location of abdominal ganglia in this or that abdominal segment is not well-fixed, as abdominal segments are able to protract backward and retract into preceding ones.

REPRODUCTIVE SYSTEM

It is usual to regard that mayfly gonads and gonoducts are paired all over their length and always open by paired gonopores. Actually, this is true for a part of mayflies only (FIGS 23:G; 59:B). In males of various non-related mayflies, left and right seminal ducts are fused one with another inside the penial base, and in the distal part of the penis diverge again, thus open by a pair of gonopores (FIG.93:C); rarely there is an unpaired gonopore. In females, left and right oviducts often unite to form a short unpaired genital chamber opened by an unpaired gonopore; in Siphlonurus/fg1 this chamber is sclerotized (FIG.18:A–C), in other mayflies membranous.

Some authors regarded the paired gonopore of mayflies to be a plesiomorphy unique among insects. This opinion is based only on a general idea about the primary nature of paired organs and secondary nature of unpaired ones, being not supported by comparison of this structure in concrete insect groups. Most probably, Hexapoda initially have an unpaired gonopore, which is present in all Entognatha, Triplura and majority of Pterygota. Paired gonopore of male mayflies can be a new formation connected with the peculiar genital structure (see above and FIG.11): As the penis is constantly articulated with a ninth abdominal tergite by a pair of penial arms, its movement should be limited by rotation around a single transverse axis; more composite movements can be made only if left and right halves of the penis are movably connected one with another; this becomes possible only if gonoducts are paired all along their length. As well as other insects, mayflies have great specific diversity in genital structure and manner of genital movement (that probably serves species reproductive isolation). Due to this, most mayfly species have paired gonoducts, and only a few species have a penis with limited mobility and unpaired gonopore. Other insects are able to combine diversity in penis structure with an unpaired gonopore, because they have no such penial arms.

Mayfly ovaria have a large indeterminate number of ovarioles (approximately from 100 to 500 in different species). Formerly it was regarded that the ovaria of mayflies have the primitive panoistic type, i. e. lack trophocytes (Soldán 1979c); however, detailed examination of a few species indicated that mayfly ovarioles belong to the meroistic telotrophic type, with linear clusters of trophocites concentrated in the apical zone of each ovariole. (Gottanka & Buning 1993). In the end of development, the trophocytes degenerate, the oocytes lost connection with them, and all ovarioles with oviduct fuse to a common sack containing numerous eggs.

Testes have a large indeterminate number of testacular follicles, each falling directly to a seminal duct.

Shape and position of ovaria and testes somewhat differs among mayfly taxa (Landa & Soldán 1985:Figs 18–20).

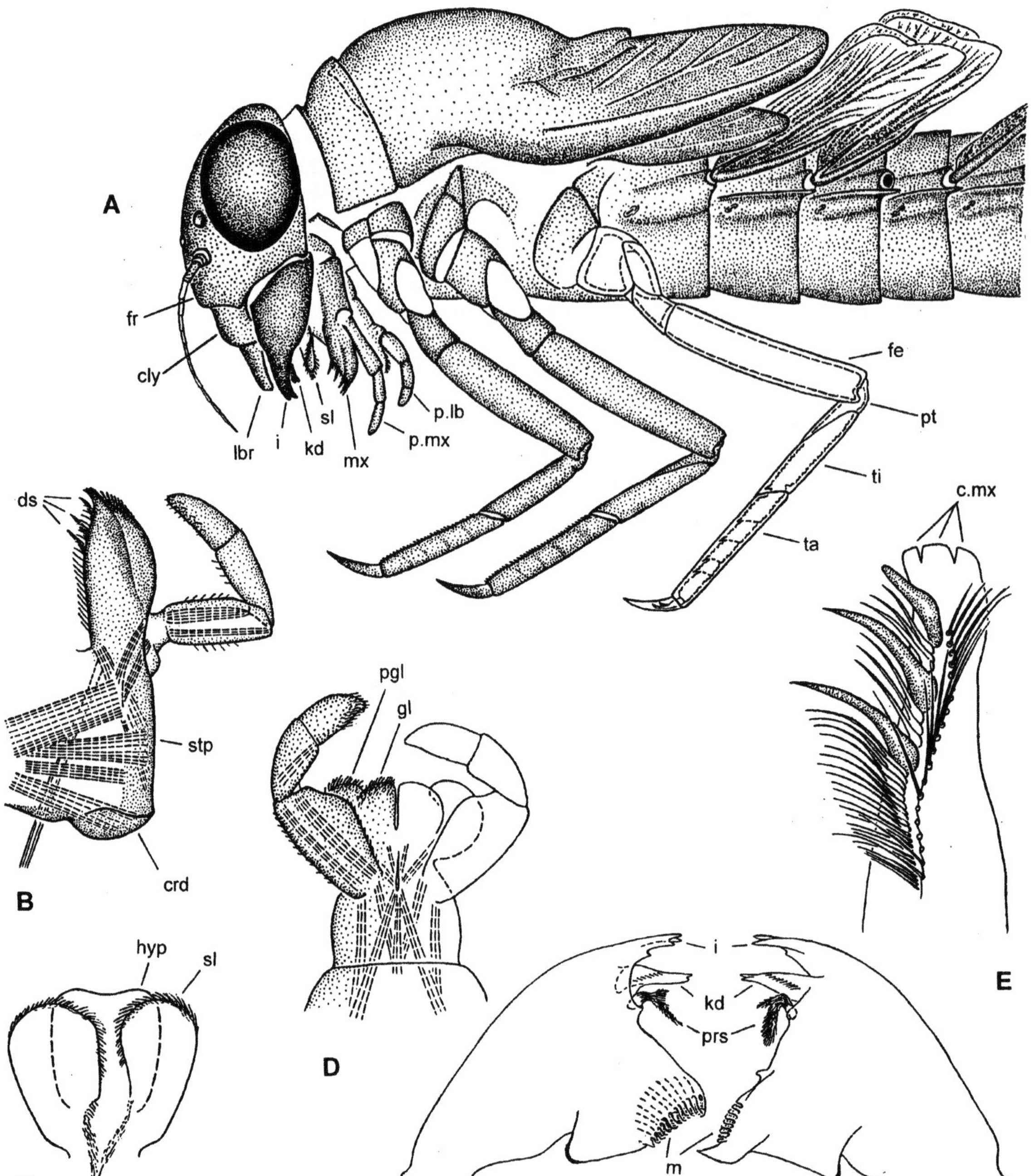

Figure 3. Larva of Siphlonurus/fg4 *aestivalis* [*Siphlurus*].
A – Anterior half of male larva of last instar, lateral view (tergalius III removed; subimaginal hind leg shown by interrupted line); **B** – left maxilla, ventral view (muscles shown by interrupted lines); **C** – hypopharynx and superlinguae, dorsal view; **D** – labium, ventral view (muscles shown by interrupted lines); **E** – apex of left maxilla, median view (dentisetae shown by dots); **F** – left mandible; **G** – right mandible. (A, B, E–G – from Kluge 1997 and 2000).

Abbreviations: **c.mx** – maxillary canines; **cly** – clypeus; **crd** – cardo; **ds** – dentisetae; **fe** – femur; **fr** – frons; **gl** – glossa; **hyp** – hypopharynx, **i** – incisor; **kd** – kinetodontium; **lbr** – labrum; **m** – mola; **mx** – maxilla; **pgl** – paraglossa; **p.lb** – labial palp; **p.mx** – maxillary palp; **prs** – prostheca; **pt** – patella, **sl** – superlingua; **stp** – stipes; **ta** – tarsus; **ti** – telotibia.

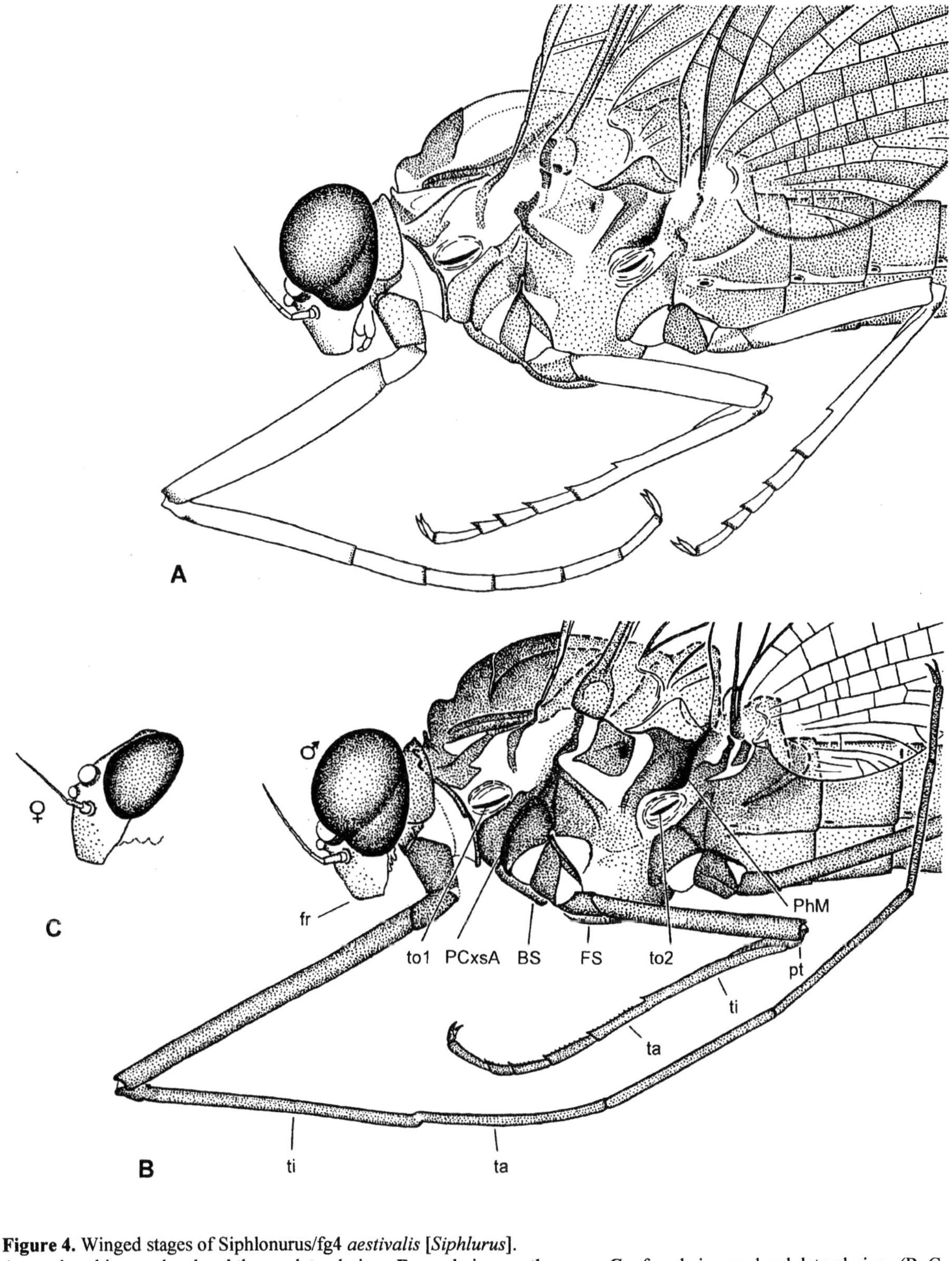

Figure 4. Winged stages of Siphlonurus/fg4 *aestivalis* [*Siphlurus*].
A – male subimago, head and thorax, lateral view; **B** – male imago, the same; **C** – female imago, head, lateral view. (B, C – from Kluge 2000)

Abbreviations: **BS** – basisternum of mesothorax; **fr** – face fold (frons); **FS** – furcasternum of mesothorax; **PCxsA** – anterior paracoxal suture of mesothorax; **PhM** – middle phragma; **pt** – patella; **ta** – tarsus; **ti** – telotibia; **to1** – stenothoracic tracheostium (first thoracic spiracle); **to2** – cryptothoracic tracheostium (second thoracic spiracle).

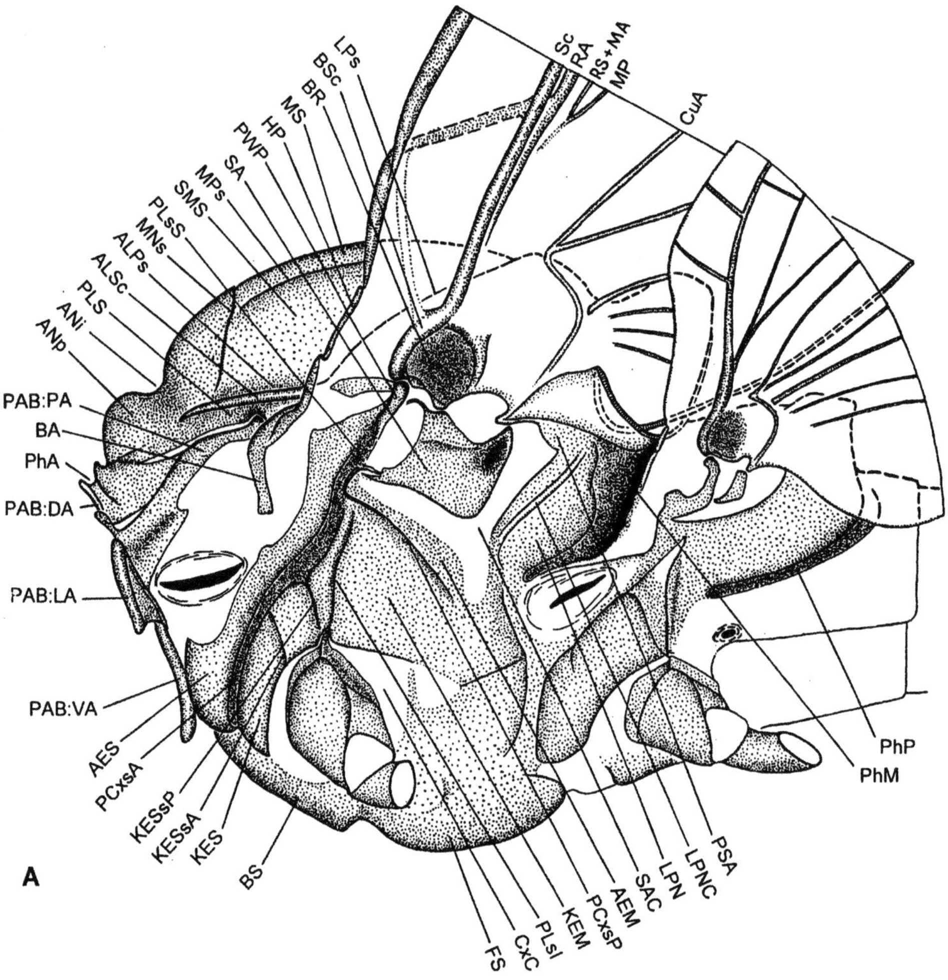

Figure 5. Pterothorax of Siphlonurus/fg4 *aestivalis* [*Siphlurus*].
A–C – imaginal pterothorax: **A–B** – lateral view; **C** – anterior view (muscles shown by interrupted lines and dotted, muscle bases outlined by dotted lines); **D** – subimaginal exuviae of left mesopleuron. (From Kluge 1994a).

Abbreviations: **AEM** – anepimeron; **AES** – anepisternum; **ALPs** – antelateroparapsidal suture; **ALSc** – anterolateral scutal costa; **ALSs** – anterolateral scutal suture; **ANi** – anteronotal transverse impression; **ANp** – anteronotal protuberance; **AN.Pm** – anteronotopleural muscle; **Ax.Fm** – axillar-furcal muscle; **Ax.PmI** – inferior axillar-pleural muscle; **BA** – basalare; **BA.Cm** – basalar-coxal muscle; **BA.SmI** – inferior basalar-sternal muscle; **BA.SmS** – superior basalar-sternal muscle; **BR** – basiradiale; **BS** – basisternum; **BSc** – basisubcostale; **CxC** – coxal conjunctiva; **Cx.Cxm** – coxo-coxal muscle; **F** – furca; **F.Cm** – furca-coxal muscle; **F.CmP** – posterior furca-coxal muscle; **F.CmS** – superior furca-coxal muscle; **Fm** – furcal muscle; **FS** – furcasternum; **FSp** – furcasternal protuberance; **HP** – humeral plate; **iFm** – intersegmental furcal muscle; **ISL** – infrascutellum; **KEM** – katepimeron; **KES** – katepisternum; **KESsA** – anterior katepisternal suture; **KESsP** – posterior katepisternal suture; **LPN** – lateropostnotum; **LPNC** – lateropostnotal crest; **LPs** – lateroparapsidal suture; **LSs** – lateral scutal suture; **MLs** – median longitudinal suture; **MNs** – mesonotal suture; **MPN** – mediopostnotum; **MPs** – medioparapsidal suture (notaulix); **MS** – medioscutum; **MTm** – median tergal muscle; **PAB:DA** – dorsal arc of prealar bridge; **PAB:LA** – lateral arc of prealar bridge; **PAB:PA** – posterior arc of prealar bridge; **PAB:VA** – ventral arc of prealar bridge; **P.Cm** – pleuro-coxal muscle; **PCxsA** – anterior paracoxal suture; **PCxsP** – posterior paracoxal suture; **PhA** – anterior phragma; **PhM** – middle phragma; **PhP** – posterior phragma; **PLS** – prelateroscutum; **PLsI** – inferior pleural suture; **PLsS** – superior pleural suture; **PSA** – postsubalar sclerite of lateropostnotum; **PSL.Cm** – parascutellar-coxal muscle; **PSp** – posterior scutal protuberance; **P.Trm** – pleuro-trochanteral muscle; **PWP** – pleural wing process; **SA** – subalare; **SAC** – subalar conjunctiva; **SA.Cm** – subalar-coxal muscle; **SA.Fm** – subalar-furcal muscle; **SA.Sm** – subalar-sternal muscle; **S.CmA** – anterior scuto-coxal muscle (tergal promotor of coxa); **S.CmP** – posterior scuto-coxal muscle (tergal remotor of coxa); **S.ESm** – scuto-episternal muscle of mesothorax; **S.ESm3** – scuto-episternal muscle of metathorax; **SL** – scutellum; **S.LPNm** – scuto-lateropostnotal muscle; **SLS** – sublateroscutum; **SMS** – submedioscutum; **S.PSLs** – scuto-parascutellar suture; **S.Trm** – scuto-trochanteral muscle; **SrA.Pm** – suralare-pleural muscle; **S.SLi** – scuto-scutellar impression; **Teg** – tegula.

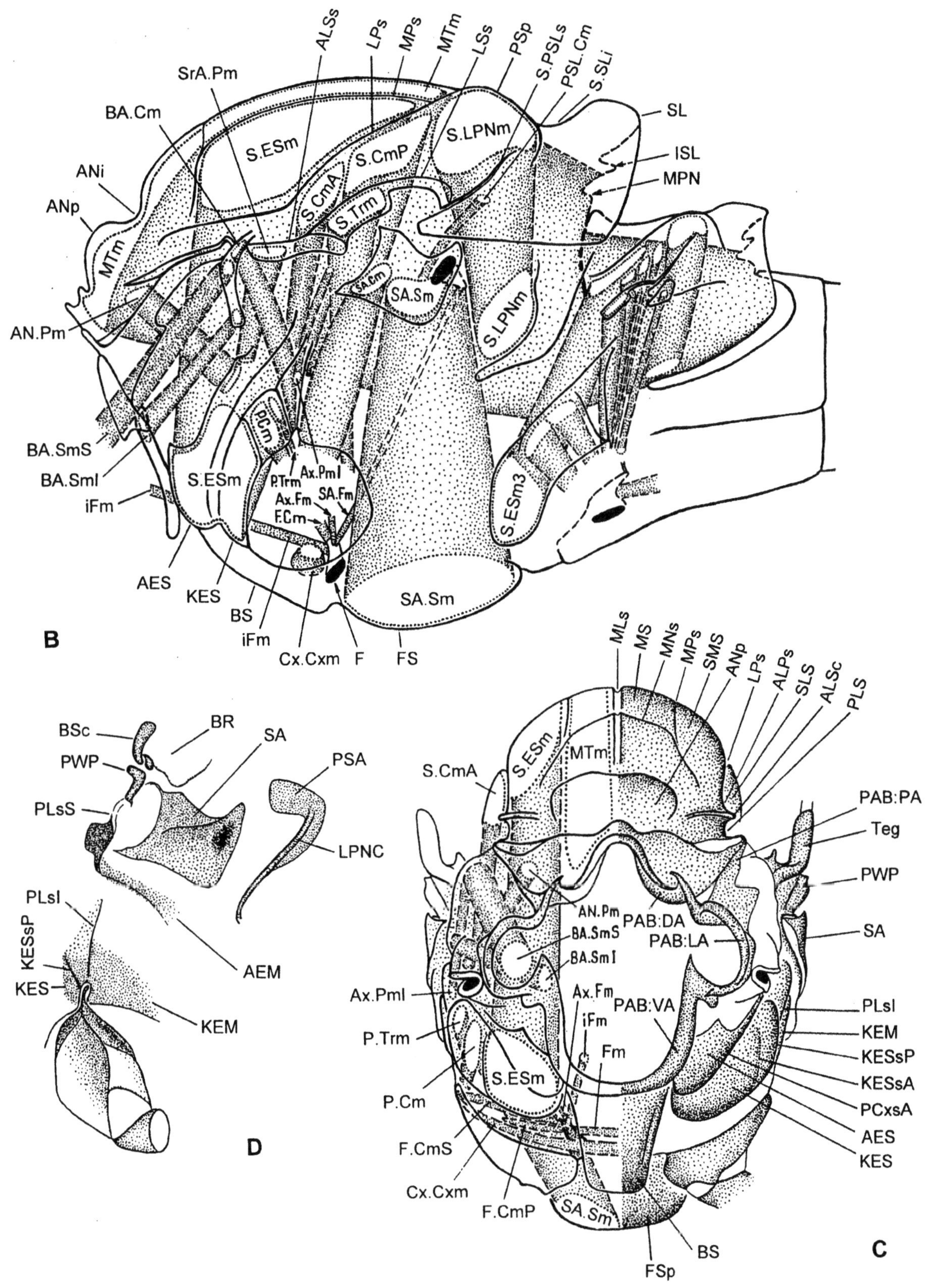

SrA.Pm
BA.Cm
ANi
ANp
MTm
AN.Pm
BA.SmS
BA.SmI
iFm
AES
KES
BS
iFm
Cx.Cxm
F
FS
ALSs
LPs
MPs
MTm
LSs
PSp
S.PSLs
PSL.Cm
S.SLi
SL
ISL
MPN
S.ESm
S.CmP
S.LPNm
S.CmA
S.Trm
SA.Cm
SA.Sm
S.LPNm
PCm
P.Trm
Ax.PmI
Ax.Fm
SA.Fm
F.Cm
S.ESm
S.ESm3
SA.Sm
B
BSc
BR
SA
PWP
PSA
PLsS
LPNC
PLsI
KESsP
KES
AEM
KEM
D
MLs
MS
MNs
MPs
SMS
ANp
LPs
ALPs
SLS
ALSc
PLS
S.CmA
S.ESm
MTm
PAB:PA
Teg
PWP
AN.Pm
PAB:DA
BA.SmS
PAB:LA
SA
BA.SmI
Ax.PmI
Ax.Fm
iFm
PAB:VA
PLsI
KEM
P.Trm
Fm
KESsP
P.Cm
S.ESm
KESsA
PCxsA
F.CmS
AES
KES
Cx.Cxm
F.CmP
SA.Sm
BS
FSp
C

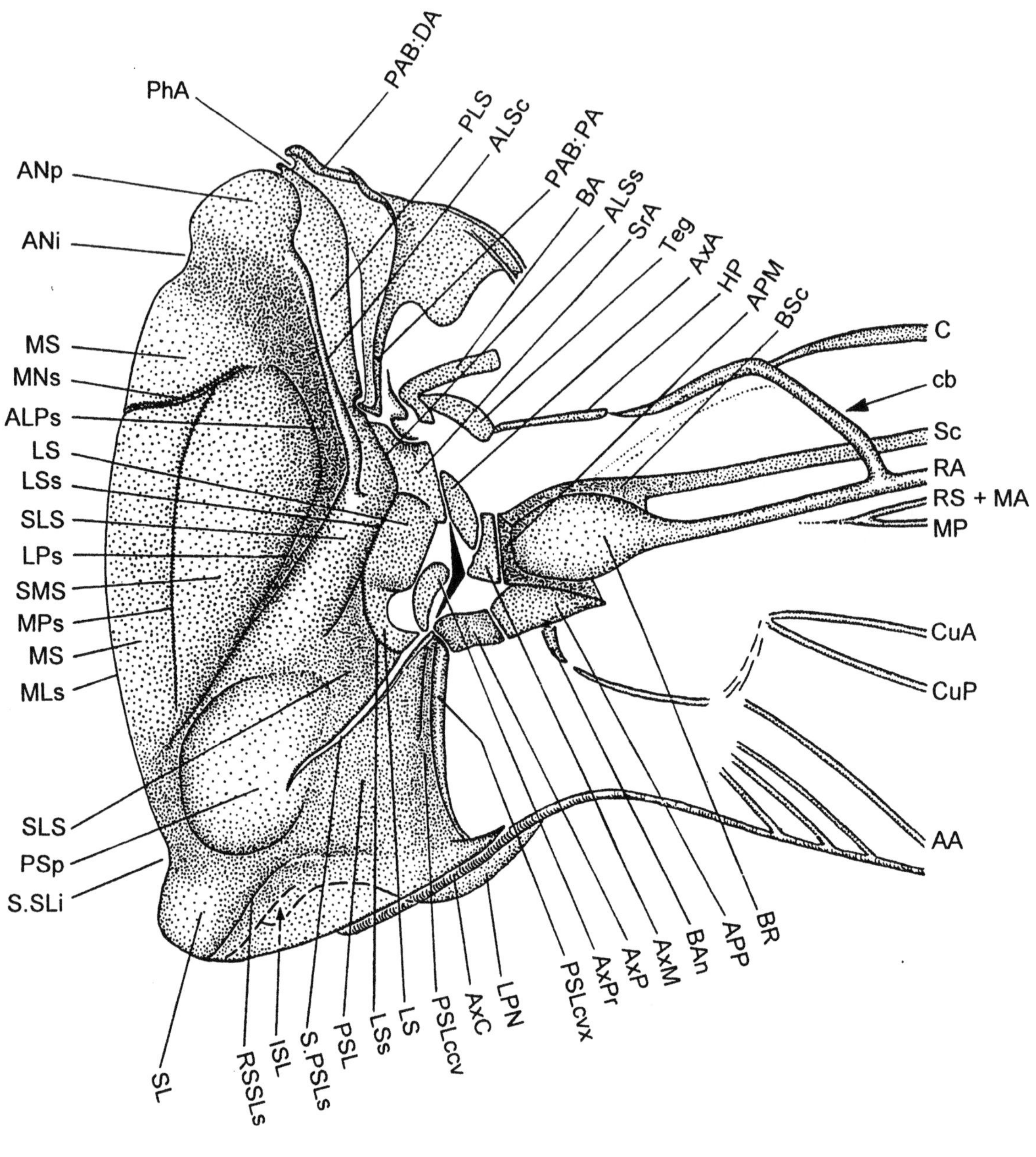

Figure 6. Right half of imaginal mesonotum and forewing base of Siphlonurus/fg4 *aestivalis* [*Siphlurus*], smoothed on slide (membrane between axillary sclerites broken, the break shown by black) (from Kluge 1994a).

Abbreviations: **ALPs** – antelateroparapsidal suture; **ALSc** – anterolateral scutal costa; **ALSs** – anterolateral scutal suture; **ANi** – anteronotal transverse impression; **ANp** – anteronotal protuberance; **APM** – middle articulatory process of wing base; **APP** – posterior articulatory process of wing base; **AxA** – anterior axillary sclerite; **AxC** – axillary cord; **AxM** – middle axillary sclerite; **AxP** – posterior axillary sclerite; **AxPr** – proximal axillary sclerite; **BA** – basalare; **BAn** – basanale; **BSc** – basisubcostale; **BR** – basiradiale; **cb** – costal brace; **HP** – humeral plate; **ISL** – infrascutellum; **LPN** – lateropostnotum; **LPs** – lateroparapsidal suture; **LS** – lateroscutum; **LSs** – lateral scutal suture; **MLs** – median longitudinal suture; **MNs** – mesonotal suture; **MS** – medioscutum; **MPs** – medioparapsidal suture (notaulix); **PAB:DA** – dorsal arc of prealar bridge; **PAB:PA** – posterior arc of prealar bridge; **PhA** – anterior phragma; **PLS** – prelateroscutum; **PSL** – parascutellum; **PSLccv** – parascutellar lateral concavity; **PSLcvx** – parascutellar lateral convexity; **PSp** – posterior scutal protuberance; **RSSLs** – recurrent scuto-scutellar suture; **SL** – scutellum; **SLS** – sublateroscutum; **SMS** – submedioscutum; **S.PSLs** – scuto-parascutellar suture; **SrA** – suralare; **S.SLi** – scuto-scutellar impression; **Teg** – tegula.

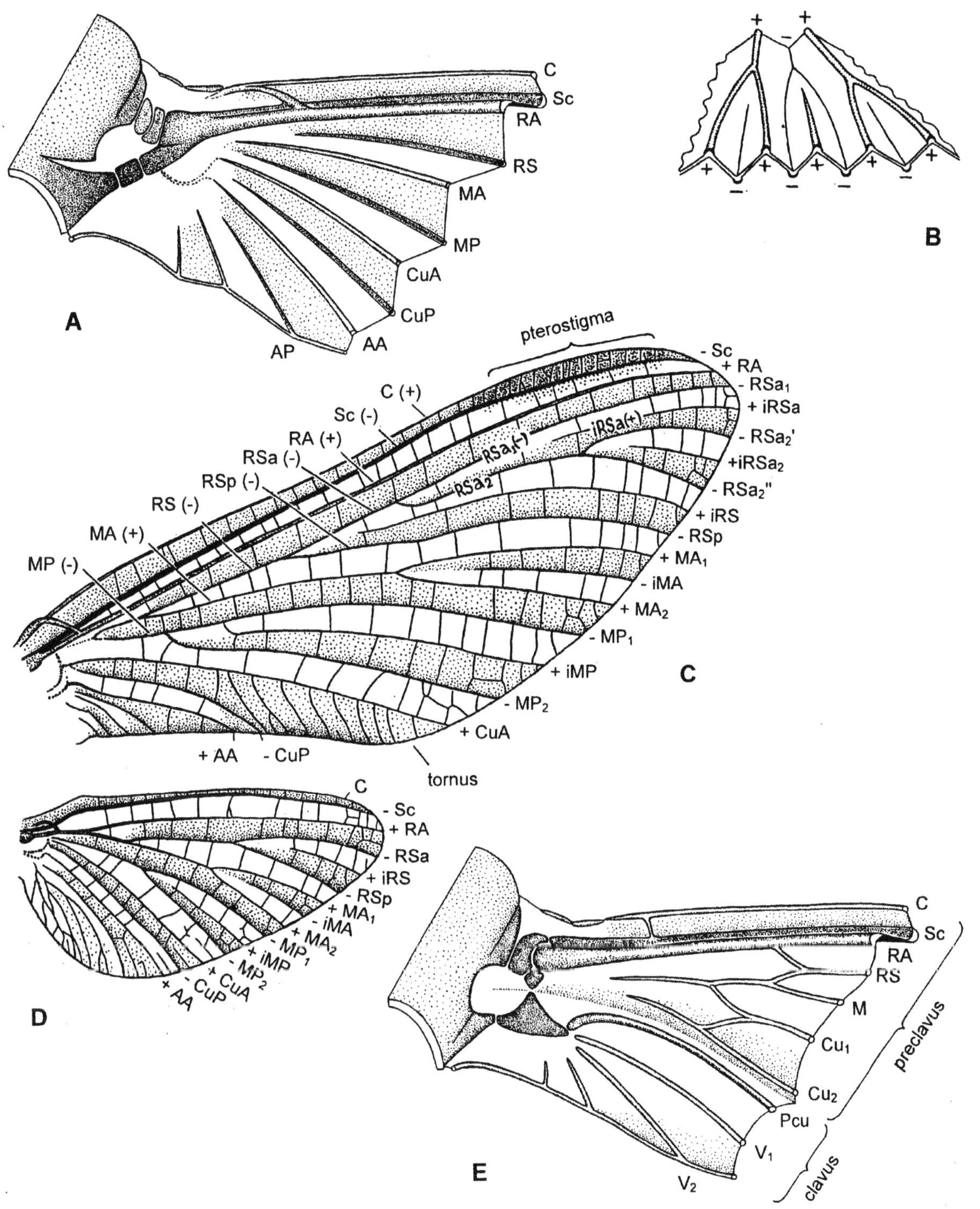

Figure 7. Wings.
A – Scheme of wing base and venation of Ephemeroptera (compare with D); **B** – scheme of triad branching of veins; **C–D** – right fore and hind wings of Siphlonurus/fg4 *aestivalis* [*Siphlurus*]; **E** – scheme of wing base and venation of Neoptera (compare with A) (B and C – from Kluge 2000).

Abbreviations: **cb** – costal brace; **pts** – pterostigma.

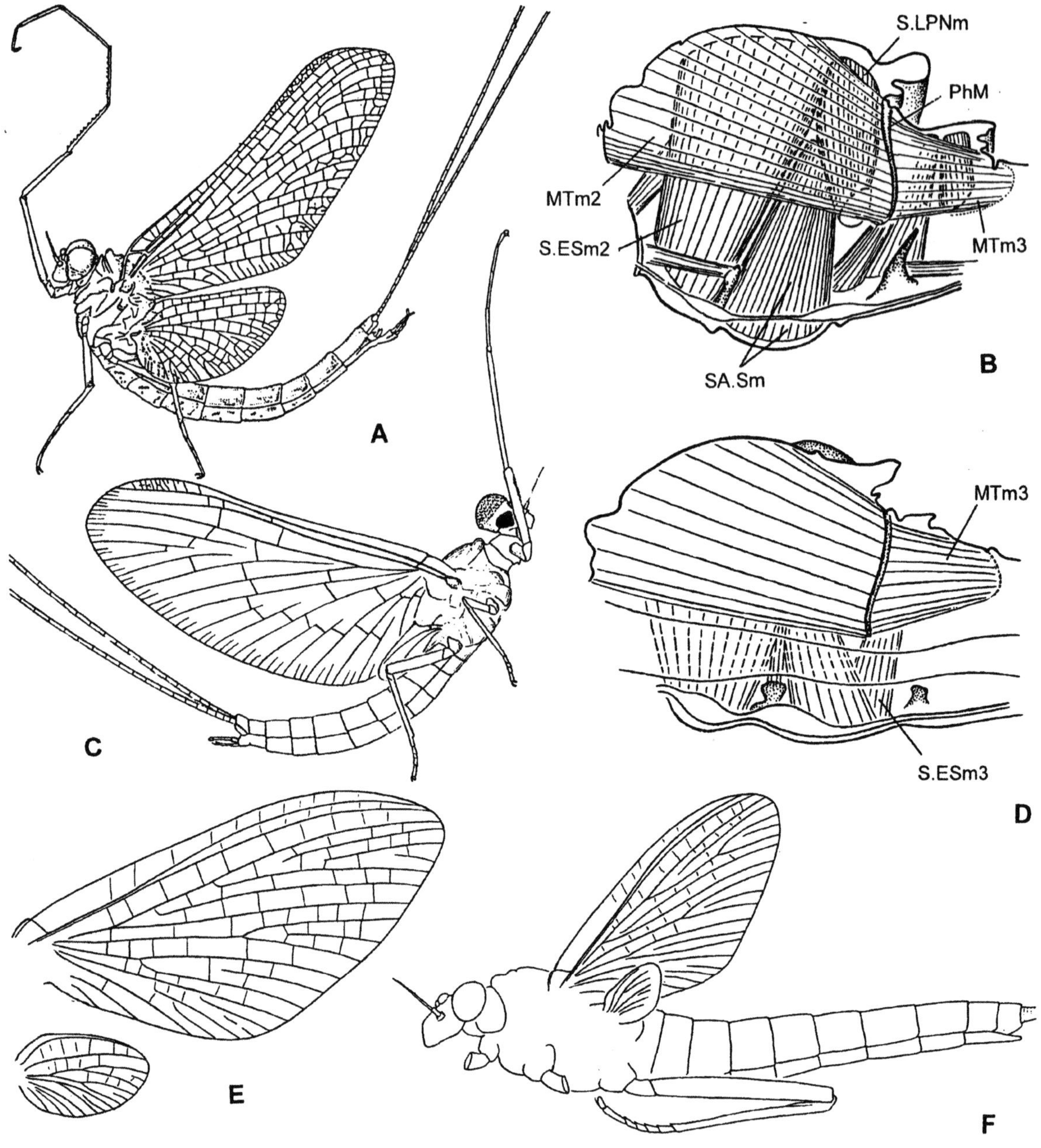

Figure 8. Reduction of wings and pterothorax among mayflies.
A–B – Siphlonurus/fg3 spp., mayflies with most usual plesiomorphic proportions: **A** – general view; **B** – median section of pterothorax. **C–D** – Turbanoculata, mayflies with reduced hind wings: **C** – *vernus* Curtis 1834 [*Baetis*], mayfly with vestigial hind wings; **D** – median section of pterothorax of Cloeon/fg* *diptera* [*Ephemera*], mayfly without hind wings. **E–F** – individual variability of wing size in an undescribed short-winged species of Rhithrogena/fg2 from Chukotka (River Amguema 12 km below meteorological station, 1 IX 1976, leg. E. Makarchenko). (A – from Tshernova 1964, B – from Kluge 2000, C – from Kluge 1997d).

Abbreviations: **MTm2** – median tergal muscle of mesothorax; **MTm3** – median tergal muscle of metathorax; **PhM** – middle phragma; **SA.Sm** – subalar-sternal muscle; **S.ESm2** – scuto-episternal muscle of mesothorax; **S.ESm3** – scuto-episternal muscle of metathorax; **S.LPNm** – scuto-lateropostnotal muscle.

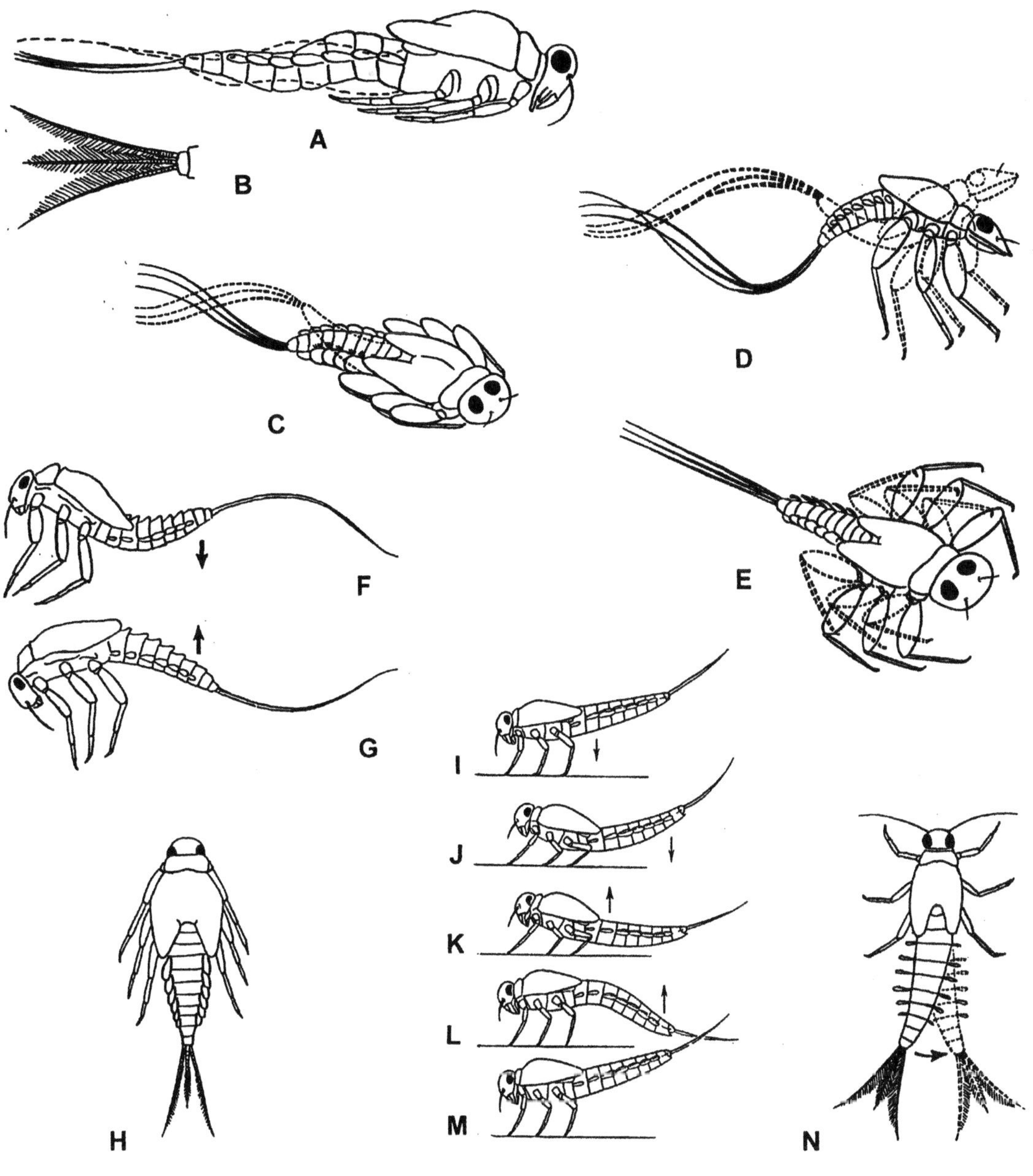

Figure 9. Movements of larvae.
A–B – Siphlonurus/fg1 (larva with primary siphlonuroid swimming specialization): **A** – lateral view of swimming larva; **B** – dorsal view of its caudalii. **C–E** – swimming larvae of various Radulapalpata (rheophilous taxon, in which swimming siphlonuroid specialization is partly or completely lost): **C** – Rhithrogena/fg3 (leg pose differs from siphlonuroid); **D** – Heptagenia/f6=g5 (legs free); **E** – Ecdyonurus/fg1 (swims by legs rather than by abdomen). **F–H** – rheophilous representatives of Turbanoculata: **F–G** – Baetiella/g1 *tuberculatum* Kazlauskas 1963 [*Pseudocloeon (Baetiella)*]; **H** – Acentrella/g1 *gnom* Kluge 1983 [*Pseudocloeon*]. **I–N** – Respiratory movements: **I–M** – respiratory movements of Ameletus/fg1 representing modified primary siphlonuroid swimming movements; **N** – special respiratory movements of Baetis/fg* different from its swimming movements. (F–N – from Kluge & Nivikova & Brodsky 1984).

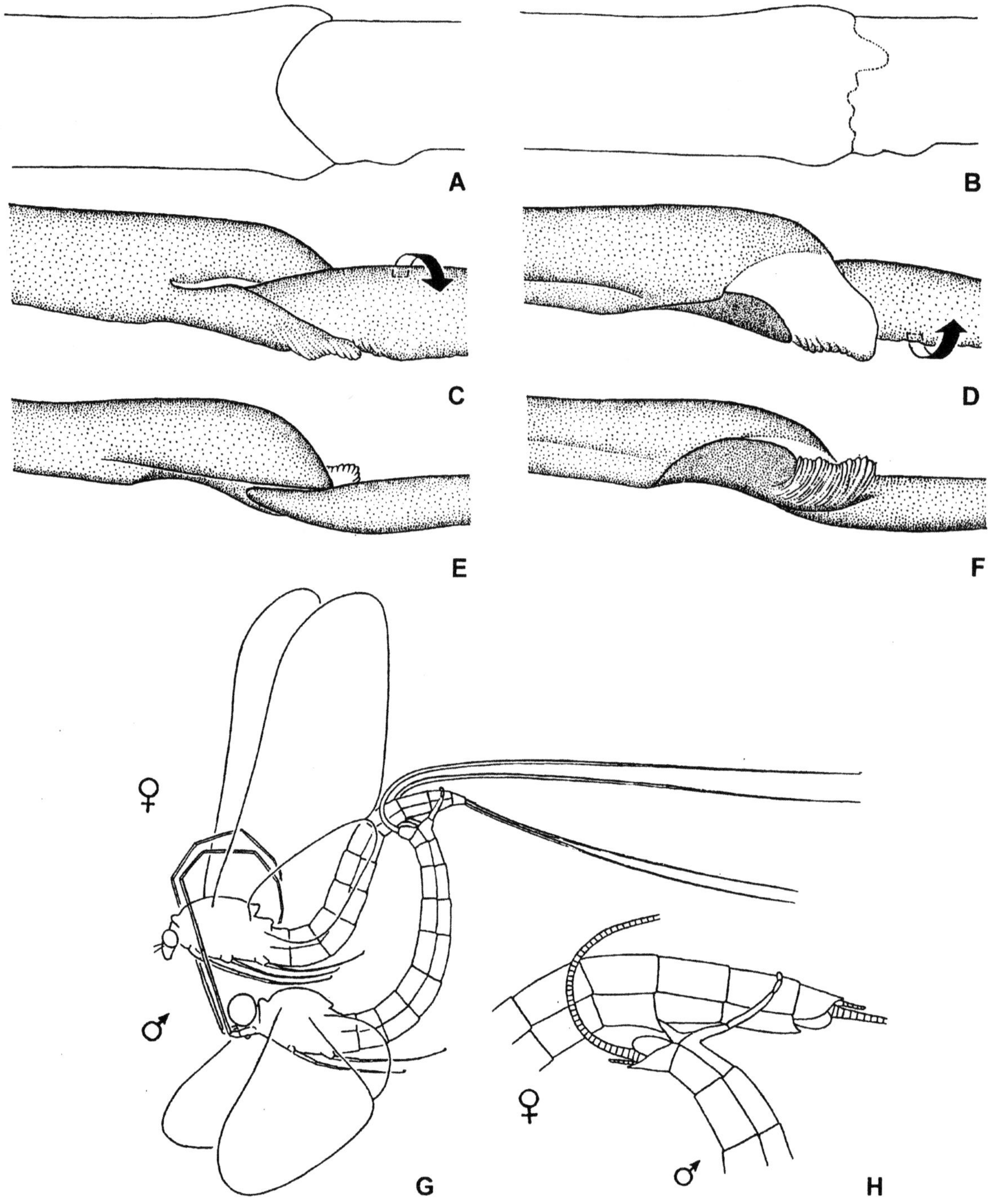

Figure 10. Copulation.
A–F – Tibia-tarsal articulation of male fore leg of Metretopus/fg* *borealis* [*Heptagenia*] (each leg lies with knee articulation directed up and tarsus directed to the right); left figures (**A**, **C**, **E**) – left leg in median view; right figures (**B**, **D**, **F**) – right leg in lateral view: **A–B** – subimago; **C–D** – imago with normal position of tarsus (arrow shows future rotation which leads to position on Figs E and F); **E–F** – imago with tarsus turned around at 180°. **G–H** – Pose of copulation of Parameletus/fg2 *chelifer* [*Parameletus*] (from Kluge 2000 based on Brinck 1957).

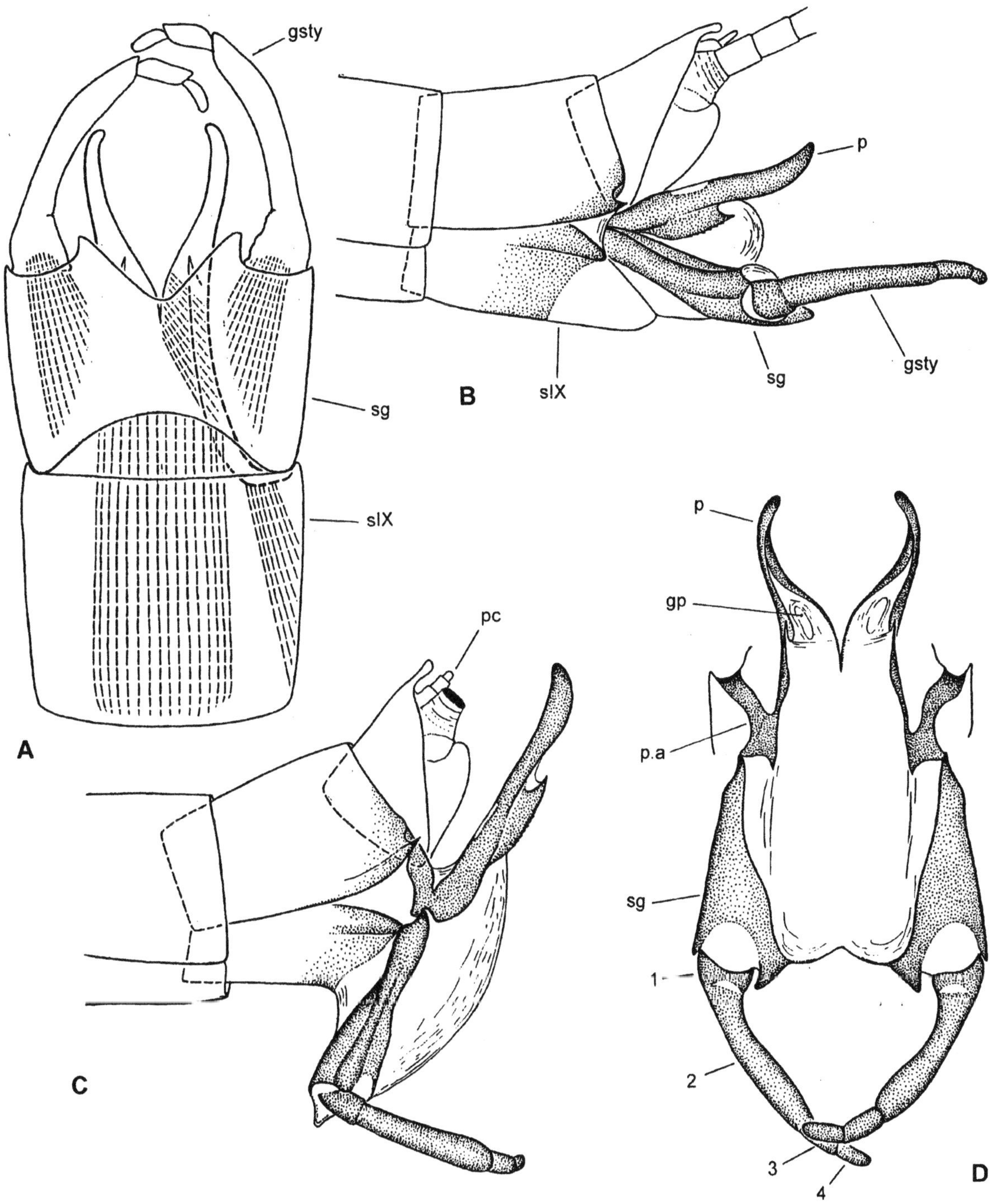

Figure 11. Genitals of male imago of Ameletus/fg2 *camtschaticus* [*Ameletus*].
A – Abdominal sternum IX with styliger, gonostyli and penis, ventral view (muscles of styliger and right half of penis are shown by interrupted lines). **B** – Apex of abdomen, lateral view. **C** – Apex of abdomen with genitals in excited condition, lateral view (left cercus removed). **D** – the same, posterior view.

Abbreviations: **1**, **2**, **3**, **4** – segments of gonostylus; **gp** – gonopore; **gsty** – gonostylus; **p** – penis; **p.a** – penial arm; **pc** – paracercus; **sIX** – abdominal sternite IX; **sg** – styliger.

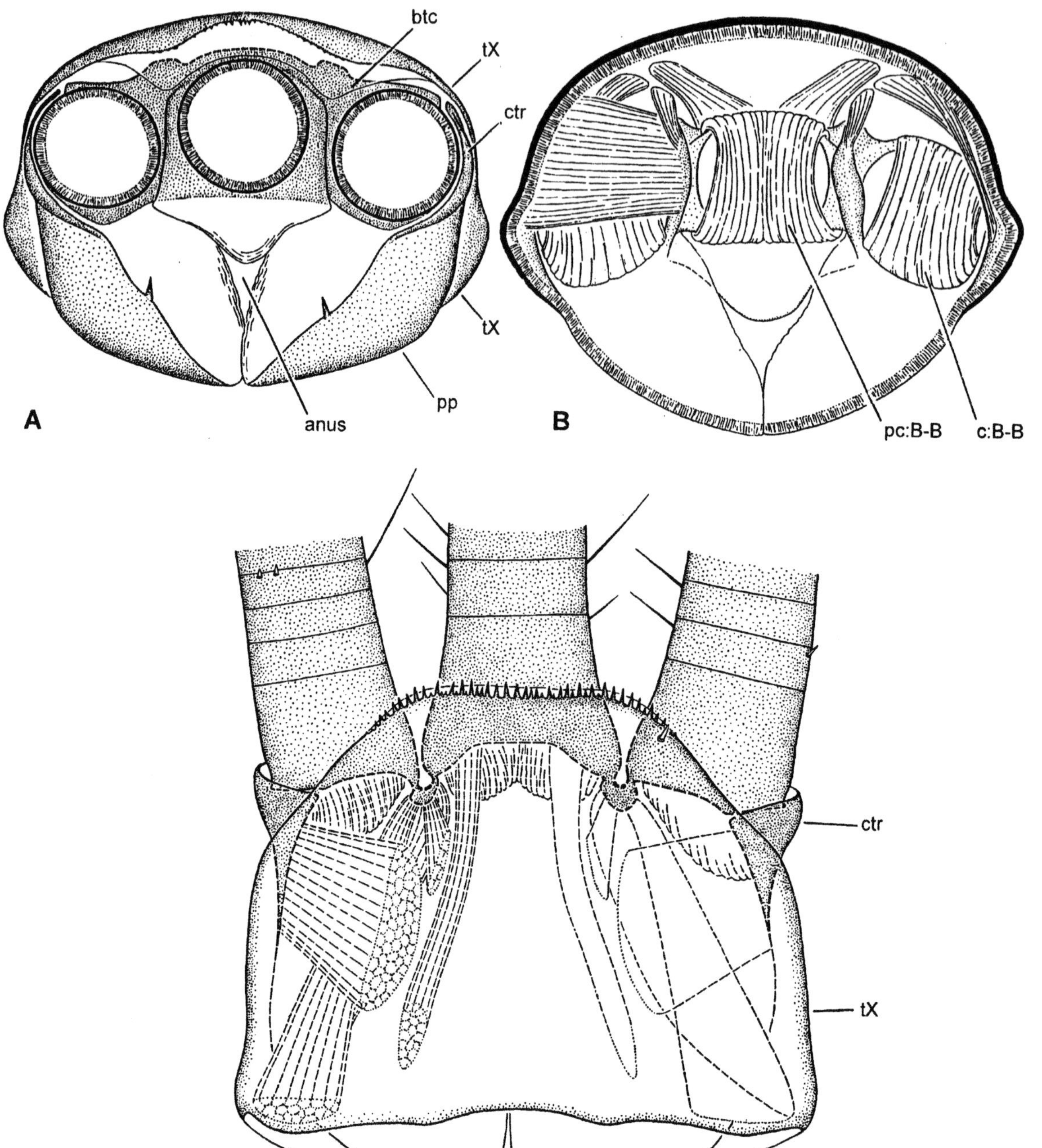

Figure 12. Larval 10th abdominal segment.
A–E – Siphlonurus/fg4 *aestivalis* [*Siphlurus*] (intestine and all muscles connecting intestine with body wall are removed): **A** – posterior view (cerci and paracercus removed, instead them three round sections are shown); **B** – anterior view, from inside; **C** – dorsal view (hidden sclerites shown by interrupted lines and dotted, muscles shown by interrupted lines, their dorsal bases – by dotted polygons); **D** – longitudinal section; **E** – lateral view (muscles shown by interrupted lines). **F** – scheme of structure usual for Ephemeroptera (corresponds to A–E); **G** – scheme of sctructutru peculiar for Geminovenata, Fossoriae and Caenotergaliae.

Abbreviations: **btc** – basitricaudale; **c** – cercus; **c:B-B** – basi-basal muscle of cercus; **ctr** – cercotractor; **pc** – paracercus; **pc:B-B** – basi-basal muscle of paracercus; **pp** – paraproct; **tX** – abdominal tergite X.

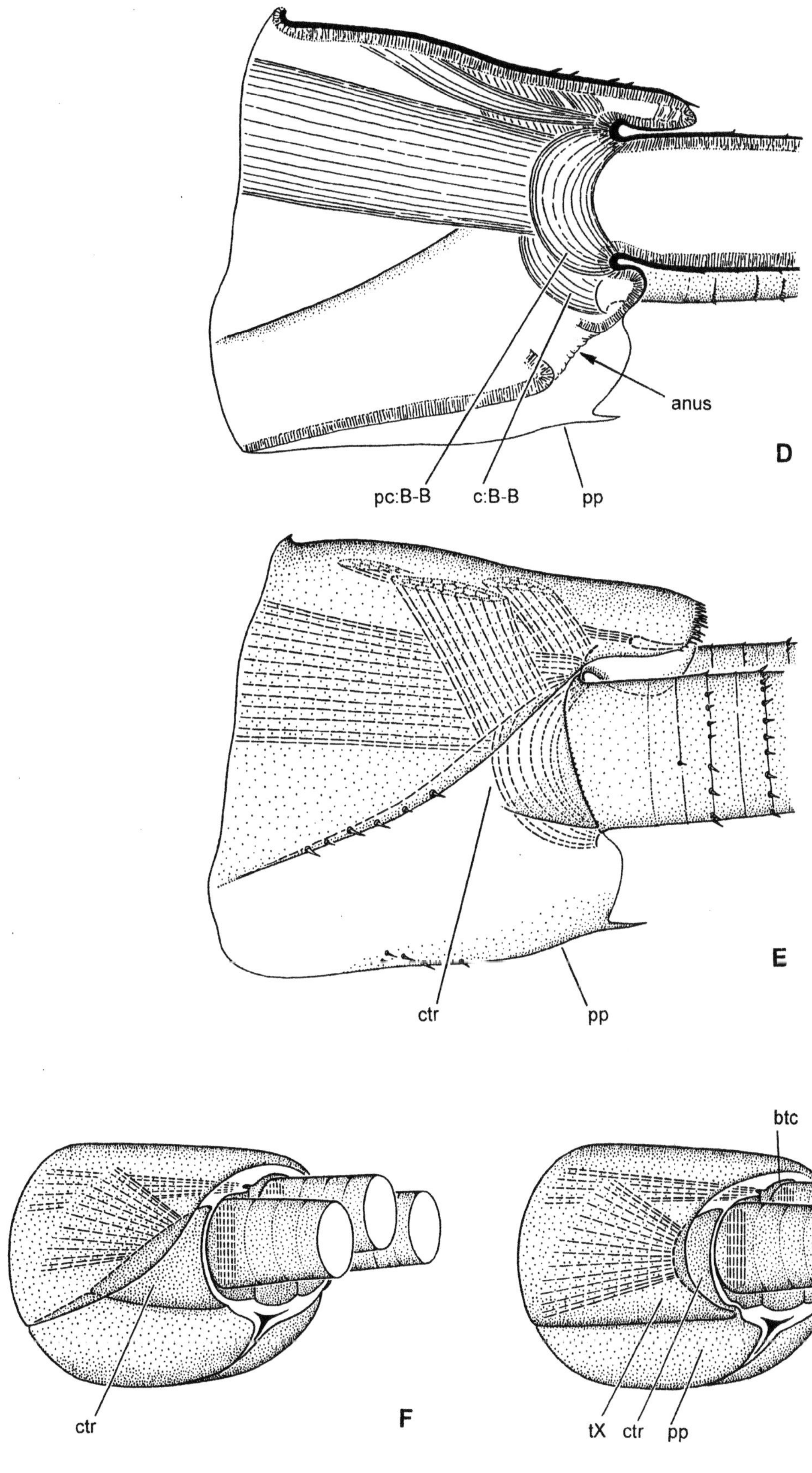
anus
pc:B-B
c:B-B
pp
D
ctr
pp
E
ctr
F
btc
pc
c
tX
ctr
pp
G

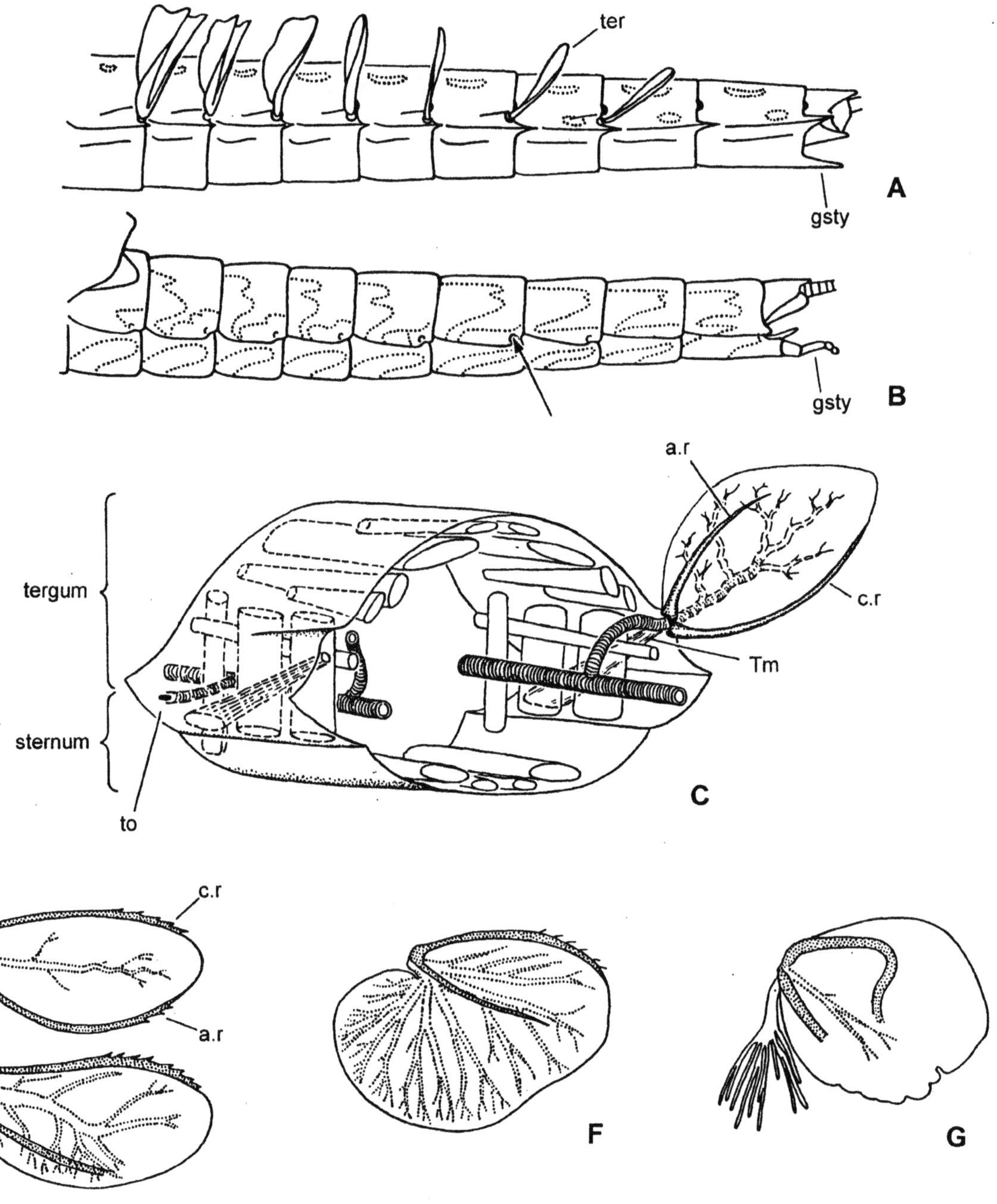

Figure 13. Tergalii.
A–B – Abdomen of male Siphlonurus/fg4 *lacustris* [*Siphlonurus*] (boundaries of pigmented spots shown by dotted lines): **A** – larva; **B** – imago (arrow shows trace of tergalia attachment). **C** – Scheme of musculature of larval abdominal segment (tergalial muscle shown by longitudinal striation, tracheal trunks – by transverse striation). **D–G** – position of ribs and tracheae in tergalii of various mayflies (ribs shown by integral lines and dotted, tracheae shown by dotted lines): **D** – *procerus* [*Ameletus*]; **E** – *montanus* [*Ameletus*]; **F** – *immanis* [*Siphlonurus*]; **G** – *eugeniae* [*Rhithrogena*]. (A–B, D–F – from Kluge 1989a; C – from Kluge 2000).

Abbreviations: **a.r** – anal rib; **c.r** – costal rib; **gsty** – gonostyli buds of larva and gonostyli of imago; **ter** – tergalius; **Tm** – tergalial muscle; **to** – tracheostium.

SPECIAL PART

Chapter III

Ephemeroptera in wider and narrower senses

If only Recent representatives are to be discussed, the taxon Ephemeroptera (mayflies) looks distinctly outlined. However, there are known several fossil Palaeozoic forms which have relationship with Recent mayflies, but do not fall into the taxon Ephemeroptera if characterize it by structure of Recent representatives. These extinct forms together with Recent mayflies form a group which can be called Ephemeroptera in widest sense, or Panephemeroptera; it includes a subordinate taxon which also can be called Ephemeroptera, or Euephemeroptera; in its turn, Euephemeroptera include a subordinate taxon Ephemeroptera in narrowest sense, or Euplectoptera, to which all Recent mayflies belong. Phylogenetic relationships of these taxa are assumed as following:

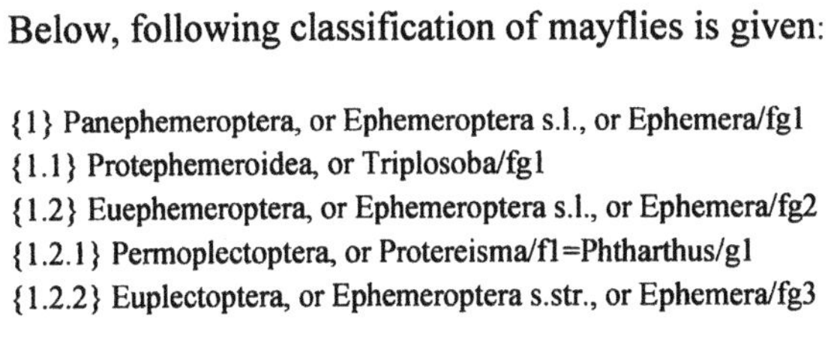

Below, following classification of mayflies is given:

{1} Panephemeroptera, or Ephemeroptera s.l., or Ephemera/fg1
{1.1} Protephemeroidea, or Triplosoba/fg1
{1.2} Euephemeroptera, or Ephemeroptera s.l., or Ephemera/fg2
{1.2.1} Permoplectoptera, or Protereisma/f1=Phtharthus/g1
{1.2.2} Euplectoptera, or Ephemeroptera s.str., or Ephemera/fg3

{1} **Panephemeroptera**, or **Ephemeroptera** sensu latissimo, or **Ephemera/fg1** (Figs 3–106)

Nomen hierarchicum: **Ephemera/fg1** (incl. *Triplosoba*) [f: Ephemerinae Latreille 1810: 273; g: *Ephemera* Linnaeus 1758: 546, typus *E. vulgata* Linnaeus 1758 (design. Latreille 1810)].
Nomina circumscribentia univoca:
— **Panephemeroptera** Crampton 1928: 85;
— **Ephemeropteria** Crampton 1938: 170;
— **Ephemeropteroidea** Rohdendorf 1968: 61.
Nomina circumscribentia non-univoca (in circumscription matching also Euephemeroptera and Euplectoptera) – see below, Euplectoptera.
In circumscription exactly matches:
— ordo Ephemeroptera: Demoulin 1956b: 8;
— ordo Ephemerida: Rohdendorf 1977: 20;
— superordo Panephemeroptera Crampton 1928: 83;
— superordo Panephemeroptera, or Ephemeropteria Crampton 1938: 170;
— superordo Ephemeroptera: Martynov 1938: 32;
— superordo Ephemeropteroidea Rohdendorf 1968: 61;
— superordo Ephemeridea: Rasnitsyn 2002: 86;
— cohors Ephemeriformes: Rohdendorf 1977: 20;
— sectio Ephemerata: Boudreaux 1979: 196;
— subclassis Ephemeroidea: Handlirsch 1906: 37;
— Panephemeroptera, or Ephemera/fg1: Kluge 2000: 241.
In circumscription non-univocally matches taxa listed below, under Euplectoptera.

References. Martynov 1938: ⊕*; – Tshernova 1962b: ⊕*; – Kluge 2000: ⊕*; – Rasnitsyn 2002: ⊕*.

Characters of unclear phylogenetic status.

(1) Wings are unable to fold on back: in rest are directed dorsally (FIG.8:A) or spread laterally (FIG. 14:A). The same in Odonatoptera Lameere 1900 (or Odonata Fabricius 1793 s. l., or Libellula/fg1) and Protorrhynchota Rohdendorf 1968 (or Dictyoneura/ /fg1), in contrast to Neoptera Martynov 1923.

(2) At least fore wing is costalized: veins Sc and RA go parallel to C nearly up to wing apex; RS begins as a common stem (FIGS 7:C–D; 14:A–B). The same in Protorrhynchota and many others.

(3) Convex and concave veins are regularly alternating (FIGS 7:A–D, 14:A–B): there are concave RS, convex MA, concave MP, convex CuA, concave CuP and others (besides concave Sc and convex RA which are common for Pterygota). The same in Protorrhynchota; Odonatoptera have less number of regularly alternating veins; Neoptera have different composition of veins behind RA (FIG.7:E) Among Panephemeroptera these veins can be lost only on vestigial hind wings of some Euplectoptera.

(4) Triad branching of veins provides regular alternating of convex and concave branches (FIGS 7: B–D, 14:A–B). Vein RS [concave – see (3)] forms a triad of concave RSa and RSp and convex iRS between them; RSa forms a triad of concave RSa_1 and RSa_2 and convex iRSa between them (the same in Odonatoptera, but in contrast to Protorrhynchota). MP [concave – see (3)] forms a triad of concave MP_1 and MP_2 and convex iMP. Euplectoptera have also triads formed by RSa_2 and MA (FIGS 7: C, 14:B). Only in extremely specialized representatives of Euplectoptera some of these veins are lost.

Plesiomorphy of Panephemeroptera. Paracercus is present, often long and multisegmented (FIGS 12, 14), sometimes reduced to a non-segmented vestige; presence of paracercus is a unique plesiomorphy among Pterygota.

Size. Fore wing length 2–40 mm.

Age and distribution. Carboniferous (see Protephemeroidea) — Recent; world-wide.

Panephemeroptera are divided into Protephemeroidea and Euephemeroptera.

{1.1} Extinct taxon **Protephemeroidea**, or **Triplosoba/fg(1)**

(Panephemeroptera Protephemeroidea)

(Fig. 14:A)

Nomen hierarchicum: **Triplosoba/fg(1)** [f: Triplosobidae Handlirsch 1906: 312; g: *Triplosoba* Handlirsch 1906: 312, typus *Blanchardia pulchella* Brongniart 1893 (monotypy); syn. obj.: ***Blanchardia*** Brongniart 1893: 325 (non *Blanchardia* Castelnau 1875)].
Nomina circumscribentia:
— **Protephemeroidea** Handlirsch 1906;
— **Protephemerida** Krausse & Wolff 1919;
— **Protephemeroptera** Crampton 1928;
— **Protephemerodea** Hamilton 1972.
In circumscription matches:
— gen. *Blanchardia* Brongniart 1893: 325 (nom. praeocc.);
— gen. *Triplosoba* Handlirsch 1906: 312;
— subfam. Triplosobinae: Demoulin 1956b: 7;
— fam. Troplosobidae Handlirsch 1906: 312;
— superfam. Triplosoboidea: Demoulin 1956b: 7;
— ordo Protephemeroidea Handlirsch 1906: 311;
— ordo Protephemerida Krausse & Wolff 1919: 156;
— ordo Protephemeroptera Crampton 1928: 83;
— ordo Protephemerodea Hamilton 1972: 146;
— Protephemeroidea, or Triplosoba/fg1: Kluge 2000: 242.
Monospecific taxon.

References. Brongniart 1893: ⊕; – Lameere 1917: ⊕; – Carpenter 1963: ⊕.

Characters of unclear phylogenetic status.

(1) On wings of both pairs RS is basally fused with RA and independent from MA (FIG.14:A) (in contrast to Euephemeroptera).

(2) RSa_2 is non-branched (probably on wings of both pairs, but for certain known for hind wings only – FIG.14:A) (in contrast to many Euephemeroptera).

(3) MA is non-branched (FIG.14:A) (in contrast to majority of Euephemeroptera).

Size. Fore wing length more than 21 mm.

Age. Carboniferous (France).

Species composition of Protephemeroidea. One species – *pulchella* Brongniart 1893 [*Blanchardia*], known as a single adult specimen.

Material examined: –.

***

{1.2} **Euephemeroptera**, or **Ephemeroptera** sensu lato, or **Ephemera/fg2**

(Panephemeroptera Euephemeroptera)

(Figs 3–13; 14:B–D; 15–106)

Nomen hierarchicum: **Ephemera/fg2** (sine *Triplosoba*; incl. *Phtharthus*).
Nomen circumscribens univocum:
— **Euephemeroptera** Kluge 2000: 242.
Nomina circumscribentia non-univoca (in circumscription matching also Panephemeroptera and Euplectoptera) – see below, Euplectoptera.
In circumscription exactly matches:
— subordo Ephemerina: Tshernova 1980: 31;
— ordo Plectoptera: Handlirsch 1906: 37;
— ordo Ephemerida: Handlirsch 1919: 63 [573];
— ordo Ephemeroptera: Krausse & Wolff 1919: 157;
— Euephemeroptera, or Ephemera/fg2: Kluge 2000: 242.
In circumscription non-univocally matches taxa listed below, under Euplectoptera.

References. Tshernova 1962b: ⊕*; – Carpenter 1963: ⏱ ⊕*; – Kluge 2000: ⏱ ⊕*; – Kluge & Sinitshenkova 2002: ⏱ ⊕*.

Characters of unclear phylogenetic status.

(1) Wings of both pairs have a **costal brace** – a short vein which goes from wing base between C and Sc, falls into Sc and at the same place is connected by a cross vein with RA. Unique character. For Protephemeroidea costal brace is not described. In Permoplectoptera costal brace is situated between C and Sc (FIG.14:B), but in Euplectoptera it is stout, convex anteriorly, and projects dorsad-anteriad of C (FIG.6); in Discoglossata it is modified (FIG.42:A).

(2) Vein RS arises not from RA, but from MA (in contrast to Protephemeroidea); the common basal stem RS+MA can be either independent from RA (FIG.7:C), or basally fused with RA (particularly on hind wings of Euplectoptera – FIG.7:D), or secondary reduced (FIG.17:A). Non-unique character, the same in some other Pterygota.

(3) Vein RSa_2 [concave – see Panephemeroptera (4)] forms a triad of concave RSa_2' and RSa_2'' and convex $iRSa_2$ between them (FIG.7:C). Unique character. This third triad of RS is present on wings of both pairs in Permoplectoptera (FIG.14:B) and fore wings of many Euplectoptera [but not on their hind wings – see Euplectoptera (1) below]. On fore wings of some Euplectoptera veins RSa_2' and $iRSa_2$ secondarily become intercalaries or lost (see Index of characters [2.2.37]).

(4) On wings of both pairs MA [convex – see Panephemeroptera (3)] forms a triad of convex MA_1 and MA_2 and concave iMA between them (FIG.7:C–D). This branching of MA secondary disappears on vestigial hind winds of some Euplectoptera and on fore wings in some groups with especially modified venation (see Index of characters [2.2.43]). Usually furcation of MA is situated approximately in middle of wing, but sometimes it can be secondarily transferred proximally (on fore wings of some specialized Euplectoptera) or toward wing margin (on vestigial hind wings of some Euplectoptera).

(5) Larva is aquatic, initially with a peculiar swimming **siphlonuroid specialization** (FIGS 9:A–B; 14:C–D; 28:A): abdomen is elongate and able to make undulate dorsoventral swimming movements; caudalii are not long (shorter than in imago), with primary swimming setae – i. e. each cercus has a row of setae on inner side, and paracercus has a pair of rows of setae of the same kind on its lateral sides; thanks to this, caudalii can function as a horizontal caudal flipper. As larvae of other Panephemeroptera are unknown, it is unclear if this specialization is an autapomorphy of Euephemeroptera or an autapomorphy of a larger taxon. In many Euplectoptera this swimming specialization is secondarily lost.

(6) Larva has tergalii on abdominal segments I–IX (FIG.14:D). As larvae of other Panephemeroptera are unknown for certain, it is unclear, if the presence of tergalii is a character of Euephemeroptera or of a larger taxon; if proceed from the assumption that tergalii are serial homologues of wings, their presence is a plesiomorphy. In various Euephemeroptera tergalii of these or that pairs are lost (see Index of characters [1.3.19]).

Size. Fore wing length 2–40 mm.

Age and distribution. Permian (see Permoplectoptera) – Recent; world-wide.

Euephemeroptera are divided into Palaeozoic plesiomorphon Permoplectoptera and Mesozoic–Recent taxon Euplectoptera. Some fossil mayflies have uncertain systematic position (see Appendix: p.359).

{1.2.1} † Plesiomorphon **Permoplectoptera**, or **Protereisma/f1=Phtharthus/g1**

(Panephemeroptera Euephemeroptera Permoplectoptera)

(Figs 14:B–D)

Nomen hierarchicum: **Protereisma/f1=Phtharthus/g1** [f: Protereismatidae (orig. Protereismidae) Lameere 1917; g: *Phtharthus* Handlirsch 1904a, typus *Ph. rossicus* Handlirsch 1904 (design. orig.; syn.subj. *Ph. netshaevi* Handlirsch 1904)].

Nomina circumscribentia:

— **Protereismephemeridae** Sellards 1907: 345;
— **Permoplectoptera** Tillyard 1932: 117.

In circumscription matches:

— fam. Protereismephemeridae Sellards 1907: 345;
— fam. Protereismidae Lameere 1917: 45;
— superfam. Protereismatoidea: Demoulin 1958: 6;
— subordo Permoplectoptera Tillyard 1932: 117;
— Permoplectoptera, or Protereisma/f1=Phtharthus/g1: Kluge 2000: 243.

References. Sellards 1907: ⊕; – Tillyard 1932: ⊕; – Carpenter 1933: ⊕; – Kukalova 1968: ℧; – Tshernova 1970: ⊕; – Carpenter 1979: ℧ ⊕.

Plesiomorphies of Permoplectoptera (in contrast to Euplectoptera). Wings are homonomous – i.e. hind wings have nearly the same size and venation as fore wings (FIG.14:B). At least in some representatives tergalii are present not only on abdominal segments I–VII, but on abdominal segments VIII–IX as well (FIG.14:D). For larva of *americana* [*Kukalova*], segmented tarsus and two claws are described, that resembles adult structure and differs from larval Euplectoptera (Kukalova 1968).

Size. Fore wing length 6–32 mm.

Age. From Early Permian to Late Jurassic (Europe and North America).

Classification of Permoplectoptera. Larvae and winged stages are associated for a single form only [see (A)]; other forms are described either as winged stages, or as larvae with uncertain characters. Thus, only artificial classification of Permoplectoptera can be used, with following groups distinguishable.

(A) On wings of both pairs CuA forms a triad; vein gemination is absent; Permian. Here belong: **Protereisma/fg** [g: *Protereisma* Sellards 1907, typus *P. permianum* Sellards 1907 (design. orig.)]; **Protechma/g** [g: *Protechma* Sellards 1907: 349, typus *P. acuminatum* Sellards 1907 (design. orig.)]; **Prodromites/g** [g: *Prodromites* Cockerell 1924: 136, typus *Prodromus rectus* Sellards 1907 (design. orig.); syn. obj.: ***Prodromus*** Sellards 1907: 349 (non *Prodromus* Distant 1904)]; **Bantisca/g** [g: *Bantisca* Sellards 1907: 349, typus *B. elongata* Sellards 1907 (design. orig.)]; **Rekter/g** [g: *Rekter* Sellards 1907: 349, typus *R. arcuatus* Sellards 1907 (design. orig.); syn. obj.: ***Recter*** Sellards 1909: 151]; **Pinctodia/g** [g: *Pinctodia* Sellards 1907: 352, typus *P. curta* Sellards 1907 (design. orig.)]; **Mecus/g** [g: *Mecus* Sellards 1907: 151, typus *Scopus gracilis* Sellards 1907 (design. orig.); syn. obj.: ***Scopus*** Sellards 1907: 352 (non *Scopus* Brisson 1760; nec Oken 1809; nec Scudder 1882)] (all these names are regarded as generic synonyms by Tillyard 1932: 244); **Kukalova/fg** [f: Kukalovidae Demoulin 1970b: 6; g: *Kukalova* Demoulin 1970b: 6, typus *K. americana* Demoulin 1970 (design. orig.)], known as larvae with well-preserved venation on wing buds (Carpenter 1979). Other species: *directum* Carpenter 1979 [*Protereisma*], *insigne* Tillyard 1932 [*Protereisma*], *latum* Sellards 1907 [*Protereisma*], *minus* Sellards 1907 [*Protereisma*], *sellardsi* Tillyard 1932 [*Protereisma*], *uralicum* Zalessky 1947 [*Protereisma*].

(B) On wings of both pairs CuA is non-branched; vein gemination is absent; Permian. Here belong: **Misthodotes/fg** [f: Misthodotidae Tillyard 1932: 260; g: *Misthodotes* Sellards 1909: 151, typus *Dromeus obtusus* Sellards 1907 (design. orig.); syn. obj.: ***Dromeus*** Sellards 1907: 257 (non *Dromeus* Reiche 1854)]; **Eudoter/fg** [f: Eudoteridae Demoulin 1954f: 553; g: *Eudoter* Tillyard 1936b: 443, typus *E. delicatulus* Tillyard 1936 (design. orig.)], regarded as a generic synonym of *Misthodotes* (Carpenter 1979: 237). Other species: *biguttatus* Tillyard 1932 [*Misthodotes*], *edmundsi* Carpenter 1979 [*Misthodotes*], *ovalis* Tillyard 1932 [*Misthodotes*], *sharovi* Tshernova 1965 [*Misthodotes*], *stapfi* Kinzelbach & Lutz 1984 [*Misthodotes*], *zalesskyi* Tshernova 1965 [*Misthodotes*]. Adults of *sharovi* [*M.*] have mayfly features (wing venation of *Misthodotes*-type and three caudalii), but at the same time have unusually elongate mouth apparatus of unclear structure resembling that of Protorrhynchota.

(C) At least on hind wing CuA forms a triad, CuP is sinuous, and veins geminate forming pairs $RSp+MA_1$ and $iMP+MP_2$; Permian. This is **Palingeniopsis/fg** [f: Palingeniopsidae Martynov 1938: 35; g: *Palingeniopsis* Martynov 1932: 10, typus *P. praecox* Martynov 1932 (design. orig.)]. This taxon known as a single wing, was regarded to be related with *Palingenia* and some other Recent taxa which also have CuP curvation and vein gemination (Martynov 1932, Demoulin 1958, Tshernova 1980). Actually such kind of gemination occurs in non-related taxa (see Index of characters [2.2.41] and [2.2.44]); CuP of *Palingeniopsis* is curved in opposite direction than in Fimbriatotergaliae.

(D) Poorly preserved winged insects.

(D.1) Permian. **Loxophlebia/g** [g: *Loxophlebia* Martynov 1928: 8, typus *L. apicalis* Martynov 1928 (design. orig.)], Permian, known as distal fragment of a wing, regarded as a generic synonym of *Protereisma* (Rohdendorf 1957: 76). Other species: *rossenrayensis* Guthorl 1967 [*Protereisma*].

(D.2) Jurassic. **Mesephemera/fg** [f: Mesephemeridae Lameere 1917: 47; g: *Mesephemera* Handlirsch 1906: 600, typus *Ephemera procera* Hagen 1862 (design. Hubbard 1981: 69)]; redescribed by Demoulin 1955i. Other formal species (according to Demoulin 1955i): *lithophilus* Germar 1842 [*Tineites*], *palaeon* Weyenbergh 1874 [*Anomalon*], *prisca* Germar 1839 [*Sciaria*], *speciosa* Oppenheim 1888 [*Ephemera*] nom. praeocc., *weyenberghi* Handlirsch 1906 [*Mesephemera*].

(E) Larvae with unknown wing venation.

(E.1) Permian. **Phtharthus/g** [see above, Protereisma/fl=Phtharthus/g1]. Originally the genus *Phtharthus* was described basing on 3 specimens deposted in Kazan' University – 2 syntypes of *Ph. netshaevi* Handlirsch 1904 (16-17 mm long) and holotype of Ph. rossicus Handlirsch 1904 (24 mm long); each specimen represents a reverse replic of dorsal side of body, lacks cuticle or pigmentation and locates at a bottom of an ovoid concavity on a separate stone. Lectotype of *Ph. netshaevi* (designated here) is a larger specimen (body 17 mm, caudalii 5 mm) (Fig.14:C). All three specimens are conspecific (Kluge & Sinitshenkova 2002). In the former descriptions and reconstruction by Handlirsch (1904, 1906-1908, 1925) it was stated that *Phtharthus* had wing buds directed posteriorly; tergalii attached ventrally, stick-like and setose; caudalii covered by setae on all sides. Basing on these descriptions, some authors concluded that initial position of insect wings is posterior rather than lateral, that mayfly tergalii originated from ventral limbs, and that most primitive recent mayflies are Pinnatitergaliae, whose larval cerci have setae on both sides. Re-examination of the fossils revealed the following: in all three specimens wing buds are not preserved at all (that is strange, because moso- and metanota of all three specimens are well-preserved, each with distinct postrior scutal protuberances and scutellum characteristic for a wing-bearing segment); tergalii are attached not ventrally, but to posterior-lateral angles of segment, and probably are lamellate and rugose (so look as being stick-like and pectinate); caudalii have primary swimming setation only (Kluge & Studemann & Landolt & Gonser 1995; Kluge & Sinitshenkova 2002). On the place where a replic of the head should be, each specimen has a sharp impression of an unusual form.

Other Permian larvae are: **Jarmila/fg** [f: Jarmilidae Demoulin 1970b: 7; g: Jarmila Demoulin 1970b: 7, typus

J. elongata Demoulin 1970 (design. orig.)]; **Oboriphlebia/fg** [f: Oboriphlebiidae Hubbard & Kukalova-Peck 1980: 29; g: *Oboriphlebia* Hubbard & Kukalova-Peck 1980: 29, typus *Kukalova moravica* Demoulin 1970 (design. orig.)]. Other species: *tertia* Hubbard & Kukalova-Peck 1980 [*Oboriphlebia*], *quarta* Hubbard & Kukalova-Peck 1980 [*Oboriphlebia*], *quinta* Hubbard & Kukalova-Peck 1980 [*Oboriphlebia*]. Other specimen: larval paratype of *sharovi* Tshernova 1965 [*Misthodotes*], possibly conspecific with the type specimens of *Phtharthus*.

(E.2) Triassic. **Mesoplectopteron/fg** [f: Mesoplectopterinae Demoulin 1955g: 345; g: *Mesoplectopteron* Handlirsch 1918, typus *M. longipes* Handlirsch 1918: 112 (design. orig.)].

Material examined: *netshaevi* [*Phtharthus*] and *rossicus* [*Ph.*]): ↺ (Kazan. Univ.); *sharovi* [*Misthodotes*]: ⊕ (Paleontol. Inst.); larva ascribed to *sharovi* [*M.*]: ↺ (Paleontol. Inst.); *zalesskyi* [*M.*]: ⊕ (Paleontol. Inst.).

***

{1.2.2} Euplectoptera, or Ephemeroptera sensu srticto, or Ephemera/fg3

(Panephemeroptera Euephemeroptera Euplectoptera)

(Figs 3–13; 15–106)

Nomen hierarchicum: **Ephemera/fg3** (sine *Phtharthus*; incl. *Prosopistoma*).

Nomen circumscribens univocum:

— **Euplectoptera** Tillyard 1932: 267 (non Euplectoptera Fischer 1853).

Nomina circumscribentia non-univoca (in circumscription matching also Panephemeroptera and Euephemeroptera):

— **Anisoptera** Stephens 1835: 53;
— **Plectoptera** Packard 1886: 808;
— **Ephemeroptera** Hyatt & Arms 1891: 13;
— **Plectopteradelphia** Crampton 1916: 305;
— **Archipterygota** Börner 1909: 121;
Prometabola Chen 1958.

In circumscription exactly matches:

— subordo Euplectoptera Tillyard 1932: 267;
— Euplectoptera, or Ephemera/fg3: Kluge 2000: 245.

In circumscription non-univocally matches following taxa (each of which also matches Euephemeroptera and Panephemeroptera):

— gen. *Ephemera* Linnaeus 1758: 546;
— natio Ephemeraedes: Billberg 1820: 97;
— tribus Ephemerides: Leach 1815: 137;
— fam. Ephemerinae Latreille 1810: 273;
— fam. Ephemerina: Burmeister 1829: 20;
— fam. Ephemeridae: Stephens 1835: 54;
— sectio Anisoptera Stephens 1835: 53;
— subordo Ephemerina: Packard 1883: 192;
— ordo Ephemerida: Haeckel 1866;
— ordo Ephemeridae: Brauer 1885: 353;
— ordo Plectoptera Packard 1886: 808;
— ordo Ephemeroptera Hyatt & Arms 1891: 13;
— ordo Agnatha: auct. (non Agnathes Cuvier 1798);
— sectio Plectopteradelphia Crampton 1916: 305;
— supersectio Archipterygota Börner 1909: 121;
— cohors Prometabola Chen 1958.

References. Tillyard 1932: ⊕*; – Tshernova 1962b: ⊕*; – Kluge 2000: ⊕*.

Characters of unclear phylogenetic status.

(1) Anteromotoric: hind wings are diminished, coupled with fore ones and modified in following manner: hind wing veins RA, RS and MA have a common stem [instead of the stem RS+MA separated from RA on fore wing – see Euephemeroptera (2)]; RS forms a single triad [instead of three successive triads of fore wing – see Euephemeroptera (3)], thus hind wing has maximum three simple triads alternating as concave and convex – RS, MA and MP (FIG.7:D). Hind wing length is subequal or less than a half of fore wing length. At flight hind and fore wings of each side couple because costal margin of hind wing is bent dorsally and basitornal margin of fore wing is bent ventrally. Thus, a functional diptery is present. In connection with this, fore wing is triangular, with tornus expressed (the same independently appeared in other anteromotoric Pterygota). As tornus of fore wing has different position in different Euplectoptera [see Posteritorna (1) and Anteritorna (1) below], its independent origin can be assumed, as well as independent shortening of hind wing. In various taxa among Euplectoptera hind wings undergone further reduction up to complete disappearance (see Index of characters [2.2.59]).

(2) Tergalii [see Euephemeroptera (4)] of abdominal segments VIII–IX are lost, thus tergalii are present only on segments I–VII or only on some of them. Pattern of this character is not quite clear, as larvae of many Permoplectoptera are unknown.

Size. Fore wing length 2–40 mm.

Age and distribution. From Jurassic (see Eusetisura INCERTAE SEDIS, Fossoriae INC. SED. and Anteritorna INC. SED.) — till Recent; world-wide. Some Triassic fossils were attributed to Euplectopteran genera *Mesoneta*, *Mesobaetis* and *Archaeobehningia* without enough ground (see below, "Other species of Euephemeroptera INCERTAE SEDIS" and "Euarthropoda INCERTAE SEDIS", pp. 360, 361). Reliable fossils of Triassic Euplectoptera are unknown.

Euplectoptera are divided into Posteritorna (Chapter IV) and Anteritorna (Chapters V–VII: p.71). Some fossil mayflies have uncertain systematic position (see Appendix: p.356).

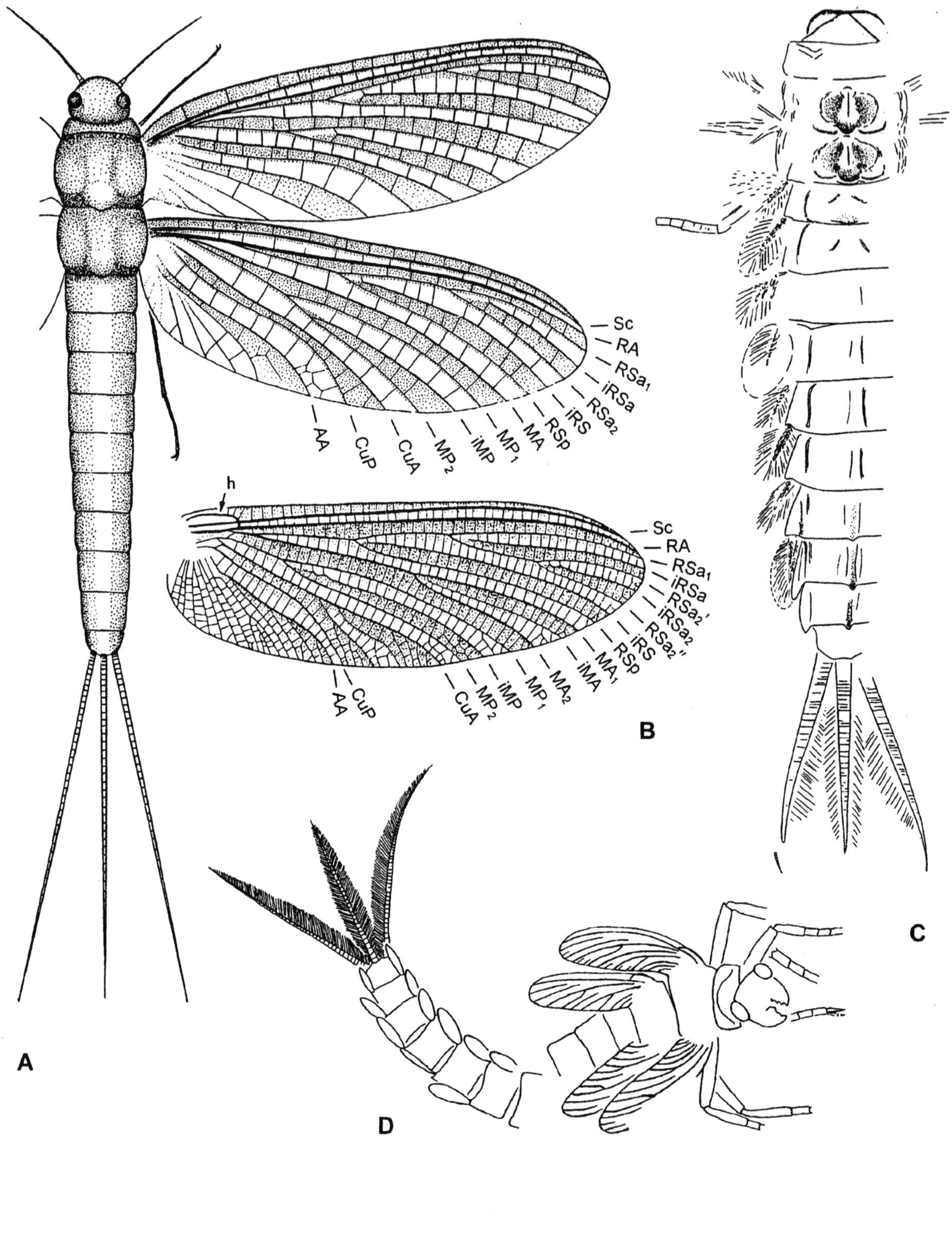

Figure 14. Palaeozoic mayflies.
A – Triplosoba/fg* *pulchella* [*Blanchardia*], reconstruction of adult (based on description by Carpenter 1963). **B** – Protereisma/fg *latum* [*Protereisma*], hind wing (based of figure and photograph by Carpenter 1933). **C** – lectotypus of *Phtharthus netshaevi*, larva. **D** – Protereisma/fg *americana* [*Kukalova*], reconstruction of larva (based on photograph of holotype in Kukalova 1968 and Carpenter 1979 and description by Carpenter 1979).

Chapter IV

Euplectoptera Posteritorna

Phylogeny of Posteritorna looks as following:

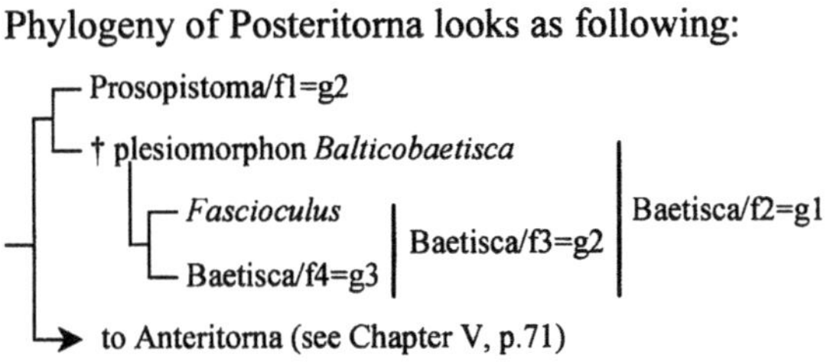

1. Posteritorna, or Baetisca/f1=Prosopistoma/g1

(Panephemeroptera Euephemeroptera Euplectoptera Posteritorna)

(Figs 15–17)

Nomen hierarchicum: **Baetisca/f1=Prosopistoma/g1** [f: Baetiscini Banks 1900: 246; g: *Prosopistoma* Latreille 1833: 33, typus *P. variegatum* Latreille 1833 (design. Eaton 1884: 150); syn. obj.: ***Chelysentomon*** Joly & Joly 1872: 438].

Nomina circumscribentia:
— **Posteritorna** Kluge & Studemann & Landolt & Gonser 1995: 105;
— **Carapacea** McCafferty 1997: 94.

In circumscription matches:
— "Larves cryptobranches": Lestage 1917: 236;
— subfam. Prosopistominae Lameere 1917: 74;
— superfam. Prosopistomatoidea: Edmunds & Traver 1954a:240;
— superfam. Baetiscoidea: Peters & Hubbard 1989: 115;
— subordo Posteritorna (as manuscript name): Kluge 1992d: 10; 1992e: 24;
— subordo Posteritorna Kluge & al. 1995: 105;
— subordo Carapacea McCafferty 1997: 94;
— Posteritorna, or Baetisca/f1=Prosopistoma/g1: Kluge 2000:247.

References. Lameere 1917: ⊕; – Edmunds & Traver 1954b (*Baetisca*, *Prosopistoma*): ⊕; – Kluge 1998: ℧* ⊕*.

Autapomorphies of Posteritorna.

(1) Tornus of fore wing is situated behind apex of CuP, usually behind apex of AA (FIGS 16:B, 17:B) [for discussion – see Anteritorna (1) below]. In connection with this, veins CuA, CuP and AA [arising independently close to wing base – see (9)] are non-branched, nearly parallel to MP_2, slightly diverge, fall to tornoapical margin of the wing.

(2) Nerve ganglia of mesothorax, metathorax and all abdominal segments are fused together, forming an integral synganglion in basisternum of mesothorax (FIGS 16:H, 17:E); at the same time, imaginal and subimaginal furcasternal protuberances and bases of subalar-sternal muscles (located inside them) are contiguous medially (FIGS 16:E, 17:E).

In contrast to Posteritorna, in all Anteritorna at lest thoracic nerve ganglia are non-fused. In that Anteritorna, whose metathoracic ganglion is more or less approximated with mesothoracic ganglion, it is shifted into furcasternum of mesothorax and separates bases of subalar-sternal muscles; thus furcasternal protuberances appear to be separated, with a median impression between them (FIG.57:A–B) (see Index of characters [2.2.23]). As in Posteritorna furcasternal protuberances retain their initial contiguous position, it seems that translocation of the metathoracic and abdominal ganglia anteriorly took place here not through that stages, which occur in certain Anteritorna, but as a saltation, i. e. in such a way, that ganglia initially situated behind mesothoracic furcasternum, appeared before it, never being inside mesothoracic furcasternum.

(3) Larva has unique structure (FIG.15:B–C): pronotum, mesonotum and fore wing buds are completely fused together and strongly enlarged, constituting an integral **notal shield**; this shield covers abdominal terga I–VI, forming a **gill chamber**, inside which all tergalii I–VI [see (4)] are located. In all larval instars notal shield has no any external or internal traces of division to pronotum and mesonotum, and no outline of margins of fore wing buds (Pescador & Peters 1974:Figs 10,11,14), thus when subimaginal wing develops, it is crumpled in free space inside notal shield (in contrast to other mayflies, where wing buds are distinctly outlined even if they are fused with notum). Lateral margins of notal shield are bent to ventral side, being pressed to sterna and forming **epipleura** (this term is taken from coleopterology, where epipleura are lateral margins of elytra bent to ventral side; epipleura of Posteritorna are partly homoplastic with the epipleura of Coleoptera, being partly formed by fore wing buds). Thoracic sterna have a pair of longitudinal ridges. At least abdominal sternum I is not expressed, abdominal segments up to VI inclusively are connected nearly immobile (in Baetisca/f2=g1) or immobile (in Prosopistoma/f1=g2).

When adult fore wings develop inside lateral-posterior parts of notal shield, a membranous process can develop inside notal shield between them; thus adult can have an unpaired **plumidium** – membranous process on apex of scutellum (FIG.16: F–G); this plumidium is present in Prosopistoma/f1= =g2, but not in Baetisca/f2=g1 (similar, but paired plumidia exist in some other mayflies with fused wing buds – see Index of characters and [2.2.13]).

(4) Tergalii [hidden in gill chamber – see (3)] strongly differ on different segments, have unique and extremely specialized structure, nearly identical in Baetisca/f2=g1 (Pescador & Peters 1974:Figs 16, 20) and Prosopistoma/f1=g2 (FIG.15:D–I).

Tergalius I is the longest, with straight costal margin and dissected anal margin (as tergalii of Posteritorna are always directed posteriorly, the costal margin is directed laterally and the anal margin – medially). Tergalius II is modified as gill operculum, covering tergalii III–V: it is enlarged, nearly square, convex. Tergalii III–V have dissected margins. Tergalius VI is widened distally, rounded, with integral margin (in some species of Prosopistoma/f1=g2 lost). Tergalii VII are lost.

Tergalial structure of Posteritorna has some common features with Caenotergaliae and Leptohyphes/fg1: tergalii II are transformed to gill opercula, tergalii III–V are dissected, tergalii VII are lost. In connection with this, hypotheses about relationship of these groups were suggested. But in other respects tergalial structure of Caenotergaliae and Leptohyphes/fg1 differs from that of Posteritorna: their tergalii I are vestigial or lost, tergalii III–VI have different shape. Each of the named here common characters in tergalial structure of Posteritorna, Caenotergaliae and Leptohyphes/fg1 occurs also in some other groups of Ephemeroptera, where it evolved independently.

As in Posteritorna all tergalii are hidden under the notal shield and thus are protected better than in any other Ephemeroptera, the transformation of tergalii II to gill opercula is not justified functionally. This leads to an assumption, that ancestors of Recent Posteritorna got specialization of tergalii and formation of gill opercula before their notal shield had appeared.

(5) Larval abdominal segment VI is enlarged, middle of posterior margin of its tergum is elevated, tergalial bases are shifted anteriorly (FIG.15:B). Such specialization of segment VI can be functionally connected with gill opercula [see (4)] or with notal shield [see (3)], because in both these cases tergum VI closes gill clamber posteriorly. Imago partly retains this larval character: abdominal segment VI is longer than others, in imaginal Baetisca/ /f3=g2 tergum VI retains convex form. Youngest larvae, which have no tergalii, have segment IV non-enlarged (Pescador & Peters 1974:Fig.10).

(6) Imaginal and subimaginal scutellum is elongate (FIGS 16:F–H, 17:F) [probably in connection with presence of notal shield in larva – see (3)].

Characters of unclear phylogenetic status.

(7) Some common characters are present in mouthpart structure (while mouth apparatus of Baetisca/f2=g1 is not specialized, and that of Prosopistoma/f1=g2 is highly specialized). On maxilla apical-ventral row of pectinate setae is absent (FIGS 15:A, 17:G) (the same in some other mayflies – see Index of characters [1.1.31]). Submentum has lateral lobes produced anteriorly (Eaton 1883–1888: Pl.43:7; Pl.52:9) (about particular structure of these lobes in Prosopistoma/f1=g2 – see below). Paraglossae are fused with mentum (the same in Ephemerella/fg1). Probably, all these are autapomorphies.

(8) Anterior paracoxal suture is transferred to anterior part of episternum, so anepisternum is much smaller than katepisternum (FIGS 16:E, G, 17:E–F). Non-unique character (see Index of characters [2.2.19]).

(9) While wing venation in Baetisca/f2=g1 and Prosopistoma/f1=g2 is quite different, their vein bases on fore wing are similar (FIGS 16:A, 17:A): common stem of MA+RS is reduced, so RS, MA, MP_1 and CuA arise from the same point; base of MP_2 lost connection with MP_1 and is transferred proximally; CuA and MP_1 diverge more strongly than MP_1 and MA; base of CuP lost connection with CuA (but at the same time, in contrast to Tetramerotarsata and Leptophlebia/fg1, its base is nearer to CuA than to AA). Other Ephemeroptera have different combinations of these characters.

(10) On hind wing MA lacks furcation; this is

clearly visible in Baetisca/f2=g1 which has primitive wing venation (FIG.17:C), but is unclear in Prosopistoma/f1=g2, where nearly all veins of hind wing are transformed to intercalaries and can not be homologized (FIG.16:C). Non-unique apomorphy (see Index of characters [2.2.67]).

(11) Patella-tibial suture (initially present on middle and hind legs) is lost on all legs of larva, but its vestige can be retained on middle and hind legs of subimago and imago. Non-unique apomorphy (see Index of characters [1.2.18]).

(12) Imaginal and subimaginal gonostylus has one distal segment only (instead of two primary segments). Non-unique apomorphy (see Index of characters [2.3.12]).

(13) Larval cerci have secondary swimming setae on lateral sides; these setae form less regular row, than primary swimming setae on median sides of cerci and lateral sides of paracercus. Non-unique apomorphy (see Index of characters [1.3.67]).

Plesiomorphies of Posteritorna. In larva: On maxilla number of dentisetae can be indeterminate and more than 3 (FIG.17:G) (in contrast to Anteritorna). Maxillary palp is 3-segmented. Labial palp is 3-segmented. Claws have no denticles. Caudalii have dense primary swimming setae [which differ from secondary swimming setae – see (13)]; distal margins of segments of cerci can be oblique (in Prosopistoma/f1=g2).

In imago and subimago: Anterior paracoxal suture [see (8)] is complete, i. e. completely crosses ventral side of episternum (FIG 16:E, 17:E) (in contrast to Furcatergaliae and some others – see Index of characters [2.2.19]). Hind wing is well-developed, as long as 0.3–0.4 of fore wing length (FIGS 16:C, 17:C) (in spite of the fact, that metathorax and its musculature are reduced in greater degree than in majority of mayflies, and in greater degree than in many mayflies lacking hind wings – FIGS 16:G–H, 17:F).

Size. Fore wing length 2–16 mm.

Age and distribution. Palaeogene (see *Balticobaetisca*) — Recent; Holarctic, Ethiopian, Oriental and Australian Regions. Late Cretaceous *Myanmarella rossi* was wrongly attributed here and actually belongs to Liberevenata INCERTAE SEDIS (see below). Larvae from Early Cretaceous of Australia described as "Siphlonuridae gen. sp." (Jell & Duncan 1986) have some similarity with Baetisca/f2=g1 and possibly belong to Posteritorna (while their tergalial structure, wing venation and other important details are unknown).

Systematic position of Posteritorna. A relationship between Posteritorna, Caenotergaliae and Ephemerella/fg1 was assumed; in connection with this different authors established following taxa: **Pannota** McCafferty & Edmunds 1979 (uniting Posteritorna, Caenotergaliae and Ephemerella/fg1); **Furcatergalia** Kluge 1989 (uniting Posteritorna and Furcatergaliae); **Rectracheata** McCafferty 1991 (uniting Posteritorna, Furcatergaliae and Vetulata). This assumption is not grounded, and the taxa Pannota, Furcatergalia and Rectracheata are actually polyphyletic (Kluge 1998).

Posteritorna are divided into Prosopistoma/f1=g2 and Baetisca/f2=g1 (p.68).

1.1. Prosopistoma/f1=g2

(Panephemeroptera Euephemeroptera Euplectoptera Posteritorna Prosopistoma/f1=g2)

(Figs 15–16)

Nomen hierarchicum: **Prosopistoma/f1=g2** (sine *Baetisca*) [f: Prosopistomatidae (orig. Prosopistomidae) Lameere 1917: 72].

In circumscription matches:
— gen. *Prosopistoma* Latreille 1833: 33;
— gen. *Chelysentomon* Joly & Joly 1872: 438;
— gen. *Binoculus*: Demoulin 1954b: 102;
— tribus Prosopistomini: Lameere 1917: 75;
— fam. Prosopistomidae: Lestage 1917: 240;
— fam. Binoculidae Demoulin 1954b: 103.

References. Vayssière 1881: ↻ ♀; – 1890: ↻; – Eaton 1883–1888: ↻ ♀; – Lestage 1917: ↻; – Lafon 1953: ↻; – Gillies 1954: ↻ ◎ ⊕; – 1956: ↻ ⊕; – Tshernova 1970: ↻ ⊕.

Autapomorphies of Prosopistoma/f1=g2.

(1) Mouth apparatus is strongly modified and specialized for carnivorism (Eaton 1883–1888:Pl.43: 4–8). Labrum is widened. Asymmetry of mandibles is lost; mandible is strongly elongate along axis of its articulation, mola is completely lost (less complete reduction of mola took place in some other carnivorous mayflies). Superlinguae are completely lost (the same in some other carnivorous mayflies). Maxilla [see Posteritorna (7)] has a single long canine (instead of initial three ones) and 3 dentisetae, which being pressed to canine, form something like an integral claw (FIG.15:A). 2nd segment of maxillary palp is curved so that its inner side is convex. Submentum is very strongly enlarged, its lateral lobes [see Posteritorna (7)] are projected forward, covering from below lateral parts of the rest of labium with labial palps, and anteriorly they reach labrum and anterior-lateral margins of head (unique apomorphy); mentum, glossae and paraglossae are fused forming an integral plate widened distally.

(2) Larva has lens-like body form (FIG.15:B–C; Eaton 1883–1888:Pl.43:1–2): head is strongly widened, with rounded anterior margin; lateral margins of notal shield [see Posteritorna (3)] are rounded, without projections (in contrast to Baetisca/f2=g1); when abdominal segments VII–X and caudalii are retracted [see (3)], larval body together with head in dorsal view has a shape of regular oval; dorsally it is evenly convex with median longitudinal keel on notal shield, ventrally – flat. Integrity of flat ventral surface is completed by expanded mentum covering mouthparts from below [see (1)]. Abdominal segments up to VI inclusively are fused immobile one with another and with thorax; while dorsally they are separated by sutures, ventrally only suture between segments V and VI is retained. Lateral margins of segment V have a pair of incisions; thus, between this segment and epipleura [see Posteritorna (3)] there is a pair of distinctly outlined apertures which lead to a gill chamber. Femora can be inserted into concavities on thorax and abdomen, and concavities for hind femora convergent posteriorly under acute angle in area of abdominal sternum IV.

(3) Abdominal segment X and caudalii have unique structure (FIG.15:B). Larval and imaginal tergite X and paraprocts are elongate, rectangular, paraprocts are expanded behind bases of caudalii. Larval (but not imaginal) caudalii can be completely retracted inside abdomen thanks to a pair of long apodemes which stretch from anterolateral angles of tergite X anteriorly and serve for attachment of the tergal-caudalial muscles.

(4) Winged stages are short-living; moult to imago takes place in males only; all legs of female and of male imago, including fore legs, are diminished and non-functional (only in male subimago legs are functional, serving moult on substrate). Non-unique apomorphy (see Index of characters [2] and [2.2.80]). In male tarsi are 2-segmented, first of these segments is fused with tibia, each leg has a single blunt claw; female legs are reduced in greater degree. Like in other short-living mayflies, subimaginal cuticle is entirely thin and colourless, without pigmented areas (FIG.16:D).

(5) Wing venation is strongly modified, unique (FIG.16:B–C). Crossveins are completely lost both on fore and hind wings. Fore wing has only veins radiating from base in a fan-form manner; RSa and RSp begin near wing base independently, i. e. have a form of intercalaries; RSa is non-branching (only sometimes female has vestige of RSa_1 arising anteriorly from RSa; MA lacks furcation, thus MA_2 is lost. In female on fore wing intercalaries are lost besides iRS and iMP (which begin near wing base and look like other longitudinal veins). In male two long intercalaries are present by sides of each of fan-form radiating veins (RSa, iRS, RSp, MA, MP_1, iMP, MP_2, CuA, CuP, AA) and several intercalaries are present before RSa and behind AA. Intercalaries bordering longitudinal veins are new formations absent in other mayflies. Hind wing has similar structure in male and female, narrow, with costal projection transferred proximally, with numerous fan-form intercalaries which can be hardly homologized.

(6) In imago [which is present in male only – see (4)] amphitornal margin of wing bears setae; these setae are smaller than in subimago or in female and in contrast to them, are present on a part of setae-bearing tubercles only (FIG.16:B). Non-unique apomorphy (see Index of characters [2.2.27]).

(7) Imaginal and subimaginal mesonotal suture is strongly curved posteriorly on each side; only its median portion is expressed as a pair of sutures, diverging from middle line backward under acute angle (FIGS 16:D, F, G). Non-unique apomorphy (see Index of characters [2.2.8]).

(8) Male imaginal eyes are not enlarged, as small as in female (non-unique apomorphy – see Index of characters [2.1.3]). In both sexes eyes are small, spherical, widely separated, with narrow bases (Gillies 1954:Fig.14).

Character of unclear phylogenetic status.

(9) In winged stages paracercus is developed, subequal to cerci (in contrast to Baetisca/f2=g1). It can be a plesiomorphy or a result of secondary development of imaginal paracercus from larval one. Caudalii (cerci and paracercus) of winged stages have no distinct segmentation, in female and in male subimago are very small (somewhat smaller than larval ones), in male imago they are several times longer, soft, with long irregular hairs (Gillies 1954:Figs 1–2).

Size. Small, fore wing length 2–7 mm (see TABLE 2).

Distribution. Eastern Hemisphere: Ethiopian, Oriental, Australian and Palaearctic Regions.

Nominal taxon included. Prosopistoma/f1=g2 includes **Binoculus/fg** (nom. praeocc.) [f: Binoculidae Demoulin 1954b: 103; g: *Binoculus* Demoulin 1954b: 102 (non *Binoculus* Geoffroy 1762 suppressed by ICZN Opinion 502, 1958)].

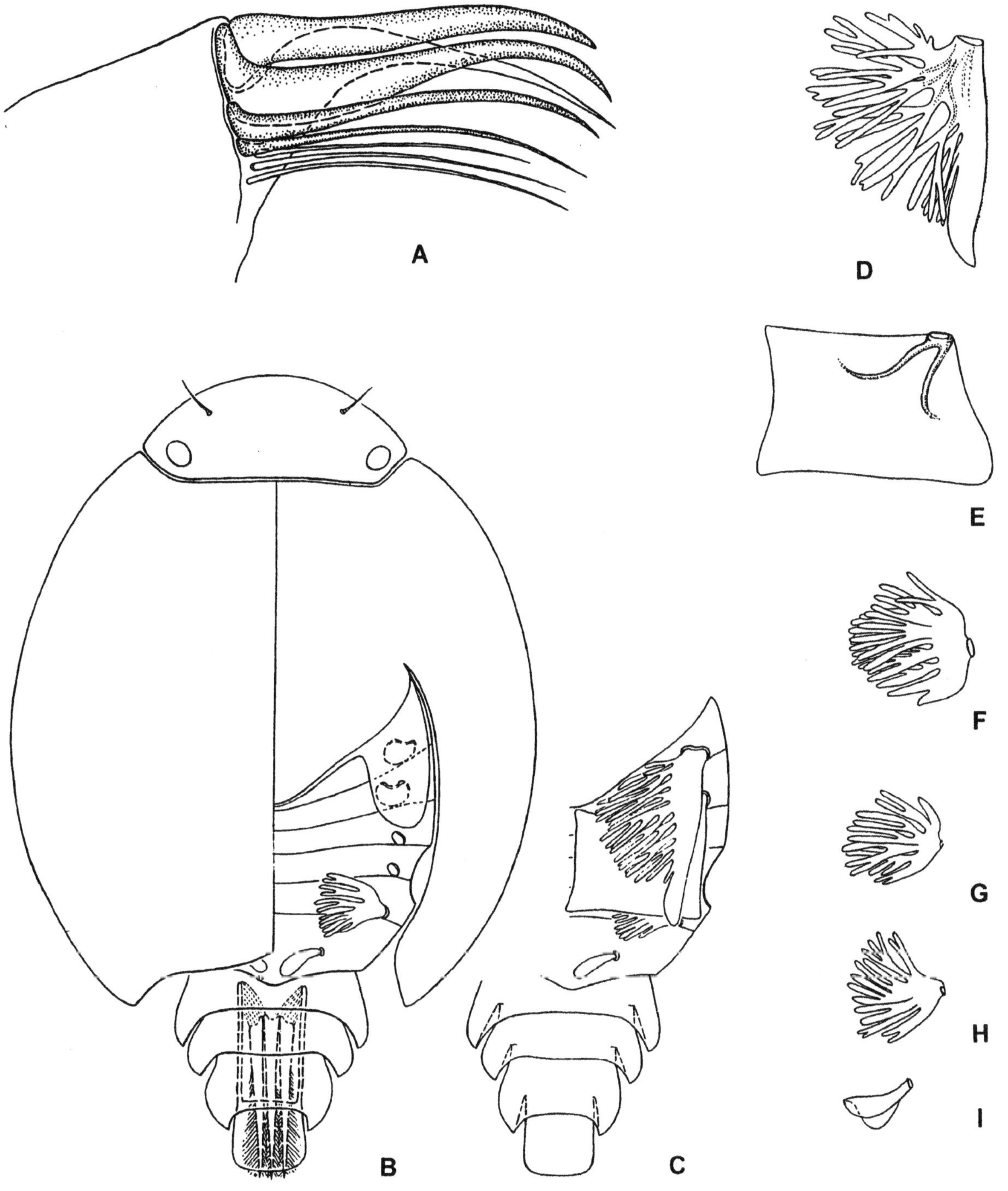

Figure 15. Prosopistoma/f1=g2 *foliaceum* [*Binoculus*], larva.
A – apex of left maxilla, dorsal view (dentisetae shown by dots). **B** – Larva of last instar, dorsal view, with abdominal segments VII–X protracted and caudalii retracted (right half of notal shield and tergalii I–IV removed; caudalii, their muscles and apodemes of 10th tergite shown by interrupted lines). **C** – Abdomen with right tergalii. **D–I** – Right tergalii I–VI.

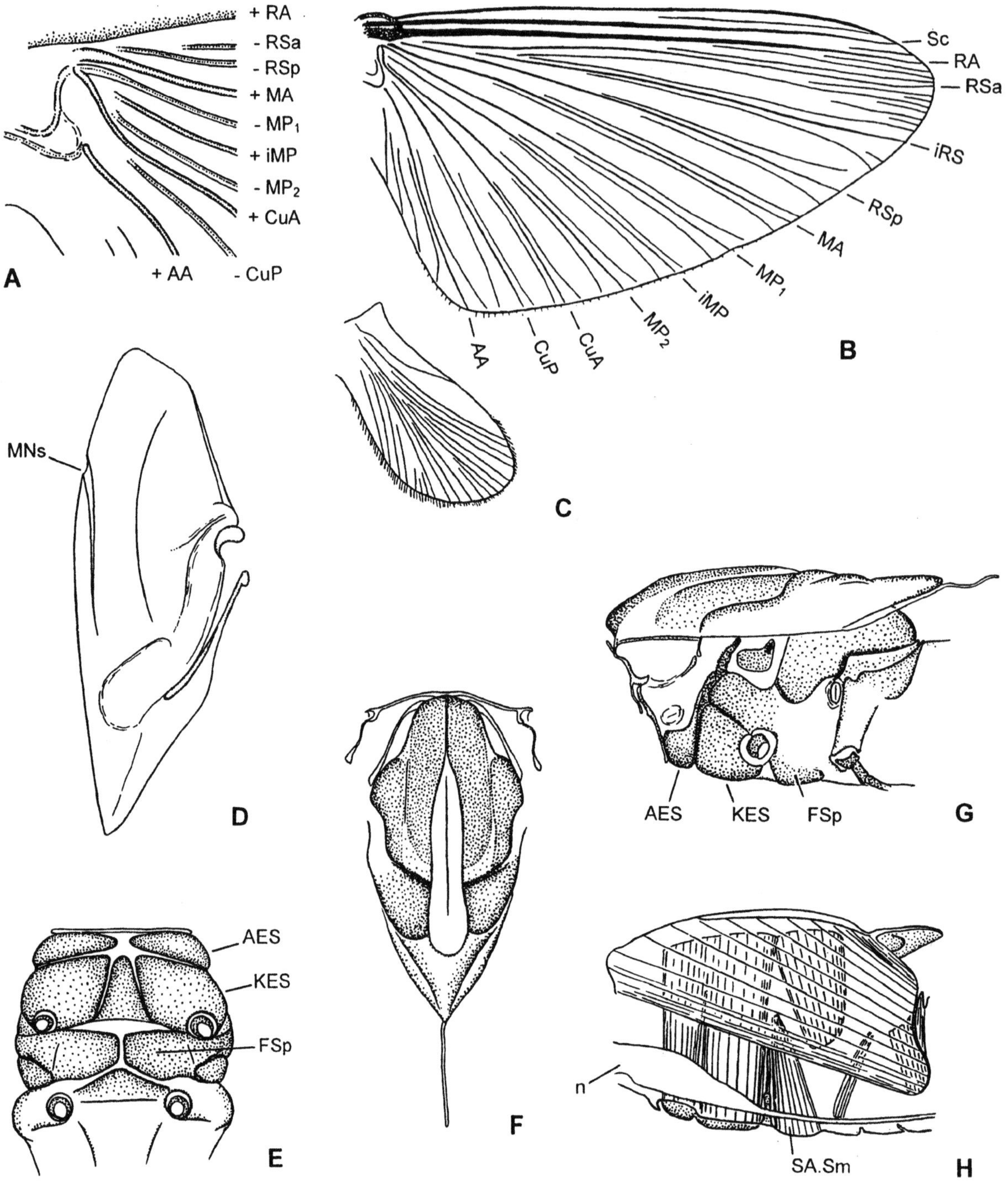

Figure 16. Prosopistoma/f1=g2 *foliaceum* [*Binoculus*], imago and subimago.
A – Base of fore wing. **B–C** – fore and hind wings of male imago. **D** – subimaginal exuviae of right half of mesonotum. **E–H** – Pterothorax of male imago: E – ventral view, F – dorsal view, **G** – lateral view, **H** – longitudinal section.

Abbreviations: **AES** – anepisternum; **FSp** – furcasternal protuberance; **KES** – katepisternum; **MNs** – mesonotal suture; **n** – nerve synganglion; **SA.Sm** – subalar-sternal muscle.

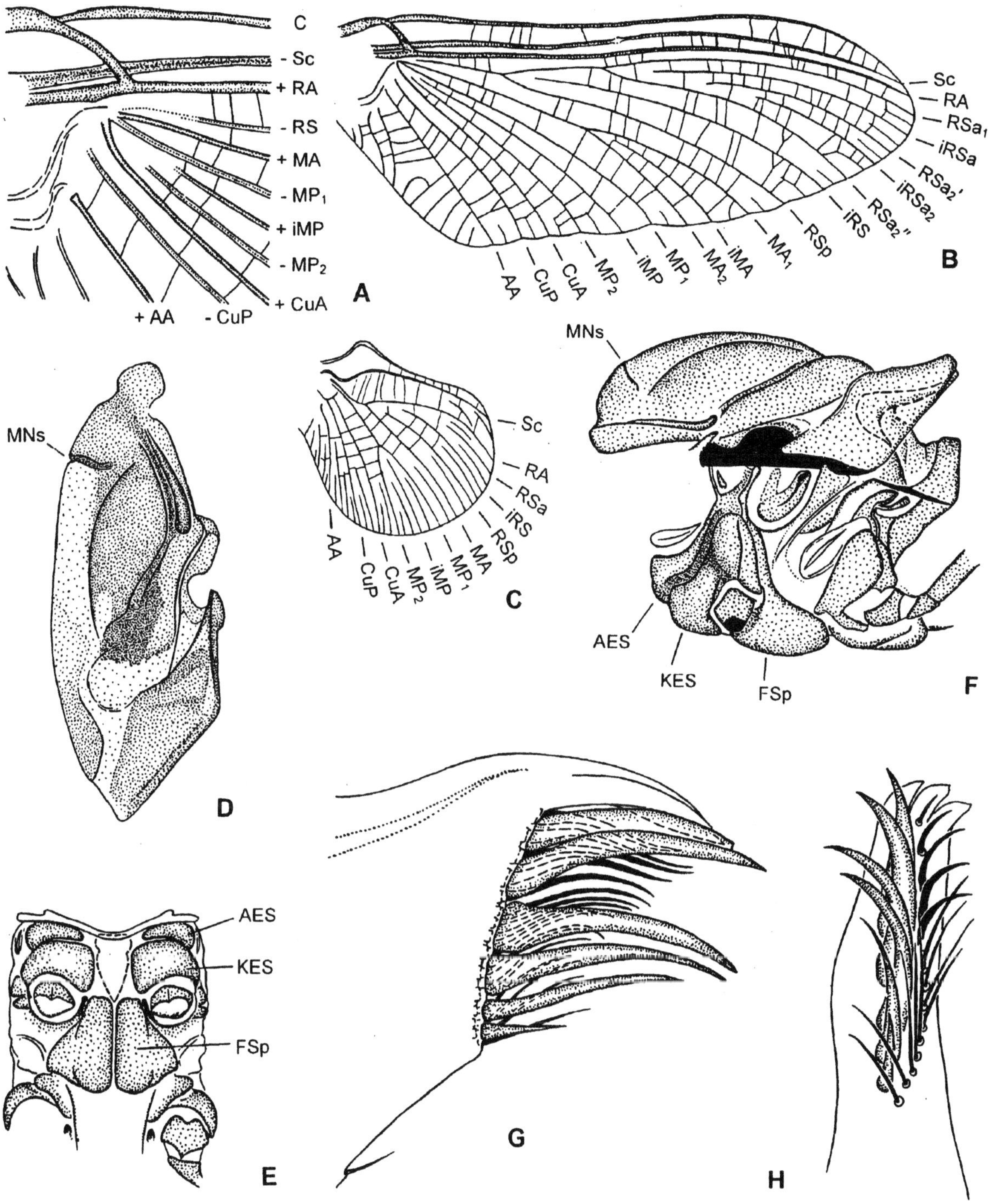

Figure 17. Baetisca/f4=g3.
A–F – Imago and subimago of *rogersi* [*Baetisca*]: **A** – base of fore wing; **B–C** – fore and hind wings of imago; **D** –subimaginal exuviae of right half of mesonotum; **E** – imaginal pterothorax, ventral view (nerve synganglion shown by interrupted line); **F** – imaginal pterothorax, lateral view. **G–H** – Larva of *carolina* [*Baetisca*]: **G** – apex of left maxilla, dorsal view; **H** – the same, medial view (dentisetae shown by dots, setae of medio-ventral row shown by black).

Abbreviations: **AES** – anepisternum; **FSp** – furcasternal protuberance; **KES** – katepisternum; **MNs** – mesonotal suture.

Species composition of Prosopistoma/f1=g2 (here [*P.*]=[*Prosopistoma*]). About 15 species are described, mainly as larvae. One **European species:** *foliaceus* Fourcroy 1785 [*Binoculus*] (synn.subj.: *pennigerus* Müller 1800 [*Limulus*], *pisciforme* Dumeril 1816 [*Binoculus*], *punctifrons* Latreille 1833 [*P.*]). **African species:** *africanum* Gillies 1954 [*P.*], *crassi* Gillies 1954 [*P.*], *deguernei* Vayssière 1893 [*P.*], *phoenicum* Alouf 1977 [*P.*] (syn. subj.: *oronti* Alouf 1977 [*P.*]), *variegatum* Latreille 1833 [*P.*]. **South Asian and Australian species:** *annamense* Soldán & Braasch 1984 [*P.*], *boreus* Peters 1967 [*P.*], *funanense* Soldán & Braasch 1984 [*P.*], *indicum* Peters 1967[*P.*], *lieftincki* Peters 1967 [*P.*], *palawana* Peters 1967 [*P.*], *pearsonorum* Campbell & Hubbard 1998 [*P.*], *sedlaceki* Peters 1967 [*P.*], *sinense* Tong & Dudgeon 2000 [*P.*], *wouterae* Lieftinck 1932 [*P.*].

Material examined: *foliaceus* [*B.*]: ʘ/◎/⊕, ⊕♀, ◎♂, ⊕♂; sp. (Iraq): ʘ; sp. (Thailand): ⊕♂.

***

1.2. **Baetisca/f2=g1**

(Panephemeroptera Euephemeroptera Euplectoptera Posteritorna Baetisca/f2=g1)

(Fig. 17)

Nomen hierarchicum: **Baetisca/f2=g1** (sine *Prosopistoma*; incl. *Balticobaetisca*) [g: *Baetisca* Walsh 1862: 378, typus *Baetis obesa* Say 1839 (monotypy)].
In circumscription matches:
— fam. Baetiscidae: Staniczek & Bechly 2002: 7.

Reference. Staniczek & Bechly 2002: ⊕♀*.

Autapomorphies of Baetisca/f2=g1.

(1) Prosternum has a bispinate projection between leg bases; this projection is prominent in imago and subimago (Staniczek & Bechly 2002:Fig. 8), but indistinct in larva.

(2) Hind wing is nearly round (FIG.17:C).

(3) Imaginal and subimaginal paracercus is vestigial, non-segmented (Pescador & Peters 1974: Fig.24D; Staniczek & Bechly 2002:Fig.10). Non-unique apomorphy (see Index of characters [2.3.22]).

Plesiomorphies of Baetisca/f2=g1 (in contrast to Prosopistoma/f1=g2). In larva (known for Baetisca/f3=g2 only): Mouth apparatus is non-specialized: labrum is not widened; mandible has normal form, with mola and prostheca; superlinguae are present; maxilla [see Posteritorna (7)] has 3 canines and 6–8 dentisetae (FIG.17:G–H) (number of dentisetae varies individually); anterior-lateral lobes of submentum [see Posteritorna (7)] are not large, situated dorsad of the rest part of labium and do not cover it; glossae and paraglossae are separated one from another (Eaton 1883–1888:Pl.52:5–10; Pescador & Berner 1981:Fig.5a–f).

In imago and subimago: Fore wing has complete venation, including all branches and intercalaries of RS (with RSa_2 arising from RSa), symmetrical bifurcation of MA near middle of wing with intercalary iMA, crossveins and intercalaries; hind wing with bifurcation of MP near middle of wing and with crossveins; costal projection is situated at a distance from base (FIG.17:B–C). Legs are functional; tarsus is 5-segmented, 1st tarsal segment is fused with tibia, longer than 2nd; claws are ephemeropteroid [except for male fore legs – see Baetisca/f3=g2 (5) below]. Moult to subimago takes place in both sexes. At least in Baetisca/f3=g2 eyes of male are large, indistinctly separated to two portions; mesonotal suture is distinct, transverse, stretches backward medially (FIG.17:D, E).

Size. Fore wing length 6–16 mm.

Age and distribution. Palaeogene (see *Balticobaetisca*) — Recent. Recently in Nearctic only (see Baetisca/f3=g2).

Baetisca/f2=g1 is divided into extinct plesiomorphon *Balticobaetisca* and Recent taxon Baetisca/f3=g2.

1.2;1. † Plesiomorphon **Balticobaetisca/g(1)**

(Euephemeroptera Euplectoptera Posteritorna Baetisca/f2=g1 *Balticobaetisca*)

Nomen hierarchicum: **Balticobaetisca/g(1)** [g: *Balticobaetisca* Staniczek & Bechly 2002: 7, typus *B. velteni* Staniczek & Bechly 2002 (design. orig.)].
In circumscription matches:
— gen. *Balticobaetisca* Staniczek & Bechly 2002: 7.
Monospecific taxon.

Reference. Staniczek & Bechly 2002: ⊕♀.

Plesiomorphies of *Balticobaetisca* (in contrast to Baetisca/f3=g2). At least in imago abdominal tergum VI [enlarged – see Posteritorna (5)] has no transverse elevation. In female imago abdominal sternum IX has even posterior margin (Staniczek & Bechly 2002:Fig.10).

Size. Fore wing length 10 mm.

Age. Palaeogene (Baltic amber).

Species composition of Balticobaetisca/g(1). 1 species – *velteni* Staniczek & Bechly 2002 [*Balticobaetisca*].

Material examined: –.

***

1.2;2. Baetisca/f3=g2

(Euephemeroptera Euplectoptera Posteritorna Baetisca/f2=g1 .../f3=g2)

(Fig. 17)

Nomen hierarchicum: **Baetisca/f3=g2** (sine *Balticobaetisca*; incl. *Fascioculus*).
In circumscription matches:
— gen. *Baetisca* Walsh 1863: 378;
— tribus Baetiscini Banks 1990: 246;
— subfam. Baetiscinae: Needham & Traver & Hsu 1935:555;
— fam. Baetiscidae: Edmunds & Traver 1954a: 240;
— superfam. Baetiscoidea: McCafferty 1991a: 349.

References. Eaton 1883–1888: ♉* ⊕*; – Needham & Traver & Hsu 1935: ♉* ⊕*; – Edmunds & Allen & Peters 1963: ♉*; – Tshernova 1970: ♉* ⊕*;– Edmunds & Jensen & Berner 1976: ♉* ⊕*; – Berner & Pescador 1980; – Pescador & Berner 1981: ♉* ⊕*; – Staniczek & Bechly 2002: ♁*.

Autapomorphies of Baetisca/f3=g2.

(1) In imago abdominal tergum VI [enlarged – see Posteritorna (5)] has mid-dorsal elevation with transverse crest, similar to that of larva.

(2) In female imago posterior margin of abdominal sternum IX produces a pair of projections with a median cleft between them (Pescador & Peters 1974:Fig.25C). Non-unique character, the same in many other mayflies (but not in *Balticobaetisca*).

Characters of unclear phylogenetic status (unknown for *Balticobaetisca*).

(3) 2nd (penultimate) segment of labial palp has apical projection on inner side (Eaton 1883–1888:Pl. 52:9; Pescador & Peters 1974:Fig.15E). Non-unique apomorphy (see Index of characters [1.1.58]); as mouth apparatus of *Prosopistoma* is strongly modified, it is not clear if this is an apomorphy of Baetisca/f2=g1 or Posteritorna in general.

(4) Larval notal shield [see Posteritorna (3)] has a pair of lateral projections which in different species have shape from small blunt protuberances to large pointed spines (Pescador & Berner 1981:Figs 10–14); they can increase from instar to instar (Pescador & Peters 1974:Figs 10,11,14).

(5) On fore legs of male imago and subimago both claws are blunt (Pescador & Peters 1974:Fig. 24C). Non-unique apomorphy (see Index of characters [2.2.77]); possibly this is an apomorphy of Posteritorna in general, because *Prosopistoma* has more simplified claws [see Prosopistoma/f1=g2 (4) above].

Size. Fore wing length 6–16 mm.

Distribution. Nearctic; dominate in Eastern Nearctic.

Baetisca/f3=g2 is divided into *Fascioculus* and Baetisca/f4=g3.

1.2;2,1. Fascioculus/g(1)

(Euephemeroptera Euplectoptera Posteritorna Baetisca/f2=g1 .../f3=g2 *Fascioculus*)

Nomen hierarchicum: **Fascioculus/g(1)** [g: *Baetisca* subgen. *Fascioculus* Pescador & Berner 1981: 168, typus *Baetisca escambiensis* Berner 1955 (design. orig.)].
In circumscription matches:
— subgen. *Fascioculus* Pescador & Berner 1981: 168.
Monospecific taxon.

Reference. Pescador & Berner 1981: ♉ ⊕.

Autapomorphy of *Fascioculus*.

(1) Larval, subimaginal and imaginal eyes with vertical alternate light and dark bands (Pescador & Berner 1981:Fig.12b).

Character of unclear phylogenetic status.

(2) Larval notal shield [see Baetisca/f3=g2 (4)] is compressed dorsoventrally (in contrast to Baetisca/ /f4=g3).

Plesiomorphy of *Fascioculus*. In contrast to Baetisca/f4=g3, larval head has no frontal projections between antennae bases; there are only genal projections latero-ventrad of antennae bases, which are more or less developed in all Baetisca/f3=g2 (Pescador & Berner 1981:Fig.12b).

Size. Fore wing length 8–12 mm.

Distribution. Florida (USA).

Species composition of Fascioculus/g(1). 1 species – *escambiensis* Berner 1955 [*Baetisca*].

Material examined: *escambiensis* [*B.*]: ♀.

***

1.2;2,2. Baetisca/f4=g3

(Euephemeroptera Euplectoptera Posteritorna Baetisca/f2=g1 .../f4=g3)

(Fig. 17)

Nomen hierarchicum: **Baetisca/f4=g3** (sine *Fascioculus*).
In circumscription matches:
— subgen. *Baetisca* s.str.: Pescador & Berner 1981: 169.

Reference. Pescador & Berner 1981: ♉* ⊕*.

Autapomorphy of Baetisca/f4=g3.

(1) Mature larva has a pair of frontal projections between antennae bases ventrad of lateral ocelli (Pescador & Peters 1974:Figs 15 I–J); in different species these projections have form from small protuberances to large denticles (Pescador & Berner 1981:Figs 10–12a, 13–14), but are absent in young larvae (Pescador & Peters 1974:Figs 10–11).

Character of unclear phylogenetic status.

(2) Larval notal shield [see Baetisca/f3=g2 (4)] is strongly convex (in contrast to *Fascioculus*).

Plesiomorphy of Baetisca/f4=g3. In contrast to *Fascioculus*, eyes are uniformly coloured.

Size. Fore wing length 6–16 mm.

Distribution. Nearctic; dominate in Eastern Nearctic.

Species composition of Baetisca/f4=g3. About 10 species – *becki* Schneider & Berner 1963 [*Baetisca*], *berneri* Tarter & Kirchner 1978 [*Baetisca*], *callosa* Traver 1931 [*Baetisca*], *carolina* Traver 1931 [*Baetisca*] (syn. subj.: *thomsenae* Traver 1937 [*Baetisca*]), *columbiana* Edmunds 1960 [*Baetisca*], *gibbera* Berner 1955 [*Baetisca*], *lacustris* McDunnough 1932 [*Baetisca*] (syn. subj.: *bajkovi* Neave 1934 [*Baetisca*]), *laurentiana* McDunnough 1932 [*Baetisca*], *obesa* Say 1839 [*Baetis*], *rogersi* Berner 1940 [*Baetisca*], *rubescens* Provancher 1878 [*Cloe*].

Material examined: *carolina* [*B.*]: ⊙; *rogersi* [*B.*]: ⊙–⊙–⊕.

*** *** *** ***

Chapter V

Euplectoptera Anteritorna: general diagnosis and systematics of Tridentiseta

Anteritorna include overwhelming majority of mayflies. Here they are divided into a larger holophyletic taxon Bidentiseta and a smaller plesiomorphon Tridentiseta; phylogenetic relationships of taxa attributed to Tridentiseta remain to be unclear:

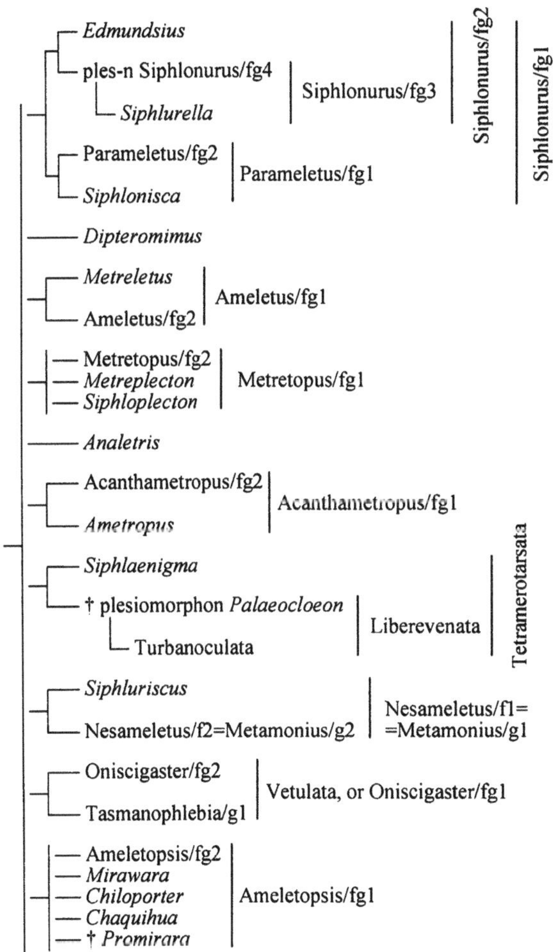

2. **Anteritorna**, or **Ephemera/fg4**

(Panephemeroptera Euephemeroptera Euplectoptera Anteritorna)

(Figs 3–13; 18–108)

Nomen hierarchicum: **Ephemera/fg4** (sine *Prosopistoma*; incl. *Baetis*).
Nomen circumscribens: **Anteritorna** Kluge 1993b: 35.
In circumscription matches:
— "Larves nudibranches": Lestage 1917: 244.
— subordo Anteritorna (as manuscript name): Kluge 1992d: 10; 1992e: 24;
— subordo Anteritorna Kluge 1993b: 35;
— Anteritorna, or Ephemera/fg4: Kluge 2000: 247.

References. Kluge 1993b: ⊕*; – Kluge & Studemann & Landolt & Gonser 1995: ⊕* – Kluge 1998: ℧* ⊕*.

Autapomorphies of Anteritorna.

(1) Tornus of fore wing is situated between apices of veins CuA and CuP; thus CuP usually terminates at basitornal margin of wing far from tornus (FIG.7:C), and only in *Ametropus* – close to tornus (FIG.24:C) (in contrast to Posteritorna, where tornus is situated behind apex of CuP – FIG.17:B).

The difference in tornus position between Posteritorna and Anteritorna does not depend on size of hind wings, as Anteritorna include species with hind wings either well-developed, or vestigial, or completely lost. In the cases when the hind wing is vestigial or lost, the tornus of fore wing is usually transferred proximally (FIG.90:A–B), becomes obtuse or disappears (FIG.101:A–B), but never changes its position relatively to terminations of the longitudinal veins, being in Anteritorna always between CuA and CuP.

In connection with presence of tornus between CuA and CuP, a field between these veins, named **cubital field**, is modified: only in proximal part it is bordered posteriorly by CuP, while in distal part its posterior border is made by basitornal wing margin; in this part, instead of usual crossveins, cubital field has peculiar veins filling the space between CuA and basitornal margin (while in Posteritorna cubital field is a simple narrow field bordered posteriorly by CuP up to tornoapical wing margin, and bearing the same venation as in fields anteriad of it).

Various Anteritorna have different venation of the cubital field. We proceed from an assumption that initial type of cubital field venation is the **siphlonuroid** one: several **sigmoid veins** go from CuA to basitornal margin (FIG.7:C). In mayflies with such venation hind wing usually retains greatest length (up to 1/2 of fore wing length); correspondingly to this, basitornal margin of fore wing is also long, CuA passes along it at some distance, so the cubital field is long and narrow all over its length. This type of cubital field is present in Siphlonurus/fg1, Ameletus/fg2, Nesameletus/f1=Metamonius/g1, Vetulata, Ameletopsis/fg1, *Rallidens*, and Coloburiscus/fg1. In the past, on the base of wing venation, these taxa were united in a family Siphlonuridae which was surely paraphyletic, because at least Ameletus/fg2 and Coloburiscus/fg1 belong to the phylogenetic branches, whose other representatives have different wing venation.

In many other mayflies cubital field venation is modified, and the same several types of its venation are repeated in non-related taxa (see Index of characters [2.2.51] and [2.2.52]). Cubital field can be wider, and veins arising from CuA can be bifurcate (FIGS 40:A; 68:A) or multibranched. These veins can be substituted by **intercalaries** going parallel to CuA. The cubital intercalaries can originate either as elongation of middle portions of the sigmoid veins (in this case the posteriormost intercalaries are the longest – FIG.53:A, C), or as anastomoses between the initial sigmoid veins (FIG.24:A, C). There are intermediate forms between the typical siphlonuroid sigmoid veins and the intercalaries, so the types of venation used in the taxa characteristics below, are rather arbitrary.

In some cases venation of cubital field provides a regular alternating of concave and convex veins, like that on the rest part of the wing; while vein alternating in general is an ancient primitive feature of Ephemeroptera, such alternating in cubital field is most probably secondary, independently evolving in several non-related groups of Anteritorna. In some taxa cubital field has intercalaries arranged in pairs (see Index of characters [2.2.52]), where each pair consists of a convex and a concave vein, alternating with convex CuA and concave CuP. Bifurcate veins arising from CuA can have hind branch convex and fore branch convex, providing the same effect. In Isonychia/fg1 two proximalmost veins going from CuA to basitornal margin, are usually non-branched (FIG.40:A), the proximalmost is convex (like CuA), and the next is concave. In some taxa CuA and the proximalmost vein arising to basitornal margin (often called "CuA_2") together form a strong bifurcate convex vein, and a concave intercalary ("iCuA") is present between its branches (FIG.75:A) (see Index of characters [2.2.50]); most probably, this structure is secondary, in spite of the fact that it resembles a triad of CuA characteristic for permoplectopteran mayflies *Protereisma* (FIG.14:B) and *Palingeniopsis* (Martynov 1932:Pl.I:2), and triads initial for many other veins of mayflies.

A hypothesis was suggested, that Posteritorna and Anteritorna independently originated from Permoplectoptera; their hind wings shortened independently, and in connection with this, their fore wings got tornus independently, at two different places (Edmunds & Traver 1954b); this idea does not contradict the known data (Kluge 1998). It was also assumed that Anteritorna originated from *Protereisma*-like ancestor which had a primitive triad of CuA; thus, some authors regard the presence of triad of CuA in Recent mayflies to be a plesiomorphy (Edmunds & Traver 1954b). However, this triad is present only in a few short-living mayflies with highly modified venation – Behningia/fg2 (but not in *Protobehningia*) and some representatives of Anagenesia/fg1 (but not other Palingenia/f2=g1), and thus, most probably, it is a secondary one. Triad of CuA is present also in extinct Hexagenites/fg1 which is regarded to be primitive only because of its Mesozoic age; however, wing venation of Hexagenites/fg1 is unique and undoubtedly modified.

(2) Maxilla has no more than 3 dentisetae; number of dentisetae is determinate, in different representatives it can be three (FIG.3:E), two, one, or none (see Index of characters [1.1.37]); a single known exception is made by Coloburiscus/fg1, where number of dentisetae can vary individually from 2 to 3.

Size. Fore wing length 2–40 mm.

Age and distribution. From Jurassic (see Eusetisura INCERTAE SEDIS, Fossoriae INC. SED. and Anteritorna INC. SED.) — till Recent; world-wide.

Anteritorna are divided into Tridentiseta and Bidentiseta (Chapters VI–VII: p.123). Some extinct taxa and some insufficiently described species have uncertain systematic position (see Appendix: p.353).

2.1. Plesiomorphon **Tridentiseta**, or **Baetis/fg1**

(Panephemeroptera Euephemeroptera Euplectoptera Anteritorna Tridentiseta)

(Figs 3–13; 18–36)

Nomen hierarchicum: **Baetis/fg1** (incl. *Siphlonurus*, *Dipteromimus*, *Ameletus*, *Metretopus*, *Acanthametropus*, *Ametropus*, *Metamonius*, *Oniscigaster*, *Ameletopsis*, *Rallidens*) [f: Baetida Leach 1815: 137; g: *Baetis* Leach 1815: 137, typus *Ephemera fuscata* Linnaeus 1761 (design. by ICZN Opinion 787, 1966); syn. obj.: ***Brachyphlebia*** Westwood 1840: 25].

Nomen circumscribens: **Tridentiseta** Kluge & Studemann & Landolt & Gonser 1995: 105.

In circumscription matches:

— superfam. Siphlonuroidea: McCafferty 1997: 95;

— infraordo Pisciforma: Kluge & al. 1995: 105 (non Pisciforma McCafferty 1991a);

— grex infraordines Tridentiseta Kluge & al. 1995: 105;

— Tridentiseta, or Baetis/fg1: Kluge 2000: 247.

References. Kluge & al. 1995: ℧* ⊕* – Kluge 1998: ℧*.

Plesiomorphies of Tridentiseta. In contrast to Bidentiseta, maxilla usually has 3 dentisetae [see Anteritorna (2)] (FIGS 3:E; 22:D; 25:D; 28:B, D; 31:A, E; 32:E, 36:A). Exceptions are made by Ameletus/fg1 (FIG.20:C) and Acanthametropus/fg1 (FIG.23:B), whose maxillae are specialized and have less number of dentisetae. Homology of maxillary parts and systematic position of Ameletopsis/fg1 (FIG.34:D) are unclear; here we conditionally attribute this taxon to Tridentiseta.

Size. Fore wing length 2–24 mm.

Age and distribution. Jurassic (see Anteritorna INCERTAE SEDIS) — Recent; world-wide.

Status and systematic position of Tridentiseta. Here we unite all Anteritorna not belonging to Bidentiseta, in a taxon Tridentiseta; probably in future paraphyly of Tridentiseta will be clarified and this taxon will be disbanded. Earlier, the groups attributed to Tridentiseta were placed into one of larger paraphyletic taxa – **Schistonota** McCafferty & Edmunds 1979 (uniting Tridentiseta, Branchitergaliae, Pinnatitergaliae and Leptophlebia/fg1) or **Costatergalia** Kluge 1989 (uniting Tridentiseta and Branchitergaliae). A taxon, whose circumscription is close to Tridentiseta, is **Pisciforma** McCafferty 1991; in contrast to Tridentiseta, Pisciforma do not include Vetulata, but include *Pseudiron*, thus being polyphyletic (Kluge 1998).

Tridentiseta are divided into 11 directly subordinate taxa: (A) **Holarctic taxa:** Siphlonurus/fg1, *Dipteromimus* (p.79), Ameletus/fg1 (p.80), Metretopus/fg1 (p.86), Acanthametropus/fg1 (p.88) and *Ametropus* (p.93); (B) **widely distributed taxa:** Tetramerotarsata (or Baetis/fg2) (p.94) and Nesameletus/fl= =Metamonius/g1 (p.105); (C) **Amphinotic taxa:** Vetulata (or Oniscigaster/fg1) (p.110), Ameletopsis/fg1 (p.114) and *Rallidens* (p.120).

2.1;1. **Siphlonurus/fg1**

(Euephemeroptera Euplectoptera Anteritorna Tridentiseta Siphlonurus/fg1)

(Figs 3–7; 9:A–B; 10:G–H; 12:A–E; 13:A–B, F; 18; 19)

Nomen hierarchicum: **Siphlonurus/fg1** (incl. *Parameletus*) [f: Siphlonuridae Ulmer 1920b: 131 (1888) (pro Siphlurines Albarda in Selys-Longchamps 1888: 150) (Peters & Hubbard 1977); g: *Siphlonurus* Eaton 1868b: 89, typus *Baetis flavida* Pictet 1865 (design. orig.); syn. obj.: ***Siphlurus*** Eaton 1871: 125].

In circumscription matches:

— *Siphlonurus*-complex: Koss & Edmunds 1974: 301;

— fam. Siphlonuridae: Kluge & Studemann & Landolt & Gonser 1995: 109.

References. Koss & Edmunds 1974: ⊙*; – Kluge & Studemann & Landolt & Gonser 1995: ⊙* ℧* ◎* ⊕*.

Autapomorphies of Siphlonurus/fg1.

(1) In female imago distal portion of seminal receptacle is sclerotized and transformed to a peculiar **copulatory pouch** situated in abdominal segment VIII (FIGS 18:A–C; 19:A–B, G). Posteriorly from copulatory pouch arises a membranous duct which opens to copulatory opening (situated at boundary of abdominal sterna VIII and IX); from postero-lateral portions of the copulatory pouch arise a pair of oviducts, and anteriorly from copulatory pouch arises a membranous seminal receptacle. Structure of copulatory pouch is species-specific. (As copulatory pouch is covered by muscles, it is invisible neither when abdomen is dissected, nor on translucent slide in Canadian balsam; it can be studied if treat abdomen by alkali to dissolve all tissues and retain cuticle only.) In female subimago sclerotized copulatory pouch is absent. Such copulatory pouch is found in all the species examined of Siphlonurus/fg3 and Parameletus/fg1 (*Edmundsius* was not examined). Unique apomorphy: in all other mayflies (including *Dipteromimus*, Ameletus/fg1, Metretopus/fg1, Acanthametropus/fg1, *Ametropus*, Nesameletus/fl=Metamonius/g1, Vetulata, Ameletopsis/fg1 and *Rallidens* which were formerly mixed with Siphlonurus/fg1) the whole female copulatory apparatus is membranous.

(2) Egg surface has convex rough spots formed by tops of threads closely pressed together (Kluge & al. 1995:Figs 61–63); probably these are vestiges of anchors. Such spots are found in all the species examined of Siphlonurus/fg3 and Parameletus/fg1; egg structure of *Edmundsius* is unknown. In other mayflies, if anchors in a form of bundle of threads are present, end of each bundle is covered by integral knob or by a regular rosette-like structure (in Ameletus/fg1, Metretopus/fg1 and some others). Probably, opened tops of bundles in Siphlonurus/fg1 were formed as a result of reduction of the knobs.

(3) Each paired penis lobe is distally split into two lobes – gonopore-bearing medial (or ventral) one and additional lateral (or dorsal) one; shape of these lobes is species-specific.

Characters of unclear phylogenetic status.

(4) Imaginal and subimaginal claws (initially ephemeropteroid) on each leg are similar and pointed (FIG.4:A–B). Non-unique apomorphy (see Index of characters [2.2.85] and [2.2.77]); among Tridentiseta the same in *Dipteromimus* and Nesameletus/fl= =Metamonius/g1.

(5) Imaginal and subimaginal paracercus is vestigial (FIG.18:A). Non-unique apomorphy (see Index of characters [2.3.22]); particularly, the same in Ameletus/fg1 and Metretopus/fg1.

(6) Larval abdominal terga and sterna with more or less developed spine-like setae (non-unique character; the same in Metretopus/fg1-*Siphloplecton*, some Ameletus/fg2 and others). Posterior margins of sterna always lack denticles, posterior margins of terga either with denticles (in Siphlonurus/fg2) or not (in Parameletus/fg1).

(7) On each tergalius I–VII anal rib is located very far from anal margin, on dorsal surface, at anterior half of tergalius, usually close to middle trachea (while costal rib has usual for mayflies position on costal margin and is developed at least in proximal part of tergalius) (FIG.18:F). The same in *Dipteromimus* and some others (see Index of characters [1.3.28]).

Plesiomorphies of Siphlonurus/fg1. Larva has primary swimming siphlonuroid specialization (FIG. 9:A): legs are able to stretch posteriorly; abdomen is large, able to make undulate swimming movements; caudalii are not long, paracercus is well-developed, primary swimming setae are dense, secondary swimming setae are absent. Larval head is hypognathous, mouth apparatus is non-specialized (FIG.3); maxilla has 3 canines and 3 dentisetae [see Anteritorna (2)] (FIG.3:B, E), setae of apical-ventral row can be from simple to pectinate (only in *Siphlonisca* absent); maxillary palp is 3-segmented; labial palp is 3-segmented. Larval (and adult) patella-tibial suture is developed on middle and hind legs only (FIGS 3–4). Larval claws are slightly curved, without denticles or with very small irregular denticles near base. Tergalii [see (7)] retain ability of rhythmical respiratory movement.

In imago and subimago: Mesonotal suture can be stretched backward medially (especially in Siphlonurus/fg3 – FIGS 6, 18:E) or nearly transverse (especially in Parameletus/fg2 – FIG.19:E). Anterior paracoxal suture is complete (FIG.5) (in contrast to *Rallidens* and some Bidentiseta). Furcasternal protuberances are contiguous (FIG.5:C) (in contrast to Amphinotic groups of Tridentiseta and some others – see Index of characters [2.2.23]). Subimaginal lateral sclerotized pigmented area of mesonotum is small, bifurcates posteriorly in such a manner, that one its branch stretches along lateroparapsidal suture, and another branch – along lateral scutal suture (FIG.18:E) (see Index of characters [2.2.14]). In cubital field of fore wing several (3–9) veins go from CuA to basitornal margin [see Anteritorna (1)]; hind wing is well-developed, as long as 0.4–0.5 of fore wing length (FIG.7:C–D). Imaginal and subimaginal tarsi are 5-segmented, 1st segment is the longest, fused with tibia (FIG.4). Gonostylus has 2 distal segments.

Size. Fore wing length 9–24 mm.

Age and distribution. Palaeogene (see Siphlonurus/fg1 INCERTAE SEDIS) — Recent; Holarctic.

Siphlonurus/fg1 is divided into Siphlonurus/fg2 and Parameletus/fg1 (p.76); one extinct species has uncertain position (p.79).

2.1;1,1. Siphlonurus/fg2

(Euephemeroptera Euplectoptera Anteritorna Tridentiseta Siphlonurus/fg1 .../fg2)

(Figs 3–7; 12:A–E; 13:A–B, F; 18)

Nomen hierarchicum: **Siphlonurus/fg2** (sine *Parameletus*; incl. *Edmundsius*).

In circumscription matches:

— subfam. Siphlonurinae: Kluge & Studemann & Landolt & Gonser 1995: 110.

Reference. Kluge & al. 1995: ⍥* ⊕*.

Autapomorphy of Siphlonurus/fg2.

(1) Tergalii I and II are bilamellate (FIG.3:A): ventral (posterior) lamella is an additional one, and represents an expanded anal-proximal lobe bent down; in Siphlonurus/fg3 this additional lamella has nearly the same shape and size as the main lamella, but in contrast to it has no ribs [see Siphlonurus/fg1 (7)] (tergalii III–VII are usually unilamellate, but in

Siphlurella became bilamellate by another way – see below).

Character of unclear phylogenetic status.

(2) Larval paraproct with a spine on inner-apical margin (FIG.18:D) (*Edmundsius* was not examined). The same in Metretopus/fg1, *Siphluriscus*, Vetulata and *Rallidens*.

Plesiomorphies of Siphlonurus/fg2. In contrast to Parameletus/fg1, larval abdominal terga (at least VI–X) have pointed denticles on posterior margins [sterna have no denticles – see Siphlonurus/fg1 (6)].

Size. Fore wing length 9–24 mm.

Distribution. Holarctic.

Siphlonurus/fg2 is divided into *Edmundsius* and Siphlonurus/fg3.

2.1;1,1-1. Edmundsius/g(1)

(Euplectoptera Anteritorna Tridentiseta Siphlonurus/fg1 .../fg2 *Edmundsius*)

Nomen hierarchicum: **Edmundsius/g(1)** [g: *Edmundsius* Day 1953: 19, typus *E. agilis* Day 1953 (design. orig.)].
In circumscription matches:
— gen. *Edmundsius* Day 1953: 19.
Monospecific taxon.

References. Day 1953: ℧ ⊕; – Edmunds & Jensen & Berner 1976: ℧ ⊕.

Plesiomorphies of *Edmundsius*. In contrast to Siphlonurus/fg3, ventral (additional) lamella of tergalii I–II [see Siphlonurus/fg2 (1)] is much smaller than dorsal (main) lamella; both lamellae are not bipointed (Day 1953:Fig.7).

Size. Fore wing length 16 mm.

Distribution. California.

Species composition of Edmundsius/g(1). 1 species *agilis* Day 1953 [*Edmundsius*].

Material examined: –.

***

2.1;1,1-2. Siphlonurus/fg3

(Euplectoptera Anteritorna Tridentiseta Siphlonurus/fg1 .../fg3)

(Figs 3–7; 12:A–E; 13:A–B, F; 18)

Nomen hierarchicum: **Siphlonurus/fg3** (sine *Edmundsius*; incl. *Siphlurella*).
In circumscription matches:
— gen. *Siphlonurus* Eaton 1868b: 89;
— gen. *Siphlurus* Eaton 1871: 125.

References. Eaton 1883–1888: ⊕*; – Needham & Traver & Hsu 1935: ℧* ⊕*; – Edmunds & Jensen & Berner 1976: ℧* ⊕*; – Studemann & Landolt 1997b: ℧ ⊕.

Autapomorphies of Siphlonurus/fg3.

(1) Dorsal and ventral lamellae of tergalii I–II [see Siphlonurus/fg2 (1)] are subequal, each lamella has a peculiar bipointed shape, like a fish tail (FIG.3:A; Kluge 1982:Fig.3; 1985:Fig.3). Unique apomorphy.

Size. Fore wing length 9–24 mm.

Distribution. Holarctic.

Siphlonurus/fg3 is divided into Siphlonurus/fg4 and *Siphlurella*.

2.1;1,1-1/1. Plesiomorphon Siphlonurus/fg4

(Euplectoptera Anteritorna Tridentiseta Siphlonurus/fg1 .../fg4)

(Figs 3–7; 12:A–E; 13:A–B, F; 18)

Nomen hierarchicum: **Siphlonurus/fg4** (sine *Siphlurella*).
In circumscription matches:
— subgen. *Siphlonurus*: Jacob 1974a: 2;
— gen. *Siphlonurus*: Bengtsson 1930: 11;
— gen. *Siphlurus*: Bengtsson 1909: 9.

References. Eaton 1883–1888 (gen. *Siphlurus*, nymph): ℧*; – Bengtsson 1930: ℧* ⊕*; – Lestage 1935a: ℧*.

Plesiomorphy of Siphlonurus/fg4. In contrast to *Siphlurella*, tergalii III–VII are unilamellate [tergalii I–II being bilamellate – see Siphlonurus/fg3 (1)].

Size. Fore wing length 9–24 mm.

Distribution. Holarctic.

Nominal taxon included. Siphlonurus/fg4 includes **Andromina/g** [g: *Andromina* Navás 1912b: 416, typus *A. grisea* Navás 1912 (design. orig.)], regarded as a generic synonym of *Siphlonurus* (Edmunds 1960: 24).

Species composition of Siphlonurus/fg4 (here [*S.*]= =[*Siphlonurus*]). About 40 species. **Eurasian species:** *aestivalis* Eaton 1903 [*Siphlurus*] (synn. subj.: *spinosus* Bengtsson 1909 [*Siphlurus*], *vernalis* Tshernova 1928 [*S.*]), *abraxas* Jacob 1968 [*S.*], *armatus* Eaton 1870 [*S.*] (syn. subj.: *latus* Bengtsson 1909 [*Siphlurus*]), *binotatus* Eaton 1892 [*Siphlurus*] (synn. subj.: *maculosus* Matsumura 1930 [*S.*], *grandiosa* Matsumura 1931 [*Siphlonisca*]), *chankae* Tshernova 1952 [*S.*], *croaticus* Ulmer 1920 [*S.*], *davidi* Navás 1932 [*Siphluriscus*], *grisea* Navás 1912 [*Andromina*], *hispanicus* Demoulin 1958 [*S.*], *flavida* Pictet 1865 [*Baetis*], *immanis* Kluge 1985 [*S.*], *ireneae* Alba-Tercedor 1990 [*S.*], *lacustris* Eaton 1870 [*S.*] (synn. subj.: *zetterstedti* Bengtsson 1909 [*Siphlurus*], *pyrenaicus* Navás 1930 [*S.*], *nuessleri* Jacob 1972 [*S.*]), *lusoensis* Puthz 1977 [*S.*], *montanus* Studemann 1992 [*S.*], *muchei* Braasch 1983 [*S.*], *palaearcticus* Tshernova 1930 [*Oniscigaster*] (syn. subj.: *brodskyi* Bajkova 1979 [*S.*]), *sanukensis* Takahashi 1929 [*Siphlurus*], *yoshinoensis* Gose 1985 [*S.*], *zhelochovtsevi* Tshernova 1952 [*S.*]. **North American species:** *autumnalis* McDunnough 1931 [*S.*], *barbaroides* McDunnough 1929 [*S.*], *barbarus*

McDunnough 1924 [*S.*], *columbianus* McDunnough 1925 [*S.*], *decorus* Traver 1932 [*S.*], *demarayi* Kondratieff & Voshell 1981 [*S.*], *luridipennis* Burmeister 1839 [*Baetis*], *marginatus* Traver 1932 [*S.*], *marshalli* Traver 1934 [*S.*], *minnoi* Provonsha & McCafferty 1982 [*S.*], *mirus* Eaton 1885 [*Siphlurus*], *occidentalis* Eaton 1885 [*Siphlurus*] (syn. subj.: *inflatus* McDunnough 1931 [*S.*]), *quebecensis* Provancher 1878 [*Cloe*] (syn. subj.: *triangularis* Clemens 1915 [*S.*]), *rapidus* McDunnough 1924 [*S.*], *securifer* McDunnough 1926 [*S.*], *spectabilis* Traver 1934 [*S.*] (syn. subj.: *maria* Mayo 1939 [*S.*]), *typicus* Eaton 1885 [*Siphlurus*], (synn. subj.: *berenice* McDunnough 1923 [*S.*], *novangliae* McDunnough 1924 [*S.*]).

Material examined: Group *palaearcticus*: *palaearcticus* [*Oniscigaster*]: ð–◎–⊕/⊙. **Group *flavida*:** *flavida* [*Baetis*]: ð, ◎♂, ⊕♀. **Group *lacustris*:** *lacustris* [*Siphlonurus*]: ð–◎–⊕/⊙; *grisea* [*Andromina*]: ð–◎–⊕/⊙; *occidentalis* [*Siphlurus*]: ð, ⊕♂. **Group *aestivalis*:** *aestivalis* [*Siphlurus*]: ð–◎–⊕/⊙; *armatus* [*Siphlonurus*]: ð, ⊕; *croaticus* [*Siphlonurus*]: ð, ◎, ⊕; *chankae* [*Siphlonurus*]: ð–◎–⊕/⊙; *zhelochovtsevi* [*Siphlonurus*]: ð–◎–⊕/⊙; *hispanicus* [*Siphlonurus*]: ð, ð♀–◎♀, ⊕♀; *lusoensis* [*Siphlonurus*]: ð, ◎, ⊕; *immanis* [*Siphlonurus*]: ð–◎–⊕/⊙; *montanus* [*Siphlonurus*]: ð, ◎; sp.S1 (= sp.n.1: Kluge 1985): ⊕♂. **Unknown group:** sp.S2 (China): ⊕♂, ⊕♀.

***

2.1;1,1-1/2. Siphlurella/g(1)

(Euplectoptera Anteritorna Tridentiseta Siphlonurus/fg1 .../fg3 *Siphlurella*)

(Fig. 8:A)

Nomen hierarchicum: **Siphlurella/g(1)** [g: *Siphlurella* Bengtsson 1909: 11, typus *S. thomsoni* Bengtsson 1909 (design. Hubbard 1979a: 412; syn. subj. *Baetis alternata* Say 1824)].

In circumscription matches:

— subgen. *Siphlurella*: Jacob 1974a: 2;

— gen. *Siphlurella* Bengtsson 1909: 11.

References. Bengtsson 1909: ð ⊕; – 1930: ð ⊕.

Autapomorphy of *Siphlurella*.

(1) Tergalii III–VII with additional dorsal (anterior) lamella which represents an expansion of anal-proximal area bent dorsally [tergalii I and II with additional ventral (posterior) lamella – see Siphlonurus/fg2 (1) and Siphlonurus/fg3 (1)]; thus all tergalii are double, but tergalii I–II and III–VII have basically different structure.

Size. Fore wing length 12–14 mm.

Distribution. Holarctic.

Species composition of Siphlurella/g(1). 2 species – *phyllis* McDunnough 1923 [*Siphlonurus*] (Nearctic) and *alternata* Say 1824 [*Baetis*] (synn. subj.: *annulata* Walker 1853 [*Baetis*], *linnaeanus* Eaton 1871 [*Siphlurus*], *thomsoni* Bengtsson 1909 [*Siphlurella*], *oblita* Bengtsson 1909 [*Siphlurella*]) (Holarctic).

Material examined: *alternata* [*B.*]: ð–◎–⊕/⊙.

*** *** ***

2.1;1,2. Parameletus/fg1

(Euplectoptera Anteritorna Tridentiseta Siphlonurus/fg1 Parameletus/fg1)

(Figs 10:G–H; 19)

Nomen hierarchicum: **Parameletus/fg1** (incl. *Siphlonisca*) [f: Parameletinae Kluge & Studemann & Landolt & Gonser 1995: 111; g: *Parameletus* Bengtsson 1908: 242, typus *P. chelifer* Bengtsson 1908 (monotypy); syn. subj.: ***Eatonia*** Bengtsson 1904: 131 (nomen nudum), typus *E. borealis* Bengtsson 1904 (monotypy, nomen nudum) (non *Eatonia* Hall 1857); syn. subj.: ***Potameis*** Bengtsson 1909: 13, typus *P. elegans* Bengtsson 1909 (design. Hubbard 1979a: 412); syn. subj.: ***Sparrea*** Esben-Petersen 1909: 554, typus *S. norvegica* Esben-Petersen 1909 (monotypy); syn. subj.: ***Palmenia*** Aro 1910 (in Lestage 1924): 28, typus *P. fennica* Aro 1910 (monotypy) (synn. subj.: *chelifer* [*P.*] = *borealis* [*E.*] = *elegans* [*P.*] = *norvegica* [*S.*] = *fennica* [*P.*])].

In circumscription matches:

— subfam. Parameletinae Kluge & Studemann & Landolt & Gonser 1995: 111.

Reference. Kluge & al. 1995: ð* ⊕*.

Autapomorphies of Parameletus/fg1.

(1) In female imago copulatory pouch [see Siphlonurus/fg1 (1)] has a pair of basal lobes projecting posteriorly from copulatory duct; pigmented area of abdominal sternum VIII anteriorly forms an unpaired projection, jutting out between the basal lobes of copulatory pouch (FIG.19:A–B, G).

(2) Penis [distally subdivided into two pairs of lobes – see Siphlonurus/fg1 (3)] in its proximal part on ventral side bears a pair of spine-like processes directed distally.

(3) Larval abdominal terga lack denticles on posterior margins [like that of sterna – see Siphlonurus/fg1 (6)]. Non-unique apomorphy.

Size. Fore wing length 10–17 mm.

Distribution. Holarctic.

Parameletus/fg1 is divided into Parameletus/fg2 and *Siphlonisca*.

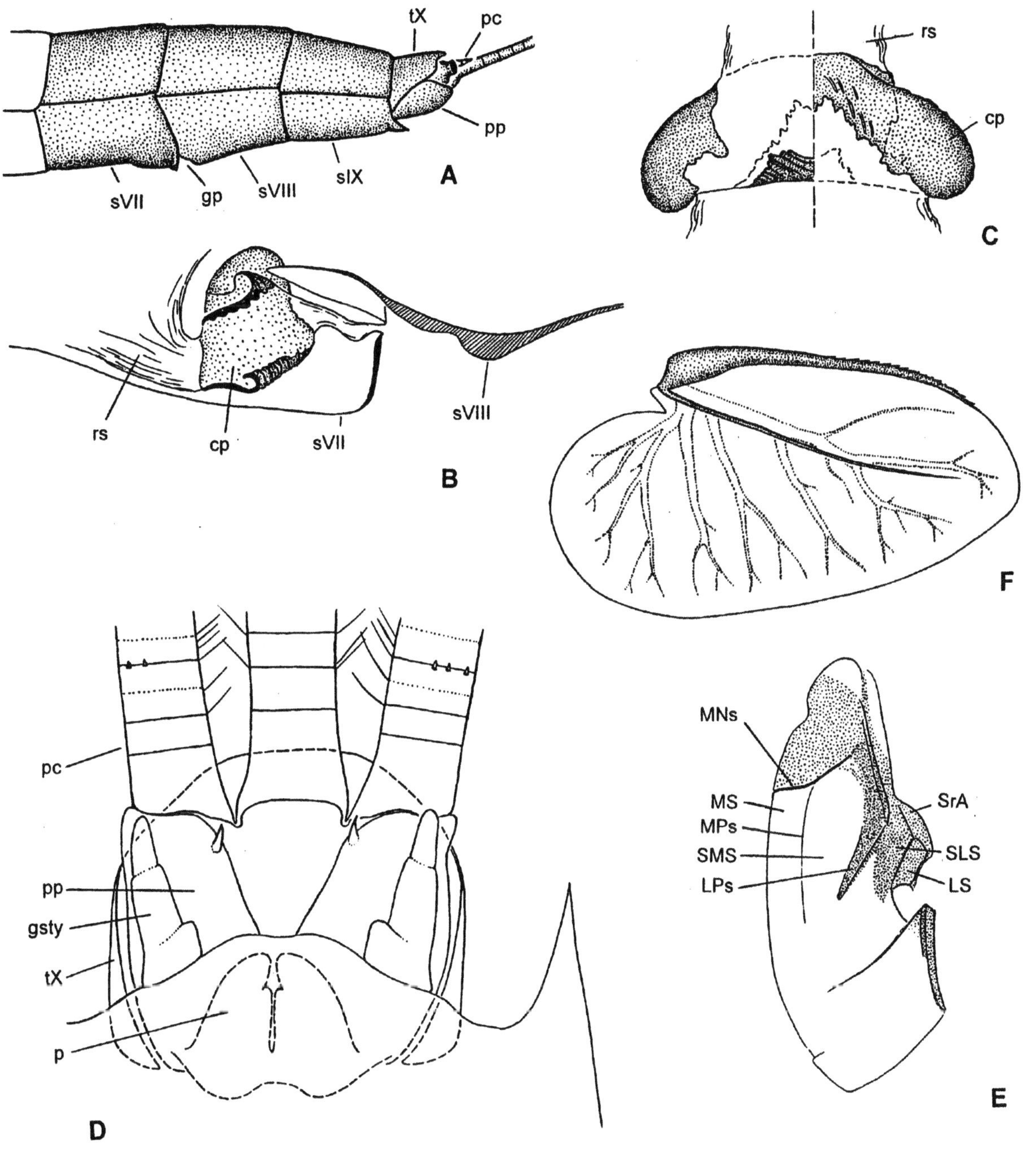

Figure 18. Siphlonurus/fg4.
A–E – Siphlonurus/fg4 *aestivalis* [*Siphlurus*]. **A–C** – Copulatory apparatus of female imago: **A** – abdominal segments VII–X, lateral view; **B** – boundary of sterna VII and VIII with copulatory pouch, median section; **C** – copulatory pouch, dorsal view (in left half) and ventral view (in right half). **D** – Apex of abdomen of mature male larva, ventral view. **E** – subimaginal exuviae of right half of mesonotum; **F** – Siphlonurus/fg4 *palaearcticus* [*Oniscigaster*]: tergalius III. (A – from Kluge 2000; B, C, E, F – from Kluge & al. 1995)

Abbreviations: **a.r** – anal rib; **c** – cercus; **cp** - copulatory pouch; **c.r** – costal rib; **gp** – gonopore; **gsty** – gonostylus; **LPs** – lateroparapsidal suture; **LS** – lateroscutum; **MNs** – mesonotal suture; **MPs** – medioparapsidal suture (notaulix); **MS** – medioscutum; **p** – penis; **pc** – paracercus; **pp** – paraproct; **rs** - receptaculum seminis; **sVII, sVIII, sIX** – abdominal sternites VII, VIII and IX; **SLS** – sublateroscutum; **SMS** – submedioscutum; **SrA** – suralare; **tX** – abdominal tergite X.

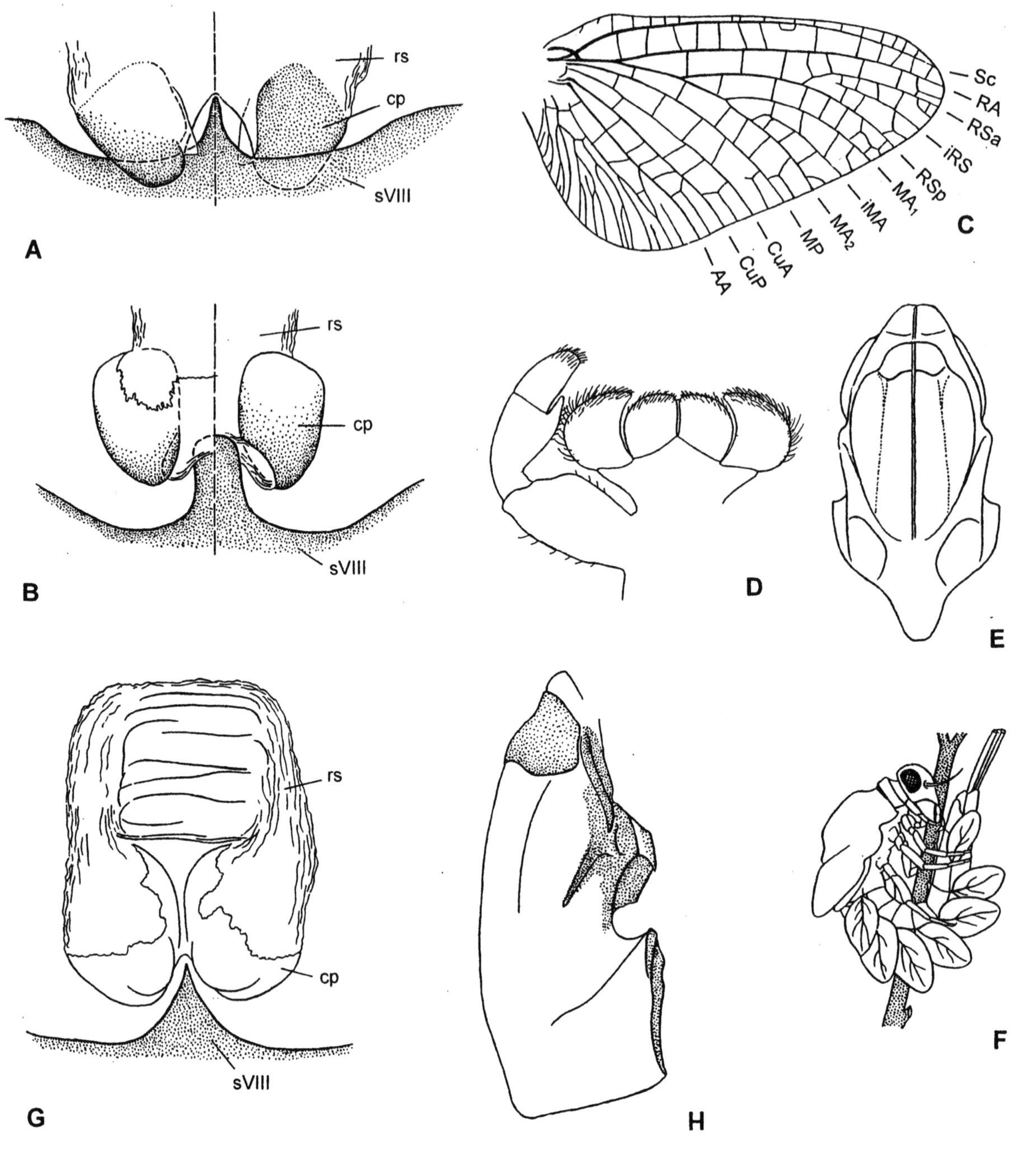

Figure 19. Siphlonurus/fg1-Parameletus/fg1.
A – Parameletus/fg2 *minor* [*Potameis*]: copulatory pouch of female imago, dorsal view (in left half) and ventral view (in right half). **B–F** – Parameletus/fg* *chelifer* [*Parameletus*]: **B** – copulatory pouch of female imago, dorsal view (in left half) and ventral view (in right half); **C** – hind wing; **D** – labium; **E** – imaginal mesonotum; **F** – larva cleaning its abdomen and tergalii. **G–H** – Siphlonisca/g* *aerodromia* [*Siphlonisca*]: **G** – copulatory pouch of female imago, dorsal view; **H** – subimaginal exuviae of right half of mesonotum. (A, B, E, G, H – from Kluge & al. 1995; D – from Kluge 1997d; F – by Novikova, from Kluge & Novikova & Brodsky 1984)

Abbreviations: **cp** - copulatory pouch; **rs** - receptaculum seminis; **sVIII** – anterior margin of abdominal sternite VIII.

2.1;1,1-1. Parameletus/fg2

(Anteritorna Tridentiseta Siphlonurus/fg1 Parameletus/fg1 ...fg2)

(Figs 10:G–H; 19A–F)

Nomen hierarchicum: **Parameletus/fg2** (sine *Siphlonisca*).
In circumscription matches:
— gen. *Parameletus* Bengtsson 1908: 242;
— gen. *Potameis* Bengtsson 1909: 15;
— gen. *Sparrea* Esben-Petersen 1909: 554;
— gen. *Palmenia* Aro 1910 (in Lestage 1924): 28;
— gen. *Siphlonuroides* McDunnough 1923: 48.

References. Needham & Traver & Hsu 1935: ℧* ⊕*; – Edmunds & Jensen & Berner 1976: ℧* ⊕*; – Söderström & Nilsson 1986: ℧ ⊕.

Autapomorphies of Parameletus/fg2.

(1) 2nd (penultimate) segment of labial palp with inner-apical projection (FIG.19:D). Non-unique apomorphy (see Index of characters [1.1.58]).

(2) On hind wing MP lacks triad, MP_2 and iMP being lost; thus only two triads are present – RS and MA [see Euplectoptera (1)] (FIG.19:C).

Plesiomorphies of Parameletus/fg2. In contrast to *Siphlonisca*: maxilla with developed apical-ventral row of slightly pectinate setae [see Plesiomorphies of Siphlonurus/fg1]; imaginal and subimaginal abdomen has no lateral lobes.

Size. Fore wing length 10–14 mm.

Distribution. Holarctic.

Nominal taxon included. Parameletus/fg2 includes **Siphlonuroides/g** [g: *Siphlonuroides* McDunnough 1923: 48, typus *S. croesus* McDunnough 1923 (design. orig.)], regarded as a generic synonym of *Parameletus* (McDunnough 1932: 81).

Species composition of Parameletus/fg2. 5 species are described – *chelifer* Bengtsson 1908 [*Parameletus*] (synn. subj.: *elegans* Bengtsson 1909 [*Potameis*], *norvegica* Esben-Petersen 1909 [*Sparrea*], *fennica* Aro 1910 [*Palmenia*]), *columbiae* McDunnough 1938 [*Parameletus*], *croesus* McDunnough 1923 [*Siphlonuroides*], *midas* McDunnough 1923 [*Siphlonuroides*], *minor* Bengtsson 1909 [*Potameis*].

Material examined: *chelifer* [*P.*]: ℧–O–⊕/⊙; *minor* [*P.*]: ℧–O–⊕/⊙; *columbiae* [*P.*]: ℧, ⊕♂; sp.S3 (Russian Far East): ⊕♂.

***

2.1;1,1-2. Siphlonisca/g(1)

(Anteritorna Tridentiseta Siphlonurus/fg1 Parameletus/fg1 *Siphlonisca*)

(Fig. 19:G–H)

Nomen hierarchicum: **Siphlonisca/g(1)** [g: *Siphlonisca* Needham 1909: 72, typus *S. aerodromia* Needham 1909 (design. orig.)].
In circumscription matches:
— gen. *Siphlonisca* Needham 1909: 72.
Monospecific taxon.

References. Needham & Traver & Hsu 1935: ℧ ⊕; – Edmunds & Jensen & Berner 1976: ℧ ⊕; – Burian & Gibbs 1988: ℧ ⊕.

Autapomorphies of *Siphlonisca*.

(1) Maxilla lacks apical-ventral row of setae (in contrast to all other Siphlonurus/fg1). Non-unique apomorphy (see Index of characters [1.1.31]).

(2) In imago and subimago of both sexes abdominal segments V–IX have large lateral lobes; non-unique apomorphy, similar lobes in selected species of Oniscigaster/fg2 and some others.

Plesiomorphies of *Siphlonisca*. In contrast to Parameletus/fg2: On hind wing MP with bifurcation proximad of wing middle (as in FIG.7:D). 2nd segment of labial palp has no projection.

Size. Fore wing length 15–17 mm.

Distribution. Eastern Nearctic.

Species composition of Siphlonisca/g(1). One species – *aerodromia* Needham 1909 [*Siphlonisca*].

Material examined: *aerodromia* [*S.*]: ℧–O–⊕/⊙.

*** ***

Siphlonurus/fg1 INCERTAE SEDIS

A single extinct species – *dubiosus* Demoulin 1968 [*Siphlonurus*] – is described from Baltic amber. Possibly, it is conspecific with the specimen examined by us (deposited in Paleontol. Inst. in Moscow) – a male imago with its subimaginal exuviae; subimaginal and imaginal thorax, wings and legs of this specimen are indistinguishable from those of Recent Siphlonurus/fg3, genitals are poorly preserved.

*** ***

2.1;2. Dipteromimus/fg(1)

(Euephemeroptera Euplectoptera Anteritorna Tridentiseta *Dipteromimus*)

(Fig. 21:A)

Nomen hierarchicum: **Dipteromimus/fg(1)** [f: Dipteromimidae Kluge & Studemann & Landolt & Gonser 1995: 111; g: *Dipteromimus* McLachlan 1875: 170, typus *D. tipuliformis* McLachlan 1875 (monotypy); syn. subj.: ***Dipteromimodes*** Matsumura 1931: 1474, typus *D. suzukii* Matsumura 1931 (monotypy) (synn. subj.: *tipuliformis* [*D.*] = *suzukii* [*D.*])].
In circumscription matches:
— gen. *Dipteromimus* McLachlan 1875: 170;
— fam. Dipteromimidae Kluge & al. 1995: 111.
Monospecific taxon.

References. McLachlan 1875: ⊕; – Eaton 1883–1888: ⊕; – Ueno 1931: ℧ ⊕; – Kluge & al. 1995: ℧ ◎ ⊕.

Autapomorphies of *Dipteromimus*.

(1) Mouth apparatus, being generally primitive, has some unique modifications (Ueno 1931:Fig.28): Glossae and paraglossae are petiolate – thick, but strongly narrowed basally. Maxilla is somewhat diminished, its apical-ventral row of setae is vestigial; maxillary palp is elongate.

(2) Hind wings are diminished, and fore wings are narrowed: Length of hind wing is about 0.1 of fore wing length; hind wing veins RS, MA and MP [which initially form triads – see Euplectoptera (1)] are shortened so strongly, that lost furcations. Fore wing lost its initial triangular shape, strongly narrowed at proximal half, thus is widest at mid-length [at the same time initial basitornal margin remains to be long (much longer than hind wing), and cubital field has 5–7 veins going from CuA to basitornal margin – see Anteritorna (1)] (Eaton 1883–1888: Pl.20:35; Pl.64:25).

Characters of unclear phylogenetic status.

(3) Subimaginal lateral pigmented area of mesonotum is strongly stretched posteriorly along sublateroscutum (FIG.21:A). By form of this area *Dipteromimus* differs from each other taxa.

(4) Imaginal and subimaginal claws (initially ephemeropteroid) on each leg are similar and pointed. Non-unique apomorphy (see Index of characters [2.2.85] and [2.2.77]); among Tridentiseta the same in Siphlonurus/fg1 and Nesameletus/f1= =Metamonius/g1.

(5) On each tergalius I-VII anal rib is located very far from anal margin, on dorsal side, near middle of tergalius, close to middle trachea (while costal rib has usual for mayflies position on costal margin) (Ueno 1931:Fig.27). The same in Siphlonurus/fg1, Nesameletus/f1=Metamonius/g1, *Rallidens* and some others (see Index of characters [1.3.28]).

(6) Larval abdominal terga and sterna lack denticles on posterior margin (probably, non-unique apomorphy) and lack stout setae (which are peculiar for Siphlonurus/fg1 and some others).

Plesiomorphies of *Dipteromimus*. Larva has primary swimming siphlonuroid specialization: legs are able to stretch posteriorly; abdomen is large, able to make undulate swimming movements; caudalii are not long, paracercus is well-developed, primary swimming setae are dense, secondary swimming setae are absent. Maxilla [see (1)] has 3 canines and 3 dentisetae [see Anteritorna (2)]; maxillary palp [see (1)] is 3-segmented; labial palp is 3-segmented. Larval (and adult) patella-tibial suture is developed on middle and hind legs only. Larval claws are slightly curved, without denticles.

In imago and subimago: Mesonotal suture is transverse (FIG.21:A); anterior paracoxal suture is complete (in contrast to *Rallidens* and some Bidentiseta); furcasternal protuberances are contiguous (in contrast to Amphinotic groups of Tridentiseta and some others – see Index of characters [2.2.23]). Imaginal and subimaginal tarsi are 5-segmented, 1st segment is the longest and fused with tibia. Paracercus is subequal to cerci. Gonostylus has 2 distal segments.

Size. Fore wing length 13–16 mm.

Distribution. Japan.

Species composition of Dipteromimus/fg(1). 1 species – *tipuliformis* McLachlan 1875 [*Dipteromimus*] (syn. subj.: *suzukii* Matsumura 1931 [*Dipteromimodes*]).

Material examined: *tipuliformis* [*D.*]: ♉, ◎, ⊕/⊙.

***

2.1;3. **Ameletus/fg1**

(Euephemeroptera Euplectoptera Anteritorna Tridentiseta Ameletus/fg1)

(Figs 9:I–M; 11; 13:D–E; 20; 21:B–C)

Nomen hierarchicum: **Ameletus/fg1** (incl. *Metreletus*) [f: Ameletidae McCafferty 1991a: 349; g: *Ameletus* Eaton 1885: 210, typus *A. subnotatus* Eaton 1885 (design. orig.)].

In circumscription matches:
— gen. *Ameletus*: Fizaine 1931: 25;
— *Ameletus*-complex: Koss & Edmunds 1974: 301;
— fam. Ameletidae: Kluge & Studemann & Landolt & Gonser 1995: 111.

References. Koss & Edmunds 1974: ⊙*; – Kluge & Studemann & Landolt & Gonser 1995: ♉* ◎* ⊕*.

Autapomorphies of Ameletus/fg1.

(1) Mouth apparatus is specialized, with mouthparts elongate and maxillae filtering (Eaton 1883–1888:Pl.49:4–9). Labrum is more or less elongate. Mandibles are elongate perpendicular to their axis of articulation, with thin weak incisor and kinetodontium, prostheca on left mandible is setiform, on right mandible is lost. Maxilla is strongly modified (FIG.20:B–C): its apical margin is widened, pectinate setae of apical-ventral row are elongate, forming a filtering apparatus; maxillary canines are completely lost, among dentisetae only one is retained, being poorly visible and directed dorsally. In contrast to other parts, labium is not elongate (FIG.20:A, D).

(2) Imaginal and subimaginal lateroparapsidal suture is elongate; subimaginal lateral pigmented area of mesonotum includes entire sublateroscutum and submedioscutum up to medioparapsidal suture (FIG.21:B). Similar form of this area in Isonychia/fg1 only (see Index of characters [2.2.14]).

(3) Imaginal and subimaginal epimeron of mesothorax has a membranous area between anepimeron and katepimeron (FIG.21:C).

(4) Tergalii (see Variable characters) lost ability of rhythmical respiratory movements. Non-unique apomorphy (see Index of characters [1.3.30]). Larvae inhabit mountain streams and lakes with high oxygen concentration, thus usually do not need in respiratory movements. Being placed into stagnant water with low oxygen concentration, larva does not move by its tergalii, but makes dorsoventral undulation by abdomen (FIG.9:I–M). Such respiratory movement is rather primitive, as it proceeds from the initial swimming movement (see Plesiomorphies), but is done slower and with legs fixed on the substrate (Kluge & Novikova & Brodsky 1984).

(5) Male imaginal styliger dorsally has a membranous area (FIG.11:B–D).

(6) Imaginal and subimaginal paracercus is vestigial (FIG.11:B–C). Non-unique apomorphy (see Index of characters [2.3.22]); particularly, the same in Siphlonurus/fg1 and Metretopus/fg1.

Plesiomorphies of Ameletus/fg1. Larva has primary swimming siphlonuroid specialization (FIG.9:I–M): legs are able to stretch posteriorly; abdomen is large, able to make undulate swimming movements; caudalii are not long, paracercus is well-developed, primary swimming setae are dense, secondary swimming setae are absent. Larval head is hypognathous [see (1)], maxillary palp is 3-segmented, labial palp is 3-segmented. Larval (and adult) patella-tibial suture is developed on middle and hind legs only. Larval claws are slightly curved.

In imago and subimago: Mesonotal suture is stretched backward medially (FIG.21:B–C). Anterior paracoxal suture is complete (FIG.21:C) (in contrast to *Rallidens* and some Bidentiseta). Furcasternal protuberances are contiguous (in contrast to Amphinotic groups of Tridentiseta and some others – see Index of characters [2.2.23]). Hind wing is well-developed, as long as 0.35–0.4 of fore wing length (as in FIG.7:D). Imaginal and subimaginal tarsi are 5-segmented, 1st segment is non-shortened and fused with tibia. All claws of imago and subimago are ephemeropteroid. Gonostylus has 2 distal segments (FIG.11).

Variable characters of Ameletus/fg1. In different species costal and anal ribs are either present on all tergalii I–VII, or lost on tergalii I–II; anal rib can be located either far from anal margin or close to anal margin; tergalius margin usually with denticles at apical part of costal rib and at the place where anal rib reaches anal margin (FIG.13:D–E; Zloty 1997:Figs 22–23).

Size. Fore wing length 7–17 mm.

Age and distribution. Palaeogene — Recent; Holarctic and Oriental Region. A single reliably determined fossil specimen is a female imago from Baltic amber (deposited in Mus. Nat. Hist. of Inst. Syst. Evol. Anim. in Krakow) which has characteristic structure of epimeron [see (3)]; its wing venation has plesiomorphic condition characteristic for Ameletus/fg2 (see below).

Ameletus/fg1 is divided into *Metreletus* and Ameletus/fg2.

2.1;3,1. Metreletus/g(1)

(Euplectoptera Anteritorna Tridentiseta Ameletus/fg1 *Metreletus*)

(Figs 20:C–D; 21:C)

Nomen hierarchicum: **Metreletus/g(1)** [g: *Metreletus* Demoulin 1951, typus *Metretopus goetghebuergi* Lestage 1938 (design. orig.; syn. subj. *Metretopus balcanicus* Ulmer 1920)].
In circumscription matches:
— gen. *Metreletus* Demoulin 1951: 10.
Monospecific taxon.

References. Demoulin 1951: ⥁ ⊕; – 1952c: ⊙ ⊕; – Ujhelyi 1960: ⥁ ⊕; – Studemann & Landolt & Tomka 1988: ⊙ ⊕.

Characters of unclear phylogenetic status.

(1) In cubital field of fore wing veins, which go from CuA to wing margin [see Anteritorna (1)], are more or less substituted by one or two intercalaries (Demoulin 1951:Figs 1a-e, 11a-d; 1952c:Fig.4; Ujhelyi 1960:Fig.5). Non-unique apomorphy (see Index of characters [2.2.51]).

(2) Larval claw on its inner side, besides irregular small denticles near base, has one row of greater denticles (Demoulin 1951:Fig.6b-d). The same in some Bidentiseta and others (see Index of characters [1.2.21]); probably non-unique autapomorphy.

Plesiomorphy of *Metreletus* (in contrast to Ameletus/fg2). Glossae are not truncate, apically-ventrally with numerous irregular pointed setae (FIG.20:D).

Size. Fore wing length 8–14 mm.

Distribution. Mountains of Europe.

Species composition of Metreletus/g(1). 1 species – *balcanicus* Ulmer 1920 [*Metretopus*] (synn. subj.: *hessei* Fizaine 1931 [*Ameletus*], *goetghebuergi* Lestage 1938 [*Metretopus*], *hungaricus* Ujhelyi 1960 [*Metreletus*]).

Material examined: *balcanicus* [*M.*]: ♉, ◎, ⊕♂, ⊕♀/⊙.

***

2.1;3,2. Ameletus/fg2

(Euplectoptera Anteritorna Tridentiseta Ameletus/fg1 .../fg2)

(Figs 11; 13:D–E; 20:A–B; 21:B)

Nomen hierarchicum: **Ameletus/fg2** (sine *Metreletus*).
In circumscription matches:
— gen. *Ameletus* Eaton 1885: 210.

References. Eaton 1883–1885: ♉* ⊕*; – Needham & Traver & Hsu 1935: ♉* ⊕*; – Edmunds & Jensen & Berner 1976: ♉* ⊕*; – Zloty 1996: ♉* ⊕*.

Autapomorphies of Ameletus/fg2.

(1) Glossae are truncate, with widened straight apical margin; all apical setae form a regular apical-ventral row and are flat, widened toward apex and rounded apically (FIG.21:A). Unique apomorphy.

Plesiomorphies of Ameletus/fg2 (in contrast to *Metreletus*). Cubital field of fore wing without intercalaries, several (4–8) veins go from CuA to basitornal margin [see Anteritorna (1)] (as in FIG.7:C). Larval claws have no denticles.

Size. Fore wing length 7–17 mm.

Age and distribution. Probably from Palaeogene (see Ameletus/fg1) — till Recent; Holarctic and Oriental Region. Dominate in Nearctic and Eastern Palaearctic; in Western Palaearctic occurs a single Holarctic species – *inopinatus* [*Ameletus*].

Nominal taxa included. Ameletus/fg2 includes: (1) **Chimura/g** [g: *Chimura* Navás 1915: 149, typus *Ch. aetherea* Navás 1915 (design. orig.)], regarded as a generic synonym of *Ameletus* (Edmunds 1960: 24); (2) **Paleoameletus/g** [g: *Paleoameletus* Lestage 1940: 124, typus *Ameletus primitivus* Traver 1939 (design. orig.)], regarded as a generic synonym of *Ameletus* (Edmunds & Traver 1954a: 237).

Species composition of Ameletus/fg2 (here [*A.*]= =[*Ameletus*]). About 50 species are described. One **Trans-Holarctic species:** *inopinatus* Eaton 1887 [*A.*] (synn.subj.: *alpinus* Bengtsson 1913 [*A.*], *eugenii* Sinitshenkova & Varykhanova 1989 [*A.*]). **Asian species:** *aetherea* Navás 1915 [*Chimura*], *alexandrae* Brodsky 1930 [*A.*] (syn.subj.: *asiacentralis* Soldán 1978 [*A.*]), *atratus* Kang & Yang 1994 [*A.*], *camtschaticus* Ulmer 1927 [*A.*] (syn. subj.: *pilatus* Sinitshenkova 1981 [*A.*]), *cedrensis* Sinitshenkova 1977 [*A.*], *cristatus* Bajkova 1976 [*A.*], *crocerus* Imanishi 1931 [*A.*], *costalis* Matsumura 1931 [*Chimura*], *dodecus* Dubey 1971 [*Ichthybotus*], *formosus* Kang & Yang 1994 [*A.*], *kyotoensis* Imanishi 1931 [*A.*], *labiatus* Sinitshenkova 1981 [*A.*], *longulus* Sinitshenkova 1981 [*A.*], *micus* Bajkova 1976 [*A.*], *montanus* Imanishi 1930 [*A.*] (syn. subj.: *procerus* Bajkova 1976 [*A.*]), *montivagus* Kang & Yang 1994 [*A.*], *parvus* Kluge 1979 [*A.*], *primitivus* Traver 1939 [*A.*]; *subalpinus* Imanishi 1931 [*A.*]. **North American species:** *amador* Mayo 1939 [*A.*], *andersoni* Zloty 1996 [*A.*], *bellus* Zloty 1996 [*A.*], *browni* McDunnough 1933 [*A.*], *celer* McDunnough 1934 [*A.*] (synn. subj.: *alticolus* McDunnough 1934 [*A.*], *celeroides* McDunnough 1934 [*A.*], *tuberculatus* McDunnough 1939 [*A.*]), *cooki* McDunnough 1929 [*A.*], *cryptostimulus* Carle 1978 [*A.*], *dissitus* Eaton 1885 [*A.*], *doddsianus* Zloty 1996 [*A.*], *edmundsi* Zloty 1996 [*A.*], *exquisitus* Eaton 1885 [*A.*], *falsus* McDunnough 1938 [*A.*], *imbellus* Day 1952 [*A.*], *lineatus* Traver 1932 [*A.*], *ludens* Needham 1905 [*A.*], *minimus* Zloty & Harper 1999 [*A.*], *majusculus* Zloty 1996 [*A.*], *oregonensis* McDunnough 1933 [*A.*], *pritchardi* Zloty 1996 [*A.*], *quadratus* Zloty & Harper 1999[*A.*], *shepherdi* Traver 1934 [*A.*] (syn. subj.: *querulus* McDunnough 1938 [*A.*]), *similor* McDunnough 1928 [*A.*] (syn. subj.: *monta* Mayo 1952 [*A.*]), *sparsatus* McDunnough 1931 [*A.*] (syn. subj.: *aequivocus* McDunnough 1934 [*A.*]), *subnotatus* Eaton 1885 [*A.*], *suffusus* McDunnough 1936 [*A.*], *tarteri* Burrows 1987 [*A.*], *tertius* McDunnough 1938 [*A.*], *tolae* Zloty 1996 [*A.*], *validus* McDunnough 1932 [*A.*], *vancouvernensis* McDunnough 1933 [*A.*] (syn. subj.: *facilis* Day 1952 [*A.*]), *velox* Dodds 1923 [*A.*] (synn. subj.: *connectus* McDunnough 1936 [*A.*], *connectina* McDunnough 1939 [*A.*]), *vernalis* McDunnough 1924 [*A.*], *walleyi* Harper 1970 [*A.*].

Material examined: *alexandrae* [*A.*]: ♉–◎–⊕/⊙; *altaicus* [*Ameletus*] (in litt.): ♉–◎–⊕/⊙; *bellus* [*A.*]: ♉♂–⊕♂; *camtschaticus* [*A.*]: ♉–◎–⊕/⊙; *cedrensis* [*A.*]: ♉–◎–⊕/⊙; *celer* [*A.*]: ♉♂–⊕♂; *costalis* [*Chimura*]: ♉, ♉♂–◎♂, ⊕/⊙; *inopinatus* [*A.*]: ♉–◎–⊕/⊙; *labiatus* [*A.*]: ♉♂–◎♂–⊕♂; *longulus* [*A.*]: ♉; *micus* [*A.*]: ♉♂/◎♂; *montanus* [*A.*]: ♉–◎–⊕/⊙; *oregonensis* [*A.*]: ♉–⊕; *parvus* [*A.*]: ♉–◎–⊕/⊙; *similor* [*A.*]: ♉, ♉/◎, ⊕; *velox* [*A.*]: ♉♂–⊕♂; sp. (China): ♉♂/◎♂.

*** ***

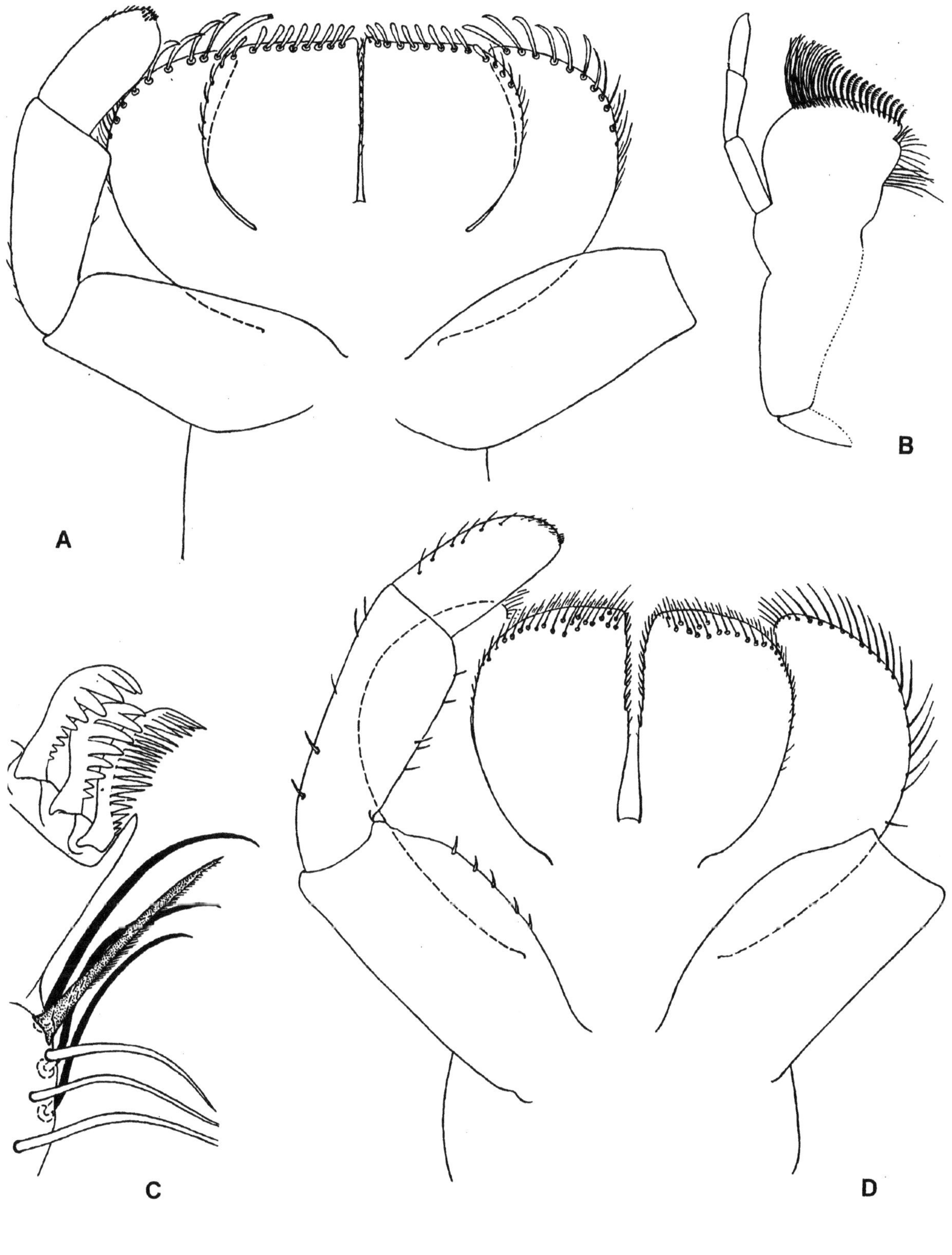

Figure 20. Ameletus/fg1.
A–B – Ameletus/fg2 *inopinatus* [*Ameletus*]: **A** – labium, ventral view; **B** – maxilla. **C–D** – Metreletus/g* *balcanicus* [*Metretopus*]: **C** – apex of left maxilla, dorsal view (dentiseta shown by dots, setae of inner-ventral row shown by black); **D** – labium, ventral view.

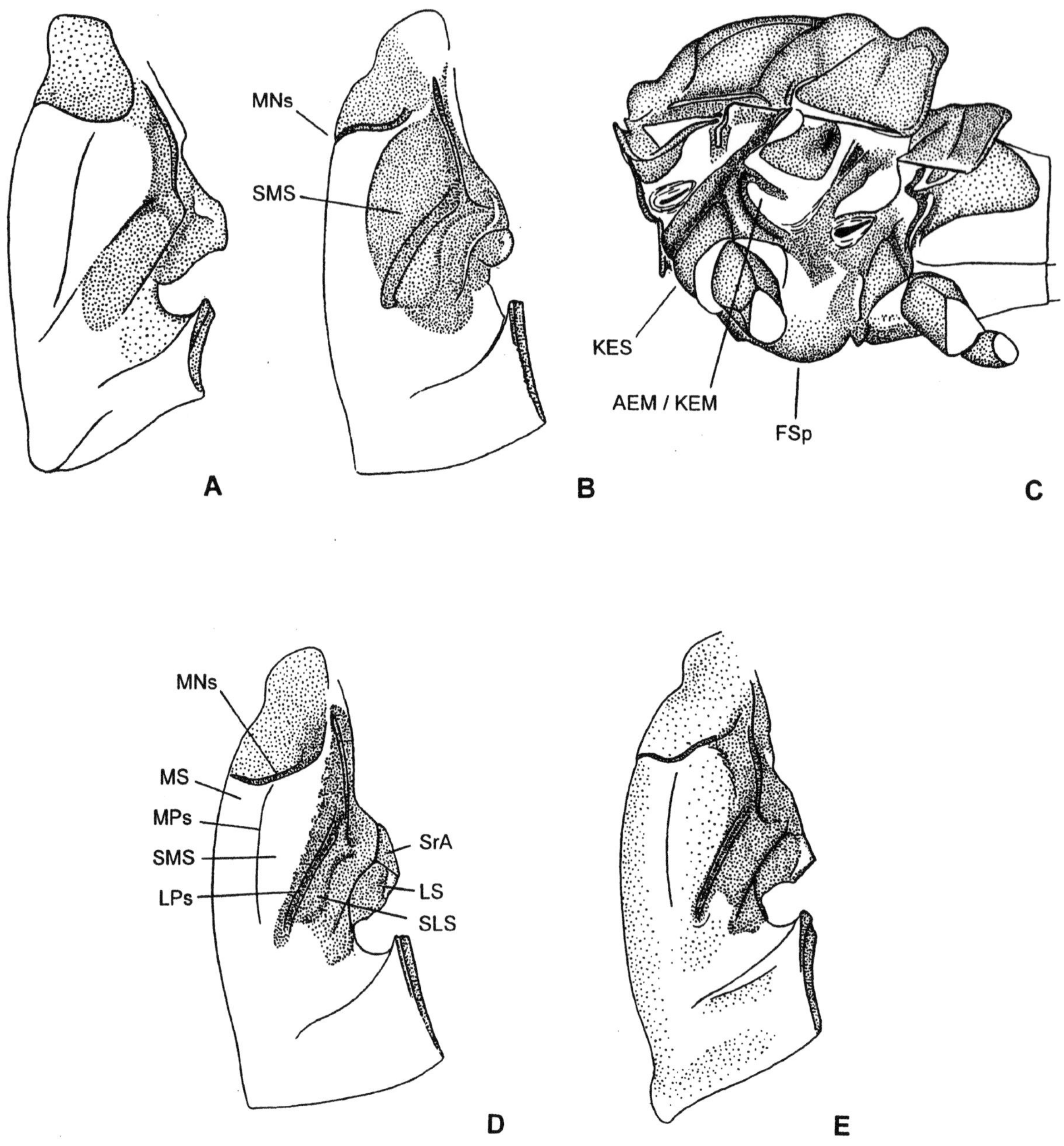

Figure 21. *Dipteromimus*, Ameletus/fg1 and Metretopus/fg1.
A – Dipteromimus/fg* *tipuliformis* [*Dipteromimus*], subimaginal exuviae of right half of mesonotum. **B–C** – Ameletus/fg1: **B** – Ameletus/fg2 *montanus* [*Ameletus*], subimaginal exuviae of right half of mesonotum; **C** – Metreletus/g* *balcanicus* [*Metretopus*], imaginal pterothorax, lateral view. **D–E** – Metretopus/fg1: **D** – Metreplecton/g* *macronyx* [*Metreplecton*], subimaginal exuviae of right half of mesonotum; **E** – Siphloplecton/fg* *basale* [*Baetis*], the same. (From Kluge & Studemann & Landolt & Gonser 1995)
Abbreviations: **AEM/KEM** – membranous area between anepimeron and katepimeron; **FSp** – furcasternal protuberance; **KES** – katepisternum; **LPs** – lateroparapsidal suture; **LS** – lateroscutum; **MNs** – mesonotal suture; **MPs** – medioparapsidal suture; **MS** – medioscutum; **SLS** – sublateroscutum; **SMS** – submedioscutum; **SrA** – suralare.

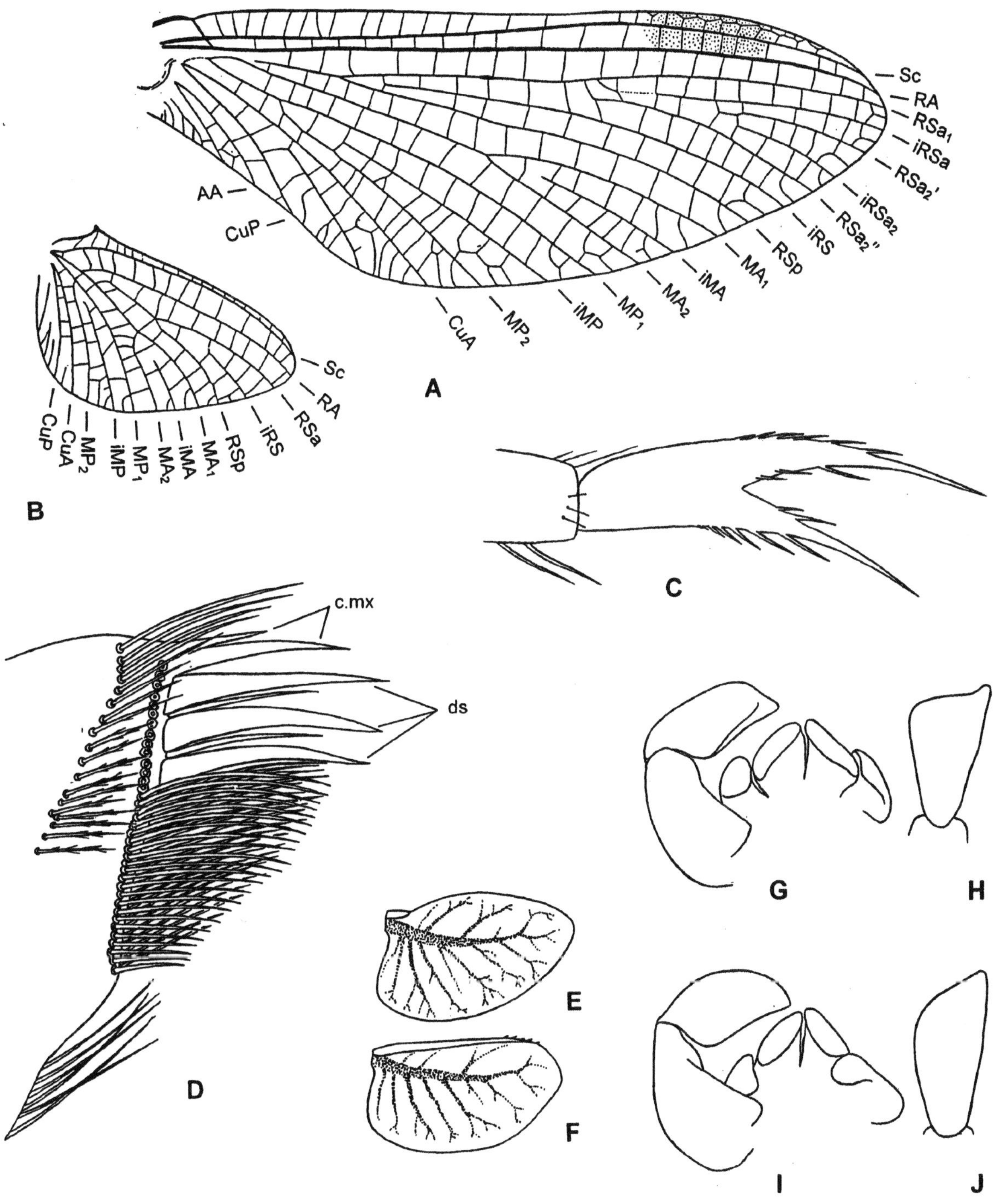

Figure 22. Metretopus/fg1.
A–H – Metreplecton/g* *macronyx* [*Metreplecton*]: **A–B** – fore and hind wings; **C** – claw of larval fore leg; **D** – apex of right maxilla, ventral view (setae of inner-ventral row partly removed to show dentisetae); **E–F** – tergalii III and IV; **G–H** – labium (ventral view) and its right palp (latero-apical view). **I–J** – Metretopus/fg* *borealis* [*Heptagenia*], the same. (From Kluge 1996b)

Abbreviations: **c.mx** – maxillary canine; **ds** – dentisetae.

2.1;4. Metretopus/fg1

(Euephemeroptera Euplectoptera Anteritorna Tridentiseta Metretopus/fg1)

(Figs 10:A–F; 21:D–E; 22)

Nomen hierarchicum: **Metretopus/fg1** (incl. *Metreplecton, Siphloplecton*) [f: Metretopodinae (orig. Metretopinae) Traver (in Needham & Traver & Hsu) 1935: 433; g: *Metretopus* Eaton 1901: 253, typus *M. norvegicus* Eaton 1901 (design. orig.; syn. subj. *Heptagenia borealis* Eaton 1871)].

In circumscription matches:

— subfam. Metretopodinae: Edmunds & Traver 1954a:237;

— fam. Metretopodidae: Tshernova 1970: 135.

References. Edmunds & Allen & Peters 1963: ℧*; – Tshernova 1970: ℧* ⊕* – Koss & Edmunds 1974: ⊙*; – Edmunds & Jensen & Berner 1976: ℧* ⊕*; – Berner 1978: ℧* ⊕*; – Kluge & Studemann & Landolt & Gonser 1995: ℧* ◎* ⊕* ⊙*.

Autapomorphies of Metretopus/fg1.

(1) Larval claw of fore leg is bifurcate (in contrast to all other Ephemeroptera); lateral margins of claw and inner margins of both its branches bear a longitudinal row of very long slender denticles resembling setae (FIG.22:C). Claws of middle and hind legs have usual structure – non-bifurcate, slender, slightly curved, without any denticles.

(2) Mouth apparatus is somewhat modified: Labrum and hypopharynx are truncate. Maxilla has 2 canines only (instead of three initial ones – see Index of characters [1.1.33]); its apical-ventral row of setae (consisted of simple non-pectinate setae) is transferred to ventral side and directed parallel to inner margin (FIG.22:D). 2nd segment of 3-segmented maxillary palp is shortened. Glossae and paraglossae are curved ventrally; labial palp is 2-segmented, its distal segment (a result of complete fusion of initial 2nd and 3rd segments) is thickened and lacks muscle initially moving 3rd segment (FIG.22:G–H, I–J).

(3) All tergalii I–VII lack anal ribs; tergalii I–III lack costal ribs also (FIG.22:E–F). In connection with this, in some species of *Siphloplecton* costal margin of tergalii I–III is expanded and bent to ventral side.

Characters of unclear phylogenetic status.

(4) Imaginal and subimaginal lateroparapsidal suture is elongate; subimaginal lateral pigmented area of mesonotum occupies most part of sublateroscutum, nearly up to posterior end of lateroparapsidal suture (FIG.21:D–E). Non-unique character (see Index of characters [2.2.14]); similar elongation of lateroparapsidal suture in Ameletus/fg1 (FIG.21: B) ; possibly synapomorphy.

(5) Cubital field of fore wing [see Anteritorna (1)] with intercalaries: usually there are two pairs of intercalaries in *Siphloplecton* (Berner 1978:Figs 1–5), one pair in Metretopus/fg2 (ibid.:Fig.6), single intercalary in *Metreplecton* (FIG.22:A). The same in Ameletus/fg1-*Metreletus* and many other mayflies (see Index of characters [2.2.52]); probably non-unique autapomorphy.

(6) Imaginal and subimaginal paracercus is vestigial. Non-unique apomorphy (see Index of characters [2.3.22]); particularly the same in Siphlonurus/fg1 and Ameletus/fg1.

(7) Larval paraproct with a spine on inner-apical margin. The same in Siphlonurus/fg2 (FIG.18:D), *Siphluriscus*, Vetulata (FIG.33:F) and *Rallidens*.

Plesiomorphies of Metretopus/fg1. Larva has primary swimming siphlonuroid specialization: legs are able to stretch posteriorly; abdomen is large, able to make undulate swimming movements; caudalii are not long, paracercus is well-developed, primary swimming setae are dense, secondary swimming setae are absent. Larval head is hypognathous. Maxilla [see (2)] with 3 dentisetae [see Anteritorna (2)] (FIG.22:D). Larval (and adult) patella-tibial suture is developed on middle and hind legs only. Tergalii [see (3)] retain ability of rhythmical respiratory movements.

In imago and subimago: Mesonotal suture can be stretched backward medially (FIG.21:E) or nearly transverse (FIG.21:D). Anterior paracoxal suture is complete (in contrast to *Rallidens* and some Bidentiseta). Furcasternal protuberances are contiguous (in contrast to Amphinotic groups of Tridentiseta and some others – see Index of characters [2.2.23]). Hind wing is well-developed, as long as 0.35–0.4 of fore wing length (FIG.22:A–B). Imaginal and subimaginal tarsi are 5-segmented, 1st segment is the longest, fused with tibia. All claws of imago and subimago are ephemeropteroid. Gonostylus has 2 distal segments.

Size. Fore wing length 9–16 mm.

Distribution. Largest part of Holarctic: North of Europe, whole Siberia, Nearctic.

Metretopus/fg1 is divided into Metretopus/fg2, *Metreplecton* and *Siphloplecton*.

2.1;4,1. Metretopus/fg2

(Euplectoptera Anteritorna Tridentiseta Metretopus/fg1 .../fg2)

(Figs 10:A–F; 22:I–J)

Nomen hierarchicum: **Metretopus/fg2** (sine *Metreplecton, Siphloplecton*).
In circumscription matches:
— gen. *Metretopus* Eaton 1901: 253.

References. Needham & Traver & Hsu 1935: ℧ ⊕; – Edmunds & Jensen & Berner 1976: ℧ ⊕; – Engblom & Lingdell & Nilsson & Savolainen 1993: ℧ ◎ ⊕; – Kluge 1996b: ℧ ⊕.

Characters of unclear phylogenetic status.

(1) Larval abdominal terga and sterna lack spine-like setae. The same in *Metreplecton*, in contrast to *Siphloplecton*.

(2) Tergalii IV–VII with denticles on costal margin [armed with rib – see Metretopus/fg1 (3)]. The same in *Metreplecton*, in contrast to *Siphloplecton*.

(3) Distal segment of labial palp [$2^{nd}+3^{rd}$ – see Metretopus/fg1 (2)] is widest near middle (FIG.22:J) (in contrast to *Metreplecton* and *Siphloplecton*).

(4) Cubital field of fore wing with 2 intercalaries [see Metretopus/fg1 (5)] (Berner 1978:Fig.6).

Size. Fore wing length 9–14 mm.

Distribution. Holarctic.

Species composition of Metretopus/fg2. 3 species – *borealis* Eaton 1871 [*Heptagenia*] (syn. subj.: *norvegicus* Eaton 1901 [*Metretopus*]), *alter* Bengtsson 1930 [*Metretopus*], *tertius* Tiunova 1999 [*Metretopus*].

Material examined: *borealis* [*H.*]: ℧–◎–⊕/⊙; *alter* [*M.*]: ℧–◎–⊕/⊙; *tertius* [*M.*]: ℧♂–◎♂–⊕♂.

***

2.1;4,2. Metreplecton/g(1)

(Euplectoptera Anteritorna Tridentiseta Metretopus/fg1 *Metreplecton*)

(Figs 21:D; 22:A–H)

Nomen hierarchicum: **Metreplecton/g(1)** [g: *Metreplecton* Kluge 1996b: 77, typus *M. macronyx* Kluge 1996 (design. orig.)].
In circumscription matches:
— gen. *Metreplecton* Kluge 1996b: 77.
Monospecific taxon.

Reference. Kluge 1996b: ℧ ⊕.

Characters of unclear phylogenetic status.

(1) Larval abdominal terga and sterna lack spine-like setae. The same in Metretopus/fg2, in contrast to *Siphloplecton*.

(2) Tergalii IV–VII have denticles on costal margin [armed with rib – see Metretopus/fg1 (3)]. The same in Metretopus/fg2, in contrast to *Siphloplecton*.

(3) Distal segment of labial palp [$2^{nd}+3^{rd}$ – see Metretopus/fg1 (2)] is widest near apex (FIG.22:H). The same in *Siphloplecton*, probably synapomorphy.

(4) Cubital field of fore wing usually with 1 intercalary (FIG.22:A), selected specimens can have up to 3 intercalaries [see Metretopus/fg1 (5)].

Size. Fore wing length 11–13 mm.

Distribution. Northern Europe, Siberia and Russian Far East.

Species composition of Metreplecton/g(1). 1 species – *macronyx* Kluge 1996 [*Metreplecton*].

Material examined: *macronyx* [*M.*]: ℧–◎–⊕/⊙.

***

2.1;4,3. Siphloplecton/fg(1)

(Euplectoptera Anteritorna Tridentiseta Metretopus/fg1 *Siphloplecton*)

(Fig. 21:E)

Nomen hierarchicum: **Siphloplecton/fg(1)** [f: Siphloplectidae (orig. Siphloplectonidae) Lestage 1938a: 180; g.: *Siphloplecton* Clemens 1915: 258, typus *Siphlurus flexus* Clemens 1913 (design. orig.; syn. subj. *Baetis basale* Walker 1853)].
In circumscription matches:
— gen. *Siphloplecton* Clemens 1915: 258;
— fam. Siphloplectonidae Lestage 1938a: 180.

References. Needham & Traver & Hsu 1935: ℧* ⊕*; – Edmunds & Jensen & Berner 1976: ℧* ⊕*; – Berner 1978: ℧* ⊕*.

Characters of unclear phylogenetic status.

(1) Larval abdominal terga and sterna have spine-like setae (in contrast to Metretopus/fg2 and *Metreplecton*).

(2) All tergalii, including IV–VII [see Metretopus/fg1 (3)], lack denticles (in contrast to Metretopus/fg2 and *Metreplecton*); only spine-like setae can be present.

(3) Distal segment of labial palp [$2^{nd}+3^{rd}$ – see Metretopus/fg1 (2)] is widest near apex. The same in *Metreplecton* (FIG.22:H), probably synapomorphy.

(4) Cubital field of fore wing usually with 4 intercalaries forming 2 pairs, among which the posterior pair arise more proximally and is the longest; in each pair the anterior intercalary (nearest to convex CuA) is concave, and the posterior intercalary (nearest to concave CuP) is convex; selected specimens have 3 intercalaries only [see Metretopus/fg1 (5)] (Berner 1978:Figs 1–5). Non-unique apomorphy; the same in Heptagennota (see Index of characters [2.2.52]).

Size. Fore wing length 9–16 mm.

Distribution. Nearctic.

Species composition of Siphloplecton/fg(1). 7 species – *basalis* Walker 1853 [*Baetis*] (synn. subj.: *flexus* Clemens 1913 [*Siphlonurus*], *signatum* Traver 1932 [*Siphloplecton*]), *brunneum* Berner 1978 [*Siphloplecton*], *costalense* Spieth 1938 [*Siphloplecton*], *fuscum* Berner 1978 [*Siphloplecton*], *interlineata* Walsh 1863 [*Baetis*], *simile* Berner 1978 [*Siphloplecton*], *speciosum* Traver 1932 [*Siphloplecton*].

Material examined: *basale* [*B.*]: ♂–♂, ♀–♀–♀/⊙; sp.(USA): ℧.

*** ***

2.1;5. Acanthametropus/fg1

(Euephemeroptera Euplectoptera Anteritorna Tridentiseta Acanthametropus/fg1)

(Fig. 23; 24:A–B)

Nomen hierarchicum: **Acanthametropus/fg1** (incl. *Analetris*) [f: Acanthametropodinae Edmunds (in Edmunds & Allen & Peters) 1963: 10; g: *Acanthametropus* Tshernova 1948: 1453, typus *A. nikolskyi* Tshernova 1948 (design. orig.)].

In circumscription matches:
— subfam. Acanthametropodinae: Hubbard 1982: 50;
— fam. Acanthametropodidae: McCafferty & Wang 1994b: 214.

References. Edmunds & Koss 1972: ℧*; – Edmunds & Jensen & Berner 1976: ℧* ◎*; – Kluge & Studemann & Landolt & Gonser 1995: ℧* ⊕*.

Autapomorphies of Acanthametropus/fg1.

(1) Mouth apparatus is specialized for carnivorism; both *Analetris* and Acanthametropus/fg2 pray on chironomid larvae. Labrum with wide median incision; asymmetry of mandibles is lost, each mandible is shortened perpendicular to axis of articulation and elongated along this axis, incisor and kinetodontium are stout, prostheca is lost, mola lacks grater; superlinguae are vestigial; (Kluge & al. 1995: Figs 23–30). Maxillae are stout, biting; apical setae are absent; canines and dentisetae are situated in one plane, only two canines are present, first and second dentisetae are canine-like, third dentiseta can be vestigial or lost (FIG.23:B) [see Plesiomorphies of Tridentiseta]. 2nd (penultimate) segment of labial palp is widened (FIG.23:C).

(2) Larval legs are specialized [see (3) and (6)], fore and middle legs differ from hind ones: femora of fore and hind legs are widened, claw of hind leg is much longer than claws of fore and middle legs (Tshernova 1948:Fig.3).

(3) On each leg tibia is shortened, tarsus is elongate; on imaginal and subimaginal middle and hind legs 1st tarsal segment is longer than tibia.

(4) Sterna with median projections: in larva projection is present at least on prosternum and mesosternum, in imago – on prosternum (FIG.23:D–E) (in larval Acanthametropus/fg2 projections are present not only on thorax, but on abdomen also).

(5) Each of tergalii I–VII consists of 3 lamellae: dorsal lamella is initial one, and 2 ventral lamellae represent 2 parts of expanded anal-proximal lobe bent down. In *Analetris* each of these 3 lobes has integral margins (Edmunds & Koss 1972:Fig.4), while in Acanthametropus/fg2 these lobes are strongly dissected and their projections are tangled (FIG.23:A). Anal rib is lost, thus only costal rib is present on initial (dorsal) lamella.

(6) Entire larval body is dorsoventrally flattened and widened, claws are strongly elongate [see (2)] (Tshernova 1948:Figs 1–3; Burks 1953:Fig.312; Edmunds & Koss 1972:Fig.1). Non-unique apomorphy – the same in some other specialized psammophils, such as *Ametropus* and *Pseudiron.*

(7) Larval abdominal terga and sterna bear stout scales – i. e. flattened setae (which nests are not widened, in contrast to scales of Turbanoculata) and lack denticles on posterior margin (Lehmkuhl 1976: Fig.13). Non-unique character.

Plesiomorphies of Acanthametropus/fg1.
Larva retains features of primary swimming siphlonuroid specialization: legs are able to stretch posteriorly [in spite of specialization – see (2), (3), (6)]; abdomen is large, able to make undulate swimming movements; caudalii are not long, paracercus is well-developed, primary swimming setae are dense, secondary swimming setae are absent. Larval head is hypognathous. Labial palp [see (1)] is 3-segmented. Larval (and adult) patella-tibial suture [on short tibia – see (3)] is developed on middle and hind legs only. Larval claws [see (2)] are slightly curved, without denticles.

In imago and subimago: Mesonotal suture is nearly transverse (at least in Acanthametropus/fg2). Anterior paracoxal suture is complete (FIG.23:E) (in contrast to *Rallidens* and some Bidentiseta). Furcasternal protuberances are contiguous (FIG.23:E) (in contrast to Amphinotic groups of Tridentiseta and some others – see Index of characters [2.2.23]). Subimaginal lateral pigmented area of mesonotum is small, stretches along lateroparapsidal suture and does not occupy sublateroscutum (FIG.23:F) (see Index of characters [2.2.14]). In cubital field of fore wing several (3–8) veins go from CuA to basitornal margin or anastomose and can be partly substituted by intercalaries [see Anteritorna (1)]; hind wing is well-developed, as long as 0.45–0.5 of fore wing length (FIG.24:A–B). Imaginal and subimaginal tarsi are 5-segmented, 1st segment [long – see (3)] is fused with tibia (FIG.23:E). Claws are ephemeropteroid [exception can be made by male fore legs only – see Acanthametropus/fg2 (6) below]. Imaginal and subimaginal paracercus is subequal to cerci.

Size. Fore wing length 12–20 mm.

Distribution. Amphipacific Sector of Holarctic: Eastern Palaearctic and Nearctic.

Acanthametropus/fg1 is divided into *Analetris* and Acanthametropus/fg2.

2.1;5,1. Analetris/fg(1)

(Euplectoptera Anteritorna Tridentiseta Acanthametropus/fg1 *Analetris*)

Nomen hierarchicum: **Analetris/fg(1)** [f: Analetridinae Demoulin 1974: 3; g: *Analetris* Edmunds (in Edmunds & Koss) 1972: 138, typus *A. eximia* Edmunds 1972 (design. orig.)].
In circumscription matches:
— gen. *Analetris* Edmunds (in Edmunds & Koss) 1972:138;
— subfam. Analetridinae Demoulin 1974: 3;
— fam. Analetrididae: McCafferty 1991a: 349.
Monospecific taxon.

References. Edmunds & Koss 1972: Ʊ ⊙; – Koss & Edmunds 1974: ⊙; – Edmunds 1975: Ʊ; – Edmunds & Jensen & Berner 1976: Ʊ ⊕; – Lehmkuhl 1976: Ʊ ⊕.

Autapomorphy of *Analetris*.

(1) Imaginal and subimaginal gonostylus has 1 distal segment (instead of two initial ones). Non-unique apomorphy (see Index of characters [2.3.12]).

Plesiomorphies of *Analetris* (in contrast to Acanthametropus/fg2). Larval and imaginal head has no projections. Maxillary palp is 3-segmented. Each tergalius has all 3 lamellae with integral margins [see Acanthametropus/fg1 (5)]. Larval abdomen has no unpaired dorsal and ventral projections. Imaginal and subimaginal foreleg claws are ephemeropteroid (as well as other claws).

Size. Fore wing length 12 mm.

Distribution. Nearctic.

Species composition of Analetris/fg(1). 1 species – *eximia* Edmunds 1972 [*Analetris*].

Material examined: *eximia* [*A.*]: Ʊ.

***

2.1;5,2. Acanthametropus/fg2

(Euplectoptera Anteritorna Tridentiseta Acanthametropus/fg1 .../fg2)

(Fig. 23; 24:A–B)

Nomen hierarchicum: **Acanthametropus/fg2** (sine *Analetris*).
In circumscription matches:
— gen. *Acanthametropus* Tshernova 1948: 1453;
— gen. *Metreturus* Burks 1953: 146;
— subfam. Acanthametropodinae: Demoulin 1974: 3;
— fam. Acanthametropodidae: McCafferty 1991a: 349.

References. Tshernova 1948: Ʊ; – 1952: Ʊ; – Burks 1953: Ʊ; – Bajkova 1970 (*Isonychia polita*): ⊕; – Edmunds & Jensen & Berner 1976: Ʊ; – Tshernova & Kluge & Sinitshenkova & Belov 1986: ⊕; – McCafferty 1991c: Ʊ; – Kluge & Studemann & Landolt & Gonser 1995: ⊙.

Autapomorphies of Acanthametropus/fg2.

(1) Larval and imaginal head mediad of antennae bases has a pair of pointed projections (FIG.23:E); in larva they are much larger than in imago.

(2) Maxillary palp is vestigial, 1-segmented.

(3) Larval abdomen has unpaired dorsal and ventral projections (McCafferty 1991:Figs 1–2).

(4) In each tergalius [3-lamellate – see Acanthametropus/fg1 (5)] both ventral (additional) lamellae and anal margin of dorsal lamella are dissected to numerous branched processes (FIG.23:A).

(5) In imago and subimago (including male) caudalii [see Plesiomorphies of Acanthametropus/fg1] are relatively short, shorter than abdomen (in contrast to overwhelming majority of mayflies, whose caudalii of male imago are strongly elongated).

(6) In male imago and subimago on fore leg both claws are blunt (while on other legs they are ephemeropteroid). Non-unique apomorphy (see Index of characters [2.2.77]).

Plesiomorphies of Acanthametropus/fg2. In contrast to *Analetris*, gonostylus has 2 distal segments (FIG.23:G).

Size. Fore wing length 14–20 mm.

Distribution. Amphipacific Sector of Holarctic: Eastern Palaearctic and Nearctic.

Nominal taxon included. Acanthametropus/fg2 includes **Metreturus/g** [g: *Metreturus* Burks 1953: 146, typus *M. pecatonica* Burks 1953], regarded as a generic synonym of *Acanthametropus* (Edmunds & Allen 1957: 318).

Species composition of Acanthametropus/fg2. 2 species – *nikolskyi* Tshernova 1948 [*Acanthametropus*] (syn. subj.: *polita* Bajkova 1970 [*Isonychia*]) (Altai-Amur Region) and *pecatonica* Burks 1953 [*Metreturus*] (Nearctic).

Material examined: *nikolskyi* [*A.*]: Ʊ–⊙–⊕/⊙.

*** ***

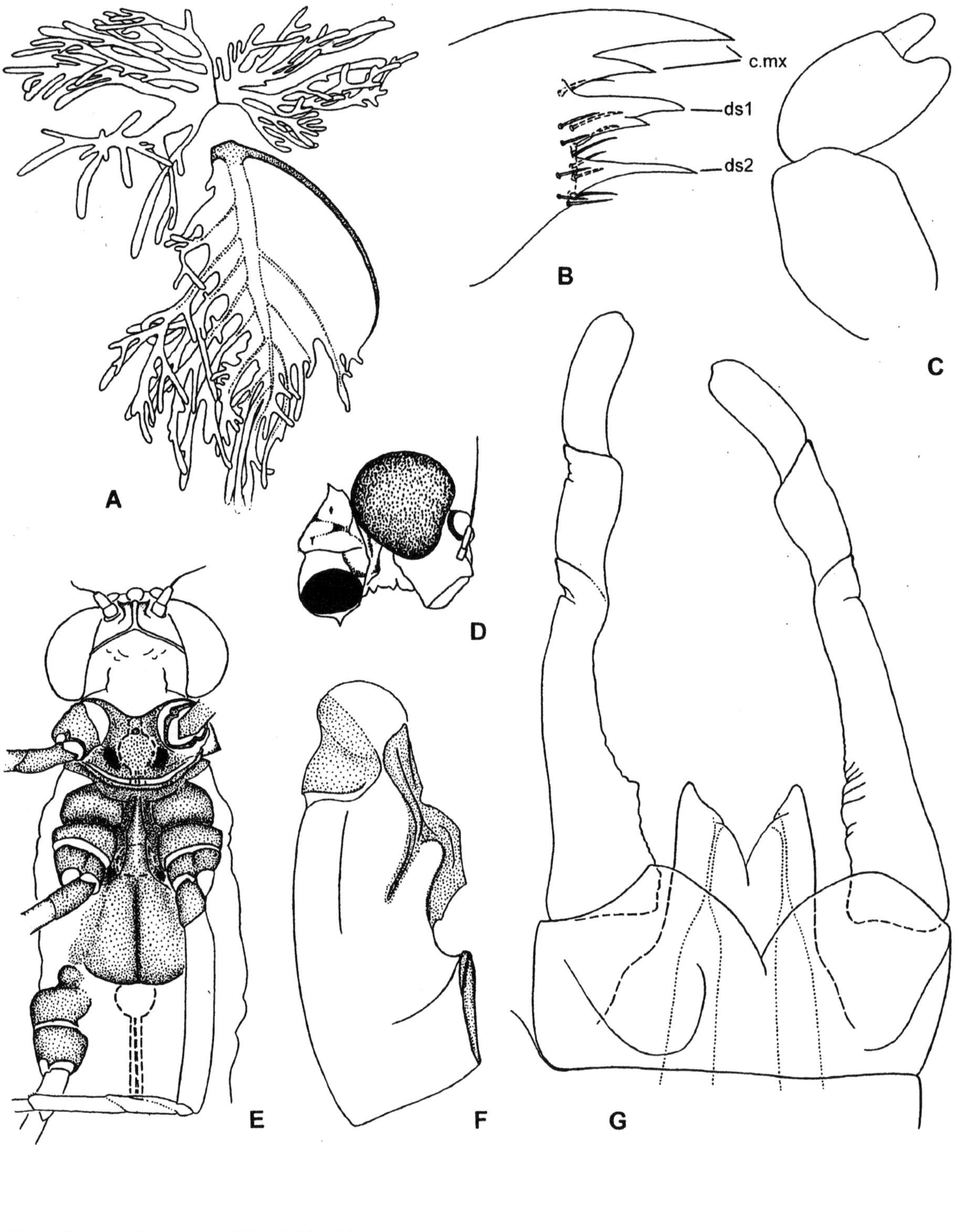

Figure 23. Acanthametropus/fg* *nikolskyi* [*Acanthametropus*].
A – tergalius VI (ventral lobe detached and spread); **B** – apex of left maxilla, dorsal view; **C** – labial palp; **D** – head and prothorax of male imago, lateral view; **E** – head and thorax of male imago, ventral view (nerve ganglions shown by interrupted lines); **F** – subimaginal exuviae of right half of meson; **G** – genitals of male imago, ventral view (gonoducts shown by dotted lines). (A–C, E, F – from Kluge & al. 1995; D – from Kluge & al. 1986)

Abbreviations: **c.mx** – maxillary canines; **ds1**, **ds2** – first and second dentisetae.

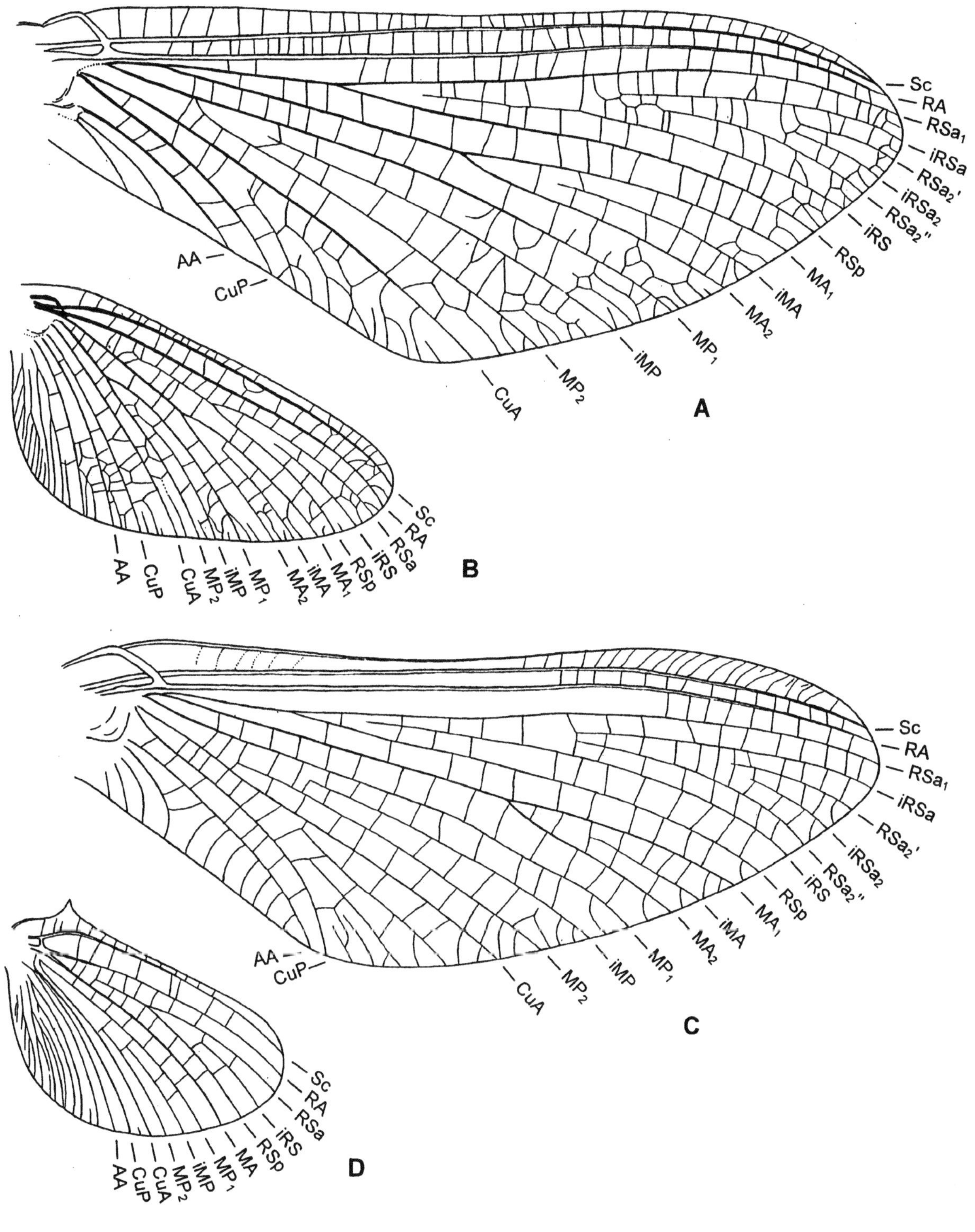

Figure 24. Wings of Acanthametropus/fg1 and Ametropus/fg1
A–B – Acanthametropus/fg* *nikolskyi* [*Acanthametropus*]. **C–D** – Ametropus/fg* *fragilis* [*Ametropus*].

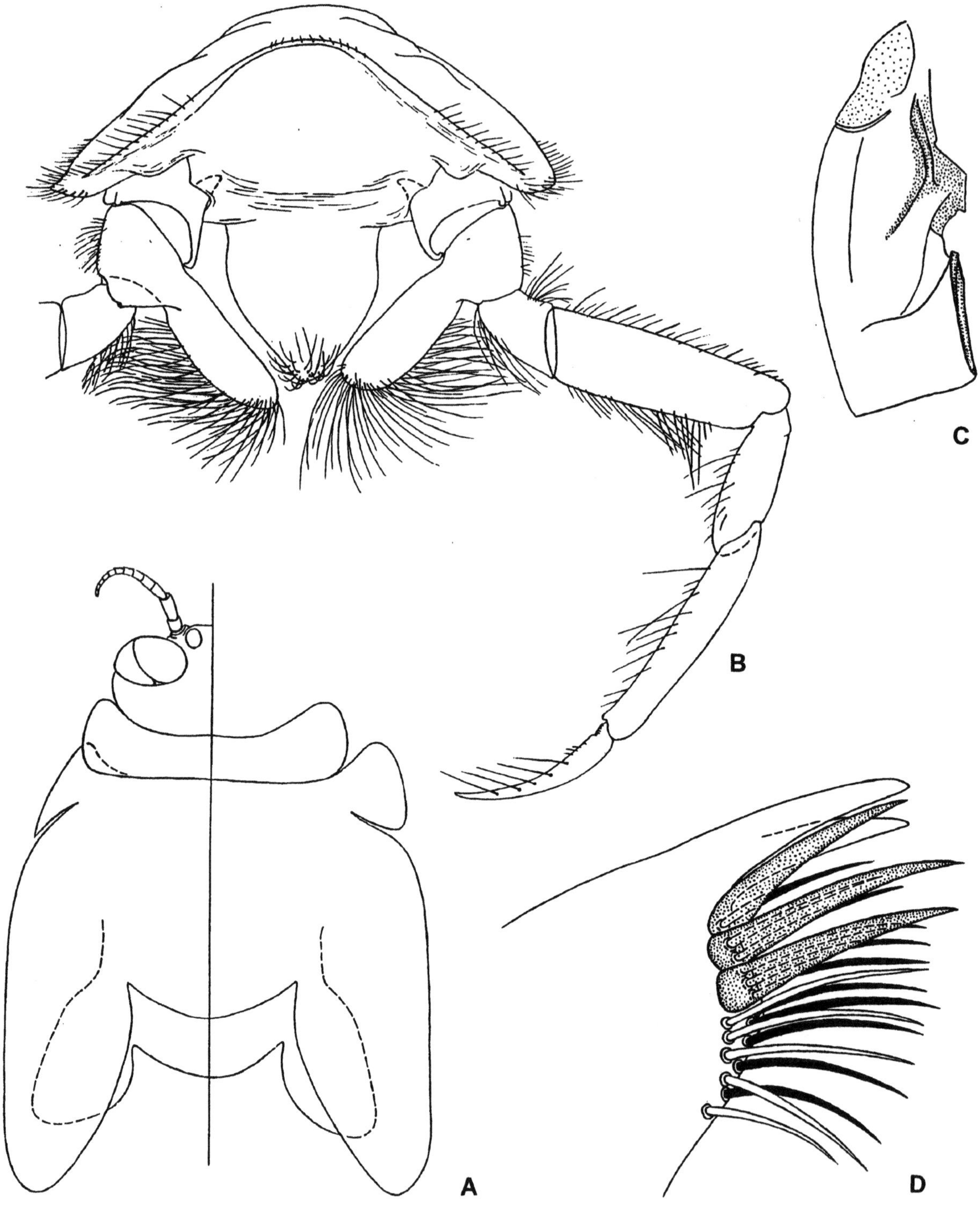

Figure 25. Ametropus/fg* *fragilis* [*Ametropus*].
A – larval head and thorax, dorsal view (in left half) and thoracic nota flattened and spread (in right half); setation not shown (numerous long fine setae are situated on margins of pronotum, mesonotum and wing buds); **B** – larval prothorax with left fore leg (anterior view); **C** – subimaginal exuviae of right half of mesonotum; **D** – apex of left maxilla, dorsal view (dentisetae shown by dots).

2.1;6. Ametropus/fg(1)

(Euephemeroptera Euplectoptera Anteritorna Tridentiseta Ametropus)

(Figs 24:C–D; 25)

Nomen hierarchicum: **Ametropus/fg(1)** [f: Ametropodidae (orig. Ametropidae) Bengtsson 1913: 305; g: *Ametropus* Albarda 1878: 129, typus *A. fragilis* Albarda 1878 (monotypy)].

In circumscription matches:

— gen. *Ametropus* Albarda 1878: 129;

— subfam. Ametropinae: Needham & Traver & Hsu 1935: 429;

— fam. Ametropodidae: Lestage 1938a: 163.

References. Needham & Traver & Hsu 1935: ↻ ⊕; – Lestage 1938a: ↻ ⊕; – Keffermüller 1959: ↻ ⊕; – Edmunds & Allen & Peters 1963 (subfam.): ↻; – Tshernova 1970: ↻ ⊕ – Koss & Edmunds 1974: ⊙; – Allen & Edmunds 1976a: ↻ ⊕; – Edmunds & Jensen & Berner 1976: ↻ ⊕; – Kluge & Studemann & Landolt & Gonser 1995: ⊙ ↻ ◎ ⊕.

Autapomorphies of *Ametropus*.

(1) Larva has peculiar body form adapted to behaviour on sandy ground in running waters (FIG. 25:A–B; Keffermüller 1959:Pl.II; Allen & Edmunds 1976:Figs 1–2): Head, while retaining hypognathous position, is relatively short, with eyes transferred anteriorly and ocelli brought together. Thorax and abdomen are flattened. Pronotum has rounded paranotal lobes. Mesonotum anteriad of fore wing buds has one more pair of paranotal lobes, separated from the wing buds by a pair of oblique incisions (FIG.25:A) (unique apomorphy). Margins of prothoracic and mesothoracic paranota are fringed with long hairs (FIG.25:B). Fore wing buds are spread, their costal margins somewhat diverge posteriorly (while in other Euplectoptera costal margins of fore wing buds never diverge, even in similarly flattened psammophilous *Pseudiron* – FIG.52:A). On all legs, especially on middle and hind ones [about fore legs see (3)] tibiae are shortened, and claws are elongate, slightly arched, without denticles (the same in other psammophilous mayflies – Acanthametropus/fg1 and *Pseudiron*).

(2) Mouth apparatus is somewhat modified (Keffermüller 1959:Pl.III; Allen & Edmunds 1976a: Figs 11–16): Labrum and hypopharynx with wide median incision apically. Biting margin of maxilla [bearing 3 dentisetae – see Anteritorna (2)] is narrow; apex with 2 canines (instead of three initial ones); apical margin lacks setae (FIG.25:D) (non-unique apomorphies – see Index of characters [1.1.33] and [1.1.31]). Labial palp (3-segmented) has 1st segment enlarged, with flexor of 2nd segment strongly widened, fan-form, proximally attached not only to base, but also to outer side of 1st segment; 2nd segment (penultimate) with more or less expressed inner-apical projection.

(3) Larval prosternum and fore legs have unique structure (FIG.25:B): Prosternum bears a lamellate process with setae on its apex (subimago and imago have similar process, but without setae). Larval fore legs are smaller than others, palp-like; fore coxa bears a long process directed medially, toward the prosternal process, and covered with long setae. Tibia-tarsal articulation is shifted to anterior side, in contrast to middle and hind legs, where it retains normal position on outer side (the same in subimago and imago). Larval fore claw bears long setae all over its length, like other leg segments (unique apomorphy).

(4) In male and female imago and subimago both claws of fore leg are blunt (unique apomorphy: in other mayflies with functional legs only male can have both fore claws blunt – see Index of characters [2.2.77] and [2.2.79]).

(5) All tergalii I–VII are large, lie in one plane; each tergalius is nearly round, entire margin is bordered by rib and bears long setae (Keffermüller 1959:Pl.II:2).

(6) Larval caudalii, besides well-developed primary swimming setae, have secondary swimming setae on lateral sides of cerci. Non-unique apomorphy (see Index of characters [1.3.67]).

Characters of unclear phylogenetic status.

(7) On fore wing CuP terminates close to tornus [see Anteritorna (1)]; AA is very long, terminates not far from tornus (FIG.24:C) (thus wing venation somewhat resembles that of Posteritorna).

(8) Cubital field of fore wing [in most part terminating anteriad of tornus – see (7)] with 2 pairs of intercalaries, among which anterior par is long, and posterior short (FIG.24:C). Non-unique apomorphy (see Index of characters [2.2.52]).

(9) On hind wing MA lacks furcation, and MP_2 has a form of intercalary, thus a single triad RS is retained [see Euplectoptera (1)] (FIG.24:D). Non-unique apomorphies (see Index of characters [2.2.67], [2.2.69] and [2.2.70]). In other respects hind wing is well developed, as long as 0.4 of fore wing length.

(10) Egg (studied for a single species *fragilis* [*A.*]) has peculiar structure: one pole bears a bunch of anchors; each anchor consists of a long thread and a terminal disc; before oviposition, each thread is folded at least three times in such a manner, that all anchors together form a cap-like polar formation externally covered by the terminal discs (Kluge & al. 1995:Fig.76). This polar formation externally resembles polar cap of Polymitarcys/f2=Ephoron/g3 (see Index of characters [3.6]), but in contrast to it, in *Ametropus* threads are not fused in to integral cap, and can be stretched becoming much longer than the initial formation.

Plesiomorphies of *Ametropus*. Maxillary and labial palps are 3-segmented. Larval (and adult) patella-tibial suture is developed on middle and hind legs only. Larval paracercus is subequal to cerci.

In imago and subimago: Mesonotal suture is transverse (FIG.25:C); anterior paracoxal suture is complete (in contrast to *Rallidens* and some Bidentiseta); furcasternal protuberances are contiguous (in contrast to Amphinotic groups of Tridentiseta and some others – see Index of characters [2.2.23]). Subimaginal lateral pigmented area of mesonotum is small, stretches along lateroparapsidal suture and does not occupy sublateroscutum (FIG.25:C) (see Index of characters [2.2.14]). Tarsi are 5-segmented, 1st segment is fused with tibia and non-shortened. Claws are ephemeropteroid [except for fore legs – see (4)]. Gonostylus with 2 distal segments. Imaginal and subimaginal paracercus is subequal to cerci.

Size. Fore wing length 10–22 mm.

Distribution. Holarctic.

Species composition of Ametropus/fg(1). Not less than 2 species – *ammophilus* Allen & Edmunds 1976 [*Ametropus*] (Western Nearctic) and a complex of hardly distinguishable forms described as species *fragilis* Albarda 1878 [*Ametropus*] (Europe), *eatoni* Brodsky 1930 [*Ametropus*] (Siberia), *neavei* McDunnough 1928 [*Ametropus*] and *albrighti* Traver 1935 [*Ametropus*] (Nearctic).

Material examined: *eatoni* [*A.*]: ♉, ◎, ⊕/⊙.

***

2.1;7. Tetramerotarsata, or Baetis/fg2

(Euephemeroptera Euplectoptera Anteritorna Tridentiseta Tetramerotarsata)

(Figs 8:C–D; 9:F–H; 26–29)

Nomen hierarchicum: **Baetis/fg2** (sine *Siphlonurus* et al.; incl. *Siphlaenigma*).

Nomen circumscribens: **Tetramerotarsata** Kluge 1997c: 528.

In circumscription matches:

— fam. Baetidae: McCafferty & Edmunds 1979: 6;

— superfam. Baetoidea: Kluge & Studemann & Landolt & Gonser 1995: 107;

— superfam. Tetramerotarsata Kluge 1997c: 528;

— Tetramerotarsata, or Baetis/fg2: Kluge 2000: 248.

References. Kluge 1997c: ♉* ⊕*; – Staniczek 1997: ⊙* ♉* ⊕*.

Autapomorphies of Tetramerotarsata.

(1) Imaginal and subimaginal tarsus of middle and hind legs consists of 4 segments only, including 1st segment fused with tibia; thus, only 3 segments are movable (FIG.27:E) (in contrast to initial for mayflies 5-segmented tarsus with 4 segments movable). This apomorphy is nearly unique among Ephemeroptera. In some other Ephemeroptera reduction of tarsal segments takes place in connection with general degeneration of legs (in Prosopistoma/f1=g2, Geminovenata, some Fossoriae); in contrast to them, in Tetramerotarsata legs are always normally developed, functional, with well-developed long tarsi. Among all mayflies only in *Machadorythus* tarsi have the same structure as in Tetramerotarsata (see Index of characters [2.2.83]).

In other respects legs of Tetramerotarsata are non-modifies or variously modified in subordinate taxa. On middle and hind legs the 1st (immobile) tarsal segment is not shortened, often the longest. Fore tarsus of male retains all 5 segments, the 1st segment being either non-shortened (in *Siphlaenigma*), or strongly shortened [see Liberevenata (3) below]. Fore tarsus of female resembles either middle and hind tarsus [see *Siphlaenigma* (1) below], or fore tarsus of male [see Liberevenata (3) below]. Claws of all legs of imago and subimago of both sexes are always ephemeropteroid.

(2) Imaginal and subimaginal mesonotum has **sublateroscutal suture** – a paired transverse suture which crosses sublateroscutum and separates dorsal bases of anterior and posterior scuto-coxal muscles. At the same time, portion of sublateroscutum behind the sublateroscutal suture is not separated from the posterior scutal protuberance (which bears base of scuto-lateropostnotal muscle) and together with it looks as an integral piece (FIGS 26:A; 29:A). Sublateroscutal suture is a new formation, not occurring in other Ephemeroptera (it should not be confused with the transverse interscutal suture of Caenoptera and some Tricorythus/fg1, which passes not anteriad, but posteriad of the base of the posterior scuto-coxal muscle).

(3) Imaginal and subimaginal anterior paracoxal suture is transferred anteriorly, thus anepisternum is smaller than katepisternum (FIGS 26:A; 27:F; 29:A). Non-unique apomorphy (see Index of characters [2.2.19]). In contrast to *Rallidens*, Heptagennota and Furcatergaliae, anterior paracoxal suture is complete.

(4) Mesonotal suture is lost because its left and right halves are curved backward, elongated and fused medially with the median longitudinal suture and laterally with lateroparapsidal suture. The conclusion that mesonotal suture had disappeared by this way, is based on the fact that subimago of *Siphlaenigma* retains a vestige of mesonotal suture: it has a form of a pair of longitudinal lines which go close to the median longitudinal suture and separate a narrow light bold area lying mediad of these lines, from pigmented microtrichia-bearing area lying laterad of them (FIG.26:A). In all other Tetramerotarsata traces of mesonotal suture are lost and subimaginal mesonotum is entirely covered with microtrichia. Non-unique apomorphy; independently from Tetramerotarsata mesonotal suture disappeared by the same way in Leptophlebia/fg1 and some Fimbriatotergaliae (see Index of characters [2.2.8]).

(5) On fore wing CuP is basally curved and strongly diverges from CuA. This curvation is retained in primitive representatives of Tetramerotarsata – *Siphlaenigma* and *Palaeocloeon* (FIG. 27:D), while in Turbanoculata the transverse basal portion of CuP is lost, so that base of CuP stands apart from base of CuA. Non-unique apomorphy; the same in Leptophlebia/fg1 only.

(6) Labrum with following features: Outer side with a transverse row of branching setae close to distal margin [in *Siphlaenigma* this row is regular in its lateral parts and irregular in median part, in Turbanoculata it is entirely regular – see Turbanoculata (4) below]; at a distance from anterior margin a pair of long submedian setae are present. In cases of strong specialization of labrum, these details can be lost.

(7) On maxilla apical-ventral row of setae is transferred to ventral side; it is well-developed in *Siphlaenigma* (FIG.26:B), being vestigial or lost in Turbanoculata.

(8) Paraglossa with a longitudinal row of setae on dorsal side and initially with a longitudinal row of setae on ventral side (FIGS 26:D; 28:C). The dorsal row is always retained, but the ventral row is secondarily lost in many Turbanoculata.

Characters of unclear phylogenetic status.

(9) On fore wing MP_2 has a form of intercalary, i. e. lost connection with MP_1 (FIG.27:A) (but at the same time begins at a distance from wing base, in contrast to Posteritorna and Caenoptera). Non-unique apomorphy, in many other taxa disjunction of MP_2 occurs as individual variability.

(10) Cubital field of fore wing [see Anteritorna (1)] initially with 2 veins arising from CuA, the proximal (posterior) of them being bifurcate; this condition can be retained *Siphlaenigma* (Penniket 1962:Fig.1); in Liberevenata these 2 veins become simple intercalaries; posterior of them can be short, in *Palaeocloeon* seems to be lost (FIG.27:A). Non-unique apomorphy (see Index of characters [2.2.52]).

(11) Larval claw on inner side with 2 longitudinal rows of denticles. Such claw structure is initial for Tetramerotarsata, being present in *Siphlaenigma* and many Turbanoculata. Non-unique apomorphy (see Index of characters [1.2.21]).

In some Tetramerotarsata denticles of that row, which is situated nearer to posterior side of the claw, are diminished or lost, thus only one row is present, being situated nearer to anterior side of the claw (anterior is the side directed anteriorly when the leg is directed by apex laterally and by knee dorsally); similar single anterior-inner row of denticles occurs also in some of that mayfly taxa which initially have one row of denticles – Pentamerotarsata, Ephemerella/fg1, Leptophlebia/fg1 (see Index of characters [1.2.21]). In selected species of Tetramerotarsata (belonging both to taxa with two rows and taxa with one row of denticles) the denticles can be secondarily lost.

(12) Tergalius has anal rib situated on anal margin (in contrast to Siphlonurus/fg1 and some others, which have anal rib at a distance from anal margin – see Index of characters [1.3.28]). Costal margin of tergalius is armed by costal rib, thus tergalius often has an integral rib entirely bordering its margins. In a few taxa (particularly, in Cloeon/ /fg1 and *Callibaetis*) anal rib is completely lost.

(13) Larval antennae are elongate, usually much longer than head; only in selected specialized species antennae are not longer than head (while in majority of other mayflies antennae are short, subequal or slightly longer than head).

Plesiomorphies of Tetramerotarsata. Larval head always retains hypognathous position (see below, Variable characters of Turbanoculata). Maxilla [see (7)] always with 3 canines and 3 dentisetae (see ibid.) (FIGS 26:B; 28:B, D).

Size. Usually small, fore wing length 2–12 mm (see TABLE 2).

Age and distribution. Late Cretaceous (see *Palaeocloeon*) — Recent; world-wide.

Tetramerotarsata are divided into *Siphlaenigma* and Liberevenata.

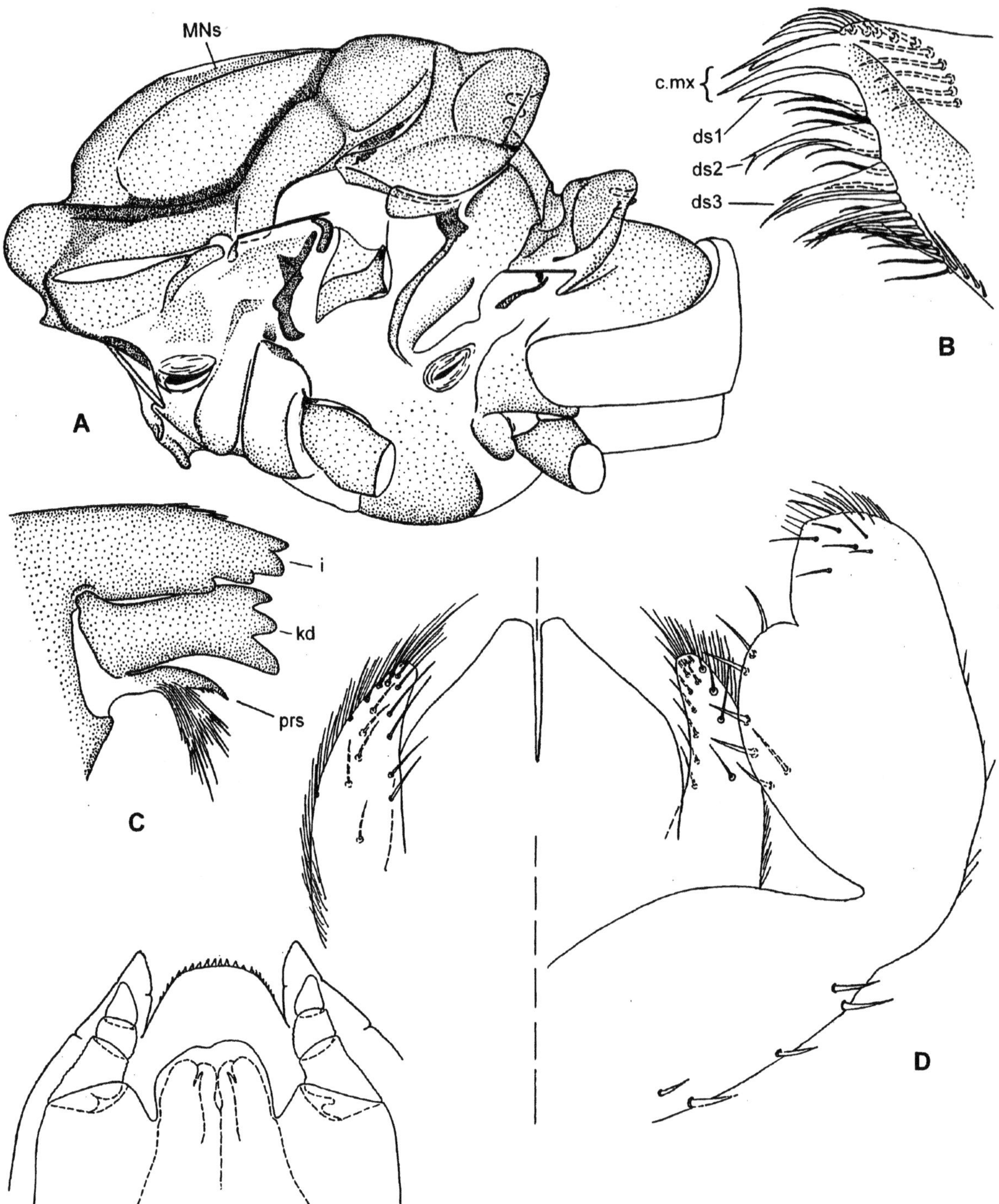

Figure 26. Tetramerotarsata-Siphlaenigma/fg* *janae* [*Siphlaenigma*].
A – subimaginal pterothorax extracted from mature larva, lateral view (compare with Fig.29:A); **B** – apex of right maxilla, dorsal view; **C** – apex of left mandible (compare with Fig.29:B); **D** – labium, dorsal view (in left half) and ventral view (in right half); **E** – genitals of male subimago under cuticle of mature larva. (From Kluge 1997c)

Abbreviations: **c.mx** – maxillary canines; **ds1**, **ds2**, **ds3** – first, second and third dentisetae; **i** – incisor; **kd** – kinetodontium; **MNs** – mesonotal suture; **prs** – prostheca.

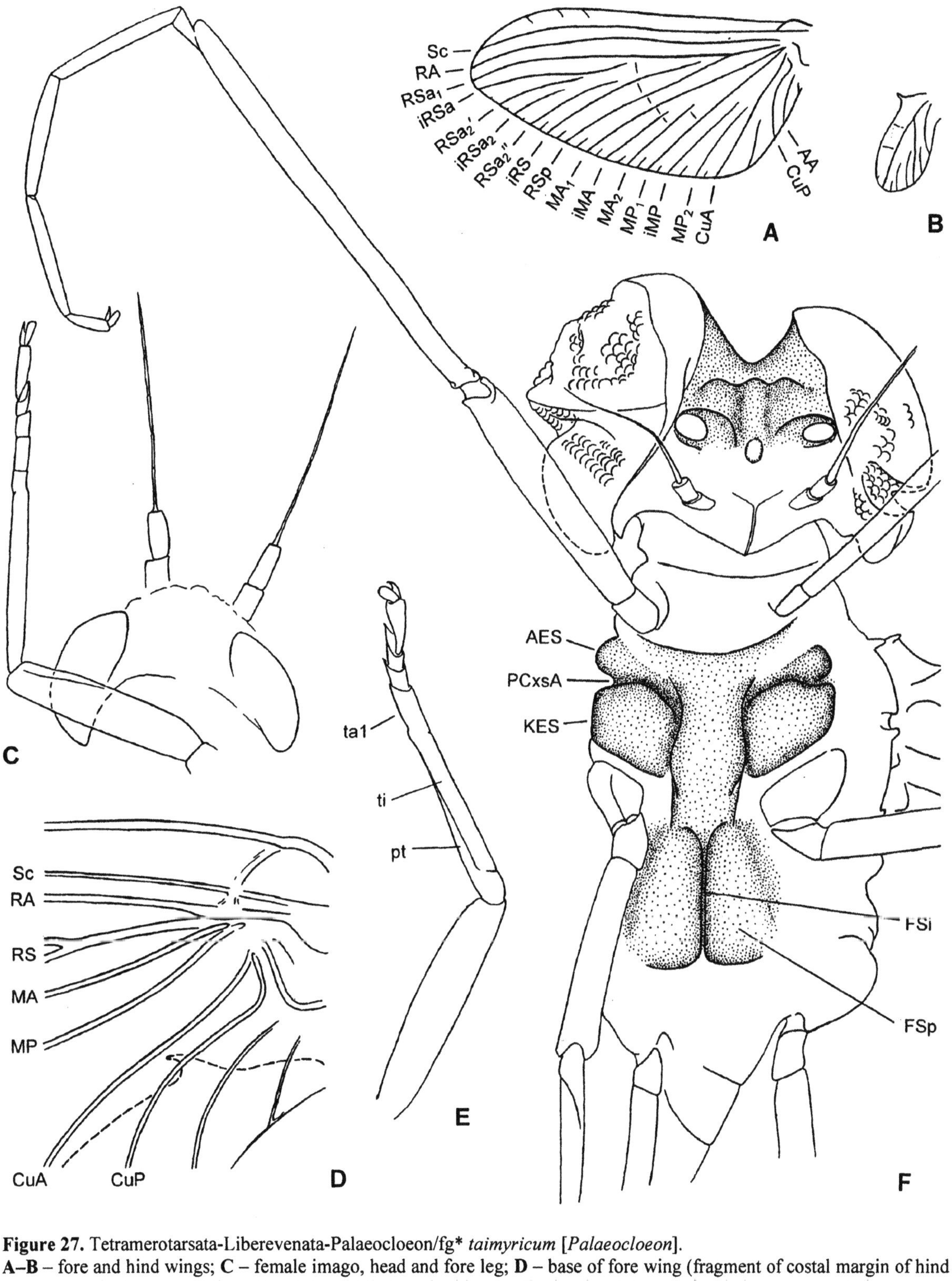

Figure 27. Tetramerotarsata-Liberevenata-Palaeocloeon/fg* *taimyricum* [*Palaeocloeon*].
A–B – fore and hind wings; **C** – female imago, head and fore leg; **D** – base of fore wing (fragment of costal margin of hind wing shown by interrupted line); **E** – right middle leg of subimago; **F** – head and thorax of male imago, ventral view. (From Kluge 1997c)

Abbreviations: **AES** – anepisternum; **FSi** – furcasternal median impression; **FSp** – furcasternal protuberance; **KES** – katepisternum; **PCxsA** – anterior paracoxal suture; **pt** – patella; **ta1** – first tarsal segment; **ti** – telotibia.

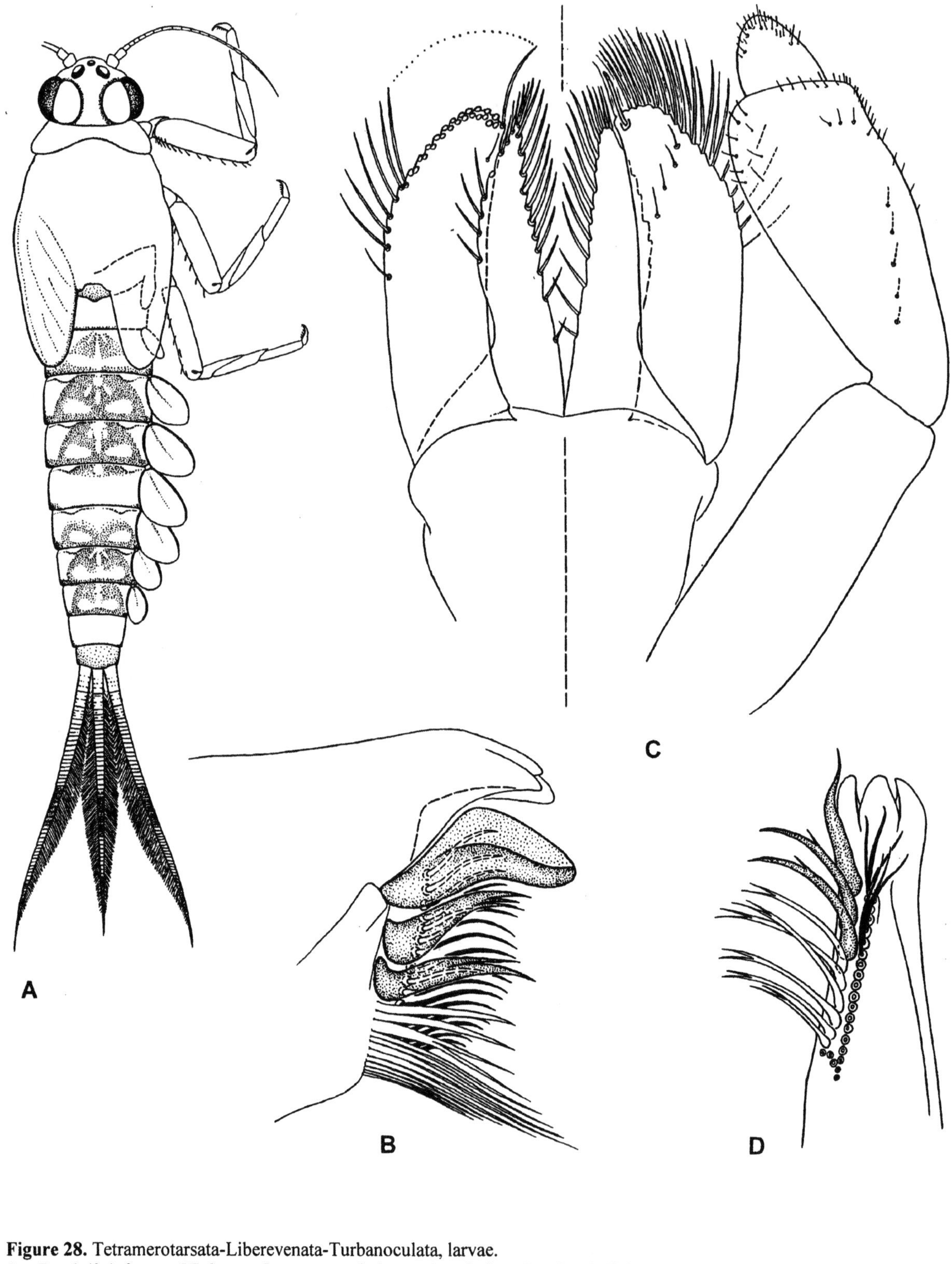

Figure 28. Tetramerotarsata-Liberevenata-Turbanoculata, larvae.
A – Baetis/fg* *fuscata* [*Ephemera*], mature male larva, dorsal view. **B** – Baetis/fg8 *vernus* Curtis 1834 [*Baetis*], apex of left maxilla, dorsal view (dentisetae shown by dots). **C** – Afroptilum/fg* *sudafricanum* [*Centroptilum*], labium, dorsal view (in left half) and ventral view (in right half). **D** – Cloeon/fg* *diptera* [*Ephemera*], apex of left maxilla, median view (dentisetae shown by dots). (A – from Kluge 1997d; B, D – from Kluge 1998)

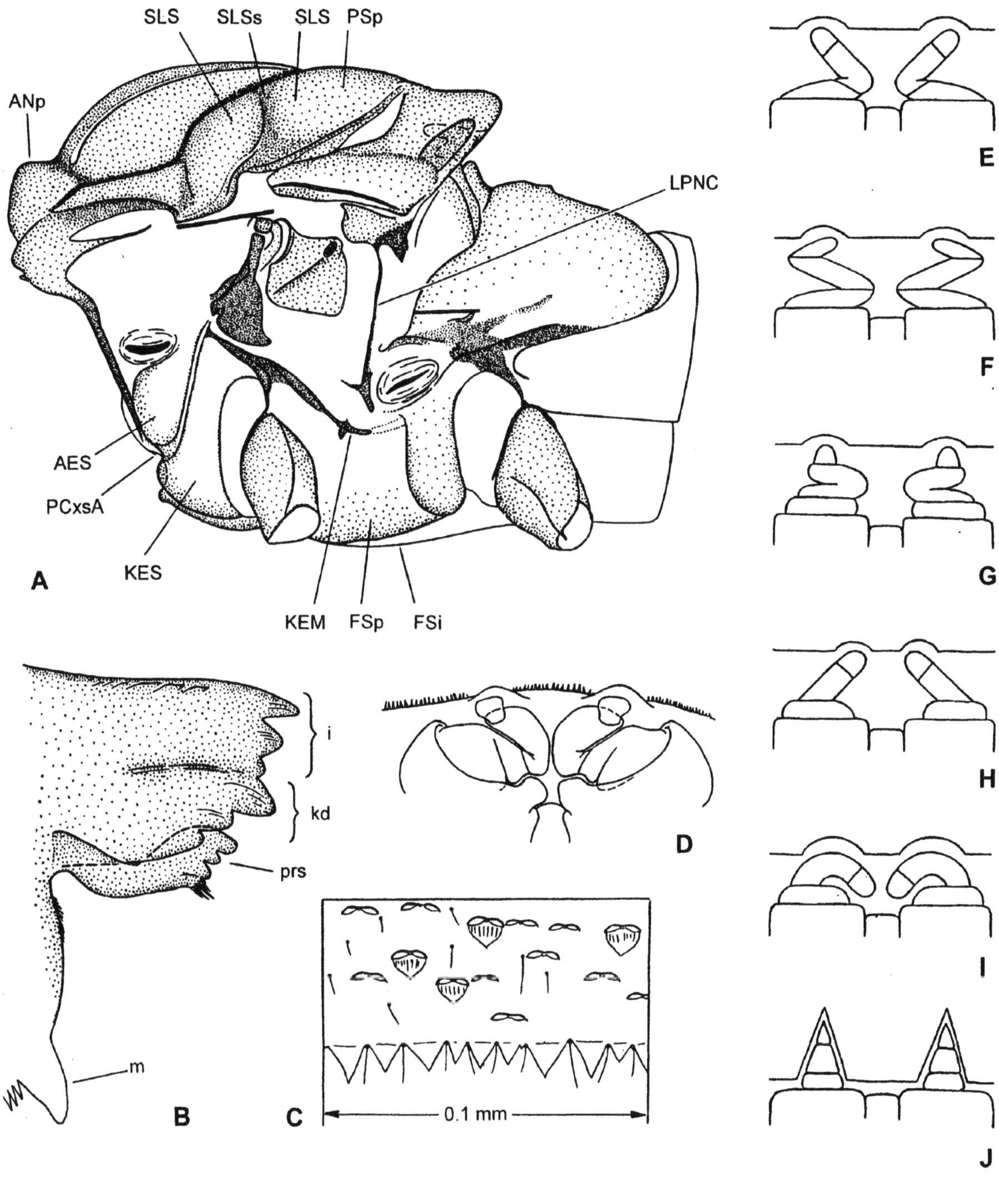

Figure 29. Turbanoculata.
A–D – Afroptilum/fg* *sudafricanum* [*Centroptilum*]: **A** – subimaginal pterothorax, lateral view (compare with Fig.26:A); **B** – apex of left mandible (compare with Fig.26:C); **C** – posterior margin of larval abdominal tergite V; **D** – subimaginal gonostyli developing under larval cuticle (correspond to scheme E); **E–J** – subimaginal gonostyli developing under larval cuticle (schematic): **E** – "*Afroptilum*-type" (see D); **F** – "*Cloeon*-type"; **G** – "*Nigrobaetis*-type"; **H** – "*Baetis*-type"; **I** – "*Labiobaetis*-type"; **J** – "*Callibaetis*-type". (From Kluge 1997c)

Abbreviations: **AES** – anepisternum; **ANp** – anteronotal protuberance; **FSi** – furcasternal median impression; **FSp** – furcasternal protuberance; **i** – mandibular incisor; **kd** – kinetodontium; **KEM** – katepimeron; **KES** – katepisternum; **LPNC** – lateropostnotal crest; **m** – mola; **PCxsA** – anterior paracoxal suture; **prs** – prostheca; **PSp** – posterior scutal protuberance; **SLS** – sublateroscutum; **SLSs** – sublateroscutal suture.

2.1;7,1. Siphlaenigma/fg(1)

(Euplectoptera Anteritorna Tridentiseta Tetramerotarsata *Siphlaenigma*)

(Fig. 26)

Nomen hierarchicum: **Siphlaenigma/fg(1)** [f: Siphlaenigmatidae Penniket 1962: 389; g: *Siphlaenigma* Penniket 1962: 389, typus *S. janae* Penniket 1962 (design. orig.)].
In circumscription matches:
— gen. *Siphlaenigma* Penniket 1962: 389;
— subfam. Siphlaenigmatinae: McCafferty & Edmunds 1979: 6;
— fam. Siphlaenigmatidae Penniket 1962: 389.
Monospecific taxon.

References. Penniket 1962: ↻ ⊕; – Edmunds & Allen & Peters 1963: ↻; – Tshernova 1970: ↻ ⊕ – Edmunds & Jensen & Berner 1976: ↻ ⊕; – Kluge 1997c: ↻ ◎ (plesiomorphies wrongly named "Apomorphies of *Siphlaenigma*"); – Staniczek 1997: ⊙ ↻ ⊕; – Lugo-Ortiz & McCafferty 1998: ↻.

Characters of unclear phylogenetic status.

(1) In female imago and subimago (in contrast to male) structure of fore tarsus is the same as of middle and hind tarsus [see Tetramerotarsata (1)]: tarsus is 4-segmented including 1st segment immobile fused with tibia. In females of other Tetramerotarsata fore tarsus is similar not to middle and hand tarsi, but to fore tarsus of male [see Liberevenata (8) below]. Polarity of this character is unclear.

(2) Maxillary palp has 3rd (distalmost) segment strongly diminished.

(3) Labial palp has 2nd segment widened inner-apically; 3rd (distalmost) segment is not large (FIG. 26:D); the same in selected taxa of Turbanoculata.

Plesiomorphies of *Siphlaenigma* (in contrast at least to Turbanoculata). In imago and subimago: Eyes of male are not divided into two portions. Subimaginal sclerites of lateropostnotal crest and katepimeron are not elongate, not convergent (FIG. 26:A). Furcasternal protuberances are contiguous. On fore wing MA forms a symmetrical fork near middle of wing; marginal intercalaries are absent; hind wing is well-developed, nearly as long as 0.3 of fore wing length, retains triads RS, MA, MP and veins CuA and CuP (Penniket 1962:Fig.1). On fore legs of male all 5 segments are long. Posterior projection of abdominal segment IX (i. e. styliger of male and postgenital plate of female) is large. Gonostylus has 2 distal segments. Penis is well-developed, with intrinsic muscles. Imaginal paracercus is well-developed, subequal to cerci. Ganglion of abdominal segment VIII is separate from that of segment VII.

In larva: Kinetodontium is jointed, prostheca of both mandibles is brush-like (FIG.26:C). Larval mesonotum has a prominent relief repeating that of winged stages (in contrast to majority of Ephemeroptera); fore wing buds are not fused with apex of scutellum. Malpighian tubes are coiled. In contrast to some specialized Turbanoculata, larva has siphlonuroid appearance – legs are able to stretch posteriorly; abdomen is able to make undulate movements; all tergalii I–VII are present; all three caudalii have dense primary swimming setae and no secondary swimming setae. In contrast to some Turbanoculata, setae proximad of prostheca are absent (FIG.26:C). Larval and adult patella-tibial suture is developed on middle and hind legs only (in contrast to larval Turbanoculata-Anteropatellata).

Size. Fore wing length 7–9 mm (see Tetramerotarsata).

Distribution. New Zealand.

Species composition of Siphlaenigma/fg(1). 1 species – *janae* Penniket 1962 [*Siphlaenigma*] (syn. subj.: *edmundsi* Lugo-Ortiz & McCafferty 1998 [*Siphlaenigma*]).

Material examined: *janae* [*S.*]: ↻, ♂/♂, ♀/♀/⊙.

***

2.1;7,2. Liberevenata, or Baetis/fg3

(Euplectoptera Anteritorna Tridentiseta Tetramerotarsata Liberevenata)

(Figs 8:C–D; 9:F–H; 27–29)

Nomen hierarchicum: **Baetis/fg3** (sine *Siphlaenigma*; incl. *Palaeocloeon*).
Nomen circumscribens: **Liberevenata** Kluge 1997c: 531.
In circumscription matches:
— fam. Baetidae: Kluge 1997c: 531;
— fam. Liberevenata Kluge 1997c: 531;
— Liberevenata, or Baetis/fg3: Kluge 2000: 248.

Reference. Kluge 1997c: ⊕.

Autapomorphies of Liberevenata.

(1) On fore wing MA_2 is free, not connected with MA (FIG.27:A). Unique apomorphy: only in Prosopistoma/f1=g2 MA is non-branched, in other Euephemeroptera MA_2 nearly always arises from MA and can be free in selected specimens only.

(2) Fore wing with free intercalaries. Non-unique apomorphy (see Index of characters [2.2.55]). Probably the initial condition for Liberevenata is presence of one marginal intercalary in each space – this is peculiar for *Palaeocloeon* (FIG.27:A), Afroptilum/f1=Centroptiloides/g1, Cloeon/fg1 and some others. In Baetis/fg6 two marginal intercalaries are present in each space (FIG.8:C).

(3) On fore tarsus of male and female imago and subimago 1st segment (fused with tibia) is strongly shortened (FIG.27:C, F). Often this 1st segment, being fused with tibia and shortened, is invisible, thus tarsus looks as 4-segmented; but in contrast to really 4-segmented tarsi of middle and hind legs [see Tetramerotarsata (1)], fore tarsus has all its 4 segments mobile.

Independently from Liberevenata, 1st tarsal segment had shortened in Furcatergaliae, but there it is shortened on all legs, while in Liberevenata the 1st segment of middle and hind leg remains to be long.

Tarsal structure of male is surely apomorphic if compare with *Siphlaenigma* (which retains all five tarsal segments non-shortened). But in female polarity of the character is unclear: in female Liberevenata structure of fore tarsus is generally the same as in male – 5 segments are retained, 1st of which is shortened, while in *Siphlaenigma* fore tarsus of female is really 4-segmented, like middle and hind tarsi.

(4) Posterior projection of abdominal sternum IX (i. e. styliger of male and postgenital plate of female) is reduced; male imago and subimago has only lateral lobes of styliger which form pedestals of gonostyli. Non-unique apomorphy.

(5) Gonostylus has 1 distal segment only (instead of two initial ones). Non-unique apomorphy (see Index of characters [2.3.12]). Among Liberevenata distal segment disappears completely in Rhithrocloeon/g1.

(6) Penis is vestigial: there can be retained a small unpaired external median projection with a pair of penial arms deeply invagined into the body by sides of it, or the external part of penis is completely lost, and only a pair of internal penial arms are retained; intrinsic penial muscles are always lost. Unique apomorphy.

(7) Imaginal and subimaginal paracercus is vestigial (in larva it can be either well-developed, or also vestigial). Non-unique apomorphy (see Index of characters [2.3.22]).

Size. Fore wing length 2–12 mm.

Age and distribution. Late Cretaceous (see *Palaeocloeon*) — Recent; world-wide.

Liberevenata are divided into extinct plesiomorphon *Palaeocloeon* and Recent taxon Turbanoculata. Extinct *Myanmarella* has uncertain systematic position (p.105).

2.1;7,2-1. † Plesiomorphon **Palaeocloeon/fg(1)**

(Anteritorna Tridentiseta Tetramerotarsata Liberevenata *Palaeocloeon*)

(Fig. 27)

Nomen hierarchicum: **Palaeocloeon/fg(1)** [f: Palaeocloeoninae Kluge 1997c: 531; g: *Palaeocloeon* Kluge 1997c: 532, typus *P. taimyricum* Kluge 1997 (design. orig.)].
In circumscription matches:
— gen. *Palaeocloeon* Kluge 1997c: 532;
— subfam. Palaeocloeoninae Kluge 1997c: 531.
Monospecific taxon.

Reference. Kluge 1997c: ⊕ (larval structure unknown).

Plesiomorphies of *Palaeocloeon*. In contrast to Turbanoculata, eyes of male are not divided into two portions; furcasternal protuberances are contiguous (FIG.27:F). On fore wing in each space only one marginal intercalary is present [see Liberevenata (2)] (FIG.27:A). At least in adult patella-tibial suture is developed on middle and hind legs only (FIG.27:E).

Size. Fore wing length 2 mm (see Tetramerotarsata).

Age. Late Cretaceous (Taimyr amber).

Species composition of Palaeocloeon/fg(1). 1 species – *taimyricum* Kluge 1997 [*Palaeocloeon*].

Material examined (Paleontol.Inst.): *taimyricum* [*P.*]: ◎♂, ⊕♂, ⊕♀.

***

2.1;7,2-2. **Turbanoculata**, or **Baetis/fg4**

(Anteritorna Tridentiseta Tetramerotarsata Liberevenata Turbanoculata)

(Figs 8:C–D; 9:F–H; 28–29)

Nomen hierarchicum: **Baetis/fg4** (sine *Palaeocloeon*; incl. *Centroptiloides*).
Nomen circumscribens: **Turbanoculata** Kluge 1997c: 532.
In circumscription matches:
— "series I of group II": Eaton 1883–1888: 153;
— "section 9 of genera": Eaton 1883–1888: 153;
— tribus Baetini: Banks 1900: 247;
— subfam. Baetidinae: Jacobson & Bianchi 1905: 875;
— grex subfamm. Turbanoculata Kluge 1997c: 532;
— fam. Baetidae: Bengtsson 1912: 109;
— Turbanoculata, or Baetis/fg4: Kluge 2000: 248.

References. Eaton 1883–1888: ℧* ⊕*; – Needham & Traver & Hsu 1935: ℧* ⊕*; – Edmunds & Allen & Peters 1963: ℧*; – Tshernova 1970: ℧* ⊕* – Edmunds & Jensen & Berner 1976: ℧* ⊕*; – Wang & McCafferty 1996b: ℧*; – Kluge 1997c: ◎* ⊕*; – Staniczek 1997: ℧* ⊕*.

Autapomorphies of Turbanoculata.

(1) In male imago dorsal part of each eye is transformed to a **turban eye**: it has cylindrical stem lacking facets (only vestiges of ommatidia with pigmented cells can be retained on stem), thus faceted surface of turban eye is separated from faceted surface of ventral eye portion (FIG.8:C; Zimmer 1897:Figs 1–13). In subimago turban eye has stem less expressed than in imago; in larva stem is completely absent and a turban eye bud has a form of flat spot distinctly separated from lower portion of eye (FIG.28:A). Exception is made by *Aturbina*, where turban eyes are secondarily reduced, thus male eyes become small and simple as in female. Besides Turbanoculata, similar turban eyes are present in some Leptophlebia/fg1.

(2) Imaginal and subimaginal furcasternal protuberances are separated (FIG.29:A), between them metathoracic nerve ganglion is situated (FIG.8:D); if furcasternal protuberances are connected, metathoracic ganglion is situated anteriad of the point of their connection. Non-unique apomorphy (see Index of characters [2.2.23]).

Characters of unclear phylogenetic status: it is unclear if these characters are autapomorphies of Turbanoculata or Liberevenata, because for *Palaeocloeon* they are unknown.

(3) Larval paired frontal sutures – i. e. anterior arms of Y-shaped epicranial suture, by which cuticle is broken at moult – pass anteriad (ventrad) of lateral ocelli or close to anterior (ventral) margins of these ocelli, thus separating the lateral ocelli from median ocellus (Wang & McCafferty 1996b:Figs 1–6). Unique apomorphy: in other Ephemeroptera the frontal suture is connected with lateral ocellus in middle or posterior part of the ocellus (ibid.:Figs 7–12).

(4) Labrum has a distinctly expressed median incision; row of branching setae near anterior margin [see Tetramerotarsata (6)] is regular all over its length. In cases of strong specialization of labrum (in predators) these details can disappear.

(5) Kinetodontium is immobile fused with mandible and more or less fused with incisor (FIG.29:B). Non-unique apomorphy.

(6) Prostheca of left mandible is elongate in a form of stout integral stick, only apically divides into several short projections (FIG.29:B) (in contrast to brush-like prostheca of other mayflies – FIG.26:C). Prostheca of right mandible either has similar structure, or is vestigial, sometimes setiform or in a form of bifurcate seta.

On inner margin of mandible proximad of prostheca, irregularly situated setae can be present (non unique character). Probably presence of these setae is initial for Turbanoculata, as they are present in such primitive groups as Afroptilum/f1= =Centroptiloides/g1 (FIG.29: B), Cloeon/fg1 and others. In selected groups inside Turbanoculata these setae are completely lost.

(7) Glossae and paraglossae have peculiar structure (FIG.28:C): bases of glossae are widened ventrad of bases of paraglossae; in their most part glossae are narrow, in middle part are situated between paraglossae, and in distal part can be projected somewhat dorsad of paraglossae; paraglossae arise from mentum apically (not laterally) and beginning from their bases have lateral margins parallel-sided or convergent distally. Such structure of labium is characteristic for overwhelming majority of Turbanoculata, but in some representatives it is secondarily changed. Somewhat similar labial structure is present only in Nesameletus/f1=Metamonius/g2 (FIG.30:F).

(8) In contrast to *Siphlaenigma*, adult mesonotal relief is not expressed in larva (as in majority of Ephemeroptera); larval fore wing buds are fused with scutellum up to apex of scutellum; thus, posterior margin of mesonotum between wing buds is straight or concave, and free portion of each wing bud narrows from its extreme base toward apex (FIG.28:A). Non-unique apomorphy (see Index of characters [1.2.6]). Usually fore wing buds are separated beginning from the apex of scutellum, but in Pseudopannota/g1 they are fused one with another behind scutellum.

(9) Subimaginal sclerites of lateral mesothoracic wall have unique structure (FIG.29:A): Sclerite of lateropostnotal crest is long and straight; the sclerite of metathoracic katepimeron is distinct, long, narrow, stretches posteriorly up to ventral end of the sclerite of lateropostnotal crest. Paired transverse postnotal sclerite, which unites mediopostnotum with posterior-dorsal angle of postsubalar sclerite, is immobile fused with postsubalar sclerite and looks as a its process. Probably initially in Turbanoculata this postnotal sclerite is large, with convex anterior-dorsal margin; such shape it has in Afroptilum/f1= =Centroptiloides/g1, Baetopus/g1, *Cheleocloeon*, Baetis/fg6; only in Cloeon/fg1 (incl. *Centroptilum*, *Procloeon* et al.) it has another shape: diminished, with concave anterior-dorsal margin.

(10) Hind wings are diminished, usually as long as 0.2 of fore wing length or shorter (rarely longer), narrowed, in many representatives lost; venation of hind wing is strongly reduced: there are no furcations of RS, MA and MP, no veins behind MP; usually only Sc and RA are present, sometimes also non-branched MP; instead of RS and MA one or several intercalaries can be present, sometimes a single vein (probably MA) arises from RA. Non-unique apomorphy (see Index of characters [2.2.59]).

Structure of hind wing of *Palaeocloeon* is not quite clear; probably some of these characters belong to Liberevenata in general.

Possibly, the most primitive for Turbanoculata is hind wing of "*Centroptilum*-type": narrow, with non-branched longitudinal veins and with long costal projection, which is situated in proximal part of wing, projected anteriorly and curved in distal direction: such form is present in species formerly placed to the artificial genus *Centroptilum* SENSU Eaton 1883–1888, and recently placed to Afroptilum/fl=Centroptiloides/g1, Cloeon/fg1 (incl. *Centroptilum* et al.) and some other taxa. In selected groups of Turbanoculata other forms of hind wing are present: wing of "*Baetis*-type" is wider, with shorter non-curved costal projection (FIG.8:C); "*Nigrobaetis*-type" has the same shape, but the second vein is bifurcate. Sometimes costal projection is lost. In many groups of Turbanoculata hind wings disappear completely, in many cases disappearance of hind wings takes place in selected species and can not be regarded as a character of supra-species taxon; formerly species without hind wings constituted artificial genera *Cloeon* SENSU Leach 1815 and *Pseudocloeon* SENSU Klapálek 1905; recently first of them are placed into Afroptilum/fl=Centroptiloides/g1, Cloeon/fg1 and other taxa, and the second ones – into Baetis/fg6. In most cases size and structure of hind wing (or its absence) is the same in both sexes, but in a few species hind wings are present in male and lost in female, or are larger in male and smaller in female.

(11) On apex of larval femur the anterior-outer projection [i. e. projection situated on anterior (dorsal) side of femur outward from incision, into which the anterior articulating condylus of tibia is inputted] is curved toward inner side of femur; it can be from small and slightly curved to very large, rounding the anterior tibial condylus from distal side and overlapping on anterior-inner apical femoral projection (Wang & McCafferty 1996b:Figs 13–16). Unique apomorphy.

(12) Larval abdominal tergal surfaces and other parts of body usually bear small (about 0.01 mm) translucent scales situated in wide semilunar nests. Probably initial form of these nests is angulate with a pair of covers at the corners (FIG.29:C) – such form occurs in Afroptilum/fl=Centroptiloides/g1 and some Anteropatellata. In other cases nests have rounded form without covers. In some Turbanoculata scales are lost. In other mayflies such scales never occur. In *Siphlaenigma* scales are absent. It is unclear if presence of scales is an autapomorphy of Turbanoculata or Tetramerotarsata, because their absence in the single species of *Siphlaenigma* may be either primary, or a result of reduction.

(13) Larval external gonostyli buds are strongly diminished, sometimes lost, thus shape of posterior margin of larval abdominal sternum IX in male is nearly the same as in female (in contrast to majority of other Ephemeroptera); only in *Callibaetis* gonostyli buds are distinctly projected (probably being enlarged secondarily). In connection with reduction of larval gonostyli buds, subimaginal gonostyli anlages, which develop in larva, are folded under larval cuticle in a special manner. Probably the initial is "*Afroptilum-Cloeon*-type", when 2nd segment of gonostylus is directed laterally as a whole; 3rd [distalmost – see Liberevenata (5)] segment can be curved medially ("*Cloeon*-type" – FIG.29:F) or also directed laterally ("*Afroptilum*-type" – FIG.29:D–E); such position occurs in Afroptilum/fl=Centroptiloides/g1, and among Anteropatellata is present in all Cloeon/fg1, Baetopus/fg1 and *Cheleocloeon*; judging by shape of gonostyli in subimago, the same position was peculiar for *Palaeocloeon* (Kluge 1997c:Fig.15). In other taxa position of subimaginal gonostyli under larval cuticle is different: 2nd segment can be directed posteriorly ("*Nigrobaetis*-type" – FIG.29:G), medio-posteriorly ("*Baetis*-type" – FIG.29:H) or curved medio-anteriorly ("*Labiobaetis*-type" – FIG.29:I).

(14) Malpighian tubes are straight, each tube lies along gut by its free end directed posteriorly; its duct arises from its anterior end posteriorly, under acute angle (Landa 1969b:Fig.12:BR) (in contrast to majority of mayflies, where Malpighian tubes are coiled and situated irregularly). Non-unique apomorphy.

(15) Ganglion of abdominal segment VIII is approximated to that of segment VII. Non-unique apomorphy.

Variable characters of Turbanoculata. Larvae of most species have primitive siphlonuroid structure (FIG.28:A); only in selected taxa these or that modifications appear.

Larval head is hypognathous and usually elongate dorsoventrally (as in Siphlonurus/fg1 et al.); in flattened representatives head is flattened dorsoventrally, but retains hypognathous condition; in extremely flattened Baetis/fg6-Acentrella/g1-Jubabaetis/g2 frons forms a large shield with setose fore margin.

Mouth apparatus [see (4)–(7)] is usually non-specialized, in rare cases specialized either for carnivorism, or for filtering, or in other manner. All three maxillary dentisetae [see Plesiomorphies of Tetramerotarsata] can be subequal and slender (FIG. 28:D) or distal dentiseta is thickened, stout and pressed to canines (FIG.28:B) (similar to Nesameletus/f1=Metamonius/g1 – FIG.31). Maxillary palp can be 3-segmented, in some groups independently becomes 2-segmented. Labial palp is usually 3-segmented, 3rd segment is usually immobile fused with 2nd, but in 2nd segment muscle-adductor of 3rd segment is usually retained; in rare cases labial palp is 2-segmented, without this muscle.

Larval legs are either slender with setae short and irregular, or have muscular widened femora with long setae forming regular row on outer margin; usually legs are able to stretch posteriorly, but in some rheophilous species lost this ability, being enlarged and widely separated (FIG.9:F–H). Usually gill-processes are absent, but selected unrelated species have gills on coxal base (some Baetiella/g1, Camelobaetidius/g1, Heterocloeon/g1 and others) or on coxal apex (some *Baetodes*).

Larval abdomen is usually long, able to make undulate swimming movements, but in some rheophilous species is shortened and unable to undulate (FIG.9:F–G). Abdominal terga are usually simple, rarely with unpaired median or paired submedian projections.

Usually all tergalii I–VII are present; only in selected species tergalii I or VII are lost, and in *Baetodes* tergalii VI–VII are lost. Usually all tergalii have normal dorsal attachment on posterior margins of segments, only in *Afrobaetodes* and *Baetodes* tergalial bases are transferred ventrally. Tergalii [with anal ribs on anal margin – see Tetramerotarsata (12)] are often oval, with entire margin bordered by costal and anal ribs; sometimes anal rib is vestigial or lost, in this case tergalius can have anal expansion; in *Callibaetis* such anal expansion is bent ventrally, in Cloeon/fg1 – dorsally, thus in both cases tergalius becomes bilamellate (the same in some other mayflies – see Index of characters [1.3.24]); in *Callibaetis* costal rib is situated far from costal margin; in *Baetodes* both tergalial ribs are lost. In some taxa [*Callibaetis*, Cloeon/fg1 (incl. *Centroptilum*, *Procloeon* et al.) and others] tergalii retain ability of rhythmical respiratory movements; in other taxa [Baetopus/g1, *Cloeodes*, *Paracloeodes*, Baetis/fg6 (incl. *Fallceon*, *Caribaetis*, *Nigrobaetis*, *Labiobaetis*, *Acentrella*, *Baetiella* et al.) and others] ability of such movements is lost (for comparison – see Index of characters [1.3.30]).

Larval caudalii are usually not long, with dense primary swimming setae and without secondary swimming setae, cerci with oblique segment boundaries; but in some rheophilous species cerci are elongate, sometimes primary swimming setae are completely lost; in other species, vice versa, besides primary ones, secondary swimming setae appear. Larval paracercus is usually well-developed [while imaginal paracercus is vestigial – see Liberevenata (7)], but in some rheophilous species larval paracercus is more or less shortened or vestigial.

Eggs have no anchors, with various chorionic sculpture; in two taxa (Cloeon/fg2 and *Callibaetis*) viviparity evolved independently.

Size. Fore wing length 2–12 mm (see Tetramerotarsata).

Age and distribution. Recently distributed world-widely and abundant everywhere. A very young taxon: being well-distinguishable in winged stages, Turbanoculata (and Liberevenata in general) are not reported from Baltic amber which contains many winged mayflies of other taxa (Siphlonurus/fg1, Ameletus/fg1, Radulapalpata, Ephemera/fg9, Ephemerella/fg1, Leptophlebia/fg1 and other Anteritorna). This allows to conclude that in Palaeogene Turbanoculata were either absent, or represented only by African Afroptilum/f1=Centroptiloides/g1.

Classification of Turbanoculata. About 800 species are described. In the past Turbanoculata were divided into Baetinae and Cloeoninae (Kazlauskas 1972). The subfamily Baetinae in such sense can be characterized by two marginal intercalaries in each space of fore wing, rheophilous larvae with one row of denticles on each claws, and convergenting subimaginal gonostyli anlages in mature larva. The subfamily Cloeoninae, or **Cloeon/fg** [f: Cloeonidae Newman 1853: 187; g: *Cloeon* Leach 1815: 137, typus *Ephemera diptera* Linnaeus 1761 (monotypy)] in such sense can be characterized by one intercalary in each space, potamophilous larvae with two rows of denticles on each claw, and divergent subimaginal gonostyli anlages in mature larva. Such classification is based on Palaearctic species only, and does not take into account mayflies of other faunas, which have various combinations of these characters. Comparison of Turbanoculata with *Palaeocloeon* testifies that one intercalary in each space and divergent gonostyli anlages are plesiomorphies (Kluge 1997c); their comparison with *Siphlaenigma* testifies that presence of two rows of denticles on larval claw is also a plesiomorphy. Thus Cloeoninae with such diagnosis is a plesiomorphon.

Recently Turbanoculata are divided into a smaller plesiomorphon Afroptilum/f1=Centroptiloides/g1 and a large holophyletic taxon Anteropatellata (Kluge 1997c).

Afroptilum/f1=Centroptiloides/g1 [f: Afroptilinae Kluge 1997c: 533; g: *Centroptiloides* Lestage 1918: 107, typus *Centroptilum bifasciatum* Esben-Petersen 1918 (design. orig.)] is characterized by a plesiomorphy – presence of patella-tibial suture on middle and hind legs only. In other respects many representatives of this taxon agree with the diagnosis of Cloeoninae SENSU Kazlauskas 1972: fore wing always with one marginal intercalary in each space, larval claw usually with two rows of denticles, subimaginal gonostyli anlages in mature larva are divergent. Larvae of various species are from typically potamophilous to typically rheophilous. Distribution is restricted by Ethiopian Region only.

Anteropatellata Kluge 1997, or **Baetis/fg5** (incl. *Cloeon* et al.) is characterized by an autapomorphy: the patella-tibial suture is equally developed on all legs of larva (but adults retain plesiomorphic condition, i. e. with vestiges of patella-tibial suture on middle and hind legs only). The same in Rhithrogena/fg3 only (see Index of characters [1.2.18]). Among the species examined, exceptions are made only by Cloeon/fg1 *kazlauskasi* Kluge 1983 [*Centroptilum*], whose larva has patella-tibial suture on middle and hind legs only, and by **Centroptella/g1** (incl. *Chopralla*) [g: *Centroptella* Braasch & Soldán 1980: 123, typus *C. longisetosa* Braasch & Soldán 1980 (design. orig.)], whose larval patella-tibial suture of fore leg is secondarily reduced in connection with peculiar leg modification. Anteropatellata have world-wide distribution and include overwhelming majority of species of Turbanoculata.

Revision of supraspecies taxa of Afroptilum/f1= =Centroptiloides/g1 and Turbanoculata is in progress (see draft versions in Internet: http://www.bio.pu.ru/win/entomol/KLUGE/sys-eph.htm). Last years a lot of new genera of Turbanoculata were described; many descriptions are insufficient, often based on a few specimens only. In order to complete the revision of this group, it will be necessary to re-examine representatives of many genera, especially pose of their subimaginal gonostyli anlages in mature larva. The revision of Turbanoculata is planned to be published as a 2nd volume of this book.

***

Liberevenata INCERTAE SEDIS

Liberevenata INCERTAE SEDIS. † **Myanmarella/g** [g: *Myanmarella* Sinitshenkova 2000b: 25, typus *M. rossi* Sinitshenkova 2000 (design. orig.)].

Reference. Sinitshenkova 2000b: ⊕.

Age. Presumably Late Cretaceous (Burmese amber).

Material examined: –.

Comment. Probably *Myanmarella* belongs to Palaeocloeon/fg1; it is impossible to state this for certain, because the only specimen is female (thus it is unclear if male had turban eyes or not), and thorax of this specimen is poorly preserved (thus furcasternum structure of mesothorax is unknown). Originally this taxon was placed to the family Prosopistomatidae basing on presence of intercalaries by sides of each longitudinal vein, as in males of Recent Prosopistoma/f2=g1 (Sinitshenkova 2000b:Figs 1d-g); actually these are not intercalaries, but traces in resin produced by longitudinal veins when insect moved (personal communication by V.V. Zherichin, who saw the holotype). 5-segmented tarsus described and figured (Sinitshenkova 2000b:Fig.1c) is non-reliable, because apical claw-bearing segment can not be so short; probably this is actually 4-segmented tarsus typical for Tetramerotarsata.

*** ***

2.1;8. Nesameletus/f1=Metamonius/g1

(Euephemeroptera Euplectoptera Anteritorna Tridentiseta Nesameletus/f1=Metamonius/g1)

(Figs 30–31)

Nomen hierarchicum: **Nesameletus/f1=Metamonius/g1** (incl. *Siphluriscus*) [f: Nesameletinae Riek 1973: 164; g: *Metamonius* Eaton 1885: 209, typus *Siphlurus anceps* Eaton 1883 (design. orig.)].

Reference. Zhou & Peters 2003: ♉ ⊕; – Hitchings & Staniczek 2003: ♉* ⊕*.

Autapomorphies of Nesameletus/f1=Metamonius/g1.

(1) Mandibular incisor is enlarged, straight, blade-like, and kinetodontium is reduced; kinetodontium can be retained as a setiform vestige (FIG.31: D, F), or lost.

(2) Maxilla is elongate, with short biting edge; all 3 canines have characteristic shape – short, with rounded outer margin; setation laterad of canines is reduced: either vestigial setae of apical-ventral row are present (in Nesameletus/f2=Metamonius/g2 – FIG.31:A–C), or all setae are lost (in *Siphluriscus* – FIG.31:E). For other character of maxilla – see (3), (4) and (7).

(3) On maxilla [see (2)] 2nd and 3rd dentisetae [among 3 dentisetae – see Anteritorna (2)] can be situated on a common base; besides dentisetae, this base bears some smaller setae (FIG.31:A, E). In other respects dentisetae are modified differently [see below – Nesameletus/f2=Metamonius/g2 (3) and Plesiomorphies of *Siphluriscus*].

(4) 3rd segment of maxillary palp is shortened (FIG.31:C).

(5) Glossae and paraglossae are elongate, narrowed, ventral surface with irregular long slender outstanding setae; glossae bear stout setae on apex (FIG.30:F); paraglossae can also bear such setae (in *Siphluriscus* – Zhou & Peters 2003:Fig.10) or not (in Nesameletus/f2=Metamonius/g2). General shape of glossae and paraglossae resembles that of Turbanoculata (FIG.28:C), but paraglossae retain lateral (not apical) attachment to mentum.

(6) Imaginal and subimaginal vertex has unpaired projection; in Nesameletus/f2=Metamonius/g2 it is well-expressed in both sexes (FIG.30:C), in *Siphluriscus* – in male only (Zhou & Peters 2003: Figs 21, 24).

Characters of unclear phylogenetic status.

(7) Maxilla [see (2)–(4)] bears a gill located laterally at articulation of cardo and stipes; in Nesameletus/f2=Metamonius/g2 this is a single small process (FIG.31:C), while in *Siphluriscus* this is a tuft of gill processes, and another tuft arises medially (Zhou & Peters 2003:Fig.8). Gill similar to that of Nesameletus/f2=Metamonius/g2 is present in *Rallidens* (FIG.36:A); possibly, synapomorphy.

(8) In imago and subimago claws (initially ephemeropteroid) on each leg are similar and pointed. Non-unique apomorphy (see Index of characters [2.2.85] and [2.2.77]); among Tridentiseta the same in Siphlonurus/fg1 and *Dipteromimus*.

(9) On each tergalius I–VII anal rib is situated on dorsal surface near middle of tergalius, just behind main trachea (the same in Siphlonurus/fg1, *Dipteromimus* and some Ameletus/fg1); tergalii II–VII have costal rib well-developed along entire costal margin, with denticles at distal part; thus, such tergalius has either 2 distinct ribs (in *Siphluriscus* – Zhou & Peters 2003:Fig.12) or 3 ribs (FIG.30:E) [see Nesameletus/f2=Metamonius/g2 (5) below]. Costal rib of tergalius I can be either lost, or small and located at a distance from costal margin (in *lacusalbinae* [*Ameletoides*]).

Plesiomorphies of Nesameletus/f1=Metamonius/g1. Larva has primary swimming siphlonuroid specialization: legs are able to stretch posteriorly; abdomen is large, able to make undulate swimming movements; caudalii are not long, paracercus is well-developed, primary swimming setae are dense, secondary swimming setae are absent. Head is hypognathous; labial palp is 3-segmented [for other features of mouthparts see (1)–(5)]. Larval (and adult) patella-tibial suture is developed on middle and hind legs only.

In imago and subimago: Mesonotal suture is nearly transverse, either not stretched backward medially (in *Siphluriscus*) or shortly stretched (in Nesameletus/f2=Metamonius/g2 – FIG.30:D). Anterior paracoxal suture is complete (FIG.30:A–B) (in contrast to *Rallidens* and some Bidentiseta). Metathoracic nerve ganglion is situated in posterior part of furcasternum; furcasternal protuberances are either contiguous except for extreme hind part (in *Siphluriscus* – Zhou & Peters 2003:Fig.25) or separated by narrow median impression widened posteriorly (in Nesameletus/f2=Metamonius/g2 – FIG.30:A). In cubital field of fore wing several branched or simple veins go from CuA to basitornal margin [see Anteritorna (1)], hind wing is well-developed; in *Siphluriscus* hind wing is especially large, a little longer than half of fore wing, with RS, MA and MP forked in proximal part, and fore wing has especially long narrow cubital field with 10–15 veins going from CuA to basitornal margin (Demoulin 1955h:Fig.1; McCafferty & Wang 1994: Figs 1–2); in Nesameletus/f2=Metamonius/g2 hind wing has more usual size, as long as 0.35–0.4 of fore wing length, with MA forked distad of middle, and fore wing has moderately long cubital field with 3–6 veins. Imaginal and subimaginal tarsi are 5-segmented, 1st segment is non-shortened and fused with tibia. Gonostylus with 2 distal segments. Imaginal and subimaginal paracercus is more or less developed: at least much longer than abdominal segment X, multisegmented, with normally developed setation; in *lacusalbinae* [*Ameletoides*] it is subequal to cerci, in other species – many times shorter than cerci.

Size. Fore wing length 12–25 mm.

Distribution. South-eastern China, Australia, New Zealand and South America.

Nesameletus/f1=Metamonius/g1 is divided into *Siphluriscus* and Nesameletus/f2=Metamonius/g2.

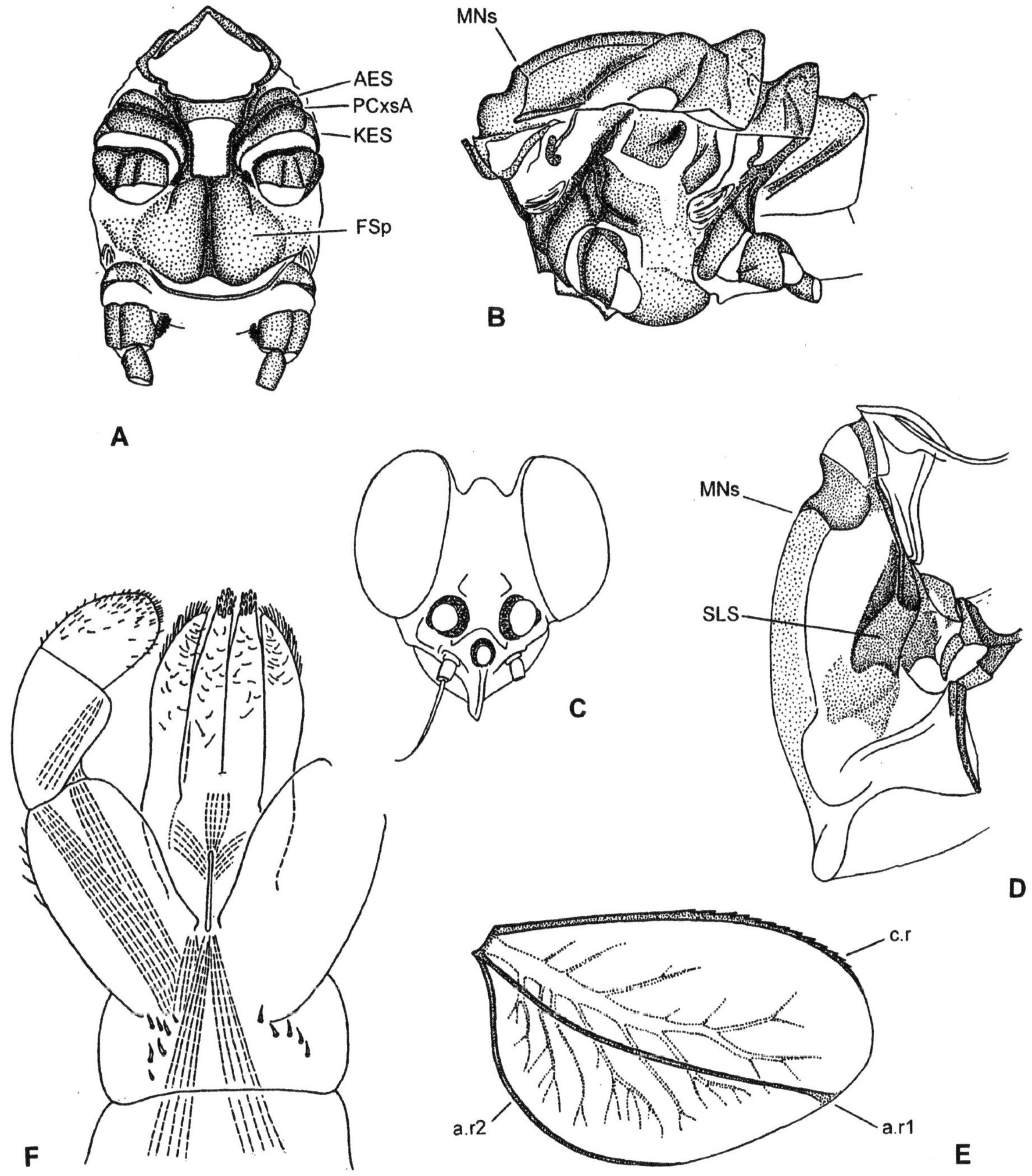

Figure 30. Nesameletus/f2=Metamonius/g2.
A–E – *flavitinctus* [*Ameletus*]: **A–B** – pterothorax of female imago: **A** – ventral view; **B** – lateral view; **C** – head of female imago; **D** – subimaginal exuviae of right half of mesonotum; **E** – tergalius. **F** – *anceps* [*Metamonius*], labium, ventral view (muscles shown by interrupted lines). (A–D – from Kluge & al., 1995).

Abbreviations: **AES** – anepisternum; **a.r1** – anterior branch of anal rib; **a.r2** – posterior branch of anal rib; **c.r** – costal rib; **FSp** – furcasternal protuberance; **KES** – katepisternum; **MNs** – mesonotal suture; **PCxsA** – anterior paracoxal suture; **SLS** – sublateroscutum.

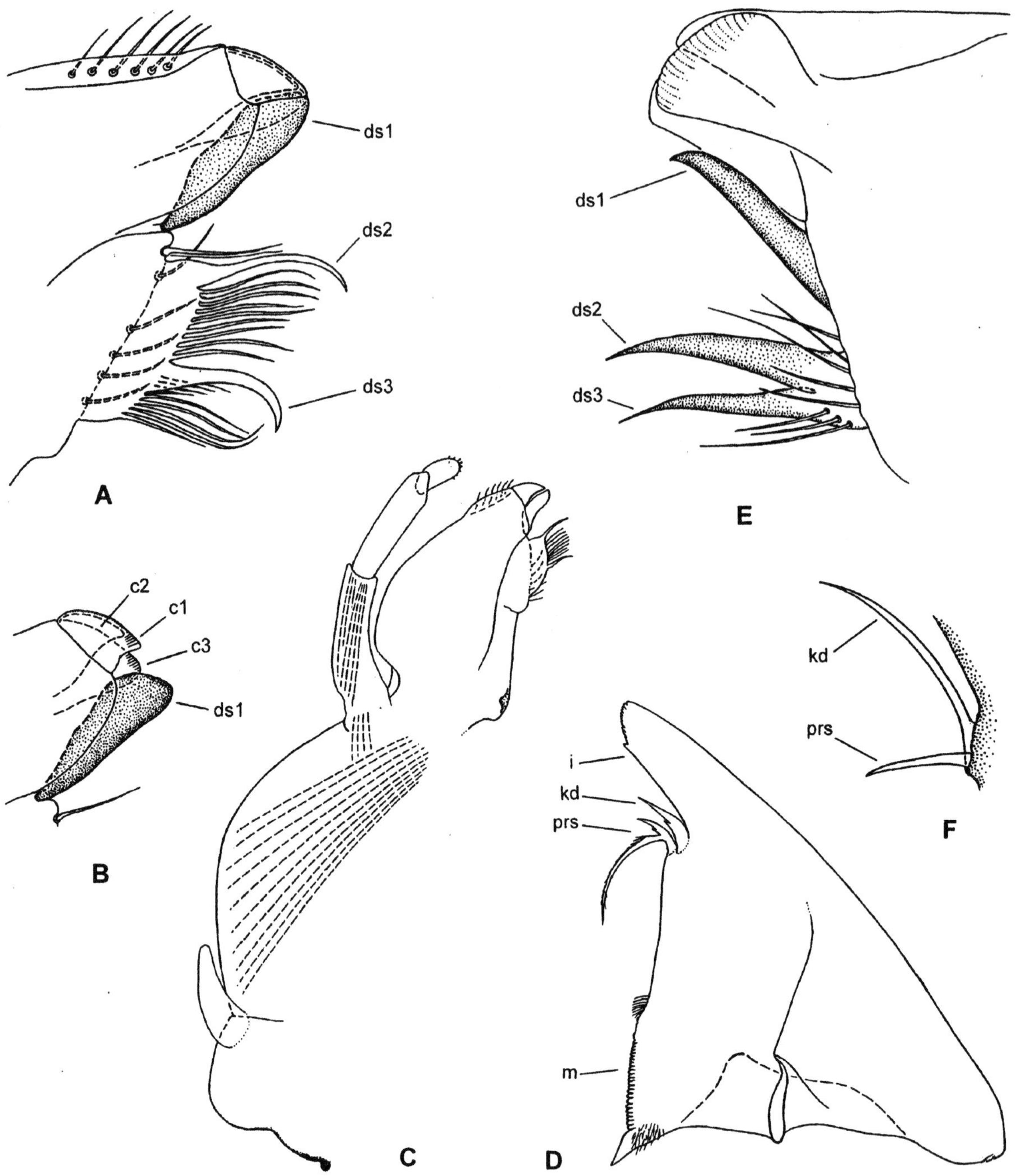

Figure 31. Nesameletus/f1=Metamonius/g1, maxillae and mandibles.
A–D – Nesameletus/f2=Metamonius/g2: **A–B** – *flavitinctus* [*Ameletus*]: **A** – apex of left maxilla, dorsal view, canines and distal dentiseta pressed together (distal dentiseta shown by dots); **B** – the same, canines and distal dentiseta not pressed together (distal dentiseta shown by dots); **C–D** – *anceps* [*Metamonius*]: **C** – left maxilla, dorsal view (muscles shown by interrupted lines); **D** – right mandible, dorsal view. **E–F** – Siphluriscus/fg* *chinensis* [*Siphluriscus*]: **E** – apex of maxilla (some bristles are not drawn; dentisetae shown by dots); **F** – kinetodontium and prostheca of right mandible, dorsal view. (A, B – from Kluge & al., 1995).

Abbreviations: **c1, c2, c3** – maxillary canines; **ds1, ds2, ds3** – first, second and third dentisetae; **i** – incisor; **kd** – kinetodontium; **m** – mola; **prs** – prostheca.

2.1;8,1. Siphluriscus/fg(1)

(Euplectoptera Anteritorna Tridentiseta Nesameletus/f1=Metamonius/g1 *Siphluriscus*)

(Fig. 31:E–F)

Nomen hierarchicum: **Siphluriscus/fg(1)** [f: Siphluriscidae Zhou & Peters 2003:345; g: *Siphluriscus* Ulmer 1920a: 61, typus *S. chinensis* Ulmer 1920 (design. orig.)].
In circumscription matches:
— gen. *Siphluriscus* Ulmer 1920a: 61;
— fam. Siphluriscidae Zhou & Peters 2003:345.
Monospecific taxon.

References. Ulmer 1920a: ⊕♂; – Demoulin 1955h: ⊕♀; – McCafferty & Wang 1994b: ◎♂ ⊕♂; – Zhou & Peters 2003: ℧ ⊕.

Autapomorphies of *Siphluriscus*.

(1) Branched filamentous gills are present on bases of mouthparts and leg: maxilla, besides lateral gill [see Nesameletus/f1=Metamonius/g1 (7)], has another multibranched gill directed medially; labium bears a pair of multibranched gills by sides of submentum; in larva, subimago and imago bases of fore and middle (but not hind) legs bear multibranched gills which arise from inner-proximal margin of coxa (Zhou & Peters 2003:Figs 8, 10, 15, 25). Similar, but differently arranged gills independently appeared in some other taxa (see Index of characters [1]).

(2) Larval claw bears an unusually large flexible projection on inner side (Zhou & Peters 2003:Fig.28). In contrast to Nesameletus/f2=Metamonius/g2, denticles are absent.

Characters of unclear phylogenetic status.

(3) Larval paraproct with a spine on inner-apical margin (Zhou & Peters 2003:Fig.20). The same in Siphlonurus/fg2, Metretopus/fg1 and Vetulata (FIG. 33:F).

Plesiomorphies of *Siphluriscus*. In larva (in contrast to Nesameletus/f2=Metamonius/g2): Mandibles [see Nesameletus/f1=Metamonius/g1 (1)] are not elongate perpendicular to axis of articulation (Zhou & Peters 2003:Figs 6–7). 1st dentiseta is not thickened, common base of 2nd and 3rd dentisetae is either stout and bears a few irregular non-fused setae (FIG.31:E) or non-expressed (Zhou & Peters 2003: Fig.9). Tergalii without additional rib on anal margin.

Subimaginal lateral pigmented area of mesonotum is not large, bifurcates posteriorly in such a manner, that one its branch stretches along lateroparapsidal suture, and another branch – along lateral scutal suture (as in FIGS 18:E and 33:D–E) (plesiomorphic condition, in contrast to Nesameletus/f2= =Metamonius/g2 and many others – see Index of characters [2.2.14]).

Size. Fore wing length 17–25 mm.

Distribution. China.

Species composition of Siphluriscus/fg(1). 1 species – *chinensis* Ulmer 1920 [*Siphluriscus*].

Material examined (Entomol. Inst.): *chinensis* [*S.*]: ◎♀, ℧ (incomplete exuviae).

***

2.1;8,2. Nesameletus/f2=Metamonius/g2

(Euplectoptera Anteritorna Tridentiseta Nesameletus/f1=Metamonius/g1 .../f2=.../g2)

(Figs 30; 31:A–D)

Nomen hierarchicum: **Nesameletus/f2=Metamonius/g2** (sine *Siphluriscus*).
In circumscription matches:
— subfam. Nesameletinae Riek 1973: 164;
— fam. Metamoniidae McCafferty 1991a: 349;
— fam. Nesameletidae: Kluge & Studemann & Landolt & Gonser 1995: 115.

References. Eaton 1833–1888 (*Metamonius*): ⊕; – Ulmer 1904 ("Nymphe des *Baetis*-Typus"): ℧; – Tillyard1933 (*Ameletoides*, *Nesameletus*): ℧ ⊕; – Riek 1955 (*Ameletoides*): ℧ ⊕; – Demoulin 1955b (*Metamonius*): ⊕♀; – 1955c (*Metamonius*): ℧; – Kimmins 1960 (*Metamonius*): ⊕♂; – Koss & Edmunds 1974: ⊙*; – Edmunds 1975: ℧*; – Kluge & Studemann & Landolt & Gonser 1995: ℧* ◎* ⊕*; – Hitchings & Staniczek 2003: ⊙* ℧* ⊕*.

Autapomorphies of Nesameletus/f2=Metamonius/g2.

(1) Mandible [see Nesameletus/f1=Metamonius/g1 (1)] is elongate perpendicular to axis of articulation (FIG.31:D).

(2) On maxilla 1st dentiseta is very massive, sclerotized, pressed to canines [shortened – see Nesameletus/f1=Metamonius/g1 (2)] (FIG.31:A–C). The same in selected taxa among Tetramerotarsata.

(3) 2nd and 3rd dentisetae [situated on a common base – see Nesameletus/f1=Metamonius/g1 (3)] are weak, their common base represents a wide thin plate (FIG.31:A).

(4) Larval claw on inner side has 2 longitudinal rows of small denticles. Non-unique apomorphy (see Index of characters [1.2.21]).

(5) On each tergalius II–VII anal rib bifurcates near base, its anterior branch is situated on dorsal surface and passes near middle of tergalius [see Nesameletus/f1=Metamonius/g1 (9)], and posterior branch is situated on anal margin and represents a 3rd (additional) rib (FIG.30:E). Similar 3 ribs are present also in *Rallidens* and Isonychia/fg1.

(6) Subimaginal lateral pigmented sclerotized area of mesonotum has peculiar form and includes whole anterior part of lateroscutum (FIG.30:D). The same in Ameletopsis/fg1.

Size. Fore wing length 12–17 mm.

Distribution. Notogea: Australia, New Zealand and Chile-Patagonian Region of South America.

Nominal taxa included. Nesameletus/f2=Metamonius/g2 includes: (1) **Nesameletus/fg** [g: *Nesameletus* Tillyard 1933: 11, typus *Chirotonetes ornatus* Eaton 1883 (design. orig.)]; (2) **Metamonius/fg** [f: Metamoniidae McCafferty 1991a: 349]; (3) **Ameletoides/g** [g: *Ameletoides* Tillyard 1933: 5, typus *A. lacusalbinae* Tillyard 1933 (design. orig.)]. Original separate generic diagnoses of South American *Metamonius*, New Zealand *Nesameletus* and Australian *Ameletoides* were based on their distribution and wrong descriptions. Hitchings & Staniczek (2003) characterized *Nesameletus*, which contains 5 New Zealand species, by structure of egg anchors, presence of spine-like setae on basal part of labium and narrowed penis base. However, no one of these characters allows to separate this group from Australian and American species.

Species composition of Nesameletus/f2=Metamonius/g2. Not less than 7 species. **South American species:** *anceps* Eaton 1885 [*Metamonius*] (syn. subj.: *fuegiensis* Lestage 1935 [*Ameletus*]). **Australian species:** *lacusalbinae* Tillyard 1933 [*Ameletoides*]. Five **New Zealand species:** *austrinus* Hitchings & Staniczek 2003 [*Nesameletus*], *flavitinctus* Tillyard 1923 [*Ameletus*], *murihiku* Hitchings & Staniczek 2003 [*Nesameletus*], *ornatus* Eaton 1883 [*Chirotonetes*], *vulcanus* Hitchings & Staniczek 2003 [*Nesameletus*].

Material examined: *anceps* [*M.*]: ♉︎; *lacusalbinae* [*A.*]: ♉︎; *ornatus* [*Ch.*]: ♉︎, ♉︎/◎, ◎/⊕/⊙.

*** ***

2.1;9. **Vetulata**, or **Oniscigaster/fg1**

(Euephemeroptera Euplectoptera Anteritorna Tridentiseta Vetulata)

(Figs 32–33)

Nomen hierarchicum: **Oniscigaster/fg1** (incl. *Tasmanophlebia*) [f: Oniscigastrina Lameere 1917: 62; g: *Oniscigaster* McLachlan 1873: 109, typus *O. wakefieldi* McLachlan 1873 (monotypy)].

Nomen circumscribens: **Vetulata** McCafferty 1991a: 349.

In circumscription matches:

— subtribus Oniscigastrina Lameere 1917: 62;

— subfam. Oniscigastrinae: Edmunds & Traver 1954a: 237;

— fam. Oniscigastridae: Landa 1973: 156.

— infraordo Vetulata McCafferty 1991a: 349.

References. Edmunds & Allen & Peters 1963: ♉︎*; – Koss & Edmunds 1974: ⊙*; – Edmunds 1975: ♉︎; – Kluge & Studemann & Landolt & Gonser 1995: ♉︎* ◎* ⊕*.

Autapomorphies of Vetulata.

(1) Larval abdominal tergum I has posterior margin elevated; tergalii I are rounded-triangular, without anal-proximal lobe [in contrast to tergalii of next pairs – see (2)] and are specialized as gill opercula. In Tasmanophlebia/fg1 they cover the rest tergalii (FIG.33A), while in Oniscigaster/fg2 they are not larger than other tergalii and do not cover them, but also have operculate specialization: anal margin, being directed anteriorly, is inserted under a hanging posterior margin of abdominal tergum I (FIG.32:A). Unique apomorphy: in some other mayfly groups gill opercula are also present, but these are tergalii not of the first pair, but of second, third or forth pair.

(2) Tergalii of next pairs (II–IV or II–VI) with projected anal-proximal lobe (having different structure in Oniscigaster/fg2 and Tasmanophlebia/g1 – see below) (FIGS 32–33). Ribs of tergalii are reduced, only weakly developed costal rib in proximal part of costal margin can be present (non-unique apomorphy – see Index of characters [1.3.27–29]).

(3) In larva at least anteriormost abdominal terga (I–IV) have median projections (FIGS 32:A; 33:A). Non-unique apomorphy (see Index of characters [1.3.3]).

(4) Larval abdomen with wide lateral lobes, thus tergalii lie dorsally (not by sides) (FIGS 32:A; 33:A). Non-unique apomorphy.

(5) In larva portion of frons between antennae bases is elevated and bordered by a pair of longitudinal keels (Eaton 1883–1888:Pl.51:1–5). Head retains hypognathous position.

Characters of unclear phylogenetic status.

(6) Imaginal and subimaginal furcasternal protuberances are not contiguous, furcasternal median impression is parallel-sided, not widening posteriorly; metathoracic nerve ganglion is situated in posterior part of furcasternum (FIG.32:D). Non-unique apomorphy (see Index of characters [2.2.23]); among Tridentiseta the same in other Amphinotic groups – Nesameletus/f2=Metamonius/g2, Ameletopsis/fg1 and *Rallidens*.

(7) Larval paraproct with a spine on inner-apical margin (FIG.33:F). The same in Siphlonurus/fg2, Metretopus/fg1, *Siphluriscus* and *Rallidens*.

Plesiomorphies of Vetulata. Larva retains features of primary swimming siphlonuroid specialization: abdomen is large; caudalii are not long, paracercus is well-developed, primary swimming setae are dense, secondary swimming setae on lateral margins of cerci are usually absent, rarely present (in Tasmanophlebia/g1 sp.V1 from Australia). Mouth apparatus is non-specialized (Eaton 1883–1888:Pl.

51:7–12); maxilla with 3 canines and 3 dentisetae [see Anteritorna (2)], setae of its apical-ventral row are from simple (FIG.32:E) to pectinate; maxillary palp is 3-segmented; labial palp is 3-segmented. Larval (and adult) patella-tibial suture is developed on middle and hind legs only. Larval claws are slightly curved, with very small irregular denticles only.

In imago and subimago: Mesonotal suture is more or less transverse (FIGS 32:B; 33:D–E). Paracoxal suture is complete (FIG.32:C–D) (in contrast to *Rallidens* and some Bidentiseta). Subimaginal lateral sclerotized pigmented area of mesonotum is small, bifurcates posteriorly in such a manner, that one its branch stretches along lateroparapsidal suture, and another branch – along lateral scutal suture (FIG. 32:B) (see Index of characters [2.2.14]). In cubital field of fore wing several (5–9) veins go from CuA to basitornal margin [see Anteritorna (1)] (as in FIG.7:C); hind wing is well-developed, as long as 0.45 of fore wing length. Imaginal and subimaginal tarsi are 5-segmented, 1st segment is non-shortened and fused with tibia. All claws of imago and subimago are ephemeropteroid. Gonostylus with 2 distal segments.

Size. Fore wing length 10–20 mm.

Distribution. Notogea: Australia, New Zealand and Chile-Patagonian Region of South America.

Systematic position of Vetulata. A relationship between Vetulata and Furcatergaliae was assumed, in connection with this there was established a suborder **Rectracheata** McCafferty 1991, which united Vetulata, Furcatergaliae and Posteritorna. This assumption is not grounded, and the taxon Rectracheata is polyphyletic (Kluge 1998).

Vetulata are divided into Oniscigaster/fg2 and Tasmanophlebia/fg1.

2.1;9,1. Oniscigaster/fg2

(Euplectoptera Anteritorna Tridentiseta Vetulata Oniscigaster/fg2)

(Fig. 32)

Nomen hierarchicum: **Oniscigaster/fg2** (sine *Tasmanophlebia*).

In circumscription matches:

— gen. *Oniscigaster* McLachlan 1873: 109.

References. Eaton 1883–1888: Ʊ* ⊕'; – Eaton 1899: ⊕* – Phillips 1930: Ʊ* ⊕*; – 1931: Ʊ*; – Lestage 1935a: Ʊ ⊕; – Edmunds 1975: Ʊ.

Characters of unclear phylogenetic status.

(1) Larva is not operculate-gilled, in spite of the fact that tergalii I strongly differ from others [see Vetulata (1)]: tergalii I are not larger than others, not covering them; anal-proximal lobe of tergalii II–VI [see Vetulata (2)] is not bent dorsally, with incised margin (FIG.32:A). Taking into account that tergalii I are specialized, it can be assumed that operculate-gilled specialization disappeared secondarily.

Plesiomorphies of Oniscigaster/fg2. Imaginal and subimaginal paracercus is developed (but much shorter than cerci). On hind wing bifurcation of MP is nearer to base than to margin of wing, more proximal than bifurcations of RS and MA (as in FIG.7:D) (in contrast to Tasmanophlebia/g1). In contrast to Tasmanophlebia/g1, all tergalii I–VII are present [about tergalii I–VI – see (1)], while tergalii VII are vestigial (FIG.32:A).

Variable characters of Oniscigaster/fg2. In imago and subimago of both sexes projecting paranota can be developed on abdominal segments VII–IX (in *wakefieldi* [*O.*]) or on segments VIII–IX only (in *intermedius* [*O.*]) or are absent (in *distans* [*O.*]).

Size. Fore wing length 15–20 mm.

Distribution. New Zealand.

Species composition of Oniscigaster/fg2. 3 species – *wakefieldi* McLachlan 1873 [*Oniscigaster*], *intermedius* Eaton 1899 [*Oniscigaster*], *distans* Eaton 1899 [*Oniscigaster*].

Material examined: *wakefieldi* [*O.*]: Ʊ; *intermedius* [*O.*]: Ʊ–⊕/⊙; *distans* [*O.*]: Ʊ, ♂/♂.

***

2.1;9,2. Tasmanophlebia/g1

(Euplectoptera Anteritorna Tridentiseta Vetulata Tasmanophlebia/g1)

(Fig. 33)

Nomen hierarchicum: **Tasmanophlebia/g1** (incl. *Siphlonella*) [g: *Tasmanophlebia* Tillyard 1921: 410, typus *T. lacustris* Tillyard 1921 (design. orig.)].

References. Tillyard 1921 (*Tasmanophlebia*): ⊕; – Needham & Murphy 1924 (*Siphlonella*): Ʊ; – Tillyard 1933 (*Tasmanophlebia*): Ʊ ⊕; – 1936a (*Tasmanophlebia*): Ʊ ⊕; – Ulmer 1938 (*Siphlonella*): ⊕; – Traver 1944 (*Siphlonella*): Ʊ ⊕; – Riek 1955 (*Tasmanophlebia*): Ʊ ⊕; – Edmunds 1975 (*Tasmanophlebia*, *Siphlonella*): Ʊ.

Autapomorphies of Tasmanophlebia/g1.

(1) On hind wing bifurcation of MP [the third triad see Euplectoptera (1)] is nearer to wing margin than to wing base, more distal than bifurcations of RS and MA (as in FIG.40:B). Non-unique apomorphy (see Index of characters [2.2.69]).

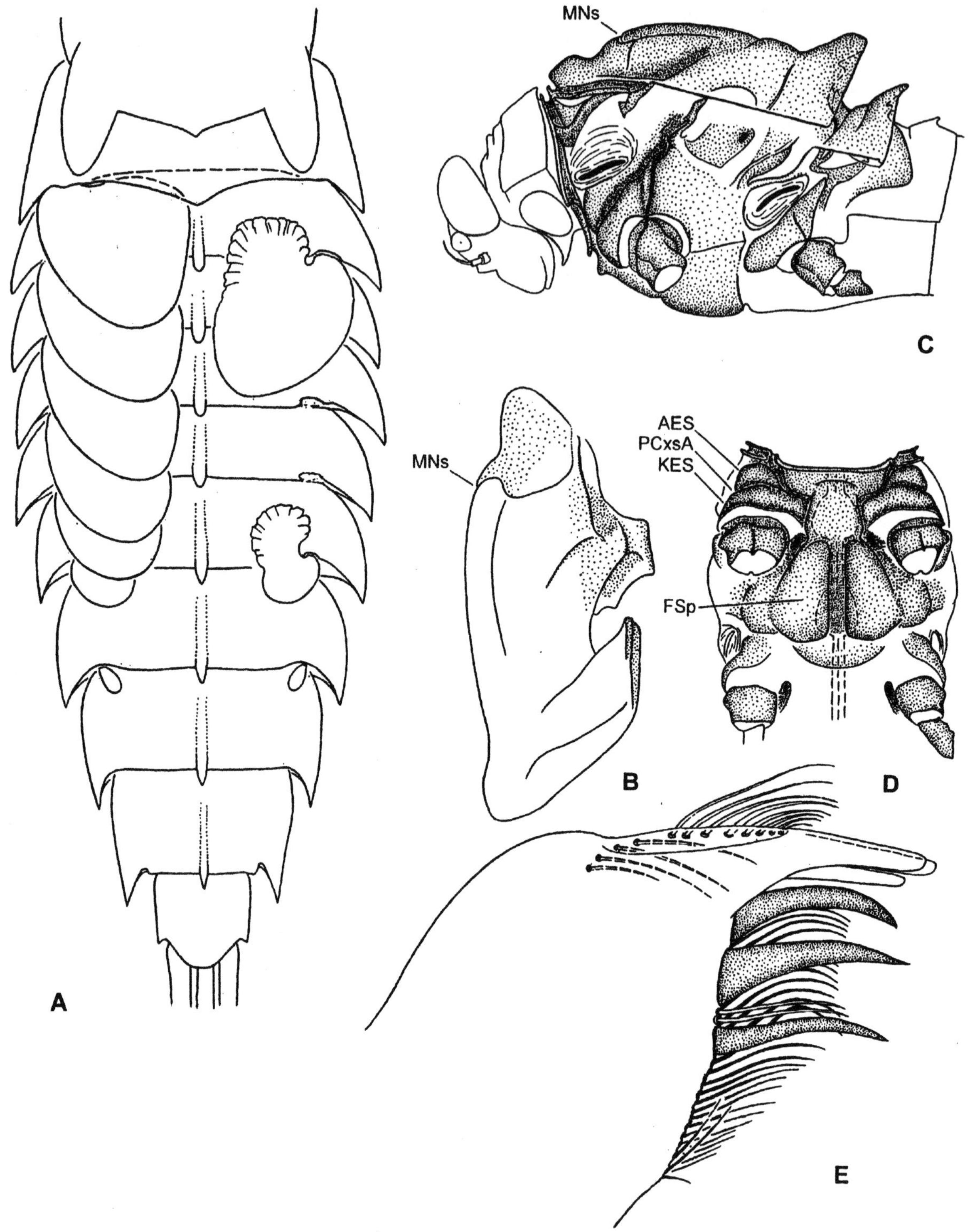

Figure 32. Vetulata-Oniscigaster/fg2.
A–B – Oniscigaster/fg2 *distans* [*Oniscigaster*]: **A** – larval metanotum and abdomen, dorsal view (on right side tergalii I, III and IV removed); **B** – subimaginal exuviae of right half of mesonotum. **C–D** – Oniscigaster/fg2 *intermedius* [*Oniscigaster*], female imago: **C** – head and thorax, lateral view; **D** – pterothorax, ventral view (nerve ganglions shown by interrupted line). **E** – Oniscigaster/fg2 sp., apex of left maxilla, dorsal view (dentisetae shown by dots). (C–E – from Kluge & al. 1995)

Abbreviations: **AES** – anepisternum; **FSp** – furcasternal protuberance; **KES** – katepisternum; **MNs** – mesonotal suture; **PCxsA** – anterior paracoxal suture.

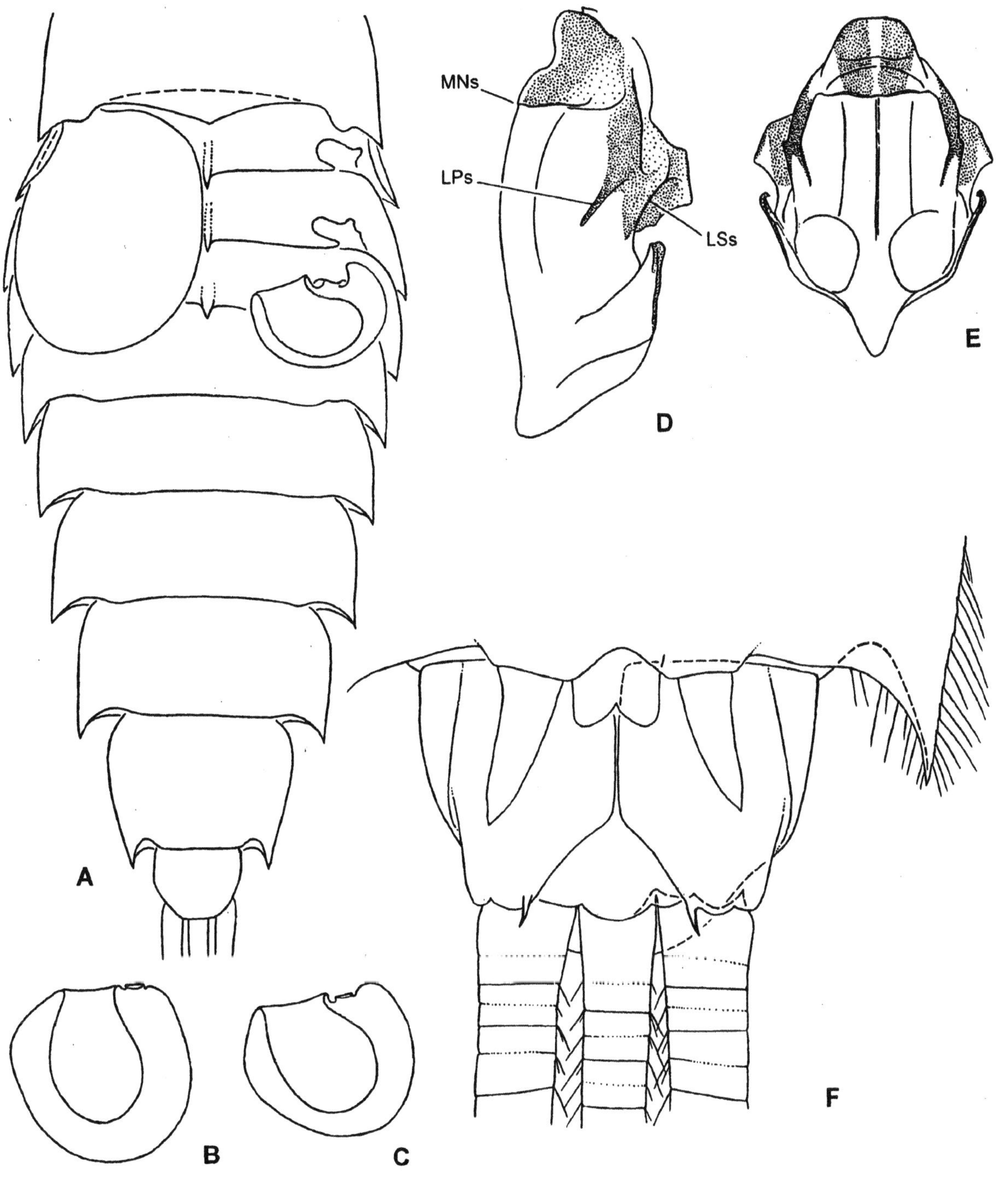

Figure 33. Vetulata-Tasmanophlebia/g1.

A–E – Tasmanophlebia/g1 *ventilans* [*Siphlonella*]: **A** – larval metanotum and abdomen, dorsal view (on right side tergalii I–III removed); **B** – its right tergalius II; **C** – its right tergalius III; **D** – subimaginal exuviae of right half of mesonotum; **E** – subimaginal mesonotum extracted from mature larva. **F** – Tasmanophlebia sp.V1 from Australia, apex of abdomen of mature male larva, ventral view. (D, E – from Kluge & al. 1995)

Abbreviations: **LPs** – lateroparapsidal suture; **LSs** – lateral scutal suture; **MNs** – mesonotal suture.

(2) Tergalii V-VII are lost, and other tergalii have true operculate structure [see Vetulata (1)] (FIG.33:A–C): tergalii I are much larger than others and completely cover them; bases of tergalii II–IV are shifted anteriorly; lateral margins of abdominal segment II are bent dorsally forming a basket for the gill-opercula. In contrast to Oniscigaster/fg2, anal-proximal lobe of each tergalius II-IV [see Vetulata (2)] is bent dorsally, with smooth margin (according to Tillyard 1933, in *lacuscoerulei* [*T.*] this lobe is developed on tergalius IV only).

(3) Imaginal and subimaginal paracercus is vestigial. Non-unique apomorphy (see Index of characters [2.3.22]).

Plesiomorphy of Tasmanophlebia/g1. In contrast to some Oniscigaster/fg2, imaginal and subimaginal abdomen has no lateral projections.

Size. Fore wing length 10–14 mm.

Distribution. Australia and Chile.

Nominal taxa included. Tasmanophlebia/g1 includes: (1) ***Tasmanophlebiodes*** Lestage 1935b: 351 – unavailable name, as type species is not designated (Hubbard 1979b: 7), regarded as a generic synonym of *Tasmanophlebia* (Riek 1955: 268); (2) **Siphlonella/g** [g: *Siphlonella* Needham & Murphy 1924: 30, typus *S. ventilans* Needham & Murphy 1924 (monotypy)], has no separate diagnosis.

Species composition of Tasmanophlebia/g1. 4 species. **Australian species:** *lacustris* Tillyard 1921 [*Tasmanophlebia*], *lacuscoerulei* Tillyard 1933 [*Tasmanophlebia*], *nigrescens* Tillyard 1933 [*Tasmanophlebia*]. **South American species:** *ventilans* Needham & Murphy 1924 [*Siphlonella*] (syn. subj.: *fazi* Navás 1931 [*Siphlonurus*]).

Material examined: *ventilans* [*S.*]: ♀/♀; sp.V1 (Australia): ♂/♂, ♀/♀.

*** ***

2.1;10. Ameletopsis/fg1

(Euephemeroptera Euplectoptera Anteritorna Tridentiseta Ameletopsis/fg1)

(Figs 34–35)

Nomen hierarchicum: **Ameletopsis/fg1** (incl. *Mirawara*, *Chiloporter*) [f: Ameletopsinae Edmunds 1957: 246; g: *Ameletopsis* Phillips 1930: 324, typus *Ameletus perscitus* Eaton 1899 (monotypy)].

In circumscription matches:

— subfam. Ameletopsinae Edmunds 1957: 246;

— fam. Ameletopsidae: Kluge & Studemann & Landolt & Gonser 1995: 120.

References. Edmunds & Allen & Peters 1963: Ꝺ*; – Koss & Edmunds 1974: ⊙*; – Edmunds 1975: Ꝺ; – Kluge & Studemann & Landolt & Gonser 1995: Ꝺ* ◎* ⊕*.

Autapomorphies of Ameletopsis/fg1.

(1) Larval head is greatly enlarged, wide, mouth apparatus is highly modified and specialized for carnivorism (FIG.35). Labrum is strongly widened, usually semicircular (only in *Chiloporter* with large median incision – see below and FIG.35:B). Mandibles are highly modified: primary asymmetry of mandibles is lost; incisor, kinetodontium and mola are elongate, mola lacks grater; prostheca is lost. Superlinguae are strongly diminished (the same in some other carnivorous mayflies – see Index of characters [1.1.27]). Maxilla is highly modified in a unique manner (FIG.34:D–E): instead of canines, dentisetae and other setae, its apex bears only 5 long stout mobile articulated appendages; these appendages are not setae, and appendages of next instar develop inside them (FIG.34:D); being pressed together, these appendages form a kind of claw which is used to grasp a prey; bases of these appendages semicircularly surround apex of sclerite, which is situated on inner margin of maxilla and to which a strongly thickened lacinia-stipital muscle is attached. Lateral sides of submentum bear a pair of lamellate lateral lobes projecting anteriorly and posteriorly. Maxillary and labial palps have unique structure: each palp retains a normal 1st segment with muscles moving the next segment; the rest part of palp (initial 2nd+3rd segment) represents a multisegmented flagellum similar to flagellum of antenna: it is narrowed toward apex, with large indeterminate number of segments increasing from age to age by division of proximalmost segments.

Structure of maxillae and mandibles are uniform in all species of Ameletopsis/fg1, while other parts have species-specific differences.

Characters of unclear phylogenetic status: non-unique apomorphies.

(2) Imaginal and subimaginal furcasternal protuberances are not contiguous at least in posterior part, with furcasternal median impression widened posteriorly; metathoracic nerve ganglion is situated in posterior part of furcasternum (FIG.34:C). Non-unique apomorphy (see Index of characters [2.2.23]); among Tridentiseta the same in other Amphinotic groups – Nesameletus/f2=Metamonius/g2, Vetulata and *Rallidens*.

(3) Subimaginal lateral pigmented sclerotized area of mesonotum has peculiar form, includes whole anterior part of lateroscutum (FIG.34:A). The same in Nesameletus/f2=Metamonius/g2.

Characters of unclear phylogenetic status: present not on all representatives.

(4) Each tergalius I–VII can have an additional

lobe arising from ventral side near base and bearing numerous branched marginal processes (TABLE 3). The same fibrillose lobe is present in *Rallidens* and Branchitergaliae – see Index of characters [1.3.25]). While costal rib is always well-developed, anal rib can be either lost, or located at anal margin, or near middle of tergalius (see TABLE 3 and Index of characters [1.3.28]).

(5) Larval caudalii, besides primary swimming setae (see Plesiomorphies) often have secondary swimming setae on lateral margins of cerci; only *Mirawara* has no them (see TABLE 3 and Index of characters [1.3.67]).

Character of unclear polarity.

(6) Larval tarsus can have 3 or 2 movably articulated segments (FIG.35:A) (adult has normal tarsal structure, with 5 segments, among which 1st one is fused with tibia). In all other Recent mayflies larval tarsus in non-segmented or with indistinct grooves corresponding to joints separating all five segments of adult tarsus. Segmentation of larval tarsus of Ameletopsis/fg1 can be either a unique plesiomorphy among Euplectoptera, or a result of secondary restoration, when adult character appears in larva.

Plesiomorphies of Ameletopsis/fg1. Larval (and adult) patella-tibial suture is developed on middle and hind legs only (FIG.35:A). Larval claws without denticles on inner margin (if not take into account proximal denticle in *Chiloporter*). Larval caudalii are not long, paracercus is equal to cerci, primary swimming setae are always dense [see (5)].

In imago and subimago: Mesonotal suture is somewhat stretched backward medially or transverse (FIG.34:A, F). Anterior paracoxal suture is complete (FIG.34:B–C) (in contrast to *Rallidens* and some Bidentiseta). In cubital field of fore wing several (4–8) veins go from CuA to basitornal margin [see Anteritorna (1)]; hind wing is well-developed, as long as 0.4–0.5 of fore wing length (as in FIG.7:C–D; Demoulin 1955c:Figs 2a, 3a,f). Imaginal and subimaginal tarsi are 5-segmented, 1st segment is non-shortened and fused with tibia. All claws of imago and subimago are ephemeropteroid.

Size. Fore wing length 13–20 mm.

Age and distribution. Probably Early Cretaceous (see *Promirara*) — Recent; Notogea: Australia, New Zealand and Chile-Patagonian Region of South America.

Ameletopsis/fg1 is divided into Recent taxa Ameletopsis/fg2, *Mirawara*, *Chiloporter* and *Chaquihua*; Mesozoic taxon *Promirara* is placed here presumably (TABLE 3).

Table 3. Characters of subordinate taxa in Ameletopsis/fg1

Taxa	Geographical distribution	Species number	Characters								
			Laval							Imaginal	
			1	2	3	4	5	6	7	8	9
Ameletopsis/fg2	New Zealand	1	–	–	–	–	–	+	+	2	++
Mirawara	Australia	3	–	–	+	–	M	+	–	3	+
Chiloporter	Chile	2	+	+	+	–	–	–	+	2	+++
Chaquihua	Chile	1	–	–	–	–	P	+	+	2	++
Promirara	extinct (Australia)	1	?	?	+	+	?	+	+	?	?

Characters:
1 – labrum with deep median incision, glossae thickened;
2 – claw with proximal denticle;
3 – tergalius with fibrillose lobe;
4 – tergalius with additional subcostal rib;
5 – tergalius with anal rib: "M" – medially, "P" – on posterior margin;
6 – posterolateral spines on abdominal segment IX;
7 – larval cerci with secondary swimming setae;
8 – number of distal segments of gonostylus;
9 – imaginal and subimaginal paracercus: "+++" – long; "++" – small; "+" – very small.

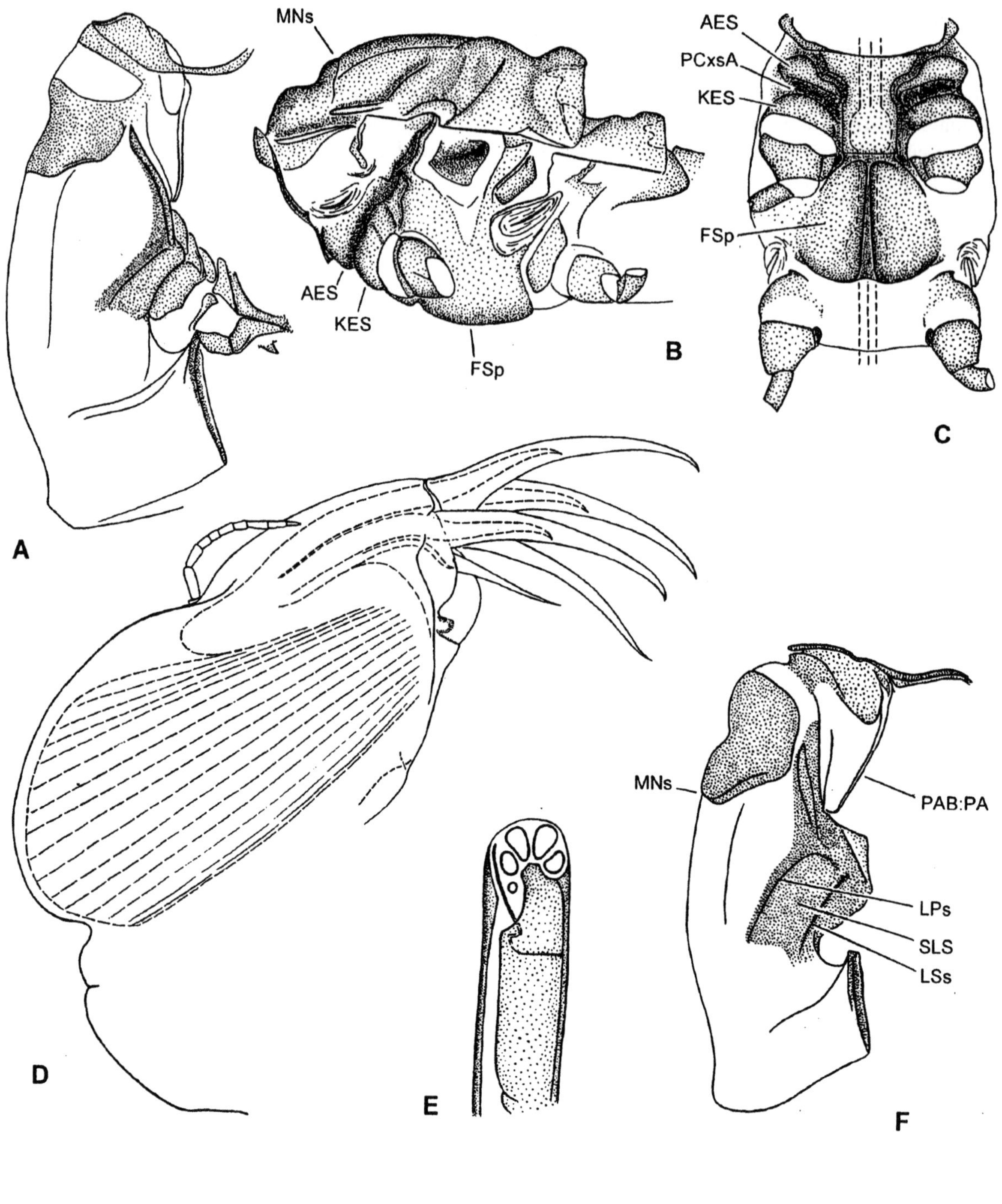

Figure 34. Ameletopsis/fg1.
A–C – Ameletopsis/fg* *perscitus* [*Ameletus*]: **A** – subimaginal exuviae of right half of mesonotum; **B** – imaginal pterothorax, lateral view; **C** – the same, ventral view (nerve ganglions shown by interrupted lines). **D–F** – Chiloporter/fg1 *penai* [*Chiloporter*]: **D** – left maxilla, dorsal view (lacinia-stipital muscle and integument of next instar shown by interrupted lines); **E** – the same, median-apical view, five movable appendages removed; **F** – subimaginal exuviae of right half of mesonotum. (From Kluge & al. 1995)

Abbreviations: **AES** – anepisternum; **FSp** – furcasternal protuberance; **KES** – katepisternum; **LPs** – lateroparapsidal suture; **LSs** – lateral scutal suture; **MNs** – mesonotal suture; **PAB:PA** – posterior arc of prealar bridge; **PCxsA** – anterior paracoxal suture; **SLS** – sublateroscutum.

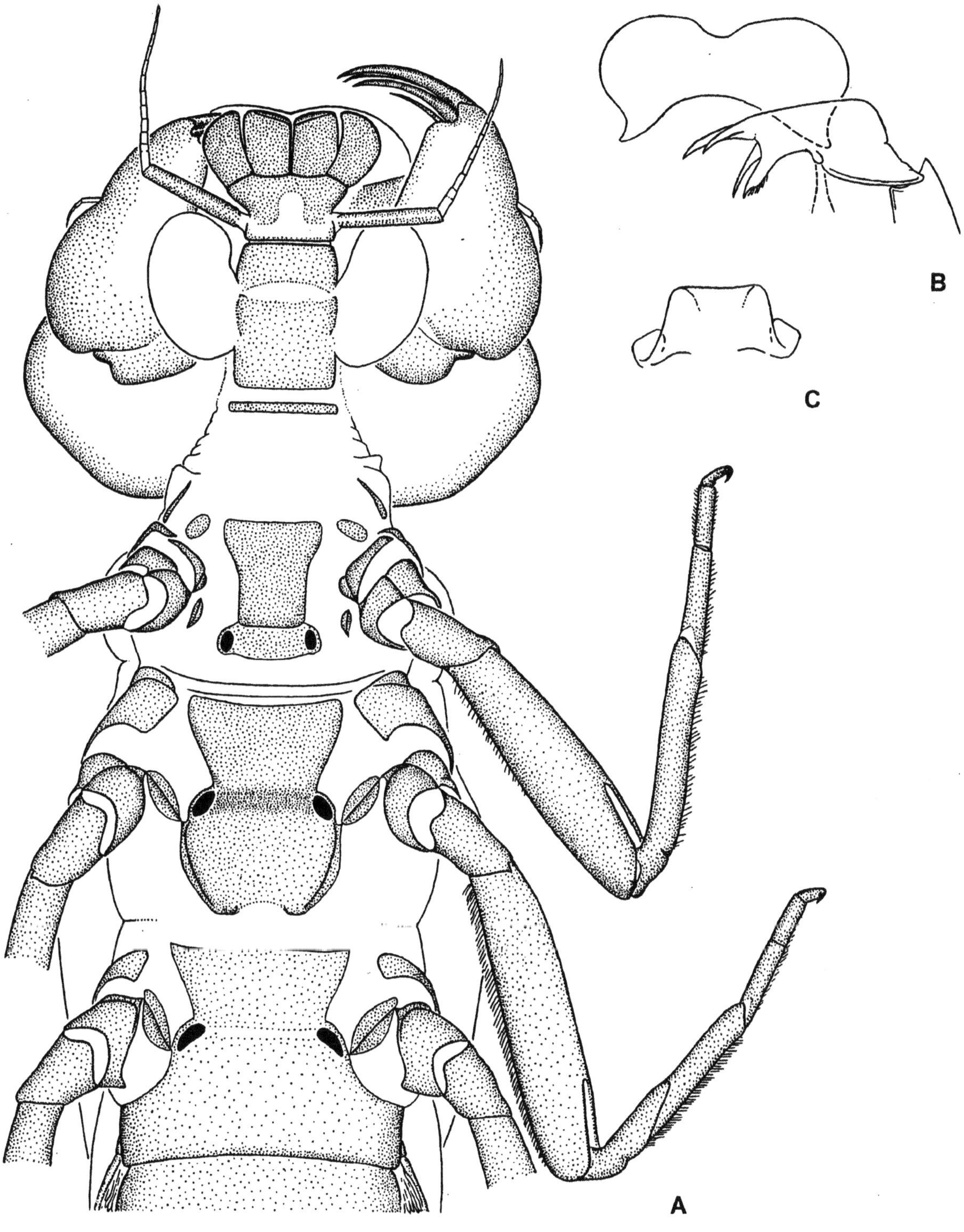

Figure 35. Ameletopsis/fg1-Chiloporter/fg1 *penai* [*Chiloporter*], larva.
A – head and thorax of mature larva, ventral view (furcal pits shown by black); **B** – labrum and left mandible, ventral view (same scale as in Figs A and C); **C** – hypopharynx with superlinguae (same scale as in Figs A and B).

2.1;10,1. **Ameletopsis/fg2**

(Euplectoptera Anteritorna Tridentiseta Ameletopsis/fg1 .../fg2)
(Fig. 34:A–C)

Nomen hierarchicum: **Ameletopsis/fg2** (sine *Mirawara*, *Chiloporter*).
In circumscription matches:
— gen. *Ameletopsis* Phillips 1930: 324.
Monospecific taxon.

References. Phillips 1930: ⟲ ⊕; – 1931: ⟲; – Lestage 1935a: ⟲ ⊕; – Edmunds 1975: ⟲.

Characters of Ameletopsis/fg2 of unclear phylogenetic status (TABLE 3).

(1) Each tergalius without fibrillose lobe [see Ameletopsis/fg1 (4)] and without anal rib.

(2) Larval cerci with secondary swimming setae [in contrast to *Mirawara* – see Ameletopsis/fg1 (5)].

(3) Imaginal and subimaginal paracercus is diminished, several times shorter than cerci (in contrast to *Chiloporter*).

Plesiomorphies of Ameletopsis/fg2. In contrast to *Chiloporter*: Labrum [widened – see Ameletopsis/fg1 (1)] is semicircular, without median incision; glossae are flat; body form is close to siphlonuroid [if not take into account strongly enlarged specialized head – see Ameletopsis/fg1 (1)]; femora are thin, claws are slightly curved, without denticles; last abdominal segments with posterolateral spines. In contrast to *Mirawara*, gonostylus with 2 distal segments.

Size. Fore wing length 17–18 mm.

Distribution. New Zealand.

Species composition of Ameletopsis/fg2. 1 species – *perscitus* Eaton 1899 [*Ameletus*].

Material examined: *perscitus* [*A.*]: ⟲, ♀/♀, ⊕.

***

2.1;10,2. **Mirawara/g(1)**

(Euplectoptera Anteritorna Tridentiseta Ameletopsis/fg1 *Mirawara*)

Nomen hierarchicum: **Mirawara/g(1)** [g: *Mirawara* Harker 1954: 261, typus *M. aapta* Harker 1954 (design. orig.)].
In circumscription matches:
— gen. *Mirawara* Harker 1954: 261.

References. Harker 1954: ⊕; – Riek 1955: ⟲ ⊕; – Edmunds 1975: ⟲; – Peters & Campbell 1991: ⟲.

Autapomorphy of *Mirawara*.

(1) Gonostylus with 3 distal segments (instead of 2 initial ones). Non-unique apomorphy (see Index of characters [2.3.12]).

Characters of *Mirawara* of unclear phylogenetic status (TABLE 3).

(2) Each tergalius I–VII has fibrillose lobe – additional ventral lamella with long marginal processes [see Ameletopsis/fg1 (4)]; anal rib is located near middle of tergalius (Riek 1973:Fig.3; Peters & Campbell 1991: Fig.16:14C) (in contrast to other Ameletopsis/fg1).

(3) Larval cerci without secondary swimming setae, with primary swimming setae only [in contrast to other Ameletopsis/fg1 – see Ameletopsis/fg1 (5)].

(4) Imaginal and subimaginal paracercus is very small (in contrast to *Chiloporter*).

Plesiomorphies of *Mirawara*. In contrast to *Chiloporter*: labrum [widened – see Ameletopsis/fg1 (1)] is semicircular, without median incision; glossae are flat; body form is close to siphlonuroid [if not take into account strongly enlarged specialized head – see Ameletopsis/fg1 (1)]; femora are thin, claws are slightly curved, without denticles; last abdominal segments have posterolateral spines.

Size. Fore wing length 18–20 mm.

Distribution. Australia.

Species composition of Mirawara/g(1). 3 species are described – *aapta* Harker 1954 [*Mirawara*], *megalopteria* Riek 1955 [*Mirawara*], *purpurea* Riek 1955 [*Mirawara*].

Material examined: –.

***

2.1;10,3. **Chiloporter/fg(1)**

(Euplectoptera Anteritorna Tridentiseta Ameletopsis/fg1 *Chiloporter*)
(Figs 34:D–F; 35)

Nomen hierarchicum: **Chiloporter/fg(1)** [f: Chiloporteridae Landa 1973: 156; g: *Chiloporter* Lestage 1931c: 50, typus *Ch. eatoni* Lestage 1931 (design. orig.)].
In circumscription matches:
— gen. *Chiloporter* Lestage 1931c: 50;
— fam. Chiloporteridae Landa 1973: 156.

References. Eaton 1883–1888 ("nameless", Pl.53): ⟲; – Lestage 1935a: ⟲; – Demoulin 1955c: ⟲ ⊕; – Edmunds 1975: ⟲.

Autapomorphies of *Chiloporter*.

(1) Labrum [widened – see Ameletopsis/fg1 (1)] with large median incision (FIG.35:B); glossae are thickened and truncate in such a manner, that their apices are inserted into the incision of labrum (FIG.35:A).

(2) Larva is rheophilous: body is somewhat flattened dorsoventrally; head is prognathous; femora are somewhat widened; claws are strongly curved, with proximal denticle on inner margin (FIG.35:A).

(3) Larval abdominal posterolateral spines are diminished, on segment IX completely lost.

Characters of *Chiloporter* of unclear phylogenetic status (TABLE 3).

(4) Each tergalius I–VII has fibrillose lobe – additional ventral lamella with long marginal processes [see Ameletopsis/fg1 (4)]; main lobe is enlarged, round, without anal rib (Demoulin 1955c: Figs 2f–i).

(5) Larval cerci with secondary swimming setae (Demoulin 1955c:Fig.2k) [in contrast to *Mirawara* – see Ameletopsis/fg1 (5)].

Plesiomorphies of *Chiloporter*. Gonostylus with 2 distal segments (in contrast to *Mirawara*). Imaginal and subimaginal paracercus is subequal to cerci (in contrast to other Ameletopsis/fg1).

Size. Fore wing length 18–19 mm.

Distribution. Chile.

Species composition of Chiloporter/fg(1). Two species are formally described – *eatoni* Lestage 1931 [*Chiloporter*] and *penai* Demoulin 1955 [*Chiloporter*].

Material examined: *penai* [*Ch.*]: Ծ, Ծ/◎, ♀, ⊕/⊙.

***

2.1;10,4. Chaquihua/g(1)

(Euplectoptera Anteritorna Tridentiseta Ameletopsis/fg1 *Chaquihua*)

Nomen hierarchicum: **Chaquihua/g(1)** [g: *Chaquihua* Demoulin 1955c: 11, typus *Ch. penai* Demoulin 1955 (design. orig.)].

References. Demoulin 1955c: ⊕ ♂; – Edmunds 1957; – 1975: Ծ (presumably associated).

Characters of *Chaquihua* of unclear phylogenetic status (TABLE 3).

(1) Each tergalius I–VII has no fibrillose lobe [see Ameletopsis/fg1 (4)]; narrow, with a short anal rib at proximal part of anal margin (in contrast to other Ameletopsis/fg1).

(2) Larval cerci with secondary swimming setae [in contrast to *Mirawara* – see Ameletopsis/fg1 (5)].

(3) Imaginal and subimaginal paracercus is diminished, several times shorter than cerci (in contrast to *Chiloporter*).

Plesiomorphies of *Chaquihua*. In contrast to *Chiloporter*: labrum [widened – see Ameletopsis/fg1 (1)] is nearly semicircular, with shallow median incision; glossae are flat; body form is close to siphlonuroid [if not take into account strongly enlarged specialized head – see Ameletopsis/fg1 (1)]; femora are thin, claws are slightly curved; last abdominal segments have posterolateral spines. In contrast to *Mirawara*, gonostylus with 2 distal segments.

Size. Fore wing length 13–18 mm.

Distribution. Chile.

Species composition of Chaquihua/g(1). 1 species – *penai* Demoulin 1955 [*Chaquihua*].

Material examined: *penai* [*Chaquihua*]: Ծ (presumably associated), ♀, ⊕♂.

***

2.1;10,5. † Promirara/g(1)

(Euplectoptera Anteritorna Tridentiseta Ameletopsis/fg1 *Promirara*)

Nomen hierarchicum: **Promirara/g(1)** [g: *Promirara* Jell & Duncan 1986: 116, typus *P. cephalota* Jell & Duncan 1986 (design. orig.)].
In circumscription matches:
— gen. *Promirara* Jell & Duncan 1986: 116.
Monospecific taxon.

Reference. Jell & Duncan 1986: Ծ.

Autapomorphies of *Promirara*.

(1) Each tergalius I–VII besides costal rib on costal margin [see Ameletopsis/fg1 (4)], has an additional strong curved **subcostal rib** at some distance from costal margin (unique apomorphy). Tergalii are strongly enlarged, tergalius I is "subelytroid" (Jell & Duncan 1986: 120), tergalius VII is the largest.

Characters of *Promirara* of unclear phylogenetic status (TABLE 3).

(2) Tergalii have "basal fibrilliform tuft" (Jell & Duncan 1986: 120) [ventral fibrillose lobe – see Ameletopsis/fg1 (4)]; the same in *Chiloporter* and *Mirawara*.

(3) Larval cerci with secondary swimming setae (Jell & Duncan 1986:Fig.2G) [in contrast to *Mirawara* – see Ameletopsis/fg1 (5)].

Plesiomorphy of *Promirara*. In contrast to *Chiloporter*, abdomen has posterolateral spines, especially large on segment IX.

Size. Larval length more than 30 mm.

Age. Early Cretaceous (Australia).

Species composition of Promirara/g(1). 1 species – *cephalota* Jell & Duncan 1986 [*Promirara*].

Material examined: –.

Comment. This taxon is attributed to Ameletopsis/fg1 presumably, basing on large larval head, conclusion by Jell & Duncan (1986) and their statement that "maxilla with long styliform terminal processes" (while these processes are not shown on figures and photographs in the original description).

There is also said that "the terminal spines on the

maxilla are shorter and the terminal tooth on the mandible is shorter than in *Mirawara*" (other Ameletopsis/fg1 were not discussed); if this is true, *Promirara* can be opposed to all other Ameletopsis/fg1 which have uniform structure of maxillae and mandibles [see Ameletopsis/fg1 (1)].

*** ***

2.1;11. **Rallidens/fg(1)**

(Euephemeroptera Euplectoptera Anteritorna Tridentiseta *Rallidens*)

(Fig. 36)

Nomen hierarchicum: **Rallidens/fg(1)** [f: Rallidentinae Penniket 1966: 169; g: *Rallidens* Penniket 1966: 164, typus *R. mcfarlanei* Penniket 1966 (design. orig.)].
In circumscription matches:
— gen. *Rallidens* Penniket 1966: 164;
— subfam. Rallidentinae Penniket 1966: 169;
— fam. Rallidentidae: Landa 1973: 156.

References. Penniket 1966: ⥀ ⊕; – Edmunds 1975: ⥀; – Koss & Edmunds 1974: ⊙; – Kluge & Studemann & Landolt & Gonser 1995: ⥀ ⊕ ⊙.

Autapomorphies of *Rallidens*.

(1) Mandible is elongate perpendicular to its axis of articulation; incisor on outer side with longitudinal plate; this plate bears a regular row of large transverse pectinate projections, forming a scraper or filter of unique structure (Penniket 1966:Figs 11–15).

(2) Maxilla [with 3 dentisetae – see Anteritorna (2)] is strongly widened and has unique structure (FIG.36:A); its initial median margin projects in such a manner that 2nd and 3rd (proximal) dentisetae are located on this projection, and 1st (distal) dentiseta is located on a straight margin directed apically; all 3 canines are flat, with unusually wide bases [Penniket (1966) erroneously wrote that canines are "movable" and "arising from the same point"]; initial apical margin is directed laterally, initial apical setae are lost .

(3) On hind wing bifurcation of MP [third triad – see Euplectoptera (1)] is nearer to wing margin than to wing base, more distal than bifurcations of RS and MA (as in FIG.40:B; Penniket 1966:Fig.1). Non-unique apomorphy (see Index of characters [2.2.69]). In other respects hind wing is well developed, as long as about 0.475 of fore wing length.

Characters of unclear phylogenetic status.

(4) Maxilla [see (1)] with a small finger-like gill located laterally at articulation of cardo and stipes (FIG.36:A). The same in Nesameletus/f1=Metamonius/g1; possibly, synapomorphy.

(5) Tergalii have peculiar structure. Each tergalius I–VII has fibrillose lobe arising from ventral side near base (FIG.36:B) (the same in Branchitergaliae and selected Ameletopsis/fg1 – see Index of characters [1.3.25]). On each tergalius III–VII anal rib bifurcates near base, thus there are 3 ribs: costal rib is situated on costal margin and in its distal part bears denticles; anterior branch of anal rib is situated on dorsal surface near middle of tergalius just behind main trachea (like anal rib of Siphlonurus/ /fg1, *Dipteromimus* and some Ameletus/fg1); posterior branch of anal rib borders proximal part of anal margin behind base of the fibrillose lobe (such 3 ribs are present also in Nesameletus/f2=Metamonius/g2 and Isonychia/fg1). Tergalii I–II have no denticles, ribs of tergalius II are vestigial, tergalius I lacks ribs.

(6) Imaginal and subimaginal anterior paracoxal suture of mesothorax is incomplete – i. e. not continued on ventral side of episternum and terminates not reaching sternum (FIG.36:C–D). In all other Tridentiseta anterior paracoxal suture is complete. Probably, this is an autapomorphy of *Rallidens*; besides *Rallidens*, the anterior paracoxal suture is incomplete in Heptagennota and Furcatergaliae.

(7) Imaginal and subimaginal furcasternal protuberances are not contiguous, furcasternal median impression is parallel-sided, not widened posteriorly; metathoracic nerve ganglion is situated in posterior part of furcasternum (FIG.36:C). Non-unique apomorphy (see Index of characters [2.2.23]); among Tridentiseta the same in other Amphinotic groups – Nesameletus/f2=Metamonius/g2, Vetulata and Ameletopsis/fg1.

(8) Larval paraproct with a spine on inner-apical margin. The same in Siphlonurus/fg2, Metretopus/ /fg1, *Siphluriscus* and Vetulata (FIG.33:F).

(9) Imaginal and subimaginal paracercus is strongly diminished. Non-unique apomorphy (see Index of characters [2.3.22]).

Plesiomorphies of *Rallidens*. Larva has primary swimming siphlonuroid specialization: legs are able to stretch posteriorly; abdomen is large, able to make undulate swimming movements; caudalii are not long, paracercus is well-developed, primary swimming setae are dense, secondary swimming setae are absent. Larval head is hypognathous. Maxillary palp is 3-segmented (FIG.36:A). Labial palp is 3-segmented. Larval (and adult) patella-tibial suture is developed on middle and hind legs only. Larval claws are slightly curved, without denticles.

In imago and subimago: Mesonotal suture is nearly transverse (FIG.36:D). In cubital field of fore

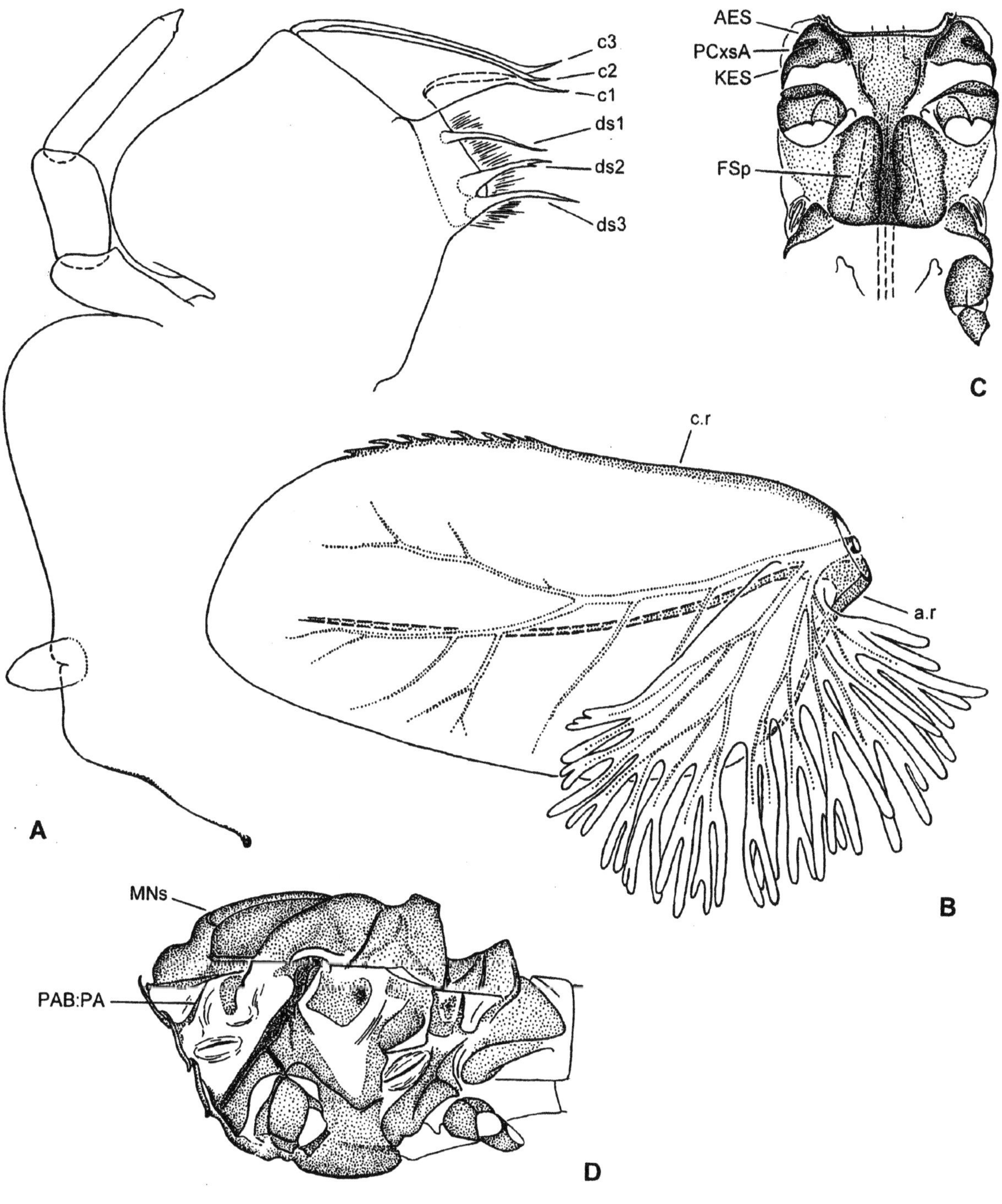

Figure 36. Rallidens/fg* *mcfarlanei* [*Rallidens*].
A – left maxilla, dorsal view; **B** – right tergalius III, ventral view; **C** – pterothorax of female imago, ventral view (nerve ganglions and inner margins of bases of subalar-sternal muscles shown by interrupted lines); **D** – the same, lateral view. (C, D – from Kluge & al., 1995)

Abbreviations: **AES** – anepisternum; **a.r** – anal rib; **c1, c2, c3** – maxillary canines; **c.r** – costal rib **ds1, ds2, ds3** – dentisetae; **FSp** – furcasternal protuberance; **KES** – katepisternum; **MNs** – mesonotal suture; **PAB:PA** – posterior arc of prealar bridge; **PCxsA** – anterior paracoxal suture;

wing several (5–6) veins go from CuA to basitornal margin [see Anteritorna (1)] (as in FIG.7:C). Imaginal and subimaginal tarsi are 5-segmented, 1st segment is non-shortened and fused with tibia. All claws of imago and subimago are ephemeropteroid. Gonostylus with 2 distal segments.

Size. Fore wing length 10–15 mm.

Distribution. New Zealand.

Species composition of Rallidens/fg(1). 2 species – *mcfarlanei* Penniket 1966 [*Rallidens*] and sp.n. (Hitchings & Staniczek, in litt).

Material examined: *mcfarlanei* [*R.*]: Ծ, ⊕/⊙.

***

Chapter VI

Anteritorna Bidentiseta: general diagnosis and systematics of Branchitergaliae

Bidentiseta include majority of mayflies and represent two large branches – Branchitergaliae and Furcatergaliae:

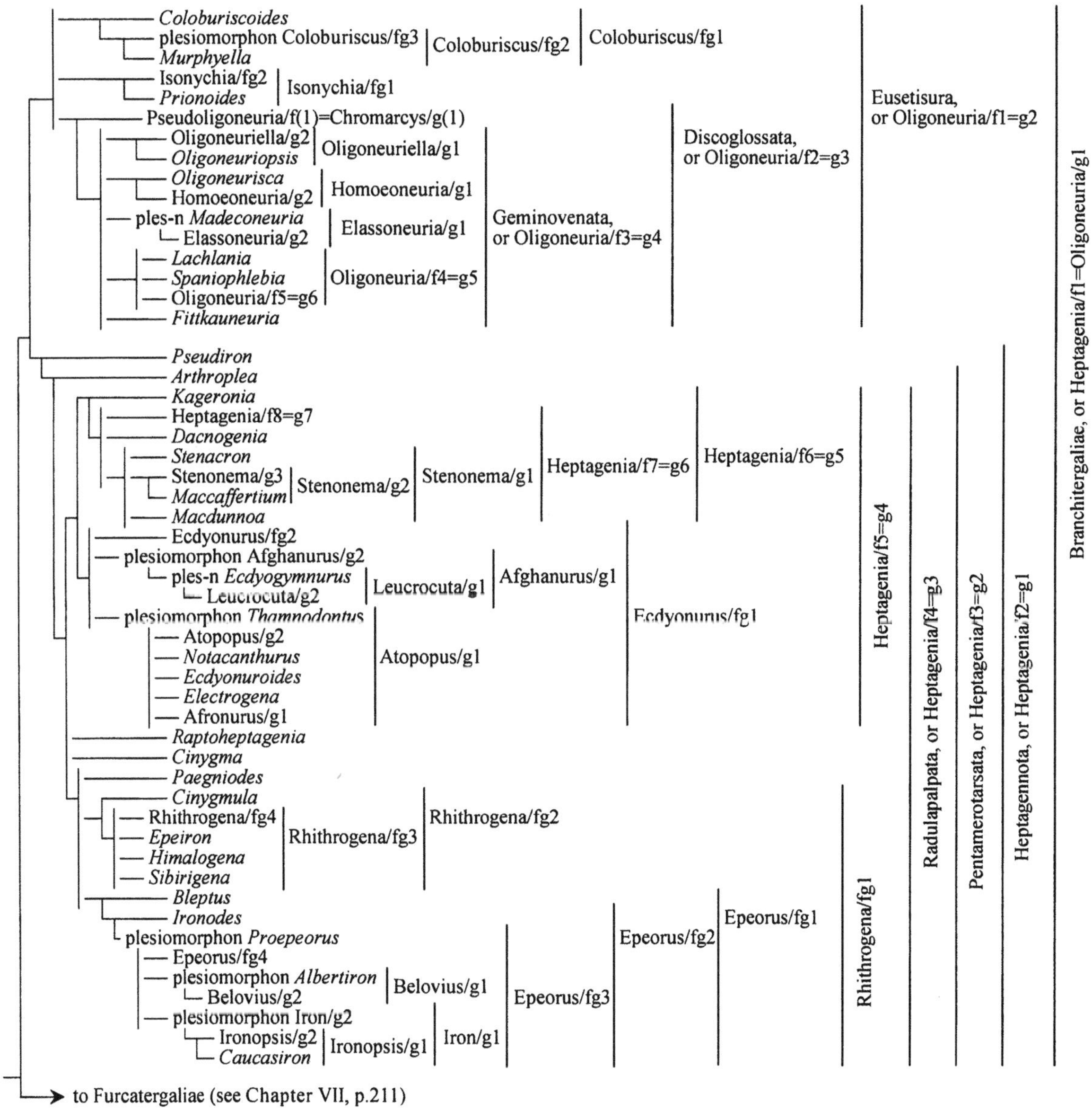

2.2. **Bidentiseta**, or **Ephemera/fg5**

(Panephemeroptera Euephemeroptera Euplectoptera Anteritorna Bidentiseta)

(Figs 8:E–F; 9:C–E; 37–106)

Nomen hierarchicum: **Ephemera/fg5** (sine *Baetis*; incl. *Oligoneuria*).
Nomen circumscribens: **Bidentiseta** Kluge 1993a: 41.
In circumscription matches:
— grex infraordines Bidentiseta: Kluge 1993a: 41;
— Bidentiseta, or Ephemera/fg5: Kluge 2000: 248.

References. Kluge 1993a: ⥁* – 1998: ⥁*.

Autapomorphy of Bidentiseta.

(1) Maxilla has 2 dentisetae only – distal and proximal ones [instead of initial 3 dentisetae – see Anteritorna (2)]. In some specialized representatives of Bidentiseta further reduction of dentisetae takes place, thus only one (proximal) dentiseta can be retained, or dentisetae disappear at all (see Index of characters [1.1.37]). Among Bidentiseta more than two dentisetae nearly newer occur; an exception is made by Coloburiscus/fg1, whose number of dentisetae is indeterminate and in specimens of the same species can be 2 or 3 (FIG.37:C).

Number of dentisetae less than three, besides Bidentiseta, is present also in some specialized taxa which we attribute to Tridentiseta: specialized filtering maxilla of Ameletus/fg1 bears a single vestigial dentiseta, and specialized for carnivorism maxilla of Acanthametropus/fg2 bears two canine-like dentisetae. Thus, a formal number of dentisetae does not allow to divide all Anteritorna into Bidentiseta and Tridentiseta.

The idea that Bidentiseta is a holophyletic taxon, is based on the fact that in each of evidently holophyletic taxa constituting the taxon Bidentiseta, the presence of two dentisetae is an initial condition: two dentisetae are present in many Eusetisura (FIG.44:D), Heptagennota (FIGS 54:M; 58:F–H; 61:H; 65:F), Fimbriatotergaliae (FIG.68:C), Ephemerella/fg1 (FIGS 92:B, F, H; 93:D; 95:C; 96:C) and primitive representatives of Leptophlebia/fg1 (FIG. 106:C, G). In many cases these maxillae with two dentisetae are non-specialized; hence, a conclusion can be made, that in common ancestor of Bidentiseta number of dentisetae was reduced to two, but maxillae remained to be non-specialized. In contrast to Bidentiseta, in Tridentiseta number of dentisetae is reduced only in the cases when maxilla is strongly specialized of as a whole.

Character of unclear phylogenetic status.

(2) Possibly, ventral fibrillose portion on tergalii of Branchitergaliae and a bilamellate structure of tergalii in Furcatergaliae have common origin [see Branchitergaliae (3) and Furcatergaliae (4) below].

Size. Fore wing length 2–40 mm.

Age and distribution. From Late Jurassic (see Eusetisura INCERTAE SEDIS and Fossoriae INCERTAE SEDIS) — till Recent; world-wide.

Bidentiseta are divided into Branchitergaliae and Furcatergaliae (Chapter VII: p.211).

2.2;1. **Branchitergaliae**, or **Heptagenia/f1=Oligoneuria/g1**

(Euephemeroptera Euplectoptera Anteritorna Bidentiseta Branchitergaliae)

(Figs 8:E–F; 9:C–E; 37–65)

Nomen hierarchicum: **Heptagenia/f1=Oligoneuria/g1** [f: Heptageniinae Needham (in Needham & Betten) 1901: 419; g: *Oligoneuria* Pictet 1843: 288, typus *O. anomala* Pictet 1843 (monotypy)].
Nomen circumscribens: **Branchitergaliae** Kluge 1998:256.
In circumscription matches:
— superfam. Heptagenioidea: McCafferty 1997: 96;
— infraordo Setisura: Kluge 1992d: 10; 1993a: 41 (non Setisura McCafferty 1991a);
— Branchitergaliae Kluge 1998: 256;
— Branchitergaliae, or Heptagenia/f1=Oligoneuria/g1: Kluge 2000: 248.

References. Kluge 1993a: ⊙* ⥁* ⊕*; – 1998: ⊙* ⥁* ⊕*.

Autapomorphies of Branchitergaliae.

(1) Maxilla on its ventral surface bears a longitudinal row of setae. This row is well-developed in Isonychia/fg1 (FIG.41:C), *Arthroplea* and Radulapalpata (FIGS 55:H; 61:D; 64:B, E, H, L; 65:F); among Radulapalpata only in Ecdyonurus/fg1 this row is transformed to a field of irregularly situated setae (FIG.58:E, H). In Discoglossata this setal row is transferred far laterally (Eaton 1883–1888:Pl.26:12), often being indistinct; in Coloburiscus/fg1 and *Pseudiron* this row is lost.

(2) Imaginal and subimaginal posterior arm of prealar bridge of mesothorax is strongly shortened and does not reach incision on ventral margin of prelateroscutum (FIGS 39:A; 40:F; 46:C; 54:J). Unique apomorphy. The same character was erroneously described for *Rallidens* (Kluge & al. 1995:128); in that paper Figure 50 has a mistake: abbreviation PA instead of the posterior arm is attributed to prelateroscutum (for correct lettering – see FIG.36:D).

(3) Tergalius [see Bidentiseta (2)] has an additional ventral (posterior) fibrillose gill lobe; anal rib is usually located on anal margin (rarely at a distance from it – see Index of characters [1.3.28]), and the fibrillose lobe arises from ventral side of tergalius close to its base between costal and anal ribs (FIGS 41:A; 44:B; 56:O; 57:C–E; 62:H–O; 65: C–E). Only in rare cases this fibrillose lobe is secondarily lost; anal rib can be also lost (see Index of characters [1.3.25] and [1.3.28]). Presence of the fibrillose lobe is probably a non-unique apomorphy: the same in *Rallidens* and certain Ameletopsis/fg1 (which are placed to Tridentiseta with some doubts). See also Furcatergaliae (4).

(4) Eggs have knob-terminated coiled threads (KCT of Koss & Edmunds 1973) – anchors, each consisting of a long cable formed by fused threads, and a small apical cap-like knob; cable is spirally coiled forming a small cylindrical papilla apically covered by the knob (FIG.43:G; Needham & Traver & Hsu 1935:Pl.15:9–18, Pl.17:42–45; Kopelke 1980: Figs 30–32; Studemann & al.1987:Figs 1–4; Belfiore & al. 1999:Figs 5–11). Such anchors, either scattered by egg surface or concentrated at certain areas, occur on eggs of various species of Pentamerotarsata, Isonychia/fg1 and Coloburiscus/fg1; Discoglossata have vestigial anchors (FIG.43:H) [see Discoglossata (11) below].

In other mayflies anchors, if present, have different structure (see Index of characters [3.4]): in Ameletus/fg1, Metretopus/fg1, Acanthametropus/ /fg1 and *Rallidens* each anchor is formed not by an integral cable, but by a bunch of threads covered by a knob, and is inserted into a chorionic crater (Kluge & al. 1995:Figs 65–75, 79–81); Siphlonurus/fg1 has bunches of threads without knobs (ibid.:Figs 61–63); in Potamanthus/fg1 and Ephemerella/fg1 the knob is surrounded by coiled thread rather than cover it (Bae & McCafferty 1991:Fig.79; Studemann & Landolt 1997a:Fig.29).

Plesiomorphies of Branchitergaliae. Mesonotal suture is never strongly curved posteriorly by sides from median suture, usually nearly transverse (FIGS 39, 40, 42, 52, 54, 56, 61, 63, 64:A); only in Geminovenata it is curved backward as a whole (FIG.46: A, C), and in Epeorus/fg1 is lost (FIG.64:D). Hind wings are well-developed, usually as long as 0.3–0.5 of fore wing length (only in *Bleptus* and selected specimens of *Paegniodes* smaller); bifurcation of MA is usually retained, but lost in *Arthroplea* and one species of *Cinygma*; bifurcation of MP is usually retained; only in Geminovenata venation is strongly modified.

Size. Fore wing length 5–27 mm.

Age and distribution. Late Jurassic (see Eusetisura INCERTAE SEDIS) — Recent; world-wide.

Branchitergaliae are divided into Eusetisura and Heptagennota (p.158).

2.2;1,1. **Eusetisura**, or **Oligoneuria/f1=g2**

(Euplectoptera Anteritorna Bidentiseta Branchitergaliae Eusetisura)

(Figs 37–51)

Nomen hierarchicum: **Oligoneuria/f1=g2** (sine *Heptagenia*; incl. *Coloburiscus*, *Isonychia*) [f: Oligoneuriidae Ulmer 1914: 97].
Nomen circumscribens: **Eusetisura** Kluge 1998: 256.
In circumscription matches:
— subtribus Oligoneuriina: Lameere 1917: 62;
— fam. Oligoneuriidae: Riek 1973: 164;
— superfam. Oligoneurioidea: Kluge & Studemann & Landolt & Gonser 1995: 105;
— Eusetisura Kluge 1998: 256;
— Eusetisura, or Oligoneuria/f1=g2: Kluge 2000: 251.

References. Lameere 1917: ♉* ⊕*; – Riek 1973: ♉*; – McCafferty & Edmunds 1979: ♉*.

Autapomorphies of Eusetisura.

(1) Larval fore legs are specialized as filtering, with femur and tibia modified (FIGS 38:B; 41:E; 44:A): On inner side of femur long filtering setae form 2 longitudinal rows which convergent proximally. Inner side of fore tibia (lacking patella-tibial suture – see Chapter II) is convex in its proximal part, thus tibia in this part is thickened or curved, in contrast to tibiae of middle and hind legs; on inner side of tibia long filtering setae also form 2 regular longitudinal rows convergenting proximally (FIG.38: B–D). Filtering setae of femur and tibia have such structure of their bases, which allow them to fold along the leg; these setae can bear thin seta-like processes, especially prominent in Oligoneuria/2=g3.

Similar setal rows on inner side of fore femur and tibia, besides Eusetisura, are present in Polymitarcys/f1=Ephoron/g2 only (FIG.77:C); in other mayflies, if fore tibia has two regular rows of filtering setae (for example, in Atalophlebia/fg1-Hagenulus/ /fg*), form of fore tibia does not differ from middle and hind ones, and femur has no two setal rows.

(2) Maxillary and labial palps are specialized for gathering food particles dispersed in water: each palp has 2nd+3rd segment massive, densely covered with long thin setae (non-unique apomorphy – see Index of characters [1.1.44] and [1.1.57]).

Degree of fusion of 2nd and 3rd segments varies: boundary between them is retained in Coloburiscus/fg1 (FIG.37:D), but lost in Isonychia/fg1 (FIG.41:B) and Discoglossata (FIG.45:B); muscle inside 2nd segment of labial palp is retained in

Coloburiscoides (FIG.37:D), Coloburiscus/fg3 and Isonychia/fg1 (FIG.41:B), but lost in *Murphyella* and Discoglossata.

(3) Base of each maxilla [see (4)] bears a tracheal gill in a form of soft process or a tuft of processes; this gill arises from articulatory membrane which connects posterior (or ventral) side of stipes and cardo with ventral side of head (FIGS 37:C; 41:D; Eaton 1883–1888:Pl.26:11). Unique apomorphy; in other mayflies similar gills evolved independently on variable parts of body, including maxillae, but not exactly on the same place as in Eusetisura (for example, in Nesameletus/f1=Metamonius/g1 and *Rallidens* a gill is present on lateral side of maxilla, in articulation of cardo with stipes – FIG.31:C). Besides the gills on maxillae, which are present in all Eusetisura, other gills can be present: on sides of mentum between paraglossae and bases of labial palps (in *Murphyella* and *Coloburiscoides* – FIG.37:D); on median side of joining of fore coxae with sternum (in *Murphyella* and Isonychia/fg1 – FIG.41:E); at the middle of prosternum, mesosternum and metasternum (in *Murphyella* only).

(4) Maxilla lacks regular apical-ventral row of setae (non-unique character – see Index of characters [1.1.31]); apical and lateral sides bear more or less dense long simple irregular setae (FIG.37:C).

(5) Larval fore wing buds are fused with scutellum up to apex of scutellum. Non-unique apomorphy (see Index of characters [1.2.6]); in Isonychia/fg1 wing buds are separated beginning from the apex of scutellum (FIG.40:C), while in other taxa they are variously fused one with another behind scutellum (FIGS 37:A; 44:A) [see below, Coloburiscus/fg1 (2) and Discoglossata (3)]. Mesal plate between fore wing buds can contain plumidia anlages: in Isonychia/fg1 such anlages occur at the beginning of last larval instar (FIG.40:C), but disappear at the end of this instar, and their space is filled by crumpled subimaginal wings; in Geminovenata plumidia anlages (FIG.44:A) develop to adult **plumidia** – a pair of membranous processes on posterior margin of scutellum (FIG.46:A, C). Non-unique apomorphy (see Index of characters [1.2.6] and [2.2.13]).

Characters of unclear phylogenetic status.

(6) Larval claws usually with one row of denticles on inner margin (only in Coloburiscus/fg2 and Homoeoneuria/g2 denticles are absent, and in *Oligoneurisca* denticles form two rows). Non-unique character (see Index of characters [1.2.21]).

(7) Tergalii lost ability of rhythmical respiratory movements (at least in the species examined). Non-unique apomorphy (see Index of characters [1.3.30]). Well-developed lamellate lobes of tergalii [bearing fibrillose branchial lobes – see Branchitergaliae (3)] are present in Isonychia/fg2 and some Geminovenata, while in Coloburiscus/fg1, *Chromarcys* and many Geminovenata the lamellate lobes are diminished, sclerotized and useless for respiration.

Plesiomorphies of Eusetisura. Larval caudalii nearly always retain well-developed primary swimming setae (these setae are lost in Coloburiscus/fg2 only) and never have secondary swimming setae.

In imago and subimago: Anterior paracoxal suture is complete (in contrast to Heptagennota and some others); furcasternal protuberances are contiguous (FIGS 39:A–B; 40:E–F; 46:C–D) (only in females of Geminovenata they are separated at posterior part). 1st tarsal segment is fused with tibia (FIG.42:G) (in contrast to Pentamerotarsata).

Size. Fore wing length 7–27 mm.

Age and distribution. Possibly, Late Jurassic (see Eusetisura INCERTAE SEDIS) — Recent; worldwide.

Eusetisura are divided into Coloburiscus/fg1, Isonychia/fg1 (p.134) and Discoglossata (p.136); some extinct Mesozoic taxa have uncertain systematic position (p.147).

2.2;1,1-1. Coloburiscus/fg1

(Anteritorna Bidentiseta Branchitergaliae Eusetisura Coloburiscus/fg1)

(Figs 37–39)

Nomen hierarchicum: **Coloburiscus/fg1** (incl. *Coloburiscoides*) [f: Coloburiscinae Edmunds (in Edmunds & Allen & Peters) 1963: 11); g: *Coloburiscus* Eaton 1888: 349, typus *Palingenia humeralis* Walker 1853 (design. orig.); syn. obj.: ***Coloburus*** Eaton 1868b: 89 (non *Coloburus* Dumeril 1851)].

In circumscription matches:

— subfam. Coloburiscinae: Edmunds & Allen & Peters 1963:11;

— fam. Coloburiscidae: Landa 1973: 156;

— Coloburiscus/fg1: Kluge 2000: 251.

References. Edmunds & Allen & Peters 1963: ♉*; – Koss & Edmunds 1974: ⊙*; – Edmunds 1975: ♉*.

Autapomorphies of Coloburiscus/fg1.

(1) Dorsal side of mandible with a field of long setae, proximally bordered by a regular arched row of the same setae (FIG.38:A). Unique apomorphy: in other mayflies, if regular rows of setae are present on mandible, they have other form.

(2) Larva has something like notal shield, as its

pronotum and mesonotum are immobile fused together all over their width, and fore wing buds are fused together (FIG.37:A): Fore wing buds [see Eusetisura (5)] convergent by their apices, in larva of last instar contiguous apically; their hind margins (except for extreme apices) are fused together by means of a triangular plate which fills an area between them and posterior margin of mesonotum. This fusion of fore wing buds differs from that of Discoglossata, where wing bud apices are divergent.

(3) Larval femur of middle leg [see (4)] bears the same 2 rows of long filtering setae as on fore leg [see Eusetisura (1)] (at the same time tibia of middle leg, as well as in other Eusetisura, has no specialization peculiar for fore legs) (FIG.38:C). Unique apomorphy.

(4) Larval femora of all legs [see (3)] bear stout spine-like setae which form transverse rows: at least a row on outer side of femur, and a more distal row on dorsal side near inner margin are present (FIG.38:B–D).

(5) On larval abdomen denticles on posterior margin of each tergum and sternum are fused forming an integral transverse sclerotized striated plate (FIG.37:B). This plate is integral beginning from tergum I or II to IX, and from sternum III or V to VIII; on sternum IX this plate is paired, being located by sides of genital buds. Selected segments of caudalii have apical tube-like striated formations of the same origin; these segments are separated one from another by 4, 8, 16 or larger number of usual segments.

(6) On each tergalius I–VII [see Eusetisura (7)] costal and anal ribs [marginal – see Branchitergaliae (3)] are unusually strongly thickened, sclerotized, covered with large stout spine-like setae; membranous part of tergalius is vestigial or lost, thus tergalius has a form of bipointed fork (Riek 1973:Fig.3) (only in *Murphyella* tergalii are completely lost).

(7) In male imago and subimago 1st segment of fore tarsus is strongly shortened, so that its length is subequal to width. Non-unique apomorphy (see Index of characters [2.2.76]).

Characters of unclear phylogenetic status.

(8) Maxilla (FIG.37:C) has indeterminate number of canines and dentisetae: canines have 3–4 apices (instead of three initial ones, in contrast to two in Isonychia/fg1 and one in Oligoneuria/f3=g4); number of dentisetae is 2 or 3, varying individually [in contrast to initial two ones – see Bidentiseta (1)].

(9) Imaginal and subimaginal paracercus is vestigial, multisegmented or 1-segmented. Non-unique apomorphy (see Index of characters [2.3.22]).

Plesiomorphies of Coloburiscus/fg1. Maxilla with nearly rectangular apical angle (in contrast to acute angle in Isonychia/fg1 and Discoglossata); 1st segment of maxillary palp is not shortened, only 2 times shorter than 2nd+3rd segment (like in Isonychia/fg1, in contrast to Discoglossata); maxillary and labial palps [see Eusetisura (2)] retain oblique suture separating 2nd and 3rd segment in such a way, that 2nd segment is shorter on inner side and longer on outer side (FIG.37:C–D). In cubital field of fore wing several (4–6) simple or branched veins go from CuA to basitornal margin [see Anteritorna (1)]. In contrast to Discoglossata, larval and adult middle and hind legs retain patella-tibial suture (FIG.38:B–D). All claws of imago and subimago are ephemeropteroid. Gonostylus has 2 distal segments.

Size. Fore wing length 12–20 mm.

Distribution. Notogea: Australia, New Zealand and Chile-Patagonian Region of South America.

Coloburiscus/fg1 is divided into *Coloburiscoides* and Coloburiscus/fg2.

2.2;1,1-1/1. Coloburiscoides/g(1)

(Bidentiseta Branchitergaliae Eusetisura Coloburiscus/fg1 *Coloburiscoides*)

(Figs 37; 38; 39:E)

Nomen hierarchicum: **Coloburiscoides/g(1)** [g: *Coloburiscoides* Lestage 1935b: 356, typus *Coloburiscus giganteus* Tillyard 1933].
In circumscription matches:
— gen. *Coloburiscoides* Lestage 1935b: 356.

References. Tillyard 1933 (*Coloburiscus*): ⊙ ⊕; – Lestage 1935b: ⊙ ⊕; – Riek 1955: ⊙ ⊕; – Peters & Campbell 1991: ⊙ ⊕.

Autapomorphy of *Coloburiscoides*.

(1) On hind wing bifurcation of MP [the third triad – see Euplectoptera (1)] is situated nearer to wing margin than to wing base, more distally than bifurcations of RS and MA (FIG.37:A). Non-unique apomorphy (see Index of characters [2.2.69]).

Plesiomorphies of *Coloburiscoides* (in contrast to Coloburiscus/fg2). Imaginal and subimaginal lateroparapsidal suture is not curved laterally; subimaginal pigmented area of sublateroscutum does not reach end of lateroparapsidal suture, and shape of lateral pigmented area of mesonotum has plesiomorphic condition (FIG.39:E) (see Index of characters [2.2.14]). All tergalii I–VII [see Coloburiscus/fg1 (6)] are present (in contrast to

Murphyella); at least each tergalius I–VI with fibrillose ventral lobe [see Branchitergaliae (3)] (in contrast to Coloburiscus/fg3); tergalius VII either also has fibrillose lobe, or not (in *giganteus* [*C.*]). Larval paracercus is subequal to cerci; primary swimming setae of caudalii are long.

Size. Fore wing length 15–20 mm.

Distribution. Australia.

Species composition of Coloburiscoides/g(1). 3 species – *haleuticus* Eaton 1871 [*Coloburus*], *giganteus* Tillyard 1933 [*Coloburiscus*], *munionga* Tillyard 1933 [*Coloburiscoides*].

Material examined: *giganteus* [*C.*]: Ɔ,♀,⊕♂; *munionga* [*C.*]: ☿/♀.

***

2.2;1,1-1/2. **Coloburiscus/fg2**

(Bidentiseta Branchitergaliae Eusetisura Coloburiscus/fg1 .../fg2)

(Fig. 39:A–D)

Nomen hierarchicum: **Coloburiscus/fg2** (sine *Coloburiscoides*; incl. *Murphyella*).

Autapomorphies of Coloburiscus/fg2.

(1) In imago and subimago posterior end of lateroparapsidal suture is strongly curved laterally; in subimago lateral pigmented area of mesonotum occupies most part of sublateroscutum and is bordered from behind by the curvation of lateroparapsidal suture (FIG.39:C–D).

(2) Swimming setae of larval caudalii are lost. Paracercus is shortened, either less than half of cerci length (in Coloburiscus/fg3), or vestigial (in *Murphyella*). Non-unique apomorphies (see Index of characters [1.3.64] and [1.3.66]).

Plesiomorphies of Coloburiscus/fg2. In contrast to *Coloburiscoides*, on hind wing fork of MP is more proximal than fork of MA.

Size. Fore wing length 12–16 mm.

Distribution. New Zealand and South America.

Coloburiscus/fg2 is divided into Coloburiscus/fg3 and *Murphyella*.

2.2;1,1-1/2.1. Plesiomorphon **Coloburiscus/fg3**

(Bidentiseta Branchitergaliae Eusetisura Coloburiscus/fg1 .../fg3)

(Fig. 39:A–C)

Nomen hierarchicum: **Coloburiscus/fg3** (sine *Murphyella*).
In circumscription matches:
— gen. *Coloburiscus*: Lestage 1935b: 357.

References. Eaton 1899: ⊕; – Phillips 1930: Ɔ ⊕; – 1931: Ɔ; – Lestage 1935a: Ɔ ⊕; – 1935b: Ɔ ⊕.

Plesiomorphies of Coloburiscus/fg3 (in contrast to *Murphyella*). All tergalii I–VII [see Coloburiscus/ /fg1 (6)] are present (but in contrast to *Coloburiscoides*, fibrillose portions of all tergalii are completely lost). Paracercus is retained [being shortened – see Coloburiscus/fg2 (2)].

Size. Fore wing length 12–16 mm.

Distribution. New Zealand.

Species composition of Coloburiscus/fg3. 3 species – *humeralis* Walker 1853 [*Palingenia*], *remota* Walker 1853 [*Baetis*], *tonnoiri* Lestage 1935 [*Coloburiscus*].

Material examined: *humeralis* [*P.*]: Ɔ, ◎, ⊕.

***

2.2;1,1-1/2.2. **Murphyella/g(1)**

(Bidentiseta Branchitergaliae Eusetisura Coloburiscus/fg1 .../fg2 *Murphyella*)

(Fig. 39:D)

Nomen hierarchicum: **Murphyella/g(1)** [g: *Murphyella* Lestage 1930: 439, typus *M. needhami* Lestage 1930 (design. orig.); syn. subj.: ***Dictyosiphlon*** Lestage 1931c: 47, typus *Heptagenia molinai* Navás 1930 (synn. subj.: *needhami* [*M.*] = *molinai* [*D.*])].
In circumscription matches:
— gen. *Murphyella* Lestage 1930: 439;
— gen. *Dictyosiphlon* Lestage 1931c: 47.
Monospecific taxon.

References. Needham & Murphy 1924 ("*Metamonius*"): Ɔ ◎; – Lestage 1930: Ɔ; – 1935a: Ɔ; – Demoulin 1955b: ⊕♂; – 1955b: Ɔ ◎♂; – Edmunds 1975: Ɔ.

Autapomorphies of *Murphyella*.

(1) Tergalii of all pairs [see Coloburiscus/fg1 (6)] are lost. Unique apomorphy, not occurring in any other Euephemeroptera.

(2) Larval, subimaginal and imaginal paracercus is vestigial, 1-segmented [see Coloburiscus/fg2 (2)]. Non-unique apomorphy (see Index of characters [1.3.64]).

Size. Fore wing length 15 mm.

Distribution. Chile.

Species composition of Murphyella/g(1). 1 species – *needhami* Lestage 1930 [*Murphyella*] (syn. subj.: *molinai* Navás 1930 [*Heptagenia*]).

Material examined: *needhami* [*M.*]: Ɔ, ☿/♀/☉.

*** *** ***

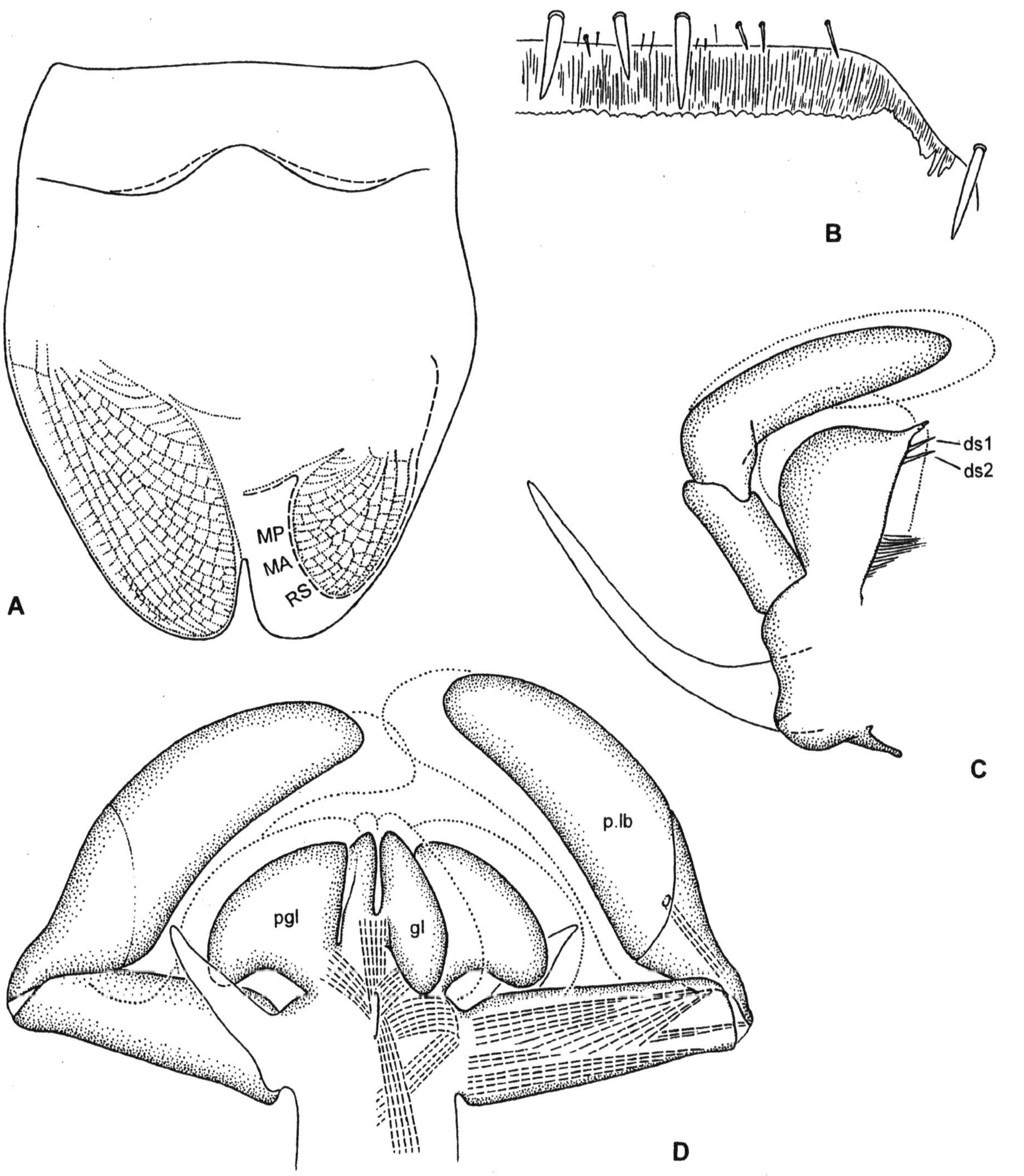

Figure 37. Coloburiscus/fg1-Coloburiscoides/g1 *munionga* [*Coloburiscoides*], larva.
A – thoracic terga of larva of beginning of last instar (larval metathorax shown by interrupted line at right half, hypodermal parts shown by dotted line); **B** – part of posterior margin of larval tergum VIII; **C** – left maxilla, dorsal view (setation not shown, areas occupied by dense irregular palpal and apical setae, and by two regular median rows of setae, shown by dotted lines); **D** – labium, dorsal view (in left half) and ventral view (in right half) (setation not shown, areas occupied by dense setae of glossae, paraglossae and palps shown by dotted lines; muscles of glossae, paraglossae and palp shown by interrupted lines) (compare with Figs 41:B and 44:B).

Abbreviations: **ds1**, **ds2** – dentisetae; **gl** – glossa; **pgl** – paraglossa; **p.lb** – 2nd+3rd segment of labial palp.

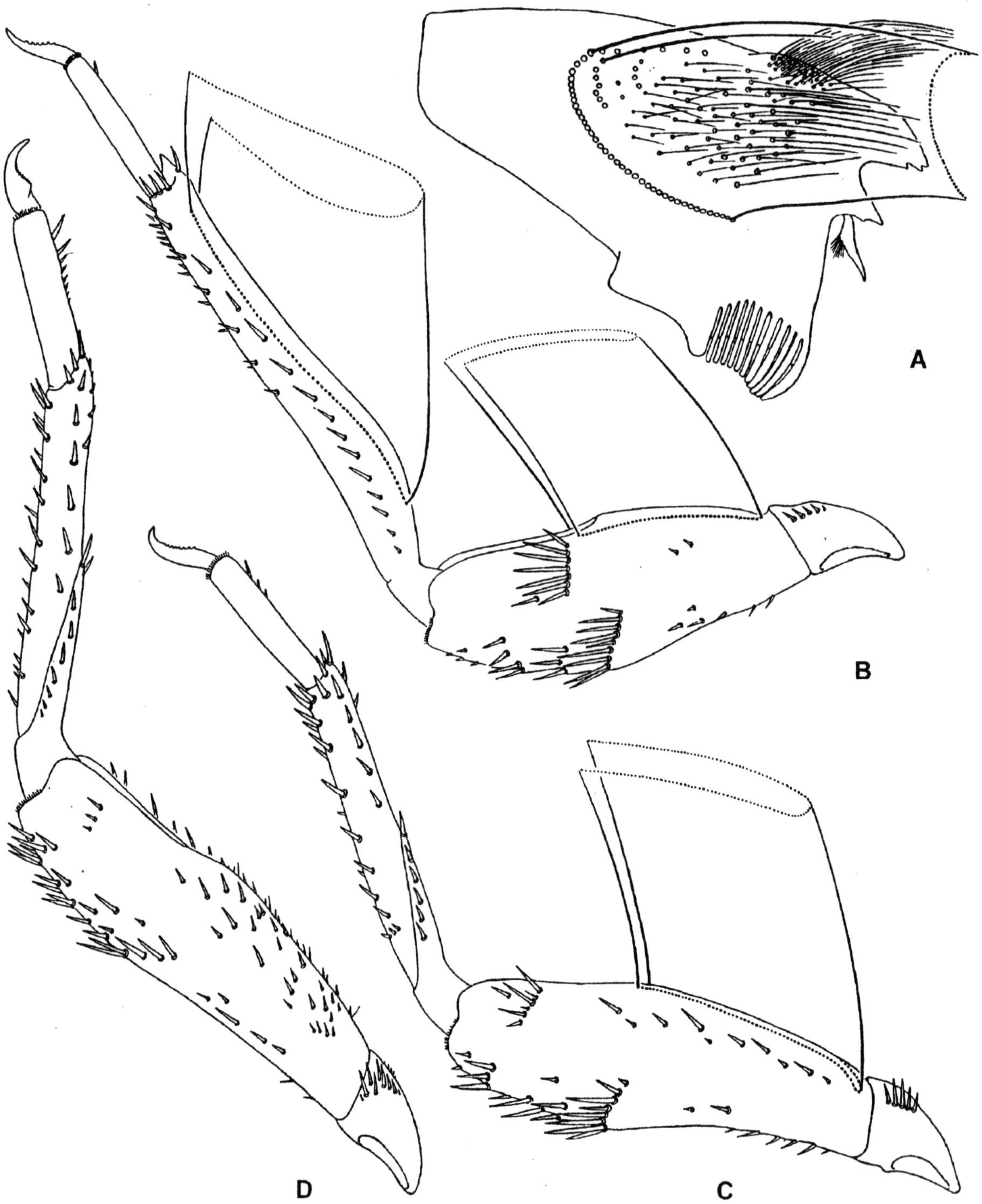

Figure 38. Coloburiscus/fg1-Coloburiscoides/g1 *munionga* [*Coloburiscoides*], larva.
A – left mandible; **B–D** – right fore, middle and hind legs, anterior (dorsal) view (coxae removed; instead of long filtering setae, their bases and occupied area are shown).

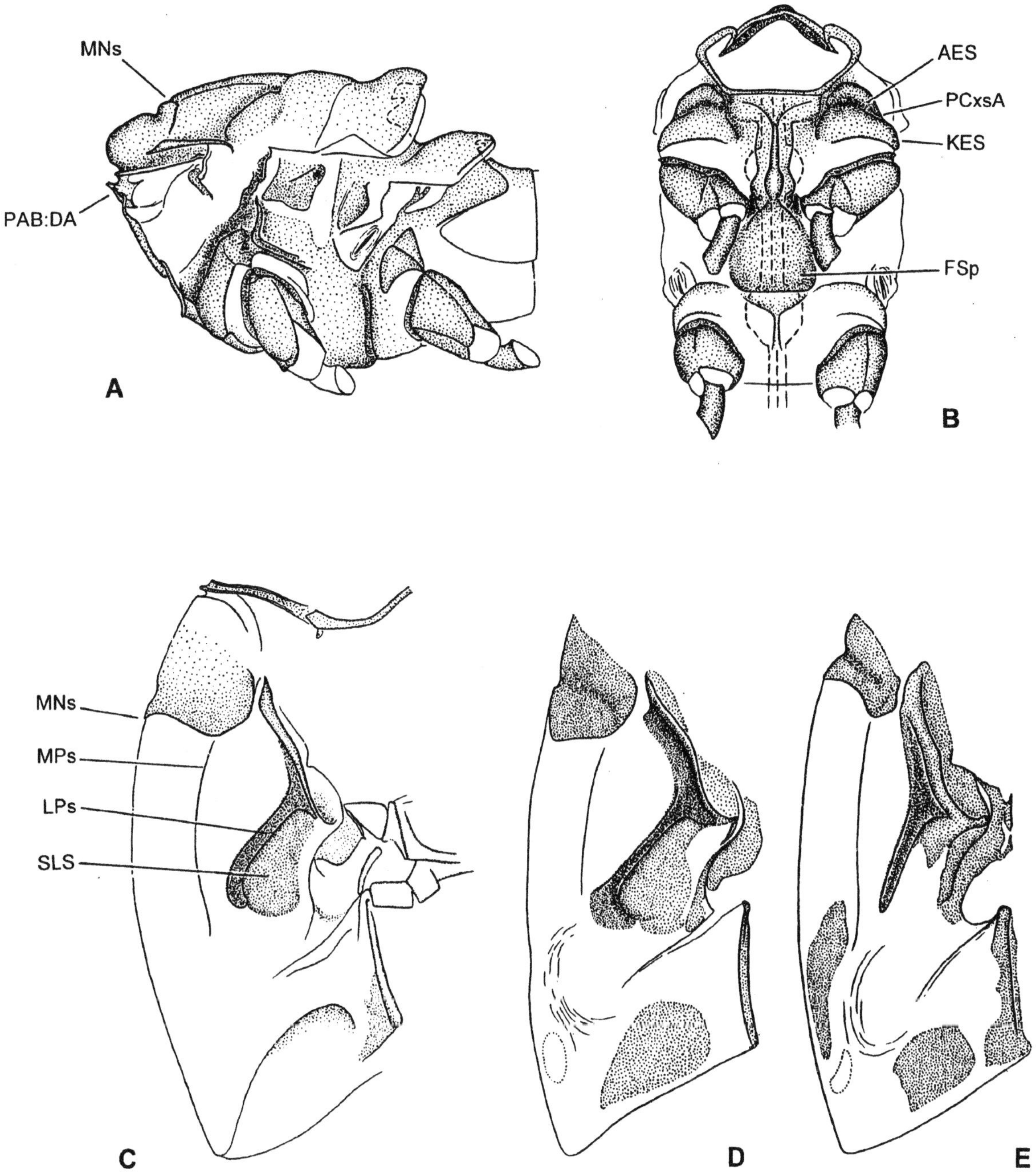

Figure 39. Coloburiscus/fg1, imago and subimago.
A–C – Coloburiscus/fg* *humeralis* [*Palingenia*]: **A** – imaginal pterothorax, lateral view; **B** – the same, ventral view (nerve ganglions shown by interrupted lines); **C** – subimaginal exuviae of right half of mesonotum. **D** – Murphyella/g* *needhami* [*Murphyella*], the same. **E** – Coloburiscoides/g1 *munionga* [*Coloburiscoides*], the same.

Abbreviations: **AES** – anepisternum; **FSp** – fused furcasternal protuberance; **KES** – katepisternum; **LPs** – lateroparapsidal suture; **MNs** – mesonotal suture; **MPs** – medioparapsidal suture; **PAB:DA** – dorsal arc of prealar bridge; **PCxsA** – anterior paracoxal suture; **SLS** – sublateroscutum.

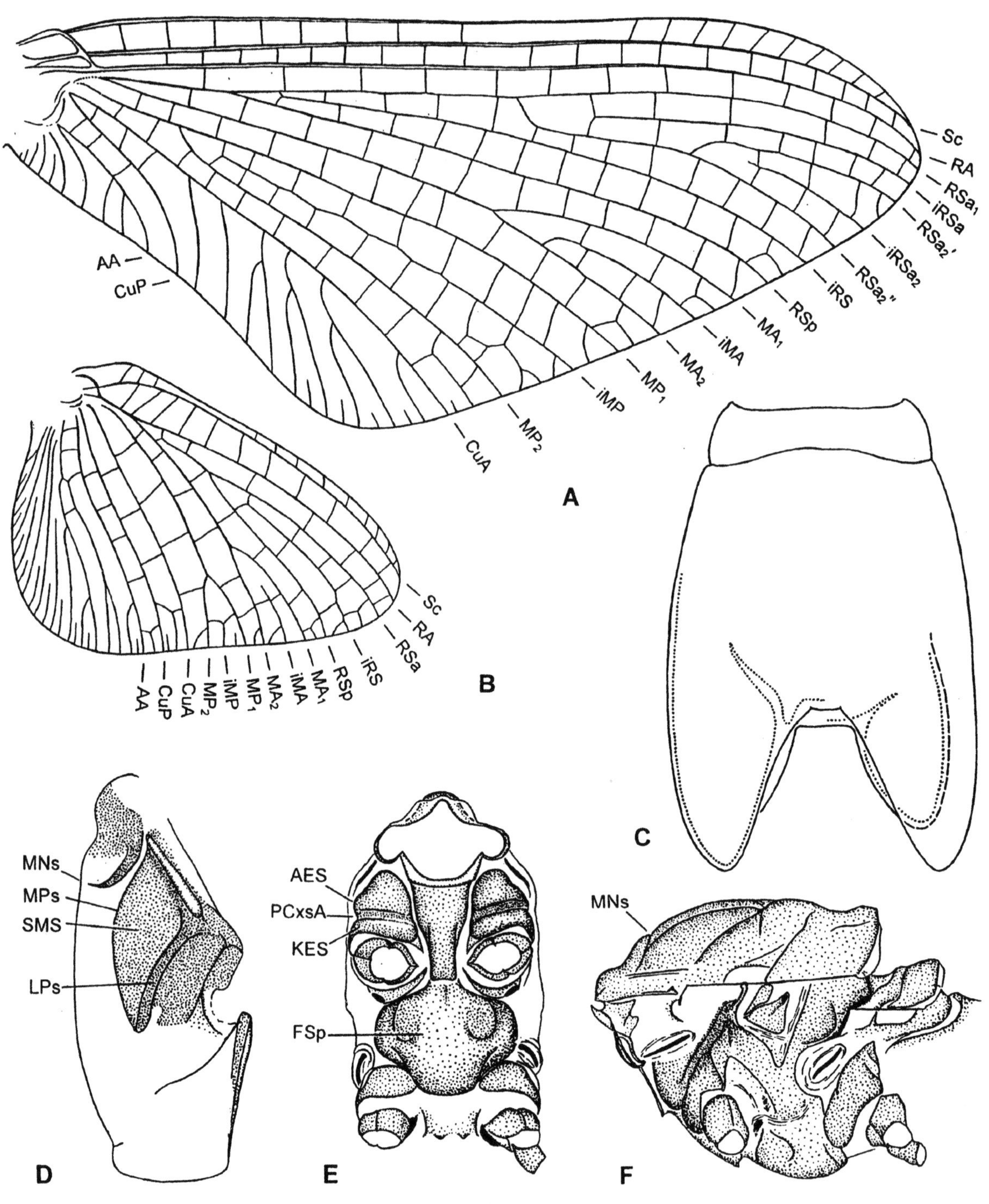

Figure 40. Isonychia/fg2 *ignota* [*Baetis*].
A–B – fore and hind wings; **C** – thoracic terga of larva of beginning of last instar (larval hind wing bud shown by interrupted line at right half, hypodermal parts shown by dotted line); **D** – subimaginal exuviae of right half of mesonotum; **E** – imaginal pterothorax, ventral view; **F** – the same, lateral view. (F – from Kluge 1998).

Abbreviations: **AES** – anepisternum; **FSp** – fused furcasternal protuberance; **KES** – katepisternum; **LPs** – lateroparapsidal suture; **MNs** – mesonotal suture; **MPs** – medioparapsidal suture; **PCxsA** – anterior paracoxal suture; **SMS** – submedioscutum.

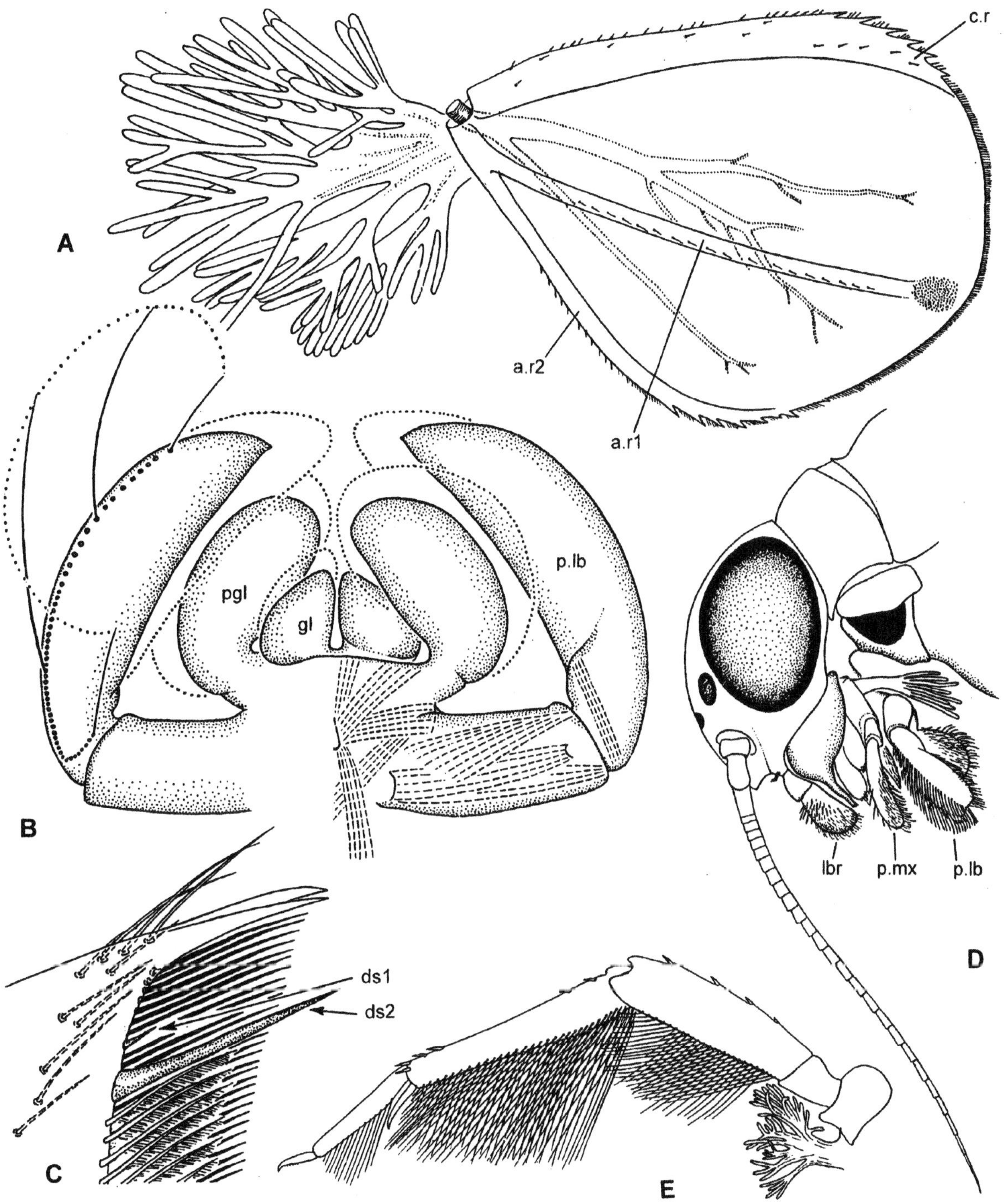

Figure 41. Isonychia/fg2

A–B – Isonychia/fg2 *crassiuscula* [*Isonychia*]: **A** – tergalius V, dorsal view (ventral fibrillose lobe turned back); **B** – labium, dorsal view (in left half) and ventral view (in right half) (setation not shown, areas occupied by dense setae of glossae, paraglossae and palps shown by dotted lines; instead of setae of regular row on palp, their bases and area occupied are shown; muscles of glossae, paraglossae and palp shown by interrupted lines in right half) (compare with Figs 37:D and 44:B); **C** – Isonychia/fg2 *ussurica* [*Isonychia*], apex of left maxilla, dorsal view; **D–E** – Isonychia/fg2 *ignota* [*Baetis*]: **D** – larval head and prothorax, lateral view (fore leg removed); **E** – larval fore leg.

Abbreviations: **a.r1** – anterior branch of anal rib; **a.r2** – posterior branch of anal rib; **c.r** – costal rib; **ds1** – vestige of distal dentiseta; **ds2** – proximal dentiseta; **gl** – glossa; **lbr** – labrum; **pgl** – paraglossa; **p.lb** – labial palp; **p.mx** – maxillary palp.

2.2;1,1-2. **Isonychia/fg1**

(Anteritorna Bidentiseta Branchitergaliae Eusetisura Isonychia/fg1)

(Figs 40–41)

Nomen hierarchicum: **Isonychia/fg1** (incl. *Prionoides*) [f: Isonychiinae Burks 1953: 108; g: *Isonychia* Eaton 1871: 134, typus *I. manca* Eaton 1871 (design. orig.; syn. subj. *Baetis sicca* Walsh 1862); syn. obj.: ***Chirotonetes*** Eaton 1881: 21)].

In circumscription matches:

— gen. *Isonychia* Eaton 1871: 134;

— gen. *Jolia* Eaton 1881: 192 (partim: ♉, non ⊕);

— gen. *Chirotonetes* Eaton 1881: 21 (partim: ⊕, non ♉);

— subfam. Isonychiinae Burks 1953: 108;

— fam. Isonychiidae: Landa 1973: 156;

— Isonychia/fg1: Kluge 2000: 251.

References. Needham & Traver & Hsu 1935: ♉* ⊕*; – Edmunds & Allen & Peters 1963: ♉*; – Koss & Edmunds 1974: ⊙*; – Edmunds & Jensen & Berner 1976: ♉* ⊕*; – Kondratieff & Voshell 1984: ♉ ⊕; – Tiunova & Kluge & Ishiwata (in press): ♉* ◎* ⊕*.

Autapomorphies of Isonychia/fg1.

(1) Larval frons between antennae bases forms a projection of unique form which narrows toward clypeus and terminates by an edge flattened from sides and hanging over clypeus (FIG.41:D).

(2) Mouth apparatus has uniform structure in all representatives, with following unique features (FIG.41:B–D):

On outer surface of labrum (densely covered with long setae, as in other Eusetisura and many other mayflies), besides slender long pointed setae, there are present more stout long stick-shape setae with truncate apices (Tiunova & al., in press.:Fig.57).

On maxilla [see (10) and Bidentiseta (1)] distal dentiseta is strongly diminished (FIG.41:C) or lost, and proximal dentiseta is long, straight, spine-like (in contrast to other Eusetisura, where both dentisetae are equally developed).

Labium has characteristic shape; $2^{nd}+3^{rd}$ segment of labial palp (integral, but with muscle retained), besides numerous irregular long setae [see Eusetisura (2)] bears a regular row of very long setae directed laterally-dorsally (FIG.41:B, D).

(3) Subimaginal lateral pigmented area of mesonotum occupies entire submedioscutum up to medioparapsidal suture (FIG.40:D). Similar form of this area is in Ameletus/fg1 only.

(4) Basisternum of mesothorax with a bifurcate projection arising of its hind margin and directed posteriorly; basisternum of metathorax with a pair of projections directed posteriorly; these projection are distinctly developed in larva, being smaller or non-expressed in imago (FIG.40:E–F).

(5) Larval fore legs [specialized as filtering – see Eusetisura (1) and bearing gills – see (12)] have unique features (FIG.41:E): apex of tibia on its inner side has a long jointed spine-like appendage (in some other mayflies on this place only non-jointed projection is present); in addition to filtering setae on femur and tibia, there are 2 longitudinal rows of smaller filtering setae on inner side of tarsus.

(6) Imaginal and subimaginal claws (initially ephemeropteroid) on all legs except for fore legs of male imago [see (11)], are similar and pointed (with this the name "*Isonychia*" is connected). This apomorphy is not unique (see Index of characters [2.2.85]), but does not occur in other Eusetisura.

(7) On hind wing bifurcation of PM is transferred to distal part of wing, distad of bifurcation of MA (FIG.40:B) (only for selected specimens of *formosus* [*Chirotonetes*] MP bifurcation proximad of MA bifurcation is reported). Non-unique apomorphy (see Index of characters [2.2.69]).

(8) Each tergalius I–VII [unable to make rhythmical movement – see Eusetisura (7)] has following structure (FIG.41:A): anal rib [arising behind base of fibrillose ventral portion – see Branchitergaliae (3)] bifurcates near base, its anterior branch passes by dorsal side near middle of tergalius up to its apex, and posterior branch borders anal margin of tergalius; thus there are three ribs (including well-developed costal rib on costal margin). Similar three ribs independently appeared in Nesameletus/f2=Metamonius/g2 and *Rallidens*. Margin of tergalius in apical part of costal rib and in area between apices of anterior and posterior branches of anal rib, bears denticles. Sometimes smaller denticles are present also on tergalius apex between these two areas (their absence was described as a character of *Prionoides*, but they are absent also in group *japonica* of Isonychia/fg2).

(9) Imaginal and subimaginal paracercus is vestigial, consists of several segments. Non-unique apomorphy (see Index of characters [2.3.22]).

Characters of unclear phylogenetic status.

(10) Maxilla [see (2)] has acute apical angle (in contrast to nearly rectangular in Coloburiscus/fg1), with 2 canines only (instead of three initial ones) (FIG.41:C). Non-unique apomorphy (see Index of characters [1.1.33]); possibly synapomorphy with Discoglossata (which has more acute angle with one canine only).

(11) On fore legs of male imago [but not subimago – see (6)] both claws are similar and blunt: each claw lacks its point and is provided with a soft plate. The same in Discoglossata (FIG.42:F); probably synapomorphy.

(12) Larval fore leg [see (5)] with a gill in joining of coxa with thorax; this gill has a form of a tuft (in majority of representatives – FIG.41:D–E) or a filament (in majority of *Prionoides*); in imago on this place a dark soft protuberance is retained. The same in *Murphyella*; possibly this gill is initial for Eusetisura [see Eusetisura (3)].

Plesiomorphies of Isonychia/fg1. Larva has the most typical siphlonuroid form among Eusetisura: legs are able to stretch posteriorly (in contrast at least to Oligoneuriella/g2, while other Eusetisura were not observed at life); abdomen is long enough and able to make undulate swimming movements; paracercus is well-developed, cerci and paracercus have dense primary swimming setae (only ends of cerci consist of elongate cylindrical segments without setation). Mouth apparatus has no autapomorphies peculiar for Discoglossata, 1st segment of maxillary palp is only 2 times shorter than distal segment [2nd+3rd – see Eusetisura (2)]. In contrast to Discoglossata, larval and adult middle and hind legs retain patella-tibial suture. Tarsus of fore leg of male imago [see (11)] is long (in contrast to Discoglossata), 1st segment is non-shortened (in contrast to Coloburiscus/fg1), equal or slightly longer than 2nd segment. In cubital field of fore wing several (4–7) veins go from CuA to basitornal and tornoapical margins: usually first 2 of these veins are non-branched and alternate as concave and convex, and next 2–4 veins are branched (FIG.40:A) [see Anteritorna (1)]. Gonostylus has 2 distal segments.

Size. Fore wing length 8–16 mm.

Distribution. Holarctic; dominate in Nearctic.

Isonychia/fg1 is divided into Isonychia/fg2 and *Prionoides*; a taxon *Borisonychia* has uncertain systematic position.

2.2;1,1-2/1. Isonychia/fg2

(Bidentiseta Branchitergaliae Eusetisura Isonychia/fg1 .../fg2)

(Figs 40–41)

Nomen hierarchicum: **Isonychia/fg2** (sine *Prionoides*).
In circumscription matches:
— subgen. *Isonychia*: McCafferty 1989: 78.

References. Kondratieff & Woshell 1983: ⊙* ♉* ⊕*; – 1984: ⊙* ♉* ⊕*; – McCafferty 1989: ⊕*; – Tiunova & Kluge & Ishiwata (in litt.): ⊙* ♉* ⊕*.

Autapomorphy of Isonychia/fg2.

(1) Styliger is strongly shortened in median part, thus gonostyli pedestals are separated one from another nearly from bases. As well as in *Prionoides* (see below), gonostyli pedestals are long, each with ventral surface flat and dorsal surface semi-cylindrically projected (Tiunova & al., in press: Figs 36–38).

Plesiomorphies of Isonychia/fg2 (in contrast to *Prionoides*). Penis with straight lateral margins, without curved projections . Eggs are spherical.

Size. Fore wing length 8–16 mm.

Distribution. Holarctic and Oriental Region; dominate in Nearctic and Eastern Palaearctic; in Western Palaearctic occurs a single Trans-Palaearctic species – *ignota* Walker 1853 [*Baetis*].

Nominal taxa included. Isonychia/fg2 includes: (1) **Jolia/g** [g: *Jolia* Eaton 1881: 192, typus *Palingenia roeselii* Joly 1870 (design. orig.; syn. subj. *Baetis ignota* Walker 1853)], regarded as a generic synonym of *Isonychia* (Needham 1905: 28); (2) **Eatonia/g** [g: *Eatonia* Ali 1970: 121, typus *E. khyberensis* Ali 1970 (monotypy), non *Eatonia* Hall 1857], regarded as generic synonym of *Isonychia* (Hubbard & Peters 1978: 31).

Species composition of Isonychia/fg2 (here [*I.*]= =[*Isonychia*]). More than 20 species are described. One **Trans-Palaearctic species:** *ignota* Walker 1853 [*Baetis*] (synn. subj.: *roeselii* Joly 1872 [*Palingenia*], *ferruginea* Albarda 1878 [*I.*], *hainanensis* She & You 1988 [*I.*]). **Asian species:** *arabica* Al-Zubaidi & Braasch & Al-Kayatt 1987 [*I.*], *crassiuscula* Tiunova & Kluge & Ishiwata [*I.*] (in litt.), *formosanus* Ulmer 1912 [*Chirotonetes*], *grandis* Ulmer 1913 [*Chirotonetes*], *guixiensis* Wu & Gui 1992 [*I.*], *japonicus* Ulmer 1920 [*Chirotonetes*] (syn. subj.: *violacea* Matsumura 1931 [*Sparrea*]), *khyberensis* Ali 1970 [*Eatonia*], *kiangsuensis* Hsu 1936 [*I.*], *sexpetala* Tiunova & Kluge & Ishiwata [*I.*] (in litt.), *sinensis* Wu & Gui 1992 [*I.*], *sumatranus* Navás 1933 [*Chirotonetes*], *ussurica* Bajkova 1970[*I.*], *vschivkovae* Tiunova & Kluge & Ishiwata [*I.*] (in litt.), *winkleri* Ulmer 1939 [*I.*]. **North American species:** *arida* Say 1839 [*Baetis*] (syn. subj.: *pictipes* Traver 1934 [*I.*]), *berneri* Kondratieff & Voshell 1984 [*I.*], *bicolor* Walker 1853 [*Palingenia*] (synn. subj.: *albomanicatus* Needham 1905 [*Chirotonetes*], *christina* Traver 1934 [*I.*], *circe* Traver 1934 [*I.*], *fattigi* Traver 1934 [*I.*], *harperi* Traver 1934 [*I.*], *matilda* Traver 1934 [*I.*], *pacoleta* Traver 1932 [*I.*], *sadleri* Traver 1934 [*I.*]), *campestris* McDunnough 1931 [*I.*], *intermedius* Eaton 1885 [*Chirotonetes*], *rufa* McDunnough 1931 [*I.*], *sicca* Walsh 1862 [*Baetis*] (synn. subj.: *manca* Eaton 1871 [*I.*], *edmundsi* Kondratieff & Voshell 1984 [*I.*]), *tusculanensis* Berner 1948 [*I.*], *velma* Needham 1932 [*I.*].

Material examined: *bicolor* [*P.*]: ♉, ♉–♂–⊕; *crassiuscula* [*I.*]: ♉–◎–⊕/⊙; *ignota* [*B.*]: ♉–◎–⊕/⊙; *japonicus* [*Ch.*]: ♉, ◎, ⊕/⊙; *sexpetala* [*I.*]: ♉/♂; *sicca* [*B.*]: ♂, ⊕;

ussurica [*I.*]: Ծ–◎–⊕/⊙; *vschivkovae* [*I.*]: Ծ–◎–⊕/⊙; *velma* [*I.*]: Ծ, ♀; spp. (Tajikistan, China, Vietnam): ⊕.

***

2.2;1,1-2/2. **Prionoides/g(1)**

(Bidentiseta Branchitergaliae Eusetisura Isonychia/fg1 *Prionoides*)

Nomen hierarchicum: **Prionoides/g(1)** [g: *Isonychia* subgen. *Prionoides* Kondratieff & Voshell 1983: 129, typus *Isonychia georgiae* McDunnough 1931 (design. orig.)].
In circumscription matches:
— subgen. *Prionoides* Kondratieff & Voshell 1983: 129.

References. Kondratieff & Woshell 1983: ⊙* Ծ* ⊕*; – 1984: ⊙* Ծ* ⊕*; – Tiunova & Kluge & Ishiwata (in litt.): ⊙* ⊕*.

Autapomorphies of *Prionoides*.

(1) Penis dorso-laterally with a pair of projections curved medially and bearing sclerotized denticles or spines (Kondratieff & Woshell 1983:Figs 1–5). Somewhat similar dorsoventral spines independently evolved in Pentamerotarsata (see below).

(2) Egg [see Branchitergaliae (4)] has shape of biconvex lens (Kondratieff & Woshell 1983:Figs 17–20) (in contrast to spherical in Isonychia/fg2). Non-unique apomorphy; the same in some species of Ameletus/fg2.

Plesiomorphy of *Prionoides*. In contrast to Isonychia/fg2, styliger is well-developed, median emargination between gonostyli pedestals being shallow (Kondratieff & Woshell 1983:Fig. 1).

Size. Fore wing length 9–16 mm.

Distribution. Eastern Nearctic and Japan.

Species composition of Prionoides/g(1). 8 species. One **Japanese species:** *shima* Matsumura 1931 [*Rhoenanthus*] (synn. subj.: *jozana* Matsumura 1931 [*Siphlonisca*], *sukashii* Matsumura 1931 [*Siphlonisca*]). **North American species:** *georgiae* McDunnough 1931 [*Isonychia*] (synn. subj.: *annulata* Traver 1932 [*Isonychia*], *thalia* Traver 1934 [*Isonychia*]), *hoffmani* Kondratieff & Voshell 1984 [*Isonychia*], *notata* Traver 1932 [*Isonychia*], *obscura* Traver 1932 [*Isonychia*], *sayi* Burks 1953 [*Isonychia*], *serrata* Traver 1932 [*Isonychia*], *similis* Traver 1932 [*Isonychia*] (syn. subj.: *aurea* Traver 1932 [*Isonychia*]).

Material examined: *georgiae* [*I.*]: Ծ; *sayi* [*I.*]: ⊕♂; *shima* [*Rh.*]: Ծ, ♂.

***

Isonychia/fg1 INCERTAE SEDIS

Isonychia/fg1 INCERTAE SEDIS. Borisonychia/g(1) [g: *Isonychia* subgen. *Borisonychia* McCafferty 1989: 78, typus *Isonychia diversa* Traver 1934 (design. orig.)].

Reference. McCafferty 1979: ⊕♂ (egg structure unknown).

Character of unclear phylogenetic status.

(1) Penis laterally with a pair of rounded projections curved toward penis base (which, in contrast to *Prionoides*, are not serrate). Probably, synapomorphy with *Prionoides*.

Size. Fore wing length 9 mm.

Distribution. South-east of USA.

Species composition. 1 species – *diversa* Traver 1934 [*Isonychia*].

Material examined: –.

*** ***

2.2;1,1-3. **Discoglossata**, or **Oligoneuria/f2=g3**

(Anteritorna Bidentiseta Branchitergaliae Eusetisura Discoglossata)

(Figs 42–51)

Nomen hierarchicum: **Oligoneuria/f2=g3** (sine *Coloburiscus*, *Isonychia*; incl. *Chromarcys*).
Nomen circumscribens: **Discoglossata** Kluge, TAXON NOV.
In circumscription matches:
— fam. Oligoneuriidae: Spieth 1943: 12;
— Oligoneuria/f2=g3: Kluge 2000: 251.

References. Demoulin 1953a: Ծ* ⊕*; – Edmunds & Allen & Peters 1963: Ծ*; – Koss & Edmunds 1974: ⊙*.

Autapomorphies of Discoglossata.

(1) Larva has a **head shield** and a peculiar modification of clypeus and labrum (FIG.44:E–F). Anterior and lateral margins of head are prominent and form a shield, covering mouthparts dorsally or anteriorly; at middle margin of head shield is narrowly rounded or pointed. Clypeus is hidden under the head shield, dorsally (or anteriorly) is projected above base of labrum, and ventrally (or posteriorly) is reduced in such a manner, that labrum is articulated directly to its base – i. e. to a pair of sclerotized portions of head capsule located at bases of anterior tentorial arms and mandibles.

Besides Discoglossata, head shield is present in Heptagennota; but in Heptagennota labrum retains its articulation on apex of clypeus, and the head shield has another form (usually more flattened and widened anteriorly), that allows to assume its independent origin.

Probably initially for Discoglossata, fore margin of the head shield bears a row of long setae directed anteriorly-medially; such setae are present at least in *Chromarcys* and Oligoneuriella/g1.

(2) Mouth apparatus is more highly specialized for gathering dispersed particles, than in other

Eusetisura, having unique structure of labium, maxillae and mandibles.

Glossae have unique structure (FIG.45): they are proximally fused together forming an integral movable semicircular **glossal disc**; this disc is so strongly widened ventrad of paraglossae that expands fare laterad of paraglossae, reaching 2nd+3rd segment of labial palp and covering all other mouthparts from ventral side. Medially, the glossal disc has a narrow longitudinal dorsal ridge with widened base, to which a pair of mentum-glossal muscles are attached. Dorsal side of the glossal disc bears numerous arched concentric rows of setae directed along these rows; between these rows there are present dense irregularly situated smaller setae of the same direction (FIG.45:A; Elpers & Tomka 1992:Figs 10–12). Paraglossae are hidden inside a pair of cavities on dorsal side of the glossal disc. As labial palps are partly covered from ventral side by glossal disc, their 1st segments are always directed laterally (but not ventrally). 2nd+3rd segment of labial palp [see Eusetisura (2)] is integral, without muscle inside; its shape is banana-like with grooved lateral side; when its densely setose ventral side overlaps the densely setose dorsal side of glossal disc, its lateral margin matches lateral margin of glossal disc.

Labium is so strongly modified, that homology of its parts is not evident: glossae, being projected fare laterally, can be taken for paraglossae, while paraglossae, being brought together, can be taken for glossae. True homology of these parts can be understood if examine muscles attachment: mentum-glossal muscles occupy median position, and mentum-paraglossal muscles are located laterad of them (FIG.45:B; compare with FIGS 37:D and 41:B).

Maxilla is strongly narrowed: it is widest in its proximal part and narrows toward apex; apically with 1 slender canine (in contrast to 3 initial canines and 2 canines in Isonychia/fg1); distalmost bristle of inner-ventral row is situated close to the canine and ventrad of it, enlarged and specialized, looks like a second canine, usually pectinate (FIG.44:D) (among the taxa examined, only in Elassoneuria/g1 it is non-pectinate). 1st segment of maxillary palp is strongly shortened, 5–6 times shorter than 2nd+3rd segment [see Eusetisura (2)] (Eaton 1883–1888:Pl.26).

Mandibles are strongly diminished, being hidden between head shield [see (1)] and glossal disc; incisor and kinetodontium are diminished, slender (Eaton 1883–1888:Pl.26:9–10).

(3) Larval fore wing buds are fused together by means of a wide plate which fills an area between their hind margins and primary posterior margin of mesonotum (FIG.44:A); fore wing buds themselves diverge posteriorly (in contrast to Coloburiscus/fg1, where they are convergent); adult plumidia can be developed [see Eusetisura (5)]. Non-unique apomorphy (see Index of characters [1.2.6] and [2.2.13]).

(4) On fore wing costal brace [see Euephemeroptera (1)] is strongly modified: from place of its proximal connection with costal vein, it passes behind C nearer to Sc up to its crossing with Sc and falling into RA (FIG.42:A). Unique apomorphy: in all other Euplectoptera costal brace at most its length passes in front of costal vein.

(5) On fore wing apex of MP_2 is approximated to apex of CuA, thus distance between apices of MP_1 and MP_2 is strongly enlarged; more or less brought together are also apices of MP_1 and MA_2, of MA_1 and RSp (FIG.42:A); besides this, apices of iRS and RSa can be brought together (in Geminovenata).

Gemination of veins (i. e. uniting in pairs) appeared independently also in Behningia/fg3 and Palingenia/f3=g2 (see Index of characters [2.2.28]); unique feature of gemination in Discoglossata is that veins iMA and iMP do not take part in gemination; thus on the place where in Behningia/fg3 and Palingenia/f3=g2 two pairs iMA+MA_2 and MP_1+iMP are present, Geminovenata have only one pair MA_2+MP_1 and two single veins iMA and iMP (which can be reduced).

(6) On fore wing RS is modified in a following manner. RSa lacks primary furcation: in *Chromarcys* branch RSa_2 lost connection with RSa-RSa_1 and arises from iRS, and iRSa arises from RSa_2 (FIG.42:A); in Geminovenata both RSa_2 and iRSa are lost (FIGS 49–51) (non-unique apomorphy). Vein iRS (which initially is intercalary) arises from RSa; in *Chromarcys* these veins widely diverge (FIG.42:A), in Geminovenata they are brought together all over their length (FIGS 50–51) (only in Elassoneuria/g1 and Homoeoneuria/g2 both are completely lost – FIG.49:A). In other mayflies iRS usually has free base, only in selected specimens being arisen from RSa.

(7) Imaginal and subimaginal fore leg is shorter than middle and hind legs (FIG.42:F–G); 1st–4th tarsal segments (i.e. all segments except for the last one bearing claws) are shortened, and their boundaries are strongly oblique; either these 4 segments are distinguishable (FIG.42:F), or only 3 of them, or only 2, or boundaries between segments are indistinct. Middle legs are the longest. These proportions are not connected with leg specialization in larva [see

Eusetisura (1)], as larval legs are specialized in the same manner as in other Eusetisura. In all other Branchitergaliae and nearly in all other Ephemeroptera, tarsi of fore legs of male imago serve to fix female during mating, and are elongate, being longer than non-specialized tarsi of middle and hind legs; only in very rare cases fore tarsus is shorter than others (see Index of characters [2.2.74]).

(8) Patella-tibial suture (initially present on middle and hind legs) is lost on all legs of larva, subimago and imago (FIGS 42:G; 44:A). Non-unique apomorphy (see Index of characters [1.2.18]).

(9) In imago and subimago at least on fore legs of male both claws are blunt (FIG.42:F). This apomorphy is not unique (see Index of characters [2.2.77]), but does not occur in other Eusetisura (in Isonychia/fg1 only in imago both fore leg claws of male are blunt).

(10) Probably the initial form of lamellate lobe of tergalius [see Branchitergaliae (3) and Eusetisura (7)] is cup-shaped (FIG.44:A–C): small, roundish; dorsal side is sclerotized, convex, without prominent ribs; margins are also sclerotized, at anterior margin sclerotization is widely spread to ventral side; most part of ventral side is non-sclerotized, concave; fibrillose portion can be partly inserted into this non-sclerotized concavity. Such tergalial structure is present in *Chromarcys* and Oligoneuriella/g1; in Oligoneuria/f4=g5 tergalii have similar convex-concave shape, but without ventral sclerotization; in other groups tergalii are secondarily enlarged up to size usual for Ephemeroptera, and have other form.

All tergalii I–VII are always present, but can be differently modified [see below, Geminovenata (1) and Homoeoneuria/g2 (3)].

(11) Each anchor on egg surface [initially consisting of a thread coiled under a knob – see Branchitergaliae (4)] has its thread reduced to a short nearly straight stem, thus anchor has mushroom-like shape (FIG.43:H; Koss & Edmunds 1974:Figs 72–78; Pescador & Peters 1980:Figs 67–70).

Character of unclear phylogenetic status.

(12) Tendency to secondary increasing of number of distal segments of gonostylus is expressed: in *Chromarcys*, *Oligoneuriopsis*, in selected specimens of Oligoneuriella/g2, and Elassoneuria/g1, instead of two initial segments, 3 or 4 segments can be present (FIGS 42:C; 47; 49:D). Non-unique apomorphy (see Index of characters [2.3.12]). In some groups of Geminovenata number of distal segments of gonostylus is, on the contrary, reduced to one.

Plesiomorphies of Discoglossata. Imaginal and subimaginal paracercus is usually well-developed (in contrast to Coloburiscus/fg1 and Isonychia/fg1); only in that representatives of Discoglossata whose paracercus is reduced in larva (*Chromarcys*, *Lachlania*, some Oligoneuriella/g2 and probably *Spaniophlebia*) paracercus of imago and subimago is also reduced.

Size. Fore wing length 7–27 mm.

Age and distribution. Early Cretaceous (see Discoglossata INCERTAE SEDIS) — Recent; nearly world-wide, except for Australian Region.

Discoglossata are divided into *Chromarcys* and Geminovenata; extinct Mesozoic taxon *Colocrus* has uncertain systematic position (p.147).

2.2;1,1-3/1.
Pseudoligoneuria/f(1)=Chromarcys/g(1)

(Bidentiseta Branchitergaliae Eusetisura Discoglossata *Chromarcys*)

(Fig. 42)

Nomen hierarchicum: **Pseudoligoneuria/f(1)=Chromarcys/g(1)** [f: Pseudoligoneuriinae Ulmer 1939: 653; g: *Chromarcys* Navás 1932: 927, typus *Ch. magnifica* Navás 1932 (design. orig.)]; syn. subj.: **Pseudoligoneuria/fg** [g: *Pseudoligoneuria* Ulmer 1939: 540, typus *P. fuernborni* Ulmer 1939 (design. orig.)]; syn. subj.: **Chromarcys/fg** [f: Chromarcyinae Demoulin 1953a:8] (synn. subj.: *magnifica* [*Ch.*]= *fuernborni* [*P.*]).

In circumscription matches:
— gen. *Chromarcys* Navás 1932: 927;
— gen. *Pseudoligoneuria* Ulmer 1939: 540;
— subfam. Pseudoligoneuriinae Ulmer 1939: 653;
— subfam. Chromarcyinae Demoulin 1953a: 8;
— fam. Chromarcidae: Tshernova 1970: 128.
Monospecific taxon.

References. Navás 1932: ⊕; – Ulmer 1939: ℧; – Demoulin 1953a: ℧/◎, ⊕; –1967a: ⊕; – Edmunds & Allen & Peters 1963: ℧; – Tshernova 1970: ℧ ⊕.

Autapomorphy of *Chromarcys*.

(1) Larval, subimaginal and imaginal paracercus is diminished, several times shorter than cerci (but not vestigial, several times longer than 10th abdominal segment) (Ulmer 1939:Fig.390; Demoulin 1967a:Fig.1d). Non-unique apomorphy (see Index of characters [2.3.22]).

Plesiomorphies of *Chromarcys* (in contrast to Geminovenata). All tergalii, including pair I, are attached dorsally and have the same structure [see Discoglossata (1)]. In imago and subimago: Mesonotal suture is normally developed, nearly transverse, not shifted posteriorly; separate sclerite forming posterior half of medioscutum is present in subimago only (FIG.42:D), while in imago the whole medioscutum is evenly sclerotized. Anterior paracoxal suture is not curved posteriorly. Middle and hind legs are functional, with ephemeropteroid claws (FIG.42:G). Wings with many intercalaries and crossveins; number of intercalaries, in contrast to Geminovenata, is not only non-reduced, but even increased if compare with other mayflies: additional intercalaries are present in widened field between MP_1 and MP_2 (FIG.42:A–B). Pedestals of gonostyli are large, with well-developed muscles (as in Isonychia/fg1 and Coloburiscus/fg1); border between 1st and 2nd segments of gonostyli is retained; penis has simple structure (FIG.42:C). Cerci are not soft, segmented, with irregular setae (Demoulin 1967a:Fig.1d) (in contrast to many Geminovenata).

Variable character of *Chromarcys*. On hind wing bifurcation of MA is shifted distally (FIG.42:B) or lost (Demoulin 1953a:Fig.1b; 1967a:Fig.1b) (it is lost also in Geminovenata; probably synapomorphy).

Size. Fore wing length 16–27 mm.

Distribution. Oriental Region.

Species composition of Pseudoligoneuria/f(1)=Chromarcys/g(1). One species is described – *magnifica* Navás 1932 [*Chromarcys*] (syn. subj.: *fuernborni* Ulmer 1939 [*Pseudoligoneuria*]).

Material examined: *magnifica* [*Ch.*]: ♂–⊕♂.

***

2.2;1,1-3/2. Geminovenata, or Oligoneuria/f3=g4

(Bidentiseta Branchitergaliae Eusetisura Discoglossata Geminovenata)

(Figs 43–51)

Nomen hierarchicum: **Oligoneuria/f3=g4** (sine *Chromarcys*; incl. *Oligoneuriella, Homoeoneuria, Elassoneuria, Fittkauneuria*).
Nomen circumscribens: **Geminovenata** Kluge, TAXON NOV.
In circumscription matches:
— gen. *Oligoneuria* Pictet 1845: 288;
— subfam. Oligoneuriinae: Demoulin 1953a: 8;
— fam. Oligoneuriidae: Ulmer 1914: 97;
— Oligoneuria/f3=g4: Kluge 2000: 251.

References. Demoulin 1953: ♉* ⊕*; – Edmunds 1961: ♉*; – Edmunds & Allen & Peters 1963: ♉*; – Tshernova 1970: ♉* ⊕* – Edmunds & Jensen & Berner 1976: ♉* ⊕*.

Autapomorphies of Geminovenata.

(1) Bases of tergalii I are transferred to ventral side of the body, while tergalii II–VII retain their dorso-lateral position (FIG.43:A). Unique apomorphy: in other mayflies in that very rare cases when tergalii bases are transferred ventrally (Behningia/ /fg2, *Raptoheptagenia* and others), tergalii of all pairs have the same position. In Geminovenata, as in all other mayflies, all tergalii are attached to tergites, but abdominal tergite I is expanded ventrally and supplied with a pair of lateral longitudinal ridges in such a manner, that its tergalii look as being attached to sternite; correct homology of parts of 1st abdominal segment can be understood if trace development of subimaginal and imaginal tergite and sternite from larval parts (FIG.43:A–C). The ventral tergalius I either retains the same structure as the dorsal tergalii II–VII, or differs [see Homoeoneuria/g2 (3) below].

(2) Winged stages are short-living; legs are non-functional, moult from subimago to imago takes place in males only (non-unique apomorphy: see Index of characters [2]). On all legs of male [including fore legs – see Discoglossata (7)] 1st–4th tarsal segments are usually shortened, with strongly inclined boundaries, can be fused together or reduced in number, all claws are blunt; legs of female are even more reduced, curved, with tarsal segmentation non-expressed, claws vestigial or lost.

(3) Wings [with gemination – see Discoglossata (5)] are highly modified: longitudinal veins are closely united in pairs, intercalaries are strongly reduced in number, thus spaces between pairs of longitudinal veins are strongly widened; crossveins demonstrate tendency to disappearance (FIGS 49–51; Demoulin 1952b:Figs 1–2).

On fore wing RSa always lacks branches and intercalaries, MA always lacks intercalary iMA; among intercalaries only iRS and iMP can be retained (sometimes they are also lost). Cubital field [see Anteritorna (1)] includes a single strong convex vein ("CuA_2") arising from CuA to basitornal margin and looking as a branch of CuA (the same in some short-living Fossoriae – see Index of characters [2.2.51]). Hence, behind RA, always following double veins are present: $RSp+MA_1$, MA_2+MP_1, MP_2+CuA and "CuA_2"+CuP; usually there is also a double vein RSa+iRS (but sometimes RSa and iRS are lost), sometimes also a singe vein iMP is present. Bifurcation of MA [initially situated near middle of wing – see Euephemeroptera (2)] is transferred close to wing base (FIGS 50:A; 51:A); only in Elassoneuria/g1 it is situated nearer to middle of wing, probably being transferred their secondarily (FIG.49:A).

On hind wing intercalaries are completely lost; MA is non-branched [see also Variable characters of *Chromarcys*] and like on fore wing, is bordered by two neighbouring veins – RSp and MP_1.

Dense cross veins can be present on larval wing bud (FIG.44:A; Demoulin 1953a:Fig.3; 1968d:Figs 1–3), but during transformation to adult they can partly or completely disappear.

(4) Imaginal mesonotal suture has a form of wide membranous stripe and is curved posteriorly as a whole, thus it crosses medioscutum not at anterior, but at middle or posterior part, dividing it to two separated sclerites (FIG.46:A). Unique apomorphy (see Index of characters [2.2.8]). At least in investigated Oligoneuriella/fg1 subimaginal cuticle of mesonotum is not sclerotized at all, being very thin, that allows to shad exuviae at flight breaking it [see (2)].

(5) Anterior paracoxal suture on ventral side of mesothoracic episternum is strongly curved posteriorly, thus terminating not at basisternum, but at anterior margin of furcasternal protuberance, and anepisternum is much larger than katepisternum (FIG.46:C–D). Unique apomorphy.

(6) Male genital apparatus has peculiar structure (FIGS 46–50). Pedestals of gonostyli are lost, and muscles moving gonostyli are diminished (in contrast to other Eusetisura which have large gonostyli pedestals with strongly developed muscles – FIG.42:C). Gonostyli themselves usually are not diminished, but soft and weak; in Homoeoneuria/g2 and *Fittkauneuria* gonostyli are lost. 1st and 2nd segments of gonostylus are fused (in contrast to *Chromarcys*, where they are separated). Left and right penis lobes are movably connected and have various structures allowing them to protract [see below, Oligoneuriella/g1 (1), Homoeoneuria/g2 (7), Elassoneuria/g1 (3) and Oligoneuria/f4=g5 (3)].

(7) Imaginal and subimaginal caudalii [cerci and paracercus – see Plesiomorphies of Discoglossata] are modified. In female [which does not moult to imago – see (2)] and male subimago caudalii are as short as in larva, more or less retaining larval structure (i.e. joinings can be oblique, primary swimming setae can be developed). In male imago caudalii are long, soft, usually lack irregularly situated setae (peculiar for majority of mayflies, including *Chromarcys*), but have whorls of long setae on apex of each segment (FIG.46:B); such structure is present at least in the species examined of Oligoneuriella/g2, *Oligoneurisca*, Homoeoneuria/g2 and Elassoneuria/ /g2; but sometimes (particularly in Lachlania/g1 *powelli* [*L.*]) cerci of male have irregularly situated setae.

(8) In larva, subimago and imago abdominal tergite X has lateral-ventral margins longitudinal, as long as the segment, reaching bases of cerci; latero-posterior angle of tergite forms a ventral condylus for cercal base and separates paraproct from cercotractor; cercotractor is transformed to a narrow semicircular sclerite exposed caudally and surrounding lateral half of cercal base (FIG.12:G), can be fused with cercal base. The same in Fossoriae and Caenotergaliae (but not in *Chromarcys* and other mayflies, whose lateral-ventral margins of tergite X are oblique, not articulating with cerci, and cercotractors are triangular, exposed laterally and widely connected with paraprocts – FIG.12:F).

Size. Fore wing length 7–22 mm.

Distribution. Holarctic, Ethiopian, Oriental and Neotropical Regions.

Geminovenata are divided into Oligoneuriella/g1, Homoeoneuria/g1, Elassoneuria/g1, Oligoneuria/f4= =g5 and *Fittkauneuria.*

2.2;1,1-3/2.1. Oligoneuriella/g1

(Branchitergaliae Eusetisura Discoglossata Geminovenata Oligoneuriella/g1)

(Figs 43:A–C; 44– 47)

Nomen hierarchicum: **Oligoneuriella/g1** (incl. *Oligoneuriopsis*) [g: *Oligoneuriella* Ulmer 1924a: 31, typus *Oligoneuria rhenana* Imhoff 1852].

Autapomorphies of Oligoneuriella/g1.

(1) Penis [see Geminovenata (6)] has peculiar structure (FIGS 46–47). Paired penis lobes are completely separated and deeply inserted into segment IX; ventral wall of each penis lobe bears a sclerotized process directed medio-ventrally; this process is always completely hidden by sternite IX and styliger (unique apomorphy); shape of these processes differs in Oligoneuriella/g2 and *Oligoneuriopsis* (see below). Distal part of each lobe (projected externally) consists of a lateral longitudinal sclerotized lobe and a median telescopic lobe with gonopore at its apex. Apices of lateral sclerotized lobes are more or less curved medially toward one another. Located between them median telescopic lobes are cylindrical, at rest have length subequal to the lateral lobes, and in everted condition can become several times longer. Inverted surface of the telescopic lobe is densely covered by microtrichia, which in everted condition are directed proximally, and in inverted condition – distally.

Besides Oligoneuriella/g1, a pair of cylindrical telescopic lobes are present in Homoeoneuria/g2 (see below), but there penis structure is different, and paired lobes have a common unpaired base.

(2) In male imago middle and hind legs [non-functional – see Geminovenata (2)] before the last (claw-bearing) segment have 2 segments only (if not take into account the first segment which is fused with tibia, shortened, and non-expressed), thus tarsus looks as 3-segmented. Non-unique apomorphy: the same in Elassoneuria/g2 (but not in Elassoneuria/g1-*Madeconeuria*).

Plesiomorphies of Oligoneuriella/g1. On fore wing RSa+iRS is present (in contrast to Elassoneuria/g1 and Homoeoneuria/g2), not approximated to RA (in contrast to *Lachlania* and Oligoneuria/f5==g6); bifurcation of MA is situated near wing base (in contrast to Elassoneuria/g1) (Demoulin 1952b: Fig.1). Gonostylus is developed (in contrast to Homoeoneuria/g2 and *Fittkauneuria*), has not less than 2 distal segments [see Discoglossata (12)] (in contrast to Oligoneuria/f4=g5). Larval head shield has no keel (in contrast to *Elassoneuria*), anterior margin of head shield with long setae directed anteriorly and medially [see Discoglossata (1)] (FIG.44:E–F).

Variable characters of Oligoneuriella/g1. Larval and imaginal paracercus is usually well-developed, but in some species vestigial – particularly in *bicaudata* [*Oligoneuriella*].

Size. Fore wing length 9–19 mm.

Distribution. Palaearctic and Ethiopian Region.

Oligoneuriella/g1 is divided into Oligoneuriella/g2 and *Oligoneuriopsis*.

2.2;1,1-3/2.1;1. Oligoneuriella/g2

(Eusetisura Discoglossata Geminovenata Oligoneuriella/g1 .../g2)

(Figs 43:A–C; 44–46; 47:A)

Nomen hierarchicum: **Oligoneuriella/g2** (sine *Oligoneuriopsis*).
In circumscription matches:
— gen. *Oligoneuriella* Ulmer 1924a: 31.

References. Hagen 1855 (*Oligoneuria rhenana*): ⊕'; – Eaton 1883–1888 (*O. rhenana*): Ꝺ' ⊕'; – Ulmer 1924a: ⊕*; – Edmunds 1961: Ꝺ*.

Autapomorphy of Oligoneuriella/g2.

(1) On each penis lobe the proximal sclerotized process directed medially-ventrally [see Oligoneuriella/g1 (1)] is saddle-shaped, curved distally and sclerotized in such a manner that its median side is membranous, and a boundary between the sclerotized and the membranous areas is dentate; thus it looks as if this process has two rows of denticles – a medial-dorsal and medial-ventral ones (FIGS 46:B; 47:A).

Size. Fore wing length 9–16 mm.

Distribution. Palaearctic; all species in Western Palaearctic, only one of them – *pallida* [*O.*] – all over Palaearctic.

Species composition of Oligoneuriella/g2. More than 10 species are described – *bicaudata* Al-Zubaidi & Braasch & Al-Kayatt 1987 [*Oligoneuriella*], *duerensis* Gonzalez & Garcia 1983 [*Oligoneuriella*], *kashmirensis* Ali 1971 [*Oligoneuria*], *keffermullerae* Sowa 1973 [*Oligoneuriella*], *marichuae* Alba-Tercedor 1983 [*Oligoneuriella*], *pallida* Hagen 1855 [*Oligoneuria*] (synn. subj.: *mikulskii* Sowa 1961 [*Oligoneuriella*], *poecile* Ikonomov 1962 [*Oligoneuriella*], *mongolica* Soldán & Landa 1977 [*Oligoneuriella*]), *orontensis* Koch 1980 [*Oligoneuriella*], *polonica* Mol 1984 [*Oligoneuriella*], *rhenana* Imhoff 1852 [*Oligoneuria*] (syn. subj.: *garumnica* Joly 1873 [*Oligoneuria*]), *skoura* Dakki & Guidicelli 1980 [*Oligoneuriella*], *tskhomelidzei* Sowa & Zosidze 1973 [*Oligoneuriella*] (synn. subj.: *baskale* Soldán & Landa 1977 [*Oligoneuriella*], *zanga* Soldán & Landa 1977 [*Oligoneuriella*]).

Material examined: *rhenana* [*Oligoneuria*]: Ꝺ, Ꝺ/⊙; *pallida* [*Oligoneuria*]: Ꝺ–⊙–⊕/⊙; *keffermullerae* [*Oligoneuriella*]: Ꝺ; *tskhomelidzei* [*Oligoneuriella*]: Ꝺ–⊙–⊕/⊙; sp. cf. *bicaudata* [*Oligoneuriella*]: Ꝺ.

***

2.2;1,1-3/2.1;2. Oligoneuriopsis/g(1)

(Eusetisura Discoglossata Geminovenata Oligoneuriella/g1 *Oligoneuriopsis*)

(Fig. 47:B)

Nomen hierarchicum: **Oligoneuriopsis/g(1)** [g: *Oligoneuriopsis* Crass 1947a: 52, typus *O. lawrencei* Crass 1947 (monotypy)]. Presumably, because description of the type species is insufficient.
In circumscription matches:
— gen. *Oligoneuriopsis* Crass 1947a: 52.

References. Crass 1947a: ⊕; – Kimmins 1960: ⊕; – Edmunds 1961: Ꝺ*; – Agnew 1973: Ꝺ; – Dakki & Giudicelli 1980: Ꝺ ⊕.

Character of unclear phylogenetic status.

(1) On each penis lobe the proximal sclerotized process directed medially-ventrally [see Oligoneuriella/g1 (1)] is narrow, curved proximally, entirely sclerotized, apically pointed or dentate (FIG.47:B) (for the type species this character is unknown).

Size. Fore wing length 11–19 mm.

Distribution. Ethiopian Region.

Species composition of Oligoneuriopsis/g(1). Not less than 2 species – *lawrencei* Crass 1947 [*Oligoneuriopsis*], *dobbsi* Eaton 1912 [*Oligoneuria*] (syn. subj.: *grandaeva* Navás 1936 [*Oligoneuriella*]), and *skhounate* Dakki & Giudicelli 1980 [*Oligoneuriopsis*] known as imagoes; *elisabethae* Agnew 1973 [*Oligoneuriopsis*] and *jessicae* Agnew 1973 [*Oligoneuriopsis*] known as larvae.

Material examined (Albany Mus.): sp.O1 (= "*Oligoneuriopsis dobbsi*" SENSU Kimmins 1960): ⥁, ⊕♂, ⊕♀ (FIG.47:B).

*** ***

2.2;1,1-3/2.2. Homoeoneuria/g1

(Branchitergaliae Eusetisura Discoglossata Geminovenata Homoeoneuria/g1)

(Figs 43: D–F)

Nomen hierarchicum: **Homoeoneuria/g1** (incl. *Oligoneurisca*) [g: *Homoeoneuria* Eaton 1881: 192, typus *H. salviniae* Eaton 1881 (design. orig.)].

Autapomorphies of Homoeoneuria/g1.

(1) Larval fore leg [specialized – see Eusetisura (1)] has tarsus vestigial, its claw is either vestigial (FIG.43:D–E), or lost (Pescador & Peters 1980:Figs 58–62).

(2) Larval middle and hind legs are highly specialized, with coxae and claws modified (Tshernova 1937:Fig.6; Pescador & Peters 1980:Fig.56). Coxa is very strongly elongate at all its sides, having a form of mobile cylindrical segment narrowly attached to the body (in contrast to elongate coxae of Coloburiscus/fg1 and some others, where only outer side of coxa is long, while inner side is short). Claws are nearly straight, slender and desclerotized, without median row of denticles on inner side [see Eusetisura (5)], either smooth (in Homoeoneuria/g2) or with two rows of protuberances by sides (in *Oligoneurisca* – FIG.43:F). Most part of the leg is covered by dense long thin setae; in contrast to other Eusetisura, without stout setae.

In winged stages coxae of middle and hind legs are also very strongly elongated and attached to the body by narrow base.

(3) Larval abdominal sterna II and III each with a median protuberance, which on its sides bears long setae directed laterally; next abdominal sterna also with long setae directed laterally-posteriorly. Unique apomorphy.

(4) In male imago tarsus of middle and hind leg [non-functional – see Geminovenata (2)] either looks as 2-segmented, or its segmentation is indistinct.

Characters of unclear phylogenetic status.

(5) Larval head is secondarily hypognathous, head shield is not large, without regular row of marginal setae [see Discoglossata (1)]: in Homoeoneuria/g2 with submarginal setae only, in *Oligoneurisca* without setae (Tshernova 1937:Fig.2; Pescador & Peters 1980:Fig.56).

(6) Tergalii II–VII are not cup-shaped [see Discoglossata (10)], flat, with dense row of long pectinate setae on anal margin (Tshernova 1937:Fig.9). The same in *Fittkauneuria*; possibly synapomorphy.

Plesiomorphies of Homoeoneuria/g1. On fore wing bifurcation of MA is situated near wing base (in contrast to Elassoneuria/g1). Larval and imaginal paracercus is well-developed (in contrast to *Lachlania*, *Spaniophlebia* and some Oligoneuriella/g2).

Size. Fore wing length 7–12 mm.

Distribution. South of Nearctic, north of Neotropical and west of Palaearctic Regions.

Homoeoneuria/g1 is divided into *Oligoneurisca* and Homoeoneuria/g2.

2.2;1,1-3/2.2;1. Oligoneurisca/g(1)

(Eusetisura Discoglossata Geminovenata Homoeoneuria/g1 *Oligoneurisca*)

(Figs 43: D–F)

Nomen hierarchicum: **Oligoneurisca/g(1)** [g: *Oligoneurisca* Lestage 1938b: 318, typus *Oligoneuriella borysthenica* Tshernova 1937 (monotypy)].
In circumscription matches:
— gen. *Oligoneurisca* Lestage 1938b: 318.
Monospecific taxon.

References. Tshernova 1937: ⥁; – Keffermüller 1957: ⥁; – Edmunds 1961: ⥁; – Keffermüller 1964: ⊙ ◎ ⊕.

Autapomorphies of *Oligoneurisca*.

(1) Larval middle and hind legs [see Homoeoneuria/g1 (2)] are modifies: tibia is much shorter than tarsus and lacks setation; tarsus is very long, its long setae are directed proximally; claw has two rows of protuberances by sides (FIG.43:F).

Plesiomorphies of *Oligoneurisca* (in contrast to Homoeoneuria/g2). In imago: On fore wing double vein RSa+iRS is present; crossveins in area of pterostigma and between RA and RSa+iRS are present. Well-developed gonostyli with both distal segments are present. Penis without retractable processes, with medio-proximal condylus [see Geminovenata (6)].

In larva: Fore legs with a roundish vestige of claw (FIG.43:D–E). Abdominal sternum I has no finger-like process. All tergalii I–VII with developed lamellate and fibrillose lobes; lamellate lobes of tergalii II–VII are not narrowed (their width is more than 1/2 of length).

Size. Fore wing length 12 mm.

Distribution. Europe.

Species composition of Oligoneurisca/g(1). 1 species – *borysthenica* Tshernova 1937 [*Oligoneuriella*].

Material examined: *borysthenica* [*O.*]: ℧, ♂.

***

2.2;1,1-3/2.2;2. Homoeoneuria/g2

(Eusetisura Discoglossata Geminovenata Homoeoneuria/g1 .../g2)

Nomen hierarchicum: **Homoeoneuria/g2** (sine *Oligoneurisca*; incl. *Notochora*).
In circumscription matches:
— gen. *Homoeoneuria* Eaton 1881: 192.

References. Eaton 1883: ♀*; – Edmunds & Berner & Traver 1958: ⊕; – Edmunds 1961: ℧*; – Edmunds & Jensen & Berner 1976: ℧* ⊕*; – Allen & Cohen 1977: ℧; – Pescador & Peters 1980: ⊙ ℧* ⊕*; – Da-Silva 1992 (*Notochora*): ℧.

Autapomorphies of Homoeoneuria/g2.

(1) Larval fore leg [see Homoeoneuria/g1 (1)] has tarsus extremely diminished and claw lost (Pescador & Peters 1980:Figs 58–62).

(2) Larval abdominal sternum I in posterior part bears an unpaired finger-like process. Unique apomorphy.

(3) Tergalii [initially consisting of lamellate and fibrillose lobes – see Branchitergaliae (3)] are differentiated: tergalius I [attached ventrally – see Geminovenata (1)] consists of the fibrillose lobe only, lamella being reduced; tergalii II–VII [see Homoeoneuria/g1 (6)] are narrowed, with width no more than 1/3 length, lack fibrillose lobe.

(4) On fore wing double vein RSa+iRS [see Geminovenata (2)] is lost, i. e. there are no veins between RA and RSp+MA (non-unique apomorphy, the same in Elassoneuria/g1); crossveins are lost (Pescador & Peters 1980:Fig.1).

(5) On hind leg of male imago femur is strongly curved, inner side of trochanter, femur and tibia with tubercles; possibly, adaptation for holding female, that is connected with loss of gonostyli (Edmunds & Berner & Traver 1958).

(6) Gonostyli are lost. Non-unique apomorphy [see Geminovenata (6)]; the same in *Fittkauneuria*.

(7) Penis [see Geminovenata (6)] with a pair of membranous retractable tubular processes, each bearing gonopore apically (Pescador & Peters 1980:Figs 15–22).

Size. Fore wing length 7–11 mm.

Distribution. America: Nearctic and Neotropical Regions.

Nominal taxon included. Homoeoneuria/g2 includes **Notochora/g** [g: *Homoeoneuria* subgen. *Notochora* Pescador & Peters 1980: 385, typus *H. (N.) fittkaui* Pescador & Peters 1980 (design. orig.)]; regarded as a subgenus, differing from other Homoeoneuria/g2 in penis shape only (Pescador & Peters 1980, Da-Silva 1992).

Species composition of Homoeoneuria/g2. 6 species – *alleni* Pescador & Peters 1980 [*Homoeoneuria*], *ammophila* Spieth 1937 [*Oligoneuria*], *cahabensis* Pescador & Peters 1980 [*Homoeoneuria*], *dolani* Edmunds & Berner & Traver 1958 [*Homoeoneuria*], *fittkaui* Pescador & Peters 1980 [*Homoeoneuria (Notochora)*], *salviniae* Eaton 1881 [*Homoeoneuria*].

Material examined: *dolani* [*H.*]: ℧.

*** ***

2.2;1,1-3/2.3. Elassoneuria/g1

(Branchitergaliae Eusetisura Discoglossata Geminovenata Elassoneuria/g1)

(Figs 48–49)

Nomen hierarchicum: **Elassoneuria/g1** (incl. *Madeconeuria*) [g: *Elassoneuria* Eaton 1881: 191, typus *Oligoneuria trimeniana* McLachlan 1868 (design. orig.)].
In circumscription matches:
— gen. *Elassoneuria* Eaton 1881: 191.

References. Demoulin 1966d: ⊕; – 1975: ⊕.

Autapomorphies of Elassoneuria/g1.

(1) On fore wing MA has a long stem and furcates nearer to middle of wing, than to its base; RSp and MP_1 are close not only to branches MA_1 and MA_2 correspondingly [see Geminovenata (3)], but to the stem of M as well; thus MA with RSp and MP_1 form a Y-shape figure (FIG.49:A; Demoulin 1952b:Fig.2). In all other Geminovenata MA bifurcates near base (i. e. MA with RSp and MP_1 form V-shaped figure – FIGS 50–51).

For Euephemeroptera in general, and particularly for Discoglossata, position of MA bifurcation near wing middle is a plesiomorphy, but in Elassoneuria/g1 elongation of stem of MA, probably, appeared secondarily [see Geminovenata (2)], that is testified by unique position of RSp and MP_1.

(2) On fore wing double vein RSa+iRS [see Geminovenata (2)] is lost (FIG.49:A). Non-unique apomorphy: this vein independently reduced in

Homoeoneuria/g2 (but not in Homoeoneuria/g1--*Oligoneurisca*).

(3) Penis has peculiar structure. Its dorsal walls of left and right lobes are movably connected proximally either by a single median condylus (FIG.48), or by a median sclerite forming a pair of articulations (FIG.49:F); in both cases penis lobes are able to protract backward turning apart. In contrast to Oligoneuria/f4=g5 (which has similar articulation – see below), penis of Elassoneuria/g1 has no muscles-retractors.

(4) Larval head shield has a median longitudinal keel and a median angulate projection on anterior margin, lacks marginal setae [see Discoglossata (1)].

Plesiomorphies of Elassoneuria/g1. Gonostylus is developed (in contrast to Homoeoneuria/g2 and *Fittkauneuria*), not less than with 2 distal segments (FIG.49:D–E) [see Discoglossata (12)] (in contrast to Oligoneuria/f4=g5). Larval and imaginal paracercus is well-developed (in contrast to *Lachlania*, *Spaniophlebia* and some Oligoneuriella/g2).

Size. Fore wing length 10–22 mm.

Distribution. Ethiopian Region.

Elassoneuria/g1 is divided into *Madeconeuria* and Elassoneuria/g2.

2.2;1,1-3/2.3;1. Plesiomorphon **Madeconeuria/g(1)**

(Eusetisura Discoglossata Geminovenata Elassoneuria/g1 *Madeconeuria*)

(Fig. 48)

Nomen hierarchicum: **Madeconeuria/g(1)** [g: *Elassoneuria* subgen. *Madeconeuria* Demoulin 1973: 8; typus *Elassoneuria insulicola* Demoulin 1966 (design. orig.)].
In circumscription matches:
— subgen. *Madeconeuria* Demoulin 1973: 8;
— gen. *Madeconeuria*: Pescador & Elouard 2001:37 (Abstract).
Monospecific taxon.

References. Demoulin 1966d: ⊕; – 1968d: Ծ; – 1973: Ծ ⊕.

Plesiomorphies of *Madeconeuria*. In male imago on tarsus of middle and hind leg [non-functional – see Geminovenata (2)] before the last (claw-bearing) segment 3 short segments are present (if not take into account the 1st segment which is fused with tibia, shortened and not expressed), thus tarsus looks as 4-segmented (Demoulin 1966d:Figs 3c-e). In contrast to Elassoneuria/g2, penis lobes are connected by median condylus [see Elassoneuria/f1 (3)] (FIG.48).

Size. Fore wing length 17 mm.

Distribution. Madagascar.

Species composition of Madeconeuria/g(1). 1 species is described – *insulicola* Demoulin 1966 [*Elassoneuria*].

Material examined: *insulicola* [*E.*]: Ծ, ♂–♂.

***

2.2;1,1-3/2.3;2. **Elassoneuria/g2**

(Eusetisura Discoglossata Geminovenata Elassoneuria/g1 .../g2)

(Fig. 49)

Nomen hierarchicum: **Elassoneuria/g2** (sine *Madeconeuria*).
In circumscription matches:
— subgen. *Elassoneuria*: [Demoulin 1973 (as autonym)];
— gen. *Elassoneuria*: Pescador & Elouard 2001 (Abstract).

References. Eaton 1883: ⊕; – Ulmer 1916: ⊕; – 1920c: Ծ; – Barnard 1932: Ծ ⊕; – Demoulin 1952b: ⊕; – Berner 1954: Ծ; – Kimmins 1960: ⊕; – Edmunds 1961: Ծ; – Demoulin 1973: ⊕; – Gillies 1974: Ծ ⊕.

Autapomorphies of Elassoneuria/g2.

(1) In male imago on middle and hind leg [non-functional – see Geminovenata (2)] before the last (claw-bearing) segment, only 2 short segments are present (if not take into account the 1st segment which is fused with tibia, shortened, and not expressed), thus tarsus looks as 3-segmented. Non-unique apomorphy, the same in Oligoneuriella/g1.

(2) Each penis lobe has two longitudinal sclerites – lateral and median ones – which are mobile connected by a membrane and move apart when penes lobes protrude backward (FIG.49:C–F); penis lobes are movably connected by a transverse sclerite [see Elassoneuria/g1 (3)] (FIG.49:F).

Size. Fore wing length 10–22 mm.

Distribution. Africa.

Species composition of Elassoneuria/g2. 6 species are described – *candida* Eaton 1913 [*Elassoneuria*], *congolana* Navás 1911 [*Elassoneuria*], *disneyi* Gillies 1974 [*Elassoneuria*], *grandis* Gillies 1974 [*Elassoneuria*], *kidahi* Gillies 1974 [*Elassoneuria*], *trimeniana* McLachlan 1868 [*Oligoneuria*].

Material examined: *congolana* [*E.*]: ⊕; sp.O2: ⊕; sp.: Ծ.

*** ***

2.2;1,1-3/2.4. **Oligoneuria/f4=g5**

(Branchitergaliae Eusetisura Discoglossata Geminovenata Oligoneuria/f4=g5)

(Figs 50–51)

Nomen hierarchicum: **Oligoneuria/f4=g5** (sine *Oligoneuriella, Homoeoneuria, Elassoneuria, Fittkauneuria*; incl. *Lachlania, Spaniophlebia*).

Autapomorphy of Oligoneuria/f4=g5.

(1) Larval head shield [see Discoglossata (1)] has a pair of distinct lateral incisions by sides of eyes (FIG.51:C–D); these incisions are retained on adult head as well (FIG.50:C).

(2) Gonostylus with one distal segment (instead of two initial ones) (FIG.50:D). Non-unique apomorphy (see Index of characters 2.3.12]).

(3) Penis has a peculiar articulatory mechanism (FIG.50:D). Left and right penis lobes are movably articulated one with another by a median condylus on proximal (i. e. anterior) margin of their dorsal wall; at rest this condylus is located proximad of lateral articulations of penial arms with tergite angles, that allows to protract penis lobes moving them apart (Peters & Pescador 1980:Figs 26–33); from this condylus toward anterior-lateral angles of sternite IX, goes a pair of strong muscles which allow to retract penis lobes back. Unique apomorphy. Among Geminovenata similar dorsal-proximal condylus occurs in *Madeconeuria* only, but no muscles are attached to it (FIG.48).

Characters of Oligoneuria/f4=g5 of unclear phylogenetic status. Among subordinate taxa of Oligoneuria/f4=g5 various combinations of following 3 characters are found (TABLE 4):

(4) Head can have anterior projection of unique shape. In this case larval head shield [with lateral incisions – see (1)] anteriorly has a huge flat projection separated by one more pair of lateral incisions from the rest part of shield (FIG.51:D); imago has a soft frontal projection of similar shape, but much smaller and curved ventrally-posteriorly (FIG.50:C). This character is present in *Spaniophlebia* and Oligoneuria/f5=g6, but absent in *Lachlania.*

(5) On fore wing double vein RSa+iRS [see Geminovenata (3)] in proximal part is closely approximated to Sc+RA, thus sometimes looks as originating not from wing base, but from Sc+RA (FIG.50:A). This character is in this or that degree expressed in *Lachlania* and Oligoneuria/f5=g6, but absent in *Spaniophlebia* (FIG.51:A).

(6) Larval and adult cerci are contiguous by their bases; paracercus can be reduced to a small sclerite (in *Lachlania*) or completely lost (in *Spaniophlebia* sp.O3). This character is present in *Lachlania* and *Spaniophlebia*, but not Oligoneuria/f5=g6. Among other Geminovenata paracercus can be strongly shortened (but not lost) in some species of Oligoneuriella/g2 (see Index of characters [1.3.64]).

Plesiomorphies of Oligoneuria/f4=g5. On fore wing bifurcation of MA is situated near wing base (FIG.50:A, 51:A) (in contrast to Elassoneuria/g1). In male imago on tarsus of middle and hind leg [nonfunctional – see Geminovenata (2)] before the last (claw-bearing) segment, 3 short segments are present (if not take into account the 1st segment which is fused with tibia, shortened and not expressed), thus tarsus looks as 4-segmented.

Size. Fore wing length 8–15 mm.

Distribution. America: Neotropical and Nearctic Regions.

Oligoneuria/f4=g5 is divided into *Lachlania, Spaniophlebia* and Oligoneuria/f5=g6.

Table 4. Distribution of derived characters among subordinate taxa of Oligoneuria/f4=g5 [for descriptions – see Oligoneuria/f4=g5 (2)–(4)]

Taxa	Apomorphies		
	1	2	3
Lachlania	+	–	+
Spaniophlebia	–	+	+
Oligoneuria/f5=g6	+	+	–

Apomorphies:
1 – proximal half of RSa+iRS brought together with RA;
2 – frontal projection;
3 – loss of paracercus.

2.2;1,1-3/2.4;1. **Lachlania/g1**

(Eusetisura Discoglossata Geminovenata Oligoneuria/f4=g5 *Lachlania*)

(Fig. 51:C)

Nomen hierarchicum: **Lachlania/g1** (incl. *Noya, Noyopsis, Alloydia*) [g: *Lachlania* Hagen 1868: 372, typus *L. abnormis* Hagen 1868 (design. orig.)].
In circumscription matches:
— gen. *Lachlania* Hagen 1868: 372.

References. Eaton 1883–1888: ⊕*; – Edmunds 1951: ☉ ⊕; – Edmunds & Berner & Traver 1958: ☉ ⊕; – Edmunds 1961: ☉*; – Koss & Edmunds 1970: ☉ ⊕; – Edmunds & Jensen & Berner 1976: ☉* ⊕*; – Allen & Cohen 1977: ☉.

Characters of *Lachlania* of unclear phylogenetic status (TABLE 4).

(1) Larval head shield without anterior projection [see Oligoneuria/f4=g5 (4)], on anterior margin instead of long setae [see Discoglossata (1)] bears dense short spatulate setae (FIG.51:C).

(2) On fore wing RSa+iRS in proximal part is closely approximated to Sc+RA. The same in Oligoneuria/f5=g6 [see Oligoneuria/f4=g5 (5)].

(3) Larval and imaginal paracercus is reduced. The same in *Spaniophlebia* [see Oligoneuria/f4=g5 (6)].

Variable characters of *Lachlania*. Larval abdominal terga with or without unpaired projections.

Size. Fore wing length 8–15 mm.

Distribution. America: Neotropical and Nearctic Regions.

Nominal taxa included. Lachlania/g1 includes: (1) **Noya/g** [g: *Noya* Navás 1912a: 746, typus *Neophlebia garciai* Navás 1912 (monotypy); syn. obj.: ***Neophlebia*** Navás 1912a: 746 (non *Neophlebia* Selys 1869)], regarded as generic synonym of *Lachlania* (Ulmer 1943: 35); (2) **Noyopsis/g** [g: *Noyopsis* Navás 1924: 70, typus *N. fusca* Navás 1924 (design. orig.)], regarded as generic synonym of *Lachlania* (Ulmer 1943: 35); (3) **Alloydia/g** [g: *Alloydia* Needham 1932: 275, typus *A. cacautana* Needham 1932 (design. orig.)], regarded as a generic synonym of *Lachlania* (Demoulin 1952d: 3).

Species composition of Lachlania/g1. About 10 species are described – *abnormis* Hagen 1868 [*Lachlania*], *boanovae* Da-Silva & Pereira 1993 [*Lachlania*], *cacautana* Needham 1932 [*Alloydia*], *dencyanna* Koss 1970 [*Lachlania*], *dominguezi* Pereira 1989 [*Lachlania*], *fusca* Navás 1924 [*Noyopsis*], *garciai* Navás 1912 [*Neophlebia*], *iops* Allen & Cohen 1977 [*Lachlania*], *lucida* Eaton 1885 [*Lachlania*], *pallipes* Eaton 1883 [*Spaniophlebia*], *radai* Navás 1926 [*Noya*], *saskatchewanensis* Ide 1941 [*Lachlania*] (syn. subj.: *powelli* Edmunds 1951 [*Lachlania*]), *talea* Allen & Cohen 1977 [*Lachlania*].

Material examined: *saskatchewanensis* [*L.*]: ⥀, ◎; spp. from Mexico, Colombia, Venezuela, Ecuador, Peru and Argentina: ⥀, ⊕ (partly deposited in Entomol. Inst.).

***

2.2;1,1-3/2.4;2. Spaniophlebia/g(1)

(Eusetisura Discoglossata Geminovenata Oligoneuria/f4=g5 *Spaniophlebia*)

(Fig. 51:A–B)

Nomen hierarchicum: **Spaniophlebia/g(1)** [g: *Spaniophlebia* Eaton 1881: 191, typus *S. trialae* Eaton 1881 (design. orig.)].
In circumscription matches:
— gen. *Spaniophlebia* Eaton 1881: 191.

Reference. Eaton 1883–1888: ⊕♂.

Characters of *Spaniophlebia* of unclear phylogenetic status (TABLE 4).

(1) Head with anterior projection (at least in the specimen examined). The same in Oligoneuria/f5==g6 [see Oligoneuria/f4=g5 (4)].

(2) On fore wing RSa+iRS is not approximated to RA (FIG.51:A). In contrast to *Lachlania* and Oligoneuria/f5=g6 [see Oligoneuria/f4=g5 (5)].

(3) At least imaginal paracercus is reduced (larvae are unknown for certain). The same in *Lachlania* [see Oligoneuria/f4=g5 (6)].

Size. Fore wing length 10 mm.

Distribution. Neotropical Region.

Species composition of Spaniophlebia/g(1). Three incompletely described species are attributed here – *assimilis* Banks 1913 [*Spaniophlebia*], *escomeli* Cockerell 1926 [*Spaniophlebia*], *trialae* Eaton 1881 [*Spaniophlebia*].

Material examined: sp.O3: ⊕♂ (single specimen from unknown locality – FIG.51:A–B).

***

2.2;1,1-3/2.4;3. Oligoneuria/f5=g6

(Eusetisura Discoglossata Geminovenata Oligoneuria/f4=g5 .../f5=g6)

(Figs 50, 51:D)

Nomen hierarchicum: **Oligoneuria/f5=g6** (sine *Lachlania*, *Spaniophlebia*).
In circumscription matches:
— gen. *Oligoneuria*: Ulmer 1924a: 32.

References. Pictet 1843–1845 (*Oligoneuria anomala*, partim): ⊕♀; – Hagen 1855 (*O. anomala*): ⊕♀; – Eaton 1883–1888: ⊕; – Ulmer 1921 (*O. anomala*, typus): ⊕♀; – Needham & Murphy 1924: ⊕♀; – Spieth 1943: ⥀; – Demoulin 1955a (*Oligoneurioides*, *Spaniophlebia* sp.): ⥀ ⊕♂; – Puthz 1973: ⊕♀.

Character of Oligoneuria/f5=g6 of unclear phylogenetic status (TABLE 4).

(1) Head with anterior projection (FIGS 50:C; 51:D). The same in *Spaniophlebia* [see Oligoneuria/ /f4=g5 (4)].

(2) On fore wing RSa+iRS in proximal part is closely approximated to Sc+RA (FIG.50:A). The same in *Lachlania* [see Oligoneuria/ /f4=g5 (5)].

(3) Larval and imaginal paracercus is well-developed. In contrast to *Lachlania* and *Spaniophlebia* [see Oligoneuria/f4=g5 (6)].

Size. Fore wing length 8–15 mm.

Distribution. Brazil.

Nominal taxon included. Oligoneuria/f5=g6 includes **Oligoneurioides/g** [g: *Oligoneurioides* Demoulin 1955a: 24, typus *O. amazonicus* Demoulin 1955 (design. orig.)].

Species composition of Oligoneuria/f5=g6. Here are attributed two incompletely described species – *amazonicus* Demoulin 1955 [*Oligoneurioides*] and *anomala* Pictet 1843 [*Oligoneuria*].

Material examined: sp.O4 (Brazil) – 1 ♉ (FIG.51:D); sp.O5 (Brazil, Amazonas State, Rio Harania, 22 I 1963, J.E. Fittkau): 1 ⊕ (FIG.50).

*** ***

2.2;1,1-3/2.5. Fittkauneuria/g(1)

(Branchitergaliae Eusetisura Discoglossata Geminovenata *Fittkauneuria*)

Nomen hierarchicum: **Fittkauneuria/g(1)** [g: *Fittkauneuria* Pescador & Edmunds 1994: 264, typus *F. carina* Pescador & Edmunds 1994 (design. orig.)].

In circumscription matches:

— gen. *Fittkauneuria* Pescador & Edmunds 1994: 264.

Reference. Pescador & Edmunds 1994: ⊙ ♉ ⊕.

Autapomorphies of *Fittkauneuria*.

(1) Gonostyli are lost. Non-unique apomorphy [see Geminovenata (6)], the same in Homoeoneuria/g2.

(2) Tergalius I [ventral – see Geminovenata (1)] lacks lamellate lobe, consists of fibrillose lobe only [see Branchitergaliae (3)]. Non-unique apomorphy.

Character of unclear phylogenetic status.

(3) Tergalii II–VII with dense row of long setae on anal margin. The same in Homoeoneuria/g1; possibly, synapomorphy.

Plesiomorphies of *Fittkauneuria*. On fore wing RSa+iRS is present (in contrast to Elassoneuria/g1 and Homoeoneuria/g2), not approximated to RA (in contrast to *Lachlania* and Oligoneuria/f5=g6); bifurcation of MA is situated near wing base (in contrast to Elassoneuria/g1). In male imago on tarsus of middle and hind leg [non-functional – see Geminovenata (2)] before the last (claw-bearing) segment, 3 short segments are present (if not take into account the 1st segment which is fused with tibia, shortened and not expressed), thus tarsus looks as 4-segmented. Paracercus is well-developed (in larva shortened, 2 times shorter than cerci, in imago only slightly shorter than cerci).

Size. Fore wing length 12 mm.

Distribution. Neotropical Region.

Species composition of Fittkauneuria/g(1). 2 species – *carina* Pescador & Edmunds 1994 [*Fittkauneuria*] and *adusta* Pescador & Edmunds 1994 [*Fittkauneuria*].

Material examined: –.

*** ***

Discoglossata INCERTAE SEDIS

Discoglossata INCERTAE SEDIS. † Colocrus/fg [f: Colocrurinae McCafferty 1990: 31; g: *Colocrus* McCafferty 1990: 31, typus *C. indivicum* McCafferty 1990, larva (design. orig.)].

Reference. McCafferty 1990: ♉ ⊕.

Age. Early Cretaceous (Brazil).

Material examined: –.

Comment. Presumably, *Colocrus* occupies an intermediate position between *Chromarcys* and Geminovenata (McCafferty 1990). Synapomorphy with Geminovenata: fore wing lacks crossveins and intercalaries besides iRS and iMP. Symplesiomorphy with *Chromarcys*: fore wing with veins going from CuA to posterior margin of wing. On the type specimen (larva) wing venation is invisible, so it is unclear, if imago and larva are correctly associated.

*** ***

Eusetisura INCERTAE SEDIS

Eusetisura INCERTAE SEDIS 1. † Clavineta/g [g: *Clavineta* Sinitshenkova 1991: 123, typus *C. cantabilis* Sinitshenkova 1991 (design. orig.)].

References. Sinitshenkova 1991: ♉; – 2000a: ♉.

Age. Late Jurassic – Early Cretaceous (Siberia and Mongolia).

Species composition: *cantabilis* Sinitshenkova 1991 [*Clavineta*], *transbaikalica* Sinitshenkova 2000 [*Clavineta*].

Material examined (Paleontol. Inst.): *cantabilis* [*C.*]: ♉; *transbaikalica* [*C.*]: ♉.

Comment. Larva of *Clavineta* retains siphlonuroid appearance, with primary setation on caudalii. As in other Eusetisura, fore leg has filtering setae; labial palp with integral banana-shape 2nd+3rd segment. Other characters are unknown.

Eusetisura INCERTAE SEDIS 2. † Cratoligoneuriella/g [g: *Cratoligoneuriella* Martins-Neto 1996: 187, typus *C. leonardi* Martins-Neto 1996 (design. orig.)].

Reference. Martins-Neto 1996: ♉.

Age. Early Cretaceous (Brazil).

Material examined: –.

Comment. Judging by larval fore leg structure, *Cratoligoneuriella* undoubtedly belongs to Eusetisura; other characters are unclear.

*** ***

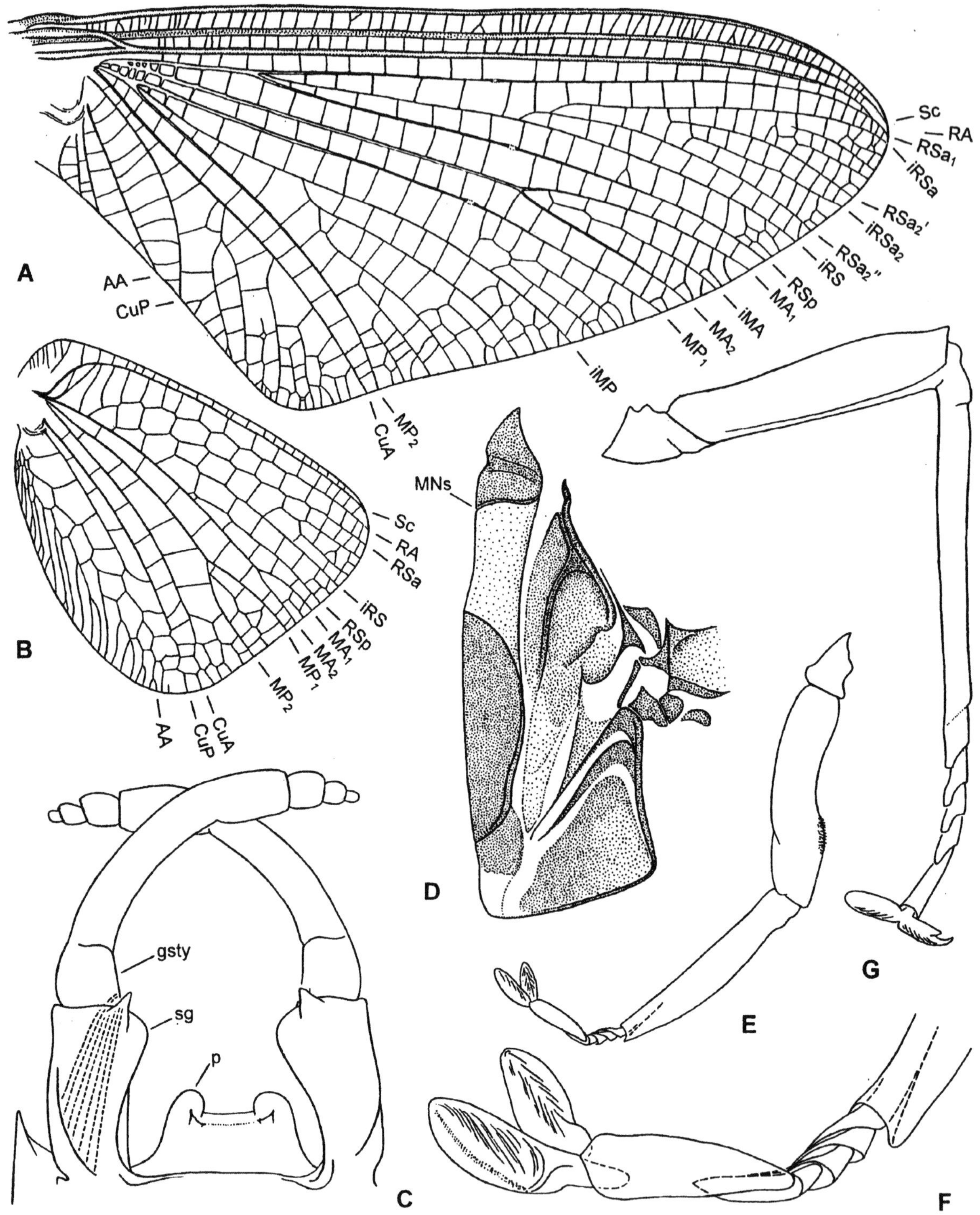

Figure 42. Discoglossata-Pseudoligoneuria/f1=Chromarcys/g1 *magnifica* [*Chromarcys*], male and subimago.
A–B – fore and hind wings; **C** – genitals, ventral view (muscle moving left gonostylus shown by interrupted lines); **D** – subimaginal exuviae of right half of mesonotum; **E** – fore leg of male imago; **F** – its tarsus enlarged; **G** – middle leg of male imago.

Abbreviations: **gsty** – gonostylus; **MNs** – mesonotal suture; **p** – penis; **sg** – pedestal of gonostylus (part of styliger).

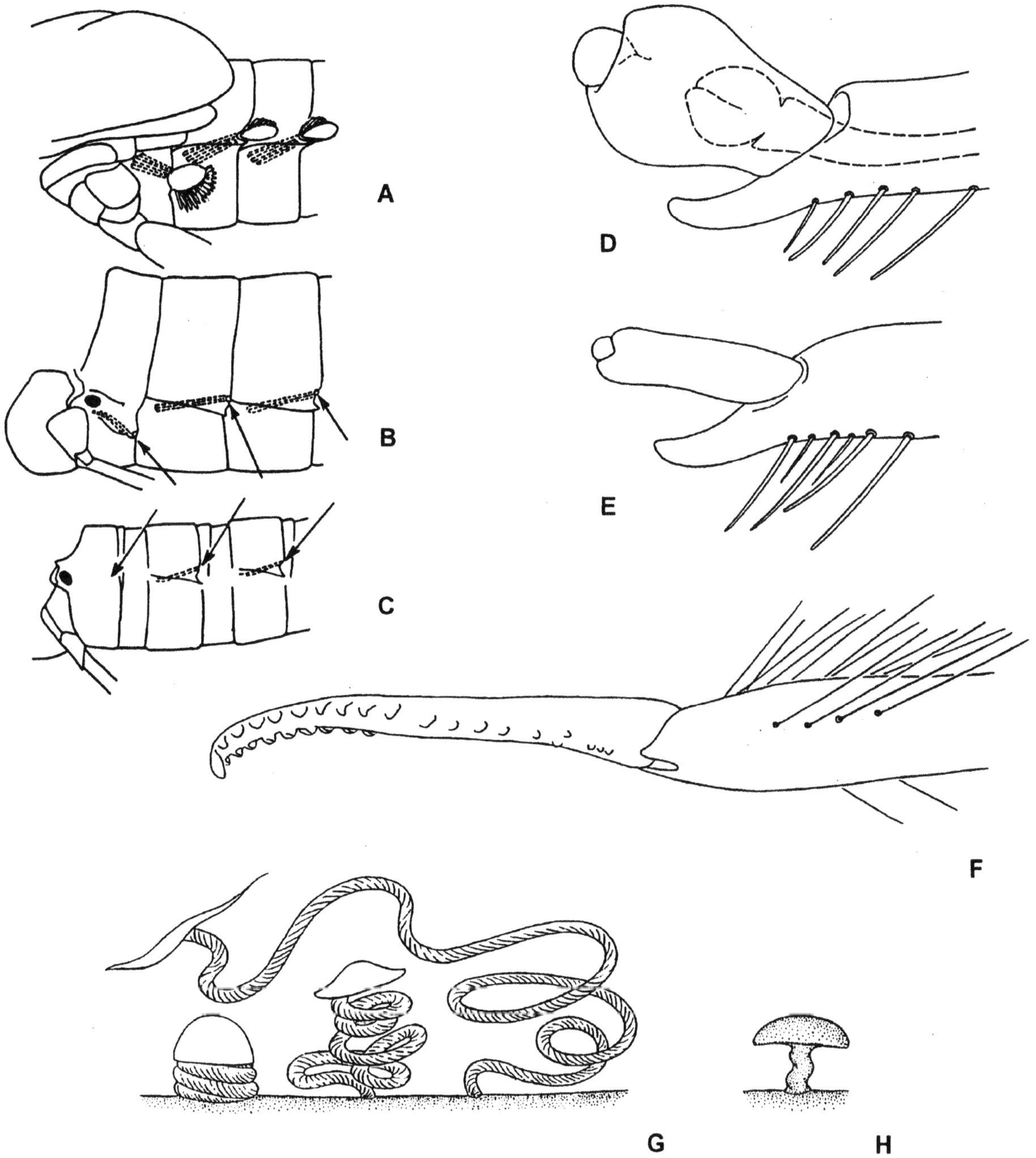

Figure 43. Discoglossata-Geminovenata and egg anchors of Branchitergaliae.
A–B – Oligoneuriella/g2 *tskhomelidzei* [*Oligoneuriella*], abdominal segments I–III, lateral view (muscles moving tergalii shown by interrupted lines, arrows show points of tergalii attachment): **A** – larva; **B** – subimago; **C** – imago. **D–F** – Oligoneurisca/g* *borysthenica* [*Oligoneuriella*], larvae (syntypi): **D** – mature male larva, tarsus of right fore leg (developing inside it adult tarsal apex shown by interrupted line); **E** – female mature larva, tarsus of right fore leg; **F** – claw of left hind leg. **G–H** – anchors on egg surface: **G** – anchor typical for Branchitergaliae, in coiled and spread position; **H** – vestigial anchor of Geminovenata (A–B – from Kluge 1989a; G–H – from Kluge 1998).

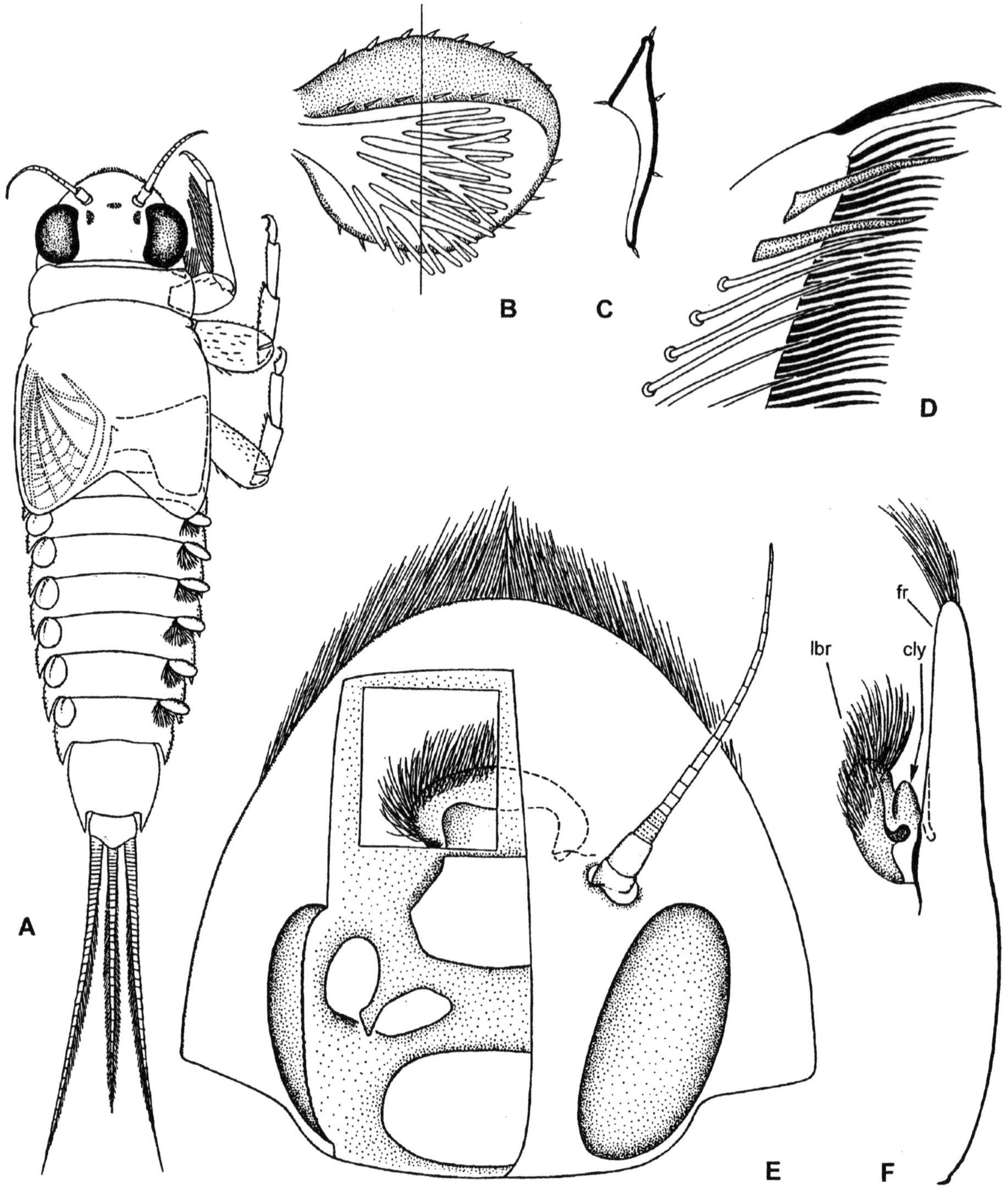

Figure 44. Discoglossata-Geminovenata-Oligoneuriella/g2, larvae.
A–D – Oligoneuriella/g2 *pallida* [*Oligoneuria*]: **A** – male larva (in left half outline of hypodermal bud of adult fore wing, its venation and plumidium are shown by dotted lines; in right half metanotum with larval hind wing bud is shown by interrupted line); **B** – tergalius, ventral view; **C** – the same, cross section by line shown in Fig.B; **D** – apex of left maxilla, dorsal view (dentisetae shown by dots, setae of inner-ventral row shown by black). **E–F** – Oligoneuriella/g2 *tskhomelidzei* [*Oligoneuriella*]: **E** – larval head, dorsal view (in left half part of dorsal wall removed to show tentorium, smaller part of ventral wall removed to show clypeus and labrum); **F** – the same, longitudinal section close to left margin of labrum. (A – from Kluge 1997, added; D – from Kluge 1998, corrected)

Abbreviations: **cly** – clypeus; **fr** – frons; **lbr** – labrum.

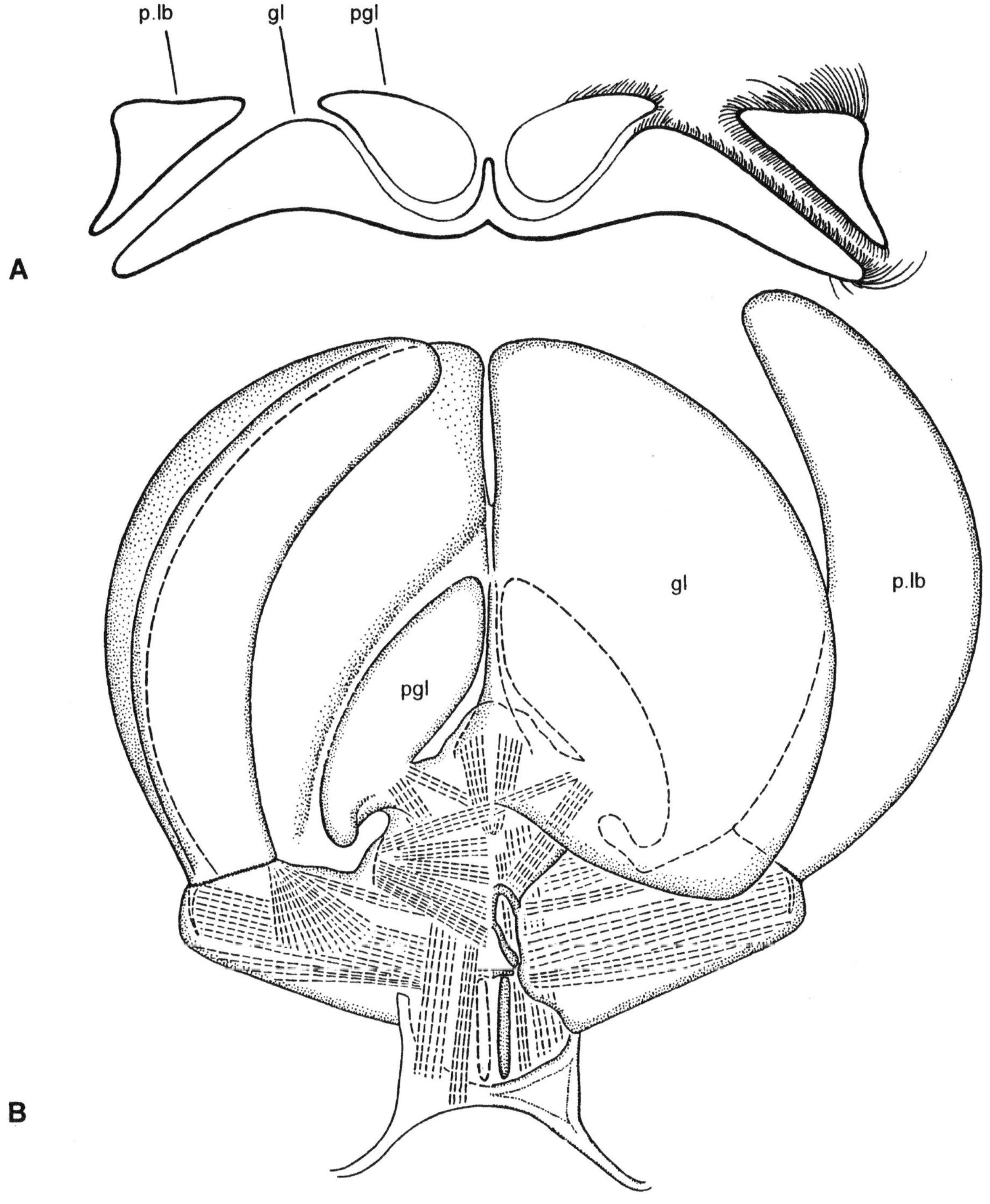

Figure 45. Oligoneuriella/g2 *tskhomelidzei* [*Oligoneuriella*], labium.
A – transverse section of labium (setae shown on right side only); **B** – tskhomelidzei [Oligoneuriella], labium, dorsal view (in left half) and ventral view (in right half) [setation not shown (actually glossae, paraglossae and palps are densely covered by setae), submentum turned at plain of glossal disc (actually it is directed perpendicular to it), muscles shown by interrupted lines]; compare with Figs 37:D and 41:B).

Abbreviations: **gl** – glossa; **pgl** – paraglossa; **p.lb** – 2nd+3rd segment of labial palp.

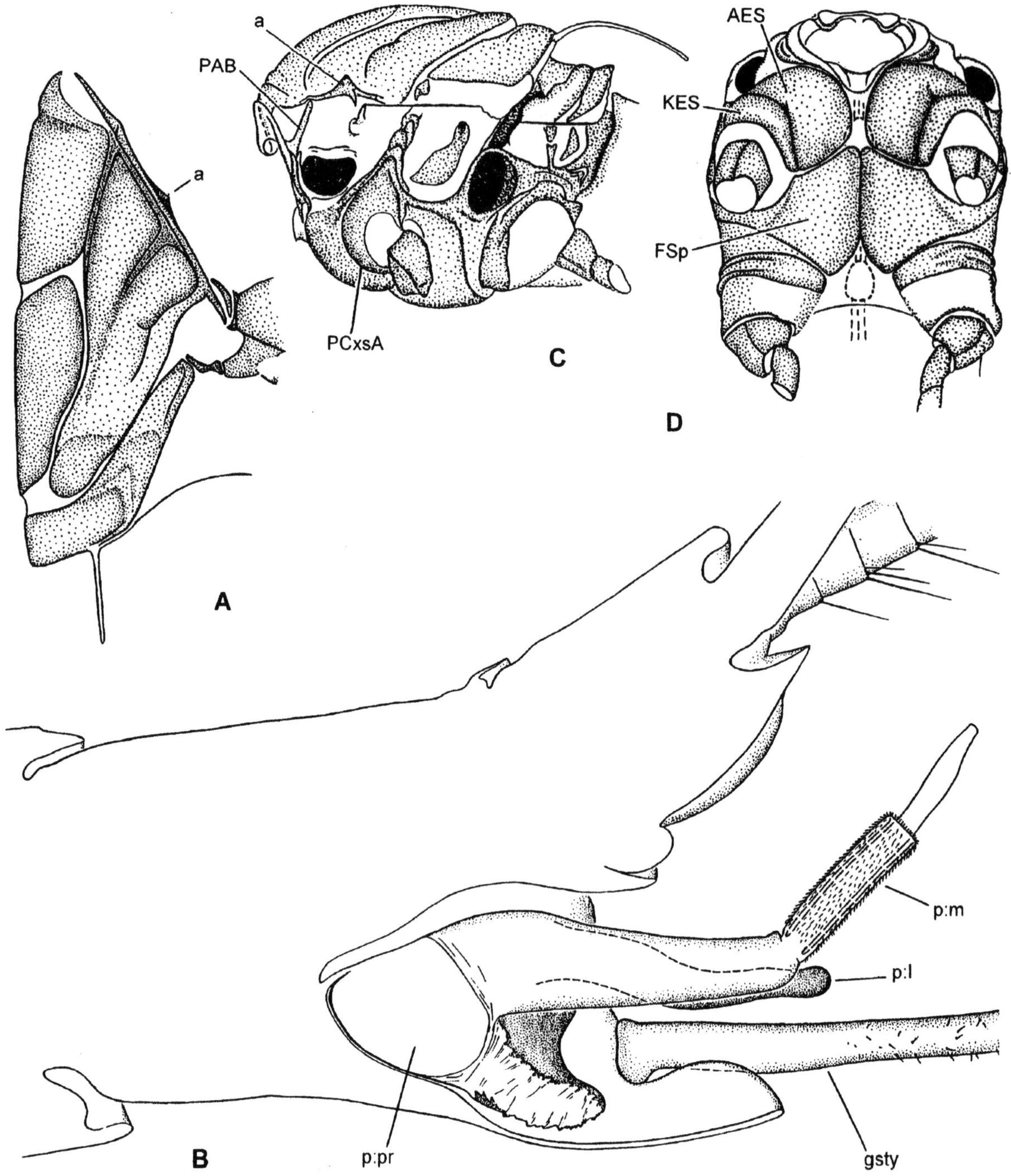

Figure 46. Discoglossata-Geminovenata-Oligoneuriella/g2
A–B – Oligoneuriella/g2 *pallida* [*Oligoneuria*]: **A** – right half of imaginal mesonotum; **B** – longitudinal median section of male imaginal abdominal segments IX and X with genitals, telescopic lobe protracted. **C–D** – Oligoneuriella/g2 *tskhomelidzei* [*Oligoneuriella*]:**C** – pterothorax, lateral view; **D** – the same, ventral view (nerve ganglions shown by interrupted lines).

Abbreviations: **a** – primary point of joining of posterior arm of prealar bridge with scutum, **PAB** – secondary dorso-posterior arm of prealar bridge; **AES** – anepisternum; **FSp** – furcasternal protuberance; **gsty** – gonostylus; **KES** – katepisternum; **p:l** – lateral longitudinal sclerotized lobe of penis; **p:m** – median telescopic lobe of penis; **p:pr** – proximal part of penis; **PCxsA** – anterior paracoxal suture.

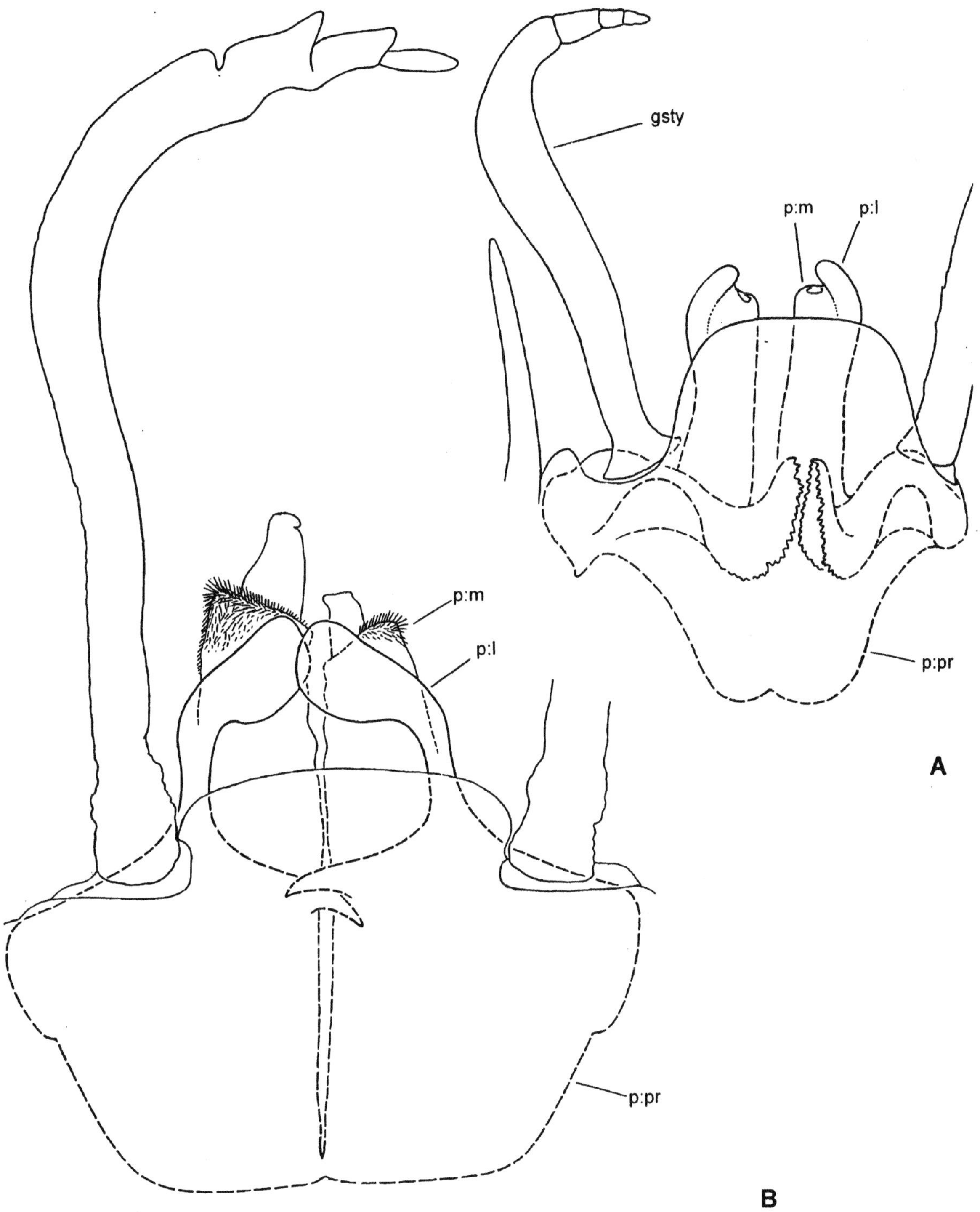

Figure 47. Oligoneuriella/g1: genitals of male imago, ventral view.
A – Oligoneuriella/g2 *pallida* [*Oligoneuria*], telescopic lobes retracted. **B** – Oligoneuriopsis/g(1) sp.O1 (="*Oligoneuriopsis dobbsi*" sensu Kimmins 1960), telescopic lobes partly protracted.

Abbreviations: **gsty** – gonostylus; **p:l** – lateral longitudinal sclerotized lobe of penis; **p:m** – median telescopic lobe of penis; **p:pr** – proximal part of penis.

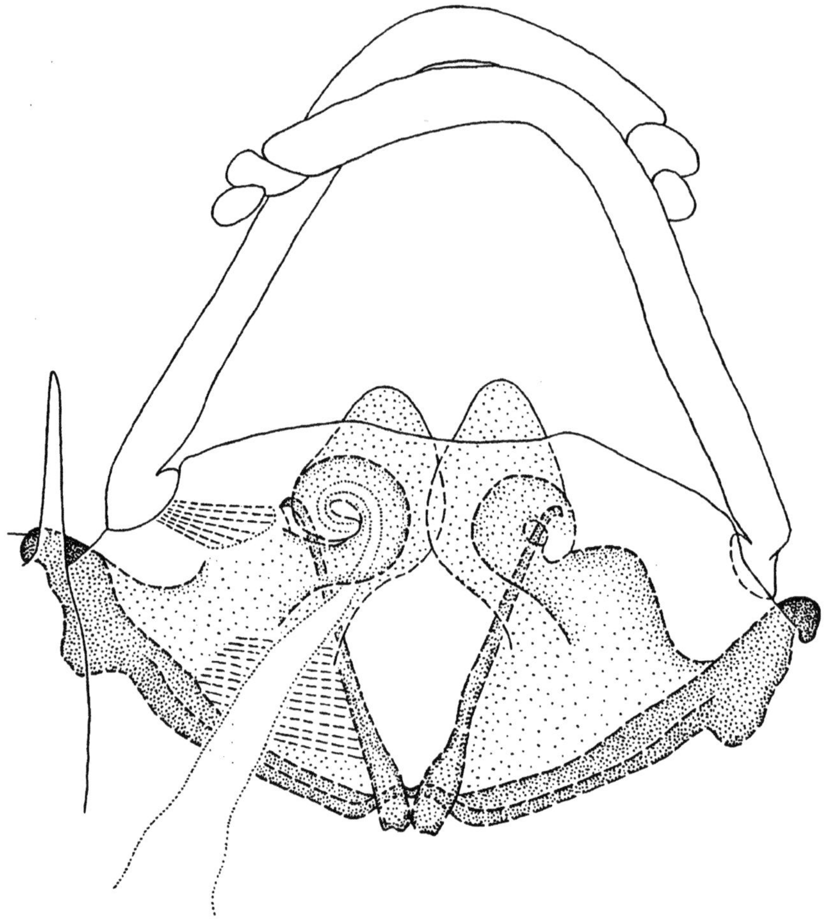

A

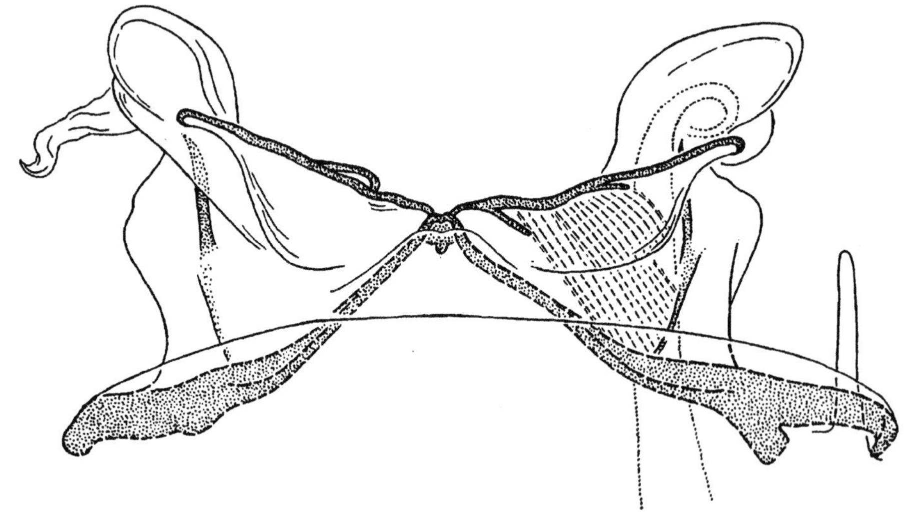

B

Figure 48. Geminovenata-Elassoneuria/g1-Madeconeuria/g* *insulicola* [*Elassoneuria*], genitals of male imago.
A – genitals with penes retracted, ventral view; **B** – penes protracted, dorsal view. In both figures hidden parts of penes shown by interrupted lines and dotted, in left half (left in Fig.A and right in Fig.B) muscles shown by interrupted lines and gonoduct shown by dotted lines.

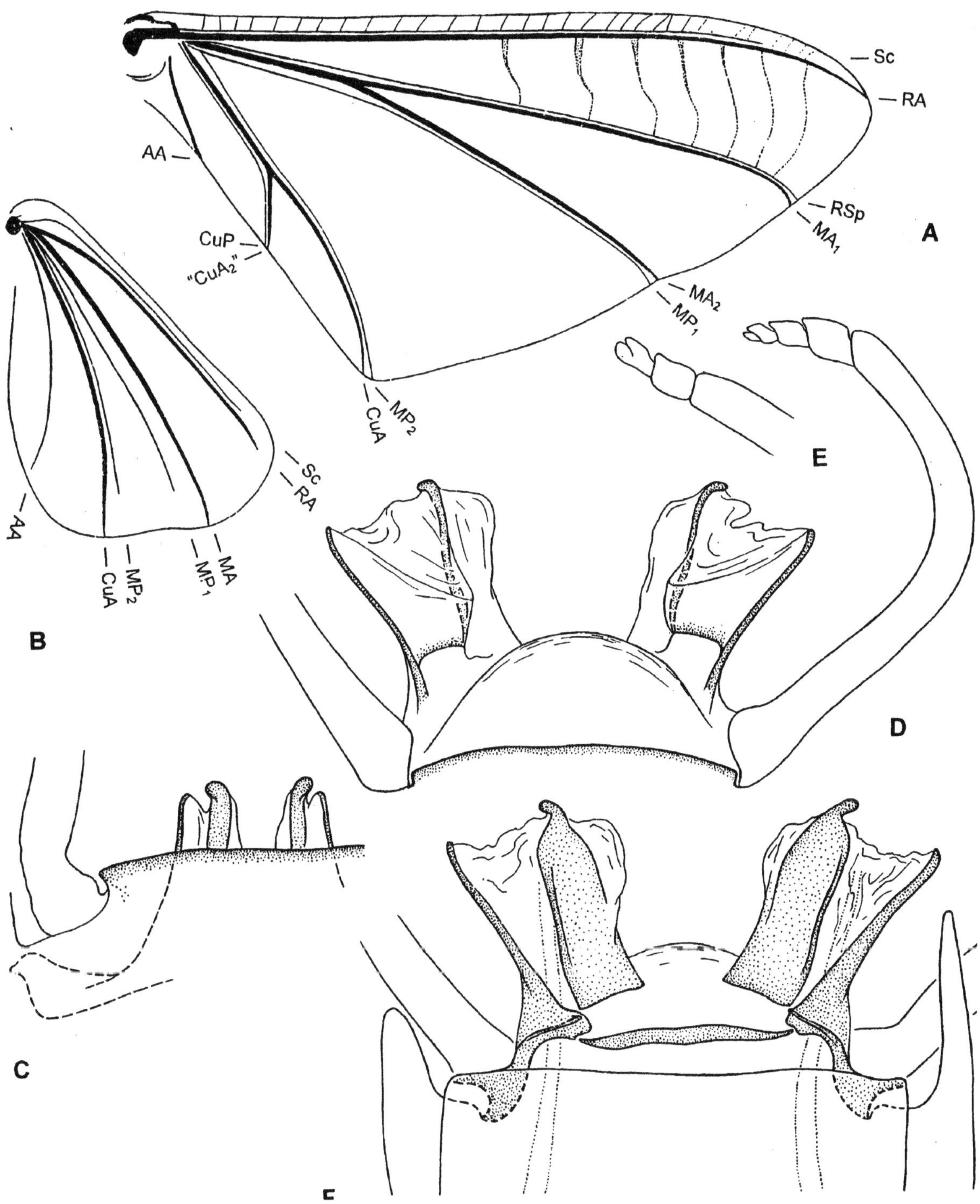

Figure 49. Elassoneuria/g2 *congolana* [*Elassoneuria*], male imagoes.
A–B – fore and hind wings (all veins conventionally shown by black, lying in one plain and spaced); **C** – genitals with penes retracted, ventral view; **D** – genitals with penes protracted, ventral view; **E** – apex of gonostylus, another specimen; **F** – genitals with penes protracted, dorsal view.

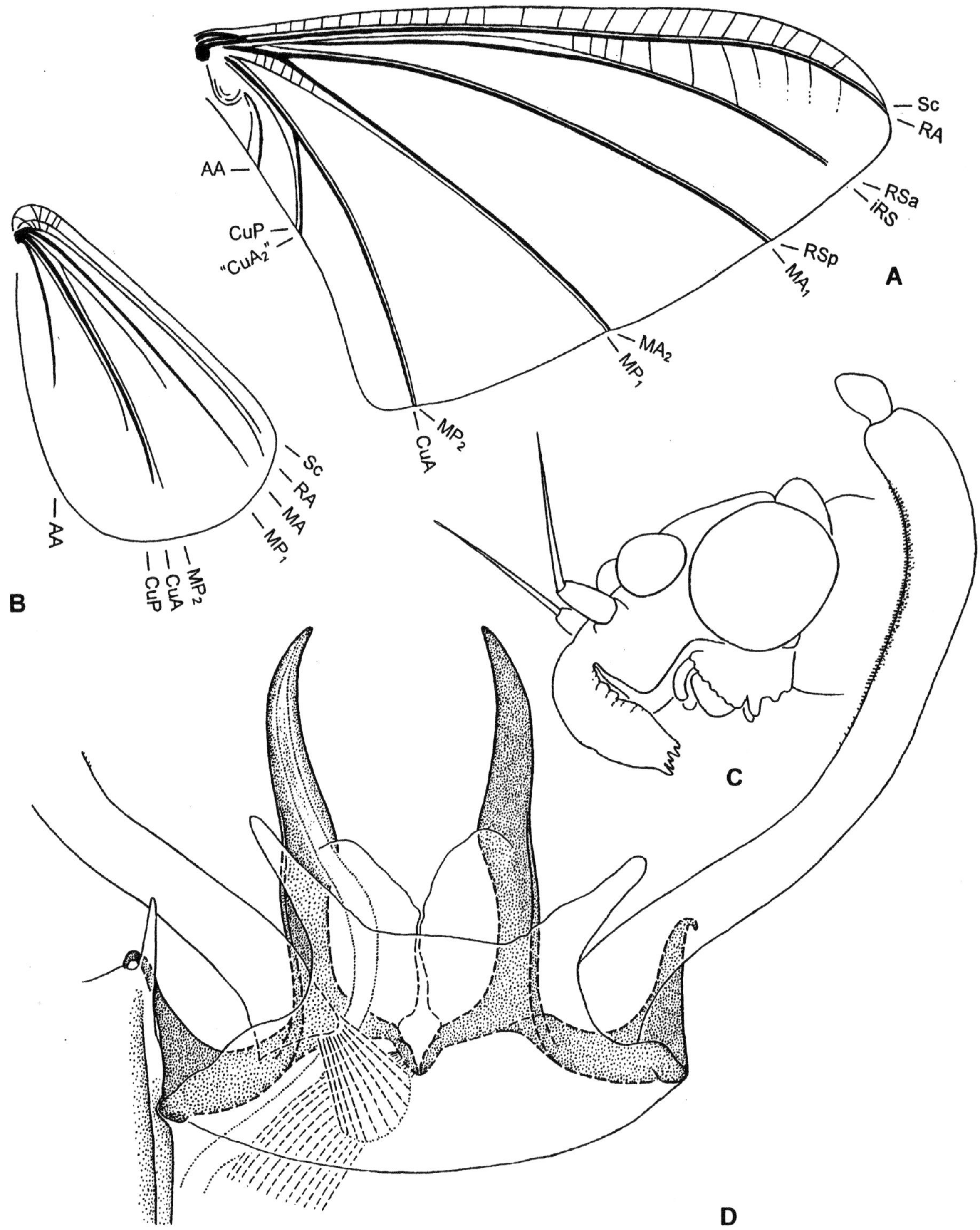

Figure 50. Discoglossata-Geminovenata-Oligoneuria/f5=g6 sp.O5, male imago.
A–B – fore and hind wings (all veins conventionally shown by black, lying in one plain and spaced); **C** – head, lateral-ventral view; D – genitals, ventral view (hidden parts of penes shown by interrupted lines and dotted, in left half muscles shown by interrupted lines and gonoduct shown by dotted lines).

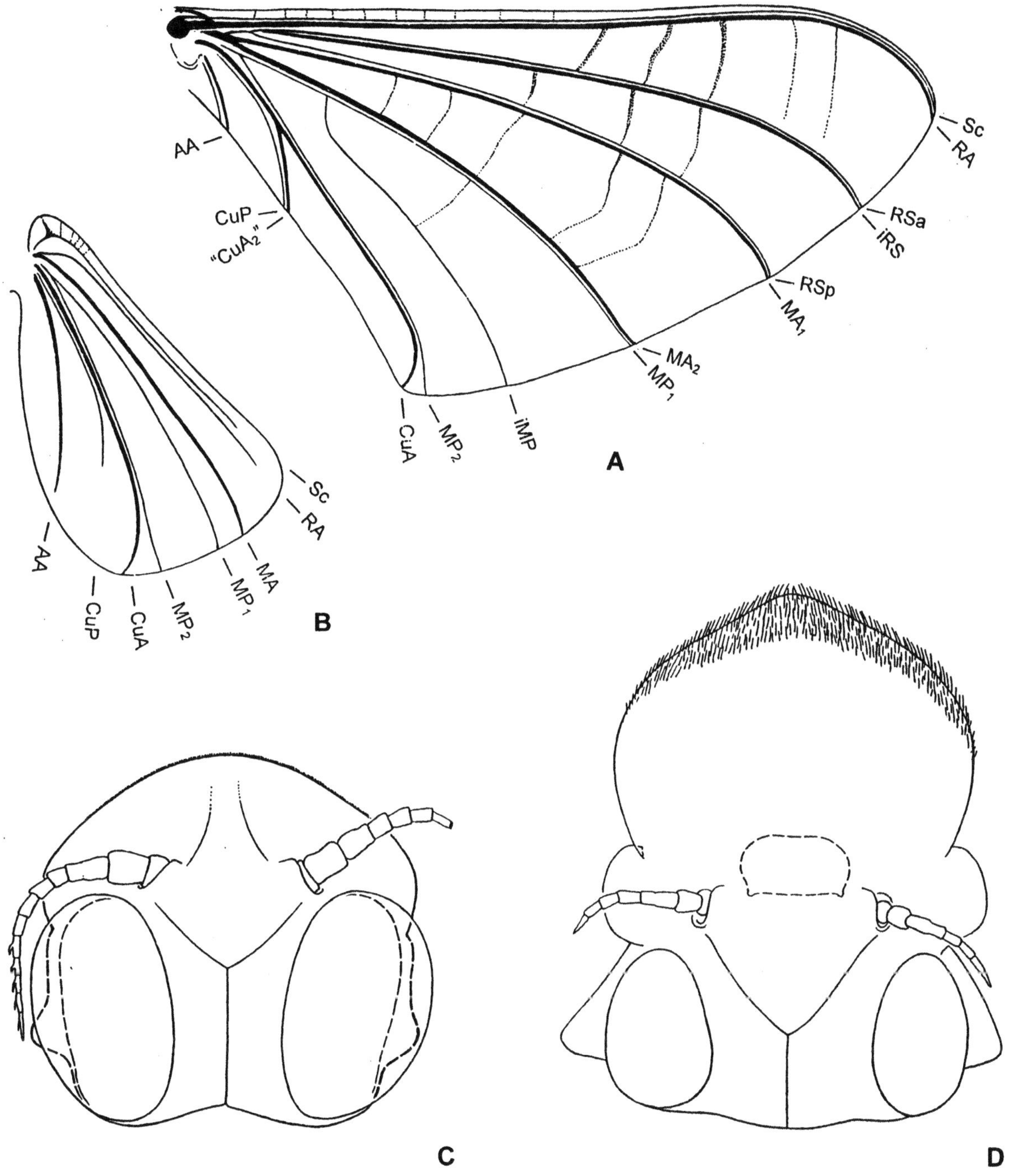

Figure 51. Oligoneuria/f4=g5.
A–B – Spaniophlebia/g1 sp.O3, fore and hind wings (all veins conventionally shown by black, lying in one plain and spaced). **C** – Lachlania/g1 *powelli* [*Lachlania*], head of mature male larva, dorsal view. D – Oligoneuria/f5=g6 sp.O4, head of young larva, dorsal view.

2.2;1,2. **Heptagennota**, or **Heptagenia/f2=g1**

(Euplectoptera Anteritorna Bidentiseta Branchitergaliae Heptagennota)

(Figs 8:E–F; 9:C–E; 52–65)

Nomen hierarchicum: **Heptagenia/f2=g1** (sine *Oligoneuria*; incl. *Pseudiron*) [g: *Heptagenia* Walsh 1863: 197, typus *Palingenia flavescens* Walsh 1863 (design. Eaton 1868b: 90)].
Nomen circumscribens: **Heptagennota** Kluge 2000: 248.
In circumscription matches:
— fam. Heptageniidae: Edmunds & Traver 1954a: 237;
— superfam. Heptagenioidea: Kluge & Studemann & Landolt & Gonser 1995: 105;
— Heptagennota, or Heptagenia/f2=g1: Kluge 2000: 248.

References. Edmunds & Allen & Peters 1963: ♉*; – Tshernova 1970: ♉* ⊕*; – Koss & Edmunds 1974: ⊙*; – Edmunds & Jensen & Berner 1976: ♉* ⊕*; – Wang & McCafferty 1995b: ♉* ⊕*.

Autapomorphies of Heptagennota.

(1) Larva is dorsoventrally flattened, with flat sterna; its head is prognathous and forms a head shield; labrum is transverse; legs are enlarged and widely separated, with femora widened and flattened (Figs 54:A; 55:A; 58:A). Most representatives inhabit current waters and use this body form to press the body to substrate.

Frons forms more or less expressed **head shield**, covering mouthparts from above; clypeus is usually membranous and hidden under the head shield, so base of labrum is also hidden under the head shield (FIG.65:I). The least expressed is head shield of *Pseudiron* (FIG.52:A) which has carnivorous specialization of mouth apparatus; but here also base of labrum is covered from above by a fold of frons (i. e. head shield) (FIG.52:D). Sclerotized clypeus is secondarily developed in some species (particularly, in Epeiron/g1 *binerve* [*Rhithrogena*] – Kluge 1987b: Figs 6–7). While in larvae of all Heptagennota head shield is present, imaginal head usually is not modified; but in selected species of Heptagennota face plate of subimago and imago is large, directed anteriorly and more or less repeats shape of larval head shield (some authors regarded this as a generic character, but it evolved repeatedly in various groups inside Heptagennota). Besides Heptagennota, larval head shield independently appeared in some other mayflies (see Index of characters [1.1.4]).

Labrum is usually transverse: width at base much exceeds length, and total width much exceeds width at base. The only exception is made by *Cinygma*, whose labrum is secondarily narrowed.

Larva lost primary swimming specialization, because its legs, being enlarged and widely separated, are unable to stretch posteriorly and press to the body when larva swims. In different taxa of Heptagennota method of swimming is different: Larva can keep its legs free and shakes them synchronously with abdominal movements, thus translational movement is served both by abdomen and legs (FIG.9:D); this is characteristic for Heptagenia/f6=g5. In some non-related taxa (among Ecdyonurus/fg1, Epeorus/fg1 and others) such method of swimming was changed to swimming by legs solely, in this case the primary abdominal dorsoventral swimming movements are lost (FIG.9:E); as femora of Heptagennota are flattened, such method of swimming is rather effective. Larvae of Rhithrogena/fg2 have another method of swimming, when primary abdominal movements are retained, but legs are folded and pressed to the body by sides (FIG.9:C).

(2) Subimaginal lateral pigmented area of mesonotum posteriorly bifurcates to a lateral portion (occupying antero-lateral part of sublateroscutum and whole lateroscutum) and a lateroparapsidal strip; the lateroparapsidal strip stretches far posteriorly, including into itself lateroparapsidal suture and being laterally limited by a **relief line** which posteriorly is curved laterally. In the case initial for Heptagennota (in *Pseudiron*, *Arthroplea*, Heptagenia/f5=g4 and *Cinygma*) the lateroparapsidal stripe is widened posteriorly, because lateroparapsidal suture posteriorly diverges with relief line and meets with medioparapsidal suture (FIGS 52:C; 54:I). In Rhithrogena/fg2 and Epeorus/fg1 lateroparapsidal stripe is curved laterally, because lateroparapsidal suture posteriorly is curved laterally, repeating curvation of the relief line (FIGS 63:A, C; 64:D). *Paegniodes* has intermediate structure (FIG.64:A).

Probably, the bifurcation of lateral pigmented area and shape of its lateral portion are plesiomorphic (see Index of characters [2.2.14]), while the enlargement of lateroparapsidal stripe and presence of the relief line are unique autapomorphies (compare with plesiomorphic condition in FIG.18:E).

(3) Imaginal and subimaginal anterior paracoxal suture of mesothorax is incomplete (i. e. does not turn to ventral side of episternum and terminates not reaching sternum) and strongly transferred posteriorly toward coxal conjunctiva, because of which katepisternum is reduced to a small piece above coxal conjunctiva and a narrow ridge stretching from this piece by anterior margin of coxal conjunctiva (FIGS 52:B; 54:J). Besides Heptagennota, paracoxal suture is incomplete in *Rallidens* and Furcatergaliae only; in most of them it is not transferred posteriorly, but in some of them (for example, Caenoptera) is more or less transferred posteriorly.

(4) Imaginal and subimaginal furcasternal protuberances are separated at least in posterior part (FIG.52:B), because metathoracic nerve ganglion is transferred into furcasternum.

Non-unique apomorphy: the same in selected groups of Tridentiseta and Furcatergaliae (see Index of characters [2.2.23]), but not in Eusetisura. In majority of Heptagennota metathoracic nerve ganglion is situated in posterior part of mesothoracic furcasternum, because of this median concavity of furcasternum is widened posteriorly (FIG.56:C–D); only in Ecdyonurus/fg1 this ganglion is transferred into anterior part of furcasternum (FIG.57:A–B).

(5) Cubital field of fore wing [see Anteritorna (1)] usually with 4 intercalaries forming 2 pairs, among which the posterior pair arise more proximally and is the longest; in each pair the anterior intercalary (nearest to convex CuA) is concave, and the posterior intercalary (nearest to concave CuP) is convex, that serves regular alternating of concave and convex veins in cubital field (FIG.53:A, C). Among Heptagennota only one species – Epeiron/g1 *binerve* [*Rhithrogena*] – has 2 intercalaries only (Kluge 1987b:Fig.2).

Non-unique apomorphy: this or that number of intercalaries in cubital field evolved independently in different groups of Ephemeroptera (see Index of characters [2.2.52]), and exactly the same two pairs are present in some Metretopus/fg1 (because of this, some authors erroneously placed them to the family Heptageniidae).

In other respects wings of Heptagennota retain normal for Euplectoptera structure and function; only in one undescribed species of Rhithrogena/fg2 from Chukotka Peninsula, fore and hind wings are proportionally diminished and probably unavailable for flight (FIG.8:E–F).

(6) Imaginal and subimaginal paracercus is always vestigial. Non-unique apomorphy (see Index of characters [2.3.22]).

Characters of unclear phylogenetic status.

(7) Prostheca of initially left mandible (with mola projected distally) is setiform or lost; prostheca of initially right mandible (with mola projected proximally) is always lost (FIGS 54:E–F; 58:B–C) [in a few representatives left and right mandibles changed places – see below, *Arthroplea* (2) and Variable characters of Radulapalpata]. Median margin of mandible between kinetodontium and mola often bears more or less long dense setae: they can either be irregular, or form a compact tuft close to prostheca and kinetodontium, or form a regular row along median margin; rarely these setae are absent (in *Pseudiron*, *Cinygma* and *Bleptus*). It is unclear if reduction of prostheca is an autapomorphy of Heptagennota in general, or it is only an autapomorphy of Pentamerotarsata, because in *Pseudiron* mandibles are strongly specialized and could lost prostheca independently.

(8) Labial palp is 2-segmented; its distal ($2^{nd}+3^{rd}$) segment lacks muscle inside (FIGS 52:F; 53:E; 57:H; 60:C). Non-unique apomorphy (see Index of characters [1.1.55]). Fusion of 2nd and 3rd segments could appear independently in connection with various palp specialization [see below, *Pseudiron* (1), *Arthroplea* (1) and Radulapalpata (1)].

Plesiomorphies of Heptagennota. Gonostylus with 2 distal segments (only in *Arthroplea* number of segments is secondary increased, but not diminished). Both dentisetae [see Bidentiseta (1)] are always present, being enlarged in *Pseudiron* (FIG.52:E) and vestigial in many Pentamerotarsata (FIGS 54:M; 58:F–H; 61:H; 65:F). All tergalii I–VII are present (tergalius VII is most greatly reduced in *Macdunnoa*).

Variable characters of Heptagennota. Imaginal and subimaginal claws of most species are ephemeropteroid; only in selected Tien-Shan–Himalayan species of *Cinygmula*, *Himalogena* and *Caucasiron* on each leg of imago and subimago both claws are pointed, and in selected species of Rhithrogena/fg2 and Epeorus/fg3 on fore legs of male both claws are blunt (but in the last case subimago has all claws ephemeropteroid). Non-unique characters (see Index of characters [2.2.77] and [2.2.85]).

Size. Fore wing length 5–21 mm.

Age and distribution. For certain known beginning from Palaeogene (see *Kageronia* and Radulapalpata INCERTAE SEDIS); reported also from Late Cretaceous (see Pentamerotarsata INCERTAE SEDIS). Recently distributed in Arctogea (Holarctic + Oriental + Ethiopian Regions) and Central America; dominate in Holarctic.

Heptagennota are divided into *Pseudiron* and Pentamerotarsata.

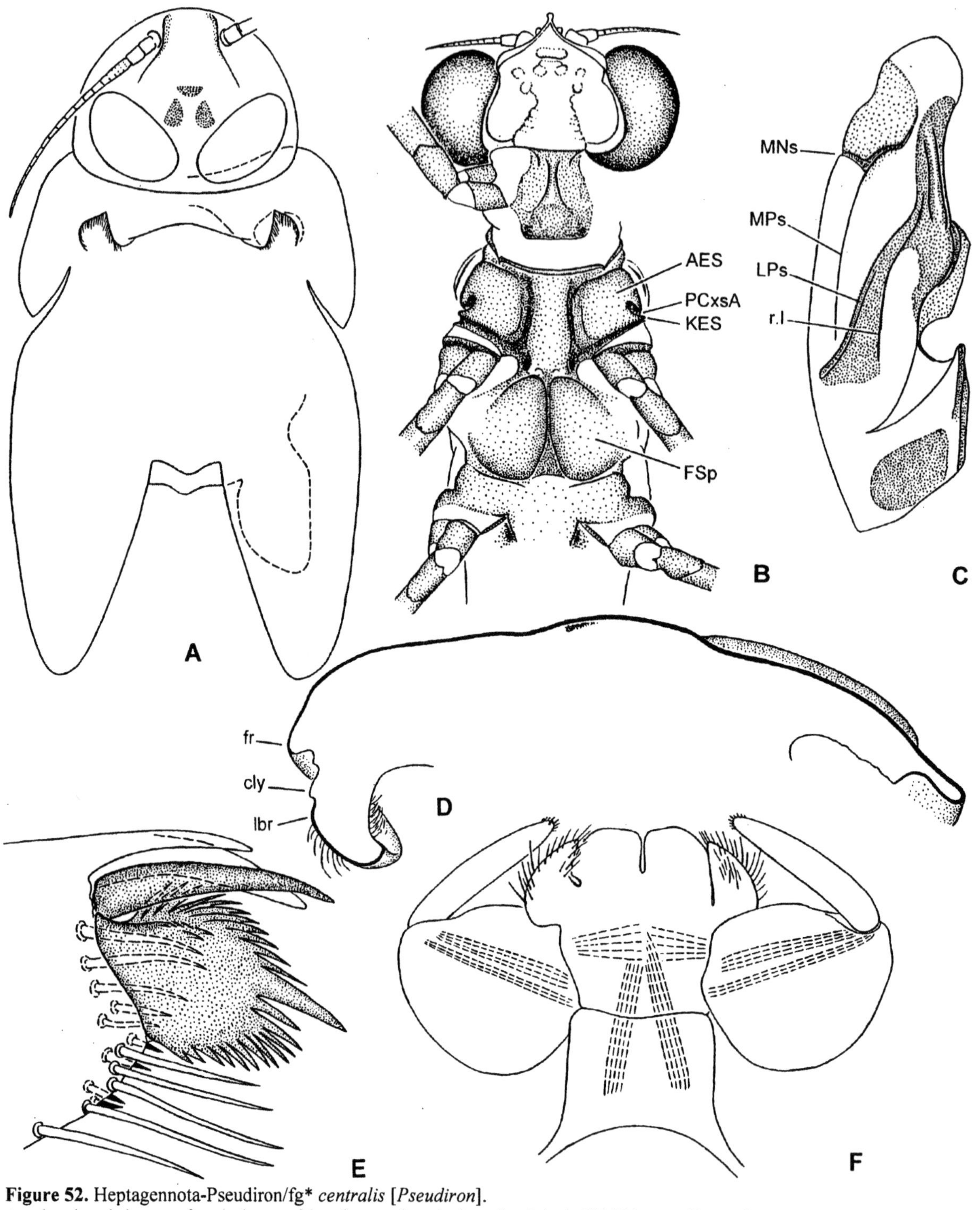

Figure 52. Heptagennota-Pseudiron/fg* *centralis* [*Pseudiron*].

A – head and thorax of male larva of last instar, dorsal view (in right half hidden outlines of pronotum, mesonotum and metanotum are shown by interrupted lines); **B** – male imaginal head and thorax, ventral view (left fore leg removed); **C** – subimaginal exuviae of right half of mesonotum; **D** – median section of larval head capsule and labrum; **E** – apex of left maxilla, dorsal view (dentisetae shown by dots); **F** – labium, dorsal view (in left half) and ventral view (in right half) (muscles shown by interrupted lines) (the same scale as in Fig.D). (E – from Kluge 1998)

Abbreviations: **AES** – anepisternum; **cly** – clypeus; **fr** – frons; **FSp** – furcasternal protuberance; **KES** – katepisternum; **lbr** – labrum; **LPs** – lateroparapsidal suture; **MNs** – mesonotal suture; **MPs** – medioparapsidal suture; **PCxsA** – anterior paracoxal suture; **r.l** – relief line.

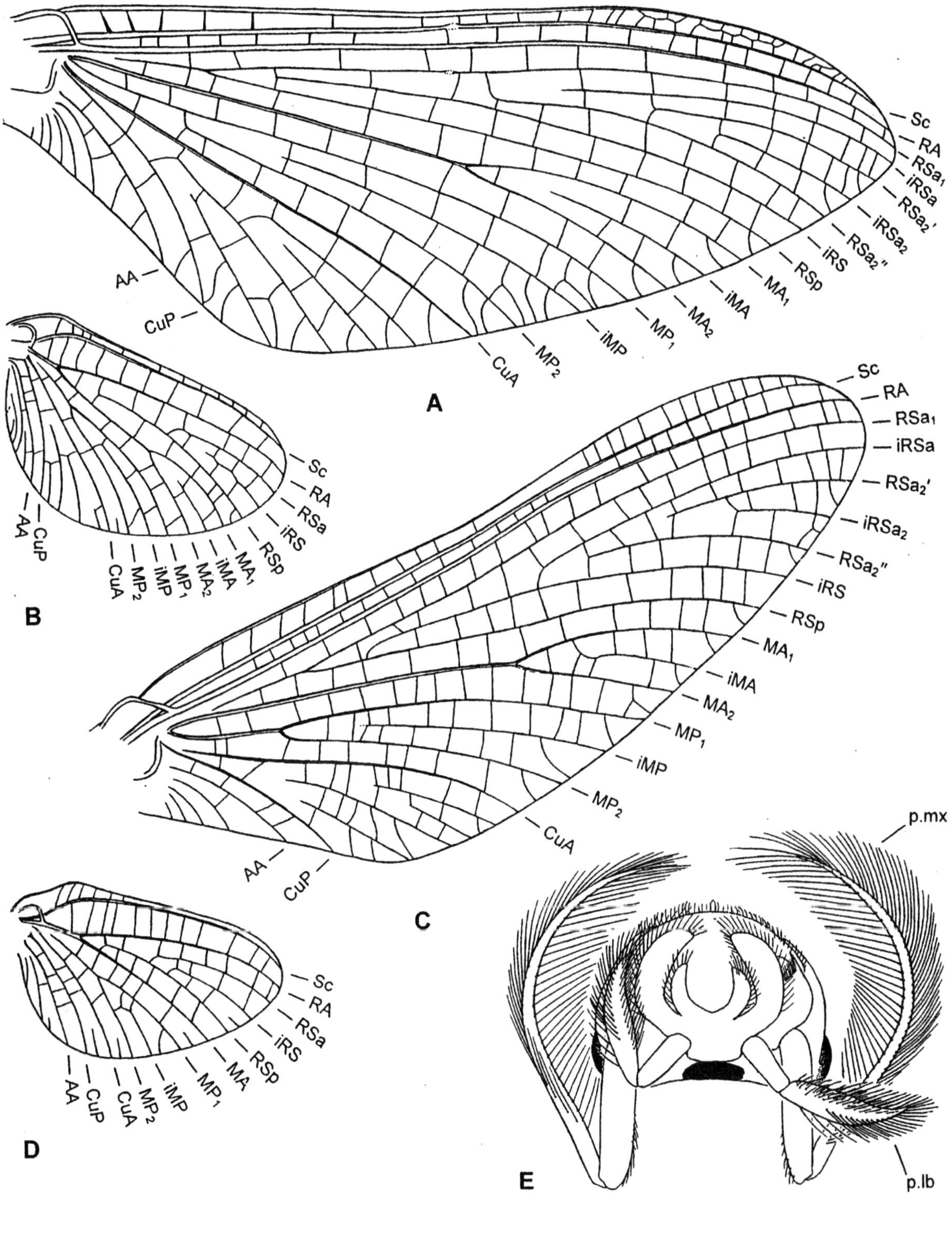

Figure 53. Heptagennota.
A–B – Pseudiron/fg(1) *centralis* [*Pseudiron*], fore and hind wings. **C–E** – Arthroplea/fg1 *congener* [*Arthroplea*]: **C–D** – fore and hind wings; **E** – larval head, ventral view. (E – from Kluge 1997d)
Abbreviations: **p.lb** – labial palp; **p.mx** – maxillary palp.

2.2;1,2-1. **Pseudiron/fg(1)**

(Anteritorna Bidentiseta Branchitergaliae Heptagennota *Pseudiron*)

(Figs 52; 53:A–B)

Nomen hierarchicum: **Pseudiron/fg(1)** [f: Pseudironinae Edmunds & Traver 1954a:237; g: *Pseudiron* McDunnough 1931a: 91, typus *P. centralis* McDunnough 1931 (design. orig.)].
In circumscription matches:
— gen. *Pseudiron* McDunnough 1931a: 91;
— subfam. Pseudironinae Edmunds & Traver 1954a: 127;
— fam. Pseudironidae: McCafferty 1991a: 349;
— Pseudiron/fg1: Kluge 2000: 251.
Monospecific taxon.

References. McDunnough 1931a: ⊕; – Needham & Traver & Hsu 1935: ⊕; – Burks 1953: ℧ ⊕; – Edmunds & Allen & Peters 1963: ℧; – Peters & Jones 1973: ℧; – Edmunds & Jensen & Berner 1976: ℧ ⊕; – Pescador 1985: ℧ ⊕; – McCafferty & Provonsha 1986: ℧.

Autapomorphies of *Pseudiron*.

(1) Larva with following peculiarities. Mouth apparatus is specialized for carnivorism: mandibles lost their asymmetry, each mandible is shortened perpendicular to axis of articulation, with stout incisor and kinetodontium, prostheca is lost [see Heptagennota (7)], mola lacks grater; superlinguae are lost; maxilla lacks apical-ventral setae, its biting margin is shortened, with 2 canines (instead of three initial ones), proximal dentiseta [see Bidentiseta (1)] is strongly enlarged and looks as a wide dentate sclerotized plate (FIG.52:E) (similar modifications of mouthparts occur in some other specialized predators). Labial palp [2-segmented – see Heptagennota (8)] with strongly widened, nearly round 1st segment (in contrast to Radulapalpata, this segment is flat, and muscles situated in it are not thickened – FIG.52:F). Larval body is flattened, pronotum has unique structure (FIG.52:A): lateral lobes are expanded far posteriorly, overlapping mesonotum and partly fused with it (that somewhat resembles Ecdyonurus/fg2); mediad of these lobes, posterior margin of pronotum forms a pair of cavities opened backward; margins of cavities bear numerous fine setae; fore wing buds are spread. Larval legs are adopted for inhabitancy on sandy ground: long; tarsus arched; claw are very long, slender, slightly arched, without denticles. Tergalii have unique structure: besides fibrillose lobe [see Branchitergaliae (3)] each tergalius II–VII has a ventral process arising near middle of lamellate lobe.

(2) Patella-tibial suture (initially present on middle and hind legs) is lost on all legs of larva [see (1)], subimago and imago (non-unique apomorphy – see Index of characters [1.2.18]).

Plesiomorphies of *Pseudiron*. Imaginal and subimaginal furcasternal protuberances are contiguous at anterior part and diverge only at posterior part, inside which metathoracic nerve ganglion is situated [see Heptagennota (4)] (FIG.52:B). 1st tarsal segment is fused with tibia (in contrast to Pentamerotarsata) and non-shortened. Penis has no sclerites peculiar for Pentamerotarsata. Maxillary palp is 3-segmented. In contrast to Pentamerotarsata, submentum is well-developed (FIG.52:F). In contrast to Pentamerotarsata, larva completely retains primary siphlonuroid swimming specialization: abdomen is large, caudalii with dense long primary swimming setae (only in distal part segments are elongate and lack setation), without secondary swimming setae. Paracercus is subequal to cerci.

Size. Fore wing length 10–12 mm.

Distribution. Nearctic.

Species composition of Pseudiron/fg(1). 1 species – *centralis* McDunnough 1931 [*Pseudiron*] (syn. subj.: *meridionalis* Traver 1935 [*Pseudiron*]).

Material examined: *centralis* [*P.*]: ℧–○–⊕/⊙.

***

2.2;1,2-2. **Pentamerotarsata**, or **Heptagenia/f3=g2**

(Anteritorna Bidentiseta Branchitergaliae Heptagennota Pentamerotarsata)

(Figs 8:E–F; 9:C–E; 53:C–E; 54–65)

Nomen hierarchicum: **Heptagenia/f3=g2** (sine *Pseudiron*; incl. *Arthroplea*).
Nomen circumscribens: **Pentamerotarsata** Kluge 2000: 251.
In circumscription matches:
— subfam. Heptageniinae Needham 1901: 419;
— fam. Heptageniidae: Lestage 1917: 266;
— grex famm. Heptageniidae: Wang & McCafferty 1995b: 252;
— Pentamerotarsata, or Heptagenia/f3=g2: Kluge 2000: 251.

References. Needham 1901: ℧* ⊕*; – Needham & Traver & Hsu 1935: ℧* ⊕*.

Autapomorphies of Pentamerotarsata.

(1) Submentum is vestigial, membranous, lacks muscles (FIGS 53:E; 57:H; 60:C) (in contrast to other Ephemeroptera which have muscles going from submentum to mentum – FIG.52:F). In other respects labium is differently specialized in *Arthroplea* and Radulapalpata (see below).

(2) Larval fore wing buds are fused with scutellum up to apex of scutellum; thus, posterior margin of mesonotum between wing buds is straight or concave, and free portion of each wing bud narrows from its extreme base to apex (FIGS 54:A; 55:A; 58:A). Non-unique apomorphy (see Index of characters [1.2.6]); in contrast to *Pseudiron* (Fig.52:A) and many Tridentiseta.

(3) In imago and subimago on all legs 1st tarsal segment is distinctly articulated with tibia – thus, all tarsi are 5-segmented (like in condition primitive for Amyocerata). This is a secondary restoration of tibia-tarsal articulation (see Chapter II); non-unique apomorphy (see Index of characters [2.2.84]).

(4) Larval claws are usually sharply curved and turned ventrally (posteriorly), perpendicular to the femur-tibia flatness; probably, primary for Pentamerotarsata is presence of one row of denticles on inner side of claw and several larger subapical denticles on anterior (dorsal) side of claw. The denticles on inner side of claw are retained in *Arthroplea* and *Kageronia* only (FIG.54:G), being lost in other Radulapalpata; the subapical denticles are present in *Arthroplea* and many Radulapalpata, being lost in selected species. Non-unique character: the same in some Ephemerella/fg1, Leptophlebia/fg1 and other rheophilous mayflies.

(5) Penis has peculiar composite structure: on apex of each penis lobe sclerotization of ventral wall terminates by a projection named **ventral sclerite**; on lateral side it is continued to a projection named **outer sclerite**; outer sclerite bears a **latero-dorsal spine** directed dorsally; on its median side each penis lobe bears a **median titillator** – sclerotized pointed appendage; from its base titillator is directed toward apex of penis, curved in median plane, and its apex is directed ventrally (FIG.59:B–C; Bogoescu & Tabacaru 1962:Fig.2). In many Pentamerotarsata, which have penis lobes fused at a long distance, the median titillators being situated between penis lobes, are dipped by their bases into deep pouches, thus in rest position only apices of titillators are projected outside; in excited condition titillators are turned out from the pouches. In other representatives of Pentamerotarsata (many Rhithrogena/fg1) penis lobes are deeply separated and median titillators are situated on their median surfaces; probably such external position is secondary for Pentamerotarsata. These details of penis structure are present in *Arthroplea* and majority of Radulapalpata, but in selected representatives some of these details are lost.

(6) Larval caudalii are usually long, primary swimming setae [see Euephemeroptera (3)] are more or less reduced. If swimming setae are present, they are simple and non-specialized: in contrast to *Pseudiron*, many Eusetisura and Tridentiseta, they are not thickened at bases and not curved distally (FIG.57:L). In spite of their simple structure, these setae can be recognized as primary swimming setae, as they form rows on lateral sides of paracercus and median sides of cerci only, but not on lateral sides of cerci. Secondary swimming setae on lateral sides of cerci are usually absent, being present in some species of Stenonema/g1 only. In many Radulapalpata swimming setae are completely lost. Non-unique apomorphy (see Index of characters [1.3.66]).

Reduction of primary swimming setae is connected with adaptation of majority (but not all) of Pentamerotarsata to inhabitancy in swift current. At the same time, in contrast to some other rheophilous mayflies, in majority of Pentamerotarsata larval paracercus is not shortened, being as long as cerci; only in Epeorus/fg1, *Acanthomola* and Anepeorus/ /fg1 larval paracercus is reduced to one segment (see Index of characters [1.3.64]); intermediate stages of shortening of paracercus in Pentamerotarsata do not occur [in imago and subimago paracercus is always vestigial – see Heptagennota (6)].

Size. Fore wing length 5–21 mm.

Age and distribution. For certain known beginning from Palaeogene (see *Kageronia* and Radulapalpata INCERTAE SEDIS); reported also from Late Cretaceous (see Pentamerotarsata INCERTAE SEDIS). Recently distributed in Arctogea (Holarctic + Oriental + Ethiopian Regions) and Central America; dominate in Holarctic.

Systematic position of Pentamerotarsata. Basing on a groundless assumption that Pentamerotarsata originated from Eusetisura, in was suggested to unite Pentamerotarsata and Eusetisura in a taxon **Setisura** McCafferty 1991 (which circumscription is close to Branchitergaliae, but does not include *Pseudiron*).

Pentamerotarsata are divided into *Arthroplea* and Radulapalpata; Anepeorus/fg1 has uncertain systematic position (p.209).

2.2;1,2-2/1. Arthroplea/fg(1)

(Bidentiseta Branchitergaliae Heptagennota Pentamerotarsata *Arthroplea*)

(Figs 53:C–E)

Nomen hierarchicum: **Arthroplea/fg(1)** [f: Arthropleidae Balthasar 1937: 204; g: *Arthroplea* Bengtsson 1908: 239, typus *A. congener* Bengtsson 1908 (monotypy); syn. subj.: ***Remipalpus*** Bengtsson 1908: 242, typus *R. elegans* Bengtsson 1908 (monotypy); syn. subj.: ***Haplogenia*** Blair 1929: 254, typus *H. southi* Blair 1929 (monotypy) (synn. subj.: *congener* [*A.*] = *elegans* [*R.*] = *southi* [*H.*])].

In circumscription matches:
— gen. *Arthroplea* Bengtsson 1908: 239;
— gen. *Remipalpus* Bengtsson 1908: 242;
— subfam. Arthropleinae: Edmunds & Allen & Peters 1963: 13;
— fam. Arthropleidae Balthasar 1937: 204;
— Arthroplea/fg1: Kluge 2000: 251.

References. Bengtsson 1909: ℧ ◯ ⊕; – 1930: ℧ ⊕; – Needham & Traver & Hsu 1935: ℧ ⊕; – Balthasar 1937: ℧ ⊕; – Burks 1953: ℧ ⊕; – Edmunds & Allen & Peters 1963: ℧; – Froechlich 1964: ℧; – Koss 1968: ⊙; – Landa 1969a: ℧ ⊕; – Edmunds & Jensen & Berner 1976: ℧ ⊕; – Soldán 1979a: ℧; – Studemann & Landolt & Tomka 1987: ℧ ⊕.

Autapomorphies of *Arthroplea*.

(1) Mouth apparatus has unique structure, being adopted to catch and filtrate food particles dispersed in water (FIG.53:E): Maxillary palp is unusually elongate, 2-segmented. Its 1st segment is elongate so greatly, that being turned posteriorly, reaches mesonotum; trachea which goes to muscles of this segment (i. e. to flexor and extensor of 2nd segment), is thicker than tracheae in other appendages. Distal (2nd+3rd) segment of maxillary palp is unusually long, about twice as long as the 1st segment, slender, arched, bears two regular longitudinal rows of long filtering setae. Working by its maxillary palps, larva can gather particles dispersed in water, and also can swim back to front. Maxilla has a single canine (instead of 3 initial ones), lacks apical-ventral row of pectinate setae, bears long simple setae on apical side. Glossae are crescent-shaped, paraglossae are much larger, crescent-shaped. Distal (2nd+3rd) segment of labial palp [see Heptagennota (8)] is large, crescent-shaped, pointed, with numerous long filtering setae, among which are present: a regular longitudinal row of setae on ventral-outer side, a regular longitudinal row of setae on inner side, a field of setae on dorsal side and other setae.

(2) Mandibles are inverted: mandible with distally projected mola is right one, and mandible with proximally projected mola is left one. Non-unique apomorphy, the same in selected species among Radulapalpata.

(3) Each tergalius I–VII lost fibrillose portion, costal and anal ribs [see Branchitergaliae (3)], has a large anterior-proximal lobe and pointed apex (Edmunds & al. 1986:Fig.54; Studemann & al. 1987: Fig.13).

(4) On fore wing RSa_2 lost its connection with RSa and has a form of intercalary which begins more distally than iRSa (FIG.53:C). This apomorphy is not unique, occurs in many Furcatergaliae and Tridentiseta, but never occurs in other Branchitergaliae.

(5) On hind wing MA lacks furcation; thus only two triads are present – RS and MP, with a single MA between them (FIG.53:D) [see Euplectoptera (1)]. Non-unique apomorphy (see Index of characters [2.2.67]); among Pentamerotarsata the same in *dimicki* [*Cinygma*].

(6) Imaginal and subimaginal gonostylus with 3 distal segments (instead of two ones initial for Ephemeroptera). Non-unique apomorphy (see Index of characters [2.3.12]).

Characters of unclear phylogenetic status.

(7) Superlinguae are curved laterally (Studemann & al. 1987:Figs 15–16). Probably apomorphy of Pentamerotarsata, as the same in Heptagenia/f5=g4 (but not in other Radulapalpata).

(8) Penis, besides the pair of median titillators [see Pentamerotarsata (5)], with a pair of ventral titillators – immovable spines, situated on ventral side proximad of the median titillators (Studemann & al. 1987:Figs 27–32).

Plesiomorphies of *Arthroplea*. Labium and labial palps completely lack that specialization which is peculiar for Radulapalpata. In contrast to Radulapalpata, posterolateral spines are present on abdominal segments II–IX and are enlarged toward segment IX (Studemann & al. 1987:Fig.11). Larval primary swimming setae [vestigial – see Pentamerotarsata (6)] are present. Larval and adult patella-tibial suture is developed on middle and hind legs only.

Size. Fore wing length 8–10 mm.

Distribution. Holarctic.

Species composition of Arthroplea/fg(1). 2 species – *bipunctata* McDunnough 1924 [*Cinygma*] (Nearctic) and *congener* Bengtsson 1908 [*Arthroplea*] (synn. subj.: *elegans* Bengtsson 1908 [*Remipalpus*], *mirabilis* Aro 1910 [*Cinygma*], *southi* Blair 1929 [*Haplogenia*], *frankenbergi* Balthasar 1937 [*Arthroplea*]) (Palaearctic).

Material examined: *congener* [*A.*]: ℧–◯–⊕/⊙; *bipunctata* [*C.*]: ℧–◯–⊕/⊙.

***

2.2;1,2-2/2. **Radulapalpata**, or **Heptagenia/f4=g3**

(Bidentiseta Branchitergaliae Heptagennota Pentamerotarsata Radulapalpata)

(Figs 8:E–F; 9:C–E; 54–65)

Nomen hierarchicum: **Heptagenia/f4=g3** (sine *Arthroplea*; incl. *Raptoheptagenia*, *Cinygma*, *Rhithrogena*).
Nomen circumscribens: **Radulapalpata** Kluge 2000: 251.
In circumscription matches:
— subfam. Heptageniinae: Kluge 1988; 1993a;
— fam. Ecdyonuridae: Balthasar 1937;
— fam. Heptageniidae: Landa 1969a: 153;
— Radulapalpata, or Heptagenia/f4=g3: Kluge 2000: 251.

References. Strenger 1953: ℧'; – Landa 1969a: ℧* ⊕*; – Kluge 1988: ℧* ⊕*; – 1993a.

Autapomorphies of Radulapalpata.

(1) Labium is strongly modified as scraping (FIG.57:F–H):

Submentum is not expressed [see Pentamerotarsata (1)]; mentum is large, transverse; most part of it is occupied by strongly thickened muscles-adductors of labial palps; these muscles are directed transversely and their immobile ends are connected together at median plane without any skeleton formations. Probably such musculature can be used for synchronous work only. Abductors of labial palps are lost.

1st segment of labial palp is very strongly thickened, with strongly convex outer margin, projecting proximally from its place of attachment; most part of the 1st segment is occupied by a strongly thickened fan-form flexor of 2nd segment, proximally this flexor is attached to the most part of the convex outer margin of the segment (somewhat similar place of attachment of this muscle in *Ametropus* – see Index of characters [1.1.56]).

Distal (2nd+3rd) segment of labial palp [see Heptagennota (8)] is massive and has following unique structure: On outer-ventral side it bears a field of densely and regularly situated small curved scraping setae. On ventral side it has a longitudinal arches sclerotized pigmented crest. On dorsal side it bears a field of irregularly situated setae; distally this field is bordered by a regular transverse row of setae. Only in one species [see *Raptoheptagenia* (1) below] 2nd+3rd segment of labial palp secondarily lost its scraping specialization and is transformed to a sclerotized claw; but on ventral side of this claw near its apex, one row of vestiges of the scraping setae is retained (FIG.60:B).

Paraglossae are widened, much wider than glossae, with divergent lateral margins; each paraglossa is roundish or with blunt apical-median angle and with wide straight or roundish apical margin (in contrast to *Arthroplea*, *Pseudiron* and majority of other mayflies, where paraglossa is crescent-shapely arched and has an acute apical-median angle). In various Radulapalpata width of paraglossa can be subequal or much larger than its length; length of paraglossa can be subequal or much less than length of glossa.

Glossae have variable shape; probably the initial for Radulapalpata is a pyramidal shape, when glossa has maximum width near its base, narrows toward its apex and has three longitudinal ridges convergenting toward apex – lateral, median and dorsal ones. Such shape of glossae occurs in non-related taxa – *Dacnogenia* (FIG.56:F–G), Rhithrogena/fg2 (FIG.62: E, G), Epeorus/fg2 (Fig.64:I) and *Paegniodes*; in other Radulapalpata glossae shape is modified: glossa can be flattened because its dorsal ridge is approximated to median ridge, can be widened apically (see Index of characters [1.1.50]).

(2) In larva posterolateral spines on abdominal segment IX are nearly always lost, and if present are much smaller than spines on previous segments: in species which have posterolateral spines on several segments, spines of segment VIII are the largest; in some species posterolateral spines are absent on all segments. The same in *Chiloporter* only. In many other mayflies posterolateral spines increase from anterior segments toward segment IX, and spines of segment IX are the largest.

Character of unclear phylogenetic status.

(3) On maxilla pectinate setae of the apical-ventral row are directed medially (i. e. toward canines) arising under acute angle, and usually have a peculiar shape: basally stout, distally widened and flattened, brush-like, with widening directed dorsally-proximally, and all processes directed medially (FIGS 58:F, H). Probably presence of a regular apical-ventral row of pectinate setae belongs to archetype of Ephemeroptera, as pectinate setae are present in such non-related groups, as Siphlonurus/fg1 (especially well-developed in Siphlonurus/ /fg4 *palaearcticus* [*Oniscigaster*]), Ameletus/fg1, Leptophlebia/fg1 and others. In Radulapalpata these setae have unique shape. In some representatives of Radulapalpata (*Raptoheptagenia*, *Cinygma*, Epeorus/fg3 and others) these setae are partly or completely reduced. It is unclear, if absence of these setae in *Arthroplea* and other Branchitergaliae is primary or secondary, so it is unclear if their peculiar structure is an autapomorphy of Radulapalpata or of a higher taxon.

Probably the initial shape of maxilla in Radulapalpata is elongate and close to rectangle – i. e. with developed apical margin, nearly straight apical-median angle and with long straight median margin which length approximately 1.5 times exceeds total width of maxilla; such maxilla shape occurs in many Heptagenia/f6=g5 (FIG.55:H), in *Cinygma* (FIG. 61:D), *Bleptus*, *Ironodes* and *Paegniodes* (FIG.64: B, E, H). In some taxa of Radulapalpata maxilla is shorter and has more acute apical-median margin [particularly, see Rhithrogena/fg2 (1) and Epeorus/ /fg3 (1) below].

Plesiomorphies of Radulapalpata. In contrast to *Arthroplea*, on fore wing RSa_2 always retains its connection with RSa (as in FIG.53:A). Gonostylus with 2 distal segments.

Variable characters of Radulapalpata. Mandibles are usually normal: mandible with distally projected mola is left one, and mandible with proximally projected mola is right one (FIG.54:E–F); only in selected species and infra-species taxa mandibles are inverted – in the single species of *Dacnogenia*, in Heptagenia/f8=g7 *sulphurea dalecarlica* [*H.*], *sulphurea albicauda* [*H.*], Leucrocuta/g1 *inversus* [*Ecdyonurus*] (the same in *Arthroplea*).

Size. Fore wing length 5–20 mm.

Age and distribution. Palaeogene (see *Kageronia* and Radulapalpata INCERTAE SEDIS) — Recent; Arctogea (Holarctic + Oriental + Ethiopian Regions) and Central America; dominate in Holarctic.

Radulapalpata are divided into Heptagenia/f5=g4, *Raptoheptagenia* (p.188), *Cinygma* (p.188) and Rhithrogena/fg1 (p.189). Some insufficiently described taxa have uncertain systematic position (p.208).

2.2;1,2-2/2.1. Heptagenia/f5=g4

(Branchitergaliae Heptagennota Pentamerotarsata Radulapalpata Heptagenia/f5=g4)
(Figs 9:D–E; 54–59)

Nomen hierarchicum: **Heptagenia/f5=g4** (sine *Raptoheptagenia*, *Cinygma*, *Rhithrogena*; incl. *Ecdyonurus*).
In circumscription matches:
— tribus Heptageniini: Kluge 1988: 292.

References. Kluge 1988: ♉*; – Tomka & Elpers 1991: ♉*; – Kluge 1993: ♉* ⊕*.

Characters of unclear phylogenetic status.

(1) Superlinguae are curved laterally. This feature is present in all Heptagenia/f5=g4, but shape of superlinguae differs in different groups (FIGS 54:B; 55:F; 56:E, H; 58:D). Non-unique apomorphy; among Heptagennota superlinguae are curved laterally also in *Arthroplea*, but have different shape.

(2) Larval caudalii with whorls of stout setae on posterior margin of each segment (FIGS 55:G; 56:K). By this character Heptagenia/f5=g4 differs from other Heptagennota; the same in many other mayflies.

(3) Abdominal ganglia VII and VIII are brought together. Non-unique apomorphy; the same in *Arthroplea* and some others, but not in Rhithrogena/ /fg1, where distance between these ganglia is much larger.

Plesiomorphies of Heptagenia/f5=g4. In contrast to *Cinygma*, all 3 maxillary canines are retained (while often are weak – FIGS 54:M; 58:F, H). Maxillary palp is not so specialized as in Rhithrogena/fg1, with a vestige of 3rd segment (FIGS 55:H; 58:E). Labial palp has no bifurcate setae (FIG.57:G) (in contrast to Rhithrogena/fg1). Tergalii retain ability of rhythmical respiratory movements (FIG.57: C–E) (in contrast to *Cinygma* and Rhithrogena/fg1) [for comment – see Rhithrogena/fg1 (6) below].

Lateroparapsidal sutures are not curved laterally, terminate near medioparapsidal sutures (FIG.56:L; as in FIG.61:A–B) (in contrast to Rhithrogena/fg1); subimaginal lateroparapsidal stripe of pigmented area of mesonotum [see Heptagennota (2)] is widened posteriorly (FIG.54:I) (in contrast to Rhithrogena/fg2 and Epeorus/fg1). As an exception, lateroparapsidal sutures can be curved laterally (in *terminata* [*Palingenia*]), but in this case they are shorter than in Rhithrogena/fg1, and do not reach bases of posterior scuto-coxal muscles (S.CmP) (FIG.56:M). Larval and adult patella-tibial suture is developed on middle and hind legs only (in contrast to larval Rhithrogena/fg3).

Size. Fore wing length 5–18 mm.

Age and distribution. Palaeogene (see *Kageronia*) — Recent; Arctogea (Holarctic + Oriental + Ethiopian Regions) and Central America; dominate in Holarctic.

Heptagenia/f5=g4 is divided into Heptagenia/f6=g5 and Ecdyonurus/fg1 (p.175).

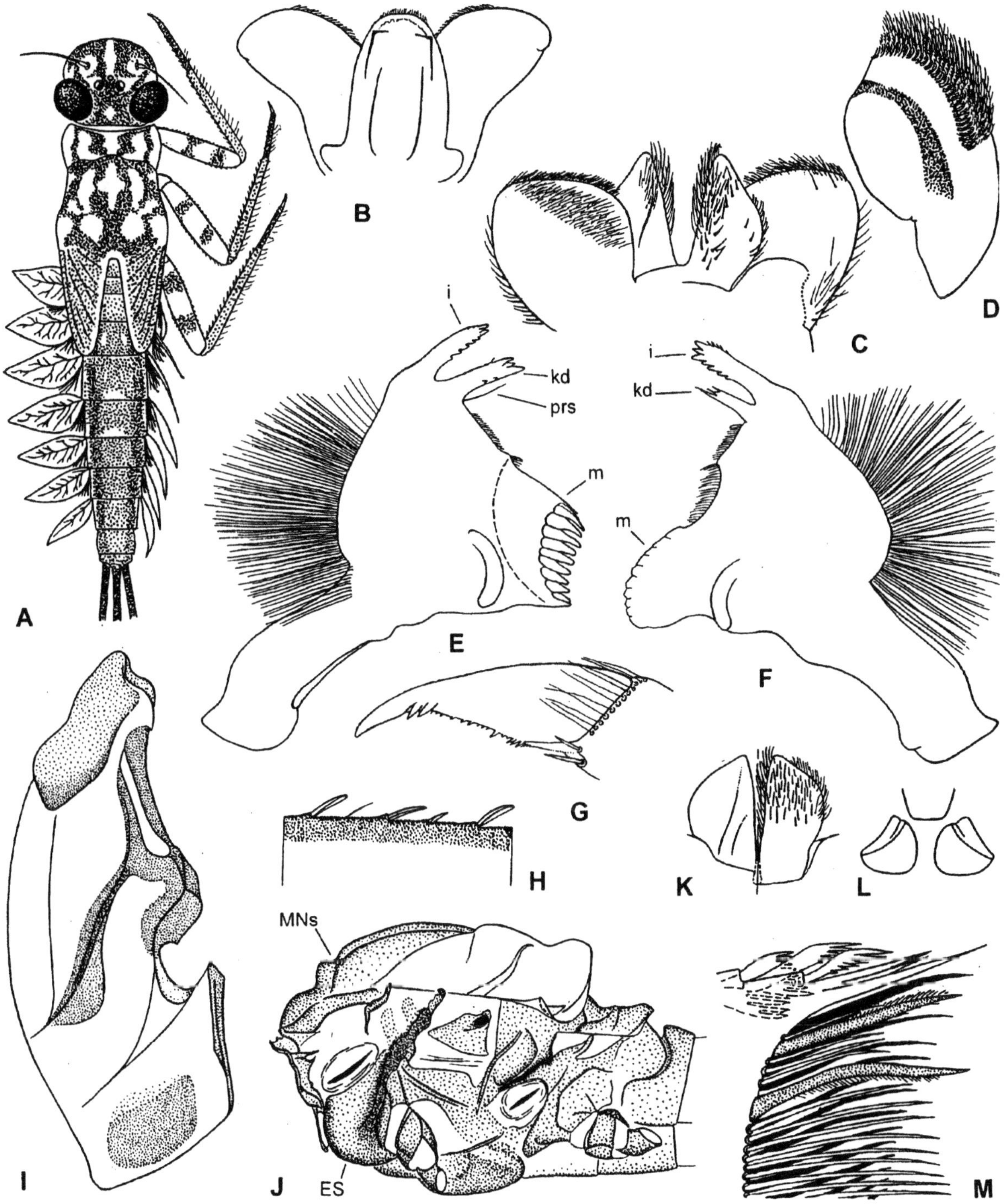

Figure 54. Radulapalpata-Heptagenia/f6=g5-Kageronia/g1.
A–H – Kageronia/g1 *orbiticola* [*Heptagenia*], larva: **A** – mature larva; **B** – hypopharynx and superlinguae, ventral view; **C** – glossae and paraglossae, dorsal view (in left half) and ventral view (in right half); **D** – distal segment of left labial palp, ventral view; **E–F** – left and right mandibles, dorsal view; **G** – claw; **H** – outer margin of femur. **I–M** – Kageronia/g1 *fuscogrisea* [*Ephemera*]: **I** – subimaginal exuviae of right half of mesonotum (see abbreviations on Fig.52:C); **J** – imaginal pterothorax, lateral view; **K** – glossae, dorsal view (in left half) and ventral view (in right half), **L** – glossae and ventral margin of hypopharynx, apical view; **M** – apex of left maxilla, dorsal view (dentisetae shown by dots, setae of inner-ventral row shown by black). (A–H – from Kluge 1987a; K–L – from Kluge 1988a; M – from Kluge 1998).

Abbreviations: **ES** – episternum; **i** – incisor; **kd** – kinetodontium; **m** – mola; **MNs** – mesonotal suture; **prs** – prostheca.

2.2;1,2-2/2.1;1. Heptagenia/f6=g5

(Heptagennota Pentamerotarsata Radulapalpata Heptagenia/f5=g4 .../f6=g5)

(Figs 9:D; 54–56)

Nomen hierarchicum: **Heptagenia/f6=g5** (sine *Ecdyonurus*; incl. *Kageronia*).
In circumscription matches:
— gen. *Heptagenia*: Kluge 1983b: 16 (Abstract); 1988: 292;
— grex genn. *Heptagenia*: Tomka 1991: 121.

References. Kluge 1986: ⊕*; – 1987a: ☋* ⊕*; – 1988: ☋* ⊕*; – Tomka 1991: ☋*; – Kluge 1993: ☋* ⊕*.

Characters of unclear phylogenetic status.

(1) Mandibles [usually with left setiform prostheca and irregular setation on median margin – see Heptagennota (7)] have uniform shape of incisors in majority of representatives: incisor of each mandible is thin, its apex bears 3 denticles more or less brought together, and inner side bears a row of smaller denticles (FIG.54:E–F). Only in rare cases, when incisor is thickened (particularly, in *Dacnogenia*) it has another form.

(2) On maxilla ventral row of setae [see Branchitergaliae (1)] is parallel to inner margin and apically is slightly curved laterally (FIG.55:H) (in contrast to *Cinygma*, *Paegniodes* and Epeorus/fg1, where this row is not curved; in contrast to Rhithrogena/fg2, where it is curved stronger; in contrast to Ecdyonurus/fg1, where this row is transformed to a field of irregular setae).

(3) Vestige of 3rd segment of maxillary palp (see Plesiomorphies of Heptagenia/f5=g4) is small, usually curved (FIG.55:H) (in contrast to Ecdyonurus/fg1, where this vestige is larger, and in contrast to other Radulapalpata, where it is not expressed).

(4) Imaginal prosternum has characteristic shape: it always has a distinct transverse ridge, and surface behind this ridge is concave (FIG.56:A–B). Among the Heptagennota examined, similar shape of prosternum occurs only in *Pseudiron*, *Bleptus* and selected Ecdyonurus/fg1; in other Heptagennota prosternum is entirely convex, without transverse ridge.

Plesiomorphies of Heptagenia/f6=g5. Larval femora [flattened – see Heptagennota (1)] have no regular row of long stout setae on outer margin: stout setae (long clavate or short spine-like) are situated on outer margin of femur irregularly (FIGS 54:H; 55:B–C). Imaginal and subimaginal median impression of mesothoracic furcasternum is strongly narrowed anteriorly, metathoracic nerve ganglion is situated in its posterior part (FIG.56:C–D) (in contrast to Ecdyonurus/fg1). Median titillators of penis [see Pentamerotarsata (5)] are always well-developed.

Size. Fore wing length 6–16 mm.

Age and distribution: Palaeogene (see *Kageronia*) – Recent; Holarctic, Oriental Region and Central America; dominate in Holarctic.

Heptagenia/f6=g5 is divided into *Kageronia* and Heptagenia/f7=g6.

2.2;1,2-2/2.1;1,1. Kageronia/g(1)

(Heptagennota Pentamerotarsata Radulapalpata Heptagenia/f5=g4 .../f6=g5 *Kageronia*)

(Fig. 54)

Nomen hierarchicum: **Kageronia/g(1)** [g: *Kageronia* Matsumura 1931: 1479, typus *K. suzukiella* Matsumura 1931 (monotypy; syn. subj. *Heptagenia kihada* Matsumura 1931)].
In circumscription matches:
—*fuscogrisea*-Gruppe: Bogoescu & Tabacaru 1962: 280;
— subgen. *Parastenacron* (manuscript name): Kluge 1983b:16;
— subgen. *Kageronia*: Kluge 1988: 293;
— gen. *Kageronia*: Jacob & Dorn & Haybach 1996: 96.

References. Kluge 1988: ☋*; – 1993: ☋*.

Characters of unclear phylogenetic status.

(1) Superlingua [curved laterally – see Heptagenia/f5=g4 (1)] with slightly concave lateral margin and rounded apex (FIG.54:B); by its form differs from any other taxon among Heptagenia/f5=g4.

(2) Glossa [see Radulapalpata (1)] with straight median margin and roundish-convex lateral margin, dorsal edge is not approximated to median edge (FIG.54:C, K–L) (in contrast to Heptagenia/f8=g7, Stenonema/g1 and Ecdyonurus/fg1).

(3) 2nd+3rd segment of labial palp [specialized as scraping – see Radulapalpata (1)] is pointed, thus the field of scraping setae terminates at a distance from the end of the sclerotized ridge (FIG.54:D). The same in Ecdyonurus/fg1 and Rhithrogena/fg1; probably symplesiomorphy. Heptagenia/f7=g6 and *Cinygma* have different shape of this segment.

(4) Larval femur on its outer margin has no long hairs and long stout setae; only short stout spine-like setae and short hairs are present (FIG.54:H) [see Plesiomorphies of Heptagenia/f6=g5]. Probably plesiomorphy.

(5) Larval claws with inner margin denticulate (FIG.54:G). Probably plesiomorphy [see Pentamerotarsata (4)] (in contrast to all other Radulapalpata). Subapical denticles on anterior (dorsal) side of claw are also present (that is usual for many Pentamerotarsata).

(6) Tergalius VII lacks fibrillose ventral lobe,

thus the fibrillose ventral lobe [see Branchitergaliae (3)] is present on tergalii I–VI only. Non-unique apomorphy (see Index of characters [1.3.59]); among Heptagenia/f5=g4 the same in Stenonema/g1 and majority of Ecdyonurus/fg1.

(7) Tergalii are pointed apically (Kluge 1987a: Figs 8–11, 17). Non-unique character.

Plesiomorphy of *Kageronia*. On tergalii costal and anal ribs are located on costal and anal margins correspondingly [see Branchitergaliae (3)] (in contrast to Stenonema/g1).

Variable characters of *Kageronia*. On larval caudalii primary swimming setae [vestigial – see Pentamerotarsata (6)] can be present (in *kihada* [*Heptagenia*] and *orbiticola* [*H.*]) or lost (in *fuscogrisea* [*Ephemera*]). In imago and subimago on fore tarsus 1st segment can be shortened (as in Heptagenia/f8=g7) (in *kihada* [*H.*] and *orbiticola* [*H.*]) or not shortened (in *fuscogrisea* [*E.*]).

Size. Fore wing length 7–16 mm.

Age and distribution. Palaeogene – Recent; the Recent species *fuscogrisea* [*E.*] is known from Baltic amber (Kluge 1986). Palaearctic; all three species occur in Eastern Palaearctic.

Nominal taxon included. Kageronia/g(1) includes ***Parastenacron*** Kluge – unavailable name, mentioned in Abstracts of 7th International Conference on Ephemeroptera (1983) and cited by Braasch & Soldán (1988).

Species composition of Kageronia/g(1). 3 species – *fuscogrisea* Retzius 1783 [*Ephemera*] (synn.subj.: *volitans* Eaton 1870 [*Heptagenia*], *convergens* Aro 1910 [*Ecdyonurus*], *rossicus* Tshernova 1928 [*Ecdyonurus*], *confinis* Tshernova 1928 [*Ecdyonurus*]), *kihada* Matsumura 1931 [*Heptagenia*] (syn. subj.: *suzukiella* Matsumura 1931 [*Kageronia*]), *orbiticola* Kluge 1987 [*Heptagenia (Kageronia)*].

Material examined: *fuscogrisea* [*E.*]: Ꝺ–◎–⊕/⊙; *kihada* [*H.*]: Ꝺ, ⊕♂, ⊕♀; *orbiticola* [*H.*]: Ꝺ–◎–⊕/⊙.

***

2.2;1,2-2/2.1;1,2. **Heptagenia/f7=g6**

(Heptagennota Pentamerotarsata Radulapalpata Heptagenia/f5=g4 .../f7=g6)

(Figs 55–56)

Nomen hierarchicum: **Heptagenia/f7=g6** (sine *Kageronia*; incl. *Dacnogenia, Stenonema*).

Reference. Kluge 1993: Ꝺ*.

Characters of unclear phylogenetic status.

(1) 2nd+3rd segment of labial palp [specialized as scraping – see Radulapalpata (1)] is blunt, thus the field of scraping setae terminates close to the end of sclerotized ridge (FIG.55:I). The same in *Cinygma* (FIG.61:J); probably, independently appeared apomorphy. In *Kageronia*, Ecdyonurus/fg1 and Rhithrogena/fg1 form of this segment is different (FIG.54:D).

(2) Whole outer margin of each larval femur always bears long hairs, and usually bears long stout blunt irregularly situated setae (FIG.55:C) (only in selected species these setae are short and pointed – FIG.55:C). Such femoral setation never occurs in other Radulapalpata [see Plesiomorphies of Heptagenia/f6=g5].

(3) Larval claws lack denticles on inner margin [see Pentamerotarsata (4)] (subapical denticles on anterior side of claw can be present or absent). The same in all other Radulapalpata except for *Kageronia.*

Plesiomorphy of Heptagenia/f7=g6. Larval caudalii retain swimming setae [vestigial – see Pentamerotarsata (6)].

Size. Fore wing length 6–16 mm.

Distribution. Holarctic, Oriental Region and Central America; dominate in Holarctic.

Heptagenia/f7=g6 is divided into Heptagenia/f8=g7, *Dacnogenia* and Stenonema/g1.

2.2;1,2-2/2.1;1,2-1. **Heptagenia/f8=g7**

(Heptagennota Pentamerotarsata Radulapalpata Heptagenia/f5=g4 .../f8=g7)

(Figs 55; 56:A–D)

Nomen hierarchicum: **Heptagenia/f8=g7** (sine *Dacnogenia, Stenonema*; incl. *Sigmoneuria, Trichogenia*).
In circumscription matches:
— subgen. *Heptagenia*: Kluge 1988: 295.

References. Kluge 1988: Ꝺ*; – 1993: Ꝺ*.

Characters of unclear phylogenetic status.

(1) Superlingua [curved laterally – see Heptagenia/f5=g4 (1)] on its lateral margin with a subapical incision and a convexity proximad of it (FIG.55:F); by its form differs from any other taxon among Heptagenia/f5=g4.

(2) Each glossa [initially pyramidal – see Radulapalpata (1)] is narrow, curved, in its distal half flattened, as dorsal ridge is approximated to median one (Fig.55:D–E). The same in Stenonema/g1; probably synapomorphy.

(3) In imago and subimago 1st tarsal segment is shortened, on fore tarsus of male several times shorter than 2nd segment. Non-unique apomorphy (see Index of characters [2.2.76]).

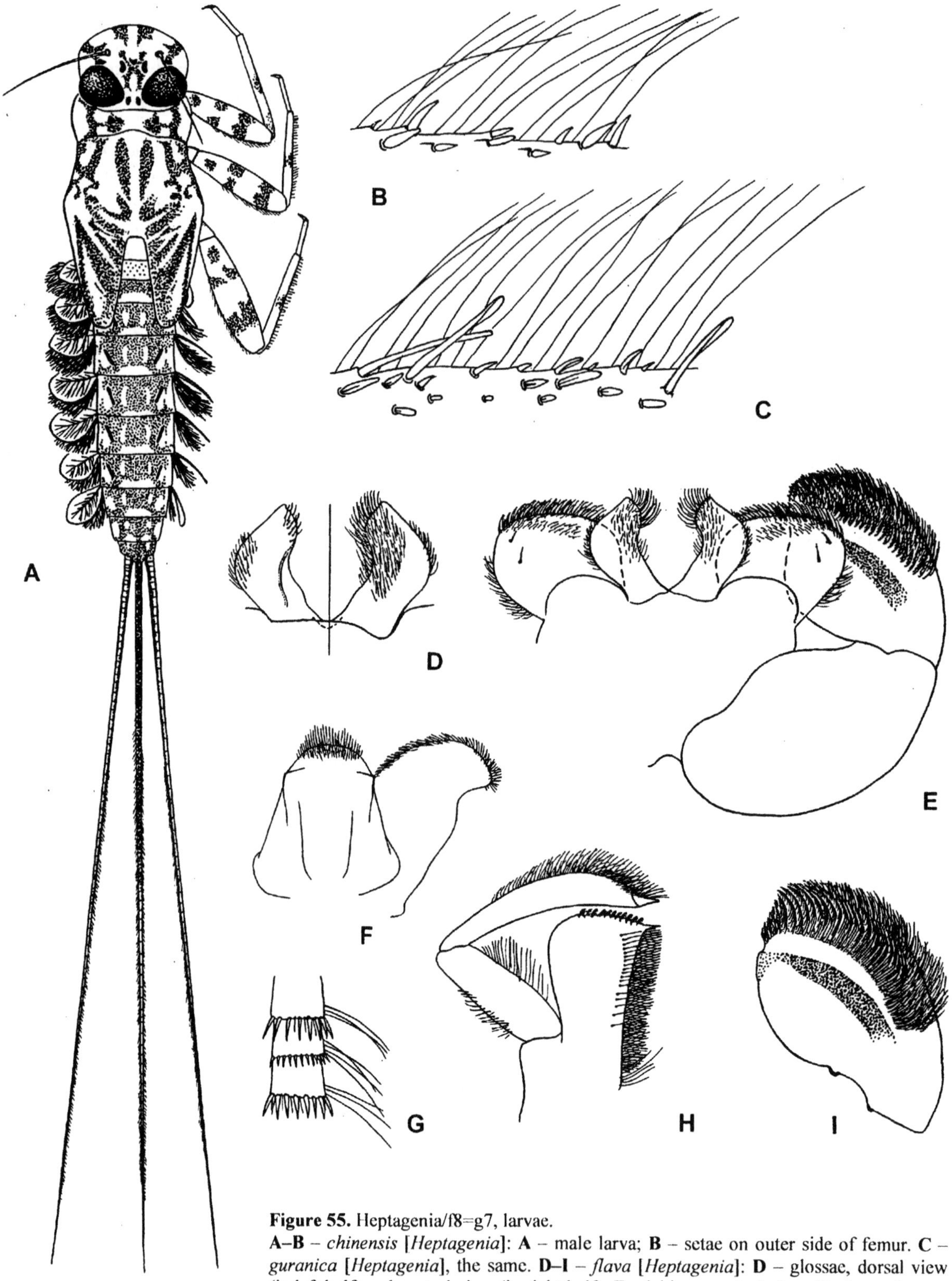

Figure 55. Heptagenia/f8=g7, larvae.
A–B – *chinensis* [*Heptagenia*]: **A** – male larva; **B** – setae on outer side of femur. **C** – *guranica* [*Heptagenia*], the same. **D–I** – *flava* [*Heptagenia*]: **D** – glossae, dorsal view (in left half) and ventral view (in right half); **E** – labium, ventral view; **F** – hypopharynx and superlinguae, ventral view; **G** – middle part of left cercus, dorsal view; **H** – right maxilla, ventral view; **I** – distal segment of left labial palp, ventral view. (A–C – from Kluge 1987a; F, G, I – from Kluge 1988a; E, F, H – from Kluge 1997d).

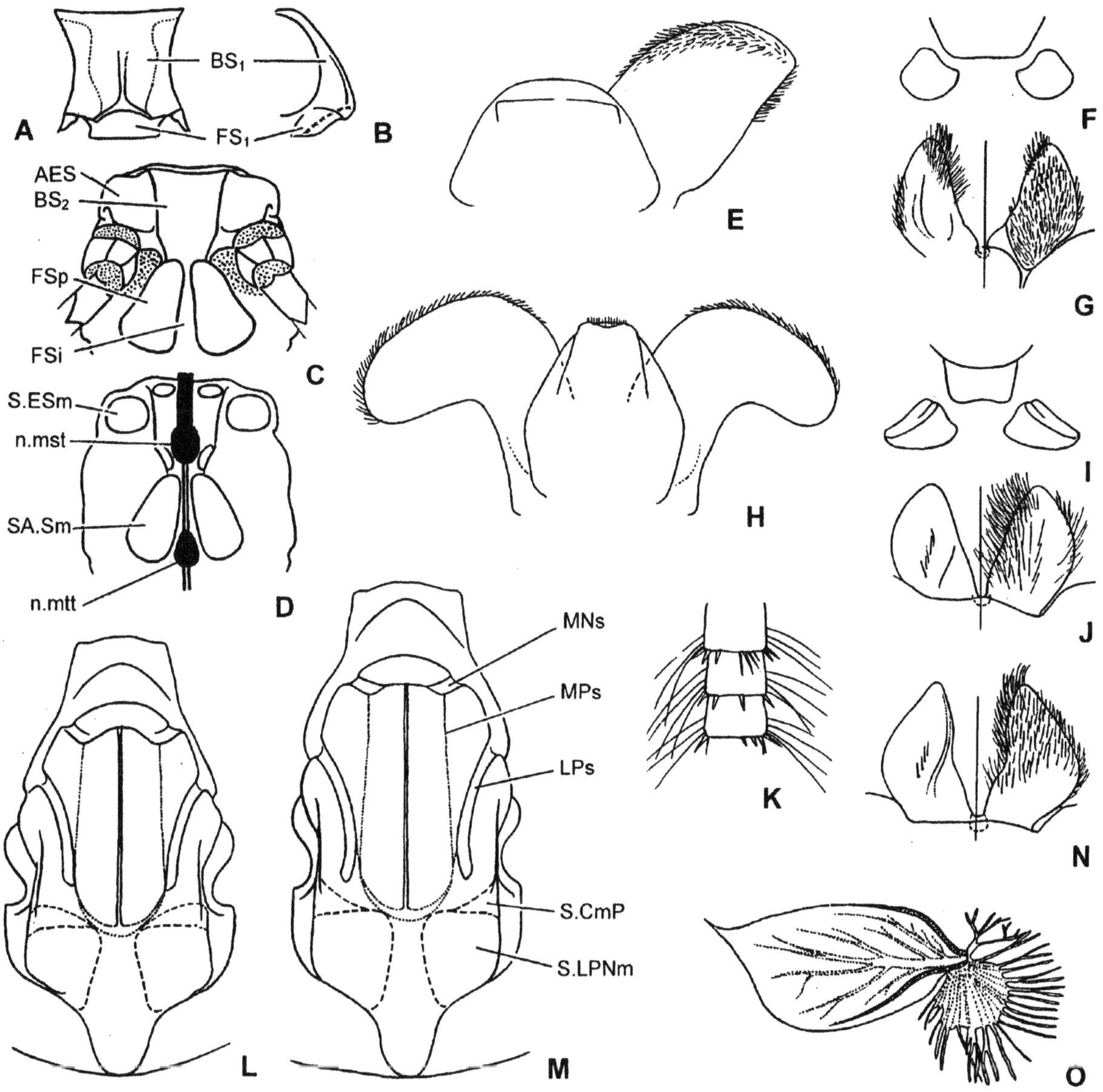

Figure 56. Heptagenia/f7=g6.
A–D – Heptagenia/f8=g7 *sulphurea* [*Ephemera*], imaginal thoracic sterna: **A–B** – prosternum: **A** – ventral view, **B** – lateral view; **C** – imaginal mesothorax, ventral view; **D** – meso- and metathoracic nerve ganglia (shown by black) and ventral areas of mesothoracic muscles attachment (compare with Fig.57:A–B). **E–G** – Dacnogenia/g* *coerulans* [*Heptagenia*]: **E** – hypopharynx and left superlingua, ventral view; **F** – glossae and ventral margin of hypopharynx, apical view; **G** – glossae, dorsal view (in left half) and ventral view (in right half). **H–O** – Stenonema/g1: **H–K** – Maccaffertium/g* *mexicana* [*Heptagenia*]: **H** – hypopharynx and superlinguae, ventral view; **I** – glossae and ventral margin of hypopharynx, apical view; **J** – glossae, dorsal view (in left half) and ventral view (in right half); **K** – middle part of larval cercus; **L** – Stenonema/g* *femorata* [*Baetis*], imaginal mesonotum, dorsal view (bases of posterior scuto-coxal and scuto-lateropostnotal muscles shown by interrupted lines) (compare with Fig.63:D–E); **M** – Maccaffertium/g1 *terminata* [*Palingenia*], the same; **N–O** – Stenacron/g1 sp.: **N** – glossae, dorsal view (in left half) and ventral view (in right half); **O** – left tergalius II, ventral portion bent out (ribs dotted, tracheae shown by dotted lines). (A–D, F–G, I–O – from Kluge 1988a).

Abbreviations: **AES** – anepisternum; **BS$_1$** – basisternum of prothorax; **BS$_2$** – basisternum of mesothorax; **FS$_1$** – furcasternum of prothorax; **FSi** – furcasternal median impression of mesothorax; **FSp** – furcasternal protuberance of mesothorax; **LPs** – lateroparapsidal suture; **MNs** – mesonotal suture; **MPs** – medioparapsidal suture; **n.mst** – mesothoracic nerve ganglion; **n.mtt** – metathoracic nerve ganglion; **SA.Sm** – ventral base of subalar-sternal muscle; **S.CmP** – dorsal base of posterior scuto-coxal muscle; **S.ESm** – ventral base of scuto-episternal muscle; **S.LPNm** – dorsal base of scuto-lateropostnotal muscle.

Plesiomorphy of Heptagenia/f8=g7. In contrast to *Dacnogenia*, hypopharynx is not shortened, has the same length as superlinguae [see (1)] (FIG.55:F).

Variable characters of Heptagenia/f8=g7. Tergalii [see Branchitergaliae (3)] are usually non-modified; usually all tergalii I–VII with fibrillose lobe; only on *maxillaris* [*T.*] tergalius VII lacks fibrillose lobe. Anal and costal ribs are either located on anal and costal margins of tergalii, or lost.

Size. Fore wing length 7–15 mm.

Distribution. Holarctic and Oriental Region.

Nominal taxa included. Heptagenia/f8=g7 includes: (1) **Sigmoneuria/g** [g: *Sigmoneuria* Demoulin 1964a: 353, typus *S. amseli* Demoulin 1964 (design. orig.; syn. subj. *Heptagenia perflava* Brodsky 1930)], regarded as a subgeneric synonym of *Heptagenia* (Kluge 1983b: 16; 1987a: 314); (2) **Trichogenia/g** [g: *Trichogenia* Braasch & Soldán 1988: 119, typus *T. maxillaris* Braasch & Soldán 1988 (design. orig.)], established for a single species known as larva only and differing from other Heptagenia/f8=g7 only by autapomorphies – branched setae on terga, differentiated tergalii, and others.

Species composition of Heptagenia/f8=g7 (here [*H.*]=[*Heptagenia*]). About 20 species. **Eurasian species:** *chinensis* Ulmer 1920 [*H.*], *flava* Rostock 1878 [*H.*] (synn. subj.: *citrina* Hummel 1825 [*Ephemera*], *bipunctata* Esben-Petersen 1916 [*H.*], *arsenjevi* Tshernova 1952 [*H.*]), *flavata* Navás 1922 [*H.*], *guranica* Belov 1981 [*H.*], *kyotoensis* Gose 1963 [*H.*], *longicauda* Stephens 1835 [*Baetis*] (synn.subj.: *flavipennis* Dufour 1841 [*Ephemera*], *cerea* Pictet 1843 [*Baetis*]), *maxillaris* Braasch & Soldán 1988 [*Trichogenia*], *perflava* Brodsky 1930 [*H.*] (synn. subj.: *tadzhikorum* Tshernova 1930 [*H.*], *platysma* Tshernova 1941 [*H.*], *amseli* Demoulin 1964 [*Sigmoneuria*]), *quadripunctata* Kluge1989 [*H.*], *samochai* Demoulin 1973 [*Sigmoneuria*] (syn. subj.: *lutea* Kluge 1987 [*H.*] nom. praeocc.), *sulphurea* Müller 1776 [*Ephemera*] (synn. subj. & subspp.: *helvola* Subz. 1776 [*Ephemera*], *leucophthalma* Ström 1783 [*Ephemera*], *ferruginea* Gmelin 1790 [*Ephemera*], *costalis* Curtis 1834 [*Baetis*], *elegans* Curtis 1834 [*Baetis*], *straminea* Curtis 1834 [*Baetis*], *cyanops* Pictet 1843 [*Baetis*], *dalecarlica* Bengtsson 1912 [*H.*], *soldatovi* Tshernova 1952 [*H.*], *albicauda* Kluge 1987 [*H.*]), *traverae* Braasch 1986 [*H.*]. **North American species** (most of them are described incompletely and placed here presumably): *adequata* McDunnough 1924 [*H.*], *culacantha* Evans & Botts & Flowers 1985 [*H.*], *diabasia* Burks 1947 [*H.*], *dolosa* Traver 1935 [*H.*], *elegantula* Eaton 1885 [*Rhithrogena*] (synn. subj.: *coxalis* Banks 1914 [*H.*], *querula* McDunnough 1924 [*H.*]), *flavescens* Walsh 1862 [*H.*], *julia* Traver 1933 [*H.*], *marginalis* Banks 1910 [*H.*], *patoka* Burks 1947 [*H.*], *pullus* Clemens 1913 [*Ecdyonurus*], *solitaria* McDunnough 1924 [*H.*], *townesi* Traver 1935 [*H.*].

Material examined: *chinensis* [*H.*]: Ꝺ–◎–⊕/⊙; *flava* [*H.*]: Ꝺ–◎–⊕/⊙; *flavata* [*H.*]: ◎♂; *guranica* [*H.*]: Ꝺ–◎–⊕/⊙; *kyotoensis* [*H.*]: Ꝺ; *longicauda* [*B.*]: Ꝺ–◎–⊕/⊙; *maxillaris* [*Trichogenia*]: Ꝺ; *perflava* [*H.*]: Ꝺ–◎–⊕/⊙; *pullus* [*E.*]: ⊕♂; *quadripunctata* [*H.*]: Ꝺ, Ꝺ♂–◎♂–⊕♂; *samochai* [*S.*]:Ꝺ–◎–⊕/⊙; *sulphurea* [*E.*]: Ꝺ–◎–⊕/⊙.

***

2.2;1,2-2/2.1;1,2-2. Dacnogenia/g(1)

(Heptagennota Pentamerotarsata Radulapalpata Heptagenia/f5=g4 .../f7=g6 *Dacnogenia*)

(Fig. 56:E–G)

Nomen hierarchicum: **Dacnogenia/g(1)** [g: *Heptagenia* subgen. *Dacnogenia* Kluge 1988: 297, typus *Heptagenia coerulans* Rostock 1878 (design. orig.)].
In circumscription matches:
— subgen. *Dacnogenia* (manuscript name): Kluge 1983b:16;
— subgen. *Dacnogenia* Kluge 1988: 297.
Monospecific taxon.

References. Eaton 1883–1888: Ꝺ ⊕; – Grandi 1953: Ꝺ; – 1960: Ꝺ ⊕; – Landa 1969a: Ꝺ ⊕; – Kluge 1988: Ꝺ; – 1989b: Ꝺ ⊕; – 1993: Ꝺ.

Autapomorphies of *Dacnogenia*.

(1) Hypopharynx is strongly shortened, twice shorter than superlinguae (FIG.56:E). In other Radulapalpata and majority of mayflies length of hypopharynx is subequal to that of superlinguae.

(2) Mouth apparatus is secondarily specialized as biting: mandibles are short and spout, maxillae are acute-angled with stout canines and dentisetae (Eaton 1883–1888:Pl.60:4–10). Non-unique apomorphy; among Radulapalpata the same in selected species of Stenonema/g1, *Himalogena* and in all Epeorus/fg3.

Characters of unclear phylogenetic status.

(3) Superlingua [curved laterally – see Heptagenia/f5=g4 (1)] with slightly concave lateral margin and angulate apex (FIG.56:E); by its form differs from any other taxon among Heptagenia/f5=g4.

(4) In imago and subimago 1st tarsal segment is shortened, on fore tarsus of male several times shorter than 2nd segment. Non-unique apomorphy (see Index of characters [2.2.76]).

Plesiomorphies of *Dacnogenia*. Glossae are pyramidal [see Radulapalpata (1)], distal edge of glossa is not approximated to median edge (FIG.56:F–G) (in contrast to Heptagenia/f8=g7, Stenonema/g1 and Ecdyonurus/fg1). All tergalii I–VII with fibrillose lobe [see Branchitergaliae (3)].

Size. Fore wing length 10–13 mm.

Distribution. Western Palaearctic.

Species composition Dacnogenia/g(1). 1 species – *coerulans* Rostock 1878 [*Heptagenia*] (syn. subj.: *gallica* Eaton 1885 [*Heptagenia*]).

Material examined: *coerulans* [*H.*]: Ɒ–O–⊕/⊙.

***

2.2;1,2-2/2.1;1,2-3. Stenonema/g1

(Heptagennota Pentamerotarsata Radulapalpata Heptagenia/f5=g4 .../f7=g6 Stenonema/g1)

(Fig. 56:H–O)

Nomen hierarchicum: **Stenonema/g1** (incl. *Stenacron*, *Macdunnoa*) [g: *Stenonema* Traver 1933: 173, typus *Heptagenia tripunctata* Banks 1910 (design. orig.; syn. subj. *Baetis femorata* Say 1823)].
In circumscription matches:
— grex subgenn. *Stenonema*: Kluge 1988: 287;
— gen. *Stenonema* Traver 1933: 173.

References. Traver 1933: Ɒ*; – Needham & Traver & Hsu 1935: Ɒ*; – Spieth 1947: Ɒ*; – Lewis 1974: Ɒ; – Tshernova 1976: Ɒ*; – Kluge 1988: Ɒ*; – 1993: Ɒ*.

Autapomorphies of Stenonema/g1.

(1) Superlingua [curved laterally – see Heptagenia/f5=g4 (1)] is especially sharply curved and stretched laterally; by its form differs from any other taxon among Heptagenia/f5=g4 (FIG.56:H).

(2) On tergalii I–VI costal rib is lost, and anal rib [initially situated on anal margin – see Branchitergaliae (3)] is situated at a distance from anal margin, being especially long and nearly straight (FIG.56:O). Unique apomorphy.

(3) Tergalii VII are vestigial, lack fibrillose ventral lobe, thus fibrillose lobe [see Branchitergaliae (3)] is present on tergalii I–VI only. Non-unique apomorphy (see Index of characters [1.3.59]); among Heptagenia/f5=g4 fibrillose lobe of tergalius VII is lost also in *Kageronia* and majority of Ecdyonurus/fg1.

Characters of unclear phylogenetic status.

(4) Glossae [initially pyramidal – see Radulapalpata (1)] with straight or slightly curved inner margin, with distal half flattened, as dorsal edge is approximated to median edge (FIG.56:I–J, N). The same in Heptagenia/f8=g7; probably, synapomorphy.

Variable characters of Stenonema/g1. In selected species of *Stenacron* and Stenonema/g3 in imago and subimago on fore tarsus 1st segment can be shortened (like in Heptagenia/f8=g7) or not shortened.

Size. Fore wing length 6–16 mm.

Distribution. North and Central America; dominate in Eastern Nearctic.

Stenonema/g1 is divided into *Stenacron*, Stenonema/g2 and *Macdunnoa*.

2.2;1,2-2/2.1;1,2-3/1. Stenacron/g(1)

(Radulapalpata Heptagenia/f5=g4 .../f7=g6 Stenonema/g1 *Stenacron*)

(Fig. 56:N–O)

Nomen hierarchicum: **Stenacron/g(1)** [g: *Stenacron* Jensen 1974: 225, typus *Baetis interpunctata* Say 1839 (design. orig.)].
In circumscription matches:
— grex spp. *interpunctata*: Spieth 1933: 332;
— grex spp. *interpunctatum*: Traver 1933: 173;
— subgen. *Stenacron*: Kluge 1983b: 16 (Abstract);
— gen. *Stenacron* Jensen 1974: 225.

References. Spieth 1933: ⊕*; – Spieth 1947: ⊙* Ɒ*, ⊕*; – Koss 1968: ⊙*; – Jensen 1974: Ɒ* ⊕*; – Edmunds & Jensen & Berner 1976: Ɒ* ⊕*.

Autapomorphy of *Stenacron*.

(1) Egg, instead of usual knob-terminated threads [see Branchitergaliae (4)], bears on each pole a ring formed by very long coiled threads, that looks as a polar cap (Koss 1968:Fig.16). Similar structure occurs in some Furcatergaliae only (see Index of characters [3.6]).

Character of unclear phylogenetic status.

(2) All tergalii I–VII have stretched pointed apices, and tergalii I–VI [which have long anal rib – see Stenonema/g1 (2)] have this pointed apex somewhat curved anteriorly (FIG.56:O). In all other Stenonema/g1 tergalii I–VI are not pointed.

Plesiomorphy of *Stenacron*. In contrast to *Macdunnoa*, tergalii VII [thin and lacking fibrillose portion – see Stenonema/g1 (3)] are not shortened.

Size. Fore wing length 8–12 mm.

Distribution. Eastern Nearctic.

Species composition of Stenacron/g(1). About 10 species – *candidum* Traver 1935 [*Stenonema*], *carolina* Banks 1914 [*Heptagenia*], *floridense* Lewis 1974 [*Stenonema*], *frontale* Banks 1910 [*Heptagenia*], *gildersleevei* Traver 1935 [*Stenonema*], *interpunctata* Say 1839 [*Baetis*] (synn. subj.: *canadensis* Walker 1853 [*Baetis*], *heterotarsalis* McDunnough 1933 [*Ecdyonurus*], *affine* Traver 1933 [*Stenonema*], *conjunctum* Traver 1935 [*Stenonema*], *majus* Traver 1935 [*Stenonema*], *ohioense* Traver 1935 [*Stenonema*], *proximum* Traver 1935 [*Stenonema*], *areion* Burks 1953 [*Stenonema*]), *minnetonka* Daggy 1945 [*Stenonema*], *pallidum* Traver 1933 [*Stenonema*].

Material examined: *interpunctata* [*B.*]: O, ⊕/⊙; sp.:Ɒ.

***

2.2;1,2-2/2.1;1,2-3/2. Stenonema/g2

(Radulapalpata Heptagenia/f5=g4 .../f7=g6
Stenonema/g1 .../g2)
(Fig. 56:H–M)

Nomen hierarchicum: **Stenonema/g2** (sine *Stenacron*, *Macdunnoa*).
In circumscription matches:
— subgen. *Stenonema*: Kluge 1983b: 16 (Abstract);
— gen. *Stenonema*: Jensen 1974: 225.

References. Edmunds & Jensen & Berner 1976: ♉* ⊕*; - Lewis 1978a: ♉*; - Bednarik & McCafferty 1979: ♉* ⊕*.

Character of unclear phylogenetic status.

(1) Tergalii I–VI are not pointed (in contrast to *Stenacron*), apically obliquely truncate or rounded, with anal margin the longest [that is connected with development of anal rib – see Stenonema/g1 (2)] (Edmunds & al. 1976:Figs 126–127).

Plesiomorphies of Stenonema/g2. In contrast to *Macdunnoa*, tergalii VII [thin and lacking fibrillose portion – see Stenonema/g1 (3)] are not shortened.

Variable characters of Stenonema/g2. In selected species of Stenonema/g2 on maxilla the apical-ventral row of pectinate setae [see Radulapalpata (3)] is vestigial; this character was erroneously regarded as an apomorphy of the whole taxon Stenonema/g2; actually in some species up to 9 normally developed pectinate setae can be present (Bednarik & McCafferty 1979: 6).

Size. Fore wing length 6–16 mm.

Distribution. North and Central America; dominate in Eastern Nearctic.

Stenonema/g2 is divided into Stenonema/g3 and *Maccaffertium*.

2.2;1,2-2/2.1;1,2-3/2.1. Stenonema/g3

(Radulapalpata Heptagenia/f5=g4 .../f7=g6
Stenonema/g1 .../g3)
(Fig. 56:L)

Nomen hierarchicum: **Stenonema/g3** (sine *Maccaffertium*).
In circumscription matches:
— subgen. *Stenonema*: Bednarik 1979: 190.
Monospecific taxon.

References. Bednarik 1979: ♉; - Bednarik & McCafferty 1979: ♉ ⊕.

Character of unclear phylogenetic status.

(1) Tergalii I–VI [see Stenonema/g2 (1)] with rounded apex (Edmunds & al. 1976:Fig.126). The same in *Macdunnoa*.

Size. Fore wing length 10–13 mm.

Distribution. Eastern Nearctic.

Species composition of Stenonema/g3. 1 species – *femorata* Say 1823 [*Baetis*] (syn. subj.: *tripunctata* Banks 1910 [*Heptagenia*], *birdi* Traver 1935 [*Stenonema*], *scitulum* Traver 1935 [*Stenonema*]).

Material examined: *femorata* [*B.*]: ♉.

***

2.2;1,2-2/2.1;1,2-3/2.2. Maccaffertium/g(1)

(Radulapalpata Heptagenia/f5=g4 .../f7=g6
Stenonema/g1 .../g2 *Maccaffertium*)
(Fig. 56:H–K, M)

Nomen hierarchicum: **Maccaffertium/g(1)** [g: *Stenonema* subgen. *Maccaffertium* Bednarik 1979: 191, typus *Heptagenia integer* McDunnough 1924 (design. orig.; syn. subj. *Heptagenia mexicana* Ulmer 1920)].
In circumscription matches:
— subgen. *Maccaffertium* Bednarik 1979: 191.

References. Bednarik 1979: ♉; - Bednarik & McCafferty 1979: ♉ ⊕.

Character of unclear phylogenetic status.

(1) Tergalii I–VI [see Stenonema/g2 (1)] with truncate apex (Edmunds & al. 1976:Fig.127).

Size. Fore wing length 6–16 mm.

Distribution. North and Central America; dominate in Eastern Nearctic.

Species composition of Maccaffertium/g(1) (here [*S.*]=[*Stenonema*]. About 15 species – *bednariki* McCafferty 1981 [*S.*], *carlsoni* Lewis 1974 [*S.*], *exiguum* Traver 1933 [*S.*] (synn. subj.: *alabamae* Traver 1937 [*S.*], *quinquespinum* Lewis 1974 [*S.*]), *ithaca* Clemens & Leonard 1924 [*Heptagenia*] (syn. subj.: *allegheniense* Carle 1977 [*S.*]), *lenati* McCafferty 1990 [*S.*], *lutea* Clemens 1913 [*Heptagenia*], *mediopunctatum* McDunnough 1926 [*Ecdyonurus*] (syn. subj.: *nepotellus* McDunnough 1933 [*Ecdyonurus*]), *meririvulanum* Carle & Lewis 1978 [*S.*], *mexicana* Ulmer 1920 [*Heptagenia*] (synn. subj.: *integer* McDunnough 1924 [*Heptagenia*], *bellum* Traver 1933 [*S.*], *wabasha* Daggy 1945 [*S.*], *metriotes* Burks 1953 [*S.*]), *modestus* Banks 1910 [*Heptagenia*] (synn. subj.: *rubromaculata* Clemens 1913 [*Heptagenia*], *ruber* McDunnough 1926 [*Ecdyonurus*], *annexum* Traver 1933 [*S.*], *varium* Traver 1933 [*S.*]), *pulchella* Walsh 1862 [*Palingenia*], *sinclairi* Lewis 1979 [*S.*], *smithae* Traver 1937 [*S.*], *terminata* Walsh 1862 [*Palingenia*] (synn. subj.: *placita* Banks 1910 [*Heptagenia*], *bipunctatus* McDunnough 1926 [*Ecdyonurus*], *lepton* Burks 1947 [*S.*], *ares* Burks 1953 [*S.*]), *vicaria* Walker 1853 [*Baetis*] (synn. subj.: *tessellata* Walker 1853 [*Baetis*], *fusca* Clemens 1913 [*Heptagenia*], *rivulicolus* McDunnough 1933 [*Ecdyonurus*]).

Material examined: *mediopunctatum* [*E.*]: ♉, ⊕'; *mexicana* [*H.*]: ♉; *terminata* [*P.*]: ♉, ⊕.

*** ***

2.2;1,2-2/2.1;1,2-3/3. **Macdunnoa/g(1)**

(Radulapalpata Heptagenia/f5=g4 .../f7=g6 Stenonema/g1 *Macdunnoa*)

Nomen hierarchicum: **Macdunnoa/g(1)** [g: *Macdunnoa* Lehmkuhl 1979b: 860, typus *M. nipawinia* Lehmkuhl 1979 (design. orig.)].
In circumscription matches:
— subgen. *Macdunnoa*: Kluge 1983b: 16 (Abstract);
— gen. *Macdunnoa* Lehmkuhl 1979b: 860.

References. Lehmkuhl 1979b: Ꝋ ⊕♀; – Flowers 1982: Ꝋ ⊕.

Autapomorphy of *Macdunnoa*.

(1) Tergalius VII [see Stenonema/g1 (3)] has a form of very small vestige (at least in *persimplex* [*Heptagenia*] and *brunnea* [*Macdunnoa*]; in original description of *nipawinia* [*M.*] is said that tergalii VII are lost). Tergalius VI is strongly diminished.

Character of unclear phylogenetic status.

(2) Tergalii I–VI with rounded apex. The same in Stenonema/g3.

Size. Fore wing length 6–9 mm.

Distribution. Nearctic.

Species composition of Macdunnoa/g(1). 3 species – *brunnea* Flowers 1982 [*Macdunnoa*], *nipawinia* Lehmkuhl 1979 [*Macdunnoa*], *persimplex* McDunnough 1929 [*Heptagenia*].

Material examined: *brunnea* [*M.*]: ⊕♂.

*** *** *** ***

2.2;1,2-2/2.1;2. **Ecdyonurus/fg1**

(Heptagennota Pentamerotarsata Radulapalpata Heptagenia/f5=g4 Ecdyonurus/fg1)

(Figs 9:E; 57–59)

Nomen hierarchicum: **Ecdyonurus/fg1** (incl. *Afghanurus*, *Atopopus*) [f: Ecdyonuridae Ulmer 1920b: 136 (1905) (pro Ecdyuridae Jacobson & Bianchi 1905: 877); g: *Ecdyonurus* Eaton 1868a: 142, typus *Ephemera venosa* Fabricius 1775 (design. orig.); syn. obj.: ***Ecdyurus*** Eaton 1868b: 90].
In circumscription matches:
— gen. *Ecdyonurus*: Kluge 1983b:16 (Abstract); 1988:298;
— grex genn. *Ecdyonurus*: Tomka 1991: 121;
— tribus Ecdyonurini: Braasch 1990: 12.

References. Bogoescu & Tabacaru 1968: Ꝋ* ⊕♂"; – Tomka 1991: Ꝋ*; – Kluge 1988: Ꝋ* ⊕*; – 1993: Ꝋ* ⊕*.

Autapomorphies of Ecdyonurus/fg1.

(1) Metathoracic nerve ganglion is strongly transferred anteriorly, nearer to mesothoracic ganglion, being connected with it by a short unpaired connective. Due to this, imaginal and subimaginal median impression of furcasternum [see Heptagennota (4)] is not narrowed in anterior part (in contrast to all other Heptagennota), but is parallel-sided or widened toward its anterior part, where the metathoracic ganglion is situated (FIG.57:A–B).

(2) Mouth apparatus has the following unique features:

Left and right mandibles [see Heptagennota (7)] have incisors dissimilar: incisor of left mandible (with mola projected distally) is pointed, with a row of denticles on inner margin; incisor of right mandible (with mola projected proximally) has a large denticle at some distance from apex and a row of smaller denticles proximad of it; denticles on outer margins of both incisors are at a distance from apex (FIG.58:B–C); in *inversus* [*Ecdyonurus*] the left and right mandibles are changed places (Kluge 1980:Figs 67–68).

Superlingua [curved laterally – see Heptagenia/ /f5=g4 (1)] has a peculiar shape uniform for all representatives: with a rounded apico-lateral projection (larger than in Heptagenia/f8=g7, more distinct than in *Kageronia* and *Dacnogenia*, and smaller than in Stenonema/g1), basally with a small rounded lateral convexity, lateral margin between this convexity and apico-lateral projection is slightly convex (FIG.58:D).

On maxilla (FIG.58:E–H) ventral row of setae [see Branchitergaliae (1)] is transformed to a field of irregularly situated setae (unique apomorphy). Proximal dentiseta [see Bidentiseta (1)] is bifurcate.

Vestige of segment 3 of maxillary palp [see Plesiomorphies of Heptagenia/f5=g4] is triangular (FIG. 58:E), larger than in Heptagenia/f6=g5.

Glossae [initially pyramidal – see Radulapalpata (1)] are rhomboid, with convex inner margin and flat apical half, as dorsal edge is approximated to median edge (FIG.57:H).

(3) Tergalii are differentiated as following (FIGS 57:D–E; 58:A): tergalius I is narrowed, with concave anal margin (i. e. has a more or less banana-like shape), that allows it to bend dorsally; tergalii II–VII are widened, with strongly convex anal margin – i. e. have more or less roundish-triangular shape. Unique apomorphy.

(4) Outer (posterior) margin of larval femur [flattened – see Heptagennota (1)] bears a regular row of long stout setae (FIG.57:J–K) [in contrast to Heptagenia/f6=g5 – see *Kageronia* (4) and Heptagenia/f7= =g6 (2)]. These setae are situated less dorsally, than similar setae in Rhithrogena/fg1 and *Cinygma*. Non-unique apomorphy: regular row of setae on outer margin of femur appeared independently in various rheophilous mayflies.

Characters of unclear phylogenetic status.

(5) 2nd+3rd segment of labial palp [specialized as scraping – see Radulapalpata (1)] is pointed, thus field of scraping setae terminates at a distance from end of sclerotized ridge (FIG.57:H). The same in *Kageronia* and Rhithrogena/fg1; probably symplesiomorphy. In Heptagenia/f7=g6 and *Cinygma* form of this segment is different.

(6) Tergalius VII [see (3)] nearly always lacks fibrillose lobe [see Branchitergaliae (3)] (thus fibrillose lobe is present only on tergalii I–VII, rarely I–V); only in selected species (particularly in Ecdyonurus/fg2 *insignis* [*H.*]) tergalius VII has fibrillose lobe (probably secondarily restored).

Size. Fore wing length 5–18 mm.

Age and distribution. Miocene (see *Pseudokageronia* below) — Recent; Arctogea (Holarctic + Oriental + Ethiopian Regions); dominate in Holarctic.

Ecdyonurus/fg1 is divided into Ecdyonurus/fg2, Afghanurus/g1 (p.180) and Atopopus/g1 (p.181). Some insufficiently described Asian and African taxa have uncertain systematic position (p.185).

2.2;1,2-2/2.1;2,1. Ecdyonurus/fg2

(Pentamerotarsata Radulapalpata Heptagenia/f5=g4 Ecdyonurus/fg1 .../fg2)
(Figs 57:F–I; 58:F)

Nomen hierarchicum: **Ecdyonurus/fg2** (sine *Afghanurus*, *Atopopus*).
In circumscription matches:
— grex spp.: Eaton 1883–1888: 281;
— subgen. *Ecdyonurus*: Kluge 1997: 200;
— gen. *Ecdyonurus*: Schoenemund 1930a: 45; 1930b: 83.

References. Eaton 1883–1888: ⊕♂*; – Schoenemund 1930a: ⊘* ⊕♂*; – Kluge 1997: ⊘*.

Autapomorphy of Ecdyonurus/fg2.

(1) In larva lateral flat projections of pronotum [which are more or less developed in all Heptagennota, in accordance with general body flattening – see Heptagennota (1)] are large, partly fused with lateral margins of mesonotum and continued posteriorly as a pair of projections situated laterad of mesonotum (FIG.57:I). Non-unique apomorphy: various posterolateral projections of pronotum independently appeared in other mayflies.

Plesiomorphies of Ecdyonurus/fg2. Distal dentiseta is either simple (FIG.58:F), or distally bifurcate [like proximal dentiseta – see Ecdyonurus/fg1 (2)] (in contrast to Atopopus/g1). Primary swimming setae [vestigial – see Pentamerotarsata (6)] are present. Median titillators of penis are well-developed, contiguous medially (in contrast to Afronurus/g1).

Variable characters of Ecdyonurus/fg2. In male imago 1st segment of fore tarsus can be relatively long (longer than 1/2 of 2nd segment) or short (several times shorter than 2nd segment, like in Heptagenia/f8=g7 and some others).

Size. Fore wing length 7–18 mm.

Distribution. Western Palaearctic.

Species composition of Ecdyonurus/fg2 (here [*E.*]= =[*Ecdyonurus*]). About 30 species are described – *alpinus* Hefti & Tomka & Zurwerra 1987 [*E.*], *angelieri* Thomas 1968 [*E.*], *androsianus* Braasch 1983 [*E.*], *asiaeminoris* Demoulin 1973 [*E.*], *aurantiaca* Burmeister 1839 [*Baetis*] (syn. subj.: *pazsiczkyi* Pongracz 1913 [*E.*]), *austriacus* Kimmins 1958 [*E.*], *autumnalis* Braasch 1980 [*E.*], *belfiorei* Haybach & Thomas 2002 [*Ecdyonurus*], *bellieri* Hagen 1860 [*Baetis*], *carpathicus* Sowa 1973 [*E.*], *codinai* Navás 1924 [*E.*], *corsicus* Esben-Petersen 1912 [*E.*], *cortensis* Belfiore 1987 [*E.*], *dispar* Curtis 1834 [*Baetis*] (syn. subj.: *fluminum* Pictet 1843 [*Baetis*]), *epeorides* Demoulin 1955 [*E.*], *forcipula* Pictet 1843 [*Baetis*], *graecus* Braasch 1984 [*E.*], *helveticus* Eaton 1887 [*Ecdyurus*] (synn. subj.: *italicus* Eaton 1883 [*E.*] NOM. NUD., *bollenganus* Navás 1933 [*E.*]), *ifranensis* Vitte & Thomas 1988 [*E.*], *insignis* Eaton 1870 [*Heptagenia*] (syn. subj.: *rhenanus* Neeracher 1910 [*E.*]), *krueperi* Stein 1863 [*Potamanthus*], *macani* Thomas & Sowa 1970 [*E.*], *moreae* Belfiore 1986 [*E.*], *ornatipennis* Tshernova 1938 [*E.*], *parahelveticus* Hefti & Tomka & Zurwerra 1986 [*E.*], *picteti* Meyer-Dür 1864 [*Baetis*] (syn. subj.: *nigrescens* Klapálek 1908 [*Heptagenia*], *puma* Jacob & Braasch 1986 [*E.*], *quaesitor* Eaton 1883 [*E.*], *rothschildi* Navás1929 [*E.*], *ruffii* Grandi 1953 [*E.*] (syn. subj.: *wautieri* Fontaine 1964 [*E.*]), *russevi* Braasch & Soldán 1985 [*E.*], *siveci* Jacob & Braasch 1984 [*E.*], *starmachi* Sowa 1971 [*E.*], *subalpinus* Klapálek 1907 [*E.*], *submontanus* Landa 1970 [*E.*], *torrentis* Kimmins 1942 [*E.*], *venosa* Fabricius 1775 [*Ephemera*] (synn. subj.: *berolinensis* Müller 1776 [*Ephemera*], *nigrimana* Dufour 1841 [*Ephemera*], *purpurascens* Pictet 1843 [*Baetis*], *subfusca* Stephens 1835 [*Baetis*]), *zelleri* Eaton 1885 [*Ecdyurus*].

Material examined: *aurantiaca* [*B.*]: ⊘–◎–⊕/⊙; *autumnalis* [*E.*]: ⊘–◎–⊕/⊙; *dispar* [*B.*]: ⊘, ⊕; *farsi* [*E.*] (in litt.): ⊘–◎–⊕/⊙; *helveticus* [*E.*]: ⊘–◎–⊕/⊙; *insignis* [*H.*]: ⊘; *ornatipennis* [*E.*]: ⊘–◎–⊕/⊙; *ornatus* [*E.*] (in litt.): ⊘–◎–⊕/⊙; *picteti* [*B.*]: ⊘, ◎, ⊕; *submontanus* [*E.*]: ⊕♂; *torrentis* [*E.*]: ⊘; *venosa* [*E.*]: ⊘, ◎, ⊕; *zelleri* [*E.*]: ⊘♂–◎♂–⊕♂; sp.H5 (Georgia): ⊘, ⊘♀–◎♀–⊕♀, ⊕♂.

***

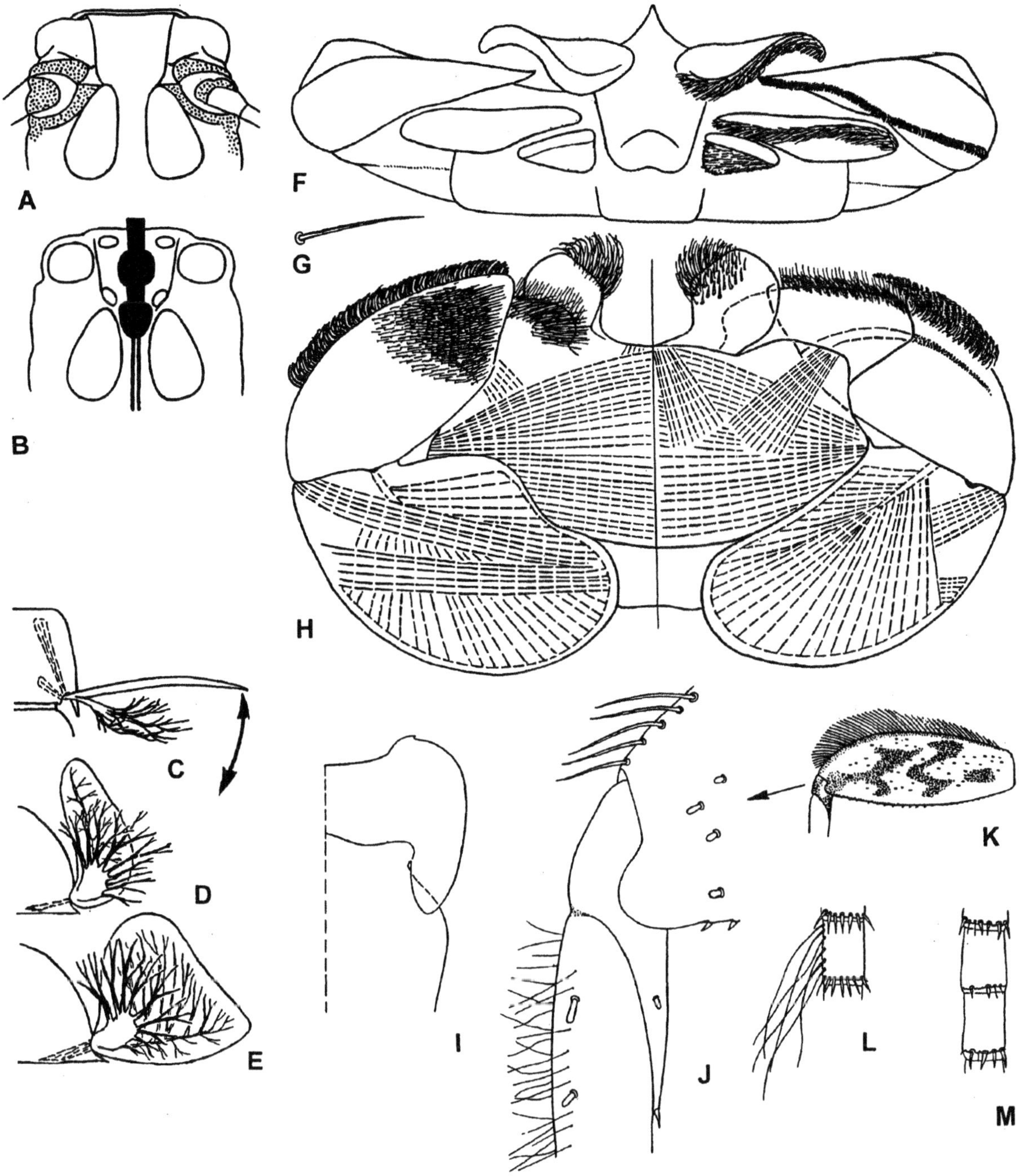

Figure 57. Ecdyonurus/fg1.
A–B – Ecdyonurus/fg1-Afronurus/g1 *levis* [*Epeorus*]: **A** – imaginal mesothorax, ventral view; **B** – meso- and metathoracic nerve ganglia (shown by black) and ventral areas of mesothoracic muscles attachment (compare with Fig.65:C–D). **C–E** – scheme of natural position of tergalii in Ecdyonurus/fg1 (muscles shown by interrupted lines) (compare with Figs 61:J–K and 62:H–I): **C** – dorsal view; **D** – tergalius I, posterior view; **E** – tergalius II, posterior view. **F–I** – Ecdyonurus/fg* *venosa* [*Ephemera*]: **F** – superlinguae, hypopharynx and labium, apical view; **G** – seta of distal row on dorsal surface of distal segment of labial palp; **H** – labium, dorsal view (in left half) and ventral view (in right half) (muscles shown by interrupted lines) (compare with Fig.62:E–G); **I** – left half of larva pronotum and anterior part of mesonotum. **J–K** – Ecdyogymnurus/g1 *scalaris* [*Ecdyonurus*], larval middle right leg, dorsal view: **J** – apex of femur, patella and base of tibia; **K** – femur. **L** – Afghanurus/g* *vicinus* [*Afghanurus*], portion of larval cercus. **M** – Ecdyogymnurus/g1 *inversus* [*Ecdyonurus*], the same. (A–H – from Kluge 1988a and 1993a; I–K – from Kluge 1997d; L–M – from Kluge 1980).

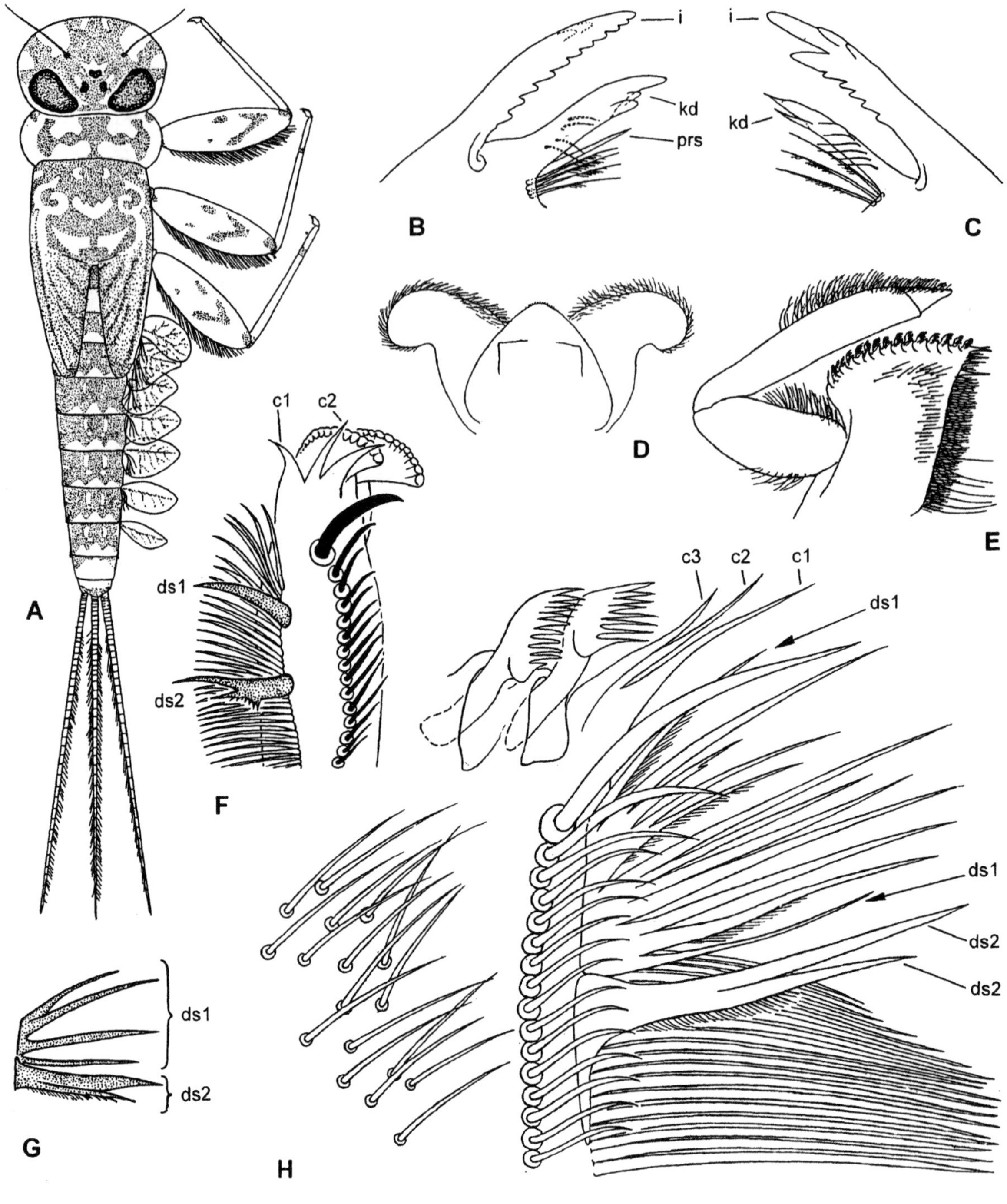

Figure 58. Ecdyonurus/fg1, larvae.
A–E – Afghanurus/fg2 *joernensis* [*Ecdyonurus*]: **A** – male larva; **B–C** – apices of left and right mandibles, dorsal view; **D** – hypopharynx and superlinguae, ventral view; **E** – right maxilla, ventral view. **F** – Ecdyonurus/fg2 *ornatus* (nomen nudum), apex of left maxilla, median view (dentisetae shown by dots, setae of medio-ventral row shown by black). **G–H** – Atopopus/g1: **G** – Afronurus/g1 *abracadabrus* [*Ecdyonurus*], dentisetae; **H** – Afronurus/g* *peringueyi* [*Ecdyonurus*], apex of right maxilla, medio-ventral view. (A–C – from Kluge 1980; from Kluge 1997d; F–G – from Kluge 1998).

Abbreviations: **c1**, **c2**, **c3** – maxillary canines; **ds1** – distal dentiseta; **ds2** – proximal dentiseta; **i** – incisor; **kd** – kinetodontium; **prs** – prostheca.

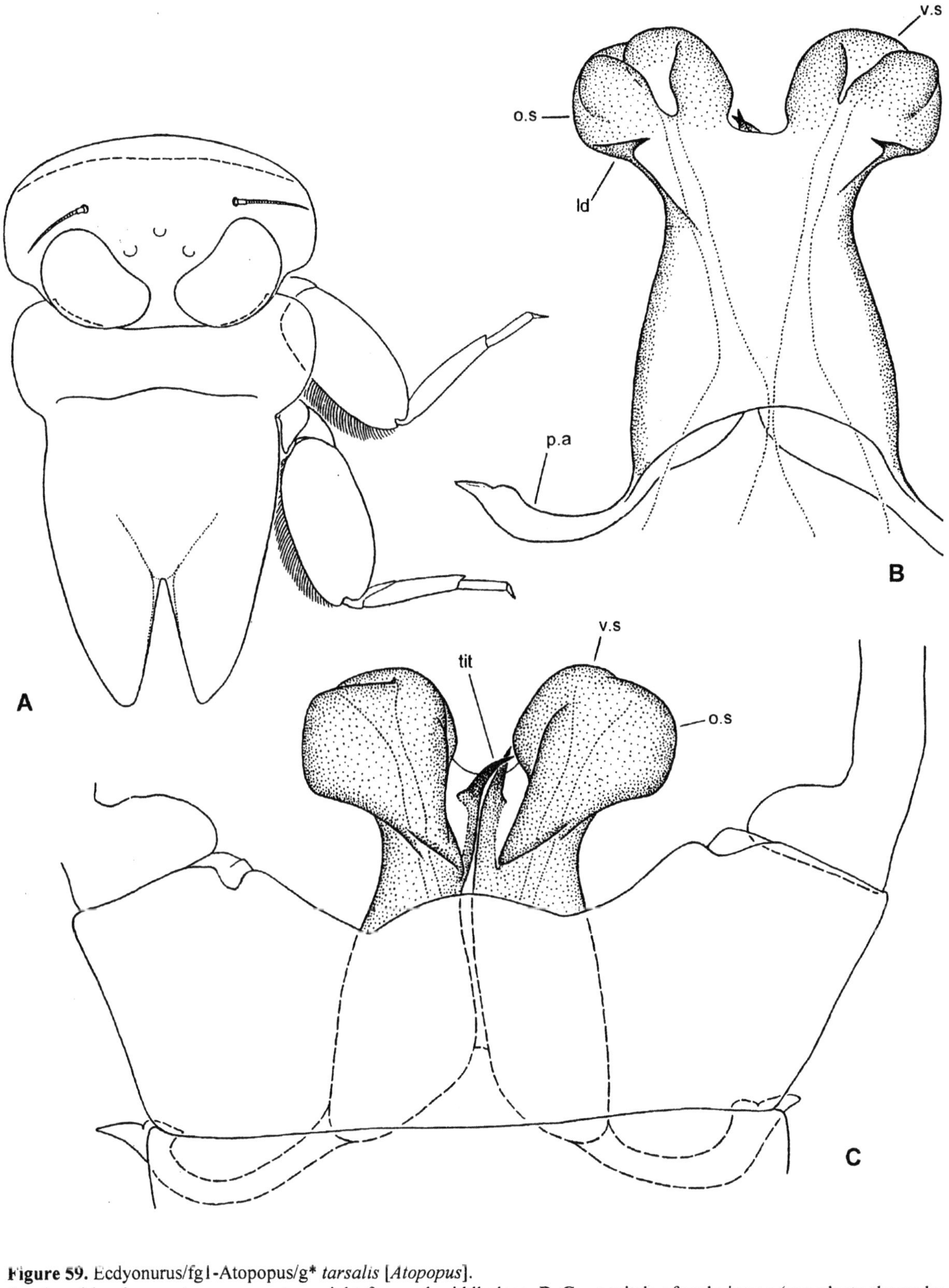

Figure 59. Ecdyonurus/fg1-Atopopus/g* *tarsalis* [*Atopopus*].
A – larval head, pronotum, mesonotum, right fore and middle legs. **B–C** – genitals of male imago (gonoducts shown by dotted lines): **B** – dorsal view; **C** –ventral view.
Abbreviations: **ld** – latero-dorsal spine; **o.s** – outer sclerite; **p.a** – penial arm; **tit** – median titillator; **v.s** – ventral sclerite.

2.2;1,2-2/2.1;2,2. **Afghanurus/g1**

(Pentamerotarsata Radulapalpata Heptagenia/f5=g4 Ecdyonurus/fg1 Afghanurus/g1)

(Figs 57:J–M; 58:A–E)

Nomen hierarchicum: **Afghanurus/g1** (incl. *Leucrocuta*) [g: *Afghanurus* Demoulin 1964a: 356, typus *A. vicinus* Demoulin 1964 (design. orig.)]. Possibly the name *Afghanurus* should be substituted by older name *Compsoneuriella* or *Notonurus* (see below, Ecdyonurus/fg1 INCERTAE SEDIS).
In circumscription matches:
— gen. *Paracinygmula*: Jacob & Dorn & Haybach 1995:99.

Autapomorphy of Afghanurus/g1.

(1) Ventral sclerites of penis dorsad-laterad of titillators [see Pentamerotarsata (5)], bear one or several pairs of denticles ("discal spines") directed distally-laterally (Kluge 1980:Figs 64,71,83,90,103). Only in a few species these "discal spines" are secondarily lost.

Character of unclear phylogenetic status.

(2) In male imago 1st segment of fore tarsus is short, several times shorter than 2nd segment. Non-unique character (see Index of characters [2.2.76]).

Plesiomorphies of Afghanurus/g1. Median titillators of penis [see (1)] are well-developed, contiguous medially (in contrast to Afronurus/g1). Larval pronotum has no posterolateral projections (in contrast to Ecdyonurus/fg2). Distal dentiseta is simple (in contrast to Atopopus/g1).

Size. Fore wing length 5–10 mm.

Distribution. Holarctic; dominate in Eastern Palaearctic and Nearctic.

Afghanurus/g1 is divided into Afghanurus/g2 and Leucrocuta/g1. Some insufficiently described species have uncertain systematic position .

2.2;1,2-2/2.1;2,2-1. Plesiomorphon **Afghanurus/g2**

(Radulapalpata Heptagenia/f5=g4 Ecdyonurus/fg1 Afghanurus/g1 .../g2)

(Figs 57:L; 58:A–E)

Nomen hierarchicum: **Afghanurus/g2** (sine *Leucrocuta*).
In circumscription matches:
— gen. *Nixe* Flowers 1980: 299;
— gen. *Afghanurus*: Braasch 1981: 128.

Reference. Flowers 1980: Ծ* ⊕*.

Plesiomorphy of Afghanurus/g2. Primarily swimming setae [vestigial – see Pentamerotarsata (6)] are present (FIGS 57:L; 58:A) (in contrast to Leucrocuta/g1).

Size. Fore wing length 5–9 mm.

Distribution. Holarctic; dominate in Eastern Palaearctic and Nearctic.

Nominal taxa included. Afghanurus/g2 includes: (1) **Paracinygmula/g** [g: *Paracinygmula* Bajkova 1975: 54, typus *P. zhiltzovae* Bajkova 1975 (design. orig.)]; (2) **Nixe/g** [g: *Nixe* Flowers 1980: 299, typus *Ecdyonurus lucidipennis* Clemens 1913 (design. orig.)]; (3) **Akkarion/g** [g: *Nixe* subgen. *Akkarion* Flowers 1980: 299, typus *Heptagenia simplicioides* McDunnough 1924 (design. orig.)] (Kluge 1988: 302).

Species composition of Afghanurus/g2. More than 20 species are described; for some North American species larvae are undescribed, so they are attributed here presumably (Flowers 1980). One **Amphipacific species:** *simplicioides* McDunnough 1924 [*Heptagenia*] (synn. subj.: *werestschagini* Tshernova 1952 [*Heptagenia*], *imanica* Bajkova 1972 [*Rhithrogena*]), One **Trans-Palaearctic species:** *joernensis* Bengtsson 1909 [*Ecdyurus*] (synn. subj.: *flavomaculatus* Aro 1928 [*Ecdyonurus*], *mongolica* Bajkova & Varykhanova 1978 [*Heptagenia*], *dentata* Braasch 1979 [*Heptagenia*], *stubbei* Braasch 1979 [*Ecdyonurus*]). **Asian species:** *littoralis* Kang & Yang 1994 [*Nixe*], *mitificus* Kang & Yang 1994 [*Nixe*], *obscurus* Kang & Yang 1994 [*Nixe*], *rubrofasciatus* Brodsky 1930 [*Ecdyonurus*] (syn. subj.: *tianshanica* Kustareva 1984 [*Stenonema*]), *vicinus* Demoulin 1964 [*Afghanurus*], *zhiltzovae* Bajkova 1975 [*Paracinygmula*] (syn. obj.: *bajkovae* Kluge 1986 [*Ecdyonurus*], syn. subj.: *subspinosa* Braasch & Soldán 1988 [*Nixe*]). **North American species:** *bella* Allen & Cohen 1977 [*Heptagenia*], *criddlei* McDunnough 1927 [*Heptagenia*], *flowersi* McCafferty 1982 [*Nixe*], *inconspicua* McDunnough 1924 [*Heptagenia*], *lucidipennis* Clemens 1913 [*Ecdyonurus*], *perfida* McDunnough 1926 [*Heptagenia*], *rusticalis* McDunnough 1931 [*Heptagenia*].

Material examined: *criddlei* [*H.*]: Ծ; *joernensis* [*E.*]: Ծ–◎–⊕/⊙; *rubrofasciatus* [*E.*]: Ծ–◎–⊕/⊙; *rusticalis* [*H.*]: Ծ; *simplicioides* [*H.*]: Ծ–◎–⊕/⊙; *vicinus* [*A.*]: Ծ–◎–⊕/⊙; *zhiltzovae* [*P.*]: Ծ–◎–⊕/⊙.

***

2.2;1,2-2/2.1;2,2-2. **Leucrocuta/g1**

(Radulapalpata Heptagenia/f5=g4 Ecdyonurus/fg1 Afghanurus/g1 Leucrocuta/g1)

(Fig. 58:J–K, M)

Nomen hierarchicum: **Leucrocuta/g1** (incl. *Ecdyogymnurus*) [g: *Leucrocuta* Flowers 1980: 297, typus *Heptagenia maculipennis* Walsh 1863 (design. orig.)].

Autapomorphy of Leucrocuta/g1.

(1) Primary swimming setae [see Pentamerotarsata (6)] are lost: larval cerci and paracercus have only whorls of setae on ends of segments [see

Heptagenia/f5=g4 (2)] (FIG.57:M). Non-unique character (see Index of characters [1.3.66]); among Ecdyonurus/fg1 the same in Atopopus/g1.

Size. Fore wing length 5–10 mm.

Distribution. Eastern Palaearctic and whole Nearctic.

Leucrocuta/g1 is divided into *Ecdyogymnurus* and Leucrocuta/g2.

2.2;1,2-2/2.1;2,2-2/1. Plesiomorphon **Ecdyogymnurus/g(1)**

(Heptagenia/f5=g4 Ecdyonurus/fg1 Afghanurus/g1 Leucrocuta/g1 *Ecdyogymnurus*)

(Fig. 58:J–K, M)

Nomen hierarchicum: **Ecdyogymnurus/g(1)** [g: *Ecdyogymnurus* Kluge, SUBGEN. NOV., typus *Ecdyonurus aspersus* Kluge 1980].

Reference. Kluge 1980 (*E. aspersus*, *E. inversus*): ʊ ⊕.

Plesiomorphy of *Ecdyogymnurus*. In contrast to Leucrocuta/g2, male eyes are not diminished and not widely separated (Kluge 1980:Figs 52, 65; 1983:Fig. 4a).

Size. Fore wing length 5–10 mm.

Distribution. Eastern Palaearctic.

Species composition of Ecdyogymnurus/g(1). Not less than 4 species are described – *aspersus* Kluge 1980 [*Ecdyonurus*], *inversus* Kluge 1980 [*Ecdyonurus*], *kibunensis* Imanishi 1936 [*Ecdyonurus*], *scalaris* Kluge 1983 [*Ecdyonurus*].

Material examined: *aspersus* [*E.*]: ʊ–ʘ–⊕/⊙; *inversus* [*E.*]: ʊ–ʘ–⊕/⊙; *kibunensis* [*E.*]: ʊ–ʘ–⊕/⊙; *scalaris* [*E.*]: ʊ–ʘ–⊕/⊙; sp.H1 (Russian Far East): ʊ, ⊕.

***

2.2;1,2-2/2.1;2,2-2/2. **Leucrocuta/g2**

(Heptagenia/f5=g4 Ecdyonurus/fg1 Afghanurus/g1 Leucrocuta/g1 .../g2)

Nomen hierarchicum: **Leucrocuta/g2** (sine *Ecdyogymnurus*).

In circumscription matches:

— gen. *Leucrocuta* Flowers 1980: 297.

Reference. Flowers 1980: ʊ* ⊕*.

Autapomorphy of Leucrocuta/g2.

(1) Male eyes are diminished and separated dorsally by 4–5 times width of median ocellus (Flowers 1980:Fig.20) (in contrast to majority of other Heptagennota and other mayflies, where male eyes are larger and separated dorsally by width of medial ocellus or less). Non-unique apomorphy: in selected species of various taxa (including some Ecdyonurus/fg1, but not *Ecdyogymnurus* or Afghanurus/g2) male eyes are also diminished and widely separated.

Size. Fore wing length 5–10 mm.

Distribution. Nearctic.

Species composition of Leucrocuta/g2. 10 species – *aphrodite* McDunnough 1926 [*Heptagenia*], *hebe* McDunnough 1924 [*Heptagenia*], *jewetti* Allen 1966 [*Heptagenia*], *juno* McDunnough 1924 [*Heptagenia*], *maculipennis* Walsh 1863 [*Heptagenia*], *minerva* McDunnough 1924 [*Heptagenia*], *petersi* Allen 1966 [*Heptagenia*], *thetis* Traver 1933 [*Heptagenia*], *umbratica* McDunnough 1931 [*Heptagenia*], *walshi* McDunnough 1926 [*Heptagenia*].

Material examined: *hebe* [*H.*]: ⊕; *maculipennis* [*H.*]: ʊ; *umbratica* [*H.*]: ⊕.

*** ***

Afghanurus/g1 INCERTAE SEDIS

The following species are known as adults only. **Asian species:** *diehli* Braasch & Soldán 1986 [*Compsoneuria*], *flowersi* Braasch & Soldán 1986 [*Compsoneuria*], *klugei* Braasch 1980 [*Ecdyonurus*], *muelleri* Braasch 1980 [*Ecdyonurus*], *naraensis* Gose 1968 [*Ecdyonurus*], *pallidus* Braasch & Soldán 1982 [*Ecdyonurus*]. **North American species:** *horrida* McDunnough 1926 [*Heptagenia*], *kennedyi* McDunnough 1924 [*Heptagenia*], *otiosa* McDunnough 1935 [*Heptagenia*], *rodocki* Traver 1935 [*Heptagenia*], *rosea* Traver 1935 [*Heptagenia*], *rubroventris* Traver 1935 [*Heptagenia*], *salvini* Kimmins 1934 [*Heptagenia*], *spinosa* Traver 1933 [*Heptagenia*].

*** ***

2.2;1,2-2/2.1;2,3. **Atopopus/g1**

(Pentamerotarsata Radulapalpata Heptagenia/f5=g4 Ecdyonurus/fg1 Atopopus/g1)

(Figs 57:A–B; 58:G–H; 59)

Nomen hierarchicum: **Atopopus/g1** (incl. *Thamnodontus*, *Notacanthurus*, *Ecdyonuroides*, *Electrogena*, *Afronurus*) [g: *Atopopus* Eaton 1881: 22, typus *A. tarsalis* Eaton 1881 (design. orig.)].

Autapomorphy of Atopopus/g1.

(1) On maxilla [whose proximal dentiseta divides into two branches – see Ecdyonurus/fg1 (2)] distal dentiseta divides into several branches divergent nearly from extreme base; thus the distal dentiseta looks as a bunch of stout setae (FIG.58: G–H). Unique apomorphy: in other Ecdyonurus/fg1 the distal dentiseta is integral, being either simple or bifurcate like the proximal one.

Among the species examined, only a few

exceptions are discovered: in Electrogena/g1 *necatii* [*Ecdyonurus*] all specimens have distal dentiseta integral, while in related species the distal dentiseta divides into branches (Belfiore & Kazanci & Tanatmis, in preparation); among Ecdyonuroides/g1 sp.H4 some specimens have distal dentiseta integral, and some – divided into branches. Possibly, these exceptions are results of secondary reversions.

Character of unclear phylogenetic status.

(2) Primary swimming setae [see Pentamerotarsata (6)] are lost; larval caudalii have only whorls of setae on ends of segments [see Heptagenia/f5=g4 (2)]. Non-unique character (see Index of characters [1.3.66]); among Ecdyonurus/fg1 the same in Leucrocuta/g1.

Size. Fore wing length 7–15 mm.

Distribution. Old World: Palaearctic, Oriental and Ethiopian Regions.

Atopopus/g1 is divided into *Thamnodontus*, Atopopus/g2, *Notacanthurus*, *Ecdyonuroides*, *Electrogena* and Afronurus/g1.

2.2;1,2-2/2.1;2,3-1. Plesiomorphon Thamnodontus/g(1)

(Radulapalpata Heptagenia/f5=g4 Ecdyonurus/fg1 Atopopus/g1 *Thamnodontus*)

Nomen hierarchicum: **Thamnodontus/g(1)** [g: *Thamnodontus* Kluge, SUBGEN. NOV., typus *Ecdyonurus aurarius* Kluge1983].

Reference. Kluge 1983 (*E. dracon*, *E. aurarius*): ℧ ⊕.

Plesiomorphies of *Thamnodontus*. Larva with normal proportions (in contrast to Atopopus/g1); larval abdomen without median ridge or spines (in contrast to *Notacanthurus*) and with posterolateral spines not enlarged (in contrast to *Ecdyonuroides*). Penis lobes [see Pentamerotarsata (5)] with distinct ventral and outer sclerites (in contrast to *Electrogena*) and with well-developed titillators contiguous medially (in contrast to Afronurus/g1).

Size. Fore wing length 7–15 mm.

Distribution. East Asia.

Species composition of Thamnodontus/g(1): Three described species can be attributed here for certain – *aurarius* Kluge 1983 [*Ecdyonurus*], *dracon* Kluge 1983 [*Ecdyonurus*], *tobiironis* Takahashi 1929 [*Ecdyonurus*] (syn. subj.: *suzukianum* Matsumura 1931 [*Cinygma*]).

Material examined: *aurarius* [*E.*]: ℧–◎–⊕/⊙; *dracon* [*E.*]: ℧–◎–⊕/⊙, *tobiironis* [*E.*]: ⊕; spp. (Nepal, China, Japan): ℧.

***

2.2;1,2-2/2.1;2,3-2. Atopopus/g2

(Radulapalpata Heptagenia/f5=g4 Ecdyonurus/fg1 Atopopus/g1 .../g2)

(Fig. 59)

Nomen hierarchicum: **Atopopus/g2** (sine *Thamnodontus*, *Notacanthurus*, *Ecdyonuroides*, *Electrogena*, *Afronurus*). In circumscription matches:

— gen. *Atopopus* Eaton 1881: 22.

References. Eaton 1883–1888: ⊕; – Wang & McCafferty 1995c: ℧ ⊕.

Autapomorphies of Atopopus/g2.

(1) Larva [flattened – see Heptagennota (1)] has very wide head and relatively short legs; head shield is so sharply expanded laterally, that behind it head margin forms a pair of postero-lateral emarginations (FIG.59:A). In other Heptagennota such emarginations are either absent, or shallower.

(2) Dorsal lamella of tergalius I [curved dorsally – see Ecdyonurus/fg1 (3)] is vestigial, much smaller than ventral fibrillose lobe [see Ecdyonurus/fg1 (6)]. Non-unique apomorphy.

(3) Wings with characteristic brown pigmentation: on fore wing – at distal part of costal area, at subcostal area and close to hind margin at cubital area; on hind wing – close to margin at distal half. Non-unique character: the same in some Oriental species of Heptagenia/f6=g5.

Plesiomorphies of Atopopus/g2. In contrast to Afronurus/g1 and *Electrogena*, penis [see Pentamerotarsata (5)] has contiguous median titillators and distinct ventral and outer sclerites with laterodorsal spines (at least in *tarsalis* [*A.*] – FIG.59:B–C).

Size. Fore wing length 11–13 mm.

Distribution. Oriental Region.

Species composition of Atopopus/g2. 3 species – *tarsalis* Eaton 1881 [*Atopopus*], *tibialis* Ulmer1920 [*Atopopus*] and *edmundsi* Wang & McCafferty 1995 [*Atopopus*].

Material examined: *tarsalis* [*Atopopus*]: Malaysia, Sabah, Kinabalu Nat. Park, Poling hot spring, Sungai Kipungit, 22 IV 1999, I. Sivec – 2 ℧♂/◎♂, 1 ⊕♂ (FIG.59).

Comment. While there are no reared specimens, larvae and imago are associated on the base of common hypodermal colour pattern: abdominal sterna dark brown, each abdominal tergum with a pair of large indistinctly outlined dark brown triangular spots and with paler median and lateral areas; on middle leg femur and base of tibia light brownish, most part of tibia pale, tarsus brown. Larval tarsus is twice shorter than tibia, while imaginal tarsus is very long, with first segment as long as tibia; in mature larva subimaginal tarsus has numerous transverse folds, especially on first segment. Other characters of subimago (dissected from larva) and imago are different: subimaginal face plate is large, imaginal – small; subimaginal wing is nearly uniformly brown, without colour pattern peculiar for imago; subimaginal titillators are swollen, imaginal – slender (FIG.59:C). Such difference is possible for imago and subimago of the same species.

***

2.2;1,2-2/2.1;2,3-3. Notacanthurus/g(1)

(Radulapalpata Heptagenia/f5=g4 Ecdyonurus/fg1 Atopopus/g1 *Notacanthurus*)

Nomen hierarchicum: **Notacanthurus/g(1)** [g: *Notacanthurus* Tshernova 1974: 812, typus *Ecdyonurus zhiltzovae* Tshernova 1972 (design. orig.)].
In circumscription matches:
— subgen. *Notacanthurus*: Kluge 1997: 202;
— gen. *Notacanthurus* Tshernova 1974: 812.

References. Tshernova 1974: ♉* ⊕'; – 1976: ♉*.

Autapomorphy of *Notacanthurus*.

(1) Larval abdominal terga have a longitudinal median ridge; on each tergite this ridge can be stretched posteriorly in a form of a spine producing behind posterior margin of tergite (Tshernova 1972: Fig.2). This apomorphy is not unique (see Index of characters [1.3.3]), but not found in other Heptagenia/f5=g4; among Heptagennota similar dorsal spines occur in selected species of Epeorus/fg1 only.

Plesiomorphies of *Notacanthurus*. In contrast to Afronurus/g1 and *Electrogena*, penis [see Pentamerotarsata (5)] has contiguous median titillators and distinct ventral and outer sclerites; in various species ventral sclerite can be enlarged and pointed (Tshernova 1972:Fig.1), or short and blunt (similar to FIG.59:B–C).

Size. Fore wing length 10–14 mm.

Distribution. Central and Eastern Asia.

Species composition of Notacanthurus/g(1). 6 species are formally described – *baekdu* Bae 1997 [*Ecdyonurus*], *cristatus* Braasch 1980 [*Ecdyonurus*], *edentatus* Braasch 1986 [*Notacanthurus*], *islamabadicus* Ali 1967 [*Ecdyonurus*], *ladakhensis* Braasch 1986 [*Notacanthurus*], *zhiltzovae* Tshernova 1972 [*Ecdyonurus*].

Material examined: *zhiltzovae* [*E.*]: ♉/⊙, ⊙, ⊕; sp.H2 (Tajikistan): ♉–⊙–⊕/⊙; sp.H3 (Kyrgyzstan): ♉–⊙–⊕/⊙; spp. (Japan, Thailand): ♉.

***

2.2;1,2-2/2.1;2,3-4. Ecdyonuroides/g(1)

(Radulapalpata Heptagenia/f5=g4 Ecdyonurus/fg1 Atopopus/g1 *Ecdyonuroides*)

Nomen hierarchicum: **Ecdyonuroides/g(1)** [g: *Ecdyonuroides* Dang 1967: 160, typus *Ecdyonurus sumatranus* Ulmer 1939 (design. orig.)].
In circumscription matches:
— gen. *Ecdyonuroides* Dang 1967: 160.

References. Ulmer 1939 (*Ecdyonurus*): ♉ ⊕; – Dang 1967: ♉; – Tshernova 1976: ♉; – Braasch & Soldán 1984 (*Thalerosphyrus*): ♉ ⊕.

Autapomorphy of *Ecdyonuroides*.

(1) In larva posterolateral projections [absent on segment IX – see Radulapalpata (2)] on segments VI–VIII are very long and pointed, exceeding segment length (Ulmer 1939:Figs 340–431; Dang 1967:Fig.8).

Size. Fore wing length 7–12 mm.

Species composition of Ecdyonuroides/g(1). Three formally described species – *flowersi* Venkataraman & Sivaramakrishnan 1987 [*Thalerosphyrus*], *sumatranus* Ulmer 1939 [*Ecdyonurus*] and *vietnamensis* Dang 1967 [*Ecdyonuroides*].

Material examined: sp.H4 (Thailand): ♉/♂, ♀/♀/⊙, ⊕.

***

2.2;1,2-2/2.1;2,3-5. Electrogena/g(1)

(Radulapalpata Heptagenia/f5=g4 Ecdyonurus/fg1 Atopopus/g1 *Electrogena*)

Nomen hierarchicum: **Electrogena/g(1)** [g: *Electrogena* Zurwerra & Tomka 1985: 102, typus *Baetis lateralis* Curtis 1834 (design. orig.)].
In circumscription matches:
— grex *lateralis*: Bogoescu & Tabacaru 1962: 281;
— subgen. *Electrogena*: Kluge 1997d: 202;
— gen. *Electrogena* Zurwerra & Tomka 1985: 102.

References. Bogoescu & Tabacaru 1962 (*lateralis*-Gruppe): ⊕*; – Zurwerra & Tomka 1985: ♉* ⊕*; – Kluge 1997d: ♉*.

Autapomorphy of *Electrogena*.

(1) Penis [see Pentamerotarsata (5)] has peculiar structure: its distal paired portion is not wider or very slightly wider than proximal integral portion; each lobe (left and right) of the distal portion lost apical incision separating ventral and outer sclerites, so that its ventral surface is integral and smooth, and apical margin is evenly roundish in ventral view; latero-dorsal spines are lost (Bogoescu & Tabacaru 1962:Fig. 5A-F). In contrast to Afronurus/g1, there are retained well-developed titillators contiguous medially.

In most other Ecdyonurus/fg1 penis has stronger widened distal part, apically with a pair of incisions between ventral and outer sclerites and/or with laterodorsal spines (FIG.59:B–C); in some Oriental taxa of Atopopus/g1 (*Ecdyonuroides* and some others) penis is similar to that of *Electrogena*, but either retains apical incisions, or has both lobes fused nearly up to apex, or has other difference from *Electrogena.*

Size. Fore wing length 7–15 mm.

Distribution. Western Palaearctic: Europe, Asia Minor and Caucasus.

Species composition of Electrogena/g(1) (here [*Ecd.*]=[*Ecdyonurus*], [*El.*]=[*Electrogena*]). About 40 species – *affinis* Eaton 1885 [*Ecd.*] (syn. subj.: *fasciioculatus* Sowa 1974 [*Ecd.*]), *anatolicus* Kazanci & Braasch 1986 [*Ecd.*], *antalyensis* Kazanci & Braasch 1986 [*Ecd.*], *armeniacus* Braasch 1983 [*Ecd.*], *aspoecki* Braasch 1984 [*Ecd.*], *azerbajdshanicus* Braasch 1978 [*Ecd.*], *boluensis* Kazanci 1990 [*El.*], *bothmeri* Braasch 1983 [*Ecd.*], *braaschi* Sowa 1984 [*Ecd.*], *calabra* Belfiore 1995 [*El.*], *dirmil* Kazanci 1990 [*El.*], *fallax* Hagen 1864 [*Baetis*] (synn. subj.: *sardoa* Costa 1882 [*Baetis*], *pseudograndiae* Zurwerra & Tomka 1986 [*El.*]), *galileae* Demoulin 1975 [*Ecd.*], *grandiae* Belfiore 1981 [*Ecd.*], *gridellii* Grandi 1954 [*Heptagenia*], *hakkaricus* Kazanci 1986 [*Ecd.*], *hellenica* Zurwerra & Tomka 1986 [*El.*] *hyblaea* Belfiore 1994 [*El.*], *kugleri* Demoulin 1973 [*Afronurus*], *kuraensis* Braasch 1978 [*Ecd.*], *lateralis* Curtis 1834 [*Baetis*] (synn. subj.: *rivulorum* Navás 1928 [*Ecd.*], *concii* Grandi 1953 [*Heptagenia*]), *lunaris* Belfiore & Scillitani & Picariello & Cataudo 1997 [*El.*], *macedonica* Ikonomov 1954 [*Heptagenia*], *madli* Kazanci 1992 [*Afronurus*], *malickyi* Braasch 1983 [*Ecd.*], *meyi* Braasch 1980 [*Ecd.*], *monticolis* Braasch 1980 [*Ecd.*], *necatii* Kazanci 1987 [*Ecd.*], *ozrensis* Tanasijevic 1975 [*Heptagenia*], *pseudaffinis* Braasch 1980 [*Ecd.*], *quadrilineata* Landa 1969 [*Heptagenia*], *ressli* Braasch 1981 [*Ecd.*], *squamatus* Braasch 1978 [*Ecd.*], *trimaculata* Ikonomov 1963 [*Heptagenia*], *ujhelyii* Sowa 1981 [*Ecd.*] (synn.subj.: *samalorum* Landa 1982 [*Ecd.*], *rivuscellana* Sartori & Landolt 1991 [*El.*]), *vipavensis* Zurwerra & Tomka 1986 [*El.*], *zebrata* Hagen 1864 [*Baetis*], *zimmermanni* Sowa 1984 [*Ecd.*].

Material examined: *affinis* [*E.*]: Ծ–◎–⊕/⊙; *antalyensis* [*E.*]: Ծ; *monticolis* [*E.*]: Ծ–◎–⊕/⊙; *necatii* [*E.*]: Ծ; *pseudaffinis* [*E.*]: Ծ–◎–⊕/⊙; *squamatus* [*E.*]: Ծ–◎–⊕/⊙; *zebrata* [*B.*]: ♂/♂, ⊕/⊙.

***

2.2;1,2-2/2.1;2,3-6. Afronurus/g1

(Radulapalpata Heptagenia/f5=g4 Ecdyonurus/fg1 Atopopus/g1 Afronurus/g1)

(Figs 57:A–B; 58:G–H)

Nomen hierarchicum: **Afronurus/g1** (incl. *Cinygmina*) [g: *Afronurus* Lestage 1924a: 349, typus *Ecdyonurus peringueyi* Esben-Petersen 1913 (design. orig.)].

Reference. Koss & Edmunds 1974: ⊙*.

Autapomorphies of Afronurus/g1.

(1) Median titillators of penis [see Pentamerotarsata (5)] are reduced: they are either widely separated and diminished (in *assamensis* [*C.*], *yoshidae* [*E.*], *levis* [*E.*]), or completely lost. This apomorphy is non-unique among Pentamerotarsata, but does not occur in other Heptagenia/f5=g4 (which have median titillators well-developed and contiguous medially).

(2) Egg [bearing small anchors – see Branchitergaliae (4)] on equator bears several additional anchors of another structure: the additional anchor is many times larger than the usual anchor, and in coiled condition represents a flat spiral (Koss & Edmunds 1974:Figs 94–95; Kopelke 1980:Figs 26–29; Flowers & Pescador 1984:Figs 11–12; Kang & Yang 1994: Figs 17–18). Unique apomorphy. Among Afronurus/g1 such additional anchors are present not in all species: among the species examined, they are present in *levis* [*E.*], spp. cf. *assamensis* [*C.*] and *peringueyi* [*E.*]; are reported for *chihpenensis* [*A.*], *hyalinus* [*E.*], *philippinensis* [*A.*] and *viridis* [*H.*]; but these additional anchors are absent in *abracadabrus* [*E.*] (at the same time *abracadabrus* [*E.*] is attributed to Afronurus/fg1 on the base of great similarity with *levis* [*E.*] in larval structure).

Size. Fore wing length 7–13 mm.

Distribution. Africa and Asia. Possibly in Miocene was distributed in Europe as well, i. e. had Arctogean distribution (see below, *Pseudokageronia* in Ecdyonurus/fg1 INCERTAE SEDIS).

Nominal taxon included. Afronurus/g1 includes **Cinygmina/g** [g: *Cinygmina* Kimmins 1937: 435, typus *C. assamensis* Kimmins 1937 (design. orig.)].

Species composition of Afronurus/g1. About 30 species are described. **African species:** *aethereus* Navás 1936 [*Ecdyonurus*], *barnardi* Schoonbee 1968 [*Afronurus*], *collarti* Navás 1930 [*Adenophlebia*] (syn. subj.: *pulcher* Ulmer 1930 [*Afronurus*]), *elongensis* Puthz 1971 [*Afronurus*], *gilliesi* Corbet 1962 [*Afronurus*], *harrisoni* Barnard 1932 [*Afronurus*], *muehlenbergi* Puthz 1971 [*Afronurus*], *negi* Corbet 1960 [*Afronurus*], *oliffi* Schoonbee 1968 [*Afronurus*], *peringueyi* Esben-Petersen 1913 [*Ecdyonurus*], *scotti* Schoonbee 1968 [*Afronurus*], *subflavus* Kopelke 1980 [*Afronurus*], *ugandanus* Kimmins 1965 [*Afronurus*]. **Asian species:** *abracadabrus* Kluge 1983 [*Ecdyonurus*], *assamensis* Kimmins 1937 [*Cinygmina*], *cervina* Braasch & Soldán 1984 [*Cinygmina*], *chihpenensis* Kang & Yang 1994 [*Afronurus*], *dama* Braasch & Soldán 1987 [*Cinygmina*], *floreus* Kang & Yang 1994 [*Afronurus*], *hunanensis* Zhang & Cai 1991 [*Cinygmina*], *hyalinus* Ulmer 1912 [*Ecdyonurus*], *keralensis* Braasch & Soldán 1987 [*Cinygmina*], *landai* Braasch & Soldán 1984 [*Cinygmina*], *levis* Navás 1912 [*Epeorus*] (syn. subj.: *zachvatkini* Tshernova 1952 [*Cinygmula*]), *nanhuensis* Kang & Yang1994 [*Afronurus*], *obliquistriata* You & Tian & Hong & Hsu 1981 [*Cinygmina*], *philip-*

pinensis Flowers & Pescador 1984 [*Afronurus*], *rangifera* Braasch & Soldán 1987 [*Cinygmina*], *rubromaculata* You & Wu & Gui & Hsu 1981 [*Cinygmina*], *viridis* Matsumura 1931 [*Heptagenia*], *yixingensis* Wu & You 1986 [*Cinygmina*], *yoshidae* Takahashi 1924 [*Ecdyonurus*] (syn. subj.: *japonicus* Ueno 1928 [*Ecdyonurus*]).

Material examined: *abracadabrus* [*E.*]: ℧–O–⊕/⊙; sp.cf. *aethereus* [*E.*] (Ethiopia): ⊕'; spp.cf. *assamensis* [*C.*] (China, Vietnam, Thailand): ℧/O/⊙, ⊕/⊙; *barnardi* [*A.*]: ℧; *levis* [*E.*]: ℧–O–⊕/⊙; *peringueyi* [*E.*]: ℧/⊙; *rubromaculata* [*C.*]: O'; sp. cf. *ugandanus* [*A.*] (Mali): ⊕; *yoshidae* [*E.*]: ℧, ⊕.

*** ***

Ecdyonurus/fg1 INCERTAE SEDIS

Ecdyonurus/fg1 INCERTAE SEDIS 1. Thalerosphyrus/g [g: *Thalerosphyrus* Eaton 1881: 22, typus *Baetis determinata* Walker 1853 (design. orig.)].

Reference. Eaton 1883–1888: ⊕.

Distribution. Oriental Region.

Material examined: –.

Comment. Larvae are not known for certain. In literature there are described imagoes and larvae of several species attributed to *Thalerosphyrus* without enough ground.

Ecdyonurus/fg1 INCERTAE SEDIS 2. Compsoneuriella/g [g: *Compsoneuriella* Ulmer 1939: 563, typus *C. thienemanni* Ulmer 1939 (design. orig.)].

Reference. Ulmer 1939: ℧ ⊕.

Distribution. Oriental Region.

Material examined: –.

Comment. Probably, the name *Compsoneuriella* is an older name for "Afghanurus/fg1" (see above).

Ecdyonurus/fg1 INCERTAE SEDIS 3. Notonurus/g [g: *Notonurus* Crass 1947b: 126, typus *N. cooperi* Crass 1947 (design. orig.; syn. subj. *Adenophlebia bequerti* Navás 1930)].

Reference. Crass 1947b: ⊕'.

Distribution. Ethiopian Region.

Species composition. 4 African species are formally attributed here – *bequerti* Navás 1930 [*Adenophlebia*] (synn. subj.: *eatoni* Navás 1931 [*Adenophlebia*], *inflexa* Navás 1932 [*Adenophlebia*], *cooperi* Crass 1947 [*Notonurus*]), *njalensis* Kimmins 1937 [*Afronurus*], *sinuosus* Navás 1931 [*Adenophlebia*], *tortinervis* Navás 1930 [*Adenophlebia*].

Material examined: sp. cf. *bequerti* Navás 1930 [*A.*] (Mali): ⊕.

Comment. Probably, the name *Notonurus* is an older name for "Afghanurus/g1" (see above).

Ecdyonurus/fg1 INCERTAE SEDIS 4. Asionurus/g [g: *Asionurus* Braasch & Soldán 1986a: 155, typus *A. primus* Braasch & Soldán 1986 (design. orig.)].

Reference. Braasch & Soldán 1986a: ℧.

Distribution. Oriental Region.

Material examined: –.

Ecdyonurus/fg1 INCERTAE SEDIS 5. † Pseudokageronia/g [g: *Pseudokageronia* Masselot & Nel 1999: 62, typus *P. thomasi* Masselot & Nel 1999 (design. orig.)].

Reference. Masselot & Nel 1999: ℧.

Age. Late Miocene (France).

Material examined: –.

Comment. The only character allowing to place *thomasi* [*P.*] to Ecdyonurus/fg1 and separate it from Heptahenia/f6=g5, is presence of a regular row of long stout setae on hind margin of femur (Masselot & Nel 1999: Photo 2); at proximal part of femur these setae are often pressed to the femur and invisible, that was erroneously taken for a generic character of *Pseudokageronia*. Other characters visible on the fossils (proportions and colour of head, thorax and legs, and pointed scales on femora) are species-specific. Wide head and wide femora of *thomasi* [*P.*] are characteristic for many species of Heptagenia/f5=g4, but not Rhithrogena/fg1 or *Cinygma* (which have similar row of setae on hind margin of femur). Characteristic femoral colour pattern (with additional subapical dark band) and other known characters of *thomasi* [*P.*] are the same as in Recent East-Asian species *abracadabrus* [*Ecdyonurus*] and *yoshidae* [*E.*], which belong to Afronurus/g1 – the taxon Recently absent in Europe.

Other species of Ecdyonurus/fg1 INCERTAE SEDIS. Following species were described without details allowing to determine their systematic position more exactly. **African species:** *ethiopicus* Soldán 1977 [*Thalerosphyrus*], *josettae* Sartori & Elouard 1996 [*Thalerosphyrus*], *matitensis* Sartori & Elouard 1996 [*Afronurus*]. **Asian species:** *apicatus* Braasch 1984 [*Ecdyonurus*], *bilineatus* Braasch 1984 [*Ecdyonurus*], *bishopi* Braasch & Soldán 1986 [*Thalerosphyrus*], *eatoni* Kimmins 1937 [*Ecdyonurus*], *flavus* Takahashi 1929 [*Ecdyonurus*], *fracta* Kang & Yang 1994 [*Electrogena*], *gilliesiana* Braasch 1990 [*Cinygmina*], *pakistanicus* Braasch 1984 [*Ecdyonurus*], *petersi* Braasch & Soldán 1986 [*Asionurus*], *rainulfiana* Braasch 1999 [*Cinygmina*], *signatus* Braasch 1984 [*Ecdyonurus*], *subfuscus* Kimmins 1937 [*Ecdyonurus*] (syn. obj.: *indicus* Hubbard 1974 [*Ecdyonurus*]), *tigris* Imanishi 1936 [*Ecdyonurus*], *tonkinensis* Soldán & Braasch 1986 [*Rhithrogenella*], *ulmeri* Braasch & Soldán 1986 [*Asionurus*], *wittmeri* Braasch 1981 [*Ecdyonurus*].

*** ***

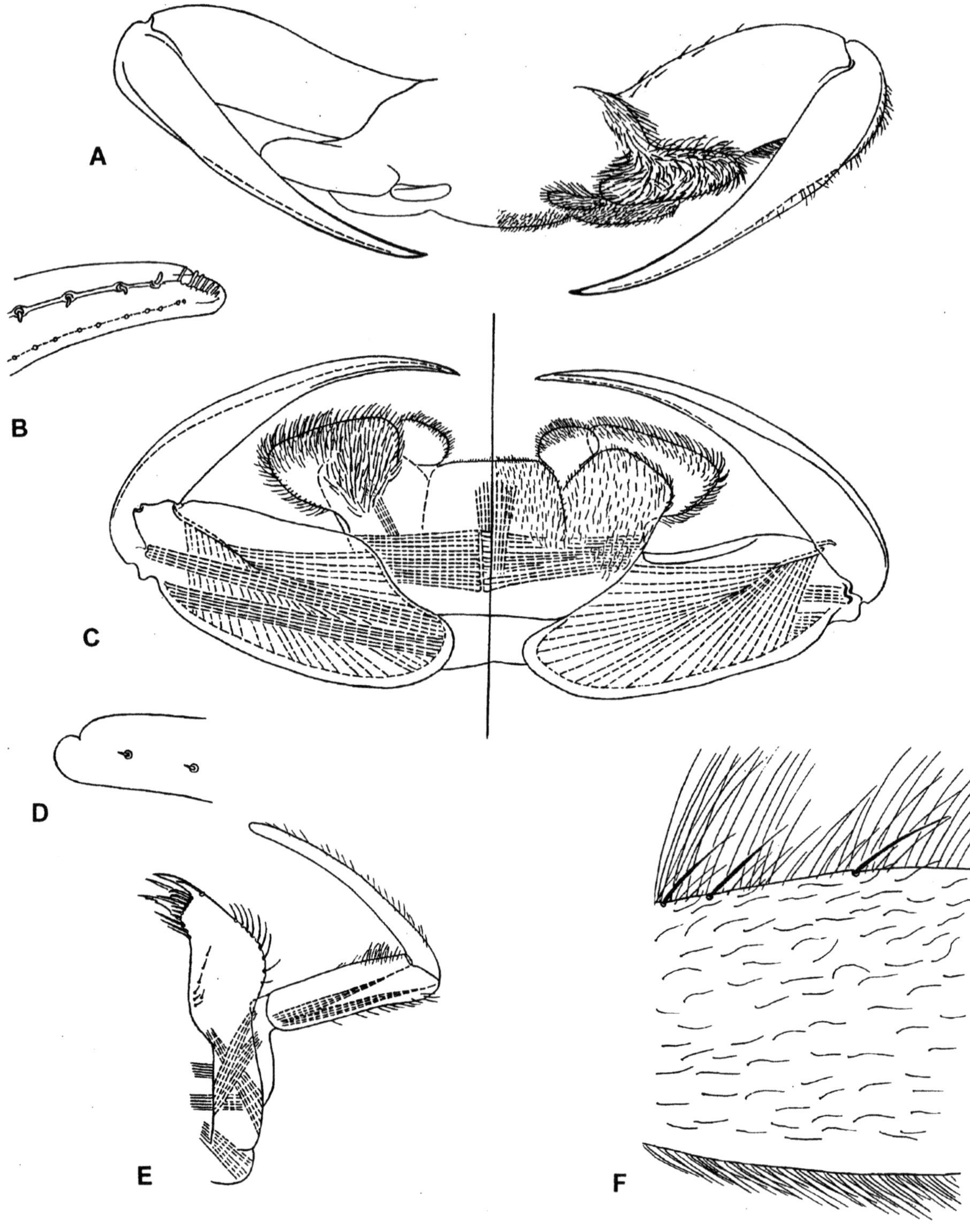

Figure 60. Radulapalpata-Raptoheptagenia/g* *cruentata* [*Heptagenia*], larva.
A – labium, apical view; **B** – apex of labial palp with vestiges of scraping setae, ventral view; **C** – labium, dorsal view (in left half) and ventral view (in right half) (muscles shown by interrupted lines); **D** – apex of maxillary palp with vestige of 3rd segment, ventral view; **E** – left maxilla, ventral view (muscles shown by interrupted lines); **F** – portion of larval femur, dorsal view. (From Kluge 1993a).

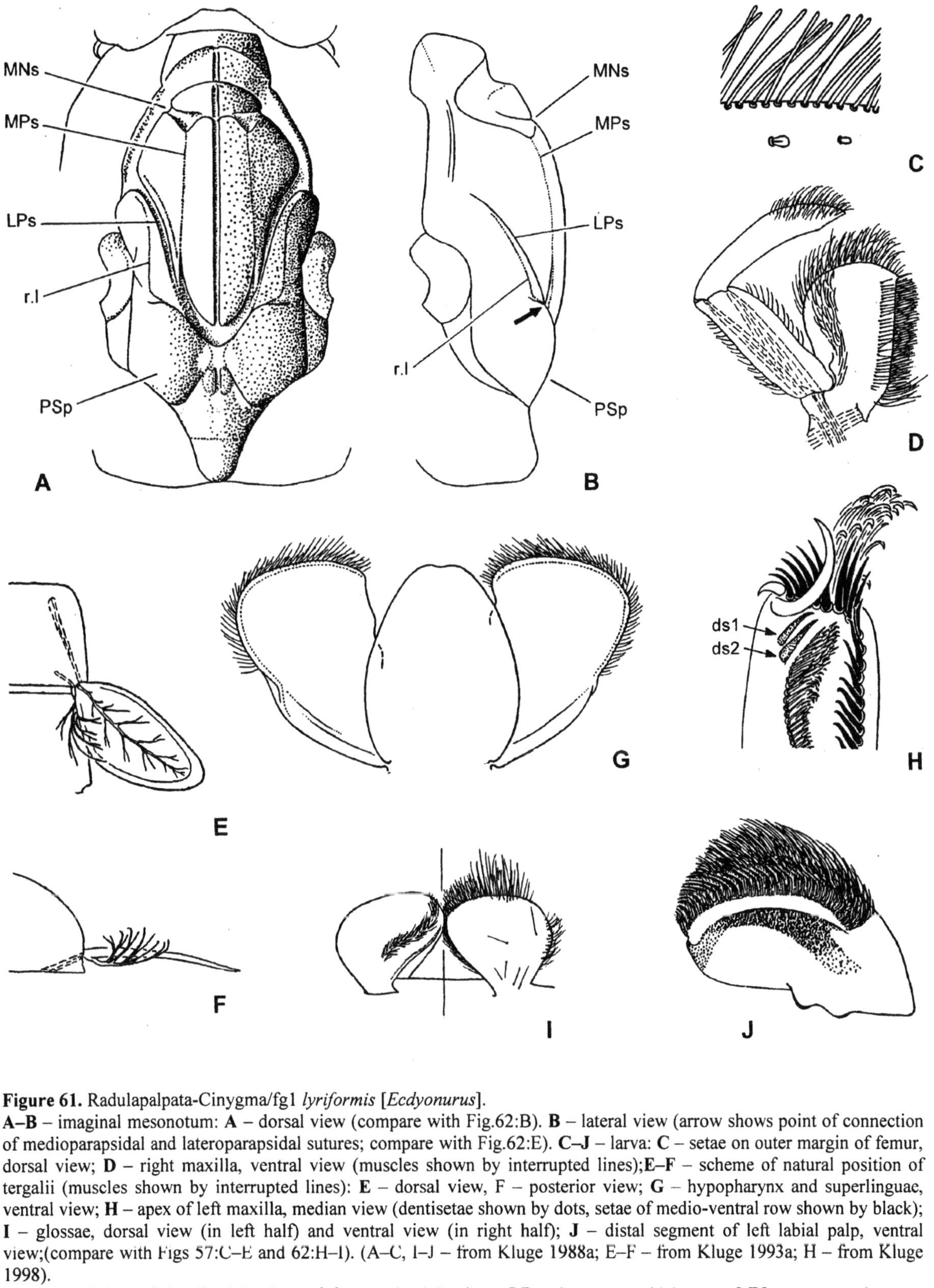

Figure 61. Radulapalpata-Cinygma/fg1 *lyriformis* [*Ecdyonurus*].
A–B – imaginal mesonotum: **A** – dorsal view (compare with Fig.62:B). **B** – lateral view (arrow shows point of connection of medioparapsidal and lateroparapsidal sutures; compare with Fig.62:E). **C–J** – larva: **C** – setae on outer margin of femur, dorsal view; **D** – right maxilla, ventral view (muscles shown by interrupted lines);**E–F** – scheme of natural position of tergalii (muscles shown by interrupted lines): **E** – dorsal view, F – posterior view; **G** – hypopharynx and superlinguae, ventral view; **H** – apex of left maxilla, median view (dentisetae shown by dots, setae of medio-ventral row shown by black); **I** – glossae, dorsal view (in left half) and ventral view (in right half); **J** – distal segment of left labial palp, ventral view;(compare with Figs 57:C–E and 62:H–I). (A–C, I–J – from Kluge 1988a; E–F – from Kluge 1993a; H – from Kluge 1998).

Abbreviations: **ds1** – distal dentiseta; **ds2** – proximal dentiseta; **LPs** – lateroparapsidal suture; **MNs** – mesonotal suture; **MPs** – medioparapsidal suture; **PSp** – posterior scutal protuberance; **r.l** – relief line.

2.2;1,2-2/2.2. Raptoheptagenia/g(1)

(Branchitergaliae Heptagennota Pentamerotarsata Radulapalpata *Raptoheptagenia*)

(Fig. 60)

Nomen hierarchicum: **Raptoheptagenia/g(1)** [g: *Raptoheptagenia* Whiting & Lehmkuhl 1987a: 405, typus *Heptagenia cruentata* Walsh 1863 (design. orig.)].
In circumscription matches:
— gen. *Raptoheptagenia* Whiting & Lehmkuhl 1987a: 405.
— gen. *Anepeorus*: Burks 1953: 198 (partim: ♉, non ⊕);
— subfam. Anepeorinae Edmunds 1962:10 (♉, non ⊕).
Monospecific taxon.

References. Burks 1953 (*Anepeorus*): ♉; – Edmunds & Allen & Peters 1963 (Anepeorinae): ♉; – Edmunds & Jensen & Berner 1976 (Anepeorinae): ♉ ⊕; – McCafferty & Provonsha 1986 (*Anepeorus*): ♉; – Whiting & Lehmkuhl 1987a: ♉ ⊕; – Kluge 1993: ♉.

Autapomorphies of *Raptoheptagenia*.

(1) Mouth apparatus is strongly modified and specialized for carnivorism (FIG.60:A–E): Labial palps, which in all other Radulapalpata (including carnivorous species) are specialized as scraping [see Radulapalpata (1)], in *Raptoheptagenia* have lost their scraping specialization and specialized as grasping: $2^{nd}+3^{rd}$ segment is sclerotized, pointed, thinned and curved in a form of hook, its setae are nearly completely reduced. Maxilla is biting, with enlarged canines and dentisetae, its apical margin is oblique, without typical for Pentamerotarsata pectinate setae [see Radulapalpata (3)] and without filtering setae. Hypopharynx is shorter and wider, and superlinguae [see (3)] are smaller than in other Radulapalpata; mandibles have shape typical for specialized predators: shortened perpendicular to axis of articulation, mola is tooth-like, without grater (McCafferty & Provonsha 1986:Figs 7, 13).

(2) Tergalii are attached to ventral side of abdomen. Non-unique apomorphy (see Index of characters [1.3.22]), but not found in other Heptagennota.

Characters of unclear phylogenetic status.

(3) Superlinguae [diminished – see (1)] are somewhat curved laterally (McCafferty & Provonsha 1986:Fig.13). Probably, a common feature with Heptagenia/f5=g4.

(4) Outer margin of larval femur [flattened – see Heptagennota (1)] bear long thin hairs and sparse stout pointed setae (FIG.60:F). Probably, a common feature with Heptagenia/f6=g5, while other Radulapalpata have a regular dense row of long stout setae.

(5) Larval caudalii lack whorls of stout setae on hind margins of segments. Among Radulapalpata the same in *Cinygma* and Rhithrogena/fg1, while in Heptagenia/f5=g4 whorls of stout setae are present.

(6) In imago and subimago 1st tarsal segment is shortened, on fore tarsus of male several times shorter than 2nd segment. Non-unique apomorphy (see Index of characters [2.2.76]).

Size. Fore wing length 9–11 mm.

Distribution. Nearctic.

Species composition of Raptoheptagenia/g(1). 1 species – *cruentata* Walsh 1863 [*Heptagenia*] (syn. subj.: *reversalis* McDunnough 1924 [*Heptagenia*]).

Material examined (Florida A & M Univ.): *cruentata* [*H.*]: ♉.

***

2.2;1,2-2/2.3. Cinygma/fg(1)

(Branchitergaliae Heptagennota Pentamerotarsata Radulapalpata *Cinygma*)

(Fig. 61)

Nomen hierarchicum: **Cinygma/fg(1)** [f: Cinygmini Kluge 1988: 303; g: *Cinygma* Eaton 1885: 247, typus *C. integrum* Eaton 1885 (design. orig.)].
In circumscription matches:
— gen. *Cinygma*: McDunnough 1933: 74;
— tribus Cinygmini Kluge 1988: 303.

References. McDunnough 1933: ♉*; – Needham & Traver & Hsu 1935: ♉*; – Tshernova 1949 (*Heptagenia abnormis*): ♉'; – Edmunds & Jensen & Berner 1976: ♉*; – Tshernova 1976: ♉*; – Lehmkuhl 1979a: ♉' ⊕'; – Kluge 1988: ♉* ⊕*; – 1993: ♉* ⊕*.

Autapomorphies of *Cinygma*.

(1) Labrum is strongly narrowed, narrowing from its base toward apex, its width is less than length (Tshernova 1949:Fig.4:2). Unique apomorphy among Heptagennota [see Heptagennota (1)].

(2) Maxilla [see also (5)] has a single canine developed (in contrast to other Radulapalpata which have all three canines subequal); on apical margin of maxilla pectinate setae [see Radulapalpata (3)] are completely lost, distal margin of maxilla is entirely covered with dense long setae (FIG.61:D, H). Non-unique apomorphy (see Index of characters [1.1.32]).

(3) Glossae [initially pyramidal – see Radulapalpata (1)] are narrow at bases and strongly widened distally, roundish-triangular (FIG.61:I). Unique apomorphy.

Characters of unclear phylogenetic status.

(4) Each superlingua is widening toward its apex (FIG.61:G), lateral margins of superlinguae are divergent (in contrast to majority of Rhithrogena/ /fg1), but apices are not curved laterally (in contrast to Heptagenia/f5=g4).

(5) On maxilla [see (2)] ventral row of setae [see Branchitergaliae (1)] is straight, at apex is not curved laterally (FIG.61:D). The same in *Paegniodes* and Epeorus/fg1; possibly, symplesiomorphy.

(6) $2^{nd}+3^{rd}$ segment of labial palp [specialized as scraping – see Radulapalpata (1)] is blunt, thus its field of scraping setae terminates close to the end of sclerotized ridge (FIG.61:J). The same in Heptagenia/f7=g6; possibly, independently appeared apomorphy. In *Kageronia*, Ecdyonurus/fg1 and Rhithrogena/fg1 form of this segment is different.

(7) Outer-dorsal margin of larval femur [flattened – see Heptagennota (1)] bears a regular row of long stout setae (FIG.61:C). Non-unique apomorphy: similar setal row occurs in some other rheophilous mayflies. Among Radulapalpata the same setal row is present in Rhithrogena/fg1; possibly synapomorphy; similar, but less dorsal row is present in Ecdyonurus/fg1.

(8) Tergalii lost ability of rhythmical respiratory movements, spread laterally (FIG.61:E–F). Non-unique apomorphy (see Index of characters [1.3.30]); among Radulapalpata the same in Rhithrogena/fg1; possibly synapomorphy [for comment – see Rhithrogena/fg1 (6) below].

(9) Larval caudalii lack whorls of stout setae on hind margins of segments. Among Radulapalpata the same in Rhithrogena/fg1 (possibly synapomorphy) and *Raptoheptagenia*, while in Heptagenia/f5=g4 whorls of stout setae are present.

(10) Primary swimming setae [see Pentamerotarsata (6)] are lost. Non-unique apomorphy (see Index of characters [1.3.66]).

Plesiomorphies of *Cinygma* (in contrast to Rhithrogena/fg1). Maxillary palp is not so strongly specialized as in Rhithrogena/fg1 (FIG.61:D). Labial palp without bifurcate setae. Imaginal and subimaginal lateroparapsidal sutures are not curved laterally, terminate near medioparapsidal sutures (FIG.61:A–B). Larval and adult patella-tibial suture is developed on middle and hind legs only (in contrast to larval Rhithrogena/fg3).

Variable characters of *Cinygma*. On hind wing bifurcation of MA is present in *integrum* [*C.*] and *lyriformis* [*E.*] (like in all other Radulapalpata), but lost in *dimicki* [*C.*] (non-unique apomorphy – see Index of characters [2.2.67]).

Size. Fore wing length 11–13 mm.

Distribution. Amphipacific Sector of Holarctic: All 3 species occur in Western Nearctic, and only one of them – *lyriformis* [*E.*] – is distributed in Eastern Palaearctic including Urals.

Species composition of Cinygma/fg(1). 3 species – *dimicki* McDunnough 1934 [*Cinygma*]; *integrum* Eaton 1885 [*Cinygma*]; *lyriformis* McDunnough 1924 [*Ecdyonurus*] (synn. subj.: *abnormis* Tshernova 1949 [*Heptagenia*], *peterseni* Lestage 1930 [*Ecdyonurus*] = *hyalinus* Esben-Petersen 1916 [*Ecdyonurus*] nom. praeocc.).

Material examined: *lyriformis* [*E.*]:Ծ,Ծ–◎♂/⊕♂,⊕♂, ⊕♀/⊙.

***

2.2;1,2-2/2.4. **Rhithrogena/fg1**

(Branchitergaliae Heptagennota Pentamerotarsata Radulapalpata Rhithrogena/fg1)

(Figs 8:E–F; 9:C; 62–65)

Nomen hierarchicum: **Rhithrogena/fg1** (incl. *Paegniodes*, *Epeorus*) [f: Rhithrogeninae Lestage 1917: 266; g: *Rhithrogena* Eaton 1881: 23, typus *Baetis semicolorata* Curtis 1834 (design. orig.)].
In circumscription matches:
— tribus Rhithrogenini: Kluge 1988: 303.

References. Kluge 1988: Ծ*; – Tomka 1991: Ծ*; – Kluge 1993: Ծ* ⊕*.

Autapomorphies of Rhithrogena/fg1. These are first of all, characters connected with adaptation of larvae to inhabitancy on stone surface in fast streams – further development of scraping specialization of mouth apparatus characteristic for Radulapalpata [see (1) and (2)], loss of ability to respiratory movements [see (6)] and presence of dorsal setae on femora [see (5)]. At the same time, unique apomorphies [see (1)–(3)] undoubtedly testify about holophyly of this taxon.

(1) Maxillary palp is specialized as scraping [in addition to the scraping labial palp – see Radulapalpata (1)] (FIGS 62:A, D; 64:B, E, H, L): 1st segment of maxillary palp is thickened; situated inside it muscle-flexor of 2nd segment is enlarged, fan-form, proximally attached not only to base of segment, but also to its outer side. 2nd and 3rd segments are fused to one; apical-ventral side of $2^{nd}+3^{rd}$ segment bears a field of densely and regularly situated small stout curved pectinate scraping setae.

(2) On dorsal side of $2^{nd}+3^{rd}$ segment of labial palp, in the field of irregularly situated setae [see Radulapalpata (1)], the most distal setae, which form a transverse row, are bifurcate (while all other setae are not bifurcate) (FIG.62:E–G). Unique apomorphy.

(3) Imaginal and subimaginal lateroparapsidal suture in its posterior part is curved laterally from medioparapsidal suture and reaches area of dorsal attachment of posterior scuto-coxal muscle S.CmP

(FIG.63). In subimago shape of the lateroparapsidal stripe of lateral pigmented area of mesonotun [see Heptagennota (2)] either repeats the shape of lateroparapsidal suture (FIGS 63:A, C; 64:D) [see Rhithrogena/fg2 (6) and Epeorus/fg1 (7) below], or is the same as in other Heptagennota (FIG.64:A) [see below, Plesiomorphies of *Paegniodes*].

Characters of unclear phylogenetic status.

(4) 2nd+3rd segment of labial palp [specialized as scraping – see (2) and Radulapalpata (1)] is pointed, thus the field of scraping setae terminates at a distance from the end of the sclerotized ridge (FIG. 62:G). The same in *Kageronia* and Ecdyonurus/fg1; probably symplesiomorphy. In Heptagenia/f7=g6 and *Cinygma* form of this segment is different.

(5) Outer-dorsal margin of larval femur [flattened – see Heptagennota (1)] bears a regular row of long stout setae. Non-unique apomorphy: similar row of setae occurs in some other rheophilous mayflies. Among Radulapalpata the same setal row is present in *Cinygma*; probably synapomorphy; similar row is present in Ecdyonurus/fg1.

(6) Tergalii lost ability of rhythmical respiratory movements. Non-unique character (see Index of characters [1.3.30]); among Radulapalpata the same in *Cinygma*; probably synapomorphy.

If ventral fibrillose lobe of tergalius [see Branchitergaliae (3)] is well-developed, it is curved in such a manner that passes around the base of tergalius from behind and projects dorsad of it (FIG.62:H–I); because of this, when tergalii are spread by sides and pressed to substrate by their ventral (posterior) sides, their fibrillose portions project above them and are used for respiration. In this respect Rhithrogena/fg1 differs from Heptagenia/f5=g4, whose tergalii are able to make rhythmical respiratory movements, at rest are directed laterally, flatness of tergalius being perpendicular to longitudinal body axis, and the fibrillose portion of each tergalius locates behind the tergalius (FIG.57:C–E). The curvation of fibrillose lobe in Rhithrogena/fg1 testifies that the spread position and immobility of tergalii is secondary (i. e. an apomorphy), while the perpendicular position and mobility of tergalii is a plesiomorphy of Heptagenia/f5=g4.

(7) Larval caudalii have no whorls of stout setae on hind margins of segments. Among Radulapalpata the same in *Cinygma* (probably synapomorphy) and *Raptoheptagenia*, while in Heptagenia/f5=g4 whorls of stout setae are present.

Plesiomorphies of Rhithrogena/fg1. In contrast to *Cinygma*, all 3 maxillary canines are retained (while sometimes being weak). In contrast to Heptagenia/f5=g4, abdominal ganglion VIII is not approximated with ganglion VII.

Size. Fore wing length 5–20 mm.

Distribution. Holarctic and Oriental Region.

Rhithrogena/fg1 is divided into *Paegniodes*, Rhithrogena/fg2 and Epeorus/fg1 (p.201).

2.2;1,2-2/2.4;1. **Paegniodes/g(1)**

(Heptagennota Pentamerotarsata Radulapalpata Rhithrogena/fg1 *Paegniodes*)
(Fig. 64: A–C)

Nomen hierarchicum: **Paegniodes/g(1)** [g: *Paegniodes* Eaton 1881: 23, typus *Heptagenia cupulata* Eaton 1871 (design. orig.)].
In circumscription matches:
— subgen. *Paegniodes*: Kluge 1988: 305;
— gen. *Paegniodes* Eaton 1881: 23.

References. Eaton 1883–1888: ⊕; – Ulmer 1925: ⊕; – 1936: ⊕; – Hsu 1932: ⊕; – 1936: ⊕; – Ueno 1969: ⊕; – Tshernova 1976: ♉; – Kluge 1988: ♉; – 1993: ♉.

Plesiomorphies of *Paegniodes*. In contrast to other Rhithrogena/fg1 (i. e. Rhithrogena/fg2 and Epeorus/fg1), subimaginal lateroparapsidal stripe of pigmented area of mesonotum retains shape initial for Heptagennota [see Heptagennota (2)], i. e. is widened toward apex and its posterior-median angle touches medioparapsidal suture [in spite of the fact that lateroparapsidal suture is curved and does not touch medioparapsidal suture – see Rhithrogena/fg1 (3)] (FIG.64:A).

In contrast to Rhithrogena/fg2: maxilla is long [see Radulapalpata (3)], ventral row of setae [see Branchitergaliae (1)] is straight, parallel to median margin (FIG.64:B) (the same in *Cinygma* and Epeorus/fg1 – see Index of characters [1.1.30]); penis without additional pair of ventral titillators, with a pair of well-developed median titillators [see Pentamerotarsata (5)]; apices of these median titillators project at ventral side of penis being directed laterally, thus have some resemblance with a single pair of ventral titillators in Rhithrogena/fg3; but in contrast to the ventral titillators, they arise from median walls of penis lobes.

In contrast to Epeorus/fg1: larval paracercus is as long as cerci, primary swimming setae [vestigial – see Pentamerotarsata (6)] are present (FIG.64:C); imaginal and subimaginal mesonotal suture is normally developed [see Plesiomorphies of Branchitergaliae] (FIG.64:A).

Each superlingua is roundish, not narrowing toward apex (Tshernova 1976:Fig.43) (in contrast to Rhithrogena/fg2 and Epeorus/fg3), but lateral margins of superlinguae are not divergent (in contrast to *Bleptus*, *Ironodes* and other Pentamerotarsata [for comment – see *Ironodes* (3) below]. Glossae are pyramidal [see Radulapalpata (1)] (Tshernova 1976: Fig.45).

Larval and adult patella-tibial suture is developed on middle and hind legs (in contrast to most of larval Epeorus/fg1), absent on fore legs (in contrast to larval Rhithrogena/fg3.

Size. Fore wing length 10–20 mm.

Distribution. Oriental Region.

Species composition of Paegniodes/g(1). Probably one species – *cupulata* Eaton 1981 [*Heptagenia*] (syn. subj.: *fukiensis* Hsu 1932 [*Paegniodes*]), with unusual variability of hind wing length – from 0.15 to 0.45 of fore wing length.

Material examined: *cupulata* [*H.*]: ℧–⊕♂, ♉/◎♂, ℧, ⊕.

***

2.2;1,2-2/2.4;2. Rhithrogena/fg2

(Heptagennota Pentamerotarsata Radulapalpata Rhithrogena/fg1 .../fg2)

(Figs 8:F; 9:C; 62–63)

Nomen hierarchicum: **Rhithrogena/fg2** (sine *Paegniodes*, *Epeorus*; incl. *Cinygmula*).

Reference. Kluge 1993: ℧*.

Autapomorphies of Rhithrogena/fg2.

(1) Maxilla [see Radulapalpata (3)] is shortened (length of median margin is subequal to maxilla width), widened apically, median margin is inclined forming acute angle with apical margin; ventral row of setae [see Branchitergaliae (1)] in its distal part is curved laterally and diverges from median margin, being perpendicular to apical margin (FIG.62:A).

In other Radulapalpata maxilla has different shape (see Index of characters [1.1.29] and [1.1.30]); only in Epeorus/fg3 median margin is inclined as strongly as in Rhithrogena/fg2, but there the ventral row of setae is straight and parallel to the median margin all over its length (FIG.64:L).

(2) Ventral side of $2^{nd}+3^{rd}$ segment of maxillary palp bears a longitudinal arched pigmented stripe which borders the field of scraping setae [see Rhithrogena/fg1 (1)] and reaches the extreme pointed apex of the palp (FIG.62:A, D) (on dorsal side, as in other Radulapalpata, a vestige of transverse suture between 2nd and 3rd segments is retained). In other Rhithrogena/fg1 such stripe is absent, or indistinct, or not reaching apex of palp.

(3) Larva has a peculiar pose of swimming: legs are folded and pressed to lateral margins of body (FIG.9:C); they are not turned ventrally (in contrast to many Heptagenia/f6=g5) and do not move when larva swims (in contrast to Epeorus/fg3 and some others) [see Heptagennota (1)]. Larvae inhabit rapid streams [see Autapomorphies of Rhithrogena/fg1], where it normally does not swim; swimming movements can be observed if put larva into stagnant water.

(4) Penis has a pair of ventral titillators – immobile spines situated on ventral side of penis; in *Cinygmula* these ventral titillators are present besides median titillators [see Pentamerotarsata (5)]; in Rhithrogena/fg3 only ventral titillators are present. In selected species of *Cinygmula* and Rhithrogena/fg3 ventral titillators are lost.

Characters of unclear phylogenetic status.

(5) Superlinguae are widest in middle part and convergent toward apex. The same in Epeorus/fg3 (FIG.64:M), but in contrast to other Radulapalpata, whose superlinguae are divergent; probably, plesiomorphy within Rhithrogena/fg1 [for comment see *Ironodes* (3) below].

(6) Subimaginal lateroparapsidal stripe of pigmented area of mesonotum in its posterior part is curved laterally, repeating curvation of lateroparapsidal suture [see Rhithrogena/fg1 (3)], thus it does not touch medioparapsidal suture (FIG.63:A, C). The same in Epeorus/fg1; possibly synapomorphy (in contrast to *Paegniodes*).

Plesiomorphies of Rhithrogena/fg2. Glossae are pyramidal [see Radulapalpata (1)] (FIG.62:E, G). In contrast to Epeorus/fg1: larval paracercus is subequal to cerci [see Pentamerotarsata (6)]; imaginal and subimaginal mesonotal suture is normally developed [see Plesiomorphies of Branchitergaliae] (FIG.63).

Size. Fore wing length 5–17 mm.

Distribution. Holarctic and Oriental Region.

Rhithrogena/fg2 is divided into *Cinygmula* and Rhithrogena/fg3.

2.2;1,2-2/2.4;2,1. Cinygmula/g1

(Heptagennota Pentamerotarsata Radulapalpata Rhithrogena/fg1 .../fg2 *Cinygmula*)

(Figs 62:A–B; 63:A–B)

Nomen hierarchicum: **Cinygmula/g1** (incl. *Ororotsia*) [g: *Cinygmula* McDunnough 1933: 75, typus *Ecdyonurus ramaleyi* Dodds 1923 (design. orig.)].
In circumscription matches:
— subgen. *Cinygmula*: Kluge 1983b:16 (Abstract); 1988:305;
— gen. *Cinygmula* McDunnough 1933: 75.

References. McDunnough 1933: ♉*; – Needham & Traver & Hsu 1935: ♉* ⊕*; – Tshernova 1974: ⊕*; – 1976: ♉*; – Edmunds & Jensen & Berner 1976: ♉* ⊕*; – Kluge 1988: ♉*; – 1993: ♉*.

Autapomorphy of *Cinygmula*.

(1) Larval caudalii have no swimming setae [see Pentamerotarsata (6)] (in contrast to Rhithrogena/ /fg3, where primary swimming setae are retained). When swimming, larvae of *Cinygmula* make the same dorsoventral movements by abdomen, as larvae with developed swimming setae [see Rhithrogena/fg2 (3) and FIG.9:C]; thus, loss of swimming setae is not connected with adaptation. Non-unique apomorphy (see Index of characters [1.3.66]).

Characters of unclear phylogenetic status.

(2) Fibrillose portions of tergalii [see Branchitergaliae (3)] are diminished or lost (in contrast to Rhithrogena/fg3, where they are always large). Non-unique apomorphy (see Index of characters [1.3.25]).

Plesiomorphies of *Cinygmula* (in contrast to Rhithrogena/fg3). On 2nd+3rd segment of maxillary palp scraping setae have no denticles, occupy about 1/2 palp length only (FIG.62:A). Larval and adult patella-tibial suture is developed on middle and hind legs only (in contrast to larval Rhithrogena/fg3). Tergalii do not form sucking disc (while in some species tergalius I is markedly widened anteriorly). Costal and anal ribs of tergalii are situated on costal and anal margin correspondingly [see Branchitergaliae (3)] (only on tergalius I costal rib can be reduced). Subimago without non-concave pigmented area behind apex of lateroparapsidal suture (FIG. 63:A). Penis has median titillators [see Rhithrogena/fg2 (4) and Pentamerotarsata (5)].

Variable characters of *Cinygmula*. Imaginal and subimaginal claws are usually ephemeropteroid, but in *hutchinsoni* [*O.*] – pointed (see Variable characters of Heptagennota and Index of characters [2.2.85]).

Size. Fore wing length 7–14 mm.

Distribution. Amphipacific Sector of Holarctic: Eastern Palaearctic and Nearctic; dominate in Eastern Palaearctic and Western Nearctic.

Nominal taxon included. Cinygmula/g1 includes **Ororotsia/g** [g: *Ororotsia* Traver 1939: 33, typus *O. hutchinsoni* Traver 1939 (design. orig.)], regarded as a subgeneric synonym of *Cinygmula* (Kluge 1983b: 16; 1988: 305). Initially the genus *Ororotsia* was established for a single Tien-Shan–Himalayan species and characterized by pointed (i.e. not ephemeropteroid) claws, that was regarded as a plesiomorphy unique among Heptagennota. Actually pointed claws evolved from ephemeropteroid claws, besides *hutchinsoni* [*O.*], in several other Tien-Shan–Himalayan species of Rhithrogena/fg1 belonging to *Himalogena* and *Caucasiron* (see below).

Species composition of Cinygmula/g1 (here [*C.*]= =[*Cinygmula*]). About 30 species. Not less than two **Tien-Shan–Himalayan species** among several formally described ones – *hutchinsoni* Traver 1939 [*Ororotsia*], *joosti* Braasch 1977 [*C.*] (syn. subj.: *oreophila* Kustareva 1978 [*C.*]), *picta* Braasch & Soldán 1979 [*C.*], *quadripunctata* Braasch & Soldán 1980 [*C.*], *rougemonti* Braasch & Soldán 1987 [*C.*], *zimmermanni* Braasch 1977 [*C.*]. **East Palaearctic species:** *adusta* Imanishi 1935 [*Cinygma*], *brunnea* Tiunova 1990 [*C.*], *cavum* Ulmer 1927 [*Cinygma*] (synn. subj.: *altaica* Tshernova 1949 [*C.*], *guentheri* Braasch 1979 [*C.*]), *dorsalis* Imanishi 1935 [*Cinygma*], *sapporensis* Matsumura 1904 [*Siphlurus*] (syn. subj.: *grandifolia* Tshernova 1952 [*C.*]), *hirasana* Imanishi 1935 [*Cinygma*] (syn. subj.: *latifrons* Tshernova & Belov 1982 [*C.*]), *inermis* Braasch 1983 [*C.*], *irina* Tshernova & Belov 1982 [*C.*], *kurenzovi* Bajkova 1962 [*Cinygma*] (*kaszabi* Landa & Soldán 1983 [*C.*]), *levanidovi* Tshernova & Belov 1982 [*C.*] (syn. subj.: *spinosa* Tshernova & Belov 1982 [*C.*]), *malaisei* Ulmer 1927 [*Cinygma*] (syn. subj.: *andrianovae* Tshernova & Belov 1982 [*C.*]), *minuta* Braasch 1980 [*C.*], *putoranica* Kluge 1980 [*C.*], *unicolorata* Tshernova 1979 [*C.*], *vernalis* Imanishi 1935 [*Cinygma*]. **North American species:** *gartrelli* McDunnough 1934 [*C.*], *kootenai* McDunnough 1943 [*C.*], *mimus* Eaton 1885 [*Cinygma*], *par* Eaton 1885 [*Cinygma*] (synn. subj.: *hyalina* McDunnough 1924 [*Cinygma*], *confusa* McDunnough 1924 [*Cinygma*]), *ramaleyi* Dodds 1923 [*Ecdyonurus*] (syn. subj.: *tollandi* Dodds 1923 [*Iron*]), *reticulata* McDunnough 1934 [*C.*], *subaequalis* Banks 1914 [*Heptagenia*] (syn. subj.: *atlantica* McDunnough 1924 [*Cinygma*]), *tarda* McDunnough 1929 [*Cinygma*], *tioga* Mayo 1952 [*C.*], *uniformis* McDunnough 1934 [*C.*].

Material examined: *brunnea* [*C.*]: ♉, ◎, ⊕; *cavum* [*Cinygma*]: ♉–◎–⊕/⊙; *grandifolia* [*C.*]: ♉–◎–⊕/⊙; *hirasana* [*Cinygma*]: ♉–◎–⊕/⊙; *hutchinsoni* [*O.*]: ♉–◎–⊕/⊙; *irina* [*C.*]: ⊕♂; *joosti* [*C.*]: ♉–◎–⊕/⊙; *kurenzovi* [*Cinygma*]: ♉–◎–⊕/⊙; *levanidovi* [*C.*]: ♉–◎–⊕/⊙; *malaisei* [*Cinygma*]: ⊕♂; *mimus* [*Cinygma*]: ⊕♂; *par* [*Cinygma*]: ♉, ◎♂–⊕♂; *putoranica* [*C.*]: ♉–◎–⊕/⊙; *ramaleyi* [*E.*]: ⊕♂; *unicolorata* [*C.*]: ⊕♂.

***

2.2;1,2-2/2.4;2,2. **Rhithrogena/fg3**

(Heptagennota Pentamerotarsata Radulapalpata Rhithrogena/fg1 .../fg3)

(Figs 9:C; 62:C–O; 63:C–E)

Nomen hierarchicum: **Rhithrogena/fg3** (sine *Cinygmula*; incl. *Epeiron, Himalogena, Sibirigena*).
In circumscription matches:
— subgen. *Rhithrogena*: Kluge 1983b: 16 (Abstract); 1988: 306;
— gen. *Rhithrogena* Eaton 1881: 23.

References. Eaton 1883–1888: ♉*; – Lestage 1917: ♉*; – Needham & Traver & Hsu 1935: ♉*; – Strenger 1953: ♉'; – Edmunds & Jensen & Berner 1976: ♉*; – Tshernova 1976: ♉*; – Kluge 1988: ♉*; – 1993: ♉*.

Autapomorphies of Rhithrogena/fg3.

(1) On 2nd+3rd segment of maxillary palp field of scraping setae [see Rhithrogena/fg1 (1) and Rhithrogena/fg2 (2)] occupies nearly whole length of the segment; denticles of these setae are enlarged (Strenger 1953:Figs 11–12). Unique apomorphy: in all other Rhithrogena/fg1 scraping specialization is not so strongly expressed, and scraping setae occupy a distal part of the segment only.

(2) Patella-tibial suture (initially present on middle and hind legs only) is equally developed on all legs of larva; this character is not due to adults, in which difference between fore tibia and other tibiae can be retained. The same in Anteropatellata only (see Index of characters [1.2.18]).

(3) Tergalii are highly specialized, forming an adhesive sucking disc of the following structure (FIG.62: C, H–O):

Tergalius I is strongly widened in such a manner that projects far anteriorly and greatly widened in front of costal rib. These anterior projections of left and right tergalii I are contiguous or closely brought together on ventral side, under thorax. Exactly anteriad of the place of tergalii attachment, ventral side of abdominal segment I has a pair of plates projected anteriorly; into a chink between this plate and sternum, a proximalmost part of anterior margin of tergalius is inputted.

Tergalii II–VI are wide, thus, being directed laterally, overlap one another by their margins.

Tergalius VII has a longitudinal fold, because of which, being directed posterior-medially, is bent under posterior end of abdomen.

Ventral fibrillose lobes of all tergalii [see Branchitergaliae (3)] are well-developed and bent to dorsal side, projecting above tergalii.

Because of their shape, all tergalii I–VII together form a united sucking disc. This specialization of tergalii appeared to be possible, because tergalii are unable to make respiratory movements [see Rhithrogena/fg1 (6)].

On each tergalius costal rib is located at a distance from anterior margin and does not reach tergalius margin; so margins of tergalii are thin and soft. This allows larva to adhere by sucking to a smooth stone surface for a short time so effectively, that if try to take it off, the larva can be only moved by stone surface but not detached (in contrast to Epeorus/fg3-Iron/g1, whose sucking disc can have the same shape, but is bordered by ribs and is not so effective). Many species of Rhithrogena/fg3 prefer to inhabit on small roundish stones with very smooth surface. Tergalius margin can be integral or incised; in all species with incised tergalii, incisions have the same form: they divide outer tergalius margin into numerous festoon-like semicircular lobes. In various species such incisions are present either on all tergalii I–VII, or on tergalii I and VII only, or on tergalius I only, or absent.

Independently from Rhithrogena/fg3, similar tergalial disc appeared in some groups of Epeorus/fg3-Iron/g1 and in some Australian Leptophlebia/fg1. In some of them outlines of the whole disc and of each tergalius are the same as in Rhithrogena/ /fg3; in Epeorus/fg3, as well as in Rhithrogena/fg3, fibrillose portions are bent up. But the structure of first abdominal sternum and the position of costal rib of tergalii II–VI are unique characters of Rhithrogena/fg3.

(4) Subimaginal lateroparapsidal stripe of pigmented area of mesonotum [non-contiguous with medioparapsidal suture – see Rhithrogena/fg2 (6)] is continued behind apex of lateroparapsidal suture; thus behind apex of lateroparapsidal suture (which as in other mayflies represents a groove-like concavity) there is present a non-concave pigmented portion whose length and width slightly exceed width of lateroparapsidal suture; often (but not always) this portion is pigmented more intensively than the whole lateroparapsidal stripe; this portion lies on posterior scutal protuberance of mesonotum, while lateroparapsidal suture terminates at anterior margin of this protuberance (FIG.63:C). Unique apomorphy: in contrast to Rhithrogena/fg3, in *Cinygmula* and Epeorus/fg1 lateroparapsidal suture reaches or nearly reaches apex of lateroparapsidal pigmented stripe, so there is no such marked non-concave pigmented area behind it (FIG.63:A).

(5) Imaginal penis lacks median titillators [see Pentamerotarsata (5)]; usually it has a pair of ventral titillators [see Rhithrogena/fg2 (4)], sometimes ventral titillators are also absent [see below, *Epeiron* (3) and *Sibirigena* (3)]. Non-unique apomorphy (see Index of characters [2.3.15]).

Plesiomorphy of Rhithrogena/fg3. Larval caudalii always retain vestigial primary swimming setae [see Pentamerotarsata (6)] at least in distal part (in contrast to *Cinygmula* and some others – see Index of characters [1.3.66]).

Size. Fore wing length 5–17 mm.

Distribution. Holarctic and Oriental Region.

Rhithrogena/fg3 is divided into Rhithrogena/fg4, *Epeiron*, *Himalogena* and *Sibirigena*; some species have uncertain systematic position.

2.2;1,2-2/2.4;2,2-1. Rhithrogena/fg4
(Heptagennota Pentamerotarsata Radulapalpata Rhithrogena/fg1 .../fg4)
(Fig. 64:D–E)

Nomen hierarchicum: **Rhithrogena/fg4** (sine *Epeiron*, *Himalogena*, *Sibirigena*).

Characters of unclear phylogenetic status.

(1) In imago and subimago 1st segment of fore tarsus is short, several times shorter than 2nd. The same in *Sibirigena*, in contrast to *Epeiron* and *Himalogena*; the same in some other taxa (see Index of characters [2.2.76]).

(2) Each penis lobe is more or less tubular, with gonopore apically; sides near apex usually bear spines or denticles, including latero-dorsal [see Pentamerotarsata (5)] and medio-dorsal ones; titillators [ventral – see Rhithrogena/fg3 (5)] are always well-developed (Sowa 1970:Figs 1–21; Tomka & Hasler 1978:Figs 1–4). The same in *Himalogena*, in contrast to *Epeiron* and *Sibirigena*).

Size. Fore wing length 5–17 mm.

Distribution. Holarctic.

Species composition of Rhithrogena/fg4 (here [*Rh.*]= =[*Rhithrogena*]). About 100 species are described. Besides species listed below, here probably belong some Asian and North American species listed as "Rhithrogena/fg3 INCERTAE SEDIS". **European, North African, Minor Asian and Caucasian species:** *adrianae* Belfiore 1983 [*Rhithrogena*], *alpestris* Eaton 1885 [*Rh.*] (synn. subj.: *alpicola* Navás 1935 [*Rh.*], *brenneriana* Klapálek 1905 [*Rh.*]), *austriaca* Sowa & Weichselbaumer 1988 [*Rh.*], *ayadi* Dakki & Thomas 1986 [*Rh.*], *beskidensis* Alba-Tercedor & Sowa 1987 [*Rh.*], *bogoescui* Sowa & Degrange 1987 [*Rh.*], *braaschi* Jacob 1974 [*Rh.*], *bulgarica* Braasch & Soldán & Sowa 1985 [*Rh.*], *buresi* Sowa 1973 [*Rh.*], *carpatoalpina* Klonowska & Olechowska & Sartori & Weichselbaumer 1987 [*Rh.*], *castellana* Navás 1927 [*Rh.*], *caucasica* Braasch 1979 [*Rh.*], *cincta* Navás 1921 [*Rh.*], *circumtatrica* Sowa & Soldán 1986 [*Rh.*], *colmarsensis* Sowa 1984 [*Rh.*], *corcontica* Sowa & Soldán 1986 [*Rh.*], *dagestanica* Braasch 1979 [*Rh.*], *daterrai* Sowa 1984 [*Rh.*], *decolorata* Sinitshenkova 1973 [*Rh.*], *degrangei* Sowa 1969 [*Rh.*], *delphinensis* Sowa & Degrange 1987 [*Rh.*], *diaphana* Navás 1917 [*Rh.*], *diensis* Sowa & Degrange 1987 [*Rh.*], *dorieri* Sowa 1970 [*Rh.*], *eatoni* Esben-Petersen 1912 [*Rh.*], *endensis* Metzler & Tomka & Zurwerra 1985 [*Rh.*], *excisa* Sinitshenkova 1979 [*Rh.*], *expectata* Braasch 1979 [*Rh.*], *ferruginea* Navás 1905 [*Rh.*], *fiorii* Grandi 1953 [*Rh.*], *fonticola* Sowa & Degrange 1987 [*Rh.*], *germanica* Eaton 1885 [*Rh.*] (synn. subj.: *fradleyi* Blair 1929 [*Rh.*], *haarupi* Esben-Petersen 1909 [*Rh.*], *ussingi* Esben-Petersen 1910 [*Rh.*]), *giudicellorum* Thomas & Bouzidi 1986 [*Rh.*], *goeldlini* Sartori & Sowa 1988 [*Rh.*], *gorganica* Klapálek 1907 [*Rh.*], *gorrizi* Navás 1913 [*Rh.*] (syn. subj.: *comitissa* Navás 1933 [*Rh.*]), *gratianopolitana* Sowa & Degrange & Sartori 1986 [*Rh.*], *grischuna* Sartori & Oswald 1988 [*Rh.*], *henschi* Klapálek 1906 [*Rh.*], *hercegovina* Tanasijevic 1985 [*Rh.*], *hercynia* Landa 1970 [*Rh.*], *hybrida* Eaton 1885 [*Rh.*], *insularis* Esben-Petersen 1913 [*Rh.*], *intermedia* Zurwerra & Metzler & Tomka 1987 [*Rh.*] (syn. subj.: *allobrogica* Sowa & Degrange 1987 [*Rh.*]), *iranica* Braasch 1983 [*Rh.*], *iridina* Kolenati 1859 [*Baetis*] (syn. subj.: *picteti* Sowa 1970 [*Rh.*]), *jacobi* Braasch & Soldán 1988 [*Rh.*], *johannis* Belfiore 1990 [*Rh.*], *joostiana* Braasch & Zimmermann 1976 [*Rh.*], *kimminsi* Thomas 1970 [*Rh.*], *klausnitzeriana* Braasch 1979 [*Rh.*], *laciniosa* Sinitshenkova 1979 [*Rh.*] *landai* Sowa & Soldán 1984 [*Rh.*], *loyolaea* Navás 1922 [*Rh.*] (syn. subj.: *tatrica* Zelinka 1953 [*Rh.*]), *lucida* Braasch 1979 [*Rh.*], *marcosi* Alba-Tercedor & Sowa 1987 [*Rh.*], *mariae* Vitte 1991 [*Rh.*], *mariaedominicae* Sowa & Degrange 1987 [*Rh.*], *marinkovici* Tanasijevic 1985 [*Rh.*], *monserrati* Alba-Tercedor & Sowa 1986 [*Rh.*], *neretvana* Tanasijevic 1985 [*Rh.*], *nivata* Eaton 1871 [*Heptagenia*], *nuragica* Belfiore 1987 [*Rh.*], *oscensis* Navás 1927 [*Rh.*], *ourika* Thomas & Mohati 1985 [*Rh.*], *podhalensis* Sowa & Soldán 1986 [*Rh.*], *pontica* Sowa & Soldán & Kazanci & Braasch 1986 [*Rh.*], *potamalis* Braasch 1979 [*Rh.*], *puthzi* Sowa 1984 [*Rh.*], *puytoraci* Sowa & Degrange 1987 [*Rh.*], *reatina* Sowa & Belfiore 1983 [*Rh.*], *rolandi* Weichselbaumer 1995 [*Rh.*], *ryszardi* Thomas & Vitte & Soldán [*Rh.*], *savoiensis* Alba-Tercedor & Sowa 1987 [*Rh.*], *semicolorata* Curtis 1834 [*Baetis*] (synn. subj.: *semitincta* Pictet 1843 [*Baetis*], *vulpecula* Klapálek 1905 [*Rh.*], *grisoculata* Bogoescu 1958 [*Rh.*]), *sibillina* Metzler & Tomka & Zurwerra 1985 [*Rh.*], *soteria* Navás 1917 [*Rh.*], *sowai* Puthz 1972 [*Rh.*], *strenua* Thomas 1982 [*Rh.*], *sublineata* Kazanci & Braasch 1988 [*Rh.*], *taurisca* Bauernfeind 1992 [*Rh.*], *teberdensis* Zimmermann 1977 [*Rh.*], *theischingeri* Braasch 1981 [*Rh.*], *thomasi* Alba-Tercedor & Sowa 1986 [*Rh.*], *thracica* Sowa & Soldán & Braasch 1988 [*Rh.*], *umbrosa* Braasch 1979 [*Rh.*], *vaillanti* Sowa & Degrange 1987 [*Rh.*], *wolosatkae* Klonowska

1987 [*Rh.*], *zelinkai* Sowa & Soldán 1984 [*Rh.*], *zernyi* Bauernfeind 1991 [*Rh.*], *zhiltzovae* Sinitshenkova 1979 [*Rh.*]. **Central and East Asian species:** *bajkovae* Sowa 1973 [*Rh.*] (syn. subj.: *quadrinotata* Sinitshenkova 1982 [*Rh.*]), *brodskyi* Kustareva1976 [*Rh.*], *japonica* Ueno 1928 [*Rh.*], *lepnevae* Brodsky 1930 [*Rh.*] (synn. subj.: *unicolor* Tshernova 1952 [*Rh.*], *binotata* Sinitshenkova 1982 [*Rh.*]); *orientalis* You 1990 [*Rh.*], *parvus* Ulmer 1912 [*Ecdyonurus*] (syn. subj.: *formosicola* Matsumura 1931 [*Rh.*]), *piechockii* Braasch 1977 [*Rh.*], *sangangensis* You 1990 [*Rh.*], *trispina* Zhou & Zheng 2000 [*Rh.*]. **North American species:** *brunneotincta* McDunnough 1933 [*Rh.*], *fasciata* Traver 1933 [*Rh.*], *robusta* Dodds 1923 [*Rh.*], *rubicunda* Traver 1937 [*Rh.*].

Material examined: *alpestris* [*Rh.*]: Ɒ; *bajkovae* [*Rh.*]: Ɒ–O–⊕/⊙; *beskidensis* [*Rh.*]: Ɒ; *brodskyi* [*Rh.*]: ⊕'; *caucasica* [*Rh.*]: Ɒ–O–⊕/⊙; *decolorata* [*Rh.*]: ⊕'; *degrangei* [*Rh.*]: Ɒ; *diaphana* [*Rh.*]: Ɒ; *dorieri* [*Rh.*]: Ɒ; *excisa* [*Rh.*]: ⊕'; *expectata* [*Rh.*]: Ɒ–O–⊕/⊙; *grischuna* [*Rh.*]: Ɒ; *intermedia* [*Rh.*]: Ɒ; *joostiana* [*Rh.*]: Ɒ–O–⊕/⊙; *laciniosa* [*Rh.*]: Ɒ; *lepnevae* [*Rh.*]: Ɒ–O–⊕/⊙; *puthzi* [*Rh.*]: Ɒ; *savoiensis* [*Rh.*]: Ɒ'/O'; *semicolorata* [*Baetis*]: Ɒ/O; *teberdensis* [*Rh.*]: Ɒ–O–⊕/⊙; *zhiltzovae* [*Rh.*]: ⊕'.

***

2.2;1,2-2/2.4;2,2-2. Epeiron/g1

(Heptagennota Pentamerotarsata Radulapalpata Rhithrogena/fg1 .../fg3 *Epeiron*)

(Fig. 62:M–O)

Nomen hierarchicum: **Epeiron/g(1)** [g: *Epeiron* Demoulin 1964a: 358, typus *E. amseli* Demoulin 1964 (design. orig.)].

In circumscription matches:

— grex *eugeniae*: Kluge 1987: 777.

— gen. *Epeiron* Demoulin 1964a: 358.

References. Demoulin 1964a: ⊕'; – Thomas & Dia 1982 (*Rhithrogena znojkoi*): ⊕'; – Kluge 1983c (*Rhithrogena eugeniae*): Ɒ ⊕; – Kluge 1987b: Ɒ* ⊕'*; – Sartori & Sowa 1992: ⊙ Ɒ ⊕.

Autapomorphies of *Epeiron*.

(1) Tergalii [sucking, with costal ribs located far from costal margin – see Rhithrogena/fg3 (3)] have peculiar shape (FIG.62:M–O): tergalii II–VI are short, with costal rib strongly curved; at least tergalius VI has incised posterior-distal margin (this is the margin, by which it overlaps anterior margin of tergalius VII).

Character of unclear phylogenetic status.

(2) In imago and subimago 1st segment of fore tarsus is relatively long, subequal or equal to 2nd segment, rarely longer. Among Rhithrogena/fg3 the same in *Himalogena*, in contrast to Rhithrogena/fg4 and *Sibirigena*; the same in many other mayflies.

(3) Penis without titillators (in contrast to Rhithrogena/fg4, *Himalogena* and part of *Sibirigena*). In most species (which probably can be united in a taxon **Epeiron/g2**) dorso-lateral spines are lost, and penis lobes have simple shape, fused proximally and divergent distally, with gonopores at a distance from tips (Thomas & Dia 1982:Figs 3–7); in one species from China (which will be described by Zhou & Peters as a new species and a new subgenus preliminary named ***Tumungula***) each penis lobe retains dorso-lateral spine [see Pentamerotarsata (5)] and has gonopore apically.

Size. Fore wing length 5–10 mm.

Distribution. Asia: from Asia Minor to Indo-China.

Species composition of Epeiron/g1. 9 species are formally described – *amseli* Demoulin 1964 [*Epeiron*], *anatolica* Kazanci 1985 [*Rhithrogena*], *binerve* Kluge 1987 [*Rhithrogena*], *eugeniae* Kluge 1983 [*Rhithrogena*], *kashmirensis* Braasch & Soldán 1982 [*Epeiron*], *paulinae* Sartori & Sowa 1992 [*Rhithrogena*], *tibiale* Ulmer 1920 [*Cinygma*], *uzbekistanicus* Braasch & Soldán 1982 [*Epeiron*], *znojkoi* Tshernova 1938 [*Ecdyonurus*] (syn. subj.: *frater* Tshernova 1938 [*Ecdyonurus*]).

Material examined: *binerve* [*Rh.*]: Ɒ–O–⊕/⊙; *eugeniae* [*Rh.*]: Ɒ–O–⊕/⊙; *uzbekistanicus* [*E.*]: Ɒ–O–⊕/⊙; *znojkoi* [*E.*]: Ɒ–O–⊕/⊙.

***

2.2;1,2-2/2.4;2,2-3. Himalogena/g(1)

(Heptagennota Pentamerotarsata Radulapalpata Rhithrogena/fg1 .../fg3 *Himalogena*)

Nomen hierarchicum: **Himalogena/g(1)** [g: *Himalogena* Kluge, SUBGEN. NOV., typus *Rhithrogena tianshanica* Brodsky 1930 (sensu Kluge 1995: 30)].

In circumscription matches:

— species *Rhithrogena tianshanica* Brodsky 1930: 713.

References. Brodsky 1930 (*Rh. tianshanica*): ⊕*; – Kluge 1995.

Characters of unclear phylogenetic status.

(1) In imago and subimago 1st segment of fore tarsus is relatively long, not less than 1/2 of 2nd segment. Among Rhithrogena/fg3 the same in *Epeiron*, in contrast to Rhithrogena/fg4 and *Sibirigena*; the same in majority of mayflies.

(2) Each penis lobe is more or less tubular, with gonopore apically; sides near apex bear spines or denticles, including latero-dorsal [see Pentamerotarsata (5)] and medio-dorsal ones; titillators [ventral – see Rhithrogena/fg3 (5)] are always well-developed. The same in Rhithrogena/fg4, in contrast to *Epeiron* and *Sibirigena*.

(3) All species have similar coloration (see original description for *Rh. tianshanica*) and incised margins of all tergalii I–VII [see Rhithrogena/fg3 (3)] (Sinitshenkova 1973:Fig.1). The same in some Rhithrogena/fg4.

Variable characters of *Himalogena*. Larvae of some species (*stackelbergi* [*Rh.*], *nepalensis* [*Rh.*], *gunti* [nom.nud.]) have mouthpart structure usual for Rhithrogena/fg3 (Sinitshenkova 1973:Fig.3:1,5), while larvae of other species (*tianshanica* [*Rh.*], *pamirica* and *carnivora*) have mouthparts highly specialized for carnivorism (Sinitshenkova 1973:Fig. 1:2–3; Kustareva 1976:Fig.6). Imaginal and subimaginal claws can be either ephemeropteroid (in *tianshanica* [*Rh.*] and *stackelbergi* [*Rh.*]), or pointed (in *nepalensis* [*Rh.*], *gunti*, *pamirica* and *carnivora*]) (see Variable characters of Heptagennota and Index of characters [2.2.85]).

Size. Fore wing length 10–15 mm.

Distribution. High mountains of Central Asia: Gissar Range, Tien-Shan and Himalayas.

Species composition of Himalogena/g(1) (here [*Rh.*]=[*Rhithrogena*]). Among species known, 3 species are described – *nepalensis* Braasch 1984 [*Rh.*], *stackelbergi* Sinitshenkova 1973 [*Rh.*] (syn. subj.: *asiatica* Sinitshenkova 1973 [*Rh.*], *angulata* Braasch 1980 [*Rhithrogena*]) and *tianshanica* Brodsky 1930 [*Rh.*].

Material examined: *carnivora* [*Rhithrogena*] (in litt.): ℧–◎–⊕/⊙; *gunti* [*Rhithrogena*] (in litt.): ℧–◎–⊕/⊙; *nepalensis* [*Rh.*]: ℧, ♀–♀–⊕; *pamirica* [*Rhithrogena*] (in litt.): ℧–◎–⊕/⊙; *stackelbergi* [*Rh.*]: ℧–◎–⊕/⊙; *tianshanica* [*Rh.*]: ℧–◎–⊕/⊙.

***

2.2;1,2-2/2.4;2,2-4. Sibirigena/g(1)

(Heptagennota Pentamerotarsata Radulapalpata Rhithrogena/fg1 .../fg3 *Sibirigena*)

(Figs 62:J–L; 63:C)

Nomen hierarchicum: **Sibirigena/g(1)** [g: *Sibirigena* Kluge, SUBGEN. NOV., typus *Rhithrogena sibirica* Brodsky 1930].

Autapomorphy of *Sibirigena*.

(1) Each tergalius II–VI [involved to sucking disk – see Rhithrogena/fg1 (3)] has a roundish projection on its posterior-dorsal margin; in various species this projection has a form from shallow convexity (FIG.62:J–K) to a sharp finger-like process.

Characters of unclear phylogenetic status.

(2) In imago and subimago 1st segment of fore tarsus is short, several times shorter than 2nd. The same in Rhithrogena/fg4, in contrast to *Epeiron* and *Himalogena*; the same in some other taxa (see Index of characters [2.2.76]).

(3) Each penis lobe is elongate, apically narrowed or roundish, with gonopore on dorsal side at some distance from tip; medio-dorsal and latero-dorsal spines are absent (in contrast to Rhithrogena/fg4 and *Himalogena*). Titillators [see Rhithrogena/fg3 (5)] can be either well-developed, or vestigial, or lost (the same in *Epeiron*).

Size. Fore wing length 7–13 mm.

Distribution. East Palaearctic (from Altai to Far East) and Nearctic.

Species composition of Sibirigena/g(1) (here [*Rh.*]= =[*Rhithrogena*]). One **East-Palaearctic** species: *sibirica* Brodsky 1930[*Rh.*]. **North American species:** *brunnea* Hagen 1875 [*Heptagenia*], *decora* Day 1954 [*Rh.*], *hageni* Eaton 1885 [*Rh.*] (syn. subj.: *doddsi* McDunnough 1926 [*Rh.*]), *impersonata* McDunnough 1925 [*Heptagenia*] (syn. subj.: *sanguinea* Ide 1954 [*Rh.*]), *jejuna* Eaton 1885 [*Rh.*], *morrisoni* Banks 1924 [*Epeorus*] (syn. subj.: *petulans* Seemann 1927 [*Iron*]), *undulatus* Banks 1924 [*Epeorus*].

Material examined: *hageni* [*Rh.*]: ⊕; *sibirica* [*Rh.*]: ℧–◎–⊕/⊙.

***

Rhithrogena/fg3 INCERTAE SEDIS

Besides species undoubtedly belonging to Rhithrogena/fg4, *Epeiron*, *Himalogena* and *Sibirigena* (see above), there are following species, whose descriptions do not allow to determine their systematic position within Rhithrogena/fg3 (here [*Rh.*]=[*Rhithrogena*]). **Asian species:** *ampla* Kang & Yang 1994 [*Rh.*], *basiri* Ali 1971 [*Rh.*], *minima* Sinitshenkova 1973 [*Rh.*], *subangulata* Braasch 1984 [*Rh.*], *tetrapunctigera* Matsumura 1931 [*Rh.*] (synn. subj.: *satsuki* Imanishi 1936 [*Rh.*], *gojoensis* Gose 1968 [*Ameletus*]). **North American species:** *amica* Traver 1935 [*Rh.*], *anomala* McDunnough 1928 [*Rh.*], *exilis* Traver 1933 [*Rh.*], *flavianula* McDunnough 1924 [*Heptagenia*], *fuscifrons* Traver 1933 [*Rh.*], *futilis* McDunnough 1934 [*Rh.*], *gaspeensis* McDunnough 1933 [*Rh.*], *manifesta* Eaton 1885 [*Rh.*] (syn. subj.: *pellucida* Daggy 1945 [*Rh.*]), *notialis* Allen & Cohen 1977 [*Rh.*], *plana* Allen & Chao 1978 [*Rh.*], *uhari* Traver 1933 [*Rh.*], *virilis* McDunnough 1934 [*Rh.*], *vitta* Allen & Chao 1978 [*Rh.*].

*** *** ***

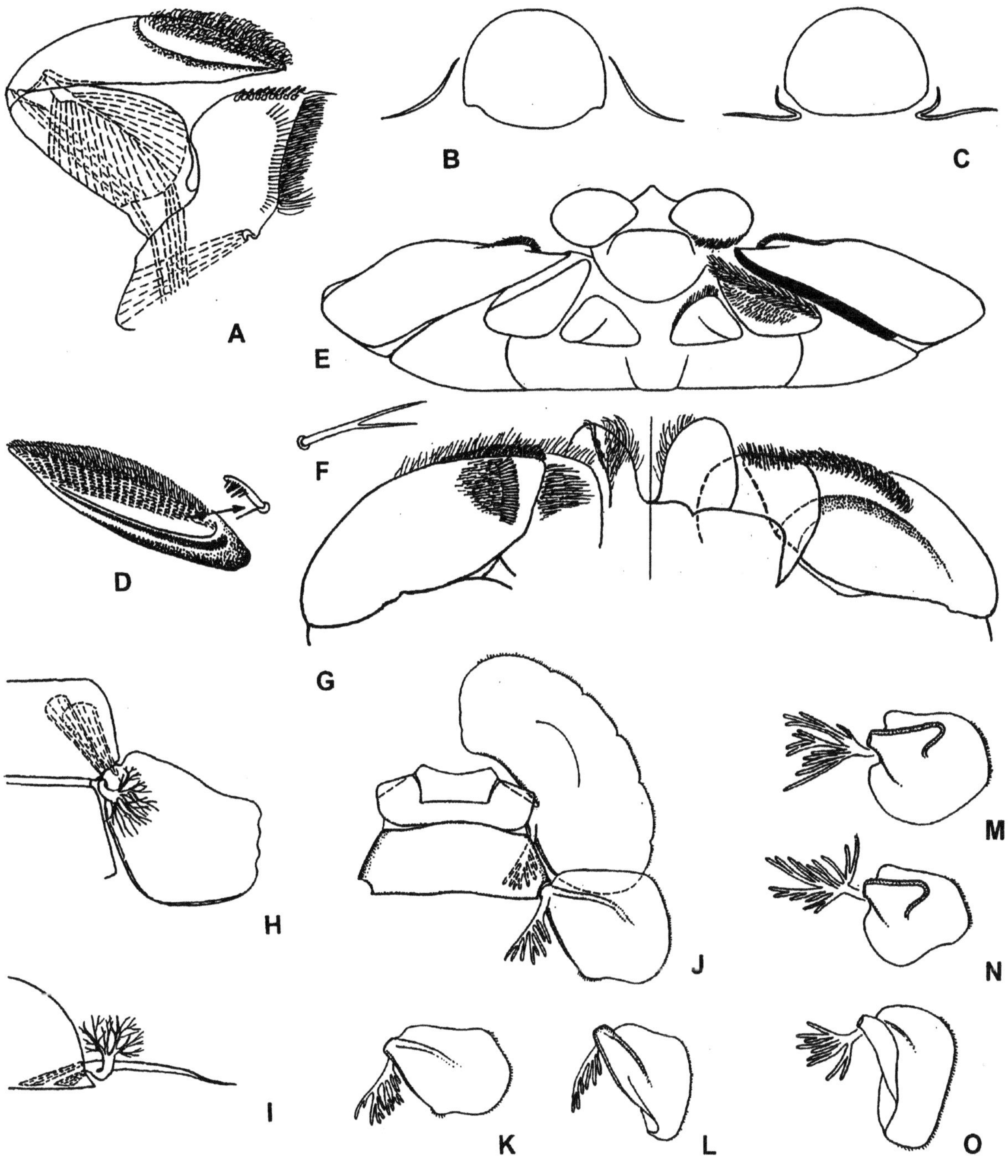

Figure 62. Rhithrogena/fg2, larvae.
A–B – Cinygmula/g1: **A** – right maxilla, ventral view (muscles shown by interrupted lines); **B** – cross section through abdominal segment VIII and tergalii VII. **C–O** – Rhithrogena/fg3: **C** – cross section through abdominal segment VIII and tergalii VII; **D** – apical (2nd+3rd) segment of maxillary palp and enlarged pectinate scraping seta; **E** – superlinguae, hypopharynx and labium, apical view; **F** – seta of distal row on dorsal surface of distal segment of labial palp; **G** – distal part of labium, dorsal view (in left half) and ventral view (in right half) (compare with Fig.57:F–H); **H–I** – scheme of natural position of tergalii (muscles shown by interrupted lines): **H** – dorsal view; **I** – posterior view (compare with Figs 57:C–E and 61:J–K); **J–L** – Rhithrogena/fg3-Sibirigena/g1* *sibirica* [*Rhithrogena*]: **J** – abdominal segments I and II with left tergalii, ventral view; **K–L** – right tergalii VI and VII, dorsal view; **M–O** – Rhithrogena/fg3-Epeiron/g1 *znojkoi* [*Ecdyonurus*]: right tergalii V–VII, dorsal view (fibrillose portions turned apart). (D–G – from Kluge 1988a; H–I – from Kluge 1993a; J–K – from Kluge 1997d).

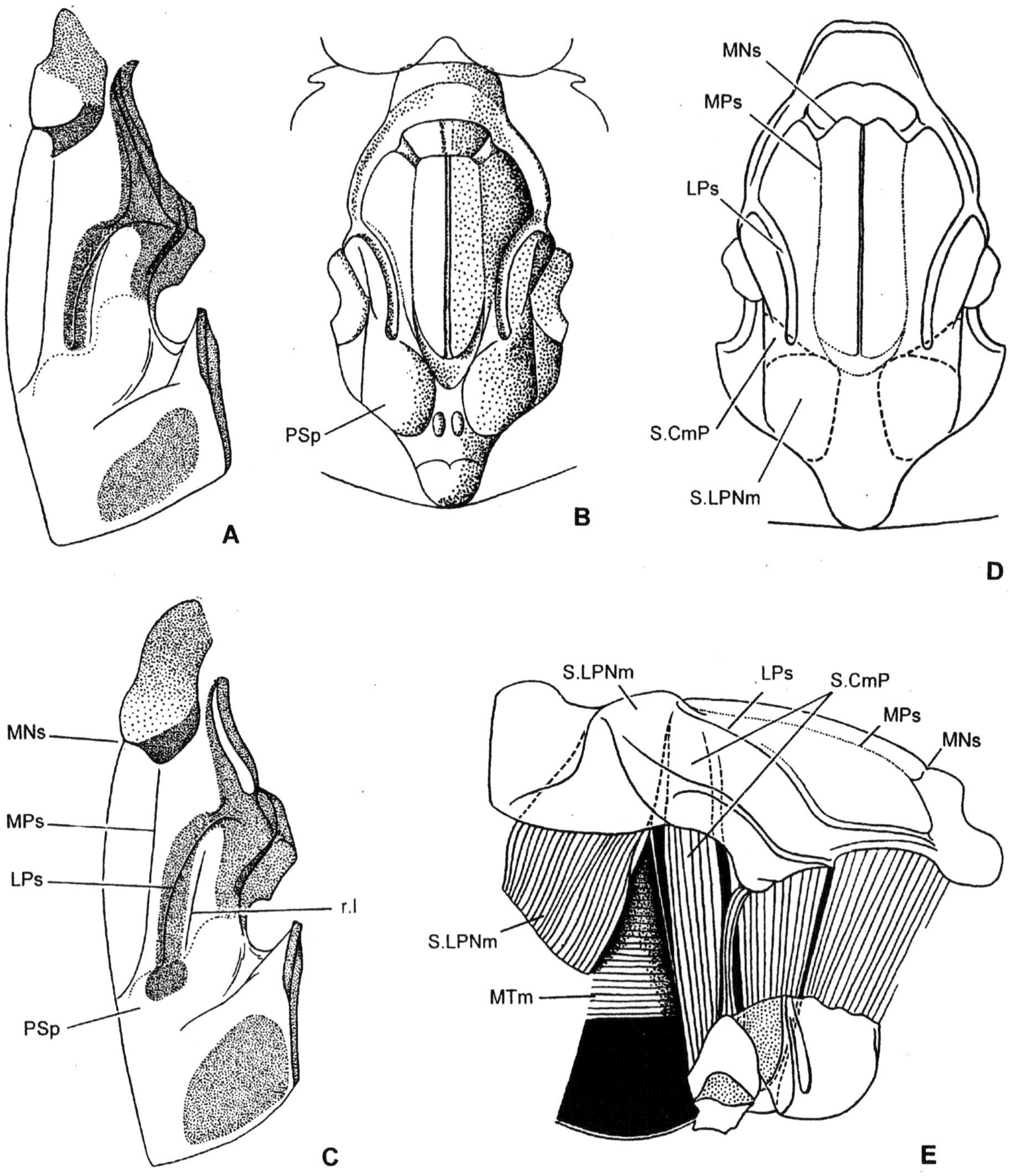

Figure 63. Rhithrogena/fg2, adult mesonotum.

A–B – Cinygmula/g1 *sapporensis* [*Siphlurus*]: **A** – subimaginal exuviae of right half of mesonotum; **B** – imaginal mesonotum, dorsal view (compare with Fig.61:A). **C–E:** – Rhithrogena/fg3: **C** – Sibirigena/g* *sibirica* [*Rhithrogena*], subimaginal exuviae of right half of mesonotum; **D–E** – Rhithrogena/fg4 *iridina* [*Baetis*], imaginal mesothorax (bases of posterior scuto-coxal and scuto-lateropostnotal muscles shown by interrupted lines) (compare with Fig.56:L–M): **D** – dorsal view, **E** – lateral view (most part of pleura removes to show muscles). (From Kluge 1988a).

Abbreviations: **LPs** – lateroparapsidal suture; **MNs** – mesonotal suture; **MPs** – medioparapsidal suture; **MTm** – median tergal muscle; **PSp** – posterior scutal protuberance; **r.l** – relief line; **S.CmP** – posterior scuto-coxal muscle; **S.LPNm** – scuto-lateropostnotal muscle.

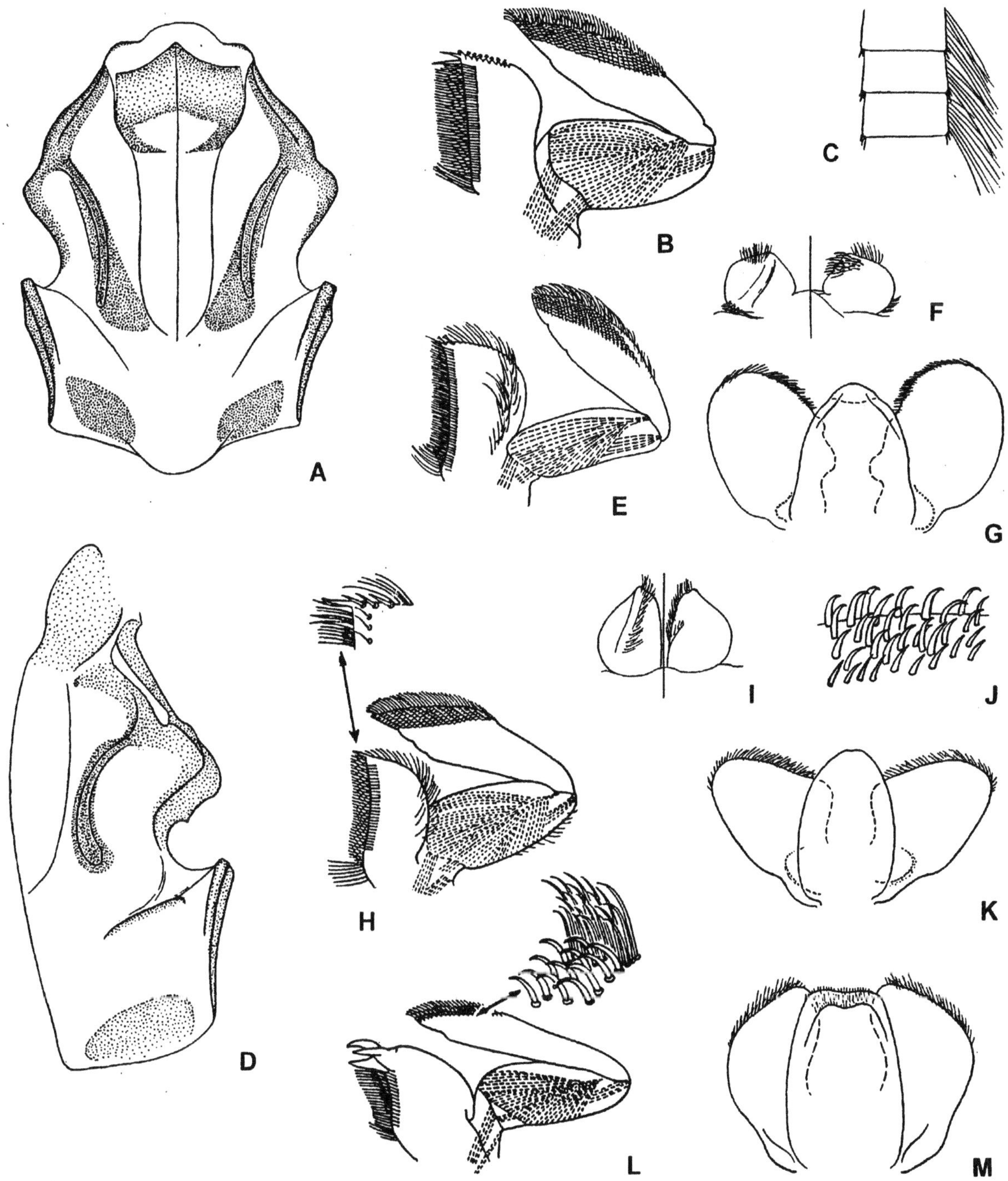

Figure 64. Rhithrogena/fg1: *Paegniodes* and Epeorus/fg1.
A–C – Paegniodes/g* *cupulata* [*Heptagenia*]: **A** – subimaginal mesonotum; **B** – left maxilla, ventral view (muscles shown by interrupted lines); **C** – middle part of larval cercus. **D–M** – Epeorus/fg1: **D–G** – Bleptus/g* *fasciatus* [*Bleptus*]: **D** – subimaginal exuviae of right half of mesonotum; **E** – left maxilla, ventral view (muscles shown by interrupted lines); **F** – glossae, dorsal view (in left half) and ventral view (in right half); **G** – hypopharynx and superlinguae, ventral view. **H–M** – Epeorus/fg2: **H–K** – Ironodes/g* *nitidus* [*Iron*]: **H** – left maxilla, ventral view, with apex enlarged (muscles shown by interrupted lines); **I** – glossae, dorsal view (in left half) and ventral view (in right half); **J** – setae on costal rib of tergalius; **K** – hypopharynx and superlinguae, ventral view; **L–M** – Epeorus/fg3 Belovius/g2 *pellucida* [*Cinygma*]: **L** – left maxilla, ventral view, with scraping setae of maxillary palp enlarged (muscles shown by interrupted lines); **M** – hypopharynx and superlinguae, ventral view) (C, E, F–J, L – from Kluge 1988a).

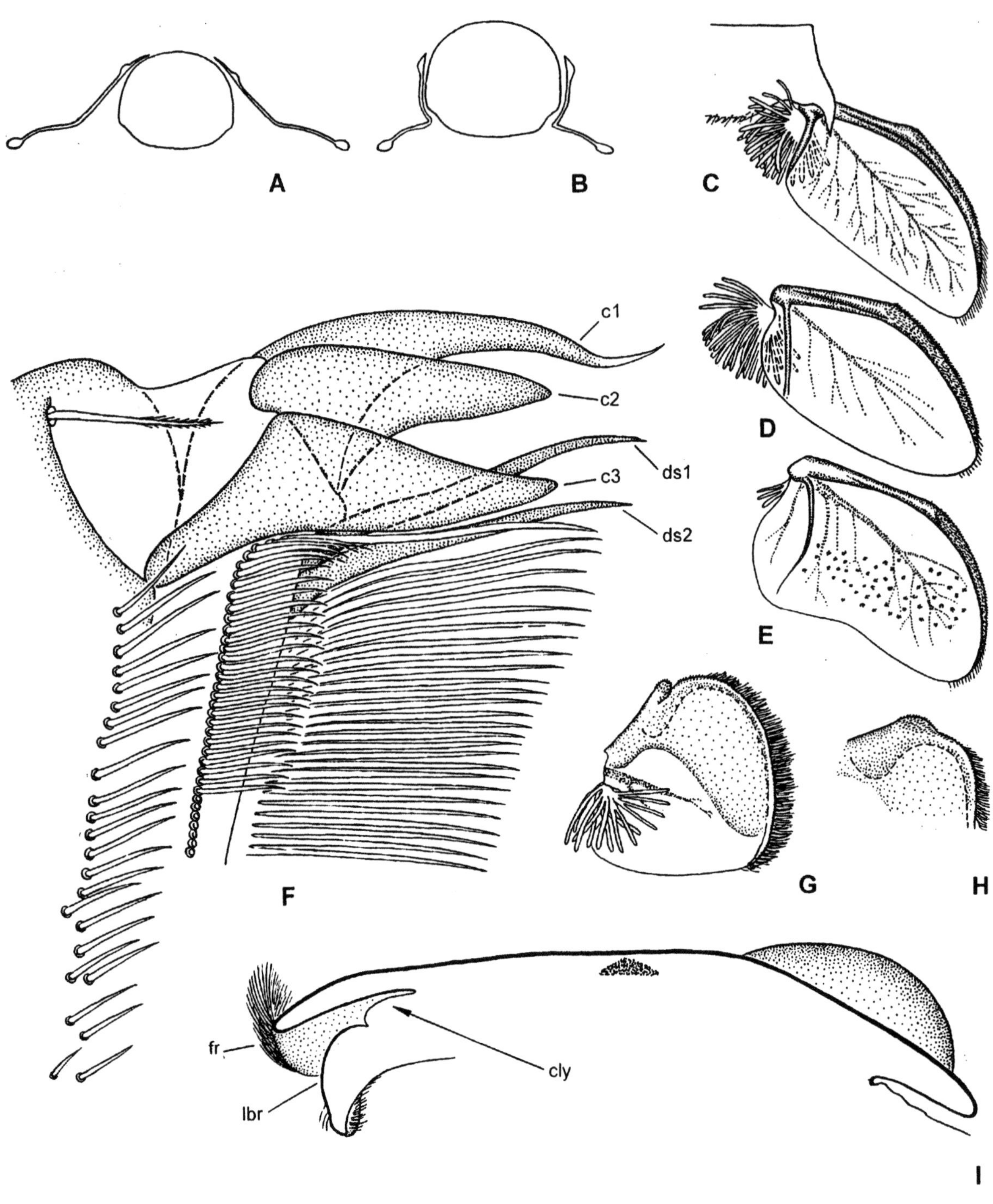

Figure 65. Epeorus/fg3, larvae.
A – Belovius/g2, cross section through abdominal segment VIII and tergalii VII; **B** – Iron/g1, the same. **C** – Epeorus/fg4 *zaitzevi* [*Epeorus*], margin of abdominal tergum IV and right tergalius IV, dorsal view. **D** – Albertiron/g* *albertae* [*Iron*], right tergalius IV, dorsal view. **E–F** – Belovius/g2 *pellucida* [*Cinygma*]: **E** – right tergalius IV, dorsal view; **F** – apex of right maxilla, ventral view (dentisetae, canines and sclerotized areas shown by dots). **G–I** – Iron/g1 Ironopsis/g1: **G** – Caucasiron/g* *caucasica* [*Cinygma*], tergalius II, dorsal view; **H–I** – Ironopsis/g2 *rheophilus* [*Iron*]: **H** – anterior margin of tergalius II, dorsal view; **I** – median section of larval head capsule and labrum. (C, G, H – from Kluge 1997d).
Abbreviations: **c1**, **c2**, **c3** – maxillary canines; **cly** – clypeus; **ds1**, **ds2** – dentisetae; **fr** – frons; **lbr** – labrum.

2.2;1,2-2/2.4;3. **Epeorus/fg1**

(Heptagennota Pentamerotarsata Radulapalpata Rhithrogena/fg1 Epeorus/fg1)

(Figs 64:D–M; 65)

Nomen hierarchicum: **Epeorus/fg1** (incl. *Bleptus*) [f: Epeorini Kluge, TRIBUS NOV.; g: *Epeorus* Eaton 1881: 26, typus *E. torrentium* Eaton 1881 (design. orig.)].
In circumscription matches:
— grex genn. *Epeorus*: Tomka 1991: 121.

References. Kluge 1988: ℧*; – Tomka 1991: ℧*; – Kluge 1993: ℧* ⊕*.

Autapomorphies of Epeorus/fg1.

(1) Imaginal and subimaginal mesonotal suture is completely lost (FIG.64:D). Loss of mesonotal suture takes place also in some other Ephemeroptera, thus, formally speaking, this apomorphy is not unique. But in Epeorus/fg1 mesonotal suture had disappeared by a peculiar way: not by gradual curvation (characteristic for Tetramerotarsata, Fimbriatotergaliae and Leptophlebia/fg1) or by gradual smoothing out (as in some Tricorythus/fg1) but by a saltation. In all other Radulapalpata mesonotal suture is distinct and has uniform shape (transverse in median part and turned posteriorly at places where it connects with medioparapsidal sutures).

(2) Larval paracercus is reduced to a 1-segmented vestige; primary swimming setae of cerci are completely lost [see Pentamerotarsata (6)]. The same in some other rheophilous mayflies (see Index of characters [1.3.64]).

(3) On fore wing [see Heptagennota (5) and Plesiomorphies of Radulapalpata] veins AA and AP proximally have a common stem (Kluge 1993:Fig. 32). In other mayflies they originate independently (FIG.53) or are connected only at extreme base.

(4) In male imago 1st segment of fore tarsus is always long, approximately equal to 2nd segment (in contrast to most other Pentamerotarsata, where it is usually equal to 1/6–7/8 of 2nd segment).

Characters of unclear phylogenetic status.

(5) On maxilla pectinate setae of apical-ventral row [see Radulapalpata (3)] are vestigial (in *Bleptus* and *Ironodes* – FIG.64:H) or lost (in Epeorus/fg3 – FIG.64:L). Non-unique apomorphy: incomplete or complete reduction of these setae takes place also in *Cinygma*, *Raptoheptagenia* and some representatives of Heptagenia/f7=g6.

(6) On maxilla ventral row of setae [see Branchitergaliae (1)] is always straight, not curved laterally (FIGS 64:E, H, L; 65:F) The same in *Cinygma* and *Paegniodes* (see Index of characters [1.1.30]); possibly symplesiomorphy. Shape of maxilla is variable, in Epeorus/fg3 being sharply different from that of *Bleptus* and *Ironodes* [see Epeorus/fg3 (1) below].

(7) Subimaginal lateroparapsidal stripe of pigmented area of mesonotum in its posterior part is curved laterally, repeating curvation of lateroparapsidal suture [see Rhithrogena/fg1 (3)], does not touch medioparapsidal suture (FIG.64:D). The same in Rhithrogena/fg2; possibly synapomorphy (in contrast to *Paegniodes*).

(8) Patella-tibial suture (initially present on middle and hind legs) is lost on all legs of larva, being retained on middle and hind legs of subimago and imago (only in *Bleptus* it is lost in subimago and imago also). Non-unique apomorphy (see Index of characters [1.2.18]). Among Epeorus/fg1 only in Belovius/g2 larva has developed patella-tibial suture on middle and hind legs; probably, in this case the suture is restored secondarily; this is possible, because the suture was retained in adult stages.

Variable characters of Epeorus/fg1. In selected species larval abdominal terga have unpaired median or paired submedian projections or spines directed posteriorly; particularly, unpaired spines are present in *Bleptus* and selected species of Iron/g1, paired spines – in *Ironodes* and selected species of Iron/g1 and Epeorus/fg4; some species have spines paired on anterior segments and unpaired on posterior segments (non-unique character – see Index of characters [1.3.3]).

Size. Fore wing length 7–20 mm.

Distribution. Holarctic and Oriental Region.

Epeorus/fg1 is divided into *Bleptus* and Epeorus/fg2.

2.2;1,2-2/2.4;3,1. **Bleptus/g(1)**

(Pentamerotarsata Radulapalpata Rhithrogena/fg1 Epeorus/fg1 *Bleptus*)

(Fig. 64:D–G)

Nomen hierarchicum: **Bleptus/g(1)** [g: *Bleptus* Eaton 1885: 243, typus *B. fasciatus* Eaton 1885 (design. orig.)].
In circumscription matches:
— gen. *Bleptus* Eaton 1885: 243.
Monospecific taxon.

References. Eaton 1883–1888: ⊕; – Ueno 1931: ℧; – Tshernova 1974: ⊕; – 1976: ℧; – Yoon & Bae 1984: ℧ ⊕; – Kluge1988: ℧ ⊕; – 1993: ℧.

Autapomorphies of *Bleptus*.

(1) Glossae [initially pyramidal – see Radulapalpata (1)] are roundish, wide (FIG.64:F) (in contrast to other Rhithrogena/fg1).

(2) Hind wing is diminished, with only 2 longitudinal veins behind triad MP (Eaton 1883–1888: Pl.65:1). Non-unique apomorphy (see Index of characters [2.2.59]).

(3) Titillators [see Pentamerotarsata (5)] are lost. Non-unique apomorphy; among Epeorus/fg1 the same in *Ironodes* and some Epeorus/fg3 (see Index of characters [2.3.15]).

Characters of unclear phylogenetic status.

(4) Superlinguae diverge toward apex (FIG.64: G). Similar form of superlinguae in *Ironodes* [for comment see *Ironodes* (3) below].

Plesiomorphies of *Bleptus*. In contrast to Epeorus/fg2, larval head is widest in its posterior part, without dense hairs on fore margin (Ueno 1931:Pl. 12:1). Maxilla is long [see Radulapalpata (3)], apical margin bears long hairs and vestiges of pectinate setae [see Epeorus/fg1 (5)], canines are not thickened (FIG.64:E) (symplesiomorphy with *Ironodes*, in contrast to Epeorus/fg3).

Size. Fore wing length 10–18 mm.

Distribution. Japan and Korea.

Species composition of Bleptus/g(1). 1 species – *fasciatus* Eaton 1885 [*Bleptus*].

Material examined: *fasciatus* [*B.*]: ℧–◎–⊕/⊙.

***

2.2;1,2-2/2.4;3,2. Epeorus/fg2

(Pentamerotarsata Radulapalpata Rhithrogena/fg1 Epeorus/fg1 .../fg2)

(Figs 64:H–M; 65)

Nomen hierarchicum: **Epeorus/fg2** (sine *Bleptus*; incl. *Ironodes*).
In circumscription matches:
— gen. *Epeorus*: Kluge 1988: 308.

References. Tshernova 1974: ⊕*; – 1976: ℧*; – Kluge 1988: ℧* ⊕*; – 1993: ℧* ⊕*.

Autapomorphies of Epeorus/fg2.

(1) Tergalius I is more or less widened anteriad of costal rib, thus the costal rib passes at a distance from margin (in contrast to *Bleptus*, where all tergalii ribs are marginal).

(2) On tergalii II–VII strongly developed costal rib (which, in contrast to tergalius I, retains its position on costal margin) is covered by dense stout curved pointed spine-like setae (FIG.64:J) (unique apomorphy). Probably, this helps larva to couple with stone, on which larva sits in a stream: tergalii are immovably spread by sides and pressed by their costal margins to the substrate. Such specialization of tergalii appeared to be possible, because tergalii are unable to make respiratory movements [see Rhithrogena/fg1 (6)].

(3) Dorsal side of larval body bears stripes of dense thin hairs: anterior margin of head shield [see Heptagennota (1)] is widened and bears a stripe of very dense thin hairs directed anteriorly or dorsally (FIG.65:I); dorsal side of each tibia bears a longitudinal stripe of dense thin hairs directed dorsally. Similar stripes of dorsally directed hairs on various parts of body (legs, head, caudalii, abdomen) evolved in various mayfly groups where larvae are specialized for inhabitancy in rapid streams. Probably these hairs have some hydrodynamic importance. In other Radulapalpata such hairs are not developed or weakly developed. Judging by uniform structure and position of hairs on head and tibiae of Epeorus/fg2, in this taxon these hairs have common origin (i. e. represent an autapomorphy of Epeorus/ /fg2). Besides this, dorsal stripes of hairs on other parts of body (cerci, abdomen) evolved independently in selected groups of Epeorus/fg3.

Plesiomorphy of Epeorus/fg2. In contrast to *Bleptus*, glossae are pyramidal [see Radulapalpata (1)] (FIG.64:I).

Size. Fore wing length 7–20 mm.

Distribution. Holarctic and Oriental Region.

Epeorus/fg2 is divided into *Ironodes* and Epeorus/fg3.

2.2;1,2-2/2.4;3,2-1. Ironodes/g(1)

(Pentamerotarsata Radulapalpata Rhithrogena/fg1 Epeorus/fg1 .../fg2 *Ironodes*)

(Fig. 64:H–K)

Nomen hierarchicum: **Ironodes/g(1)** [g: *Ironodes* Traver 1935: 32, typus *Iron nitidus* Eaton 1885 (design. orig.)].
In circumscription matches:
— subgen. *Ironodes*: Kluge 1988: 308;
— gen. *Ironodes* Traver 1935: 32.

References. Traver 1935: ℧* ⊕*; – Needham & Traver & Hsu 1935: ℧* ⊕*; – Day 1952: ℧'; – Edmunds & Jensen & Berner 1976: ℧* ⊕*; – Tshernova 1976: ℧*; – Kluge 1988: ℧*.

Autapomorphies of *Ironodes*.

(1) Tergalius VII lacks fibrillose lobe [see Branchitergaliae (3)]. Non-unique apomorphy; among Radulapalpata the same in some Heptagenia/f5=g4 and some *Cinygmula*.

(2) Titillators [see Pentamerotarsata (5)] are lost. Non-unique apomorphy; among Epeorus/fg1 the same in *Bleptus* and some Epeorus/fg3 (see Index of characters [2.3.15]).

Character of unclear phylogenetic status.

(3) Superlinguae diverge toward apex (FIG.64:K). The same in *Bleptus* (FIG.64:G). Probably, this is a symplesiomorphy of *Bleptus* and *Ironodes* within Epeorus/fg1, but in this case similarity of Rhithrogena/fg2 and Epeorus/fg3 (whose superlinguae are widest in middle part and convergent toward apex – FIG.64:M) can be explained only as a convergention. Possibly, vice verse, convergention is the similarity of *Bleptus* and *Ironodes*, while similarity of Rhithrogena/fg2 and Epeorus/fg3 is a symplesiomorphy within Rhithrogena/fg1.

Plesiomorphies of *Ironodes*. Mandible is longer than in Epeorus/fg2, with incisor and kinetodontium slender. Maxilla is long [see Radulapalpata (3)], apical margin bears long hairs and vestiges of pectinate setae [see Epeorus/fg1 (5)], canines are not thickened (FIG.64:H) (symplesiomorphy with *Bleptus*, in contrast to Epeorus/fg3). Larval cerci have no dorsal hairs (in contrast to Epeorus/fg3).

Size. Fore wing length 8–14 mm.

Distribution. Western Nearctic.

Species composition of Ironodes/g(1). 6 species – *arctus* Traver 1935 [*Ironodes*], *californicus* Banks 1910 [*Heptagenia (Epeorus)*], *flavipennis* Traver 1935 [*Ironodes*], *geminatum* Eaton 1885 [*Cinygma*], *lepidus* Traver 1935 [*Ironodes*], *nitidus* Eaton 1885 [*Iron*].

Material examined: *nitidus* [*Iron*]: ℧.

***

2.2;1,2-2/2.4;3,2-2. Epeorus/fg3

(Pentamerotarsata Radulapalpata Rhithrogena/fg1 Epeorus/fg1 .../fg3)

(Figs 64:L–M; 65)

Nomen hierarchicum: **Epeorus/fg3** (sine *Ironodes*; incl. *Proepeorus*, *Belovius*, *Iron*).

In circumscription matches:

— grex subgenn. *Epeorus*: Kluge 1988: 308;

— gen. *Epeorus*: Edmunds & Jensen & Berner 1976: 190.

References: Edmunds & Jensen & Berner 1976: ℧* ⊕*; - Kluge 1988: ℧*; - 1993: ℧*.

Autapomorphies of Epeorus/fg3.

(1) Mouth apparatus, retaining the scraping specialization of labial and maxillary palps [see Radulapalpata (1) and Rhithrogena/fg1 (1)], has a peculiar biting specialization which is expressed in modification of maxillae and mandibles.

Maxilla is shortened, inclined, with acute apical-median margin [but ventral row of setae is straight and goes parallel to median margin – see Epeorus/fg1 (6)] (FIG.64:L). Apical margin of maxilla is bare, vestiges of pectinate setae [see Epeorus/fg1 (5)] and hairs are completely lost. All 3 maxillary canines are very massive, strongly sclerotized; at the same time canines are not pressed together, and both dentisetae [see Bidentiseta (1)] are not thickened (FIG.65:F) (in contrast to Nesameletus/f2=Metamonius/g2 and some Turbanoculata whose maxillary canines are also thickened).

Mandible is shorter (in relation to perpendicular to axis of articulation) and more stout, than in other Pentamerotarsata, incisor and kinetodontium are stout; median margin is short, its setae and setiform prostheca form a compact tuft close to kinetodontium [see Heptagennota (7)] (Eaton 1883–1888:Pl. 55–56). Mandibles with similar proportions but of somewhat different shape occur in selected carnivorous species of Heptagenia/f7=g6 and *Himalogena*.

(2) On tergalii II–VII apical part of costal rib is covered by long slender soft hairs; they form a common band with curved spin-like setae which cover more proximal part of the costal rib [see Epeorus/fg2 (2)] (FIG.65:C–E, G–H) (in contrast to *Ironodes*, whose apex of costal rib is glabrous).

Characters of unclear phylogenetic status.

(3) Superlinguae are widest at middle part, convergent and narrowed toward apex (FIG.64:M). The same in Rhithrogena/fg2, in contrast to *Ironodes* and *Bleptus* [for comment – see *Ironodes* (3) above].

(4) Larval abdominal segments can have more or less prominent projections close to tergalial articulatory membranes: besides a small rounded projection ventrad of the tergalius base (corresponding to posterolateral spine of many mayflies), there can be developed a larger **supra-tergalial projection** just dorsad of the tergalius base. Supra-tergalial projection is flat, can be either small and rounded (in Iron/fg1, some *Proepeorus* and some Epeorus/fg4), or long, pointed and curved postero-medially (in Belovius/fg1, some *Proepeorus* and some Epeorus/fg4 – FIG.65:C).

Plesiomorphy of Epeorus/fg3. Each tergalius I–VII retains fibrillose lobe [see Branchitergaliae (3)] (in contrast to *Ironodes*).

Size. Fore wing length 7–20 mm.

Distribution. Holarctic and Oriental Region.

Epeorus/fg3 is divided into *Proepeorus*, Epeorus//fg4, Belovius/g1 and Iron/g1. Some insufficiently described species have uncertain systematic position (p.207).

2.2;1,2-2/2.4;3,2-2/1. Plesiomorphon **Proepeorus/g(1)**

(Pentamerotarsata Radulapalpata Rhithrogena/fg1 Epeorus/fg1 .../fg3 *Proepeorus*)

Nomen hierarchicum: **Proepeorus/g(1)** [g: *Proepeorus* Kluge, SUBGEN. NOV., typus *Epeorus anatolii* Sinitshenkova 1981].

References. Sinitshenkova 1981 (*E. anatolii*): Ծ'; – 1982 (*E. rautiani*): Ծ' ⊕'.

Plesiomorphies of *Proepeorus*. In contrast to Epeorus/fg4, small median titillators [see Pentamerotarsata (5)] are present. In contrast to Belovius/g1, tergalii II–VII have anal rib on anal margin (as in FIG.65:C). In contrast to Iron/fg1, tergalius I [see Epeorus/fg2 (1)] is widened not so strongly, tergalii II–VII are long, and tergalius VII usually without longitudinal fold, only with smooth longitudinal concavity, which is not enough to allow tergalius to turn under abdomen (as in FIG.65:A) (among the species examined, only *dispar* [*Iron*] has distinct longitudinal fold).

Size. Fore wing length 12–15 mm.

Distribution. Eastern Palaearctic and Nearctic.

Species composition of Proepeorus/g(1). Several species are described, among which **Asian species:** *anatolii* Sinitshenkova 1981 [*Epeorus*] (syn.subj.: *rautiani* Sinitshenkova 1982 [*Epeorus*]), *nipponicus* Ueno 1931 [*Iron*]; **North American species:** *dispar* Traver 1933 [*Iron*], *vitrea* Walker 1853 [*Palingenia*] (syn. subj.: *humeralis* Morgan 1911 [*Epeorus*]).

Material examined: *anatolii* [*E.*]: Ծ♂–◎♂–⊕♂; *dispar* [*I.*]:Ծ; *nipponicus* [*I.*]: Ծ, ⊕♂; *vitrea* [*P.*]: Ծ♂/◎♂.

***

2.2;1,2-2/2.4;3,2-2/2. **Epeorus/fg4**

(Pentamerotarsata Radulapalpata Rhithrogena/fg1 Epeorus/fg1 .../fg4)

(Fig. 65:C)

Nomen hierarchicum: **Epeorus/fg4** (sine *Proepeorus*, *Belovius*, *Iron*).

References. Grandi 1960 (*Epeorus assimilis*, *E. torrentium*): Ծ ⊕; – Berthelemy & Thomas 1967: ⊕♂.

Autapomorphies of Epeorus/fg4.

(1) Penis has characteristic structure: titillators [see Pentamerotarsata (5)] are lost; left and right penis lobes are immobile fused basally and divergent apically. Among Epeorus/fg3 titillators are lost also in Ironopsis/g2 and some Belovius/g1 (see Index of characters [2.3.15]).

Plesiomorphies of Epeorus/fg4. Tergalius I [see Epeorus/fg2 (1)] is widened not so strongly as in majority of Iron/g1; tergalii II–VII have anal rib on anal margin (FIG.65:C) (in contrast to Belovius/g1); tergalius VII without longitudinal fold (in contrast to Iron/g1), only with smooth longitudinal concavity, which is not enough to allow tergalius to turn under abdomen (as in FIG.65:A).

Size. Fore wing length 12–18 mm.

Distribution. Palaearctic and Oriental Region.

Species composition of Epeorus/fg4 (here [*E.*]= =[*Epeorus*]). Three **European species:** *assimilis* Eaton 1885 [*E.*], *sylvicola* Pictet 1865 [*Baetis*] (syn. subj.: *geminus* Eaton 1885 [*E.*]), *torrentium* Eaton 1881 [*E.*] (syn. subj.: *bernardezi* Navás 1924 [*E.*]). Several **Asian species:** *bispinosus* Braasch 1980 [*E.*], *hieroglyphicus* Braasch & Soldán 1984 [*E.*], *zaitzevi* Tshernova 1981 [*E.*]; probably here also belong *bifurcatus* Braasch & Soldán [*E.*] and *aculeatus* Braasch 1990 [*E.*] (related to *bispinosus* [*E.*]) and some other species whose male imagoes are unknown.

Material examined: *aculeatus* [*E.*]: Ծ♀/◎♀/⊙; *bispinosus* [*E.*]:Ծ–◎–⊕/⊙; *sylvicola* [*B.*]: Ծ, ◎, ⊕/⊙; *zaitzevi* [*E.*]: Ծ–◎–⊕/⊙.

***

2.2;1,2-2/2.4;3,2-2/3. **Belovius/g1**

(Pentamerotarsata Radulapalpata Rhithrogena/fg1 Epeorus/fg1 .../fg3 Belovius/g1)

(Figs 64:L–M; 65:A, D–F)

Nomen hierarchicum: **Belovius/g1** (incl. *Albertiron*) [g: *Epeorus* subgen. *Belovius* Tshernova 1981: 326, typus *Epeorus latifolium* Ueno 1928 (design. orig.)].

Autapomorphy of Belovius/g1.

(1) On tergalii II–VII [see Epeorus/fg3 (2)] anal rib [which initially is situated on anal margin – see Branchitergaliae (3)] is situated at a distance from anal margin (FIG.65:D–E). Non-unique apomorphy.

Plesiomorphies of Belovius/g1. In contrast to Iron/fg1, tergalius I [see Epeorus/fg2 (1)] is widened not so strongly, and tergalius VII has no longitudinal fold, only with smooth longitudinal concavity, which is not enough to allow tergalius to turn under abdomen (FIG.65:A).

Variable characters of Belovius/g1. Titillators are either well-developed, or vestigial, or lost.

Size. Fore wing length 10–15 mm.

Distribution. Amphipacific: Eastern Palaearctic, Oriental Region and Western Nearctic.

Belovius/g1 is divided into *Albertiron* and Belovius/g2.

2.2;1,2-2/2.4;3,2-2/3.1. Plesiomorphon **Albertiron/g(1)**

(Radulapalpata Rhithrogena/fg1 Epeorus/fg1 .../fg3 Belovius/g1 .../g2 *Albertiron*)

(Fig. 65:D)

Nomen hierarchicum: **Albertiron/g(1)** [g: *Albertiron* Kluge, SUBGEN. NOV., typus *Iron albertae* McDunnough 1924].

Plesiomorphies of *Albertiron* (in contrast to Belovius/g2). Tergalii II–VII are non-enlarged, without anal-proximal projection, anal rib [situated at a distance from anal margin – see Belovius/g1 (1)] is not curved (FIG.65:D). Larval patella-tibial suture is absent on all legs [see Epeorus/fg1 (8)].

Size. Fore wing length 10–11 mm.

Distribution. Nearctic (possibly, only Western).

Species composition of Albertiron/g(1). Probably several species (unknown exactly, because Nearctic species of Epeorus/fg3 are described insufficiently); one species is attributed here for certain – *albertae* McDunnough 1924 [*Iron*] (syn. subj.: *youngi* Traver 1935 [*Iron*]).

Material examined: *albertae* [*I.*]: ℧, ⊕̄.

***

2.2;1,2-2/2.4;3,2-2/3.2. **Belovius/g2**

(Radulapalpata Rhithrogena/fg1 Epeorus/fg1 .../fg3 Belovius/g1 .../g2)

(Figs 64:L–M; 65: A, E–F)

Nomen hierarchicum: **Belovius/g2** (sine *Albertiron*).
In circumscription matches:
— subgen. *Belovius*: Sinitshenkova 1981: 813.

References. Sinitshenkova 1981:℧*; – Kluge 1997d:℧*.

Autapomorphies of Belovius/g2.

(1) Tergalii of middle pairs are large, with wide rounded anal-proximal projection; on tergalii II–VI anal rib [situated at a distance from anal margin – see Belovius/g1 (1)], is arched, directed by its apex toward anal margin and supports the anal-proximal projection of tergalius (FIG.65:E) (unique apomorphy). On tergalius VII anal rib is also situated at a distance from anal margin, but is nearly straight. On tergalius I this rib is lost, as well as in many other Epeorus/fg3.

(2) Patella-tibial suture [initially lost in larva but retained on middle and hind legs of adult – see Epeorus/fg1 (8)] is restored on larval middle and hind legs. This condition coincides with plesiomorphic for other mayflies (see Index of characters [1.2.18]).

Size. Fore wing length 10–15 mm.

Distribution. Eastern Palaearctic and Oriental Region.

Species composition of Belovius/g2 (here [*E.*]= =[*Epeorus*] and [*B.*]=[*Belovius*]. 8 species are formally described – *carinatus* Braasch & Soldán 1984 [*E.*], *ermolenkoi* Tshernova 1981 [*E.(B.)*] (syn. subj.: *daedaleus* Sinitshenkova 1981 [*E.(B.)*]), *frolenkoi* Sinitshenkova 1981 [*E.(B.)*], *gornostajevi* Tshernova 1981 [*E.(B.)*], *latifolium* Ueno 1928 [*E.*] (syn.subj.: *L-nigrum* Matsumura 1931 [*Epeorus*]), *pellucida* Brodsky 1930 [*Cinygma*] (synn. subj.: *smirnovi* Tshernova 1978 [*Cinygmula*], *tshernovae* Braasch 1979 [*Epeorus*]), *rubeus* Tiunova 1991 [*E.(B.)*], *sinitshenkovae* Tshernova 1981 [*E.(B.)*] (syn. obj.: *ninae* Kluge 1995 [*E.(B.)*].

Material examined: *ermolenkoi* [*E.*]: ℧, O, ⊕; *frolenkoi* [*E.*]: ℧; *gornostajevi* [*E.*]: ℧–O–⊕/⊙; *latifolium* [*E.*]: ℧; *pellucida* [*C.*]: ℧–O–⊕/⊙; *rubeus* [*E.*]: ℧̄–Ō–⊕̄, *sinitshenkovae* [*E.*]: ⊕̄.

*** ***

2.2;1,2-2/2.4;3,2-2/4. **Iron/g1**

(Pentamerotarsata Radulapalpata Rhithrogena/fg1 Epeorus/fg1 .../fg3 Iron/g1)

(Figs 65:B, G–I)

Nomen hierarchicum: **Iron/g1** (incl. *Ironopsis*) [g: *Iron* Eaton 1883: Pl.24, typus *Epeorus longimanus* Eaton 1883: Pl.23 (monotypy)].
In circumscription matches:
— subgen. *Iron*: Kluge 1983b: 16 (Abstract); 1988: 309;
— gen. *Iron*: Sinitshenkova 1976b: 853.

Reference. Kluge 1997d: ℧'.

Character of unclear phylogenetic status.

(1) Tergalii [see Epeorus/fg3 (2)] form an adhesive disk, resembling sucking disk of Rhithrogena/fg3 [with this is connected the name "*Iron*" which means "a dissembler" (Eaton 1883–1888: 245)]: tergalius I [which has costal rib at a distance from costal margin – see Epeorus/fg2 (1)] is more or less widened anteriorly and turned under thorax, thus apices of left and right tergalii I are brought together or contiguous; tergalius VII has a sharp longitudinal fold, because of which it is bent ventrally under last abdominal segments (FIG.65:B). In contrast to Rhithrogena/fg3, the tergalial disc of Iron/g1 does not function as true sucker, because its margins are rigid and armed with hook-like setae [see Epeorus/fg2 (2)]. The anterior widening of tergalius I can be widened distally (like in Rhithrogena/fg3) or narrowed distally, sometimes it can be expressed as poorly as in Epeorus/fg4; the fold of tergalius I is always expressed. It can be assumed that the tergalial disk appeared as a convergention in several non-related groups of Epeorus/fg3 (in the same manner as it appeared independently in Rhithrogena/fg3), because no other apomorphies of Iron/g1 are known.

Plesiomorphy of Iron/g1. In contrast to Belovius/g1, tergalii II–VII have anal rib on anal margin.

Size. Fore wing length 7–20 mm.

Distribution. Holarctic.

Iron/g1 is divided into Iron/g2 and Ironopsis/g1.

2.2;1,2-2/2.4;3,2-2/4.1. Plesiomorphon **Iron/g2**

(Radulapalpata Rhithrogena/fg1 Epeorus/fg1 .../fg3 Iron/g1 .../g2)

Nomen hierarchicum: **Iron/g2** (sine *Ironopsis*).
In circumscription matches:
— subgen. *Iron*: Kluge 1997d: 206.

Reference. Kluge 1997d: ♉'.

Plesiomorphies of Iron/g2. In larva hairs on anterior margin of head shield [see Epeorus/fg2 (2)] are directed anteriorly (the same in Belovius/g1 and Epeorus/fg4, in contrast to Ironopsis/g1). Penis [see Pentamerotarsata (5)] has median titillators (in contrast to Ironopsis/g2 and Epeorus/fg4) and well-developed latero-dorsal spines.

Size. Fore wing length 7–13 mm.

Distribution. Holarctic.

Species composition of Iron/g2. About 20 species are formally described. **Asian species:** *aesculus* Imanishi 1934 [*Epeorus*] (synn. subj.: *grunini* Sinitshenkova 1982 [*Iron*], *levanidovae* Sinitshenkova 1982 [*Iron*], *koreanicus* Braasch & Soldán 1988 [*Iron*]), *alexandri* Kluge & Tiunova 1989 [*Epeorus (Iron)*], *curvatulus* Matsumura 1931 [*Epeorus*] (="*Iron* sp.?": Ueno 1928), *gultsha* Kustareva 1984 [*Iron*], *gurvitshi* Kustareva 1984 [*Iron*], *inaequalis* Braasch & Soldán 1980 [*Iron*], *intermedia* Brodsky 1930 [*Iron montanus*], *kirgisicus* Kustareva 1984 [*Iron*], *maculatus* Tshernova 1949 [*Iron*] (synn. subj.: *latericius* Sinitshenkova 1982 [*Iron*], *tshernovae* Sinitshenkova 1982 [*Iron*]), *minuta* Brodsky 1930 [*Iron montanus*], *montanus* Brodsky 1930 [*Iron*], *nigromaculata* Brodsky 1930 [*Iron montanus*], *pamirensis* Kustareva 1984 [*Iron*], *sinespinosus* Braasch 1978 [*Iron*], *siveci* Braasch 1980 [*Iron*], *tianshanica* Kustareva 1984 [*Iron*]. **North American species:** *fragilis* Morgan 1911 [*Iron*] (syn. subj.: *tenuis* Traver 1935 [*Iron*]), *longimanus* Eaton 1883 [*Epeorus*] (syn. subj.: *proprius* Traver 1935 [*Iron*]), *pleuralis* Banks 1910 [*Heptagenia (Epeorus)*] (synn.subj.: *confusus* Traver 1935 [*Iron*], *fraudator* Traver 1935 [*Iron*]).

Material examined: Asian species: group *aesculus*: *aesculus* [*E.*]: ♉–◎–⊕/⊙; *alexandri* [*E.*]: ♉–◎–⊕/⊙; *maculatus* [*I.*]: ♉–◎–⊕/⊙; **group *montanus*:** *montanus* [*I.*]: ♉–◎–⊕/⊙; *sinespinosus* [*I.*]: ♉–◎–⊕/⊙; *inaequalis* [*I.*]: ♉–◎–⊕/⊙; spp. (incl. *gultsha* [*I.*], *gurvitshi* [*I.*], *intermedia* [*I.*], *kirgisicus* [*I.*], *minuta* [*I.*], *nigromaculata* [*I.*], *pamirensis* [*I.*], *tianshanica* [*I.*]): ♉–◎–⊕/⊙. **North American species:** *longimanus* [*E.*]: ♉/♂; *fragilis* [*I.*]: ♉/♂.

***

2.2;1,2-2/2.4;3,2-2/4.2. **Ironopsis/g1**

(Radulapalpata Rhithrogena/fg1 Epeorus/fg1 .../fg3 Iron/g1 Ironopsis/g1)

(Fig. 65: G–I)

Nomen hierarchicum: **Ironopsis/g1** (incl. *Caucasiron*) [g: *Ironopsis* Traver 1935: 36, typus *Iron grandis* McDunnough 1924 (design. orig.)].

Autapomorphies of Ironopsis/g1.

(1) In larva long dense hairs of anterior margin of head shield [see Epeorus/fg2 (2)] are directed not anteriorly as in other Epeorus/fg2, but medially-dorsally, pressed to the head and one to another (FIG.65:I).

(2) Penis lacks latero-dorsal spines [see Pentamerotarsata (5)]. Non-unique apomorphy: these spines independently disappear in some other Radulapalpata, including some Epeorus/fg4.

Character of unclear phylogenetic status.

(3) On tergalius I anterior widening [see Iron/g1 (1)] is always widened distally (the same in some Iron/g2).

Size. Fore wing length 10–20 mm.

Distribution. Holarctic.

Ironopsis/g1 is divided into Ironopsis/g2 and *Caucasiron.*

2.2;1,2-2/2.4;3,2-2/4.2;1. **Ironopsis/g2**

(Rhithrogena/fg1 Epeorus/fg1 .../fg3 Iron/g1 Ironopsis/g1 .../g2)

(Fig. 65:H–I)

Nomen hierarchicum: **Ironopsis/g2** (sine *Caucasiron*).
In circumscription matches:
— subgen. *Ironopsis*: Edmunds & Jensen & Berner 1976:193;
— gen. *Ironopsis* Traver 1935: 36.

References. Traver 1935: ♉* ⊕*; – Edmunds & Jensen & Berner 1976: ♉* ⊕*; – Kluge 1997d: ♉*.

Character of unclear phylogenetic status.

(1) Titillators [see Pentamerotarsata (5)] are lost. Non-unique apomorphy (see Index of characters [2.3.15]); among Epeorus/fg3 the same in Epeorus/fg4.

Plesiomorphy of Ironopsis/g2. In contrast to *Caucasiron*, tergalii II–VII have no such sharp projection on costal margin (on its place only a shallow bulge can be present – FIG.65:H).

Size. Fore wing length 13–20 mm.

Distribution. Holarctic.

Species composition of Ironopsis/g2. Not less than 5 species. Two **European species:** *alpicola* Eaton 1871 [*Heptagenia*] (synn. subj.: *alpinus* Hubault 1927 [*Iron*], *steinmanni* Leger 1926 [*Iron*]), *yougoslavicus* Šamal 1935 [*Iron*] (syn. obj.: *jugoslavicus* Šamal 1939 [*Iron*]). One **Tien-Shan–Himalayan species:** *rheophilus* Brodsky 1930 [*Iron*]. Two **West Nearctic species:** *grandis* McDunnough 1924 [*Iron*], *permagnus* Traver 1935 [*Ironopsis*].

Material examined: *alpicola* [*H.*]: ʊ, ○; *permagnus* [*I.*]: ʊ; *rheophilus* [*I.*]: ʊ–○–⊕/⊙.

***

2.2;1,2-2/2.4;3,2-2/4.2;2. Caucasiron/g(1)

(Rhithrogena/fg1 Epeorus/fg1 .../fg3 Iron/g1 Ironopsis/g1 *Caucasiron*)

(Fig. 65:G)

Nomen hierarchicum: **Caucasiron/g(1)** [g: *Epeorus* subgen. *Caucasiron* Kluge 1997a: 233, typus *Cinygma caucasica* Tshernova 1938 (design. orig.)].
In circumscription matches:
— subgen. *Caucasiron* Kluge 1997a: 233.

References. Tshernova 1938 (*Cinygma caucasica*): ⊕'; – Sinitshenkova 1976 (*Iron*): ʊ' ⊕'; – Braasch 1978 (*Iron*): ʊ'; – 1979 (*Iron*): ʊ' ⊕'; – Braasch & Soldán 1979 (*Iron*): ʊ'; – Kluge 1997a: ʊ* ⊕*; – 1997d: ʊ*.

Autapomorphy of *Caucasiron*.

(1) Each tergalius II–VII on its costal rib [which is situated on costal margin, bears spines and setae – see Epeorus/fg3 (2)] has a sharp projection covered with spines and directed anteriorly-dorsally (FIG.65:G); this projection can be well-developed on all tergalii II–VII, or at least on anteriormost of them. Unique apomorphy.

Plesiomorphy of *Caucasiron*. Penis has a pair of well-developed median titillators [see Pentamerotarsata (5)] (in contrast to Ironopsis/g2, Epeorus/fg4 and some others).

Variable characters of *Caucasiron*. Imaginal and subimaginal claws are usually ephemeropteroid, but in *guttatus* [*Iron*] – pointed (see Variable characters of Heptagennota and Index of characters [2.2.85]).

Size. Fore wing length 10–20 mm.

Distribution. Greece, Caucasus and Tien-Shan–Himalayan Region.

Species composition of Caucasiron/g(1). 12 species are formally described. One **European species:** *insularis* Braasch 1983 [*Iron znojkoi*]. **Caucasian species:** *alpestris* Braasch 1979 [*Iron*], *caucasica* Tshernova 1938 [*Cinygma*], *fuscus* Sinitshenkova 1976 [*Iron*], *longimaculatus* Braasch 1980 [*Iron*], *magnus* Braasch 1978 [*Iron*], *nigripilosus* Sinitshenkova 1976 [*Iron*], *sinitshenkovae* Braasch & Zimmermann 1979 [*Iron*], *soldani* Braasch 1979 [*Iron*], *znojkoi* Tshernova 1938 [*Iron*]. **Tien-Shan–Himalayan species:** *guttatus* Braasch & Soldán 1979 [*Iron*], *paraguttatus* Braasch 1983 [*Iron*].

Material examined: *caucasica* [*C.*]: ʊ–○–⊕/⊙; *guttatus* [*I.*]: ʊ–○–⊕/⊙; *magnus* [*I.*]: ʊ–○–⊕/⊙; *znojkoi* [*I.*]: ʊ–○–⊕/⊙; spp. (Caucasus, incl. *alpestris* [*I.*], *fuscus* [*I.*], *nigripilosus* [*I.*], *sinitshenkovae* [*I.*], *soldani* [*I.*]): ʊ–○–⊕/⊙; sp. (Nepal): ʊ, ○.

*** *** ***

Epeorus/fg3 INCERTAE SEDIS

Descriptions of the following species are not sufficient enough to determine there exact systematic position. **Asian species:** *gilliesi* Braasch 1981 [*Epeorus*], *hiemalis* Imanishi 1934 [*Epeorus*], *ikanonis* Takahashi 1924 [*Epeorus*] (syn. subj.: *hanazononis* Matsumura 1931 [*Epeorus*]), *jacobi* Braasch 1978 [*Iron*], *kapurkripalanorum* Braasch 1983 [*Iron*], *L-nigrum* Matsumura 1931 [*Epeorus*], *martensi* Braasch 1981 [*Iron*], *martinus* Braasch & Soldán 1984 [*Iron*], *napaeus* Imanishi 1934 [*Epeorus*], *rhithralis* Braasch 1980 [*Epeorus*], *tiberius* Braasch & Soldán 1984 [*Epeorus*], *uenoi* Matsumura 1933 [*Iron*], *unispinosus* Braasch 1980 [*Epeorus*]. **North American species:** *deceptiva* McDunnough 1924 [*Cinygma*], *dulciana* McDunnough 1935 [*Iron*], *frisoni* Burks 1947 [*Iron*], *hespera* Banks 1924 [*Rhithrogena*], *lagunitas* Traver 1935 [*Iron*], *margarita* Edmunds & Allen 1964 [*Epeorus (Iron)*], *metlacensis* Traver 1965 [*Iron*], *namatus* Burks 1946 [*Iron*], *packeri* Allen & Cohen 1977 [*Epeorus (Iron)*], *punctatus* McDunnough 1925 [*Iron*], *rubidus* Traver 1933 [*Iron*], *sanctagabriel* Traver 1935 [*Iron*], *subpallidus* Traver 1937 [*Iron*], *suffusus* McDunnough 1925 [*Iron*].

*** *** *** ***

Radulapalpata INCERTAE SEDIS 1. Acanthomola/g(1)

Nomen hierarchicum: **Acanthomola/g(1)** [g: *Acanthomola* Whiting & Lehmkuhl 1987b: 410, typus *A. pubescens* Whiting & Lehmkuhl 1987 (design. orig.)].
In circumscription matches:
— gen. *Acanthomola* Whiting & Lehmkuhl 1987b: 410.
Monospecific taxon.

Reference. Whiting & Lehmkuhl 1987b: ℧.

Autapomorphies of *Acanthomola*.

(1) Mouth apparatus is adapted for carnivorism: mandibular mola lacks grater; maxilla without pectinate setae [see Radulapalpata (3)] (Whiting & Lehmkuhl 1987b: Figs 21 a–g).

Character of *Acanthomola* of unclear phylogenetic status.

(2) Larval paracercus is vestigial. Non-unique apomorphy (see Index of characters [1./3.64]); among Heptagennota the same in Epeorus/fg1 and Anepeorus/fg1 only.

Size. Larval body length 9 mm.

Distribution. Canada.

Species composition of Acanthomola/g(1). 1 species – *pubescens* Whiting & Lehmkuhl 1987 [*Acanthomola*].

Material examined: –.

***

Other Radulapalpata INCERTAE SEDIS

Radulapalpata INCERTAE SEDIS 2. Compsoneuria/g [g: *Compsoneuria* Eaton 1881: 23, typus *C. spectabilis* Eaton 1881 (design. orig.)].

References. Eaton 1883–1888: ⊕; – Ulmer 1939: ℧ ⊕; – Braasch & Soldán 1986: ⊕.

Distribution. Oriental Region.

Material examined: –.

Comment. Larva ascribed to the type species (Ulmer 1939) belongs to Heptagenia/f8=g7 (sine *Dacnogenia, Stenonema*); another larva ascribed to the genus *Compsoneuria* (Braasch & Soldán 1986) belongs to Ecdyonurus/fg1.

Radulapalpata INCERTAE SEDIS 3. Cinygmoides/g [g: *Cinygmoides* Matsumura 1931: 1474, typus *C. hekachii* Matsumura 1931 (monotypy)].

References. Matsumura 1931: ⊕.

Distribution. Japan.

Material examined: –.

Radulapalpata INCERTAE SEDIS 4. Rhithrogenella/g [g: *Rhithrogenella* Ulmer 1939: 575, typus *Rh. ornata* Ulmer 1939 (design. orig.)].

Reference. Ulmer 1939: ⊕.

Distribution. Oriental Region.

Material examined: –.

Radulapalpata INCERTAE SEDIS 5. Epeorella/g [g: *Epeorella* Ulmer 1939: 578, typus *E. borneonia* Ulmer 1939 (design. orig.)].

Reference. Ulmer 1939: ⊕.

Distribution. Oriental Region.

Material examined: –.

Radulapalpata INCERTAE SEDIS 6. † Miocenogenia/g [g: *Miocenogenia* Tshernova 1962a: 944, typus *M. gorbunovi* Tshernova 1962 (design. orig.)].

Reference. Tshernova 1962a: ℧.

Age. Miocene (Siberia).

Material examined (Paleontol. Inst.): *gorbunovi* [*M.*]: ℧.

Comment. Probably belongs to one of Recent taxa among Heptagenia/f6=g5.

Radulapalpata INCERTAE SEDIS 7. † Succinogenia/g [g: *Succinogenia* Demoulin 1965a: 151, typus *S. larssoni* Demoulin 1965 (design. orig.)].

Reference. Demoulin 1965a: ℧.

Age. Eocene (Baltic amber).

Material examined: –.

Other species of Radulapalpata INCERTAE SEDIS. Following species were described as adults only; their thorax structure is undescribed or described insufficiently. **Extinct species from Baltic amber:** *atypica* Demoulin 1968 [*Heptagenia*], *bachofeni* Demoulin 1968 [*Heptagenia*], *baltica* Demoulin 1968 [*Cinygma*], *gleissi* Demoulin 1968 [*Heptagenia*], *ligata* Demoulin 1968 [*Heptagenia*], *sepulta* Demoulin 1968 [*Rhithrogena*], *sinex* Demoulin 1968 [*Heptagenia*]. **Recent European species** (probably wrongly described and known under other names): *catalaunica* Navás 1917 [*Rhithrogena*], *diversus* Navás 1923 [*Ecdyonurus*], *excelsus* Navás 1927 [*Ecdyonurus*], *flavimanus* Klapálek 1905 [*Ecdyonurus*]. **Recent Asian species:** *annulifera* Walker 1860 [*Palingenia*], *asiaticum* Ulmer 1924 [*Cinygma*], *bengalensis* Ulmer 1920 [*Ecdyonurus*], *bifasciatus* Navás 1933 [*Ecdyonurus*], *cingulatus* Navás 1933 [*Thalerosphyrus*], *costata* Navás 1936 [*Heptagenia*], *curtus* Dubey 1971 [*Afronurus*], *davidi* Navás 1934 [*Ecdyonurus*], *dayongensis* Gui & Zhang 1992 [*Epeorus (Iron)*], *diehliana* Braasch & Soldán 1986 [*Rhithrogena*], *dominans* Navás 1936 [*Iron*], *erratus* Braasch 1981 [*Epeorus*], *famulans* Navás 1936 [*Iron*], *hazaraensis* Ali 1970 [*Heptagenia*], *hummeli* Ulmer 1936 [*Anepeorus*], *illotus* Navás 1933 [*Ecdyonurus*], *javanicus* Ulmer 1939 [*Afronurus*], *kerklotsi* Hsu 1933 [*Ecdyonurus*], *kiangsuensis* Puthz 1971 [*Ecdyonurus*] (syn. obj.: *hyalinus* Navás 1936 [*Ecdyonurus*] nom. praeocc.), *kurobensis* Imanishi 1935 [*Cinygma*], *lahaulensis* Kapur & Kripalani 1961 [*Epeorus*], *lieftincki* Ulmer 1939 [*Heptagenia*], *limbata* Navás 1936 [*Heptagenia*], *lobatus* Ulmer 1924 [*Ecdyonurus*], *magawana* Imanishi 1936 [*Rhithrogena*], *marhieus*

Dubey 1970 [*Thraulodes*], *melli* Ulmer 1925 [*Thalerosphyrus*], *minazuki* Imanishi 1936 [*Rhithrogena*], *minor* You & Gui 1995 [*Heptagenia*], *minor* Hsu 1936 [*Iron*], *nasuta* Ulmer 1939 [*Heptagenia*], *ngi* Hsu 1936 [*Heptagenia*], *nubila* Kimmins 1937 [*Heptagenia*], *pallescens* Navás 1936 [*Ecdyonurus*], *pectoralis* Matsumura 1931 [*Heptagenia*], *pichoni* Navás 1934 [*Ecdyonurus*], *pingguoyuanensis* You 1987 [*Epeorus*], *psi* Eaton 1885 [*Epeorus*], *radialis* Navás 1936 [*Ecdyonurus*], *sibuyanensis* Mol 1987 [*Afronurus*], *sinensis* Ulmer 1925 [*Iron*], *sinuosus* Navás 1933 [*Ecdyonurus*], *sangangensis* You & Su & Hsu 1982 [*Afronurus*], *solangensis* Dubey 1971 [*Afronurus*], *solangensis* Dubey 1971 [*Heptagenia*], *tateyamana* Imanishi 1936 [*Rhithrogena*], *torrida* Walker 1853 [*Baetis*], *xiasimaensis* You 1987 [*Cinygma*], *yadohglinensis* You 1987 [*Cinygmula*], *yadongxiasimaensis* You 1987 [*Afronurus*].

*** ***

Pentamerotarsata INCERTAE SEDIS 1
Anepeorus/fg1

Nomen hierarchicum: **Anepeorus/fg1** (incl. *Spinadis*) [f: Anepeorinae Edmunds 1962: 10; g: *Anepeorus* McDunnough 1925: 190, typus *A. rusticus* McDunnough 1925 (design. orig.)].

In circumscription matches:
— gen. *Anepeorus* McDunnough 1925: 190;
— gen. *Spinadis* Edmunds & Jensen 1974: 495;
— subfam. Spinadinae Edmunds & Jensen 1974: 495.

References. McDunnough 1925: ⊕; – Needham & Traver & Hsu 1935: ⊕; – Burks 1953: ⊕ (non Ꝺ); – Edmunds & Jensen 1974 (*Spinadis*): Ꝺ; – Edmunds & Jensen & Berner 1976 (*Anepeorus*): ⊕ (non Ꝺ); – ibid. (*Spinadis*): Ꝺ; – McCafferty & Provonsha 1984: ♁; – 1985: ⊕; – 1986 (*Spinadis*): Ꝺ.

Characters of unclear phylogenetic status.

(1) Mouth apparatus is adapted for carnivorism: mandible is shortened perpendicular to axis of articulation, mola lacks grater; superlinguae are vestigial; maxilla lacks apical-ventral setae, its biting margin is shortened and both dentisetae are enlarged (McCafferty & Provonsha 1986:Figs 8, 11, 14). Non-unique apomorphy (see Index of characters [1.1.12]).

(2) Maxillary palp is 2-segmented (as in majority of other Pentamerotarsata, in contrast to *Pseudiron*).

(3) Larval paracercus is vestigial. Non-unique apomorphy (see Index of characters [1.3.64]); among Pentamerotarsata the same in Epeorus/fg1 and *Acanthomola* only.

Plesiomorphy of Anepeorus/fg1. Paraglossae are not widened (in contrast to Radulapalpata).

Size. Fore wing length 7–10 mm.

Distribution. Nearctic.

Nominal taxon included. Anepeorus/fg1 includes **Spinadis/fg** [f: Spinadinae Edmunds & Jensen 1974: 495; g: *Spinadis* Edmunds & Jensen 1974: 495, typus *S. wallacei* Edmunds & Jensen 1974 (design. orig.; syn. subj. *Heptagenia simplex* Walsh 1863)].

Species composition of Anepeorus/fg1. 2 species – *simplex* Walsh 1863 [*Heptagenia*] (syn. subj.: *wallacei* Edmunds & Jensen 1974 [*Spinadis*]) and *rusticus* McDunnough 1925 [*Anepeorus*].

Material examined: –.

***

Other Pentamerotarsata INCERTAE SEDIS

Pentamerotarsata INCERTAE SEDIS 2. † **Amerogenia/g** [g: *Amerogenia* Sinitshenkova 2000c: 116, typus *A. macrops* Sinitshenkova 2000 (design. orig.)].

Reference. Sinitshenkova 2000c: ♁.

Age. Late Cretaceous (New Jersey amber).

Material examined: –.

*** *** *** ***

Chapter VII

Bidentiseta Furcatergaliae

Phylogeny of Furcatergaliae is rather intricate (see discussion below and Table 5); following taxa belong here:

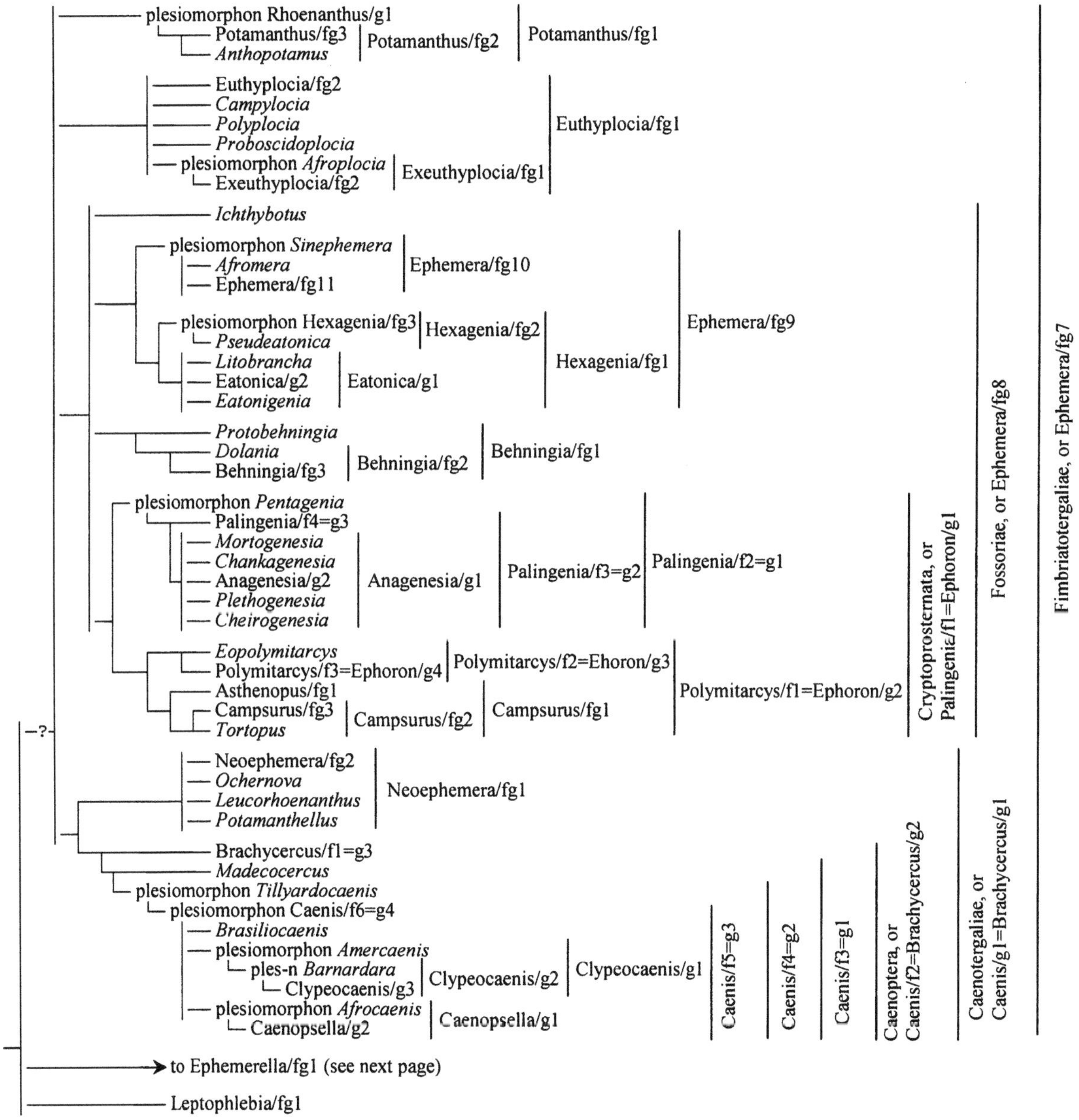

Phylogeny of Furcatergaliae-Ephemerella/fg1 (continuation of the scheme from previous page):

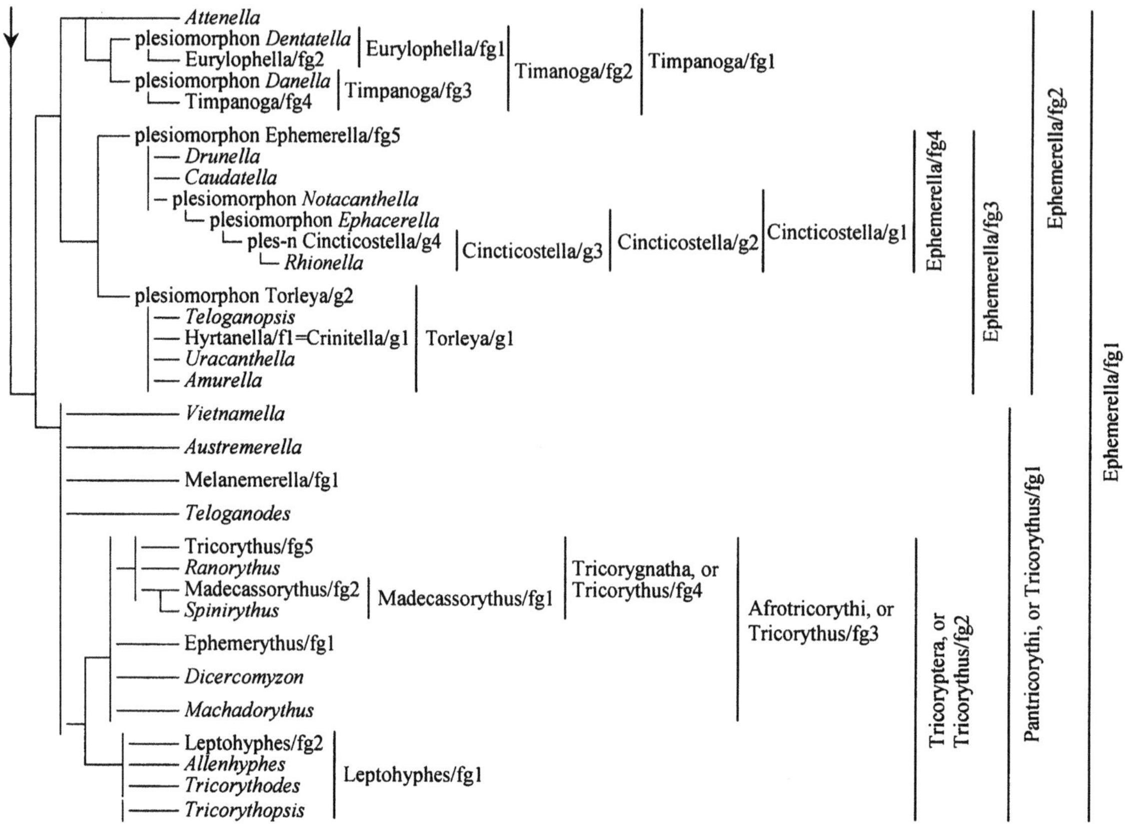

2.2;2. **Furcatergaliae**, or **Ephemera/fg6**

(Euephemeroptera Euplectoptera Anteritorna Bidentiseta Furcatergaliae)

(Figs 66–106)

Nomen hierarchicum: **Ephemera/fg6** (sine *Oligoneuria*; incl. *Ephemerella*, *Leptophlebia*).
Nomen circumscribens: **Furcatergaliae** Kluge 1998: 256.
In circumscription matches:
— infraordo Furcatergalia: Kluge 1993: 41; Kluge & al. 1995: 105 [non Furcatergalia Kluge 1989];
— Furcatergaliae Kluge 1998: 256;
— Furcatergaliae, or Ephemera/fg6: Kluge 2000: 251.

Reference. Kluge 1997b; – 1998: ⊕*.

Autapomorphies of Furcatergaliae.

(1) Pleura of prothorax are modified (FIG.106:A): external wall of pleurite is very low at least at area of dorsal coxal articulation, thus fore coxa articulates close to lateral margin of tergite; pleural suture is extremely shortened, indistinct or lost (in other mayflies, as in majority of other Pterygota, pleural suture stretches from dorsal coxal articulation to dorsal margin of pleurite); pleurite area dorsad of coxal articulation is transformed to a slender apodeme (**cryptopleurite**) deeply penetrating into the body and invisible from outside; being a continuation of the dorsal margin of pleurite, cryptopleurite is flattened dorsoventrally. This apodeme serves for attachment of pleural muscles – tergo-pleural, pleura-coxal and cranio-pleural ones. A part of pleurite posteriad of this apodeme is variable, in different representatives can be either low and vestigial, or rather high and well visible externally. This pleurite structure is characteristic for larva, and is retained in subimagoes and imagoes (whose prothorax sclerotization is less distinct than in larva).

This structure has no direct correspondence with larval specialization: among Furcatergaliae there are larvae with thorax either cylindrical (in many Fossoriae) or flattened (in many others), with fore legs non-specialized or highly specialized as burrowing (in Fossoriae).

Sometimes larval pleurites undergo further reduction and are immovably fused with tergite, so that fore legs are directly articulated with the tergite [see Campsurus/fg1 (3) below].

In most other mayflies pleurite is completely or largely exposed externally and retains distinct pleural suture (FIG.3:A). Pleurite structure similar to that of Furcatergaliae is found only in some Heptagennota (especially Heptagenia/f5=g4): their larvae, being flattened dorsoventrally, have prothorax flattened, with wide paranotal lobes, and pleurites are folded under them in a form of flat apodemes; sometimes these apodemes are narrowed, as in Furcatergaliae. Other mayfly larvae with distinctly flattened body or flattened ventral side either have normally developed external prothoracic pleurites (Discoglossata, Baetisca/f2=g1, Acanthametropus/ /fg1, *Ametropus* – FIG.25:B), or have pleurites hidden, but not in a form of so narrow apodemes (Prosopistoma/f1=g2).

(2) In imago and subimago on all legs 1st tarsal segment (fused with tibia) is strongly shortened, its length usually does not exceed its thickness (FIG.83:G). In other mayfly taxa 1st tarsal segment is either long, or more or less shortened in selected species only.

In some representatives of Furcatergaliae tarsal structure is secondarily changed (see Index of characters [2.2.84]): rarely 1st tarsal segment is secondarily separated from tibia (FIG.81:C), and/or secondarily elongate (FIG.86:G), and/or fused with the 2nd segment (FIG.98:A–B). Particularly, *Machadorythus* has unusually modified tarsi which have 4 segments only, with the proximalmost segments long and fused with tibia (FIG.99:F), like that of Tetramerotarsata. In representatives with non-functional legs, tarsi can have less number of segments, or lack segments, or are lost (see Index of characters [2.2.83]).

Characters of unclear phylogenetic status.

(3) On imaginal and subimaginal mesothorax anterior paracoxal suture is incomplete: on ventral side of episternum it becomes shallow and indistinct, or absent, does not reach basisternum (FIG.68:D–E). Non-unique apomorphy, the same in *Rallidens* and Heptagennota (see Index of characters [2.2.19]).

(4) On hind wing MA lost its bifurcation; thus hind wing has maximum two primary triads – RS and MP, with a simple MA between them (FIG. 68:B). Non-unique apomorphy (see Index of characters [2.2.67]). In Furcatergaliae, as well as in other groups of Ephemeroptera, hind wing can be strongly reduced or disappears completely, in this case this character loses its actuality. Rarely secondary triads can be added [see below, *Vietnamella* (4) and FIG.95:B].

(5) Tergalii are initially bifurcate or bilamellate, without marginal ribs (with this the name "Furcatergaliae" is connected). All or at least part of tergalii are bifurcate or bilamellate at least in primitive representatives of each subordinate taxon [see below, Fimbriatotergaliae (6), Ephemerella/fg1 (7) and Leptophlebia/fg1 (7)] (FIGS 67:A; 82:E; 91:B–F). In some Furcatergaliae one of the two lamellae is secondarily lost, thus tergalius becomes single-lamellate. Costal and anal ribs are either lost (in majority of representatives), or located at a distance from costal and anal margins correspondingly (in some Fimbriatotergaliae and Pantricorythi only – see Index of characters [1.3.27] and [1.3.28]); thus tergalial margins are never armed with ribs.

Besides Furcatergaliae, tergalii independently got bilamellate structure in some other mayfly groups: in some Metretopus/fg1 – because of ventral curvation of costal lobe; in Siphlonurus/fg2 and some others – because of ventral curvation of anal lobe; in Tasmanophlebia/fg1 and some others – because of dorsal curvation of anal lobe; in Branchitergaliae and some others – because of development of secondary ventral fibrillose portion (see Index of characters [1.3.24]). Origin of bilamellate tergalial structure in Furcatergaliae is unclear. Possibly, this is a synapomorphy with Branchitergaliae [see Bidentiseta (2)].

(6) Larva lost primary swimming siphlonuroid specialization: legs are unable to stretch posteriorly, and caudalii have no siphlonuroid setation; at the same time ability to make dorsoventral abdominal movements is retained; in some taxa larva retains long abdomen able to make effective undulate movements; in others larval abdomen is shortened.

When larva swims, its legs are either folded under the body by knees directed posteriorly, or pressed to the body in a special manner [see Fossoriae (1) below].

Larval caudalii never have primitive siphlonuroid structure with swimming setae on paracercus and median sides of cerci only: either cerci have swimming setae both on median and lateral sides, or swimming setae are lost. Among Furcatergaliae only Fossoriae retain difference between the secondary swimming setae on lateral sides of cerci and the primary swimming setae on median sides of cerci and lateral sides of paracercus; some of them also retain oblique boundaries of cercal segments. In other representatives swimming setae, if present, have equal structure and density on both lateral and median sides of cerci and lateral sides of paracercus. Such swimming setae are present in all Potamanthus/fg1, all Euthyplocia/fg1, some Caenotergaliae and some Ephemerella/fg1. In many representatives of Caenotergaliae, Ephemerella/fg1 and Leptophlebia/fg1 caudalii lack swimming setae at all (but can have whorls of long setae on segment boundaries).

Size. Fore wing length 2–40 mm.

Age and distribution. Late Jurassic (see Fossoriae INCERTAE SEDIS) — Recent; world-wide.

Classifications of Furcatergaliae. Besides the accepted here division of Furcatergaliae into Fimbriatotergaliae, Ephemerella/fg1 and Leptophlebia/fg1, the following classifications are possible (TABLE 5).

I. Taxon uniting Potamanthus/fg1 + Caenotergaliae + Ephemerella/fg1. In this case taxon Fimbriatotergaliae should not be recognized.

Potamanthus/fg1, Caenotergaliae and Ephemerella/fg1 have the following common apomorphies.

(1) Tergalius I is vestigial, in a form of cylindrical sclerotized stick (instead of initial lamellate), covered with thin standing setae; it arises not from posterior-lateral angle of segment, but from a peculiar prominent cylindrical **pedestal** usually located far anteriad of hind margin of the segment (FIGS 67:A; 82:B–C; 88:A; 91:G–H; 95:D–E). Unique apomorphy (see Index of characters [1.3.33]).

(2) In larva on dorsal (anterior) side of fore femur a transverse row of setae can be developed (FIG.89:H–J). Unique apomorphy (see Index of characters [1.2.29]).

(3) Larval mesonotum has a **collar** – transverse concave band on anterior margin which is separated from the rest part of mesonotum by a transverse crest (FIGS 67:B–C; 82:F; 88:B; 92:C–D) (see Index of characters [1.2.5]). The same crest separating the collar is present in some Euthyplocia/fg1 (distinct in *inaccessibile* [*E.*], nearly absent in the *Polyplocia* examined); some Fossoriae have a concave transverse setose area at the same place.

(4) Egg can have one or two polar caps; anchors of the same structure are present in Potamanthus/fg1 and many Ephemerella/fg1 (see Index of characters [3.4]–[3.6]).

(5) Venation of cubital field of fore wing [see Anteritorna (1)] can be similar or even identical in certain specimens of Potamanthus/fg1, Neoephemera/fg1, Ephemerella/fg2, *Vietnamella* and *Austremerella*: proximalmost vein arising from CuA to wing margin is bifurcate (x_1 and x_2 in FIGS 68:A; 81:A; 90:A; 95:A); more distally from CuA can arise one or several bifurcate or simple veins more [see Potamanthus/fg1 (6), Caenotergaliae (9) and Ephemerella/fg1 (15)]. Similar venation occurs in some other taxa (see Index of characters [2.2.51]).

II. Taxon uniting Caenoptera + Tricoryptera. In the past this taxon was recognized as a "section 7 of the genera – type of *Caenis*" (Eaton 1883–1888: 137) or a family Brachycercidae SENSU Lestage 1931a:52.

From the one hand, Caenoptera are attributed to Caenotergaliae on the base of unique apomorphies in structure of tergalii [see Caenotergaliae (1)]; from the other hand, Tricoryptera are attributed to Ephemerella/fg1 on the base of a number of unique autapomorphies including that in tergalial structure [see Ephemerella/fg1 (1)–(8)]; from this, an assumption can be made that Caenoptera and Tricoryptera can not have close relationship (TABLE 5). However, some characters in structure of imaginal pterothorax and wings, as well as in structure of larval caudalii, have deep similarity in Caenoptera and all or selected representatives of Tricoryptera. This similarity can not be explained as occasional coincidence or as adaptations for a similar mode of life. Both Caenoptera and Tricoryptera have small size, their adults can be shortly-moulting, actively flying, with mesothorax enlarged: In Caenoptera and Tricoryptera size ranges from 2 to 8–10 mm; representatives of other Caenotergaliae and Ephemerella/fg1 are usually larger, their size ranges from 5 to 20 mm (see TABLE 2). In Caenoptera proportion of mesonotum length / fore wing length is 0.4–0.5, in Tricoryptera – 0.3–0.4, in other mayflies it is usually less and ranges from 0.2 to 0.35.

Among Tricoryptera the most similar to Caenoptera in imaginal characters is *Tricorythodes* attributed to Leptohyphes/fg1.

II.a. Apomorphies common for Caenoptera and all Tricoryptera:

(1) Infrascutellum of mesothorax is interrupted medially, scutellum can be large with enlarged lateral impressions (FIGS 66:A, C; 89:B). The same in Teloganodes/fg1 (see Index of characters [2.2.17]).

(2) On fore wing vein CuP is not connected with CuA and looks as arising from AA; cubital field with bifurcate vein iCu [see Caenoptera (10) and Tricoryptera (1)] (FIGS 66:A, C).

II.b. Apomorphies common for Caenoptera and selected groups of Tricoryptera – Tricorygnatha and Leptohyphes/fg1:

(3) Sublateroscutum of mesothorax has a **transverse interscutal suture** (FIGS 66:A, C; 87; 89; 102; 104) (see Index of characters [2.2.10]). This secondary suture has a form of sharp transverse furrow, separating sublateroscutum (with areas of attachment of scuto-coxal muscles – S.CmA and S.CmP) from posterior scutal protuberance (with area of attachment of scuto-lateropostnotal muscle – S.LPNm). In other mayflies sublateroscutum is separated from posterior scutal protuberance by indistinct concavity only (FIG.66:B, D).

(4) Anterior paracoxal suture of mesothorax is transferred posteriorly, close to anterior margin of coxal conjunctiva, so that episternum is integral, consisting of anepisternum only (FIG.66:A, C). Non-unique apomorphy (see Index of characters [2.2.19]).

(5) On fore wing [see (2)] marginal intercalaries are absent (FIG.66:A, C). Non-unique character (see Index of characters [2.2.55]).

(6) In imago amphitornal margin of wing with setae as in subimago (FIG.66:A, C). Non-unique, but rare apomorphy (see Index of characters [2.2.27]).

(7) Caudalii (cerci and paracercus) of male larva are thickened at proximal part, in contrast to caudalii of female larva which have usual form (FIG.103:E–F) (see Index of characters [1.3.63]). Thickening of male larval caudalii is connected with their strong elongation in male imago: when larva is ready to moult to subimago, inside its thickened caudalius locates a crumpled anlage of adult caudalius. In other respects larval structure retains sharp difference from imaginal one. In other mayflies sexual dimorphism of caudalii is usually expressed in imagoes and subimagoes only, but not in larva; only in some Fossoriae, whose larval abdomen has imaginal features (such as developed genitals), larval caudalii have sexual dimorphism similar to that of Caenoptera and Tricoryptera.

(8) Adults are shortly-moulting but not short-living (see Index of characters [2]).

II.c. Apomorphies common for Caenoptera and *Tricorythodes* only (FIG.66:A, C):

(9) In imago and subimago scuto-lateropostnotal muscle (S.LPNm) is strongly enlarged; in connection with this, lateropostnotum is also strongly enlarged, fused with epimeron and reaches coxal conjunctiva (FIG.66:A, C; 87; 104). Unique apomorphy (see Index of characters [2.2.16]).

(10) Metathorax is strongly reduced (FIG.66:A, C; 87; 104); hind wings are lost in both sexes (the same in Tricorygnatha and some others); in female metathoracic wing muscles, including scuto-episternal muscle (S.ESm) and median tergal muscle (MTm), are lost (see Index of characters [2.2.26]). Unique apomorphy; in other mayflies, even if hind wings are completely lost, in metathorax S.ESm and MTm are retained in both sexes.

(11) Mesonotal suture is lost (FIG.66:A, C). In different mayflies mesonotal suture disappears by different ways (see Index of characters [2.2.8]). In *Tricorythodes* mesonotal suture initially was transverse [see Ephemerella/fg1 (9)] and than smoothed; about this testifies the fact that in different representatives of Tricoryptera mesonotal suture is expressed in different degree. Caenoptera have no traces of mesonotal suture; if proceed from the assumption about relationship of Caenoptera and Neoephemera/fg1, we should assume that in Caenoptera mesonotal suture disappeared because was stretched posteriorly at each side and fused with median and lateroparapsidal sutures [see Neoephemera/fg1 (5)]. In this case, absence of mesonotal suture in *Tricorythodes* and Caenoptera is not their common character.

(12) Female imaginal and subimaginal caudalii are strongly diminished [see Caenoptera (6) and *Tricorythodes* (6)].

III. Lanceolata = Potamanthus/fg1 + Euthyplocia/fg1 + Fossoriae + Leptophlebia/fg1. In this case the taxon Fimbriatotergaliae is not recognized. The taxon **Lanceolata** McCafferty 1991 united Potamanthus/fg1, Euthyplocia/fg1, Fossoriae and Leptophlebia/fg1 and was characterized by a single character – elongate double tergalii. Probably this is a plesiomorphy among Furcatergaliae [see Furcatergaliae (5)].

IV. Taxon uniting Caenotergaliae + Ephemerella/fg1. Some authors unite Caenotergaliae and Ephemerella/fg1 in a taxon called **Pannota** auct. (non Pannota McCafferty & Edmunds 1979). Actually Caenotergaliae and Ephemerella/fg1 have a number of common characters, but no one of them is characteristic for these two taxa only (TABLE 5).

Here we divide Furcatergaliae into Fimbriatotergaliae, Ephemerella/fg1 (p.293) and Leptophlebia/fg1 (p.349). Some fossil taxa have uncertain systematic position (p.351).

Table 5. The paradoxical ring: synapomorphies of subordinate taxa in Furcatergaliae.

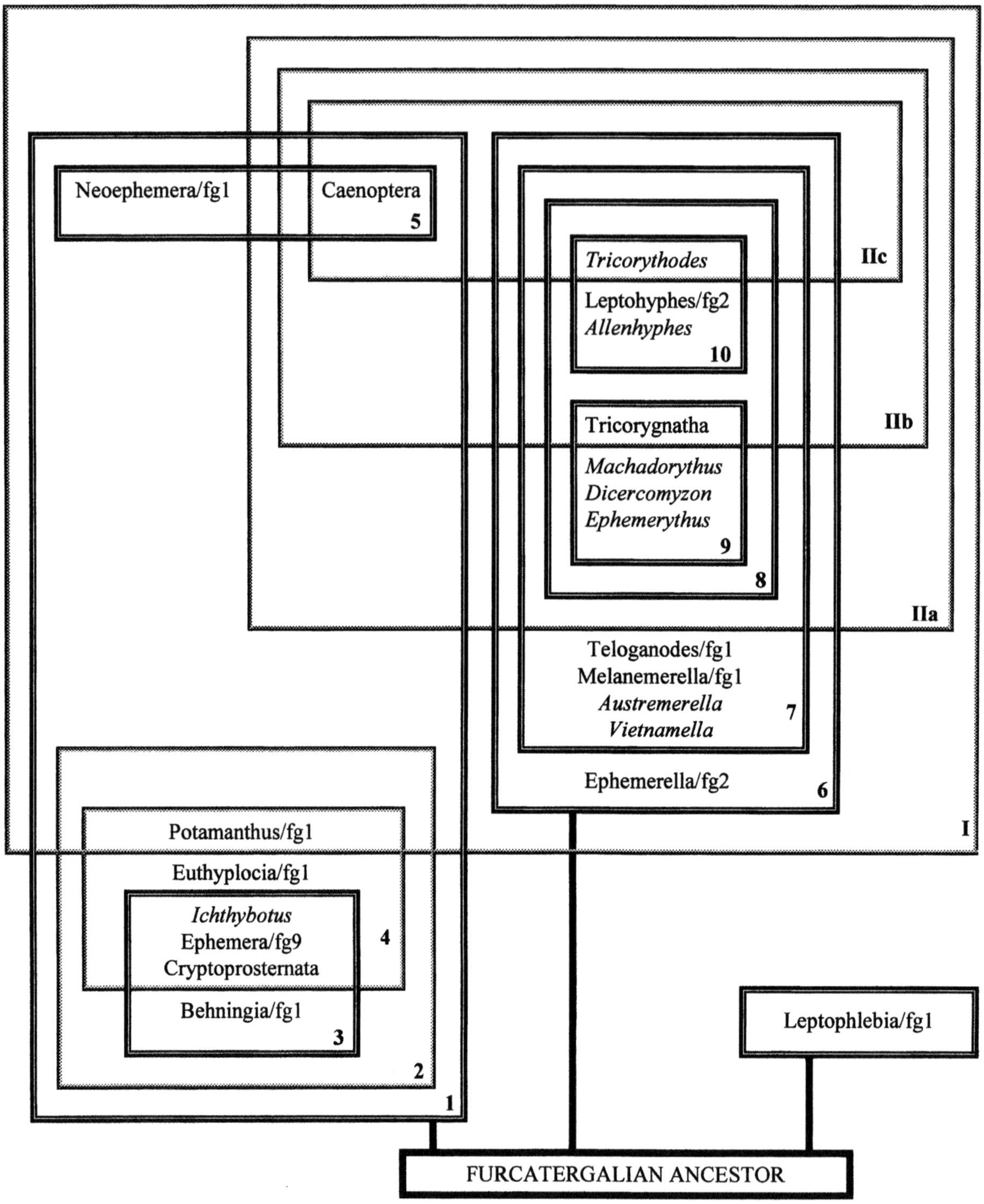

By the double frames there are outlined taxa accepted in this book, by single frames – taxa under discussion: **1** – Fimbriatotergaliae, **2** – Pinnatitergaliae, **3** – Fossoriae, **4** – Scapphodonta, **5** – Caenotergaliae, **6** – Ephemerella/fg1, **7** – Pantricorythi, **8** – Tricoryptera, **9** – Afrotricorythi, **10** – Leptohyphes/fg1;
I, IIa, IIb, IIc – groups with common apomorphies discussed in the division "Classifications of Furcatergaliae"

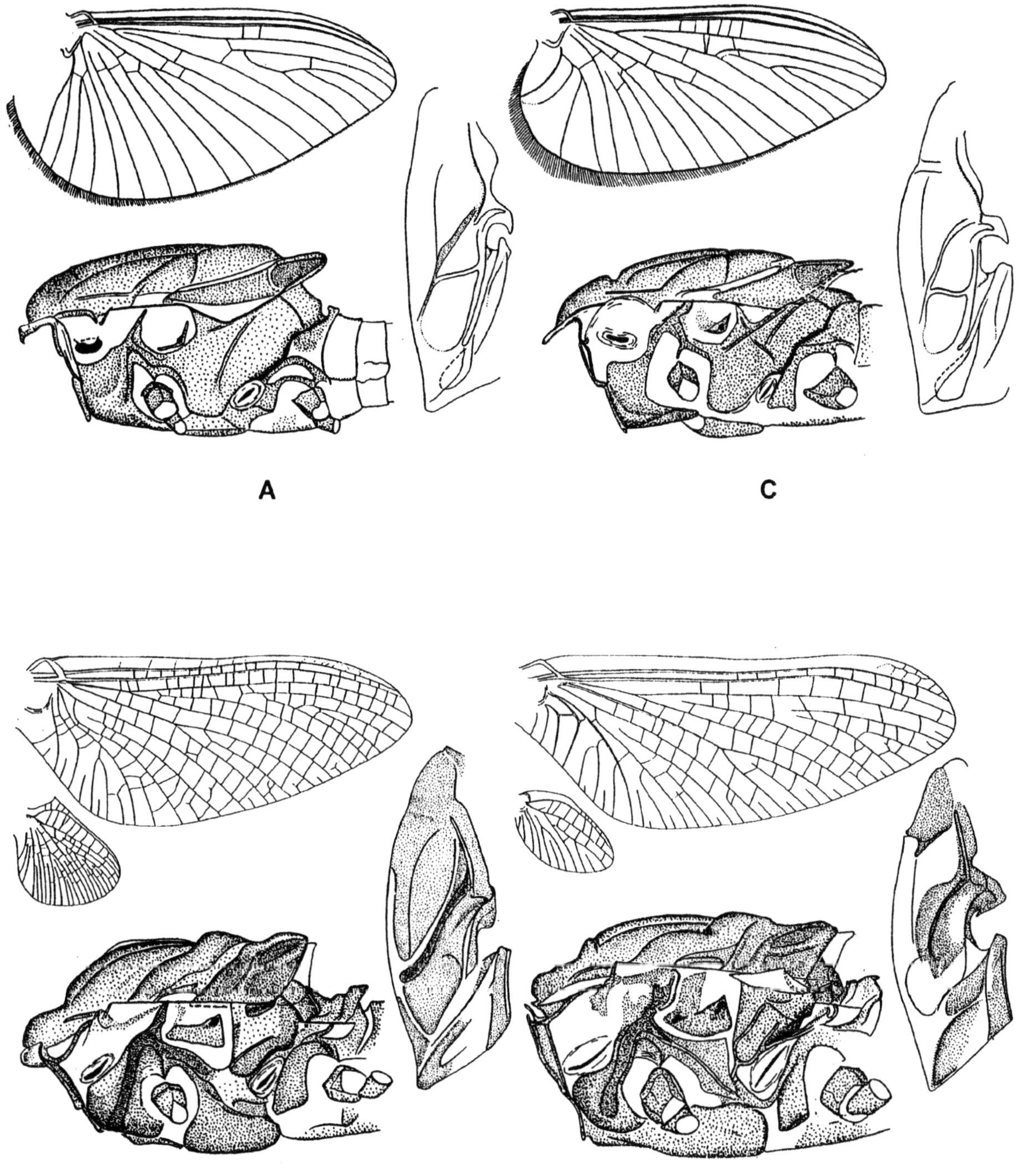

A C

B D

Figure 66. Similarity of Caenoptera and *Tricorythodes* compared with related taxa – Neoephemera/fg1 and Ephemerella/fg2. Wings, pterothorax and subimaginal exuviae of right half of mesonotum of male imago: **A–B** – Caenotergaliae: **A** – Caenoptera *macrura* [*Caenis*]; **B** – Neoephemera/fg1 spp.; **C–D** – Ephemerella/fg1: **C** – Tricorythodes/fg1 *cubensis* [*Tricorythodes*]; **D** – Ephemerella/fg2 spp.

2.2;2,1. **Fimbriatotergaliae**, or **Ephemera/fg7**

(Euplectoptera Anteritorna Bidentiseta Furcatergaliae Fimbriatotergaliae)

(Figs 67–88)

Nomen hierarchicum: **Ephemera/fg7** (sine *Ephemerella*, *Leptophlebia*; incl. *Potamanthus*, *Euthyplocia*, *Brachycercus*).
Nomen circumscribens: **Fimbriatotergaliae** Kluge, TAXON NOV.
In circumscription matches:
— superfam. Ephemeroidea: Riek 1970: 231.

Reference: Riek 1970: ♁* ⊕*.

Notes. Here are attributed such different mayflies as Fossoriae and Caenoptera; they have a single common feature [see (1)] that is not enough to prove their relationship; however, the holophyly of Fimbriatotergaliae is proven by synapomorphies of Caenoptera with Neoephemera/fg1 [see below, Autapomorphies of Caenotergaliae] and synapomorphies of Neoephemera/fg1 with Fossoriae, Potamanthus/fg1 and Euthyplocia/fg1 listed here [see (2)–(6)]. Imago of Caenoptera, being sharply different from other Fimbriatotergaliae, has hardly explainable similarity with certain Ephemerella/fg1 (see Classifications of Furcatergaliae II). Most probably, the characters (8) and (9) are also autapomorphies of Fimbriatotergaliae, but they are never present in Caenotergaliae, thus can not be used to argue the holophyly of Fimbriatotergaliae.

Autapomorphies of Fimbriatotergaliae.

(1) Larval pronotum has a **collar** – a concave band at anterior margin, separated from the rest part of pronotum by a transverse crest; laterally this crest is bent anteriad and terminates near anterior-lateral angle of the pronotum. At this point margin of pronotum can be simple (in some Potamanthus/fg1 and Caenotergaliae – FIGS 67:B–C; 82:F; 88:A–B) or produced to a denticle (in some Potamanthus/fg1, Euthyplocia/fg1 and Fossoriae – FIGS 69:G; 70:C); in Behningia/fg1 this denticle is transformed to a setose projection (FIG.73:A, E). Possibly, the pronotal collar is homologous with the area separated by V-shaped impression [see Ephemerella/fg1 (4) below and FIG.92:C–D].

(2) Adult mesonotal suture is strongly curved posteriorly by sides of median line; both its halves can be stretched posteriorly either moderately (FIGS 68:F; 71:E), or strongly (FIG.71:E); medially they can approximate with median suture, and laterally either approximate with lateroparapsidal sutures (FIG.83:F), or disappear (FIGS 78:D–F; 87:G–H). Non-unique apomorphy: the same in Leptophlebia/fg1 (FIG.106:D) and some others, but not in Ephemerella/fg1 (see Index of characters [2.2.8]). Extinct taxon *Baba* has the same wing venation as Fimbriatotergaliae [see (5)], but its mesonotal suture has plesiomorphic transverse shape (see below, Furcatergaliae INCERTAE SEDIS).

In more primitive representatives of Fimbriatotergaliae – Potamanthus/fg1, *Ichthybotus*, most Ephemera/fg9 (except for Afromera/g1), *Pentagenia* among Palingenia/f2=g1 and Neoephemera/fg1 among Caenotergaliae – mesonotal suture is well visible in subimago: subimaginal cuticle of medioscutum and submedioscutum anteriad of mesonotal suture is covered with microtrichia, sclerotized and usually more or less pigmented, and cuticle posteriad of mesonotal suture is smooth and non-pigmented. In these representatives lateral pigmented area of subimaginal mesonotum occupies the whole lateroparapsidal suture and area laterad and behind it (i. e. sublateroscutum, lateroscutum and posterior scutal protuberance); thus subimaginal mesonotal pigmentation has a characteristic pattern (FIGS 68:F; 71:E; 83:F). In imago mesonotal suture can be either also well-expressed (in Neoephemera/ /fg1 – FIG.83:E), or poorly visible, as the whole mesonotal cuticle is smooth and can be evenly sclerotized and pigmented.

In various shortly-moulting representatives with modified adult structure – Polymitarcys/f1=Ephoron/g2, Palingenia/f3=g2 among Palingenia/f2=g1 and Caenoptera among Caenotergaliae – subimaginal mesonotal pigmentation is lost, but mesonotal suture is well-expressed in imago: imaginal cuticle behind-mediad of mesonotal suture is weakly sclerotized and non-pigmented, forming a **medionotal membrane** (FIG.78:D). Median portions of left and right halves of the mesonotal suture can stretch longitudinally parallel to the median suture, in this case medionotal membrane is opened posteriorly (FIG.78:E); in some taxa (Caenoptera and selected Polymitarcys/f1=Ephoron/g2) these portions convergent posteriorly, and medionotal membrane is closed (FIGS 78:F; 87:G). Imaginal cuticular pigmentation can be masked by colour of internal tissues, in this case outline of medionotal membrane is visible only after treating by alkali; but in all Caenoptera and some others it is distinctly visible on intact specimens.

(3) Lateroscutal suture stretches from suralare posteromedially, being nearly straight; its posterior end is not curved laterally. This character is expressed in Potamanthus/fg1 (FIG.68:F), Euthyplocia/fg1, Fossoriae (FIGS 71:E–F; 74:A; 78:E–F) and Neoephemera/fg1 (FIG.83:F), being especially well visible in subimago, but is not expressed in Caenoptera whose mesonotum is highly modified (FIG.87: H). In other mayflies posterior end of lateral scutal suture is curved laterally (FIGS 6; 90:D; 106: D–E).

(4) Subimaginal lateropostnotal crest begins from anterior margin of postsubalar sclerite and stretches in ventral direction by margin of subalar conjunctiva (FIGS 71:D; 83:C). This is visible on representatives whose subimaginal cuticle has distinctly pigmented sclerites (Potamanthus/fg1, *Ichthybotus*, Ephemera/fg9, *Pentagenia*, Neoephemera/fg1), but it is non-expressed in representatives with colourless cuticle (Palingenia/f3=g2 and Polymitarcys/f1=Ephoron/g2). In Caenoptera lateropostnotum is strongly modified and lacks this character (FIG.87:A) [see Caenoptera (1) and (8) below]. In majority of other mayflies lateropostnotal crest begins from postsubalar sclerite at a distance from its anterior margin and passes at a distance from margin of subalar conjunctiva, thus subimaginal sclerotization of lateropostnotum has an axe-like shape (FIGS 5:D; 90:C).

(5) In proximal part of fore wing MP_2 and CuA are curved and divergent from MP_1 (FIGS 68, 69, 71, 75, 78–81). This curvation is more or less expressed in all Fimbriatotergaliae except for Caenoptera, where it is completely absent (FIG.86:F). In place of curvation MP_2 and CuA can be brought together, sometimes they are fused at some distance, and in this case MP_2 looks as originating from CuA, but not from MP_1 (FIGS 75:A; 78:A).

On hind wing MP_2 either retains straight shape, or becomes similar to that of fore wing – basally curved and approximated to CuA or originates from CuA (FIG.79:B).

The origin of MP_2 (from MP_1 or CuA) on fore and hind wings was erroneously regarded to be characteristic for certain supra-species taxa; however this character varies individually (Lestage 1922:Figs 1A–C).

The same vein curvation as in Fimbriatotergaliae, is present in extinct taxon *Baba* which differs from true Fimbriatotergaliae by primitive mesonotum structure (see Fimbriatotergaliae INCERTAE SEDIS).

(6) Tergalii II–VII [bilamellate, without marginal ribs – see Furcatergaliae (5)] have numerous marginal trachea-bearing processes. In Potamanthus/fg1, Euthyplocia/fg1 and Fossoriae (formerly united in a taxon Pinnatitergaliae) tergalii have a uniform structure (FIGS 67:A; 70:C; Eaton 1883–1888:Pl.28–31): each tergalius II–VII has both lamellae well-developed and elongate (length not less than 4 times exceeds its width, if not take into account marginal processes), with numerous (several dozens) non-branched trachea-bearing processes on both sides. Only in *lankensis* [*Ephemera*] tergalius VII is non-bifurcate and has processes on one side only. Ribs are usually absent, but sometimes one or two ribs can be present by sides of central trachea as poorly sclerotized semi-cylindrical convexities (especially thick in Palingenia/f3=g2); possibly, these ribs are secondary.

In Caenotergaliae tergalial structure is strongly modified [see Caenotergaliae (1) below]; primitively there are also two lamellae with numerous marginal trachea-bearing processes (FIG.82:E), but in Caenoptera ventral lamella is lost.

Tergalius I always strongly differs from others: in Behningia/fg1 it is non-bifurcate (FIG.73:A, E); in other Fossoriae and Euthyplocia/fg1 it is diminished and lacks processes, being bifurcate (FIG.70:C) or non-bifurcate (see Index of characters [1.3.33]), but retaining normal attachment and musculature; in Potamanthus/fg1 and Caenotergaliae it is transformed to a stick-like vestige on a pedestal at anterior part of the segment (FIG.67:A), which has the same structure as in Ephemerella/fg1, that is hardly explainable (see Classifications of Furcatergaliae I).

(7) Chromosome sex determination is XX in female and X0 in male (non-unique character). According to Soldán & Puthz (2000), such sex determination is found in three species of Ephemera/fg10 – *vulgata* [*E.*], *danica* [*E.*] and *lineata* [*E.*], one species of Potamanthus/fg3 – *lutea* [*E.*], and two species of Caenis/f3=g1 – *horaria* [*E.*] and *macrura* [*C.*], while chromosomes of other Fimbriatotergaliae are not examined. In all other mayfly species examined (about 20 species), sex determination is XX in female and XY in male.

Probable autapomorphies of Fimbriatotergaliae, absent in all Caenotergaliae.

(8) Mandible bears a **tusk** – a process on outer (anterior, or dorsal) side of mandible laterad of incisor. Tusk can be much longer than mandible (in majority of Potamanthus/fg1, Euthyplocia/fg1 and Fossoriae – FIGS 67:E; 69; 70; 75; 76; 79), or in a form of small denticle (in some Potamanthus/fg1 – FIG.67:D), or reduced to a thin vestige (in *Afromera*). Among Fimbriatotergaliae only Behningia/fg1 and Caenotergaliae completely lack tusks (see below, Classifications I and II).

Among other mayflies, tusks occur only in one

Australian species of Leptophlebia/fg1 – *doantrange* Campbell 1993 [*Kalbaybaria*]. In another species of Leptophlebia/fg1 – Paraleptophlebia/g1 *packii* Needham 1927 [*Leptophlebia*] – mandibular incisor is strongly enlarged and can be erroneously taken for a tusk.

(9) On larval hind leg (but not on other legs) inner side of femur and inner side of tibia (patella and/or telotibia) has dense protruding short stout pointed setae (FIG.70:D). Unique character. Among Fimbriatotergaliae these setae are present in Potamanthus/fg1, Euthyplocia/fg1 and most Fossoriae, but absent in Palingenia/f2=g1, Behningia/fg1, and all Caenotergaliae.

(10) Larval abdominal segments have characteristic shape, with lateral margins convex, rounded, and not produced to posterolateral spines (FIG.70:C). Among Fimbriatotergaliae such abdominal shape is present in Potamanthus/fg1, Euthyplocia/fg1 and most Fossoriae, but not in Behningia/fg1 and Caenotergaliae which can have well-developed posterolateral spines on segment IX and some previous ones (FIGS 73:A–B; 88:A) (as in many other mayflies).

Character of unclear phylogenetic status.

(11) Maxilla lacks apical-ventral row of setae (in contrast to many Leptophlebia/fg1, Radulapalpata and Tridentiseta – see Index of characters [1.1.31]).

Plesiomorphies of Fimbriatotergaliae. In most taxa attributed here larval claws have no denticles: subapical denticles or two rows are never present (in contrast to many Ephemerella/fg1 and Leptophlebia/fg1); one row is present in some Caenis/f4=g2 and on specialized fore legs of Asthenopus/fg1 only. Larval paracercus has the same length as cerci or only slightly shorter (adult paracercus can be developed or vestigial – see Index of characters [2.3.22]).

Size. Fore wing length 5–40 mm.

Age and distribution. Late Jurassic (see Fossoriae INCERTAE SEDIS) — Recent; world-wide.

Classifications of Fimbriatotergaliae. Besides the taxa accepted here, following taxa were suggested:

I. **Pinnatitergaliae** Kluge 2000

In circumscription matches:
— superfam. Ephemeroidea: Edmunds&Traver 1954a:239;
— subordo Ephemeroidea: Lestage 1924a: 319;
— Pinnatitergaliae, or Ephemera/fg7: Kluge 2000: 251.

This taxon is widely accepted and characterized by uniform structure of tergalii II–VII [bilamellate with marginal processes – see Fimbriatotergaliae (6)]: no one pair is reduced, each lamella is elongate, with marginal processes numerous and non-branched. In other mayflies tergalii can be bifurcate and elongate, but without processes (majority of Leptophlebia/fg1), or with less number of processes (Habrophlebia/fg2 and some other Leptophlebia/ /fg1), or bifurcate and with numerous non-branched processes, but shorter than in Pinnatitergaliae (*Thraulus* and some other Leptophlebia/fg1). Thus, structure of tergalii II–VII is a reliable diagnostic character of Pinnatitergaliae. Besides tergalial structure, larvae of Pinnatitergaliae are characterized by presence of dense swimming setae on caudalii [including lateral sides of cerci – see Furcatergaliae (6)].

However, both characters are plesiomorphies in comparison with Caenotergaliae. Other characters of Pinnatitergaliae are either synapomorphic with Caenotergaliae [see Fimbriatotergaliae (1)–(7), or present not in all representatives [see ibid., (8)–(9)]. Thus, most probably Pinnatitergaliae is a paraphyletic taxon ancestral for Caenotergaliae.

II. **Scapphodonta** McCafferty

In circumscription matches:
— superfam. Ephemeroidea: McCafferty 1991: 349;
— infraordo Scapphodonta McCafferty (Internet publication).

This taxon unites all Pinnatitergaliae (see I) except for Behningia/fg1 – i. e. Potamanthus/fg2, Euthyplocia/fg1, *Ichthybotus*, Ephemera/fg9 and Cryptoprosternata. An assumption about holophyly of Scapphodonta contradicts the assumption about holophyly of Fossoriae. The following characters could be taken into account as possible autapomorphies of Scapphodonta.

(1) Presence of tusks [see Fimbriatotergaliae (8)].

(2) Lateral margins of larval abdominal segments are rounded, without posterolateral spines [see Fimbriatotergaliae (10)].

(3) Non-pectinate tergalius I [see Fimbriatotergaliae (6)].

However, Behningia/fg1 has apomorphies of Fossoriae (see below); larvae of Behningia/fg1 are deeply specialized, that allows to assume that their mandibular tusks disappeared secondarily. Structure of tergalius I is not uniform among Scapphodonta: in this respect Potamanthus/fg1 is similar to Caenotergaliae and principally differs from other Fimbriatotergaliae (see above, Classifications of Furcatergaliae I). Most probably, the taxon Scapphodonta is paraphyletic.

Here Fimbriatotergaliae are divided into Potamanthus/fg1, Euthyplocia/fg1 (p.226), Fossoriae (p.232) and Caenotergaliae (p.269); some extinct Mesozoic taxa have uncertain systematic position (p.292).

2.2;2,1-1. **Potamanthus/fg1**

(Anteritorna Bidentiseta Furcatergaliae Fimbriatotergaliae Potamanthus/fg1)

(Figs 67–68)

Nomen hierarchicum: **Potamanthus/fg1** (incl. *Rhoenanthus*) [f: Potamanthines Albarda in Selys-Longchamps 1888: 148 (Peters & Hubbard 1977); g: *Potamanthus* Pictet 1843: 197, typus *Ephemera lutea* Linnaeus 1764 (design. Eaton 1868b: 86); syn.subj.: ***Eucharidis*** Joly & Joly 1877: 314, typus *E. reaumuri* Joly & Joly 1877 (monotypy) (synn. subj.: *lutea* [*E.*] = *reaumuri* [*E.*])].

In circumscription matches:

— "series I of group II:" Eaton 1883–1888: 77;
— "section 4 of genera": Eaton 1883–1888: 78;
— tribus Potamanthini: Lameere 1917: 68;
— subfam. Potamanthinae: Jacobson & Bianchi 1905: 873;
— fam. Potamanthidae: Klapálek 1909: 9;
— Potamanthus/fg1: Kluge 2000: 251.

References. Eaton 1883–1888: ♉* ⊕*; – Edmunds & Allen & Peters 1963: ♉*; – Tshernova 1970: ♉* ⊕*; – Koss & Edmunds 1974: ⊙; – Bae & McCafferty 1991: ⊙* ♉* ⊕*.

Autapomorphy of Potamanthus/fg1.

(1) On fore wing AA forms a fork which is either symmetric, or has an anterior branch arising forward from the main stem; no one branch arises backward from the main stem (FIG.68:A). In other mayflies AA is either non-branched or has one or several branches arising backward from the main stem (FIGS 71:A; 81:A).

Apomorphies of Potamanthus/fg1 common with Caenotergaliae and Ephemerella/fg1 (see Classifications of Furcatergaliae I).

(2) In larva [which has collar on pronotum – see Fimbriatotergaliae (1)] mesonotum also has a **collar** – a concave band at anterior margin, separated from the rest part of mesonotum by a transverse crest (FIG.67:B–C). The same in Euthyplocia/fg1 and Caenotergaliae; the same mesonotal collar in Ephemerella/fg1.

(3) Dorsal (initially anterior) surface of larval fore femur can bear a transverse row of setae (as in FIG.89:H): regular row is present in some Rhoenanthus/g1, vestigial row – in selected specimens of Potamanthus/fg3 *lutea* [*E.*]; in other Potamanthus/ /fg2 this row is absent. The same in Caenotergaliae and Ephemerella/fg1 only.

(4) Tergalius I [different from others – see Fimbriatotergaliae (6)] is stick-like, setose, attached on a prominent cylindrical pedestal arisen from anterior part of the abdominal segment I close to metathorax (the same in Caenotergaliae and Ephemerella/fg1 only); pedestal is sclerotized (the same in Caenotergaliae); tergalius itself is subdivided to 2 segments (FIG.67:A) (the same occurs in Neoephemera/fg1 – FIG.82:B).

(5) Egg with 2 polar caps and several (8–12) anchors; each anchor consists of a terminal knob and a skein of threads which surround the knob in a form of regular ring (Bae & McCafferty 1991:Figs 76–79). Similar caps and anchors are present in Ephemerella/ /fg1, and caps – also in some Caenotergaliae.

(6) In cubital field of fore wing [see Anteritorna (1)] several (2–5) veins go from CuA to basitornal and tornoapical margins; they are usually bifurcate (FIG.68:A: x, y), rarely simple. The same in Caenotergaliae and Ephemerella/fg1 (which initially have 1–3 bifurcate veins), as well as in selected specimens of *Ichthybotus*, Ephemera/fg9, and *Pentagenia* (which have 3–6 bifurcate or simple veins) and some others (see Index of characters [2.2.51]).

Non-unique characters of unclear phylogenetic status.

(7) Mandibular tusks [see Fimbriatotergaliae (8)] are curved medially (FIG.67:D–E). The same in Euthyplocia/fg1, *Ichthybotus* and Polymitarcys/f1= =Ephoron/g2; probably symplesiomorphy.

(8) Maxilla [see Fimbriatotergaliae (10)] on apical-ventral side close to base of canines has a field of densely and regularly situated long setae (FIG.68:C; Elpers & Tomka 1994:Figs 4a–d). The same in Euthyplocia/fg1 and Ephemerella/fg1, and a similar field – in Leptophlebia/fg1 (see Index of characters [1.1.32]).

(9) In imago and subimago proximal part of gonostylus is integral, boundary between 1st and 2nd segments being non-expressed (the same in Euthyplocia/fg1 and some others). Both distal segments are always retained, thus gonostylus is 3-segmented.

Plesiomorphies of Potamanthus/fg1. In larva: Maxilla with 3 canines and 2 dentisetae [see Bidentiseta (1)] (FIG.68:C). Maxillary palp is 3-segmented; labial palp is 3-segmented (in contrast to Cryptoprosternata in Fossoriae). Larval (and adult) patella-tibial suture is developed on middle and hind legs only. Larva retains long abdomen able to make undulate swimming movements; caudalii with dense secondary swimming setae equally developed on lateral and median sides of cerci and lateral sides of paracercus [see Furcatergaliae (6)]. All tergalii I–VII [modified – see Fimbriatotergaliae (6)] are present; tergalii II–VII are spread by sides of abdomen, not bent dorsally (the same in Euthyplocia/fg1, in contrast to Fossoriae and Caenotergaliae).

In imago and subimago: Mesonotal suture is moderately stretched posteriorly, and subimaginal mesonotal cuticle has pigmented areas of characteristic shape [see Fimbriatotergaliae (2)] (FIG.68:F). Furcasternal protuberances are contiguous (FIG.68: D) (in contrast to Campsurus/fg1, Caenotergaliae and some others – see Index of characters [2.2.23]). Hind wing is well-developed, as long as 0.35–0.4 of fore wing length (FIG.68:B). Imaginal and subimaginal claws are ephemeropteroid (exception can be made by male imaginal fore legs only – see below).

Variable characters of Potamanthus/fg1. Eyes of male from large (as in majority of Ephemeroptera) to small as in female. In male imago on fore legs claws can be ephemeropteroid (in Rhoenanthus/g1 *speciosus* [*Rh.*]), or both claws blunt (while in subimago they are always ephemeropteroid).

Size. Fore wing length 7–19 mm.

Distribution. Holarctic and Oriental Region; dominate in Oriental Region.

Systematic position of Potamanthus/fg1. About possibility to unite Potamanthus/fg1 with Caenotergaliae and Ephemerella/fg1 – see above, Classifications of Furcatergaliae I.

Potamanthus/fg1 is divided into Rhoenanthus/g1 and Potamanthus/fg2.

2.2;2,1-1/1. Plesiomorphon **Rhoenanthus/g1**

(Bidentiseta Furcatergaliae Fimbriatotergaliae Potamanthus/fg1 Rhoenanthus/g1)

(Fig. 67:E)

Nomen hierarchicum: **Rhoenanthus/g1** [g: *Rhoenanthus* Eaton 1881: 192, typus *Rh. speciosus* Eaton 1881 (design. orig.)].

References. Eaton 1883–1888 (*Rhoenanthus*): ⊕'; – Ulmer 1920a (*Potamanthodes*): ⊕'; – Lestage 1931b: ⊕'; – You 1984 (*Potamanthodes*): ⊕'; – Yoon & Bae 1985 (*Potamanthodes*): ℧' ⊕'; – Bae & McCafferty & Edmunds 1990 (*Stygifloris*): ℧' ⊕'; – Bae & McCafferty 1991: ℧' ⊕'.

Plesiomorphies and variable characters of Rhoenanthus/g1. Larvae of some species (*speciosus* [*Rhoenanthus*], *obscurus* [*Rh.*], *coreanus* [*Potamanthus*] and *youi* [*Neopotamanthus*]) have long tusks [see Potamanthus/fg1 (7)] (FIG.67:E) and elongated fore legs without transverse row of setae on femur (as in *Anthopotamus*). Larvae of other species (*formosus* [*Potamanthus*], *sabahensis* [*Stygifloris*] and some others) have short tusks (as in Potamanthus/fg3 or somewhat longer) and a distinctly expressed transverse row of setae on fore femur [see Potamanthus/fg1 (3)].

Radial vein of hind wing from slightly arched (as in Potamanthus/fg2) to sharply arched basally. In various species imaginal and subimaginal paracercus is either completely developed in both sexes, or vestigial in male and developed in female, or vestigial in both sexes (non-unique character – see Index of characters [2.3.22]).

Shape of penis is various, but not as in Potamanthus/fg2: penis lobes are either narrow without widenings, with blunt or pointed apices (in *formosus* [*Potamanthus*] and others), or wide with blunt apices (in *speciosus* [*Rhoenanthus*] and others).

Size. Fore wing length 7–19 mm.

Distribution. South-East Asia (to the north up to Primorye Territory of Russia).

Nominal taxa included. Rhoenanthus/g1 includes: (1) **Potamanthodes/g** [g: *Potamanthodes* Ulmer 1920a: 11, typus *Potamanthus formosus* Eaton 1892 (design. orig.)]; (2) **Potamanthindus/g** [g: *Potamanthindus* Lestage 1931b: 123, typus *P. auratus* Lestage 1930 (design. orig.; syn. subj. *Rhoenanthus obscurus* Navás 1922)]; (3) **Rhoenanthopsis/g** [g: *Rhoenanthopsis* Ulmer 1932: 212, typus *Rhoenanthus magnificus* Ulmer 1920 (design. orig.); syn.subj.: ***Neopotamanthodes*** Hsu 1938: 221, typus *N. lanchi* Hsu 1938 (design. orig.; syn. subj. *Rhoenanthus magnificus* Ulmer 1920)]; (4) **Neopotamanthus/g** [g: *Neopotamanthus* Wu & You 1986: 401, typus *N. youi* Wu & You 1986 (design. orig.)]; (5) **Stygifloris/g(1)** [g: *Stygifloris* Bae & McCafferty & Edmunds 1990: 887, typus *S. sabahensis* Bae & McCafferty & Edmunds 1990 (design. orig.)]. Different authors regarded these taxa as genera or subgenera. The most species are usually attributed to taxa *Rhoenanthus* and *Potamanthodes*.

The genus *Rhoenanthus* initially was established for a single species. Different authors characterized the genus *Rhoenanthus* by one of the following characters: imaginal paracercus is vestigial (Eaton 1883–1888); on fore wing MP_2 originates not from MP_1, but from CuA (Lestage 1931b); on hind wing MP_2 originates not from MP_1, but from CuA, larva has long tusks and has no transverse row of setae on fore femur (Bae & McCafferty 1991). Circumscription of the taxon *Rhoenanthus* depends upon choice of character. Places of originating of MP_2 on fore and hind wings are non-unique and non-reliable characters [see Fimbriatotergaliae (5)]; in Ephemera/fg10 the same characters vary individually.

The genus *Potamanthodes* initially was established for a single species and was characterized as following: on hind wing RA near its base is sharply arched repeating curvation of Sc (which as in other Potamanthus/fg1 repeats curvation of costal margin with its costal projection – FIG.68:B). In various species curvation of RA is expressed in various degree, hence this character does not allow to make boundary between *Potamanthodes* and other taxa.

The genera *Potamanthindus*, *Rhoenanthopsis*, *Neopotamanthus* and *Stygifloris* were established for selected species which have various combinations of characters ascribed to the genera *Rhoenanthus* and *Potamanthodes*.

Species composition of Rhoenanthus/g1. About 20 species – *coreanus* Yoon & Bae 1985 [*Potamanthus*] (syn. subj.: *rohdendorfi* Tshernova 1985 [*Rhoenanthus*]), *distafurcus* Bae & McCafferty 1991 [*Rhoenanthus*], *formosus* Eaton 1892 [*Potamanthus*] (synn. subj.: *iyonis* Matsumura 1931 [*Potamanthus*], *kamonis* Imanishi 1940 [*Potamanthus*]), *hunanensis* You & Gui 1995 [*Neopotamanthus*], *idiocerus* Bae & McCafferty 1991 [*Potamanthus (Potamanthodes)*], *kwangsuensis* Hsu 1937 [*Potamanthodes*] (syn. subj.: *fujianensis* You & Wu & Gui & Hsu 1980 [*Potamanthodes*]), *longitibius* Bae & McCafferty 1991 [*Potamanthus (Potamanthodes)*], *magnificus* Ulmer 1920 [*Rhoenanthus*] (synn. subj.: *ferrugineus* Navás 1930 [*Rhoenanthus*], *lanchi* Hsu 1938 [*Neopotamanthodes*], *vitalisi* Navás 1922 [*Rhoenanthus*]), *macrophthalmus* You 1984 [*Potamanthodes*], *nanchangi* Hsu 1936 [*Potamanthodes*], *obscurus* Navás 1922 [*Rhoenanthus*] (syn. subj.: *auratus* Lestage 1930 [*Potamanthindus*]), *sabahensis* Bae & McCafferty 1990 [*Stygifloris*], *sangangensis* You 1984 [*Potamanthodes*], *speciosus* Eaton 1881 [*Rhoenanthus*], *subcostalis* Navás 1932 [*Potamanthus*], *yooni* Bae & McCafferty 1991 [*Potamanthus (Potamanthodes)*], *youi* Wu & You 1986 [*Neopotamanthus*], *yunnanensis* You & Wu & Gui & Hsu 1982 [*Potamanthodes*].

Material examined: *formosus* [*P.*]: Ȣ, ◎, ⊕; *magnificus* [*Rh.*]: ♂, ⊕; *obscurus* [*Rh.*]: Ȣ; *coreanus* [*P.*]: Ȣ, ⊕; sp.P1 (China): ♂; sp.P2 (China): ⊕; sp.P3 (China): ♂; spp.: Ȣ.

***

2.2;2,1-1/2. **Potamanthus/fg2**

(Bidentiseta Furcatergaliae Fimbriatotergaliae Potamanthus/fg1 .../fg2)

(Figs 67:A–D; 68)

Nomen hierarchicum: **Potamanthus/fg2** (sine *Rhoenanthus*; incl. *Anthopotamus*).
In circumscription matches:
— gen. *Potamanthus*: Eaton 1871.

References. Eaton 1883–1888: ⊕*; – Needham & Traver & Hsu 1935: Ȣ* ⊕*; – Edmunds & Jensen & Berner 1976: Ȣ* ⊕*.

Autapomorphy of Potamanthus/fg2.

(1) Penis has a characteristic shape similar in all representatives: each penis lobe has a stretched pointed apex and a flat rounded lateral lobe (Bae & McCafferty 1991:Figs 124–126).

Character of unclear phylogenetic status.

(2) In larva on fore femur transverse row of setae [see Potamanthus/fg1 (3)] is not expressed or poorly expressed (in contrast to some Rhoenanthus/g1).

Plesiomorphy of Potamanthus/fg2. Imaginal and subimaginal paracercus is not shortened (in contrast to some Rhoenanthus/g1).

Size. Fore wing length 7–15 mm.

Distribution. Holarctic and Oriental Region.

Potamanthus/fg2 is divided into Potamanthus/fg3 and *Anthopotamus* (p.226).

2.2;2,1-1/2.1. **Potamanthus/fg3**

(Bidentiseta Furcatergaliae Fimbriatotergaliae Potamanthus/fg1 .../fg3)

(Figs 67:A–D; 68)

Nomen hierarchicum: **Potamanthus/fg3** (sine *Anthopotamus*).
In circumscription matches:
— subgen. *Potamanthus*: Bae & McCafferty 1991: 48;
— gen. *Potamanthus*: McCafferty & Bae 1990: 201.

References. Eaton 1883–1888: Ȣ ⊕; – Ueno 1928: Ȣ; – Grandi 1960: Ȣ ⊕; – Landa 1969a: Ȣ ⊕; – McCafferty & Bae 1990: Ȣ ⊕; – Bae & McCafferty 1991: Ȣ ⊕.

Character of unclear phylogenetic status.

(1) Mandibular tusks [see Potamanthus/fg1 (7)] are short, shorter than mandibles themselves (FIG. 67:D). The same in some Rhoenanthus/g1.

Size. Fore wing length 8–15 mm.

Distribution. Eurasia.

Species composition of Potamanthus/fg3. 2 close species – *huoshanensis* Wu 1987 [*Potamanthus*] and *lutea* Linnaeus 1767 [*Ephemera*] (synn. subj.: *reticulata* Fourcroy 1785 [*Ephemera*], *hyalina* Panzer 1804 [*Ephemera*], *mellea* Curtis 1834 [*Baetis*], *marginalis* Burmeister 1839 [*Baetis*], *flavicans* Rambur 1842 [*Ephemera*], *chlorotica* Rambur 1842 [*Ephemera*], *ferreri* Pictet 1843 [*Potamanthus*], *reaumuri* Joly & Joly 1876 [*Eucharidis*], *oriens* Bae & McCafferty 1991 [*Potamanthus luteus*]).

Material examined: *lutea* [*E.*]: Ȣ–◎–⊕/⊙.

***

2.2;2,1-1/2.2. **Anthopotamus/g(1)**

(Bidentiseta Furcatergaliae Fimbriatotergaliae Potamanthus/fg1 .../fg2 *Anthopotamus*)

Nomen hierarchicum: **Anthopotamus/g(1)** [g: *Anthopotamus* McCafferty & Bae 1990: 201, typus *Baetis verticis* Say 1839 (design. orig.)].
In circumscription matches:
— gen. *Anthopotamus* McCafferty & Bae 1990: 201.

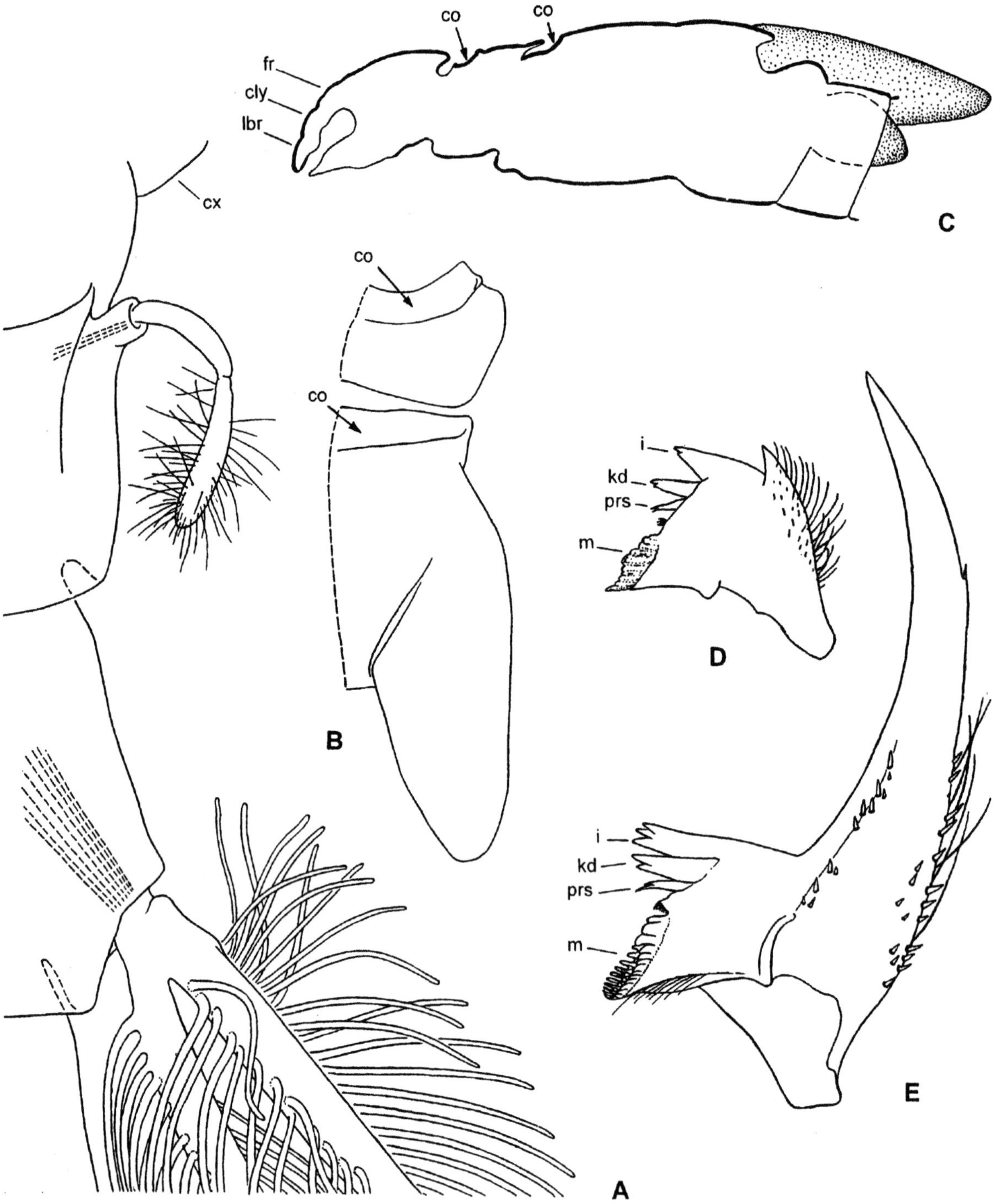

Figure 67. Potamanthus/fg1, larvae.
A–D – Potamanthus/fg* *lutea* [*Ephemera*]: **A** – right margins of abdominal segments I and II with tergalius I and base of tergalius II, dorsal view (tergalial muscles shown by interrupted lines, other muscles not shown); **B** – larval exuviae of right half of pronotum and mesonotum; **C** – median section of larval head and thorax; **D** – right mandible, dorsal view. **E** – Rhoenanthus/g1 *coreanus* [*Potamanthus*], right mandible, dorsal view. (B, C – from Kluge 1997b).

Abbreviations: **cly** – clypeus; **co** – collar; **cx** – margin of hind coxa; **fr** – frons; **i** – incisor; **kd** – kinetodontium; **lbr** – labrum; **m** – mola; **prs** – prostheca.

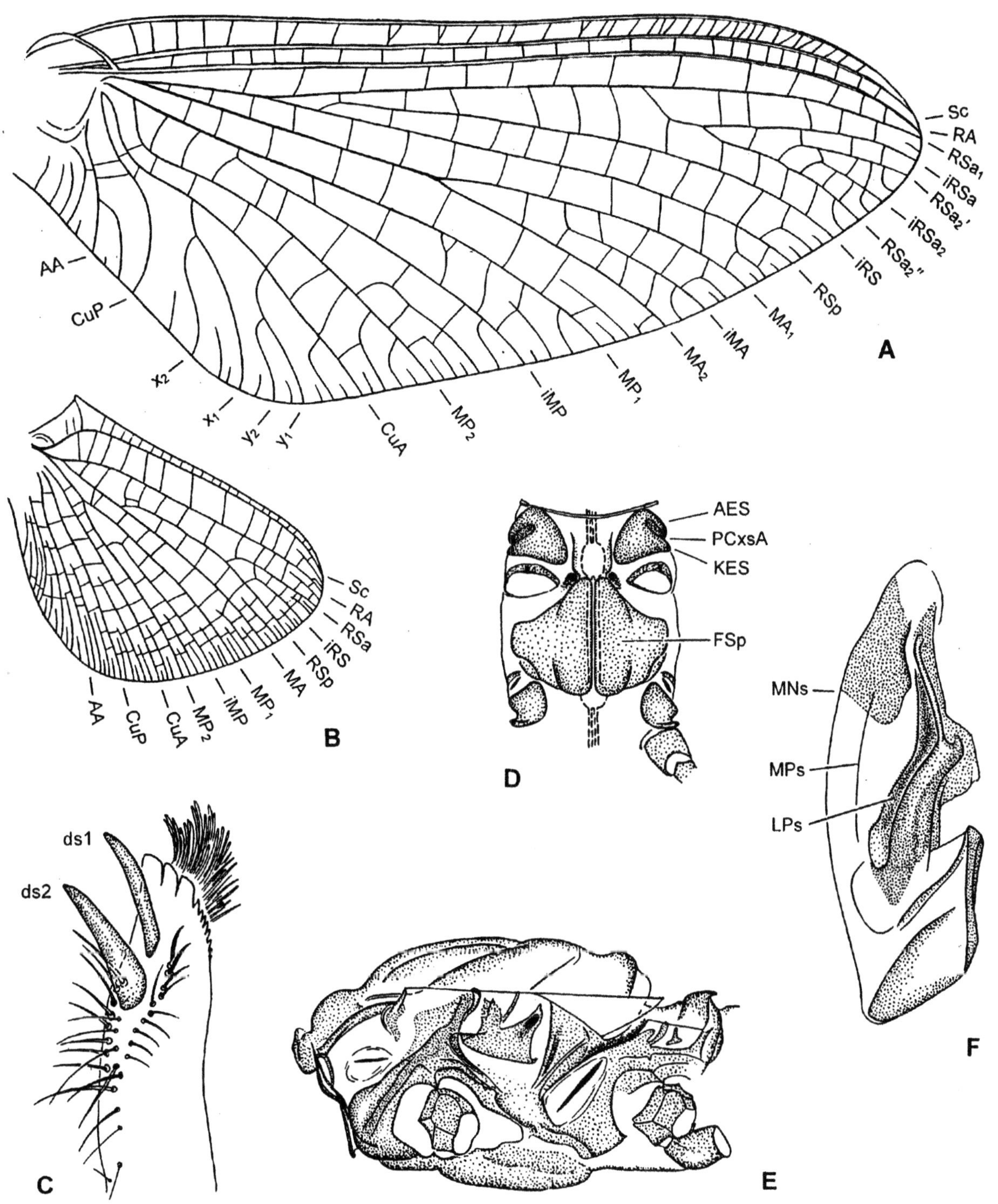

Figure 68. Potamanthus/fg* *lutea* [*Ephemera*].
A–B – fore and hind wings. **C** – apex of left maxilla, median view (dentisetae shown by dots, setae of medio-dorsal row shown by black); **D–E** – imaginal pterothorax: **D** – ventral view (nerve chain shown by interrupted line); **E** – lateral view. **F** – subimaginal exuviae of right half of mesonotum. (C, E – from Kluge 1998).

Abbreviations: **AES** – anepisternum; **ds1** – distal dentiseta; **ds2** – proximal dentiseta; **FSp** – furcasternal protuberance; **KES** – katepisternum; **LPs** – lateroparapsidal suture; **MNs** – mesonotal suture; **MPs** – medioparapsidal suture; **PCxsA** – anterior paracoxal suture.

References: McCafferty & Bae 1990: ℧* ⊕*; – Bae & McCafferty 1991: ℧* ⊕*.

Character of *Anthopotamus* of unclear phylogenetic status.

(1) Mandibular tusks [see Potamanthus/fg1 (7)] are long, much longer than mandibles themselves. The same in some Rhoenanthus/g1; probably symplesiomorphy.

Size. Fore wing length 7–15 mm.

Distribution. Eastern Nearctic.

Species composition of Anthopotamus/g(1). 4 species – *distinctus* Traver 1935 [*Potamanthus*], *myops* Walsh 1863 [*Ephemera*] (synn. subj.: *medius* Banks 1908 [*Potamanthus*], *inaequalis* Needham 1908 [*Potamanthus*], *rufous* Argo 1927 [*Potamanthus*]), *neglectus* Traver 1935 [*Potamanthus*], *verticis* Say 1839 [*Baetis*] (synn. subj.: *flaveola* Walsh 1862 [*Ephemera*], *diaphanus* Needham 1907 [*Potamanthus*], *betteni* Morgan 1913 [*Potamanthus*], *walkeri* Ide 1935 [*Potamanthus*]).

Material examined: *verticis* [*B.*]: ○, ⊕/⊙.

*** *** ***

2.2;2,1-2. Euthyplocia/fg1

(Anteritorna Bidentiseta Furcatergaliae Fimbriatotergaliae Euthyplocia/fg1)

(Fig. 69)

Nomen hierarchicum: **Euthyplocia/fg1** (incl. *Campylocia, Polyplocia, Proboscidoplocia, Exeuthyplocia*) [f: Euthyplociinae Lestage 1921: 213; g: *Euthyplocia* Eaton 1871: 67, typus *Palingenia hecuba* Hagen 1861 (monotypy)].

In circumscription matches:

— subfam. Euthyplociinae: Demoulin 1952e: 2;
— fam. Euthyplociidae: Edmunds & Traver 1954a: 239;
— Euthyplocia/fg1: Kluge 2000: 251.

References. Demoulin 1952e: ⊕; – Edmunds & Allen & Peters 1963: ℧*; – Tshernova 1970: ⊕*; – Edmunds & Jensen & Berner 1976: ℧* ⊕*.

Autapomorphies of Euthyplocia/fg1.

(1) Larval clypeus projects forwards as a shelf above base of labrum (FIG.69:F, I). Unique apomorphy. By its form this clypeal shelf resembles frontal projection of Fossoriae, but is not homologous to it: clypeal shelf of Euthyplocia/fg1 is located ventrad of anterior tentorial pits, while frontal projection of Fossoriae is located dorsad of them (FIG.76:A).

(2) Mandibular tusks [see Fimbriatotergaliae (8)] are covered by dense irregular long slender setae (FIG.69:G); on inner side of tusk these setae are directed nearly perpendicular to the tusk and probably are used for filtering. Unique apomorphy: in Potamanthus/fg1 and Fossoriae setae on tusks are not so dense and long. Tusks are curved by their apices medially and besides the slender setae, bear stout spine-like setae (that is probably a plesiomorphy within Fimbriatotergaliae).

(3) Larval fore tibia on inner-apical corner bears a pointed projection (FIG.69:D–E). Non-unique apomorphy (see Index of characters [1.2.35]).

(4) On fore wing bifurcation of MA [initially situated at middle of MA – see Euephemeroptera (2)] is transferred proximally (FIG.69:A). Non-unique apomorphy; among Fimbriatotergaliae the same in Behningia/fg1, Anagenesia/g1 and Polymitarcys/f1=Ephoron/g2.

(5) Gonostylus has no more than 1 distal segment (instead of two initial ones) (non-unique apomorphy – see Index of characters [2.3.12]); proximal part of gonostylus is integral, boundary between 1st and 2nd segments is absent (the same in Potamanthus/fg1 and some other mayflies); thus gonostylus has no more than 2 segments.

Characters of Euthyplocia/fg1 common with Potamanthus/fg1.

(6) Maxilla [see Fimbriatotergaliae (10)] on apical-ventral side close to base of maxillary canines has a field of densely and regularly situated long setae. The same in Potamanthus/fg1 and Ephemerella/fg1, a similar field – in Leptophlebia/fg1 (see Index of characters [1.1.32]).

(7) Larva has collar not only on pronotum [see Fimbriatotergaliae (1)], but on mesonotum as well (FIG.69:G). The same in Potamanthus/fg1 and Caenotergaliae, and the same mesonotal collar in Ephemerella/fg1.

Characters of Euthyplocia/fg1 common with Fossoriae.

(8) On imaginal and subimaginal mesothorax epimeron and lateropostnotum [with crest on anterior margin – see Fimbriatotergaliae (4)] are fused by their ends, forming an integral framework (FIG.69:C). The same in Fossoriae (FIG.71:C–D), while in Potamanthus/fg1 and Neoephemera/fg1 ends of epimeron and lateropostnotum are separated by membranous concavity (FIGS 68:E; 83:C).

(9) Tergalius I [see Fimbriatotergaliae (6)] is vestigial, bilamellate, without processes. The same in majority of Fossoriae except for Behningia/fg1 (see Index of characters [1.3.33]); probably, symplesiomorphy of Euthyplocia/fg1 and Fossoriae.

Other characters of unclear phylogenetic status.

(10) In male imago (but not in subimago) on fore leg both claws are similar and blunt. Non-unique apomorphy (see Index of characters [2.2.77]). Imaginal and subimaginal claws of middle and hind legs can be either ephemeropteroid, or pointed (TABLE 6); subimaginal claws of male fore leg (examined in *Polyplocia* only) are similar to that of other legs.

(11) In some species, whose females are much larger than males, fore wing of female (but not male) has vein gemination: apices of certain longitudinal veins are brought together forming following pairs: RSa_2"+iRS; RSp+MA_1; iRS+MA_2; sometimes also MP_1+iMP; each space between these pairs contains one or several additional intercalaries, thus spaces between all longitudinal veins have subequal width. Such gemination is present in the females examined of *Campylocia*, *Polyplocia* and *Proboscidoplocia*. In males of the same species and in both sexes of other species (belonging to Euthyplocia/fg2 and Exeuthyplocia/fg2) veins are not brought in pairs, and additional intercalaries are absent. Possibly this character varies among species or individually, as gemination is also absent in some females of *Proboscidoplocia* (Sartori & al. 1999:Fig.1). The same gemination occurs in some other mayfly taxa (see Index of characters [2.2.28]).

Plesiomorphies of Euthyplocia/fg1. In larva: Maxilla has 3 canines and 2 dentisetae [see Bidentiseta (1)] (distal dentiseta can be slender, pressed to canines and poorly visible). Maxillary palp is 3-segmented; labial palp is 3-segmented (in contrast to Cryptoprosternata in Fossoriae). Larva retains long abdomen able to make undulate swimming movements; caudalii with secondary swimming setae equally developed on lateral and median sides of cerci and lateral sides of paracercus [see Furcatergaliae (6)]. All tergalii I–VII [modified – see Fimbriatotergaliae (6)] are present; tergalii II–VII are spread by sides of abdomen, not bent dorsally (the same in Potamanthus/fg1, in contrast to Fossoriae and Caenotergaliae).

In imago: Furcasternal protuberances are contiguous (FIG.69:H) (in contrast to Campsurus/fg1, Caenotergaliae and some others – see Index of characters [2.2.23]). In cubital field of fore wing sigmoid veins going toward basitornal margin [see Anteritorna (1)] are always developed; in various representatives they arise either directly from CuA, or from a secondary intercalary (see Table 6 and characteristics below). Hind wing is usually well-developed (as long as 0.3–0.5 of fore wing length), rarely diminished (FIG.69:B).

Variable characters of Euthyplocia/fg1. Patella-tibial suture (initially present on middle and hind legs) can be developed on middle leg only (in larvae of the Polyplocia/g1 examined), or slightly traced on middle leg only (in larva of Exeuthyplocia/g1 sp.), or lost an all legs (in larva and imago of *inaccessible* [*Euthyplocia*]).

In imago 1st segment of middle and hind leg can be fused with tibia and shortened [see Furcatergaliae (2)] (in *Polyplocia*), or is secondarily separated from tibia. In *Campylocia* it is said to be separated on middle leg, remaining to be fused on hind leg (Berner & Thew 1961:Figs 8c–f); in *Proboscidoplocia* it is separated on middle and hind legs (Demoulin 1966e:Figs 1e–d; Fontaine 1968:Figs 3–5); both in *Campylocia* and *Proboscidoplocia* it remains to be shortened. In Euthyplocia/fg2 (at least in *inaccessibile* [*E.*] and *intermedia* [*Mesoplocia*]) first segment of middle and hind legs is separated from tibia and elongated, being a little longer than second segment (Demoulin 1952e:Figs 1c–d); non-unique apomorphy (see Index of characters [2.2.84]). First tarsal segment of male fore leg is always short, as in other Furcatergaliae.

Egg structure, being known for a few representatives only, is quite variable. In all species of *Proboscidoplocia* egg has a form of mammalian erythrocyte and lacks any anchors or polar caps (Sartori & al. 1999:Figs 8–11). In Campylocia/fg1 *incepts* [*E.*] egg has two polar anchors, each consisting of lamellate round cap and a bunch of very long numerous threads which connect the cap with egg pole; threads are coiled as a regular spiral under the cap, and when stretched exceed egg length many times. Exeuthyplocia/fg2 *minima* [*E.*] has on one pole a similar lamellate round cup attached to the egg by a short integral stem, without threads.

Size. Fore wing length 5–40 mm.

Distribution. Neotropical, Oriental and Ethiopian Regions.

Euthyplocia/fg1 is divided into taxa Euthyplocia//fg2, *Campylocia*, *Polyplocia*, *Proboscidoplocia* and Exeuthyplocia/fg1, which differ one from another by combinations of non-unique characters (TABLE 6).

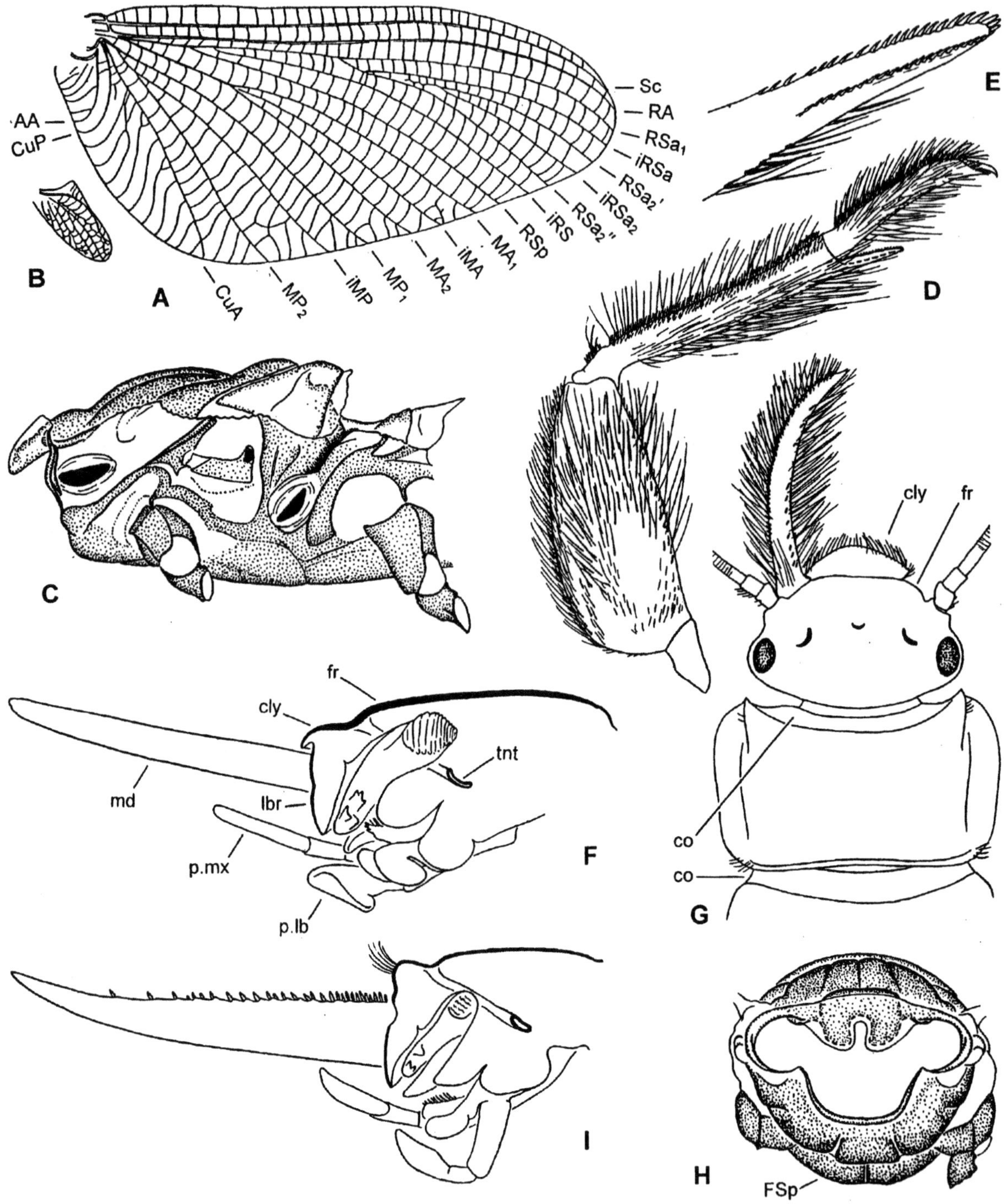

Figure 69. Euthyplocia/fg1.
A–H – Euthyplocia/fg2 *inaccessibile* [*Euthyplocia*]: **A–B** – fore and hind wings; **C** – imaginal pterothorax, lateral view; **D** – larval left fore leg, dorsal view; **E** – apical projection of fore tibia; **F** – median section of larval head; **G** – larval head and prothorax (right mandible not shown); **H** – imaginal pterothorax, front view. **I** – Exeuthyplocia/fg* *minima* [*Euthyplocia*], median section of larval head (compare with Figs 67:C and 76:A). (A, B, D–F – from Kluge & Naranjo 1994).

Abbreviations: **cly** – clypeus; **co** – collar; **fr** – frons; **FSp** – furcasternal protuberance; **lbr** – labrum; **md** – mandibular tusk; **p.lb** – labial palp; **p.mx** – maxillary palp; **tnt** – tentorium.

Table 6. Non-unique characters of subordinate taxa in Euthyplocia/fg1

Taxon	Geographical distribution	Species number	Characters				
			Imaginal				Larval
			1	2	3	4	5
Euthyplocia/fg2	Neotropical Region	4	–	1	))	±	+
Campylocia	Neotropical Region	2	1–3	–	))	+	–
Polyplocia	Oriental Region	2	2	–	))	+	+
Proboscidoplocia	Madagascar	10	–	1	0)	+	–
Exeuthyplocia/fg1	Ethiopian Region	2	–	–	0)	–	–

Characters:
1 – number of intercalaries in cubital field of fore wing;
2 – number of distal segments of gonostylus;
3 – claws of imago (except for fore legs of male): "0)" – ephemeropteroid, "))" – both pointed;
4 – paracercus of male imago: "+" – developed, "–" – vestigial;
5 – projection on apex of larval tarsus: "+" – present, "–" – absent.

2.2;2,1-2/1. Euthyplocia/fg2

(Bidentiseta Furcatergaliae Fimbriatotergaliae Euthyplocia/fg1 .../fg2)

(Fig. 69:A–H)

Nomen hierarchicum: **Euthyplocia/fg2** (sine *Campylocia, Polyplocia, Proboscidoplocia, Exeuthyplocia*; incl. *Mesoplocia*).
In circumscription matches:
— gen. *Euthyplocia*: Needham & Murphy 1924: 24.

References. Ulmer 1920c: Ծ; – Needham & Murphy 1924: ⊕; – Gros & Lestage 1927: Ծ ⊕; – Ulmer 1942: ⊕; – Demoulin 1952e: ⊕; – Roback 1966: Ծ; – Edmunds & Jensen & Berner 1976: Ծ* ⊕*; – Kluge & Naranjo 1994: Ծ* ⊕*.

Characters of Euthyplocia/fg2 of unclear phylogenetic status (TABLE 6).

(1) Imaginal claws [except for blunt claws on fore legs of male – see Euthyplocia/fg1 (10)] are similar, pointed. Non-unique apomorphy (see Index of characters [2.2.85]); among Euthyplocia/fg1 the same in *Campylocia* and *Polyplocia*, in contrast to *Proboscidoplocia* and *Afroplocia*.

(2) In larva apex of fore tarsus has a projection dorsad of claw (FIG.69:D). The same in *Polyplocia*.

Plesiomorphies of Euthyplocia/fg2. In cubital field of fore wing several (6–14) sigmoid veins arise directly from CuA (FIG.69:A) [see Plesiomorphies of Euthyplocia/fg1] (in contrast to *Campylocia* and *Polyplocia*). Gonostylus with distal segment [single – see Euthyplocia/fg1 (5)].

Variable characters of Euthyplocia/fg2. Imaginal paracercus can be well-developed in both sexes (in *hecuba* [*Palingenia*], *haenschi* [*Euthyplocia*]), or well-developed in female and vestigial in male (in *inaccessibile* [*E.*]).

Size. Fore wing length 20–40 mm.

Distribution. Neotropical Region.

Nominal taxon included. Euthyplocia/fg2 includes **Mesoplocia/g** [g: *Mesoplocia* Demoulin 1952e: 11, typus *M. intermedia* Demoulin 1952], regarded as a generic synonym of *Euthyplocia* (Kluge & Naranjo 1994: 778).

Species composition of Euthyplocia/fg2. 4 species – *haenschi* Ulmer 1942 [*Euthyplocia*], *hecuba* Hagen 1861 [*Palingenia*], *inaccessibile* Kluge & Naranjo 1994 [*Euthyplocia*], *intermedia* Demoulin 1952 [*Mesoplocia*].

Material examined: *inaccessibile* [*E.*]: Ծ♂, Ծ♀, ⊕♂; spp. (Costa Rica, Peru; deposited in Entomol. Inst.): Ծ, ⊕♂, ⊕♀.

***

2.2;2,1-2/2. Campylocia/g(1)

(Bidentiseta Furcatergaliae Fimbriatotergaliae Euthyplocia/fg1 *Campylocia*)

Nomen hierarchicum: **Campylocia/g(1)** [g: *Campylocia* Needham & Murphy 1924: 25, typus *Euthyplocia anceps* Eaton 1883 (design. orig.); syn. subj.: ***Longinella*** Gros & Lestage 1927: 158, typus *Euthyplocia guntheri* Navás 1920 (design. orig.) (synn. subj.: *anceps* [*E.*] = *guntheri* [*E.*])].
In circumscription matches:
— gen. *Campylocia* Needhan & Murphy 1924: 25.

References. Needham & Murphy 1924: Ծ ⊕; – Ulmer 1920c: Ծ; – Gros & Lestage 1927: Ծ ⊕; – Berner & Thew 1961: ⊕; – Edmunds & Jensen & Berner 1976: Ծ ⊕; – Pereira & Da-Silva 1990: Ծ ⊕.

Characters of *Campylocia* of unclear phylogenetic status (TABLE 6).

(1) On fore wing directly from CuA arise only 1–4 veins going to basitornal margin; there are 1–3 longitudinal intercalaries which go parallel to CuA toward tornoapical margin; from the posteriormost intercalary arise other 5–10 sigmoid veins going to basitornal margin [see Plesiomorphies of Euthyplocia/fg1] (Edmunds & al. 1996:Fig.201; Berner & Thew 1961:Figs 4–5; Pereira & Da-Silva 1990: Fig.1). Non-unique character (see Index of characters [2.2.51]); among Euthyplocia/fg1 intercalaries in cubital field are present also in *Polyplocia*.

(2) Imaginal claws [except for blunt claws on fore legs of male – see Euthyplocia/fg1 (10)] are similar, pointed. Non-unique apomorphy (see Index of characters [2.2.85]); among Euthyplocia/fg1 the same in Euthyplocia/fg2 and *Polyplocia*, in contrast to *Proboscidoplocia* and *Afroplocia*.

(3) Gonostylus without distal segment [i.e. 1-segmented – see Euthyplocia/fg1 (5)]. Non-unique apomorphy (see Index of characters [2.3.12]); among Euthyplocia/fg1 the same in *Polyplocia* and Exeuthyplocia/fg1; possibly, synapomorphy.

Plesiomorphies of *Campylocia*. Imaginal paracercus is well-developed in both sexes (in contrast to Exeuthyplocia/fg1 and some Euthyplocia/fg2). Larval fore tarsus without projection (in contrast to Euthyplocia/fg2 and *Polyplocia*).

Size. Fore wing length 10–40 mm.

Distribution. Neotropical Region.

Species composition of Campylocia/g(1). 3 species are described – *anceps* Eaton 1883 [*Euthyplocia*] (synn. subj.: *ampla* Needham & Murphy 1924 [*Campylocia*], *burmeisteri* Hagen 1888 [*Euthyplocia*], *guntheri* Navás 1920 [*Euthyplocia*], *intercalata* Banks 1918 [*Euthyplocia*]), *bocainensis* Pereira & Da-Silva 1990 [*Campylocia*], *dochmia* Berner & Thew 1961 [*Campylocia*].

Material examined: *anceps* [*E.*]: ʘ (Entomol. Inst.), ⊕♂, ⊕♀.

***

2.2;2,1-2/3. Polyplocia/g(1)

(Bidentiseta Furcatergaliae Fimbriatotergaliae Euthyplocia/fg1 *Polyplocia*)

Nomen hierarchicum: **Polyplocia/g(1)** [g: *Polyplocia* Lestage 1921: 212, typus *P. vitalisi* Lestage 1921 (design. orig.)].
In circumscription matches:
— gen. *Polyplocia* Lestage 1921: 212.

References. Ulmer 1939: ⊕; – Demoulin 1952e: ⊕; – 1953b: ⊕; – 1966c: ʘ.

Characters of *Polyplocia* of unclear phylogenetic status (TABLE 6).

(1) On fore wing just behind CuA there are 2 longitudinal intercalaries which begin as a common stem and go parallel to CuA toward tornoapical margin; thus sigmoid veins going to basitornal margin of wing [see Plesiomorphies of Euthyplocia/fg1) arise not from CuA, but from the posteriormost of these intercalary veins (Demoulin 1952e:Figs 3–4; 1953b:Fig.1). Non-unique character (see Index of characters [2.2.51]); among Euthyplocia/fg1 intercalaries in cubital field are present also in *Campylocia*.

(2) Imaginal and subimaginal claws [except for blunt claws on fore legs of male imago – see Euthyplocia/fg1 (10)] are similar, pointed. Non-unique apomorphy (see Index of characters [2.2.85]); among Euthyplocia/fg1 the same in Euthyplocia/fg2 and *Campylocia*, in contrast to *Proboscidoplocia* and *Afroplocia*.

(3) Gonostylus without distal segment [i.e. 1-segmented – see Euthyplocia/fg1 (5)]. Non-unique apomorphy (see Index of characters [2.3.12]); among Euthyplocia/fg1 the same in *Campylocia* and Exeuthyplocia/fg1; probably, synapomorphy.

(4) In larva apex of fore tarsus with a projection dorsad of claw (Demoulin 1966c:Fig.1h). The same in Euthyplocia/fg2 (FIG.69:D).

Plesiomorphy of *Polyplocia*. Imaginal paracercus is well-developed in both sexes (in contrast to Exeuthyplocia/fg1 and some Euthyplocia/fg2).

Size. Fore wing length 12–20 mm.

Distribution. Oriental Region.

Species composition of Polyplocia/g(1). 2 species are described – *vitalisi* Lestage 1921 [*Polyplocia*] and *campylociella* Ulmer 1939 [*Polyplocia*] (syn. subj.: *crassinervis* Ulmer 1939 [*Polyplocia*]).

Material examined: sp. (Thailand): ♀; sp. (East Malaysia): ♀, ♂/♂.

***

2.2;2,1-2/4. Proboscidoplocia/g(1)

(Bidentiseta Furcatergaliae Fimbriatotergaliae Euthyplocia/fg1 *Proboscidoplocia*)

Nomen hierarchicum: **Proboscidoplocia/g(1)** [g: *Proboscidoplocia* Demoulin 1966e: 946, typus *Euthyplocia sikorai* Vayssière 1891 (design. orig.)].
In circumscription matches:
— gen. *Proboscidoplocia* Demoulin 1966e: 954.

References. Ulmer 1920c: ʘ; – Gros & Lestage 1926 (*Campylocia sikorai*): ʘ; – Demoulin 1966e: ʘ ⊕; – Fontaine 1968: ʘ ⊕; – Elouard & Sartori 1997: I; – Sartori & Elouard & Ruffieux & L'Eplattenier 1999: ⊕ ⊙.

Plesiomorphies of *Proboscidoplocia*. In imago and subimago: in cubital field of fore wing several (3–6) simple or branched sigmoid veins arise directly from CuA [see Plesiomorphies of Euthyplocia/fg1] (in contrast to *Campylocia* and *Polyplocia*)

(Demoulin 1966e:Fig.1; Elouard & Sartori 1997:Figs 1–11); claws [except for blunt claws on fore legs of male imago – see Euthyplocia/fg1 (10)] are ephemeropteroid (in contrast to Euthyplocia/fg2, *Campylocia* and *Polyplocia*); gonostylus with distal segment [single – see Euthyplocia/fg1 (5)]; paracercus is well-developed in both sexes (in contrast to Exeuthyplocia/fg1 and some Euthyplocia/fg2). Larval fore tarsus without projection (in contrast to Euthyplocia/fg2 and *Polyplocia*).

Size. Fore wing length 12–36 mm.

Distribution. Madagascar.

Species composition of Proboscidoplocia/g(1) (here [*P.*]=[*Proboscidoplocia*]. 8 species – *auberti* Elouard & Sartori 1997 [*P.*], *billi* Elouard & Sartori 1997 [*P.*], *leplattenierae* Elouard & Sartori 1997 [*P.*], *magdaleinae* Elouard & Sartori 1997 [*P.*], *mccaffertyi* Elouard & Sartori 1999 [*P.*], *ruffieuxae* Elouard & Sartori 1997 [*P.*], *sikorai* Vayssière 1891 [*Euthyplocia*], *vayssiere* Elouard & Sartori 1997 [*P.*].

Material examined: sp.: ♉, ⊕ (Entomol. Inst.); *billi* [*P.*]: ⊕/⊙.

***

2.2;2,1-2/2. **Exeuthyplocia/fg1**

(Bidentiseta Furcatergaliae Fimbriatotergaliae Euthyplocia/fg1 Exeuthyplocia/fg1)

(Fig. 69:I)

Nomen hierarchicum: **Exeuthyplocia/fg1** (incl. *Afroplocia*) [f: Exeuthyplociinae Gillies 1980a: 218; g: *Exeuthyplocia* Lestage 1918: 74, typus *Euthyplocia minima* Ulmer 1916 (monotypy)].
In circumscription matches:
— gen. *Exeuthyplocia*: Barnard 1937: 275;
— subfam. Exeuthyplociinae Gillies 1980a: 218.

References. Barnard 1940: ♉; – Gillies 1980a: ♉ ⊕.

Autapomorphies of Exeuthyplocia/fg1.

(1) Imago is short-living, middle and hind legs are non-functional. Non-unique apomorphy (see Index of characters [2] and [2.2.80]).

(2) In imago amphitornal margin of wings with setae similar to that of subimago. Non-unique apomorphy (see Index of characters [2.2.27]).

(3) Larvae (attributed to Exeuthyplocia/fg1 only presumably) have burrowing specialization. Tusks [see Euthyplocia/fg1 (2)] are more stout than in other Euthyplocia/fg1, thick at base and narrowed toward apex (FIG.69:I). Fore leg has femur and tibia thicker than in other Euthyplocia/fg1. Femur of middle leg is shortened, sharply widened and somewhat curved in such a manner, that middle leg can be directed anteriorly; its form is similar to middle femur of Fossoriae, but in contrast to it, retains dense setation on initially anterior (dorsal) side (directed anteriorly and inward), but has no dense setation on initially posterior (ventral) side (directed posteriorly and outside). Abdominal terga with long dense setae directed dorsally; these setae form a pair of widely separated longitudinal stripes on segments III–VII; tergalii are able to fold, being pressed from sides to these stripes of setae (in contrast to Fossoriae, whose tergalii are fold on abdomen, being brought together at median line).

An assumption was suggested, that burrowing specialization is a synapomorphy of *Exeuthyplocia* and Polymitarcys/f1=Ephoron/g2 (Gillies 1980a). This assumption contradicts the ideas about holophyly of Euthyplocia, holophyly of Cryptoprosternata and holophyly of Fossoriae (see below). Actually, the similarity between larvae of Exeuthyplocia/fg1 and Polymitarcys/f1=Ephoron/g2 is superficial; particularly, similar projection in front of head in Exeuthyplocia is formed by clypeus ventrad of tentorial pits (FIG.69:I) [see Euthyplocia/fg1 (1)], while in Polymitarcys/f1=Ephoron/g2 it is formed by frons dorsad of tentorial pits [see Fossoriae (2)] (FIG.76:A).

Characters of Exeuthyplocia/fg1 of unclear phylogenetic status (TABLE 6).

(4) Gonostylus without distal segment [i.e. 1-segmented – see Euthyplocia/fg1 (5)]. Non-unique apomorphy (see Index of characters [2.3.12]); among Euthyplocia/fg1 the same in *Campylocia* and *Polyplocia*; probably, synapomorphy.

(5) Imaginal paracercus of male is vestigial (while in female paracercus is subequal to cerci). Non-unique apomorphy (see Index of characters [2.3.22]); among Euthyplocia/fg1 the same in one species of Euthyplocia/fg2.

Plesiomorphies of Exeuthyplocia/fg1. In cubital field of fore wing several (5–7) simple or branched sigmoid veins arise directly from CuA [see Plesiomorphies of Euthyplocia/fg1] (in contrast to *Campylocia* and *Polyplocia*). Larval fore tarsus [see (3)] without projection (in contrast to Euthyplocia/fg2 and *Polyplocia*).

Size. Fore wing length 5–20 mm.

Distribution. Ethiopian Region.

Exeuthyplocia/fg1 is divided into *Afroplocia* and Exeuthyplocia/fg2.

2.2;2,1-2/5.1. Plesiomorphon **Afroplocia/g(1)**

(Furcatergaliae Fimbriatotergaliae Euthyplocia/fg1 Exeuthyplocia/fg1 *Afroplocia*)

Nomen hierarchicum: **Afroplocia/g(1)** [g: *Afroplocia* Lestage 1939: 134, typus *Exeuthyplocia sampsoni* Barnard 1937 (design. orig.)].
In circumscription matches:
— gen. *Afroplocia* Lestage 1939: 134.
Monospecific taxon.

References. Barnard 1937: ⊕; – Lestage 1939: ⊕.

Plesiomorphies of *Afroplocia*. In imago of both sexes hind legs [non-functional – see Exeuthyplocia/fg1 (1)] are developed (in contrast to Exeuthyplocia/fg2); claws [except for blunt claws on fore legs of male – see Euthyplocia/fg1 (10)] are ephemeropteroid (in contrast to Euthyplocia/fg2, *Campylocia* and *Polyplocia*).

Size. Fore wing length 15–20 mm.

Distribution. Ethiopian Region.

Species composition of Afroplocia/g(1). 1 species – *sampsoni* Barnard 1937 [*Exeuthyplocia*].

Material examined: –.

***

2.2;2,1-2/5.2. **Exeuthyplocia/fg2**

(Furcatergaliae Fimbriatotergaliae Euthyplocia/fg1 Exeuthyplocia/fg1 .../fg2)

(Fig. 69:I)

Nomen hierarchicum: **Exeuthyplocia/fg2** (sine *Afroplocia*).
In circumscription matches:
— gen. *Exeuthyplocia*: Lestage 1918: 74.
Monospecific taxon.

References. Ulmer 1916 (*Euthyplocia minima*): ⊕; – Lestage 1918: ⊕; – Demoulin 1952e: ⊕; – Hideux 1987: ⊕; – Gillies 1980a: ♉ ⊕.

Autapomorphies of Exeuthyplocia/fg2.

(1) In imago of both sexes hind leg [non-functional – see Exeuthyplocia/fg1 (1)] is strongly reduced, being composed only of coxa, trochanter and very short pointed vestige of femur (at the same time all segments of middle leg are retained, though very weak). Unique apomorphy: in other short-living mayflies middle and hind legs, if reduced (in Campsurus/fg2 and others), are reduced in the same degree.

(2) On developed legs – middle leg of male imago and on fore and middle leg of female imago [see (1)] – claws are vestigial and blunt. Non-unique apomorphy (see Index of characters [2.2.85]).

Size. Fore wing length 6–17 mm.

Distribution. Ethiopian Region.

Species composition of Exeuthyplocia/fg2. 1 species – *minima* Ulmer 1916 [*Euthyplocia*].

Material examined: *minima* [*E.*]: ♉ (presumably associated), ⊕♂, ⊕♀/⊙.

*** *** ***

2.2;2,1-3. **Fossoriae**, or **Ephemera/fg8**

(Anteritorna Bidentiseta Furcatergaliae Fimbriatotergaliae Fossoriae)

(Figs 70–80)

Nomen hierarchicum: **Ephemera/fg8** (sine *Potamanthus*, *Euthyplocia*; incl. *Ichthybotus*, *Behningia*, *Ephoron*).
Nomen circumscribens: **Fossoriae** Kluge 2000: 252.
In circumscription matches:
— Fossoriae, or Ephemera/fg8: Kluge 2000: 252.

Autapomorphies of Fossoriae.

(1) Larval legs are initially specialized as burrowing. Fore leg [articulated to vestigial apodeme-bearing pleurite – see Furcatergaliae (1)] is usually strengthened, its tibia is thickened (only in Behningia/fg2 fore legs secondarily lost burrowing specialization and become palp-like). In all representatives (including Behningia/fg2) on femora of fore and middle legs anterior side (which in other mayflies is directed more or less dorsally) is concave, bare, most time directed inward, while posterior side (which in other mayflies is directed more or less ventrally) is convex, setose, most time directed outward; femur of hind leg, vice versa, has anterior side convex, setose and most time directed outward, and posterior side concave, bare, most time directed inward (FIGS 70:C; 73:E). This difference in femoral structure is present not only in those representatives of Fossoriae that use their legs for burrowing, but also in Behningia/fg2, whose legs strongly changed their specialization; in Behningia/fg2 the opposite curvations of middle and hind femora are retained in spite of the fact that deeply specialized middle and hind legs have similar function (FIG.73:A–B).

In all other mayfly larvae the outer (usually most convex and most setose) side of femur of all legs is their anterior (dorsal) side. This is true also for Exeuthyplocia/fg1, which have burrowing specialization evolved independently from Fossoriae (see above).

Probably, the following structure of fore tibia is initial for Fossoriae: apical margin projects posteriad of tarsus base and forms two projections – outer-apical and inner-apical ones, with incision between them (FIG.74:D–E). Such tibia shape is present in *Ichthybotus*, Hexagenia/fg1 (among Ephemera/fg9) and Palingenia/f2=g1 (among Cryptoprosternata); presence of the same tibial structure in these non-related groups testifies in favour of its presence in the common Fossoriae ancestor. However, in other Fossoriae fore tibia has more simple shape and looks as more "primitive" (FIGS 70:C; 77).

(2) Larval frons forms a shelf-like projection with flat dorsal surface. Usually this frontal projection hangs over clypeus and mandibular bases (FIGS 70:C, E; 75:C; 76:A–C), but in some representatives it is poorly expressed or non-expressed. Similar projection independently appeared in some other mayflies (particularly, in Drunella/g1). In contrast to Fossoriae, in Euthyplocia/fg1 projection of similar shape is formed not by frons, but by clypeus (FIG.69:F, I).

(3) All tergalii I–VII [modified – see Fimbriatotergaliae (6)], while retain lateral attachment, at rest are bent dorsally-medially-posteriorly (FIG.70:C) (with exception for Behningia/fg2, whose tergalii are attached ventrally because of expanded lateral tergal projections – FIG.73:A–B).

Characters of unclear phylogenetic status.

(4) On imaginal and subimaginal mesothorax epimeron and lateropostnotum [with crest on anterior margin – see Fimbriatotergaliae (4)] are fused by their ends, forming an integral framework (FIGS 71:D; 78:C). The same in Euthyplocia/fg1, while in Potamanthus/fg1 and Neoephemera/fg1 ends of epimeron and lateropostnotum are separated by membranous concavity (FIG.83:C).

(5) In larva, subimago and imago abdominal tergite X has lateral-ventral margins longitudinal, as long as the segment, reaching bases of cerci; latero-posterior angle of tergite forms a ventral condylus for cercal base and separates paraproct from cercotractor; cercotractor is either transformed to a narrow semicircular sclerite exposed caudally and surrounding lateral half of cercal base (FIG.12:G), or completely fused with cercal base. The same in Caenotergaliae and Geminovenata (but not in Potamanthus/fg1, Euthyplocia/fg1 and other mayflies, whose lateral-ventral margins of tergite X are oblique, not articulating with cerci, and cercotractors are triangular, exposed laterally and widely connected with paraprocts – FIG.12:F).

Plesiomorphies of Fossoriae. Larva retains long abdomen able to make undulate swimming movements (FIG.70:C); caudalii retain dense secondary swimming setae on outer sides of cerci [see Furcatergaliae (6)] and more regular longitudinal rows of more stout primary swimming setae on median sides of cerci and lateral sides of paracercus (only in Campsurus/fg1 longitudinal rows of primary setae are substituted by transverse rows). Hind wing is always well-developed, as long as 0.3–0.5 of fore wing length (FIGS 71; 75; 78–80).

Size. Fore wing length 5–30 mm.

Age and distribution. Late Jurassic (see Fossoriae INCERTAE SEDIS) — Recent; nearly world-wide, except for Australia.

Fossoriae are divided into *Ichthybotus*, Ephemera/ /fg9 (p.234), Behningia/fg1 (p.243) and Cryptoprosternata (p.248); some extinct Mesozoic taxa have uncertain systematic position (p.268).

2.2;2,1-3/1. Ichthybotus/fg(1)

(Bidentiseta Furcatergaliae Fimbriatotergaliae Fossoriae *Ichthybotus*)

Nomen hierarchicum: **Ichthybotus/fg(1)** [f: Ichthybotidae Demoulin 1957a: 336; g: *Ichthybotus* Eaton 1899: 285, typus *Ephemera hudsoni* McLachlan 1894 (monotypy)].

In circumscription matches:
— gen. *Ichthybotus* Eaton 1899: 285;
— subfam. Ichthybotinae: Koss & Edmunds 1974: 295;
— fam. Ichthybotidae Demoulin 1957a: 336;
— Ichthybotus/fg1: Kluge 2000: 252.

References. Eaton 1899: ⊕; – Tillyard 1923: ⊕; – Phillips 1930: ⊙ ⟲ ⊕; – 1931: ⟲; – Koss & Edmunds 1974: ⊙.

Characters of unclear phylogenetic status.

(1) Mandibular tusks [see Fimbriatotergaliae (8)] are curved medially (probably plesiomorphy, as the same in Potamanthus/fg1, Euthyplocia/fg1, Polymitarcys/f1=Ephoron/g2); dorsal and lateral sides of tusks bear numerous tubercles (the same in Polymitarcys/f2=Ephoron/g3).

(2) Apex of larval fore tibia projects posteriad of tarsus base and forms two projections – outer-apical and inner-apical ones, with incision between them. Similar in Hexagenia/fg1 and Palingenia/f2=g1 (FIG.74:D–E); possibly, symplesiomorphy [see Fossoriae (1)].

(3) Tergalius I [see Fimbriatotergaliae (6)] is vestigial, bilamellate, without processes. The same in majority of Fossoriae (except for Behningia/fg1) and in Euthyplocia/fg1; possibly, symplesiomorphy.

(4) Imaginal and subimaginal paracercus of male is vestigial, while paracercus of female is subequal to cerci. Non-unique apomorphy (see Index of characters [2.3.22]); among Fossoriae the same in Cryptoprosternata.

Plesiomorphies of *Ichthybotus*. In larva: Maxilla with 3 canines and 2 dentisetae [see Bidentiseta (1)]. Maxillary palp is 3-segmented; labial palp is 3-segmented (in contrast to Cryptoprosternata). Larval (and adult) patella-tibial suture is developed on middle and hind legs only. In contrast to Ephemera/fg9, all tergalii are attached to posterior part of segment.

In imago and subimago: Mesonotal suture is moderately stretched posteriorly, and subimaginal mesonotal cuticle has pigmented areas of characteristic shape [see Fimbriatotergaliae (2)] (as in FIG.71:E). Furcasternal protuberances are contiguous (in contrast to Campsurus/fg1, Caenotergaliae and some others – see Index of characters [2.2.23]). In cubital field of fore wing several (3–4) branched veins go from CuA to basitornal margin [see Anteritorna (1)]. Claws of all legs (including fore legs of male) are ephemeropteroid. Gonostylus is 4-segmented (i. e. with expressed 1st segment and with 2 distal segments).

Size. Fore wing length 15–22 mm.

Distribution. New Zealand.

Species composition of Ichthybotus/fg(1). 2 species – *hudsoni* McLachlan 1894 [*Ephemera*] and *bicolor* Tillyard 1923 [*Ichthybotus*].

Material examined: sp.: ☉, ☿/♀, ♀/⊙.

***

2.2;2,1-3/2. Ephemera/fg9

(Bidentiseta Furcatergaliae Fimbriatotergaliae Fossoriae Ephemera/fg9)

(Figs 70–72)

Nomen hierarchicum: **Ephemera/fg9** (sine *Ichthybotus*, *Behningia*, *Ephoron*; incl. *Hexagenia*).
In circumscription matches:
— subfam. Ephemerinae: Koss & Edmunds 1974: 295;
— Ephemera/fg9: Kluge 2000: 252.

Autapomorphy of Ephemera/fg9.

(1) Abdominal segments VII–IX are elongate owing to elongation of their posterior portions, thus bases of tergalii VII are located at midlength or at anterior half of segment VII (FIG.70:C). This character can be used as diagnostic not only for larvae, but also for winged stages, because they have visible traces of tergalial bases. In all other Fossoriae, as well as in majority of mayflies, at least tergalii II–VII are attached to posterior parts of abdominal segments.

Characters of unclear phylogenetic status.

(2) Mandibular tusk [see Fimbriatotergaliae (8)] is curved laterally-dorsally, round in section, without denticles or protuberances (FIG.70:C, E). Such structure of tusk is unique; lateral-dorsal curvation of the tusk is present also in Palingenia/f2=g1.

(3) Larval hind tibia with a prominent inner-distal angle (FIG.70:C). Non-unique apomorphy, the same in Palingenia/f2=g1.

(4) Tergalius I [see Fimbriatotergaliae (6)] is vestigial, without processes, initially bilamellate (FIG.70:C) (in *Litobrancha* unilamellate). The same in majority of Fossoriae (except for Behningia/fg1) and in Euthyplocia/fg1; probably, symplesiomorphy.

(5) In male imago on fore leg both claws are blunt (in subimago ephemeropteroid). Non-unique apomorphy (see Index of characters [2.2.77]).

(6) On fore wing from AA to basitornal margin go several (3–10) veins (FIG.71:A). Non-unique character.

Plesiomorphies of Ephemera/fg9. In larva: Maxilla with 2 dentisetae [see Bidentiseta (1)] (in Hexagenia/g1 distal dentiseta is vestigial). In contrast to Cryptoprosternata, maxillary palp is normally 3-segmented (FIG.70:C); in selected specimens suture separating 2nd and 3rd segments is indistinct or lost, in this case palp is 2-segmented. In contrast to Cryptoprosternata, labial palp is 3-segmented (FIG.70:C); sometimes suture between 2nd and 3rd segments is non-expressed, but in this case both segments retain shape peculiar for each of them; in 2nd segment muscle-adductor of 3rd segment can be retained or lost (particularly among individuals of Hexagenia/fg3 *limbata* [*E.*] all variants can be found – from wide fan-formed muscle to its absence). Larval (and adult) patella-tibial suture is developed on middle and hind legs only.

In imago and subimago: Furcasternal protuberances are contiguous (in contrast to Campsurus/fg1, Caenotergaliae and some others – see Index of characters [2.2.23]). In cubital field of fore wing several (3–6) branched or simple veins go from CuA to basitornal and tornoapical margins (FIG.71:A) [see Anteritorna (1)].

Imaginal and subimaginal claws are ephemeropteroid [except for fore legs of male imago – see (5)]. 1st segment of gonostylus is distinct from 2nd (in contrast to Behningia/fg2, Euthyplocia/fg1, Potamanthus/fg1 and some others).

Size. Fore wing length 8–30 mm.

Age and distribution. Palaeogene (see Ephemera/fg9 INCERTAE SEDIS) — Recent; Holarctic, Ethiopian, Oriental and Neotropical Regions.

Ephemera/fg9 is divided into Ephemera/fg10 and

Hexagenia/fg1 (p.237). An extinct taxon *Denina* has uncertain systematic position (p.243).

2.2;2,1-3/2.1. **Ephemera/fg10**

(Furcatergaliae Fimbriatotergaliae Fossoriae Ephemera/fg9 .../fg10)

(Figs 70:B–E; 71; 72)

Nomen hierarchicum: **Ephemera/fg10** (sine *Hexagenia*; incl. *Sinephemera, Afromera*).
In circumscription matches:
— gen. *Ephemera*: Eaton 1883–1888: 58;
— subfam. Ephemerinae: McCafferty 1991a: 354.

References. Eaton 1883–1888: ℧* ⊕*; – Needham & Traver & Hsu 1935: ℧* ⊕*; – McCafferty & Edmunds 1973: ℧* ⊕*; – McCafferty 1975: ℧; – Edmunds & Jensen & Berner 1976: ℧* ⊕*.

Autapomorphy of Ephemera/fg10.

(1) Larval frontal projection [see Fossoriae (2)] is bipointed (FIG.70:E).

Characters of unclear phylogenetic status.

(2) Larval antennal flagellum dorsally with whorls of long dense setae (FIG.70:C, E). Among Hexagenia/fg1 the same in Hexagenia/fg2, but not in Eatonica/g1.

(3) Larval fore tibia [burrowing – see Fossoriae (1)] on distal margin has no incision and denticles (FIG.70:C) (in contrast to *Ichthybotus*, Hexagenia/ /fg1 and Palingenia/f1=g2). Probably a secondary simplification.

Plesiomorphies of Ephemera/fg10. Maxilla with 3 canines, both dentisetae [see Plesiomorphies of Ephemera/fg9] are well-developed; paraglossae are not stretched proximally (in contrast to Hexagenia/fg1). Vestige of tergalius I [see Ephemera/fg9 (4)] is bilamellate (FIG.70:C) (in contrast to *Litobrancha*). In imago and subimago of both sexes paracercus is not shortened or slightly shortened, being subequal to cerci (in contrast to Hexagenia/ /fg1); gonostylus with 2 distal segments [i. e. 4-segmented – see Plesiomorphies of Ephemera/fg9].

Size. Fore wing length 8–20 mm.

Distribution. Holarctic, Oriental and Ethiopian Regions; dominate in Oriental and Palaearctic Regions.

Ephemera/fg10 is divided into *Sinephemera, Afromera* and Ephemera/fg11. A number of insufficiently described Asian species have uncertain systematic position (p.237).

2.2;2,1-3/2.1;1. Plesiomorphon **Sinephemera/g(1)**

(Furcatergaliae Fimbriatotergaliae Fossoriae Ephemera/fg9 .../fg10 *Sinephemera*)

(Figs 71:D–E; 72:A)

Nomen hierarchicum: **Sinephemera/g(1)** [g: *Sinephemera* Kluge, SUBGEN. NOV., typus *Ephemera strigata* Eaton 1892].

Plesiomorphies and variable characters of *Sinephemera*. In contrast to *Afromera*, mesonotal suture is moderately stretched posteriorly, and subimaginal mesonotal cuticle has pigmented areas of characteristic shape [see Fimbriatotergaliae (2)]: pigmented microtrichia-bearing area anteriad of mesonotal suture occupies only a part of submedioscutum and is rounded posteriorly (FIG.71:E). In contrast to Ephemera/fg11, penis has no titillators (FIG.72:A).

Size. Fore wing length 8–20 mm.

Distribution. Eurasia: Palaearctic and Oriental Regions.

Species composition of Sinephemera/g(1) (here [*E.*]=[*Ephemera*]). Not less than 5 described species. One **European species:** *glaucops* Pictet 1843 [*E.*]. **Asian species:** *japonica* McLachlan 1875 [*E.*], *separigata* Bae 1995 [*E.*], *shengmi* Hsu 1937 [*E.*], *strigata* Eaton 1892 [*E.*] (synn.subj.: *kuwamayai* Navás 1920 [*E.*], *iwatensis* Matsumura 1931 [*E.*], *jezonica* Matsumura 1931 [*E.*]).

Material examined: *glaucops* [*E.*]: ⊕̄; *japonica* [*E.*]: ℧, ⊕̄, ⊕/⊙; *separigata* [*E.*]: ⊕̄; *shengmi* [*E.*]: ℧, ⊙, ⊕/⊙; *strigata* [*E.*]: ℧, ⊙, ⊕/⊙; sp.E1 (China): ⊕̄, sp.E2 (China): ⊕̄.

***

2.2;2,1-3/2.1;2. **Afromera/g1**

(Furcatergaliae Fimbriatotergaliae Fossoriae Ephemera/fg9 .../fg10 *Afromera*)

(Figs 71:F; 72:C–D)

Nomen hierarchicum: **Afromera/g1** (incl. *Dicrephemera*) [g: *Afromera* Demoulin 1955f: 292, typus *A. congolana* Demoulin 1955 (design. orig.)].
In circumscription matches:
— gen. *Afromera*: McCafferty & Gillies 1979: 170.

References. Demoulin 1955f: ⊕̄'; – McCafferty & Edmunds 1973 (*Dicrephemera*): ℧' ⊕'; – McCafferty & Gillies 1979: ℧*; – Elouard 1986: ⊕̄'.

Autapomorphies of *Afromera*.

(1) Mandibular tusks [see Ephemera/fg9 (2)] are vestigial, subequal to incisor (McCafferty & Edmunds 1973:Fig.11; McCafferty & Gillies 1979: Fig.8). This character is true at least for *siamensis*

[*E.*] and *evae* [*A.*], whose larvae are described, while larvae of other species are unknown.

(2) Mesonotal suture [stretched backward by sides of median line – see Fimbriatotergaliae (2)] is modified: its lateral portion is approximated to lateroparapsidal suture, and median portion stretches along medioparapsidal suture up to its end; thus, in subimago the microtrichia-bearing sclerotized area (located anteriad of mesonotal suture) occupies the whole submedioscutum and is pointed posteriorly (FIG.71:F). Such shape of this area is unique (see Index of characters [2.2.8]). This character is know only for the species examined, while for others mesonotal structure is not described.

(3) Hind wing is smaller, and basitornal margin of fore wing is shorter than in other Ephemera/fg10 (Demoulin 1955f:Fig.1a; McCafferty & Edmunds 1973:Figs 7–8; McCafferty & Gillies 1979:Figs 1–2; Elouard 1986b:Fig.2).

Variable characters of *Afromera*. Penis has no such rolled curved titillators as in Ephemera/fg11; titillators of other form can be present or absent; particularly in *siamensis* [*E.*] titillators are blunt and flat in subimago and absent in imago (FIG.72:C–D).

Size. Fore wing length 8–14 mm.

Distribution. Oriental and Ethiopian Regions.

Nominal taxon included. Afromera/g1 includes **Dicrephemera/g** [g: *Ephemera* subgen. *Dicrephemera* McCafferty & Edmunds 1973: 302, typus *Ephemera siamensis* Ueno 1969 (design. orig.)], regarded as a generic synonym of *Afromera* (McCafferty & Gillies 1979: 170).

Species composition of Afromera/g1. 7 species are formally described, most African species being poorly known and attributed here presumably. One **Asian species:** *siamensis* Ueno 1969 [*Ephemera*]. **African species:** *aequatorialis* Kimmins 1956 [*Ephemera*], *congolana* Demoulin 1955 [*Afromera*], *evae* McCafferty & Gillies 1979 [*Afromera*], *gilliesi* Elouard 1986 [*Afromera*], *natalensis* Barnard 1932 [*Ephemera*], *troubati* Elouard 1986 [*Afromera*].

Material examined: *siamensis* [*E.*]: ♂/⊕, ♂–⊕, ⊕, ♀/⊕/⊙; sp. (Mali): ♀/⊙.

***

2.2;2,1-3/2.1;3. Ephemera/fg11

(Furcatergaliae Fimbriatotergaliae Fossoriae Ephemera/fg9 .../fg11)

(Figs 70:B–E; 71:A–C; 72:B)

Nomen hierarchicum: **Ephemera/fg11** (sine *Sinephemera*, *Afromera*).

Autapomorphy of Ephemera/fg11.

(1) Penis has a pair of titillators of unique structure: each titillator arises ventrally near penis base and is directed distally, along membranous concavity of ventral penis surface; titillator represents a sclerotized lamella rolled forming a tube; thus, titillator looks as a smooth spine round in cross section; it is arched medially, with pointed apex (FIG.72:B). Larval penis bud has no any external titillator buds; subimaginal penis is identical to imaginal one, with just the same rolled titillators. In other Ephemera/fg10 penis also can have projections mediad of ventral sclerotized wall, but they are not rolled in such manner (FIG.72:C).

Plesiomorphies of Ephemera/fg11. In contrast to *Afromera*, mesonotal suture is moderately stretched posteriorly, and subimaginal mesonotal cuticle has pigmented areas of characteristic shape [see Fimbriatotergaliae (2)]: pigmented microtrichia-bearing area anteriad of mesonotal suture occupies only a part of submedioscutum and is rounded posteriorly (as in FIG.71:E).

Size. Fore wing length 8–20 mm.

Distribution. Holarctic, Oriental and Ethiopian Regions; dominate in eastern Asia.

Nominal taxon included. Ephemera/fg11 includes **Nirvius/g** [g: *Nirvius* Navás 1922a: 56, typus *N. punctatus* Navás 1922 (design. orig.; syn. subj. *Ephemera duporti* Lestage 1921)], regarded as a generic synonym of *Ephemera* (Lestage 1922: 56).

Species composition of Ephemera/fg11 (here [*E.*]= =[*Ephemera*]). More than 30 species. One **African species:** *mooiana* McCafferty 1971 [*E.*]. **European and Caucasian species:** *danica* Müller 1764 [*E.*] (synn. subj.: *maculata* Villers 1789 [*Ephemera*], *cognata* Stephens 1835 [*E.*], *hispanica* Rambur 1842 [*E.*], *parnassiana* Demoulin 1958 [*E.*]), *lineata* Eaton 1870 [*E.*], *romantzovi* Kluge 1988 [*E.*], *vulgata* Linnaeus 1758 [*E.*] (synn. subj.: *communis* Retzius 1783 [*E.*]), *zettana* Kimmins 1937 [*E.*] (syn. subj.: *paulae* Grandi 1956 [*E.*]). **Asian species:** *annandalei* Chopra 1937 [*E.*], *brunnea* Hubbard & Peters 1978 [*E.*] (syn. obj.: *striatum* Ali 1970 [*Ephemera*] nom. praeocc.), *consors* Eaton 1892 [*E.*], *diffusa* Chopra 1937 [*E.*], *duporti* Lestage 1921 [*E.*] (syn. subj.: *punctatus* Navás 1922 [*Nirvius*]), *hainanensis* You & Gui 1995 [*E.*], *hongjiangensis* You & Gui 1995 [*E.*], *hunanensis* You & Gui 1995 [*E.*], *immaculata* Eaton 1871 [*E.*], *jianfengensis* You & Gui 1995 [*E.*], *kirinensis* Hsu 1937 [*E.*], *orientalis* McLachlan 1875 [*E.*] (synn. subj.: *amurensis* Navás 1912 [*E.*], *modesta* Brodsky 1930 [*E.*], *kuccharonis* Matsumura 1931 [*E.*]), *pramodi* Dubey 1972 [*E.*], *pulcherrima* Eaton 1892 [*E.*], *remensa* Eaton 1892 [*E.*], *sachalinensis* Matsumura 1931 [*E.*], *serica* Eaton 1871 [*E.*], *transbajkalica* Tshernova 1973 [*E.*], *wanquanensis* You & Gui 1995 [*E.*], *wuchowensis* Hsu 1937 [*E.*], *yaoshani* Hsu 1937 [*E.*], *zhangjiajiensis* You & Gui 1995 [*E.*]. **North**

American species: *blanda* Traver 1932 [*E.*], *compar* Hagen 1875 [*E.*], *guttulata* Pictet 1843 [*E.*], *simulans* Walker 1853 [*E.*] (synn. subj.: *decora* Walker 1853 [*E.*], *natata* Walker 1853 [*Palingenia*]), *traverae* Spieth 1938 [*E.*], *triplex* Traver 1935 [*E.*], *varia* Eaton 1883 [*E.*].

Material examined: *danica* [*E.*]: Ծ, **O**, ⊕/⊙; *hainanensis* [*E.*]: **O**, ⊕; *lineata* [*E.*]: Ծ, ⊕; *orientalis* [*E.*]: Ծ, **O**, ⊕/⊙; *sachalinensis* [*E.*]: Ծ–**O**–⊕/⊙; *romantzovi* [*E.*]: Ծ–**O**–⊕/⊙; *serica* [*E.*]: ⊕;, *transbajkalica* [*E.*]: Ծ, Ծ–**O**–⊕, **O**, ⊕/⊙; *vulgata* [*E.*]: Ծ, **O**, ⊕/⊙; spp. (China, Vietnam, Thailand): Ծ, **O**, ⊕/⊙.

***

Ephemera/fg10 INCERTAE SEDIS

Ephemera/fg10 INCERTAE SEDIS. Aethephemera/g [g: *Ephemera* subgen. *Aethephemera* McCafferty & Edmunds 1973: 306, typus *Ephemera (Aethephemera) nadinae* McCafferty & Edmunds 1973 (design. orig.)].

References. McCafferty & Edmunds 1973: Ծ; – Balasubramanian & Venkataraman & Sivaramakrishnan 1991: ⊕.

Distribution. India.

Material examined. –

Comment. The subgenus *Aethephemera* was established for a single species; larva differs from other known species by shortened tusk of right mandible; penis has no titillators (in contrast to Ephemera/fg11); mesonotal structure is not described.

Other species of Ephemera/fg10 INCERTAE SEDIS (here [*E.*]=[*Ephemera*]). For the following Asian species genital and/or mesonotal structure is unknown or non-described: *axillaris* Navás 1930 [*E.*], *distincta* Hubbard 1982 [*E.*], *expectans* Walker 1860 [*Potamanthus*] (syn. subj.: *vedana* Banks 1913 [*E.*]), *formosana* Ulmer 1920 [*E.*], *fulvata* Navás 1935 [*E.*], *hasalakensis* Hubbard 1983 [*E.*], *hellenica* Demoulin 1955 [*E.*], *hsui* You & Gui 1995 [*E.*], *innotata* Navás 1922 [*E.*], *javana* Navás 1930 [*E.*], *koshunensis* Matsumura 1931 [*E.*], *lankensis* Hubbard 1983 [*E.*], *longiventris* Navás 1917 [*E.*], *lota* Navás 1934 [*E.*], *maoyangensis* You & Gui 1995 [*E.*], *media* Ulmer 1936 [*E.*], *nathani* Hubbard 1982 [*E.*], *nigroptera* Zhou & Gui & Su 1998 [*E.*], *pictipennis* Ulmer 1924 [*E.*], *pictiventris* McLachlan 1904 [*E.*], *pieli* Navás 1934 [*E.*], *posticus* Banks 1914 [*Rhoenanthus*], *purpurata* Ulmer 1920 [*E.*], *quadriguttata* Lestage 1927 [*E.*], *sauteri* Ulmer 1912 [*E.*], *soanica* Ali 1967 [*E.*] (syn. obj.: *soanensis* Ali 1971 [*E.*]), *spilosa* Navás 1936 [*E.*], *supposita* Eaton 1883 [*E.*].

*** ***

2.2;2,1-3/2.2. **Hexagenia/fg1**

(Furcatergaliae Fimbriatotergaliae Fossoriae Ephemera/fg9 Hexagenia/fg1)

(Fig. 70:A)

Nomen hierarchicum: **Hexagenia/fg1** (incl. *Eatonica*) [f: Hexageniinae McCafferty 1991a: 354; g: *Hexagenia* Walsh 1863: 197, typus *Ephemera limbata* Serville 1829 (design. Eaton 1868b: 85)].

In circumscription matches:

— *Hexagenia*-complex: Koss & Edmunds 1974: 295;

— subfam. Hexageniinae McCafferty 1991a: 354.

Reference. Koss & Edmunds 1974: ⊙*.

Autapomorphies of Hexagenia/fg1.

(1) Paraglossae are stretched proximally-laterally in such a manner, that their stretched parts lie laterad of mentum and dorsad of palps (Eaton 1883–1888: Pl.63: 6–7).

(2) Maxilla has 2 canines only (in contrast to three initial ones in Ephemera/fg10); distal dentiseta [see Plesiomorphies of Ephemera/fg9] is vestigial, nearly invisible between proximal dentiseta and canines (only Hexagenia/g3 *limbata* [*E.*] and Eatonigenia/g1 *chinei* [*H.*] are examined).

(3) In imago and subimago of both sexes paracercus is vestigial. Non-unique apomorphy (see Index of characters [2.3.22]).

Character of unclear phylogenetic status.

(4) Apex of larval fore tibia projects posteriad of tarsus base and forms outer-apical denticle and inner-apical projection, with incision between them (Eaton 1883–1888:Pl.63:9–10). Similar in Palingenia/f2=g1 (FIG.74:D–E) and *Ichthybotus*; possibly, symplesiomorphy [see Fossoriae (1)].

Plesiomorphies of Hexagenia/fg1. Larval frontal projection is not bipointed (in contrast to Ephemera/fg10): semicircular, conic or non-expressed. In imago and subimago: Mesonotal suture is moderately stretched posteriorly, and subimaginal mesonotal cuticle has pigmented areas of characteristic shape [see Fimbriatotergaliae (2)] (as in FIG.71:E).

Size. Fore wing length 10–30 mm.

Distribution. Nearctic, Neotropical, Ethiopian and Oriental Regions; probably, also Palaeogene of Palaearctic (see *Denina* in Ephemera/fg9 INCERTAE SEDIS).

Hexagenia/fg1 is divided into Hexagenia/fg2 and Eatonica/g1.

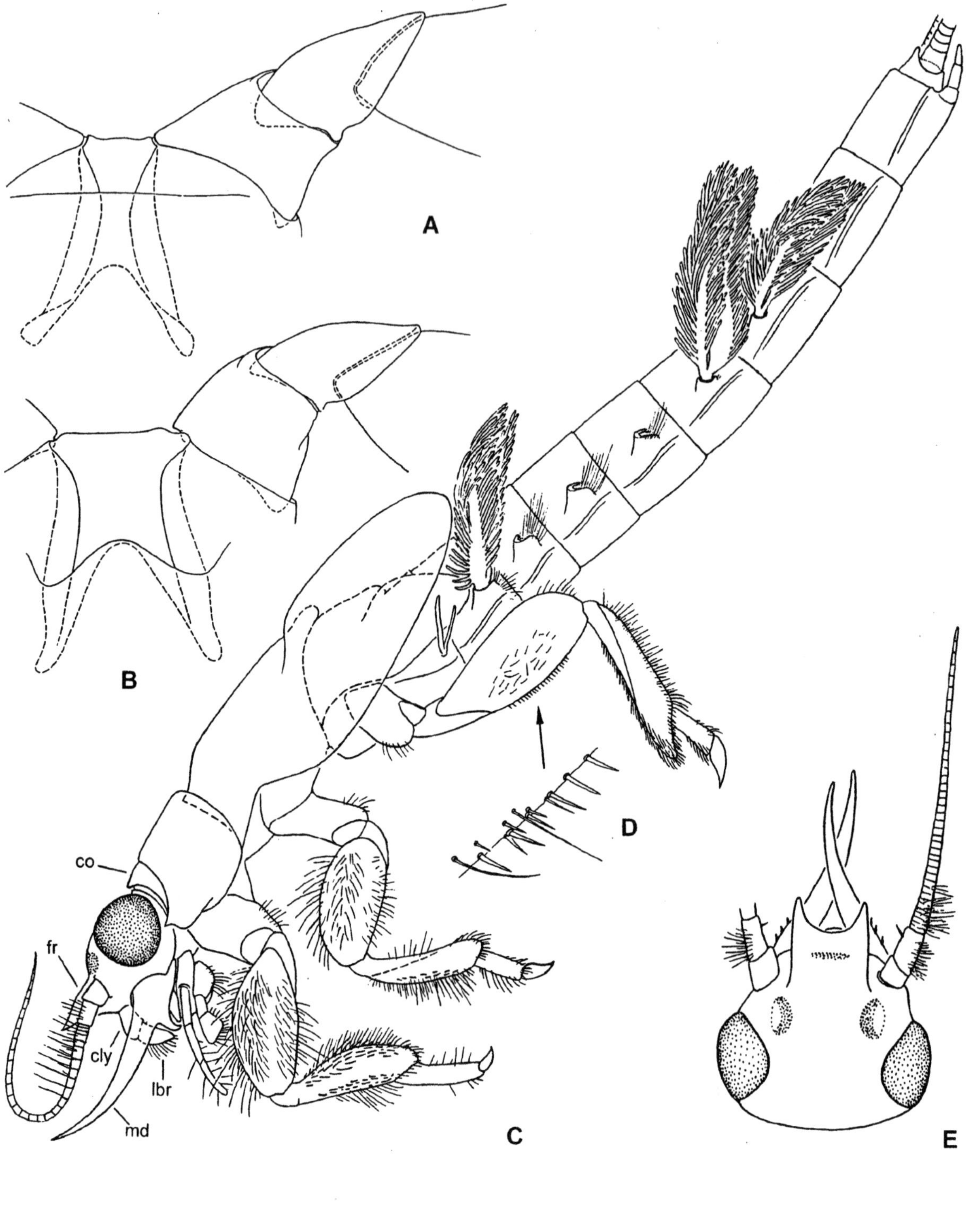

Figure 70. Ephemera/fg9.
A – Hexagenia/fg* *limbata* [*Ephemera*], larval prosternum and base of left fore leg, view from behind (compare with Figs 74:C and 77:B). **B–E** – Ephemera/fg* *vulgata* [*Ephemera*]: **B** – larval prosternum and base of left fore leg, view from behind; **C** – male larva of last instar, lateral view (tergalii III-V removed); **D** – setae on inner side of hind femur; **E** – head, dorsal view.

Abbreviations: **cly** – clypeus; **co** – collar; **fr** – frontal projection; **lbr** – labrum; **md** – mandibular tusk.

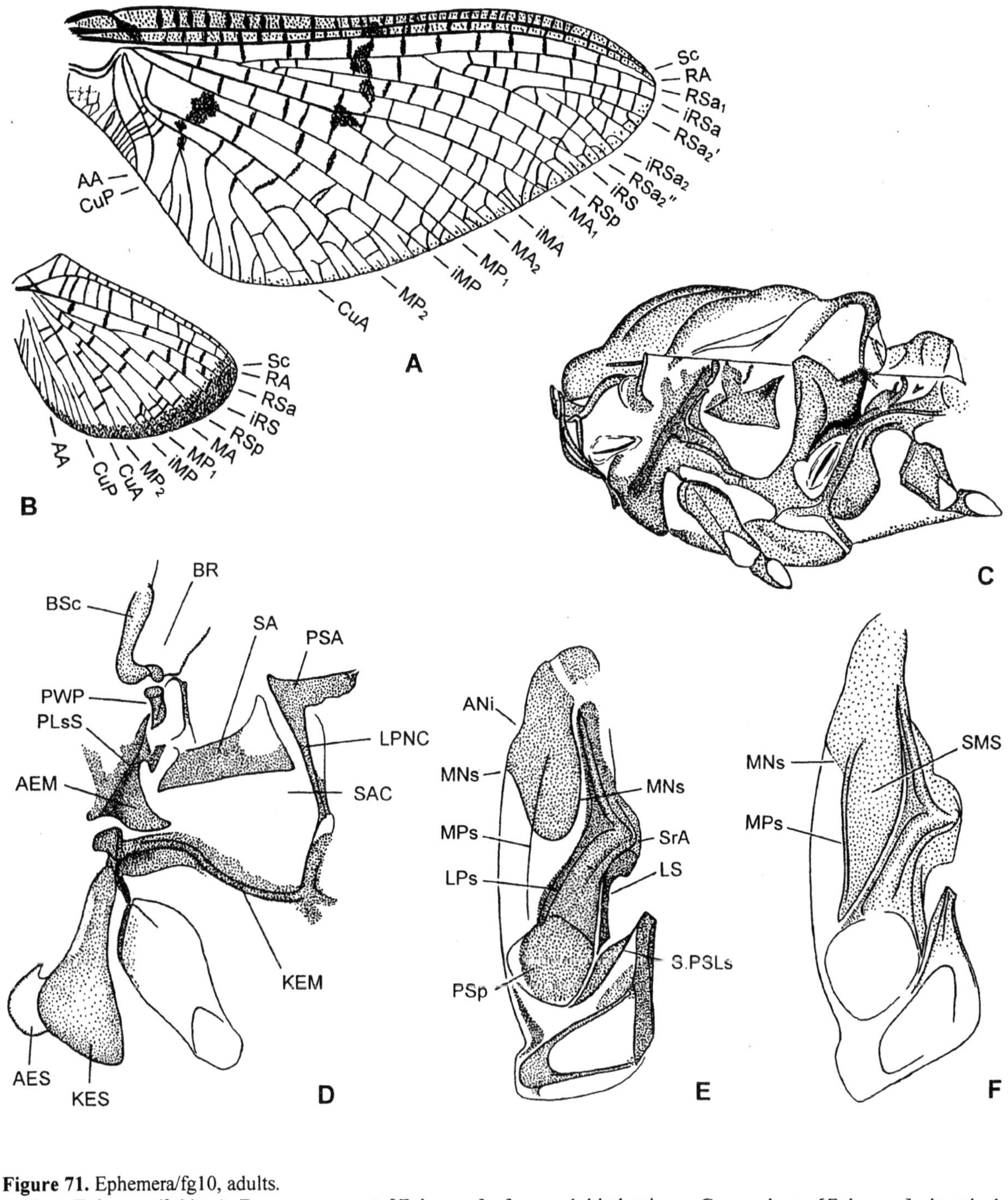

Figure 71. Ephemera/fg10, adults.

A–C – Ephemera/fg11: **A–B** – *romantzovi* [*Ephemera*], fore and hind wings; **C** – *vulgata* [*Ephemera*], imaginal pterothorax, lateral view. **D–E** – Sinephemera/g* *strigata* [*Ephemera*]: **D** – subimaginal exuviae of left mesopleuron (compare with Fig.5:D); **E** – subimaginal exuviae of right half of mesonotum. **F** – Afromera/g1 *siamensis* [*Ephemera*], subimaginal exuviae of left mesopleuron (here dotes show sclerotized ares, but not pigmentation; cuticle is nearly entirely colourless) (A–B – from Kluge 1988b; D–E – from Kluge 1997b).

Abbreviations: **AEM** – anepimeron; **AES** – anepisternum; **ANi** – anteronotal transverse impression; **BR** – basiradiale; **BSc** – basisubcostale; **KEM** – katepimeron; **KES** – katepisternum; **LPNC** – lateropostnotal crest; **LPs** – lateroparapsidal suture; **LS** – lateroscutum; **MNs** – mesonotal suture; **MPs** – medioparapsidal suture; **PLsS** – superior pleural suture; **PSA** – postsubalar sclerite of lateropostnotum; **PSp** – posterior scutal protuberance; **PWP** – pleural wing process; **SA** – subalare; **SAC** – subalar conjunctiva; **SMS** – submedioscutum; **S.PSLs** – scuto-parascutellar suture; **SrA** – suralare.

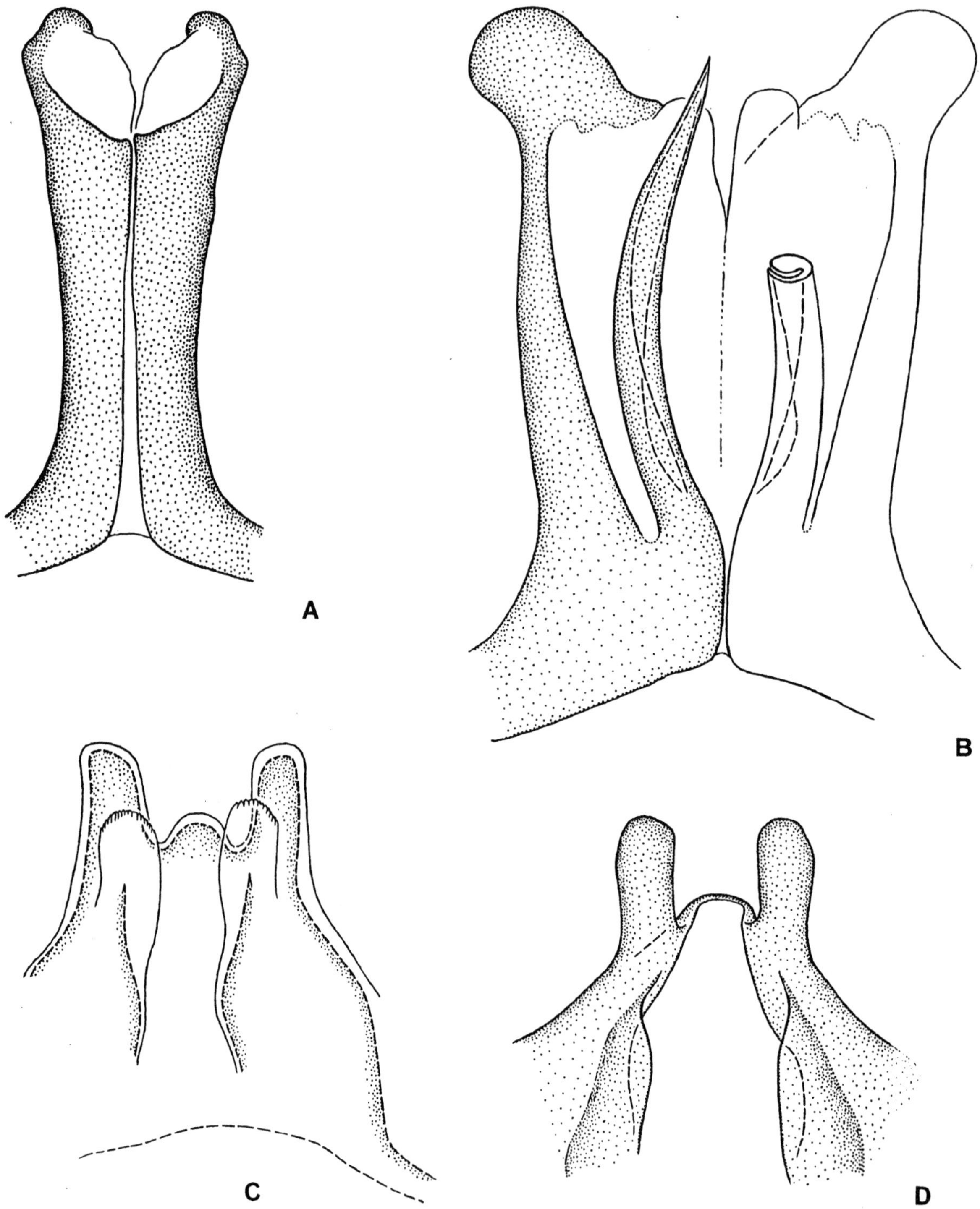

Figure 72. Ephemera/fg10, penis structure.
A – Sinephemera/g* *strigata* [*Ephemera*], penis of imago, ventral view; **B** – Ephemera/fg11 *vulgata* [*Ephemera*], penis of imago, ventral view (left titillator is cut off to show transverse section). **C–D** – Afromera/g1 *siamensis* [*Ephemera*]: **C** – penis of subimago, ventral view (internal subimaginal parts shown by interrupted lines); **D** – penis of imago, ventral view.

2.2;2,1-3/2.2;1. **Hexagenia/fg2**

(Fimbriatotergaliae Fossoriae Ephemera/fg9 Hexagenia/fg1 .../fg2)

(Fig. 70:A)

Nomen hierarchicum: **Hexagenia/fg2** (sine *Eatonica*; incl. *Pseudeatonica*).
In circumscription matches:
— gen. *Hexagenia*: Eaton 1883–1888: 48.

References. Eaton 1883–1888: Ꝺ ⊕; – Edmunds & Jensen & Berner 1976: Ꝺ* ⊕*.

Character of unclear phylogenetic status.

(1) Larval flagellum of antenna with whorls of long dense setae (Eaton 1883–1888:Pl.63:1). The same in Ephemera/fg10 (FIG.70:C, E).

Plesiomorphies of Hexagenia/fg2. Vestige of tergalius I [see Ephemera/fg9 (4)] is bilamellate (in contrast to Eatonica/g1-*Litobrancha*). Gonostylus without projection peculiar for Eatonica/g1.

Size. Fore wing length 10–25 mm.

Distribution. America: Nearctic and Neotropical Regions; probably in Palaeogene was distributed in Palaearctic as well (see *Denina* in Ephemera/fg9 INCERTAE SEDIS).

Hexagenia/fg2 is divided into Hexagenia/fg3 and *Pseudeatonica.*

2.2;2,1-3/2.2;1,1. Plesiomorphon **Hexagenia/fg3**

(Fimbriatotergaliae Fossoriae Ephemera/fg9 Hexagenia/fg1 .../fg3)

(Fig. 70:A)

Nomen hierarchicum: **Hexagenia/fg3** (sine *Pseudeatonica*).
In circumscription matches:
— subgen. *Hexagenia*: McCafferty 1970: 225;
— gen. *Hexagenia*: Kimmins 1960: 309.

References. Needham & Traver & Hsu 1935: Ꝺ* ⊕*; – Edmunds & Jensen & Berner 1976 (subgen.): Ꝺ* ⊕*; – McCafferty 1970: ⊕*; – 1975: Ꝺ*; – 1979b: ⊕*.

Plesiomorphy of Hexagenia/fg3. Imaginal and subimaginal gonostylus with 2 distal segments (in contrast to *Pseudeatonica* and some of Eatonica/ /fg1). Only in selected specimens of Hexagenia/fg3 distal segments of gonostyli are separated indistinctly or fused one with another.

Size. Fore wing length 10–25 mm.

Distribution. Nearctic; probably in Palaeogene was distributed in Palaearctic as well (see *Denina* in Ephemera/fg9 INCERTAE SEDIS).

Species composition of Hexagenia/fg3 (here [*H.*]= =[*Hexagenia*]). Not less than 4 species – *atrocaudata* McDunnough 1924 [*H.*], *bilineata* Say 1824 [*Baetis*], *rigida* McDunnough 1924 [*H.*] and *limbata* Serville (in Guerin) 1829 [*Ephemera*] (synn. subj.: *affiliata* McDunnough 1927 [*H.*], *angulata* Walker 1953 [*Baetis*], *californica* Upholt 1937 [*H.*], *carolina* Traver 1931 [*H.*], *elegans* Traver 1931 [*H.*], *kanuga* Traver 1937 [*H.*], *marilandica* Traver 1931 [*H.*], *mingo* Traver 1931 [*H.*], *munda* Eaton 1883 [*H.*], *occulta* Walker 1853 [*Palingenia*], *orlando* Traver 1931 [*H.*], *pallens* Traver 1935 [*H.*], *rosacea* Traver 1931 [*H.*], *variabilis* Eaton 1885 [*H.*] *venusta* Eaton 1883 [*H.*], *viridescens* Walker 1853 [*Palingenia*], *weewa* Traver 1931 [*H.*]).

Material examined: *limbata* [*E.*]: Ꝺ, ⊙, ⊕, ⊕/⊙.

***

2.2;2,1-3/2.2;1,2. **Pseudeatonica/g(1)**

(Fimbriatotergaliae Fossoriae Ephemera/fg9 Hexagenia/fg1 .../fg2 *Pseudeatonica*)

Nomen hierarchicum: **Pseudeatonica/g(1)** [g: *Hexagenia* subgen. *Pseudeatonica* Spieth 1941: 269, typus *Hexagenia mexicana* Eaton 1883 (design. orig.)].
In circumscription matches:
— subgen. *Pseudeatonica* Spieth 1941: 269;
— gen. *Pseudeatonica*: Kimmins 1960: 312.

References. Eaton 1883–1888 (*Hexagenia mexicana*): ⊕; – 1892a (*H. mexicana*): ⊕; – Needham & Murphy 1924 (*H. mexicana*): ⊕; – Kimmins 1934 (*H. mexicana*): ⊕; – Kimmins 1960: ⊕; – McCafferty 1970: Ꝺ ⊕*; – Edmunds & Jensen & Berner 1976: Ꝺ ⊕*.

Character of unclear phylogenetic status.

(1) Imaginal and subimaginal gonostylus with 1 distal segment only (instead of two initial ones). Non-unique apomorphy (see Index of characters [2.3.12]); among Ephemera/fg9 the same in Eatonica/g2 and *Eatonigenia.*

Size. Fore wing length 13 mm.

Distribution. Neotropical Region.

Species composition Pseudeatonica/g(1). 3 species – *callineura* Banks 1914 [*Hexagenia*], *mexicana* Eaton 1883 [*Hexagenia*] and *albivitta* Walker 1853 [*Baetis*] (synn. subj.: *continua* Walker 1860 [*Palingenia*], *dorsigera* Hagen 1861 [*Palingenia*], *benedicta* Navás 1922 [*Hexagenia*], *dominans* Navás 1937 [*Hexagenia*]).

Material examined: –.

*** ***

2.2;2,1-3/2.2;2. **Eatonica/g1**

(Fimbriatotergaliae Fossoriae Ephemera/fg9 Hexagenia/fg1 Eatonica/g1)

Nomen hierarchicum: **Eatonica/g1** (incl. *Eatonigenia*, *Litobrancha*) [g: *Eatonica* Navás 1913: 181, typus *Ephemera schoutedeni* Navás 1911 (design. orig.)].

Autapomorphy of Eatonica/fg1.

(1) In imago and subimago 1st segment of gonostylus ventrally with apical rounded projection which projects distad of place of attachment of 2nd segment (McCafferty 1971a:Fig.7; 1971b:Figs 1C, 2C; 1973:Figs 9–12).

Character of unclear phylogenetic status.

(2) Larval flagellum of antenna without long setae, with short sparse setae only (in contrast to Hexagenia/fg2 and Ephemera/fg10).

Size. Fore wing length 10–24 mm.

Distribution. Nearctic, Ethiopian an Oriental Regions.

Eatonica/fg1 is divided into *Litobrancha*, Eatonica/g2 and *Eatonigenia.*

2.2;2,1-3/2.2;2,1. **Litobrancha/g(1)**
(Fossoriae Ephemera/fg9 Hexagenia/fg1 Eatonica/g1 *Litobrancha*)

Nomen hierarchicum: **Litobrancha/g(1)** [g: *Litobrancha* McCafferty 1971a: 45, typus *Hexagenia recurvata* Morgan 1913 (design. orig.)].
In circumscription matches:
— gen. *Litobrancha* McCafferty 1971a: 45.
Monospecific taxon.

References. Burks 1953 (*Hexagenia recurvata*): Ꝺ ⊕; – McCafferty 1971a: Ꝺ ⊕; – 1975: Ꝺ; – Edmunds & Jensen & Berner 1976: Ꝺ* ⊕*.

Autapomorphy of *Litobrancha*.

(1) Vestige of tergalius I [see Ephemera/fg9 (4)] is unilamellate. Non-unique apomorphy (see Index of characters [1.3.33]).

Plesiomorphies of *Litobrancha*. Larval claw of fore leg is normally developed (in contrast to *Eatonigenia*). Imaginal and subimaginal gonostylus with 2 distal segments (in contrast to Eatonica/g2 and *Eatonigenia*). Penis has no shape peculiar for *Eatonigenia.*

Size. Fore wing length 15–24 mm.

Distribution. Eastern Nearctic.

Species composition of Litobrancha/g(1). 1 species – *recurvata* Morgan 1913 [*Hexagenia*].

Material examined: *recurvata* [*H.*]: Ꝺ.

***

2.2;2,1-3/2.2;2,2. **Eatonica/g2**
(Fossoriae Ephemera/fg9 Hexagenia/fg1 Eatonica/g1 .../g2)

Nomen hierarchicum: **Eatonica/g2** (sine *Litobrancha*, *Eatonigenia*).
In circumscription matches:
— gen. *Eatonica* Navás 1913: 181.

References. Barnard 1932: ⊕; – Demoulin 1968b: Ꝺ; – 1969: ⊕; – McCafferty 1971b: ⊕; – McCafferty & Gillies 1979: Ꝺ ⊕; – Elouard 1986a: ⊕; – Elouard & Oliarinony & Sartori 1998: ⊕'.

Character of unclear phylogenetic status.

(1) Imaginal and subimaginal gonostylus with 1 distal segment only (instead of two initial ones). Non-unique apomorphy (see Index of characters [2.3.12]); among Ephemera/fg9 the same in *Eatonigenia* (possibly synapomorphy) and *Pseudeatonica.*

Plesiomorphies of Eatonica/g2. Larval claw of fore leg is normally developed (in contrast to *Eatonigenia*). Vestige of tergalius I [see Ephemera/ /fg9 (4)] is bilamellate (in contrast to *Litobrancha*). Penis has no shape peculiar for *Eatonigenia.*

Size. Fore wing length 13–30 mm.

Distribution. Ethiopian Region.

Species composition of Eatonica/g2. 6 species – *crassi* McCafferty 1971 [*Eatonica*], *denysae* Elouard & Sartori 1998 [*Eatonica*], *josettae* Demoulin 1969 [*Eatonica*], *luciennae* Elouard & Oliarinony 1998 [*Eatonica*], *patriciae* Elouard 1986 [*Eatonica*], *schoutedeni* Navás 1911 [*Ephemera*] (synn. subj.: *fulva* Esben-Petersen 1913 [*Hexagenia*], *illustris* Eaton 1913 [*Hexagenia*], *reticulata* Navás 1929 [*Hexagenia*]).

Material examined: *crassi* [*E.*]: ♀, ⊕♂.

***

2.2;2,1-3/2.2;2,3. **Eatonigenia/g1**
(Fossoriae Ephemera/fg9 Hexagenia/fg1 Eatonica/g1 *Eatonigenia*)

Nomen hierarchicum: **Eatonigenia/g1** [g: *Eatonigenia* Ulmer 1939: 477, typus *Hexagenia chaperi* Navás 1935 (design. orig.)].
In circumscription matches:
— gen. *Eatonigenia* Ulmer 1939: 477;
— gen. *Heterogenesia* Dang 1967: 159.

References. Ulmer 1939: ⊕; – Dang 1967: Ꝺ; – McCafferty 1973: Ꝺ* ⊕*; – Zhang 1988: ⊕.

Autapomorphies of *Eatonigenia*.

(1) On larval fore leg [burrowing – see Fossoriae (1)] claw is vestigial, completely dipped into an incision on apex of tibia (McCafferty 1973:Figs 23–24). Unique apomorphy.

(2) Penis lobes are fused all along their length and have a characteristic shape – trapezoid, widened apically, with gonopores located on a pair of small ventral projections in proximal-median part (McCafferty 1973:Figs 9–12).

Character of unclear phylogenetic status.

(3) Imaginal and subimaginal gonostylus with 1 distal segment only (instead of two initial ones).

Non-unique apomorphy (see Index of characters [2.3.12]); among Ephemera/fg9 the same in Eatonica/g2 (possibly synapomorphy) and *Pseudeatonica*.

Plesiomorphy of *Eatonigenia*. Vestige of tergalius I [see Ephemera/fg9 (4)] is bilamellate (in contrast to *Litobrancha*).

Size. Fore wing length 10–17 mm.

Distribution. Oriental Region.

Nominal taxon included. Eatonigenia/g1 includes **Heterogenesia/g** [g: *Heterogenesia* Dang 1967:159, typus *H. chinei* Dang 1967 (design. orig.)], regarded as a generic synonym of *Eatonigenia* (McCafferty 1991d:180).

Species composition of Eatonigenia/g1. 6 species – *chaperi* Navás 1935 [*Hexagenia*], *chinei* Dang 1967 [*Heterogenesia*], *indica* Chopra 1924 [*Hexagenia*], *philippina* Navás 1933 [*Hexagenia*], *seca* McCafferty 1973 [*Eatonigenia*], *trirama* McCafferty 1973 [*Eatonigenia*].

Material examined: *chinei* Dang 1967 [*Heterogenesia*]: Ѽ.

*** *** ***

Ephemera/fg9 INCERTAE SEDIS

Ephemera/fg9 INCERTAE SEDIS 1. † Denina/g [g: *Denina* McCafferty 1987: 472, typus *D. dubiloca* McCafferty 1987 (design. orig.)].

Reference. McCafferty 1987: ⊕'.

Age. Eocene (Baltic amber).

Material examined: –.

Comment. In contrast to Ephemera/fg10, paracercus is shortened (nearly 2 times longer than abdominal tergum X), but not as strongly reduced as in Hexagenia/fg1. Imago has no apomorphies peculiar for *Pseudeatonica* and Eatonica/g1. Shape of penis is similar to that of Hexagenia/fg2.

Other Ephemera/fg9 INCERTAE SEDIS. Single fossil larva from Oligocene (USA) was described as "*Hexagenia* sp." (Lewis 1978).

*** ***

2.2;2,1-3/3. Behningia/fg1

(Bidentiseta Furcatergaliae Fimbriatotergaliae Fossoriae Behningia/fg1)

(Fig. 73)

Nomen hierarchicum: **Behningia/fg1** (incl. *Protobehningia*) [f: Behningiidae Motas & Bacesco 1937: 29; g: *Behningia* Lestage 1930: 436, typus *B. ulmeri* Lestage 1930 (design. orig.)].

In circumscription matches:

— fam. Behningiidae: Tshernova & Bajkova 1960: 410;

— superfam. Behningioidea: McCafferty 1991a: 349;

— Behningia/fg1: Kluge 2000: 252.

References. Tshernova 1970: Ѽ* ⊕*; – Peters & Gillies 1991: Ѽ* ⊕*; – Elpers & Tomka 1994b: Ѽ*; – Kluge 1997d: Ѽ*.

Autapomorphies of Behningia/fg1.

(1) Larva [burrowing – see Fossoriae (1)] has on its head, thorax and abdomen projections of unique shape which are covered by stout setae directed posteriorly, that helps to move in sand (FIG.73:A, E):

Larval frons anteriorly bears a pair of wide rounded projections which cover from above not only clypeus [see Fossoriae (2)], but also antennae bases; these projections are densely covered with stout setae directed posteriorly. Due to this, imaginal antennae are located on a pair of projections directed anteriorly (Tshernova 1938:Fig.1).

On larval pronotum ends of the crest separating collar [see Fimbriatotergaliae (1)] are transformed to a pair of large flat antero-lateral projections, densely covered with stout setae directed posteriorly.

Larval mesothoracic episterna are more or less projected anteriorly in a form of bald sclerotized projections (especially remarkable in Behningia/ /fg2), and epimera are densely covered with stout setae directed posteriorly.

Larval abdominal terga bear a pair of longitudinal ridges covered with setae; in *Protobehningia* these ridges are shallow (FIG.73:E), while in Behningia/fg2 they are projected laterally and cover tergalii from above (FIG.73:A–B).

In *Protobehningia* burrowing leg specialization is retained, but in Behningia/fg2 body shape and method of movement in sand are strongly modified, and leg specialization is changed (see below).

(2) Mouth apparatus [located at a distance from anterior margin of head because of frontal projections – see (1)] is specialized for carnivorism:

Mandible is shortened perpendicular to its axis of articulation, mola without grater, incisor and kinetodontium are stout; initial asymmetry of mandibles is either lost (in Behningia/fg3), or nearly lost (in *Protobehningia* and *Dolania*) (Elpers & Tomka 1994b:Figs 4, 5, 19, 20).

Labium has glossae and paraglossae more or less diminished and palps enlarged, massive, without muscle inside 2nd segment (border between 2nd and 3rd segments is distinct in *Dolania*, and indistinct in *Protobehningia* and Behningia/fg3).

(3) Winged stages are short-living; moult from subimago to imago takes place in males only; middle and hind legs of male and all legs of female are non-functional (non-unique apomorphy – see Index of characters [2]). In female all legs are vestigial, tarsi are non-segmented, claws are lost. In male imago middle and hind legs lack tarsi; at least in *Dolania* (according to Peters & Peters 1986) middle and hind legs of male subimago with 2-segmented tarsus and one claw.

(4) RSa_2 of fore wing simple, not forming triad. The same in some other short-living mayflies [see (3)] (see Index of characters [2.2.37]).

(5) On fore wing bifurcation of MA is transferred to wing base. The same in some other short-living mayflies [see (3)] (see Index of characters [2.2.43]).

(6) Cubital field of fore wing [see Anteritorna (1)] with 1–8 intercalaries sometimes alternating as concave and convex, the anteriormost intercalary being the longest (non-unique apomorphy – see Index of characters [2.2.52]). CuA is either simple (in *Protobehningia*), or bifurcate [see Behningia/fg2 (3) below], thus no more than one vein arises from CuA.

(7) Amphitornal margin of imaginal wing bears setae similar to that of subimago. Non-unique apomorphy (see Index of characters [2.2.27]).

(8) In male imago on fore legs (elongate and specialized for grasping female, as in other mayflies) tarsus is non-segmented, with one blunt claw only (Keffermüller 1959:Pl.V; Peters & Peters 1986:Fig.1; Peters & Gillies 1991:Fig.2). Unique apomorphy.

(9) Penis lobes are very long, much longer than gonostyli [at the same time gonostyli are normally developed, not shortened – see (10)]. Unique apomorphy: so long penes are present also in some Geminovenata, but there penes are protractible, while penes of Behningia/fg1 are rigid.

(10) Gonostylus [see (9)] lacks distal segments. Non-unique apomorphy (see Index of characters [2.3.12]).

(11) Imaginal and subimaginal caudalii [see (12)] lack distinct segmentation. Non-unique apomorphy, the same in some Palingenia/f3=g2 and others.

(12) In male imago paracercus [see (11)] is shortened, much shorter than cerci: 1/2 of cerci length in *Protobehningia* and strongly shortened in Behningia/fg2 (but in female paracercus is subequal to cerci). Non-unique apomorphy (see Index of characters [2.3.22]).

Characters of Behningia/fg1 of doubtful polarity (see Classifications of Fimbriatotergaliae II).

(13) Mandibles lack tusks. Probably result of reduction [see Fimbriatotergaliae (8)].

(14) On inner side of femur and tibia of larval hind leg spiny setae are absent. Probably result of reduction [see Fimbriatotergaliae (9)].

(15) Tergalius I [different from others – see Fimbriatotergaliae (6)] has a form of a single large (as long as tergalii of next pairs) lamella with numerous marginal processes (FIG.73:A, E). As in other Fimbriatotergaliae tergalius I is strongly diminished and always lacks processes, we can assume that in Behningia/fg1 vestige of this tergalius has been secondarily enlarged and got the same processes as on tergalii II–VII, but remains to be unilamellate.

Plesiomorphies of Behningia/fg1. Maxillary palp is always 3-segmented [in spite of specialization of mouth apparatus – see (2)] (in contrast to Cryptoprosternata). Furcasternal protuberances are contiguous (in contrast to Campsurus/fg1, Caenotergaliae and some others – see Index of characters [2.2.23]).

Size. Fore wing length 6–24 mm.

Distribution. Holarctic and Oriental Region.

Behningia/fg1 is divided into *Protobehningia* and Behningia/fg2.

2.2;2,1-3/3.1. Protobehningia/g(1)

(Furcatergaliae Fimbriatotergaliae Fossoriae Behningia/fg1 *Protobehningia*)

(Fig. 73:E)

Nomen hierarchicum: **Protobehningia/g(1)** [g: *Protobehningia* Tshernova & Bajkova 1960: 410, typus *P. asiatica* Tshernova & Bajkova 1960 (design. orig.)].
In circumscription matches:
— gen. *Protobehningia* Tshernova & Bajkova 1960: 410.

References. Tshernova & Bajkova 1960: ⥀; – Peters & Gillies 1991: ⥀ ⊕; – Elpers & Tomka 1994b: ⥀.

Autapomorphies of *Protobehningia*.

(1) On larval fore leg [burrowing – see Fossoriae (1)] tarsus is fused with tibia (FIG.73:E; Tshernova & Bajkova 1960:Fig.2). Non-unique apomorphy, similar fusion in Campsurus/fg1.

(2) Patella-tibial suture (initially present on middle and hind legs) is lost on all legs of larva (FIG. 73:E; Tshernova & Bajkova 1960:Fig.2). Non-unique apomorphy (see Index of characters [1.2.18]).

(3) On maxilla both dentisetae [see Bidentiseta (1)] are lost (Tshernova & Bajkova 1960:Fig.2; Elpers & Tomka 1994b:Figs 2a–d).

Plesiomorphies of *Protobehningia* (in contrast to Behningia/fg2). Maxilla is not diminished, with 2 canines [without dentisetae – see (3)]; maxillary palp is not large, its muscles are not thickened; labium is normally developed, with elongate glossae and paraglossae; labial palp of moderate size, its muscles are not thickened; on all legs well-developed pointed claws are present (Tshernova & Bajkova 1960:Fig.2; Elpers & Tomka 1994b:Figs 1–6). Longitudinal ridges of larval abdomen [see Behningia/fg1 (1)] are not widened, thus tergalial bases are visible from above (FIG.73:E; Tshernova & Bajkova 1960:Fig.1). On fore wing CuA [see Behningia/fg1 (6) and Anteritorna (1)] without furcation (Tshernova & Bajkova 1960:Fig.3; Peters & Gillies 1991:Fig.4). Gonostylus [without distal segments – see Behningia/fg1 (10)] is 2-segmented, as its proximal segment is expressed, being very short (Peters & Gillies 1991:Fig.3).

Size. Fore wing length 6–8 mm.

Distribution. Eastern Asia.

Species composition of Protobehningia/g(1). One or two species – *asiatica* Tshernova & Bajkova 1960 [*Protobehningia*] and *merga* Peters & Gillies 1991 [*Protobehningia*].

Material examined: *asiatica* [*P.*]: Ꝺ; *merga* [*P.*]: Ꝺ, ⊕♂.

***

2.2;2,1-3/3.2. Behningia/fg2

(Furcatergaliae Fimbriatotergaliae Fossoriae Behningia/fg1 .../fg2)

(Fig. 73:A–D)

Nomen hierarchicum: **Behningia/fg2** (sine *Protobehningia*; incl. *Dolania*).
In circumscription matches:
— fam. Behningiidae Motas & Bacesco 1937: 29.

References. Edmunds & Traver 1959: Ꝺ* ⊕*; – Edmunds & Allen & Peters 1963: Ꝺ*.

Autapomorphies of Behningia/fg2.

(1) Larval body and legs are strongly modified in such a manner that allow larva to push itself through sand not burrowing it [see Behningia/fg1 (1)] (FIG. 73:A–D):

Segments of thorax and abdomen, except for posteriormost abdominal segments, are shortened and widened. Pronotum, besides the pair of flat anterolateral projections covered with setae [see Behningia/fg1 (1)], with a pair of flat bald posterolateral projections.

On all legs claws are lost. Fore legs [initially burrowing – see Fossoriae (1)] are diminished, palp-like. Middle and hind legs are modified in such a manner, that tibia and tarsus of middle leg and femur of hind leg are directed by their apices posteriorly, widened, covered with setae and are used for pushing through sand.

On middle leg femur [whose initial posterior side is external, convex and setose – see Fossoriae (1)] is shortened, on outer side bears stout setae directed posteriorly; tibia is shortened and strongly widened, with large apical projection inward from tarsus (but at the same time there is retained a patella-tibial suture in a form of long arc on wide flat bare inner side of the tibia – FIG.73:C); tarsus is wide, narrowed toward apex, curved by its convexity outwards; on the side directed outwards (corresponding to posterior or ventral side of non-specialized leg) tibia and tarsus are densely covered with stout setae directed posteriorly.

Hind leg has coxa enlarged, movable; femur is more or less widened; tibia is very strongly diminished; tarsus is strongly elongate, straight.

The same proportions of leg segments are retained on vestigial legs of subimago [see Behningia/fg1 (3)].

Tergalii are transferred to ventral side because the paired longitudinal ridges bearing setae [see Behningia/fg1 (1)] at least on abdominal segments III–VII are strongly expanded laterally and cover tergalii from above (in *Dolania* such distinct lateral projections are present on abdominal segments III–VII, in Behningia/fg3 – on segments III–IX).

(2) Mouth apparatus [carnivorous – see Behningia/fg1 (2)] is strongly modified (Keffermüller 1959:Pl.VII; Edmunds & Traver 1959:Figs 8–22; Elpers & Tomka 1994b:Figs 8–21):

Maxilla is diminished, with a single slender canine (instead of three initial canines and two canines in *Protobehningia*) (the same in Palingenia/f2=g1 and some others – see Index of characters [1.1.33]). Maxillary palp [3-segmented – see Plesiomorphies of Behningia/fg1] is strongly enlarged; 1st segment of palp is much thicker than maxilla, sharply widened just from place of its attachment; flexor and extensor of 2nd segment (situated in the 1st segment) are strongly thickened, being proximally attached to widened base of 1st segment, correspondingly inward and outward from place of attachment of 1st segment.

Labium is diminished, glossae and paraglossae are strongly shortened, labial palp [3-segmented – see Plesiomorphies of Behningia/fg1] is strongly enlarged; flexor of 2nd segment (situated in 1st segment) is strongly thickened, fan-form, proximally attached not only to base of 1st segment, but also to its inner side.

(3) Cubital field of fore wing [see Anteritorna (1)] includes a strong convex vein "CuA_2" arising from CuA to basitornal margin and looking as a branch of CuA; one or several intercalaries [see

Behningia/fg1 (6)] are situated between "CuA_2" and the distal part of CuA (as in FIG.75:A; Edmunds & Traver 1959:Figs 24, 31). The same in some other short-living mayflies – Geminovenata and certain Anagenesia/g1 (see Index of characters [2.2.51]).

(4) Gonostylus [which has no distal segments – see Behningia/fg1 (10)] is 1-segmented, because proximal segment is not expressed. Non-unique apomorphy (see Index of characters [2.3.10]).

Plesiomorphies of Behningia/fg2 (in contrast to *Protobehningia*). On maxilla [see (2)] dentisetae are retained (but can be slender and hardly distinguishable among setae): Behningia/fg3 has both dentisetae [see Bidentiseta (1)], *Dolania* – a single proximal dentiseta only (Elpers & Tomka 1994b:Fig.15b). Larval tarsus of fore leg is not fused with tibia (FIG.73:A–B).

Size. Fore wing length 13–24 mm.

Distribution. Holarctic.

Behningia/fg2 is divided into *Dolania* and Behningia/fg3.

2.2;2,1-3/3.2;1. Dolania/g(1)

(Furcatergaliae Fimbriatotergaliae Fossoriae Behningia/fg1 .../fg2 *Dolania*)

(Fig. 73:D)

Nomen hierarchicum: **Dolania/g(1)** [g: *Dolania* Edmunds & Traver 1959: 46, typus *D. americana* Edmunds & Traver 1959 (design. orig.)].
In circumscription matches:
— gen. *Dolania* Edmunds & Traver 1959: 45.
Monospecific taxon.

References. Edmunds & Traver 1959: ℧; – Edmunds & Jensen & Berner 1976: ℧ ⊕; – Tsui & Hubbard 1979: ℧; – Soldán 1979b: ℧; – Peters J.G. & Peters W.L. 1986: ○ ⊕; – Elpers & Tomka 1994: ℧.

Plesiomorphies of *Dolania* (in contrast to Behningia/fg3). Wings without gemination. Larval hind leg (FIG.73:D) has femur moderately widened, tibia not folded into femur, tarsus not bent relative to tibia; thus, femur, tibia and tarsus are stretched to one line; in spite of very strong shortening of hind tibia [see Behningia/fg2 (1)] its patella-tibial suture is retained (as well as on middle leg).

Size. Fore wing length 13–16 mm.

Distribution. Eastern Nearctic.

Species composition of Dolania/g(1). 1 species – *americana* Edmunds & Traver 1959 [*Dolania*].

Material examined: *americana* [*D.*]: ℧, ⊕♂, ⊕♀/⊙.

***

2.2;2,1-3/3.2;2. Behningia/fg3

(Furcatergaliae Fimbriatotergaliae Fossoriae Behningia/fg1 .../fg3)

(Fig. 73:A–C)

Nomen hierarchicum: **Behningia/fg3** (sine *Dolania*).
In circumscription matches:
— gen. *Behningia* Lestage 1930: 436.

References. Ulmer 1924d ("Ephemeriden-Nymphe"): ℧; – Motas & Bacesco 1937: ℧; – Tshernova 1938: ⊕; – 1952: ℧; – Keffermüller 1959: ℧ ⊕; – Edmunds & Traver 1959: ℧ ⊕; – Elpers & Tomka 1994b: ℧.

Autapomorphies of Behningia/fg3.

(1) Wings with strongly expressed gemination. On fore wing [which has non-branched RSa_2 – see Behningia/fg1 (4), and bifurcate CuA – see Behningia/fg2 (3)] following pairs of veins are present: RSa_2+iRS, RSp+MA_1, iMA+MA_2, MP_1+iMP, MP_2+CuA. Between these vein pairs intercalaries are lost and crossveins are weak; tornoapical margin is undulate, with concavities between apices of the vein pairs. Non-unique apomorphy (see Index of characters [2.2.28]): similar vein pairs (except for MP_2+CuA) are present in Palingenia/f3=g2 (FIG. 75:A) (being especially expressed in *Cheirogenesia*) and in some female Euthyplocia/fg1; Discoglossata also have vein pair MP_2+CuA, but veins anteriad of it are united in pairs in another manner.

(2) Larval hind leg [strongly modified – see Behningia/fg2 (1)] has femur strongly widened, with concavity, into which can be inputted tibia [strongly shortened – see Behningia/fg2 (1)]. Tarsus is articulated to tibia movably, thus can bent at acute angle to tibia (in such a way that if leg would be non-specialized, tarsus would be directed outward). As a result, in folded condition femur, tibia and tarsus form a Z-figure, in which femur and tarsus are directed posteriorly. Patella-tibial suture of hind tibia is lost [but is retained on middle leg – see Behningia/fg2 (1)] (FIG.73:A–B).

Size. Fore wing length 15–24 mm.

Distribution. Palaearctic.

Species composition of Behningia/fg3. Not less than 2 forms, described under 3 species names – *ulmeri* Lestage 1929 [*Behningia*] (locus typicus – Volga River), *lestagei* Motas & Bacesco 1937 [*B.*] (locus typicus – Dnestr River), *tshernovae* Edmunds & Traver 1959 [*B.*] (locus typicus – Amur River).

Material examined: *ulmeri* [*B.*]: ℧ (Ural River and Belaya River); *lestagei* [*B.*]: ℧, ⊕♀, ⊕♂ (Warta River); *tshernovae* [*B.*]: ℧, ⊕♀, ⊕♂ (Amur River).

*** *** ***

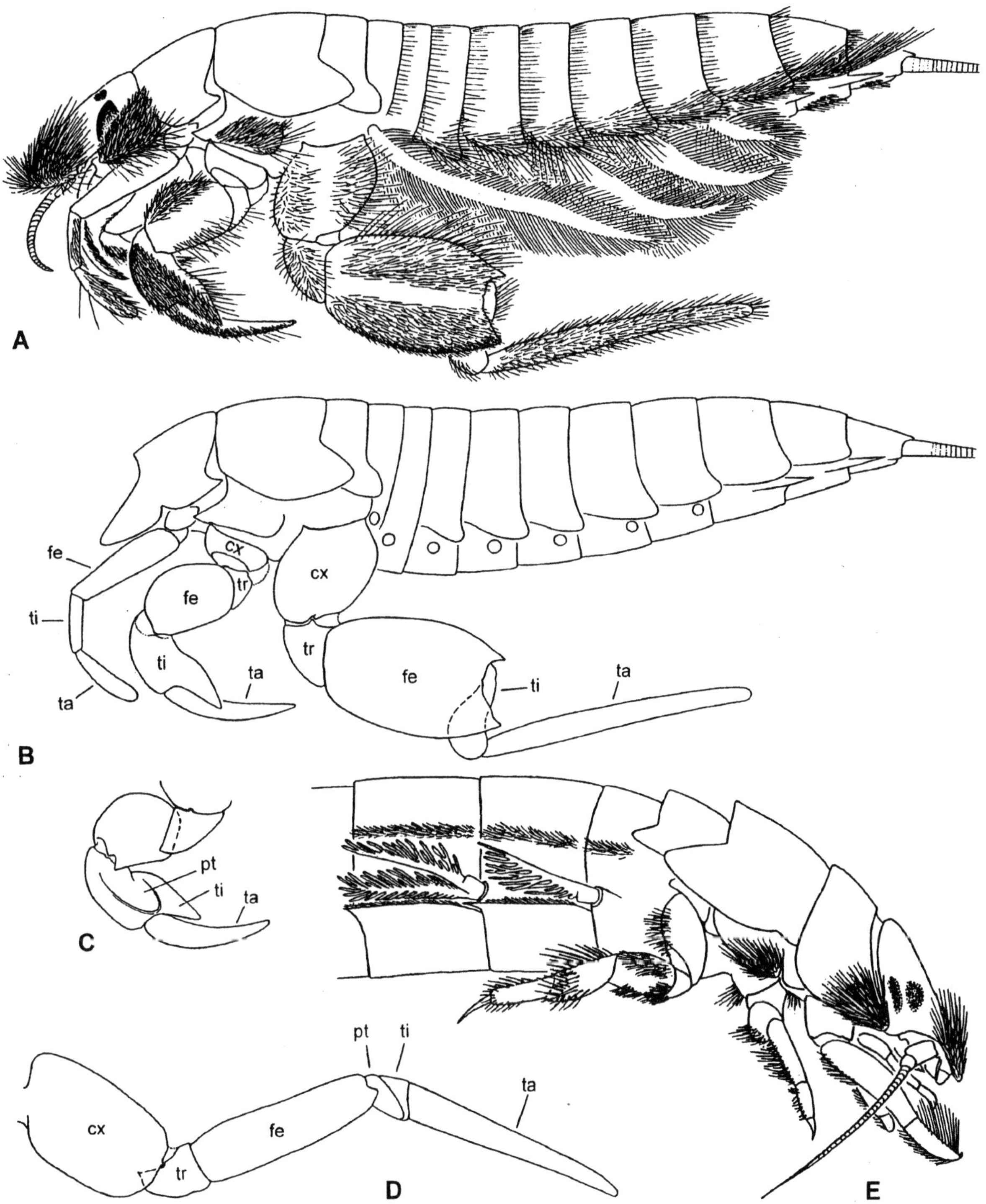

Figure 73. Behningia/fg1, larvae.
A–C – Behningia/fg3 *tshernovae* [*Behningia*], young larva: **A** – lateral view; **B** – thorax and abdomen, lateral view (setae and tergalii removed, round spots indicate places of tergalial attachment); **C** – larval right middle leg, median view. **D** – Dolania/g* *americana* [*Dolania*], larval hind leg. **E** – Protobehningia/g* *asiatica* [*Protobehningia*], anterior part of young larva, lateral view.

Abbreviations: **cx** – coxa; **fe** – femur; **pt** – patella; **ta** – tarsus; **ti** – tibia or telotibia; **tr** – trochanter.

2.2;2,1-3/4. **Cryptoprosternata, or Palingenia/f1=Ephoron/g1**

(Bidentiseta Furcatergaliae Fimbriatotergaliae Fossoriae Cryptoprosternata)

(Figs 74–80)

Nomen hierarchicum: **Palingenia/f1=Ephoron/g1** [f: Palingenines Albarda in Selys-Longchamps 1888: 147 (Peters & Hubbard 1977); g: *Ephoron* Williamson 1802: 71, typus *E. leucon* Williamson 1802 (monotypy)].
Nomen circumscribens: **Cryptoprosternata** Kluge 2000: 252.
In circumscription matches:
— gen. *Palingenia* Burmeister 1839: 802;
— subfam. Palingenines: Selys-Longchamps 1888: 147.
— Cryptoprosternata, or Palingenia/f1=Ephoron/g1: Kluge 2000: 252.

Autapomorphies of Cryptoprosternata. Autapomorphies (1)–(3) are connected with deepening of the burrowing specialization of mandibles and fore legs in larva [see Fossoriae (1)].

(1) Larval, subimaginal and imaginal prosternum is narrowed, bases of fore coxae are brought together; larval fore coxae are contiguous or nearly contiguous at one point where they are articulated with the prosternum; a part of the prosternum behind coxae bases is dipped into the body, thus furca represents a fork-like hollow formation, opened outside by a small unpaired opening exactly behind the place of connection of leg coxae (FIGS 74:C; 77:B–C). At the same time, furcae of mesothorax and metathorax are not modified (i. e. each consists of a pair of widely separated apodemes). In other Ephemeroptera median coxal condyli are widely separated (FIG.70:B); among the species examined, Hexagenia/fg3 *limbata* [*Ephemera*] has the narrowest unpaired prosternal furca, but its coxae are not brought together (FIG.70:A).

(2) In larva on fore leg [specialized as burrowing – see Fossoriae (1)] inner side of femur is proximally convex; distal edge of trochanter which bears a condylus, wedges itself into anterior side of the femur (FIGS 74:C–D; 77:B–C). In other mayflies, including those which have specialized burrowing fore legs (Ephemera/fg9 and *Ichthybotus*), inner side of fore femur is straight, and trochanter is longest not on anterior, but on its inner side (FIG.70:A–B).

(3) Larval mandibles and fore legs have the following regularly situated long setae. **(A)** Ventral side of mandible with an arched row of setae, this row is directed by its convex side laterally-dorsally; in Polymitarcys/f1=Ephoron/g2 this is a U-shape row (FIGS 76:B; 79:C–D), in Palingenia/f2=g1 this is a widely-arched stripe of irregularly situated setae (FIG.75:C). **(B)** The proximal convexity of inner side of fore femur [see (2)] with a U-shape row of setae: curvation of this row is situated at extreme femur base, and two its arms are directed distally; in Polymitarcys/f1=Ephoron/g2 this row is regular (FIG.77:C), and in Palingenia/f2=g1 it has a form of stripe (FIG.74:D). **(C)** Anterior (dorsal) side of fore tibia has at least a proximal oblique row of setae; in Polymitarcys/f1=Ephoron/g2 this row is regular, both its ends are sharply curved distally and continued as two regular longitudinal rows (FIG.77:C); in Palingenia/f2=g1 this row is double and has no continuations (FIG.74:D). **(D)** Inner side of fore tibia has a regular row of setae, whose shape is different in Polymitarcys/f1=Ephoron/g2 and Palingenia/f2==g1 (FIGS 74:D; 77:C). At least presence of the rows (B) and (C) is a unique apomorphy.

(4) Maxillary and labial palps are 2-segmented; palps of both pairs have 1st segment shortened, and 2nd+3rd segment thickened, of a simple shape – oval, oval-conic, banana-shaped, etc. (FIGS 75:C; 76:B). Correspondingly, labial palp has no muscle in 2nd segment (as well as maxillary palp – see Chapter II).

In contrast to Cryptoprosternata, in various other Fimbriatotergaliae – in Ephemera/fg9, *Ichthybotus*, Potamanthus/fg1 and Euthyplocia/fg1 – maxillary palp is long and slender, and labial palp often has widened 3rd segment; if 2nd and 3rd segments of labial palp are fused, they retain general composite shape and sometimes muscle of 2nd segment (see Index of characters [1.1.55] and [1.1.58]).

(5) Imaginal and subimaginal paracercus is reduced at least in male; in *Pentagenia* and Polymitarcys/f2=Ephoron/g3 paracercus is well-developed in female, while in Palingenia/f3=g2 and Campsurus/fg1 it is reduced in both sexes. Non-unique apomorphy (see Index of characters [2.3.22]).

Character of unclear phylogenetic status.

(6) Tergalius I [see Fimbriatotergaliae (6)] is vestigial, without processes, initially bilamellate (in Anagenesia/g1, Polymitarcys/f2=Ephoron/g3 and *Tortopus* – unilamellate). The same in majority of Fossoriae (except for Behningia/fg1) and in Euthyplocia/fg1; probably, symplesiomorphy.

Size. Fore wing length 5–30 mm.

Distribution. Holarctic, Oriental, Ethiopian and Neotropical Regions.

Cryptoprosternata are divided into Palingenia/f2=g1 and Polymitarcys/f1=Ephoron/g2 (p.257).

2.2;2,1-3/4.1. Palingenia/f2=g1

(Furcatergaliae Fimbriatotergaliae Fossoriae Cryptoprosternata Palingenia/f2=g1)

(Figs 74; 75; 76:D–E)

Nomen hierarchicum: **Palingenia/f2=g1** (sine *Ephoron*; incl. *Pentagenia*) [g: *Palingenia* Burmeister 1839: 802, typus *Ephemera longicauda* Olivier 1791 (design. Hagen in Walsh 1863: 173)].
In circumscription matches:
— fam. Palingeniidae: McCafferty & Edmunds 1976a:488;
— Palingenia/f2=g1: Kluge 2000: 252.

Reference. McCafferty & Edmunds 1976a: ♂* ⊕*.

Autapomorphies of Palingenia/f2=g1.

(1) Larval frontal projection [see Fossoriae (2)] anteriorly is more or less denticulate, the lateralmost denticles are projected anteriorly more than others (thus, sometimes the frontal projection is bipointed like in Ephemera/fg10). Laterad of the frontal projection, there is a pair of denticles above antennae bases; laterad of them one more pair of denticles are present (FIG.75:C).

(2) Lateral edge of mandibular tusk [see (8) and Fimbriatotergaliae (8)] is flattened in a form of ridge and bears denticles; each denticle terminates by a stout spine-like seta (FIGS 75:C; 76:D). Only in Palingenia/f4=g3, *Mortogenesia* and *Plethogenesia* these denticles are integral, without setae (FIG.76:E) (TABLE 7).

(3) Maxilla with 1 canine only (instead of initial three) and with 1 dentiseta only [instead of initial two – see Bidentiseta (1)] (FIG.74:B). Non-unique apomorphy (see Index of characters [1.1.33] and [1.1.37]).

(4) In larva anterior side of fore tibia [specialized as burrowing – see Fossoriae (1)] proximally with 2 oblique rows of long setae (FIG.74:D): a regular double row close to outer margin of tibia [see Cryptoprosternata (3C)]; proximad of it begins a regular single row which continues as an irregular row on inner margin of tibia [see Cryptoprosternata (3D)] [see also (5) and (9)].

(5) Outer margin of larval fore tibia [see (4) and (9)] with protuberances bearing spine-like setae (FIG.74:E); these protuberances are present in *Pentagenia* and majority of Anagenesia/g1, but in Palingenia/f4=g3 and Anagenesia/g1-*Mortogenesia* they are substituted by large denticles lacking setae and similar to denticles on mandibular tusks [see (2)] and frons [see (1)].

(6) In larva each of abdominal segments III–VII ventrad of tergalial bases has a pair of long soft lateral processes; these processes are directed latero-posteriorly, curved dorsally and covered with long dense hairs. Unique apomorphy.

(7) In male imago and subimago fore leg has both claws blunt (non-unique character – see Index of characters [2.2.77]). Fore leg of male imago is only slightly longer than that of female, less than 1/2 of body length (i. e. shorter than in most mayflies).

Characters of unclear phylogenetic status.

(8) Mandibular tusks [see (2)] are curved laterally (FIG.76:D). Non-unique character, similar curvation in Ephemera/fg9.

(9) Apex of larval fore tibia [see (4) and (5)] projects posteriad of tarsus base and forms outer-apical denticle and inner-apical projection, with incision between them (FIG.74:D–E). Similar in Hexagenia/fg1 and *Ichthybotus*; possibly, symplesiomorphy [see Fossoriae (1)].

(10) Larval hind tibia [specialized – see Fimbriatotergaliae (9)] with projected inner-distal corner. Non-unique character, the same in Ephemera/fg9 (FIG.70:C).

(11) On inner side of femur and tibia of larval hind leg spiny setae are absent. Probably a result of reduction [see Fimbriatotergaliae (9)].

Plesiomorphies of Palingenia/f2=g1. Larval and adult patella-tibial suture is developed on middle and hind legs (in contrast to Campsurus/fg1 and some others). Furcasternal protuberances are contiguous (in contrast to Campsurus/fg1, Caenotergaliae and some others – see Index of characters [2.2.23]).

Size. Fore wing length 11–30 mm.

Distribution. Holarctic, Oriental and Ethiopian Regions.

Systematic position of Palingenia/f2=g1. Besides the accepted here attributing of Palingenia/f2=g1 to Cryptoprosternata, it was suggested to unite it with Ephemera/fg9 in a taxon **Ephemera/fg** (incl. *Palingenia*). Assumed apomorphies of this taxon are the following: tusks are curved by their apices latero-dorsally [see Ephemera/fg9 (2) and Palingenia/f2=g1 (8)]; larval hind tibia [specialized – see Fimbriatotergaliae (9) and Fossoriae (1)] with oblique apex and projected inner-distal angle [see Ephemera/fg9 (3) and Palingenia/f2=g1 (10)]. The idea about holophyly of this group contradicts the idea about holophyly of Cryptoprosternata. Possibly these characters are symplesiomorphies or independently originated apomorphies.

Palingenia/f2=g1 is divided into *Pentagenia* and Palingenia/f3=g2.

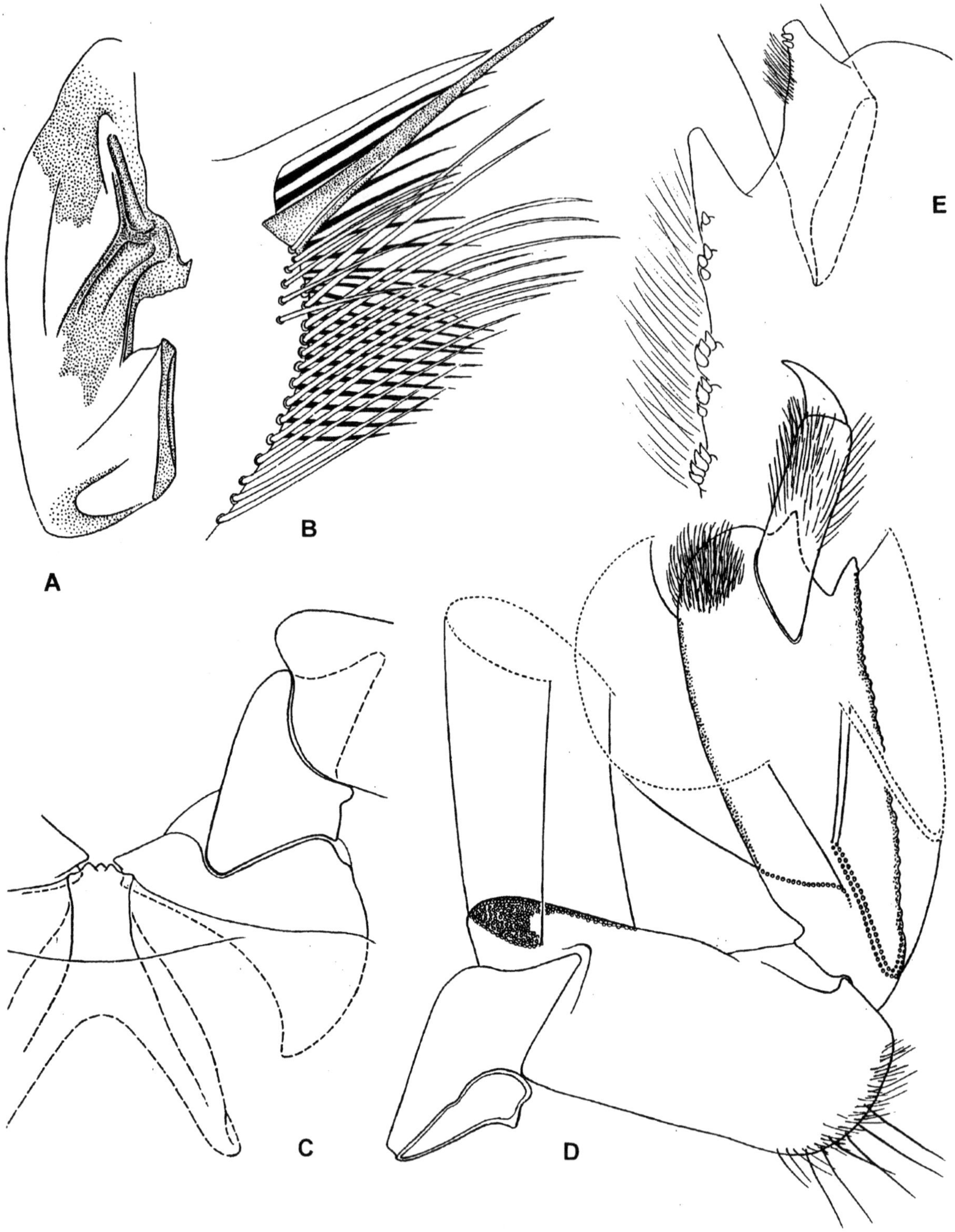

Figure 74. Cryptoprosternata-Palingenia/f2=g1-Pentagenia/g* *vittigera* [*Palingenia*].
A – subimaginal exuviae of right half of mesonotum; **B** – apex of maxilla, dorsal view (dentiseta shown by dots, setae of medio-dorsal shown by black); **C** – larval prosternum and base of left fore leg, view from behind; **D** – larval right fore leg, front view (coxa removed, instead of long setae, their bases and occupied area are shown); **E** – outer-apical angle of larval right fore tibia with tarsal base, view from behind.

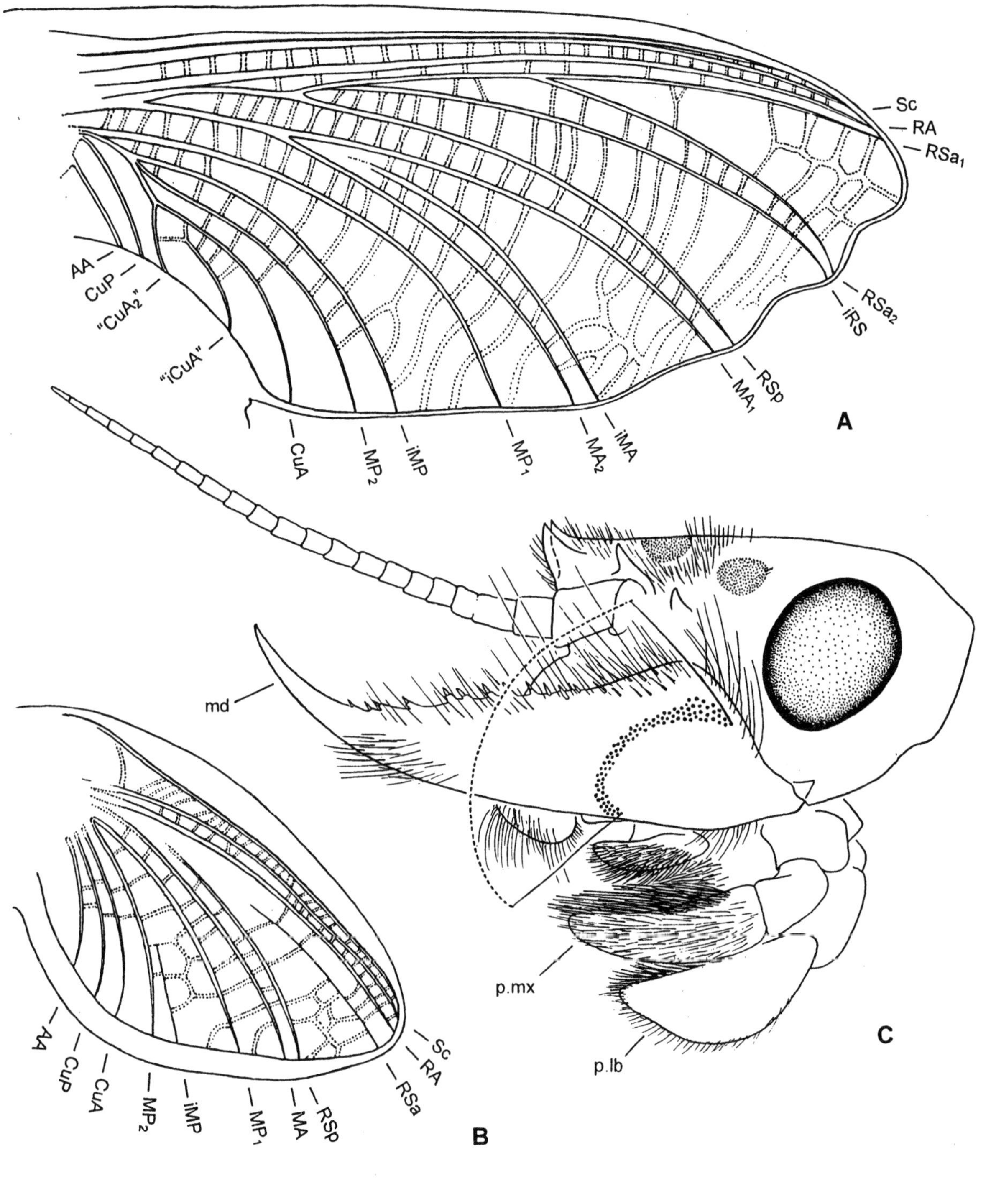

Figure 75. Palingenia/f3=g2-Anagenesia/g2 *paradoxa* [*Anagenesia*].
A–B – larval buds of fore and hind wings (hypoderm with venation shown by integral and dotted lines); **C** – larval head, lateral view (instead of long setae, their bases and occupied area are shown).
Abbreviations: **md** – mandibular tusk; **p.lb** – labial palp; **p.mx** – maxillary palp.

2.2;2,1-3/4.1;1. Plesiomorphon **Pentagenia/fg(1)**

(Fimbriatotergaliae Fossoriae Cryptoprosternata Palingenia/f2=g1 *Pentagenia*)

(Fig. 74)

Nomen hierarchicum: **Pentagenia/fg(1)** [f: Pentageniidae McCafferty 1972: 51; g: *Pentagenia* Walsh 1863: 196, typus *Palingenia vittigera* Walsh 1862 (design. Eaton 1868b: 85)].

In circumscription matches:

— gen. *Pentagenia* Walsh 1863: 196;

— subfam. Pentageniinae: McCafferty & Edmunds 1976a: 489;

— fam. Pentageniidae McCafferty 1972: 51.

Monospecific taxon.

References. Eaton 1883–1888: ⊕; – Needham & Traver & Hsu 1935: ℧ ⊕; – Burks 1953: ℧ ⊕; – McCafferty 1972; – 1975: ℧; – McCafferty & Edmunds 1976a: ℧ ⊕; – Edmunds & Jensen & Berner 1976: ℧ ⊕; – Keltner & McCafferty 1986: ℧.

Plesiomorphies of *Pentagenia*. On outer margin of mandibular tusk each denticle terminates by a spine-like seta [see Palingenia/f2=g1 (2)]. Larval outer margin of fore tibia without denticles, with spine-like setae [see Palingenia/f2=g1 (5)] (FIG.74: D–E). Vestige of tergalius I [see Cryptoprosternata (6)] is bilamellate (in contrast to Anagenesia/g1).

In contrast to other Cryptoprosternata, imago and subimago are not short-living: mesonotal suture is moderately stretched posteriorly, and subimaginal mesonotum has pigmented areas of characteristic shape [see Fimbriatotergaliae (2)] (FIG.74:A); imaginal and subimaginal legs are normally developed, claws are ephemeropteroid [except for male fore legs – see Palingenia/f2=g1 (7)]; wing structure is similar to that of Ephemera/fg9, without vein gemination; in cubital field of fore wing several (3–6) branched or simple veins go from CuA to basitornal margin [see Anteritorna (1)] (in contrast to Palingenia/f3=g2). Gonostylus is 4-segmented: with distinct long 1st segment and 2 distal segments (in contrast to some Palingenia/f3=g2 which have increased number of segments). In female imago and subimago paracercus is subequal to cerci [while in male paracercus is vestigial – see Cryptoprosternata (5)].

Size. Fore wing length 14–19 mm.

Distribution. Nearctic.

Species composition of Pentagenia/fg(1). Probably one species – *vittigera* Walsh 1862 [*Palingenia*] (synn. subj.: *quadripunctata* Walsh 1863 [*Pentagenia*], *robusta* McDunnough 1926 [*Pentagenia*]).

Material examined: *vittigera* [*P.*]: ℧, ♂/⊕♂, ♀/⊙.

***

2.2;2,1-3/4.1;2. **Palingenia/f3=g2**

(Fimbriatotergaliae Fossoriae Cryptoprosternata Palingenia/f2=g1 .../f3=g2)

(Figs 75; 76:D–E)

Nomen hierarchicum: **Palingenia/f3=g2** (sine *Pentagenia*; incl. *Anagenesia*).

In circumscription matches:

— gen. *Palingenia*: Eaton 1868b: 84;

— subfam. Palingeniinae: Lestage 1917: 254;

— fam. Palingeniidae: Edmunds & Traver 1954a: 239.

References. Eaton 1883–1888: ℧* ⊕*; – Edmunds & Allen & Peters 1963: ℧*; – Demoulin 1965b: ℧ ⊕; – Tshernova 1970: ℧* ⊕*; – McCafferty & Edmunds 1976a: ℧* ⊕*; – Edmunds & Jensen & Berner 1976: ℧* ⊕*.

Autapomorphies of Palingenia/f3=g2.

(1) Winged stages are short-living: Legs are non-functional (except for fore legs of male). Moult from subimago to imago takes place in males only. Pterothorax is modified, mesonotal suture is extremely stretched posteriorly up to disappearance (as in FIG.78:E); subimaginal cuticle lost pigmentation [see Fimbriatotergaliae (2)]. Non-unique apomorphies (see Index of characters [2], [2.2.8] and [2.2.14]); among Cryptoprosternata the same in Polymitarcys/f1=Ephoron/g2.

(2) Wings with gemination: on fore wing apices of veins are brought together in pairs RSa_2+iRS, RSp+MA_1, iMA+MA_2; main longitudinal veins are firm, membrane and crossveins are soft (FIG.75:A–B; Eaton 1883–1888:Pl.1–2; Demoulin 1965b:Figs 1–8). Non-unique apomorphy (see Index of characters [2.2.28]): similar vein pairs are present in Behningia/fg3 and some female Euthyplocia/fg1; in Discoglossata veins are united in pairs in another manner.

(3) In male subimago fore leg [short – see Palingenia/f2=g1 (7)] is shortened and thickened in a special manner: tibia is thick; tarsus is short, thick and conically narrowed apically. In male imago of Palingenia/f4=g3 tarsus has normal form – long and non-conic (Demoulin 1965b:Figs 1–2); in male imago of Anagenesia/g1 tarsus has the same form as in subimago (ibid.:Figs 3–8).

(4) In female [non-moulting – see (1)] fore leg is vestigial, with a single blunt claw, sometimes with blunt vestige of second claw [in male imago and subimago fore leg has two blunt claws – see Palingenia/f2=g1 (7)]; on middle and hind legs claws can be either ephemeropteroid [see below, Plesiomorphies of Palingenia/f4=g3], or vestigial [see below, Anagenesia/g1 (2)].

Some species have sexual dimorphism in shape of larval claws [lacking denticles – see Plesiomorphies of Fimbriatotergaliae], which is connected with strong reduction of adult claws in female: claws of all legs of male mature larva can be swollen in proximal part, inside which adult claw develops, while claws of female mature larva and young larvae of both sexes are entirely thin (Soldán 1981). In other mayflies larval claws have no sexual dimorphism, independently of sexual dimorphism in adults (mature larvae of *Pentagenia* were not examined).

(5) Gonostyli have peculiar structure. 1st segment is not separated by suture from second; inner side of gonostylus has a more or less expressed projection near base (possibly corresponding to apex of strongly shortened 1st segment). Usually 2 distal segments are present, but tendency to increasing of their number exists (TABLE 7): Anagenesia/g2 *paradoxa* [*A.*] has 3–4 distal segments, Palingenia/f4=g3 and *Mortogenesia* have 4–6 distal segments; selected specimens have 1 distal segment.

(6) Imaginal and subimaginal paracercus is vestigial not only in male [see Cryptoprosternata (5)], but in female as well. Non-unique apomorphy (see Index of characters [2.3.22]).

(7) Egg has a form of biconvex round lens; thread-like anchors are absent; at least in some representatives egg can be attached to substrate by a thin lamella which covers egg surface (Kosova 1967:Figs 1–2; Landolt & al. 1995:Figs 11–14). Egg of *Pentagenia* has the same kind of anchor, but egg form is irregular with convexity. Lens-like eggs are found in Palingenia/f4=g3, Anagenesia/g2, *Chankagenesia* and *Cheirogenesia*. Non-unique apomorphy: the same in *Prionoides*; ovoid (but not round) lens-like eggs occurs also in some Ameletus/fg2.

Size. Fore wing length 11–30 mm.

Distribution. Old World: Palaearctic, Oriental and Ethiopian Regions.

Palingenia/f3=g2 is divided into Palingenia/f4=g3 and Anagenesia/g1.

2.2;2,1-3/4.1;2,1. **Palingenia/f4=g3**

(Fimbriatotergaliae Fossoriae Cryptoprosternata Palingenia/f2=g1 .../f4=g3)

(Fig. 76:E)

Nomen hierarchicum: **Palingenia/f4=g3** (sine *Anagenesia*).

In circumscription matches:

— subgen. *Palingenia*: Eaton 1883: 23;

— gen. *Palingenia*: Lestage 1917: 254.

References. Eaton 1883–1885: ⭮ ⊕; – Demoulin 1965b: ⊕; – Soldán 1978b: ⭮ ⊕.

Characters of Palingenia/f4=g3 of unclear phylogenetic status (TABLE 7).

(1) Denticles on outer margin of mandibular tusk [see Palingenia/f2=g1 (2)] are large and integral, without terminal spine-like setae (FIG.76:E). The same in *Mortogenesia* and *Plethogenesia.*

(2) In larva on outer margin of fore tibia spine-like setae [see Palingenia/f2-g1 (5)] are lost and substituted by large denticles. The same in *Mortogenesia.*

Plesiomorphies of Palingenia/f4=g3. Vestige of tergalius I [see Cryptoprosternata (6)] is bilamellate (Soldán 1978b:Figs 20–22) (in contrast to Anagenesia/g1).

In imago and subimago: On fore wing bifurcation of MA is situated near middle; RSa_2' and $iRSa_2$ are present; many additional intercalary veins are present between paired veins RSa_2+iRS, RSp+MA_1 and iMA+MA_2 [see Palingenia/f3=g2 (2)] and other longitudinal and intercalary veins; tornoapical wing margin has no incisions (in contrast to Anagenesia/g1); CuA without furcation (in contrast to some Anagenesia/g1 – see TABLE 7) [see Anteritorna (1)]; MP_1 and iMP are not brought together (in contrast to *Cheirogenesia*) (Eaton 1883–1888:Pl.1; Demoulin 1965b:Figs 1–2; Soldán 1978b:Figs 1–2). On fore leg of male tarsus is not shortened [in contrast to subimago – see Palingenia/f3=g2 (3)]. In female tarsi of all legs are segmented; on middle and hind legs of male and female claws are ephemeropteroid (in contrast to Anagenesia/g1); male subimago moults to imago on bank attaching itself to some support by middle and hind legs (Landolt & al. 1995) [females do not moult – see Palingenia/f3=g2 (1)].

Size. Fore wing length 23–25 mm.

Distribution. West Palaearctic (including West Siberia) and northern Africa (to the south up to Liberia).

Species composition of Palingenia/f4=g3. 6 species – *anatolica* Jacob 1977 [*Palingenia*], *apatris* Demoulin 1965 [*Palingenia*], *fuliginosa* Georgi 1802 [*Ephemera*], *longicauda* Olivier 1791 [*Ephemera*] (synn. subj.: *flosaquae* Illiger 1802 [*Ephemera*], *marginata* Panzer 1804 [*Semblis*], *swammerdiana* Latreille 1805 [*Ephemera*], *swammerdamiana* Shaw 1806 [*Ephemera*]), *orientalis* Chopra 1927 [*Palingenia*], *sublongicauda* Tshernova 1949 [*Palingenia*].

Material examined: *fuliginosa* [*E.*]: ⊕♂, ⊕♀; *longicauda* [*E.*]: ⭮, ⊕♂, ⊕♀; *sublongicauda* [*P.*]: ⊕♂, ⊕♀.

***

2.2;2,1-3/4.1;2,2. **Anagenesia/g1**

(Fimbriatotergaliae Fossoriae Cryptoprosternata Palingenia/f2=g1 .../f3=g2 Anagenesia/g1)

(Figs 75; 76:D)

Nomen hierarchicum: **Anagenesia/g1** (incl. *Mortogenesia, Chankagenesia, Plethogenesia, Cheirogenesia*) [g: *Anagenesia* Eaton 1883: 25, typus *Palingenia lata* Walker 1853 (design. orig.)].

In circumscription matches:

— subgen. *Anagenesia* Eaton 1883–1888: 25;

— gen. *Anagenesia*: Lestage 1917: 256.

Reference. Eaton 1883–1888: ⊕*.

Autapomorphies of Anagenesia/g1.

(1) Fore wing is modified in following manner: RSa_2 [approximated to iRS – see Palingenia/f3=g2 (2)] does not form a triad (non-unique apomorphy – see Index of characters [2.2.37]); bifurcation of MA is transferred to proximal part of wing (non-unique apomorphy – see ibid. [2.2.43]); additional intercalaries are absent or weekly developed, thus anterior-distal part of wing is armed only by paired veins RSa_2+iRS, RSp+MA_1 and iMA+MA_2 [see Palingenia/f3=g2 (2)], wing margin between their apices can be incised (FIG.75:A; Demoulin 1965b:Figs 3–8). This form of wing has convergent similarity with Behningia/fg3. Some representatives of Anagenesia/ /g1 retain ability to fly, while some of them (*Chankagenesia, Cheirogenesia*, partly Anagenesia/g2) are unable to fly and can only skim the water surface (Buldovsky 1935a, 1935b; Tshernova 1952; Sartori & Elouard 1999).

(2) In male and female imago legs [non-functional – see Palingenia/f3=g2 (1)] are strongly modified. In male imago fore leg (elongate in majority of mayflies) retains subimaginal structure [see Palingenia/f3=g2 (3)]: it is short, with tarsus strongly shortened, thickened, usually arched (Demoulin 1965b: Figs 3-8). In both sexes on middle and hind legs all claws are blunt; in male posterior claw (i. e. the claw corresponding to pointed claw when claw structure is ephemeropteroid) can be more or less reduced (TABLE 7) [claws of fore leg are also blunt – see Palingenia/f2=g1 (7) and Palingenia/f3=g2 (4)]. Female legs are reduced in greater degree; not only fore tarsus [see Palingenia/f3=g2 (4)], but all tarsi are non-segmented, each with 1 blunt claw only. Male subimago [skimming the water surface – see (1)] can moult to imago on the water surface immediately after moult from larva to subimago, some specimens do not moult at all [females never moult – see Palingenia/f3=g2 (1)] (Buldovsky 1935a, 1935b). Some of these features independently appeared in some other short-living mayflies.

(3) Vestige of tergalius I [initially bilamellate – see Cryptoprosternata (6)] is unilamellate. Non-unique apomorphy (see Index of characters [1.3.33]).

Size. Fore wing length 11–30 mm.

Distribution. Asia, New Guinea, Madagascar.

Anagenesia/g1 is divided into *Mortogenesia, Chankagenesia*, Anagenesia/g2, *Plethogenesia* and *Cheirogenesia*; these taxa differ one from another by combinations of non-unique characters (TABLE 7).

Table 7. Non-unique characters of subordinate taxa in Anagenesia/g1, in comparison with Palingenia/f4=g3

Taxa	Geographical distribution	Species number	Characters						
			Imaginal				Larval		
			1	2	3	4	5	6	7
Palingenia/f4=g3	West Palaearctic	6	I	I I	0)	4–6	Δ	Δ	2
Anagenesia/g1:									
Mortogenesia	Iraq	1	I	I I	00	4–6	Δ	Δ	1
Chankagenesia	Eastern Asia	1	I	I I	0	2	I	I	1
Anagenesia/g2	Asia	14	Y	I I	0	2–4	I	I	1
Plethogenesia	New Guinea	4	Y	I I	00	2	Δ	I	1
Cheirogenesia	Madagascar	1	Y	II	00	2	I	I	1

1 – vein CuA of fore wing: "I" – simple, "Y" – with bifurcation;

2 – veins MP_1 and iMP of fore wing: "II" – brought together, "I I" – not brought together;

3 – claws of middle and hind legs of male imago: "0)" – two ephemeropteroid, "00" – two blunt, "0" – single blunt;

4 –number of segments of gonostylus;

5 – tusk denticles: " I " – denticle with articulated apical spine-like seta; "Δ" – integral denticle;

6 – outer margin of fore tibia of larva: " I " – with spine-like setae; "Δ" – with denticles;

7 – number of lamellae of first tergalius.

2.2;2,1-3/4.1;2,2-1. **Mortogenesia/g(1)**

(Fossoriae Cryptoprosternata Palingenia/f2=g1 .../f3=g2 Anagenesia/g1 *Mortogenesia*)

Nomen hierarchicum: **Mortogenesia/g(1)** [g: *Mortogenesia* Lestage 1923a: 110, typus *Palingenia mesopotamica* Morton 1921 (design. orig.)].
In circumscription matches:
— gen. *Mortogenesia* Lestage 1923a: 110.
Monospecific taxon.

References. Morton 1921: ⊕; – Demoulin 1965b: ⊕.

Characters of *Mortogenesia* of unclear phylogenetic status (TABLE 7).

(1) Denticles on outer margin of mandibular tusk [see Palingenia/f2=g1 (2)] are large and integral, without terminal spine-like setae; these denticles are retained on tusk vestiges of subimago and imago. The same in Palingenia/f4=g3 and *Plethogenesia.*

(2) In larva on outer margin of fore tibia spine-like setae [see Palingenia/f2=g1 (5)] are lost and substituted by large denticles. The same in Palingenia/f4=g3.

Plesiomorphies of *Mortogenesia* (TABLE 7). On fore wing MP_1 and iMP are not brought together (in contrast to *Cheirogenesia*); CuA without furcation. At least in male imago all legs with 2 claws [blunt – see Anagenesia/g1 (2)].

Size. Fore wing length 17–22 mm.

Distribution. Iraq.

Species composition of Mortogenesia/g(1). 1 species – *mesopotamica* Morton 1921 [*Palingenia*].

Material examined: *mesopotamica* [*P.*]: ℧ (exuviae, deposited in Entomol. Inst.), ♂, ♀.

***

2.2;2,1-3/4.1;2,2-2. **Chankagenesia/g(1)**

(Fossoriae Cryptoprosternata Palingenia/f2=g1 .../f3=g2 Anagenesia/g1 *Chankagenesia*)

(Fig. 76:D)

Nomen hierarchicum: **Chankagenesia/g(1)** [g: *Chankagenesia* Buldovsky 1935a: 831; 1935b: 152, typus *Ch. natans* Buldovsky 1935 (design. orig.)].
In circumscription matches:
— subgen. *Chankagenesia*: Tshernova 1952: 247;
— gen. *Chankagenesia* Buldovsky 1935a:831; 1935b:152.
Monospecific taxon.

References. Buldovsky 1935a: ⊕; – 1935b: ℧ ⊕; – Tshernova 1952: ℧ ⊕; – Demoulin 1952a: ⊕; – 1965b: ⊕; – Tshernova & Kluge & Sinitshenkova & Belov 1986: ⊕.

Character of *Chankagenesia* of unclear phylogenetic status (TABLE 7).

(1) In male imago middle and hind legs with one (anterior) claw on each, thus only on fore legs of male two claws are retained [all claws are blunt – see Anagenesia/g1 (2)]. The same in Anagenesia/g2; possibly, synapomorphy.

Plesiomorphies of *Chankagenesia* (TABLE 7). On fore wing MP_1 and iMP are not brought together (in contrast to *Cheirogenesia*); CuA without furcation (in contrast to Anagenesia/fg2, *Plethogenesia* and *Cheirogenesia*). In larva: on outer margin of mandibular tusk each denticle terminates by a spine-like seta [see Palingenia/f2=g1 (2)] (FIG.76: D); outer margin of fore tibia without denticles, with spine-like setae [see Palingenia/f2=g1 (5)].

Size. Fore wing length 10–19 mm, body length 17–27 mm.

Distribution. Eastern Asia.

Species composition of Chankagenesia/g(1). 3 species are formally described – *natans* Buldovsky 1935 [*Chankagenesia*], *sibirica* McLachlan 1872 [*Palingenia*] (nomen dubium), *yangi* Hsu 1937 [*Anagenesia*].

Material examined: *natans* [*Ch.*]: ℧, ♂, ♀.

***

2.2;2,1-3/4.1;2,2-3. **Anagenesia/g2**

(Fossoriae Cryptoprosternata Palingenia/f2=g1 .../f3=g2 Anagenesia/g1 .../g2)

(Fig. 75)

Nomen hierarchicum: **Anagenesia/g2** (sine *Mortogenesia, Chankagenesia, Plethogenesia, Cheirogenesia*).
In circumscription matches:
— subgen. *Anagenesia*: Tshernova 1952: 247;
— gen. *Anagenesia*: Ulmer 1920b: 102.

References. Gravely 1920: ℧ ⊕; – Chopra 1927: ⊙ ⊕; – Tshernova 1952: ℧ ⊕; – Demoulin 1965b: ⊕.

Characters of Anagenesia/g2 of unclear phylogenetic status (TABLE 7).

(1) Cubital field of fore wing [see Anteritorna (1)] includes a strong convex vein "CuA_2" arising from CuA to basitornal margin and looking as a branch of CuA; a single intercalary "iCuA" is situated between "CuA_2" and the distal part of CuA (FIG.75:A; Demoulin 1965b:Figs 3–6). Non-unique apomorphy (see Index of characters [2.2.51]); among Anagenesia/g1 the same in *Plethogenesia* and *Cheirogenesia*; probably synapomorphy.

(2) In male imago each middle and hind leg with 1 (anterior) claw only, thus only on fore legs of male 2 claws are retained [all claws are blunt – see Anagenesia/g1 (2)]. The same in *Chankagenesia*; possibly, synapomorphy.

Plesiomorphies of Anagenesia/g2 (TABLE 7). On fore wing MP_1 and iMP are not brought together

(FIG.75:A) (in contrast to *Cheirogenesia*). In larva: on outer margin of mandibular tusk each denticle terminates by a spine-like seta [see Palingenia/f2=g1 (2)]; outer margin of fore tibia without denticles, with spine-like setae [see Palingenia/f2=g1 (5)].

Size. Fore wing length 11–27 mm.

Distribution. Eastern and Southern Asia.

Species composition of Anagenesia/g2 (here [*A.*]= =[*Anagenesia*]). 15 species (Demoulin 1965b) – *albescens* Demoulin 1965 [*A.*], *ampla* Eaton 1885 [*Palingenia (A.)*] (syn. subj.: *chaperi* Navás 1932 [*A.*]), *birmanica* Navás 1932 [*A.*], *javanica* Eaton 1885 [*Palingenia (A.)*], *lata* Walker 1871 [*Palingenia*], *leucoptera* Navás 1919 [*A.*], *lontona* Hafiz 1937 [*A.*], *minor* Eaton 1892 [*Palingenia*], *nana* Demoulin 1965 [*A.*], *nanoides* Demoulin 1965 [*A.*], *paradoxa* Buldovsky 1935 [*A.*], *picta* Gravely 1920 [*Palingenia (A.)*], *robusta* Eaton 1892 [*Palingenia*], *spodiocephala* Demoulin 1965 [*A.*], *tenera* Eaton 1885 [*Palingenia (A.)*].

Material examined: *paradoxa* [*A.*]: Ꝺ, ⊕♂, ⊕♀; sp. (India): ⊕♂.

***

2.2;2,1-3/4.1;2,2-4. Plethogenesia/g1

(Fossoriae Cryptoprosternata Palingenia/f2=g1 .../f3=g2 Anagenesia/g1 *Plethogenesia*)

Nomen hierarchicum: **Plethogenesia/g1** [g: *Plethogenesia* Ulmer 1920b: 102, typus *Palingenia papuana* Eaton 1879 (monotypy)].
In circumscription matches:
— gen. *Plethogenesia* Ulmer 1920b: 102.

References. Lestage 1923a: ⊕; – Ulmer 1939: Ꝺ ⊕; – Demoulin 1965b: ⊕; – 1966b: ⊕.

Characters of *Plethogenesia* of unclear phylogenetic status (TABLE 7).

(1) Cubital field of fore wing [see Anteritorna (1)] includes a strong convex vein "CuA_2" arising from CuA to basitornal margin and looking as a branch of CuA; a single intercalary "iCuA" is situated between "CuA_2" and the distal part of CuA (Demoulin 1965b:Figs 7–8). Non-unique apomorphy (see Index of characters [2.2.51]); among Anagenesia/g1 the same in Anagenesia/g2 (FIG.75:A) and *Cheirogenesia*; probably synapomorphy.

(2) Denticles on outer margin of mandibular tusk [see Palingenia/f2=g1 (2)] are large and integral, without terminal spine-like setae; sometimes these denticles are retained on tusk vestiges of imago (Demoulin 1965b:Fig.7p). The same in Palingenia/ /f4=g3 and *Mortogenesia*.

Plesiomorphies of *Plethogenesia* (TABLE 7). On fore wing MP_1 and iMP are not brought together (in contrast to *Cheirogenesia*). At least in male imago all legs with 2 claws [blunt – see Anagenesia/g1 (2)]. In larva outer margin of fore tibia without denticles, with spine-like setae [see Palingenia/f2=g1 (5)].

Size. Fore wing length 16–29 mm.

Distribution. New Guinea.

Nominal taxon included. Plethogenesia/g1 includes **Tritogenesia/g** [g: *Tritogenesia* Lestage 1923a:111, typus *T. bibisica* Lestage 1923], regarded as a generic synonym of *Plethogenesia* (Demoulin 1965b: 330).

Species composition of Plethogenesia/g1. 4 species (Demoulin 1965b) – *bibisica* Lestage 1923 [*Tritogenesia*], *delicata* Demoulin 1965 [*Plethogenesia*], *pallida* Demoulin 1965 [*Plethogenesia*], *papuana* Eaton 1879 [*Palingenia*].

Material examined (Entomol. Inst.): sp.: ⊕♂.

***

2.2;2,1-3/4.1;2,2-5. Cheirogenesia/g(1)

(Fossoriae Cryptoprosternata Palingenia/f2=g1 .../f3=g2 Anagenesia/g1 *Cheirogenesia*)

Nomen hierarchicum: **Cheirogenesia/g(1)** [g: *Cheirogenesia* Demoulin 1952a: 10, typus *Anagenesia decaryi* Navás 1926 (design. orig.); syn. subj.: ***Fontainica*** McCafferty 1968: 293, typus *F. josettae* McCafferty 1968 (design. orig.) (synn. subj.: *decaryi* [*A.*] = *josettae* [*F.*])].
In circumscription matches:
— gen. *Cheirogenesia* Demoulin 1952a: 10;
— gen. *Fontainica* McCafferty 1968: 293.

References. Demoulin 1952a: ⊕♂; – 1965b: ⊕♂; – Fontaine 1968 ["Ephemeridae (larve)" & "*Cheirogenesia* sp."]: Ꝺ ⊕♂; – McCafferty 1968 (*Fontainica*): Ꝺ; – McCafferty & Edmunds 1976b: Ꝺ; – Sartori & Elouard 1999: ⊙ Ꝺ ⊕♂ ⊕♀.

Autapomorphy of *Cheirogenesia*.

(1) On fore wing [see Anagenesia/g1 (1)] MP_1 and iMP are brought together [besides the vein pairs situated anteriad of them – see Palingenia/f3=g2 (2)]. This character is especially well expressed in male, where distance between apices of MP_1 and iMP is as small as between apices of previous geminating veins (RSa_2+iRS, RSp+MA_1, iMA+ +MA_2) (Demoulin 1952a:Fig.4); in female distance between MP_1 and iMP can be larger. Non-unique apomorphy: the same type of gemination is present in Behningia/fg3.

Character of *Cheirogenesia* of unclear phylogenetic status (TABLE 7).

(2) Cubital field of fore wing [see Anteritorna (1)] includes a strong convex vein "CuA_2" arising from CuA to basitornal margin and looking as a branch of CuA; a single intercalary "iCuA" is situated between "CuA_2" and the distal part of CuA (Demoulin 1952a:Fig.4). Non-unique apomorphy (see Index of characters [2.2.51]); among Anagenesia/g1 the same in Anagenesia/g2 (FIG.75:A) and *Plethogenesia*; probably synapomorphy.

Plesiomorphies of *Cheirogenesia* (TABLE 7). At least in male imago all legs with 2 claws [blunt – see Anagenesia/g1 (2)]. In larva: on outer margin of mandibular tusk each denticle terminates by a spine-like seta [see Palingenia/f2=g1 (2)] (McCafferty & Edmunds 1976:Fig.4); outer margin of fore tibia without denticles, with spine-like setae [see Palingenia/f2=g1 (5)] (Fontaine 1968:Pl.V–VI).

Size. Fore wing length 18 mm.

Distribution. Madagascar.

Species composition of Cheirogenesia/g(1). 3 species are described (Sartori & Elouard 1999) – *decaryi* Navás 1926 [*Anagenesia*] (syn. subj.: *josettae* McCafferty 1968 [*Fontainica*]), *edmundsi* Sartori & Elouard 1999 [*Cheirogenesia*], *laurencae* Sartori & Elouard 1999 [*Cheirogenesia*].

Material examined (Entomol. Inst.): sp.: ⊕, ⊕.

*** *** *** ***

2.2;2,1-3/4.2. Polymitarcys/f1=Ephoron/g2

(Furcatergaliae Fimbriatotergaliae Fossoriae Cryptoprosternata Polymitarcys/f1=Ephoron/g2)

(Figs 76:A–C; 77–80)

Nomen hierarchicum: **Polymitarcys/f1=Ephoron/g2** (sine *Palingenia*; incl. *Campsurus*) [f: Polymitarcini Banks 1900: 246].

In circumscription matches:

— fam. Polymitarcidae: Edmunds & Traver 1954a: 239;

— Polymitarcys/f1=Ephoron/g2: Kluge 2000: 252.

References. Edmunds & Allen & Peters 1963: ℧*; – Tshernova 1970: ℧* ⊕*; – Edmunds & Jensen & Berner 1976: ℧* ⊕*.

Autapomorphies of Polymitarcys/f1=Ephoron/g2.

(1) In larva rows of long setae on mandibles and fore legs [see Cryptoprosternata (3A–D)] are regular, U-shaped, and have a peculiar form (FIGS 76:B; 77:A, C; 79:C–D). **(A)** Mandible has a regular U-shaped row of setae: curvation of this row is situated on lateral side of mandible, and its two arms go parallel one another toward ventral side of mandible across tusk base. **(B)** Inner side of fore femur in its proximal part has a regular U-shaped row of setae: curvation of this row is situated close to base of femur, and its two arms do longitudinally in distal direction. **(C)** On fore side of fore tibia the regular proximal oblique row of setae is strongly curved at both its ends and is continued distally in a form of two regular longitudinal rows. **(D)** Inner side of tibia has a regular U-shaped row of setae, curvation of this row is situated near base of tibia, and its two arms go longitudinally in distal direction. Shape of setal rows (A) and (C) is a unique apomorphy, while setal rows (B) and (D) have some similarity with Eusetisura.

(2) Maxilla is angulate: its apical side (laterad of canines) forms an obtuse or right angle with lateral side; apical side is wide, straight, and bears a field of very densely situated long filtering setae (Eaton 1883–1888: Pls.25, 28); in contrast to Potamanthus/fg1, Euthyplocia/fg1 and some others (see Index of characters [1.1.32]), this field of setae is not so closely adjacent to canines. In other Fossoriae apical side of maxilla also bears numerous long setae, but they do not form such dens and distinctly outlined field. Canines and dentisetae are slender, straight and brought together at apical-inner angle; all 3 canines and 2 dentisetae [see Bidentiseta (1)] can be developed (in Polymitarcys/f2=Ephoron/g3), sometimes middle canine is lost (in some Campsurus/fg1), or both dentisetae are lost (in *adusta* [*Povilla*]).

(3) Winged stages are short-living: At least imaginal middle and hind legs are always non-functional (but functional subimaginal middle and hind legs are retained in male Polymitarcys/f1=Ephoron/g2). Pterothorax is modified, mesonotal suture is extremely stretched posteriorly up to disappearance (FIG.78:D–F); subimaginal cuticle lost its colour pattern [see Fimbriatotergaliae (2)], being colourless or evenly pigmented. Non-unique apomorphies (see Index of characters [2], [2.2.8] and [2.2.14]); among Cryptoprosternata the same in Palingenia/f3=g2. Moult from subimago to imago either takes place in male only (in Polymitarcys/f2=Ephoron/g3), or is retained in both sexes (in Asthenopus/fg1).

(4) On fore wing bifurcation of MA [initially situated in middle of MA – see Euephemeroptera (2)] is transferred proximally (FIGS 78–80). Non-unique apomorphy (see Index of characters [2.2.43]); among Fimbriatotergaliae the same in Euthyplocia/ /fg1, Behningia/fg1 and Anagenesia/g1.

(5) Cubital field of fore wing with one or several intercalaries alternating as concave and convex (FIGS 78–80); sigmoid veins going to basitornal margin [see Anteritorna (1)], can arise from the most posterior of these intercalaries (the same in some Euthyplocia/fg1), or absent (the same in some other Ephemeroptera – see Index of characters [2.2.51]).

(6) In male imago claws of fore leg of male are

blunt, more or less elongate [in subimago they are either ephemeropteroid (in Polymitarcys/f2=Ephoron/f3), or also blunt (in Campsurus/fg1)].

(7) 1st segment of gonostylus is not expressed, being fused with second; thus gonostylus is 3-segmented (in Polymitarcys/f2=Ephoron/g3) or 1-segmented (in Campsurus/fg3). Non-unique apomorphy (see Index of characters [2.3.10]). Some authors erroneously take for 1st segment a pedestal of gonostylus (which contains muscles); in most Polymitarcys/f1=Ephoron/g2 gonostylus pedestal is long, separated from remainder part of styliger and resembles a segment. In Campsurus/fg2 genitals undergone further modification (see below).

Character of unclear phylogenetic status.

(8) Mandibular tusks [see Fimbriatotergaliae (8)] are curved medially (FIGS 76:C; 79:C–D). The same in *Ichthybotus*, Potamanthus/fg1 and Euthyplocia/ /fg1; probably symplesiomorphy.

Variable character of Polymitarcys/f1=Ephoron/g2. In imago on amphitornal margin of wings can be more or less developed setae (FIG.78: A–B) similar to that of subimago; non-unique apomorphy (see Index of characters [2.2.27]).

Size. Fore wing length 5–20 mm.

Distribution. Holarctic, Oriental, Neotropical and Ethiopian Regions.

Polymitarcys/f1=Ephoron/g2 is divided into Polymitarcys/f2=Ephoron/g3 and Campsurus/fg1 (p.265).

2.2;2,1-3/4.2;1. Polymitarcys/f2=Ephoron/g3

(Fimbriatotergaliae Fossoriae Cryptoprosternata Polymitarcys/f1=Ephoron/g2 .../f2=.../g3)
(Figs 76:A–C; 77; 78:A–E)

Nomen hierarchicum: **Polymitarcys/f2=Ephoron/g3** (sine *Campsurus*; incl. *Eopolymitarcys*).
In circumscription matches:
— gen. *Polymitarcys*: Tshernova 1952: 239;
— gen. *Ephoron*: Edmunds & Traver 1954a: 239;
— subfam. Polymitarcinae: Lestage 1917: 258;
— subfam. Ephoroninae: Needham & Traver & Hsu 1935:241;
— fam. Polymitarcidae: Tshernova 1980: 32.

References. Eaton 1883–1888: Ծ ⊕; – McCafferty 1975: Ծ; – Needham & Traver & Hsu 1935: Ծ* ⊕*; – Edmunds & Allen & Peters 1963: Ծ*; – Koss 1968: ⊙; – Koss & Edmunds 1974: ⊙; – Edmunds & Jensen & Berner 1976: Ծ* ⊕*; – Ishiwata 1996: ⊙ Ծ ⊕.

Autapomorphies of Polymitarcys/f2=Ephoron/g3.

(1) On inner side of right mandible between incisor and kinetodontium, a denticle-shape or finger-shape process is present (Ishiwata 1996:Figs 12–15).

(2) Larval fore leg [specialized as burrowing – see Polymitarcys/f1=Ephoron/g2 (1)] has rows of roundish strongly sclerotized wart-shape tubercles: a short row at proximal part of inner side of femur and two long longitudinal rows on posterior side of tibia – one nearer to inner margin, another nearer to outer margin (FIG.77).

(3) Vestige of tergalius I [initially bilamellate – see Cryptoprosternata (6)] is unilamellate. Non-unique apomorphy (see Index of characters [1.3.33]).

(4) On hind wing between RA and RS [which forms a triad – see Euplectoptera (1)] is present a long additional intercalary vein (FIG.78:B); this vein is convex like RA, thus it breaks the regularity in alternating of convex and concave veins typical for mayfly wings. This additional intercalary vein is distinct, integral and very long in male, while in female it can be interrupted or absent. Unique apomorphy: in other mayflies such additional intercalaries are usually absent, rarely present in all spaces between longitudinal and primary intercalary veins (FIG.81:B).

(5) Egg with one or two polar caps of unique structure: external surface of such cap bears numerous anchors, each anchor consists of a short slightly coiled thread terminating by a knob; these knobs can be pressed one to another covering the cap from outside; the cap is situated on a short stem formed by numerous fibrils fused together; base of this stem is surrounded by a collar – a crater-like elevation of chorion (Ishiwata 1996:Figs 22–29). In Polymitarcys/f3=Ephoron/g4 only one pole bears such cap, and in *Eopolymitarcys* each of two poles has a cap. Externally the polar cap of Polymitarcys/f2=Ephoron/g3 resembles a bunch of polar anchors in *Ametropus* (which are not fused in a cap) and differs from simple polar caps of some Euthyplocia/fg1, Potamanthus/fg1, Caenotergaliae and Ephemerella/ /fg1 (see Index of characters [3.6]).

(6) Moult from subimago to imago takes place in male only; female does not moult, and lies eggs being retained in subimaginal stage [imago and subimago of both sexes are short-living with non-functional legs – see Polymitarcys/f1=Ephoron/g2 (3)]. Non-unique apomorphy (see Index of characters [2]).

Character of unclear phylogenetic status.

(7) Mandibular tusks [curved medially – see Polymitarcys/f1=Ephoron/g2 (8)] bear numerous tubercles on dorsal and lateral side (FIG.76:B–C; in contrast to Campsurus/fg1, there are no denticles on inner side. The same in *Ichthybotus*; possibly plesiomorphy within Fossoriae.

Plesiomorphies of Polymitarcys/f2=Ephoron/g3 (in contrast to Campsurus/fg1). Larval (and adult) patella-tibial suture is developed on middle and hind legs. Larval propleura [vestigial – see Furcatergaliae (1)] retain movable articulation with pronotum. Larval fore tarsus is normally developed and articulated with tibia (FIG.77:A, C).

In imago and subimago: Furcasternal protuberances of mesothorax are contiguous (FIG.78:D). On fore wing RS with complete set of veins, bifurcation of MA [proximal – see Polymitarcys/f1=Ephoron/g2 (4)] is situated at a distance from wing base; on hind wing in costal field a series of crossveins of equal thickness are present, MP with bifurcation (FIG.78: A–B). While in imago legs are non-functional [see Polymitarcys/f1=Ephoron/g2 (3)], in male subimago middle and hind legs retain ability to function; because of this male moults on the bank, keeping by its legs on substrate, and exuviae represent an integral case [female does not moult and its subimaginal legs are non-functional – see (6)]. Claws are ephemeropteroid [except for male imaginal fore legs – see Polymitarcys/f1=Ephoron/g2 (6)]. Gonostylus with 2 distal segments [i. e. 3-segmented – see Polymitarcys/f1=Ephoron/g2 (7)]. Female imaginal and subimaginal paracercus is subequal to cerci [while male paracercus is vestigial – see Cryptoprosternata (5)].

Size. Fore wing length 10–18 mm.

Distribution. Holarctic, Oriental and Ethiopian Regions.

Polymitarcys/f2=Ephoron/g3 is divided into *Eopolymitarcys* and Polymitarcys/f3=Ephoron/g4.

2.2;2,1-3/4.2;1,1. Eopolymitarcys/g(1)

(Fimbriatotergaliae Fossoriae Cryptoprosternata Polymitarcys/f1=Ephoron/g2 .../f2=.../g3 *Eopolymitarcys*)

(Figs 76:A–C; 77; 78:A–E)

Nomen hierarchicum: **Eopolymitarcys/g(1)** [g: *Eopolymitarcys* Tshernova 1934: 240, typus *E. nigridorsum* Tshernova 1934 (design. orig.)].
In circumscription matches:
— gen. *Eopolymitarcys* Tshernova 1934: 240.
Monospecific taxon.

References. Tshernova 1934: ⊕; – Tshernova & Kluge & Sinitshenkova & Belov 1986 (*Ephoron nigridorsum*): ⊕; – Kluge 1997 (*E. nigridorsum*): ♂.

Characters of unclear phylogenetic status.

(1) In cubital field of fore wing [see Polymitarcys/f1=Ephoron/g2 (5)] usually only one concave intercalary is present; it goes parallel to CuA toward tornoapical margin; sigmoid veins arise from this intercalary and directly from CuA (FIG.78:A). In selected specimens there are two or none intercalaries (in the last case all sigmoid veins arise from CuA). Non-unique character (see Index of characters [2.2.51] and [2.2.52]).

(2) Egg has 2 polar caps [see Polymitarcys/f2= =Ephoron/g3 (5)].

Size. Fore wing length 10–14 mm.

Distribution. Palaearctic.

Species composition of Eopolymitarcys/g(1). 1 species – *nigridorsum* Tshernova 1934 [*Eopolymitarcys*]. (syn. subj.: *ladogensis* Tiensuu 1935 [*Polymitarcys*]).

Material examined: *nigridorsum* [*Eopolymitarcys*]: ♂–♂–⊕, ♀–⊕/⊙.

***

2.2;2,1-3/4.2;1,2. Polymitarcys/f3=Ephoron/g4

(Fimbriatotergaliae Fossoriae Cryptoprosternata Polymitarcys/f1=Ephoron/g2 .../f3=.../g4)

Nomen hierarchicum: **Polymitarcys/f3=Ephoron/g4** (sine *Eopolymitarcys*).
In circumscription matches:
— gen. *Ephoron* Williamson 1802: 71;
— gen. *Polymitarcys* Eaton 1868b: 84.

References. Eaton 1883–1888: ⊕; – Needham & Traver & Hsu 1935: ⊕; – Koss 1968: ⊙; – Edmunds & Jensen & Berner 1976: ⊕; – Tshernova & al. 1986: ⊕; – Ishiwata 1996: ⊙ ♂ ⊕.

Autapomorphies of Polymitarcys/f3=Ephoron/g4.

(1) Cubital field of fore wing [see Polymitarcys/ /f1=Ephoron/g2 (5)] contains 3–5 intercalaries alternating as concave and convex; they go parallel to CuA toward tornoapical margin and tornus (Eaton 1883–1888:Pl.6:10a-c; Ishiwata 1996:Figs 16–21) (in contrast to all other Polymitarcys/f1=Ephoron/g2 which have no more than two such intercalaries – see Index of characters [2.2.51] and [2.2.52]).

Character of unclear phylogenetic status.

(2) Egg has 1 polar cap [see Polymitarcys/f2= =Ephoron/g3 (5)] (Ishiwata 1996:Figs 22–29).

Size. Fore wing length 10–18 mm.

Distribution. Holarctic, Oriental and Ethiopian Regions.

Nominal taxa included. Polymitarcys/f3=Ephoron/g4 includes: (1) **Ephoron/fg** [f: Ephoroninae Traver (in Needham & Traver & Hsu) 1935: 241] and (2) **Polymitarcys/fg** [g: *Polymitarcys* Eaton 1868b: 84, typus *Ephemera virgo* Olivier 1791 (design. orig.)], regarded as a generic synonym of *Ephoron* (Spieth 1940: 109).

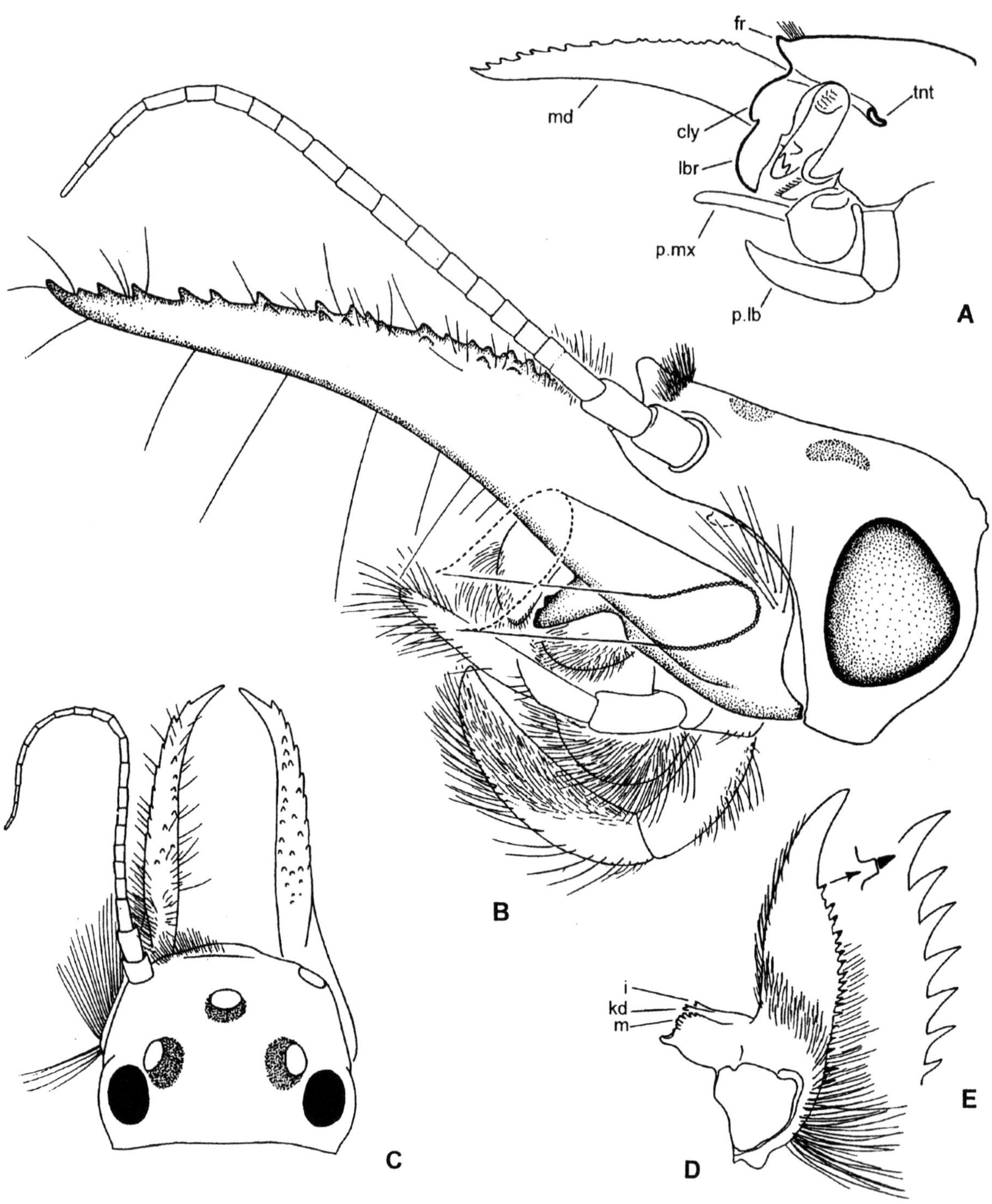

Figure 76. Cryptoprosternata, larval head.
A–C – Polymitarcys/f2=Ephoron/g3 Eopolymitarcys/g* *nigridorsum* [*Eopolymitarcys*]: **A** – median section of head (compare with Figs 69:G, I, 67:C); **B** – head, lateral view (instead of long setae, their bases and occupied area are shown); **C** – head, dorsal view (on right setae are not shown). **D–E** – Palingenia/f3=g2: **D** – Anagenesia/g1 Chankagenesia/g* *natans* [*Chankagenesia*], right mandible, dorsal view, separately enlarged denticle terminated by a spine-like seta; **E** – Palingenia/f4=g3 *longicauda* [*Ephemera*], lateral margin of mandibular tusk. (C–E – from Kluge 1997d).

Abbreviations: **cly** – clypeus; **fr** – frons; **i** – incisor; **kd** – kinetodontium; **lbr** – labrum; **m** – mola; **md** – mandibular tusk; **p.lb** – labial palp; **p.mx** – maxillary palp; **tnt** – tentorium.

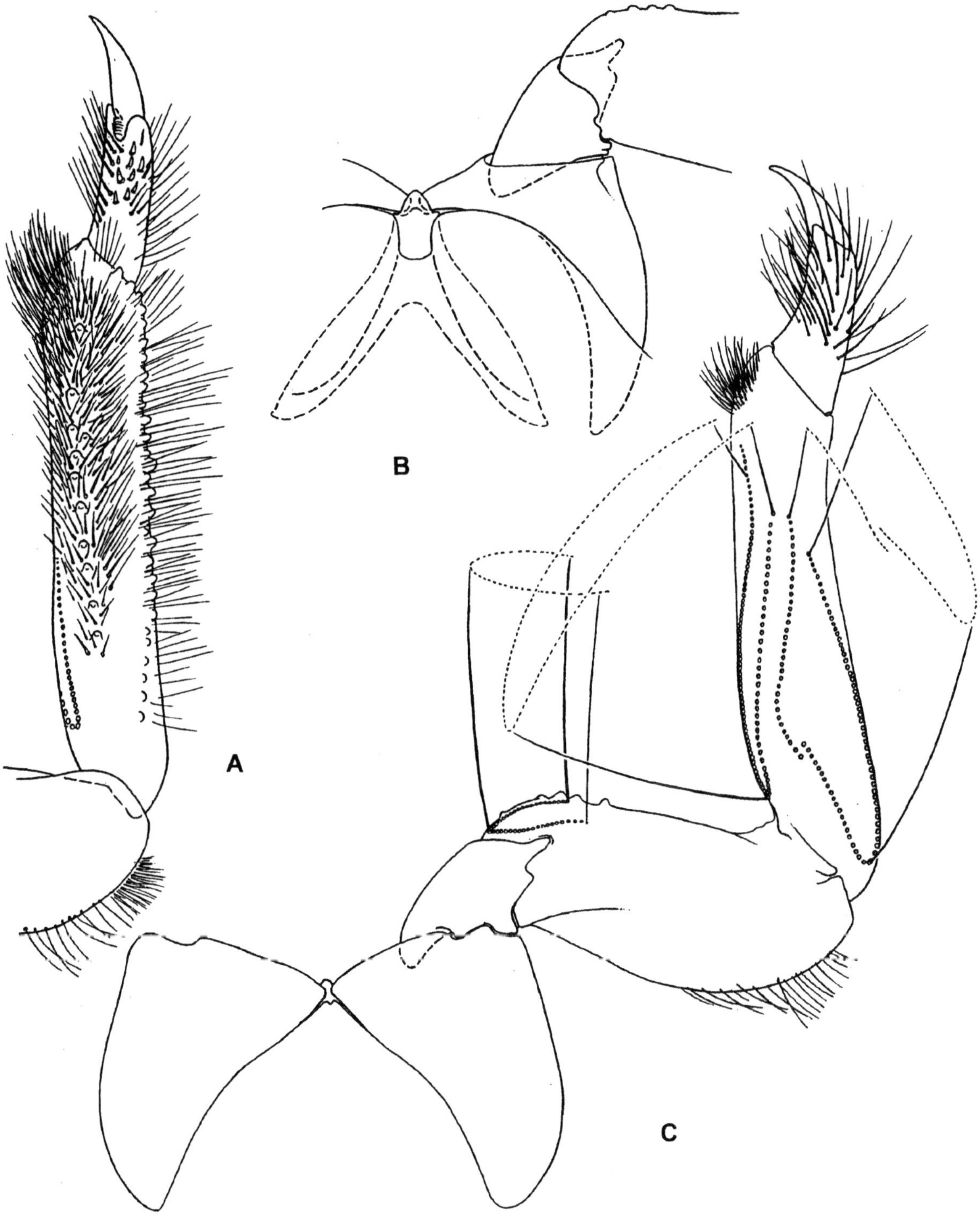

Figure 77. Eopolymitarcys/g* *nigridorsum* [*Eopolymitarcys*], larval fore leg.
A – tibia and tarsus of left fore leg, view from behind (instead of long setae, only their bases are shown; **B** – prosternum and base of left fore leg, view from behind; **C** – right fore leg, front view (instead of long setae, their bases and occupied area are shown).

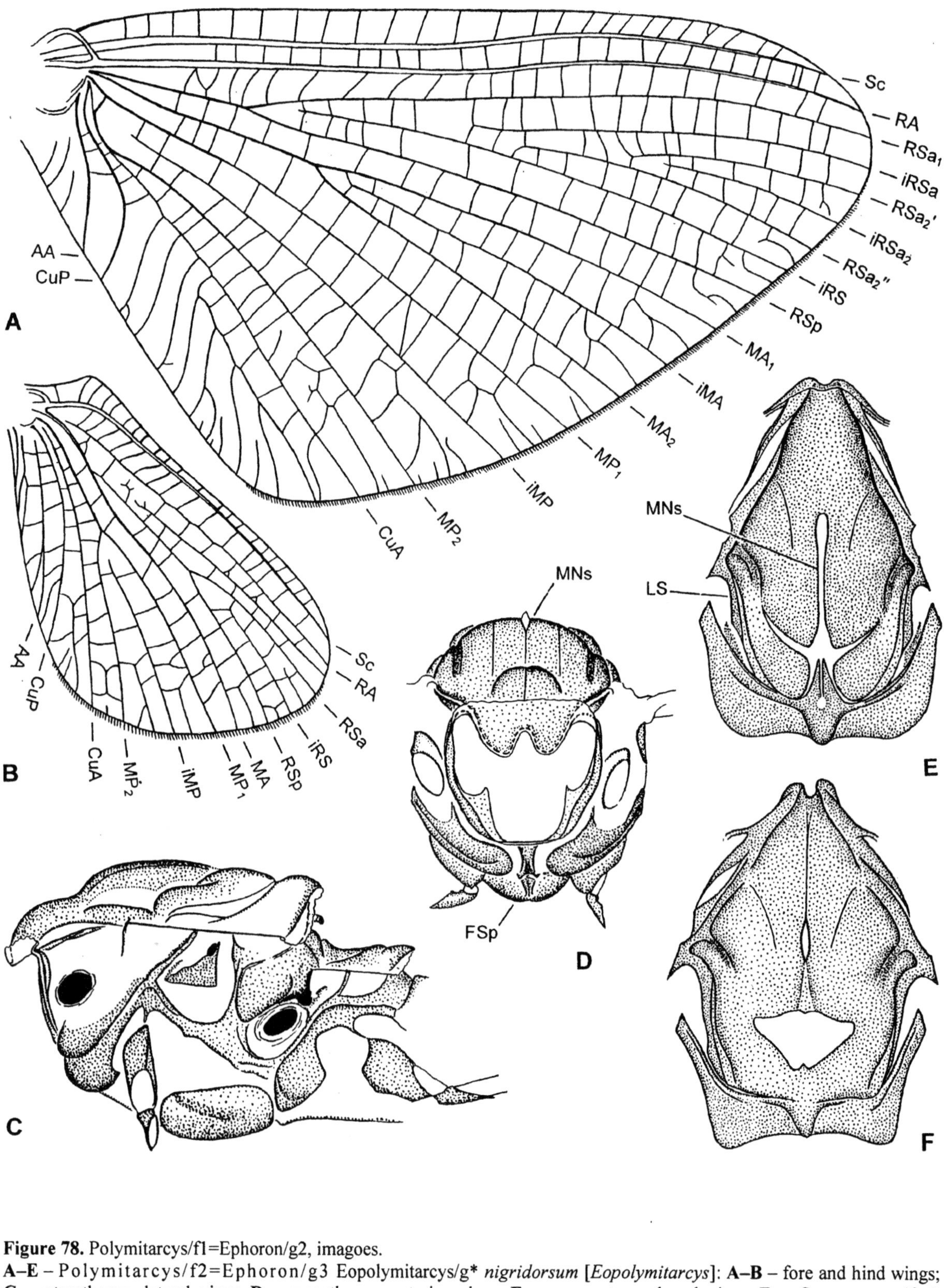

Figure 78. Polymitarcys/f1=Ephoron/g2, imagoes.
A–E – Polymitarcys/f2=Ephoron/g3 Eopolymitarcys/g* *nigridorsum* [*Eopolymitarcys*]: **A–B** – fore and hind wings; **C** – pterothorax, lateral view; **D** – mesothorax, anterior view; **E** – mesonotum, dorsal view. **F** – Campsurus/fg2 Tortopus/g1 *puella* [*Palingenia*], mesonotum, dorsal view.
Abbreviations: **FSp** – furcasternal protuberance; **LS** – lateroscutum; **MNs** – mesonotal suture.

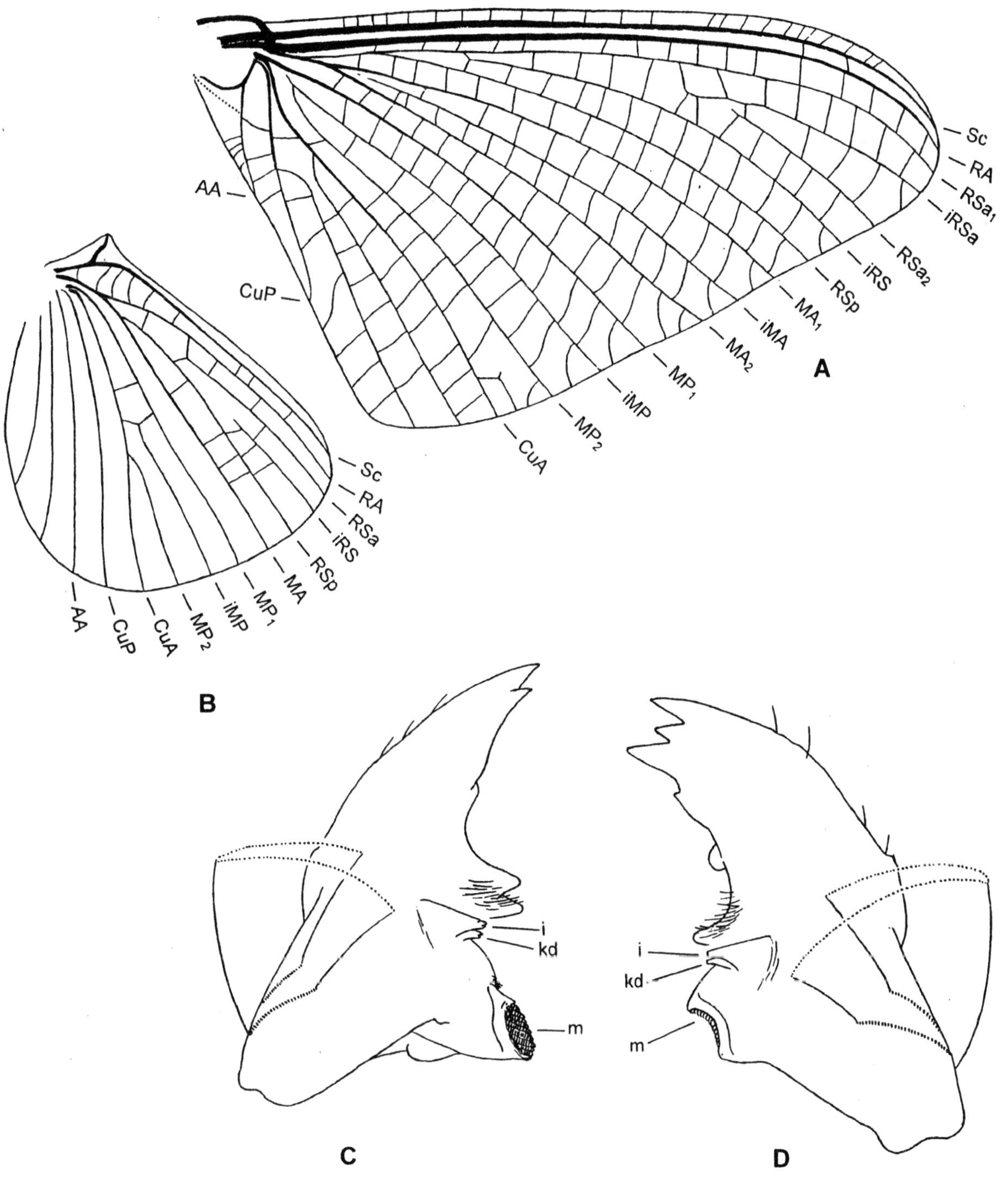

Figure 79. Campsurus/fg1-Asthenopus/fg1 *adusta* [*Povilla*], imago.
A–B – fore and hind wings (trachea passing out of veins shown by dotted line); **C–D** – right and left mandibles, ventral view (instead of long setae, their bases and occupied area are shown).
Abbreviations: **i** – incisor, **kd** – kinetodontium.

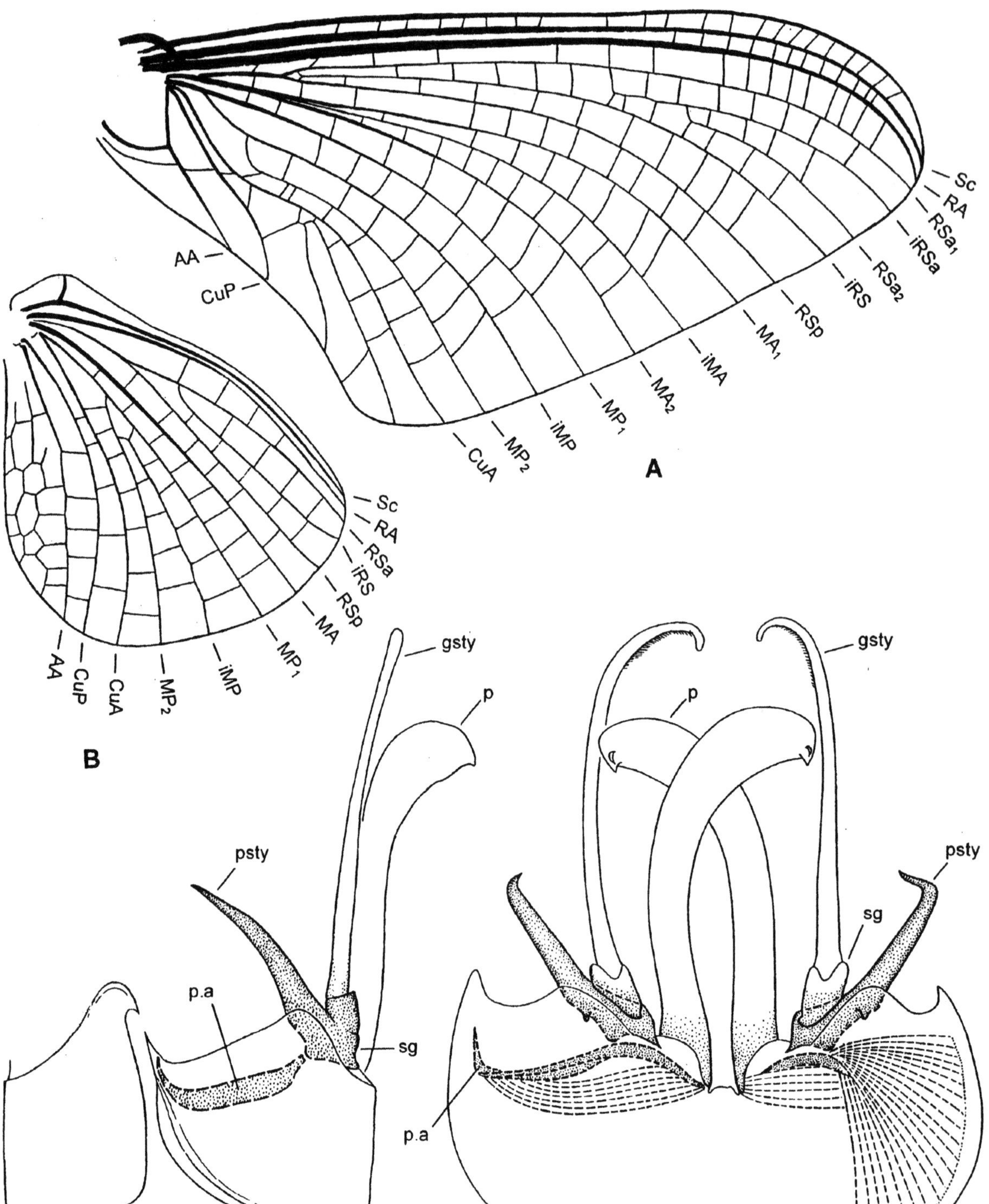

Figure 80. Campsurus/fg2-Tortopus/g1 *puella* [*Palingenia*], male imago.
A–B – fore and hind wings. **C–D** – abdominal segment IX of with genitals (penis arm and pedestal of gonostylus shown by dots): **C** – lateral view; **D** – ventral view (penial muscles shown by interrupted lines, left penial-styliger muscle not shown).

Abbreviations: **gsty** – gonostylus; **p** – penis; **p.a** – penis arm; **psty** – parastylus, **sg** – paired separated portion of styliger (pedestal of gonostylus).

Species composition of Polymitarcys/f3=Ephoron/g4. About 10 species are described. Two **North American species:** *alba* Say 1823 [*Baetis*], *leucon* Williamson 1802 [*Ephoron*]. One **African species:** *savignyi* Pictet 1843 [*Palingenia*] (synn. subj.: *capensis* Petersen 1913 [*Polymitarcys*], *temerata* Navás 1916 [*Polymitarcys*]). **Eurasian species:** *annandalei* Chopra 1927 [*Polymitarcys*], *birmanus* Navás 1933 [*Polymitarcys*], *eophilum* Ishiwata 1996 [*Ephoron*], *indica* Pictet 1843 [*Palingenia*] (synn. subj.: *australis* Hagen 1888 [*Polymitarcys*], *greeni* Banks 1914 [*Anagenesia*]), *limnobium* Ishiwata 1996 [*Ephoron*], *nanchangi* Hsu 1937 [*Polymitarcys*], *punensis* Dubey 1971 [*Euthyplocia*], *shigae* Takahashi 1924 [*Polymitarcys*], *virgo* Olivier 1791 [*Ephemera*] (synn. subj.: *marocana* Fabricius 1793 [*Ephemera*], *horaria* Burmeister 1839 [*Palingenia*]).

Material examined: *alba* [*B.*]: ⊕♂, ⊕♀/⊙; *savignyi* [*P.*]: ⊕♀/⊙; *shigae* [*P.*]: ტ/◎/⊕, ♂–⊕♂, ⊕♀/⊙; *virgo* [*E.*]: ტ, ⊕; spp. (Thailand, China): ⊕♂, ⊕♀/⊙.

***   ***

2.2;2,1-3/4.2;2. Campsurus/fg1

(Fimbriatotergaliae Fossoriae Cryptoprosternata Polymitarcys/f1=Ephoron/g2 Campsurus/fg1)

(Figs 78:F; 79; 80)

Nomen hierarchicum: **Campsurus/fg1** (incl. *Asthenopus*) [f: Campsurinae Traver (in Needham & Traver & Hsu) 1935: 284; g: *Campsurus* Eaton 1868b: 83, typus *Palingenia latipennis* Walker 1853 (design. orig.)].

In circumscription matches:

— gen. *Campsurus* Eaton 1868b: 83;

— subfam. Campsurinae: Needham & Traver & Hsu 1935:284;

— fam. Campsuridae: Tshernova 1980: 32.

References. Eaton 1883–1888: ⊕*; – Needham & Traver & Hsu 1935: ⊕*.

Autapomorphies of Campsurus/fg1.

(1) Mandibular tusks [curved medially – see Polymitarcys/f1=Ephoron/g2 (8)] with denticles on inner side; incisor and kinetodontium are extremely diminished (FIG79:C–D) [about tusk modifications – see below, Asthenopus/fg1 (1), Campsurus/fg3 (2) and *Tortopus* (2)]. In contrast to Polymitarcys/f2= =Ephoron/g3, dorsal and outer sides of tusks without tubercles.

(2) Paraglossae are crescent-shaped, with apices stretched and pointed; in normal position paraglossae are directed ventrally, and their pointed apices are directed dorsally (Eaton 1883–1888:Pl.25:11); if put labium in one flatness and spread paraglossae laterally, their apices become convergent. In contrast to it, in Polymitarcys/f2=Ephoron/g3 paraglossae have more usual semicircular shape with rectangular apices (Eaton 1883–1888:Pl.28:14).

(3) On larval prothorax suture separating collar from the rest part of pronotum [see Fimbriatotergaliae (1)] does not bent forward by sides, but has a form of transverse groove terminating at dorsal condyli of coxae; propleura [vestigial – see Furcatergaliae (1)] are fused with notum. This structure of pronotum is especially well expressed in larva with its strong burrowing fore legs [see (6)], and is retained in subimago and imago as well.

(4) Imaginal furcasternal protuberances of mesothorax are separated. Non-unique apomorphy (see Index of characters [2.2.23]), but not occurring in other Fossoriae.

(5) Wing venation is modified in the following manner (FIGS 79:A–B; 80:A–B):

Fore wing:

Number of RS branches is diminished: usually RSa_2 does not form a triad (non-unique apomorphy – see Index of characters [2.2.37]).

Bifurcation of MA [transferred proximally – see Polymitarcys/f1=Ephoron/g2 (4)] is transferred to extreme wing base (the same in some other taxa – see Index of characters [2.2.43]).

In cubital field number of intercalaries [see Polymitarcys/f1=Ephoron/g2 (5)] is constant and equal to 2 – concave and convex ones.

From AA backward toward wing base goes an arched vein. Trachea penetrating into CuA and MP_2, goes from wing base posteriad of this arched vein, crossing AA and CuP; a chain of crossveins can enclose this trachea partly (FIG.79:A) or completely (FIG.80:A) (in other mayflies this trachea passes anteriorly, in a common bunch with tracheae of RA, RS and MA).

Hind wing:

In costal field the proximalmost crossvein, which supports costal projection, is thickened (crossveins distad of it can be developed or absent) (the same in some Geminovenata)

Normal bifurcation of MP is lost: MP_2 either arises not from MP_1, but from CuA, or has a form of intercalary, or is lost.

(6) On larval fore leg [which is specialized as burrowing – see (3) and Polymitarcys/f1=Ephoron/ /g2 (1)] tarsus is strongly shortened, widened and fused with tibia, thus it seems that claw is directly jointed to tibia (Edmunds & al. 1976:Figs 33, 39). Non-unique apomorphy, similar fusion in *Protobehningia.* In connection with shortening of larval fore tarsus and great length of fore tarsus in male imago (as in many mayflies), in male subimago fore tarsus is moniliform, with strongly thickened segments (Eaton 1883–1888: Pl.5).

(7) Patella-tibial suture (initially present on middle and hind legs) is lost on all legs. Non-unique apomorphy (see Index of characters [1.2.18]).

(8) Male fore leg claws are blunt not only in imago [see Polymitarcys/f1=Ephoron/g2 (6)], but in subimago as well; they are strongly elongate, stick-like (Eaton 1883–1888:Pl.5).

(9) In imago and subimago middle and hind legs of male and all legs of female [non-functional – see Polymitarcys/f1=Ephoron/g2 (3)] are vestigial, tarsal segmentation and claws are non-expressed (in Campsurus/fg2 they are greatly reduced).

(10) Gonostylus without distal segments [i. e. 1-segmented – see Polymitarcys/f1=Ephoron/g2 (7)]. Non-unique apomorphy (see Index of characters [2.3.12]).

(11) Imaginal and subimaginal paracercus is reduced not only in male [see Cryptoprosternata (5)], but in female as well. Non-unique apomorphy (see Index of characters [2.3.22]).

(12) Larval cercus [with secondary swimming setae on lateral side – see Furcatergaliae (6)] lost longitudinal row of primary swimming setae on median side; instead of it, apex of each cercal segment on median side bears a transverse row of long setae which are thicker and shorter than secondary swimming setae on lateral side of cercus. In contrast to Campsurus/fg1, other Fossoriae (including Polymitarcys/f2=Ephoron/g3) retain longitudinal row or stripe of primary swimming setae on median side of cercus.

Plesiomorphies and variable characters of Campsurus/fg1 (in contrast to Polymitarcys/f2= =Ephoron/g3). Larval fore legs without rows of tubercles. Hind wing of both sexes without additional vein between RA and RS. Eggs without anchors-bearing polar caps: they either have no polar structures (in some Asthenopus/fg1 and some Campsurus/fg2), or have long threads on one or both poles, which being coiled to a large spiral, can form a polar cap [see below, Campsurus/fg2 (3) and Plesiomorphies of Asthenopus/fg1].

Size. Fore wing length 5–20 mm.

Distribution. Ethiopian, Oriental, Neotropical and Nearctic Regions.

Campsurus/fg1 is divided into *Asthenopus* and Campsurus/fg2.

2.2;2,1-3/4.2;2,1. Asthenopus/fg1

(Fossoriae Cryptoprosternata Polymitarcys/f1=Ephoron/g2 Campsurus/fg1 *Asthenopus*)

(Fig. 79)

Nomen hierarchicum: **Asthenopus/fg1** (incl. *Povilla*) [f: Asthenopodinae Edmunds & Traver 1954a: 239; g: *Asthenopus* Eaton 1871: 59, typus *Palingenia curta* Hagen 1861]

In circumscription matches:

— subfam. Asthenopodinae Edmunds & Traver 1954a:239.

References. Ulmer 1916 (*Povilla*): ⊕'; – Needham & Murphy 1924: ⊕'; – Ulmer 1939 (*Povilla*): ℧'; – 1942: ⊕; – Spieth 1943: ⊕; – Traver 1950: ⊕; – Demoulin 1956c (*Povilla*): ℧'; – 1966a: ℧; – Edmunds & Allen & Peters 1963: ℧*; – Roback 1966: ℧; – Hubbard 1984 (*Povilla*): ℧', ⊕'; – Domínguez 1988: ⊕; – Hubbard & Domínguez 1988: ⊕.

Autapomorphies of Asthenopus/fg1.

(1) Mandibular tusks [see Campsurus/fg1 (1)] are specialized as biting: not long, very thick and stout, with serrate inner margin; these tusks resemble mandibular incisors of other biting insects (FIG.79:C–D).

(2) Larval claw of fore leg [see Campsurus/fg1 (6)] has a row of denticles on inner margin (while claws of middle and hind legs have no denticles). Unique apomorphy: in all other Fossoriae denticles on all claws are absent; in that Ephemeroptera where denticles are present, they are developed on all legs.

Plesiomorphies of Asthenopus/fg1. On fore wing CuA is not so strongly curved as in Campsurus/fg2, thus both intercalaries of cubital field [see Campsurus/fg1 (5)] go nearly parallel to basitornal margin and terminate near or anteriad of tornus (FIG.79:A). In contrast to Campsurus/fg2, genitals retain small median styliger which is articulated to posterior margin of sternite IX and bears immobile gonostyli pedestals [narrow basally – see Polymitarcys/f1=Ephoron/g2 (7)]; penial arms retain lateral articulations with posterior-lateral angles of tergite IX; gonostyli pedestals retain muscles which move gonostyli [non-segmented – see Campsurus/fg1 (10)]. In spite of short-living specialization of adult [see Polymitarcys/f1=Ephoron/g2 (3)], moult from subimago to imago takes place not only in male, but in female as well. In contrast to Campsurus/fg2, egg has a usual ellipsoid shape; long threads being coiled, can either form two polar caps, or cover the whole egg surface (Koss & Edmunds 1974:Figs 211–214). Vestige of tergalius I [see Cryptoprosternata (6)] is bilamellate.

Size. Fore wing length 8–20 mm.

Distribution. Ethiopian, Oriental and Neotropical Regions.

Nominal taxa included. Asthenopus/fg1 includes: (1) **Povilla/g** [g: *Povilla* Navás 1912c: 402, typus *P. adusta* Navás 1912 (design. orig.)]; (2) **Asthenopodes/g** [g: *Asthenopodes* Ulmer 1924a: 26, typus *A. picteti* Hubbard 1975 (= *Palingenia albicans*: Pictet 1843, non *Ephemera albicans* Percheron 1838, design. orig.)]; (3) **Languidipes/g** [g: *Povilla* subgen. *Languidipes* Hubbard 1984: 29, typus *Asthenopus corporaali* Lestage 1922 (design. orig.)].

Usually 2 genera are accepted – a genus *Asthenopus* for South American species and a genus *Povilla* for African and Asian species; difference between these genera is not described. Within the genus *Povilla*, a subgenus *Languidipes* was established, whose difference from the genus *Asthenopus* is not described. *Asthenopodes* is regarded to be a generic synonym of *Asthenopus* (Hubbard & Domínguez 1988: 107).

Correctness of the recently used names *Povilla* and *Povilla adusta* is doubtful: in the original description (Navás 1912c) in cubital field of fore wing behind two long intercalaries [see Plesiomorphies], two more short intercalaries are figured; known mayflies attributed to Asthenopus/fg1, and particularly to *adusta* [*Povilla*], have no such veins (FIG.79:A).

Species composition of Asthenopus/fg1. 9 species are described. One **African species:** *adusta* Navás 1912 [*Povilla*]. **Asian species:** *andamanensis* Hubbard 1984 [*Povilla*], *cambodjensis* Ulmer 1920 [*Povilla*], *corporaali* Lestage 1922 [*Asthenopus*], *heardi* Hubbard 1984 [*Povilla*] (synn. subj.: *junki* Hubbard 1984 [*Povilla*], *ulmeri* Hubbard 1984 [*Povilla*]), *taprobanes* Hubbard 1984 [*Povilla*]. **South American species:** *curta* Hagen 1861 [*Palingenia*] (syn. subj.: *amazonicus* Hagen 1888 [*Campsurus*]), *gilliesi* Domínguez 1988 [*Asthenopus*], *paraguarius* Navás 1920 [*Campsurus*], *picteti* Hubbard 1975 [*Asthenopodes*].

Material examined: *adusta* [*P.*] (see above): ♉, ⊕, ⊕; *heardi* [*P.*]: ⊕, ⊕; sp. (Colombia): ⊙, ♉/♀/⊕/⊙; sp. (Ecuador): ♂/⊕, ⊕.

***

2.2;2,1-3/4.2;2,2. Campsurus/fg2

(Fossoriae Cryptoprosternata Polymitarcys/f1=Ephoron/g2 Campsurus/fg1 .../fg2)

(Figs 78:F; 80)

Nomen hierarchicum: **Campsurus/fg2** (sine *Asthenopus*; incl. *Tortopus*).
In circumscription matches:
— subfam. Campsurinae: Edmunds & Traver 1954a: 239.

References. Needham & Traver & Hsu 1935: ♉* ⊕*; – Edmunds & Allen & Peters 1963: ♉*; – Koss & Edmunds 1974: ⊙*; – Edmunds & Jensen & Berner 1976: ♉* ⊕*.

Autapomorphies of Campsurus/fg2.

(1) On fore wing vein CuA [proximally arched by its convexity posteriorly – see Fimbriatotergaliae (5)] distally is strongly arched by its convexity anteriorly, thus terminating close to tornus, and two intercalaries of cubital field [see Campsurus/fg1 (5)] are turned in such a manner, that arise from CuA and go posteriorly: the anteriormost of them terminates close to tornus, and the posteriormost – on basitornal margin (FIG.80:A). Unique apomorphy.

(2) Genitals have unique structure (FIG.80:C–D). Median part of styliger is completely lost, thus gonostyli pedestals [initially well-separated – see Polymitarcys/f1=Ephoron/g2 (7)] are movably articulated to membrane which connects posterior margin of sternite IX with base of penis. Penial arms lost articulations with tergite IX; each penial arm is laterally articulated with posterior-lateral angle of sternite IX and ventrally – with base of gonostylus pedestal; muscles going from penial arms to sternite IX, are enlarged. Muscles of gonostyli (initially located inside gonostyli pedestals) are lost. Thus gonostyli [non-segmented – see Campsurus/fg1 (10)] are either immobile fused with gonostyli pedestals, or have only passive mobility, while gonostyli pedestals have active mobility served by the muscles going from penial arms to sternite IX.

(3) Egg is hemispherical, with one side convex and another flattened or concave. In other respects eggs are variable: either without polar structures (in *Tortopus* and some Campsurus/fg3), or with long threads on one pole, which being coiled to a large spiral, form a cap with smooth external surface (in some Campsurus/fg3).

Plesiomorphies of Campsurus/fg2. Mandibular tusks have no that specialization which is peculiar for Asthenopus/fg1. In contrast to Asthenopus/fg1, all larval claws without denticles.

Size. Fore wing length 5–20 mm.

Distribution. America; dominate in Neotropical Region.

Campsurus/fg2 is divided into Campsurus/fg3 and *Tortopus*.

2.2;2,1-3/4.2;2,2-1. Campsurus/fg3

(Fossoriae Cryptoprosternata Polymitarcys/f1=Ephoron/g2 Campsurus/fg1 .../fg3)

Nomen hierarchicum: **Campsurus/fg3** (sine *Tortopus*).
In circumscription matches:
— gen. *Campsurus*: Eaton 1871.

References. Lestage 1923b: ⊕; – McCafferty 1975: ♉; – Edmunds & Jensen & Berner 1976: ♉* ⊕*.

Autapomorphies of Campsurus/fg3.

(1) In male and female imago middle and hind legs [vestigial – see Campsurus/g1 (6)] consist of coxa and trochanter only.

(2) Mandibular tusk [see Campsurus/fg1 (1)] on median margin with a denticle near base and a small denticles distad of it.

Plesiomorphies of Campsurus/fg3 (in contrast to *Tortopus*). Vestige of tergalius I [see Cryptoprosternata (6)] is bilamellate. Pedestal of gonostylus without pointed lateral process.

Size. Fore wing length 5–20 mm.

Distribution. America; dominate in Neotropical Region.

Species composition of Campsurus/fg3 (here [*C.*]= =[*Campsurus*]). About 40 species are formally described, most of them being poorly known – *albicans* Percheron 1838 [*Ephemera*], *albifilum* Walker 1853 [*Palingenia*], *argentinus* Esben-Petersen 1912 [*C.*], *assimilis* Traver 1944 [*C.*], *brasilianus* Traver 1944 [*C.*], *burmeisteri* Ulmer 1942 [*C.*], *claudus* Needham & Murphy 1924 [*C.*], *corumbanus* Needham & Murphy 1924 [*C.*], *cuspidatus* Eaton 1871 [*C.*], *cuyuniensis* Traver 1947[*C.*], *dallasi* Navás 1927 [*C.*], *decolorata* Hagen 1861 [*Palingenia*], *dorsalis* Burmeister 1839 [*Palingenia*], *duplicatus* Spieth 1943 [*C.*], *emersoni* Traver 1947 [*C.*], *essequibo* Traver 1947 [*C.*], *evanidus* Needham & Murphy 1924 [*C.*], *holmbergii* Weyenbergh 1883 [*Ephemera*], *indivisus* Ulmer 1942 [*C.*], *joergenseni* Esben-Petersen 1912 [*C.*], *juradinus* Navás 1930 [*C.*], *latipennis* Walker 1853 [*Palingenia*], *litaninensis* Spieth 1943 [*C.*], *longicauda* Navás 1931 [*C.*], *lucidus* Needham & Murphy 1924 [*C.*], *mahunki* Puthz 1973 [*C.*], *major* Needham & Murphy [*C.*], *melanocephalus* Pereira & Da-Silva 1991 [*C.*], *meyeri* Navás 1934 [*C.*], *mutilus* Needham & Murphy 1924 [*C.*], *nappii* Weyenbergh 1883 [*Palingenia*], *notatus* Needham & Murphy 1924 [*C.*], *pallidus* Needham & Murphy [*C.*], *paranensis* Navás 1934 [*C.*], *pedicellarius* Spieth 1943 [*C.*], *pfeifferi* Navás 1931 [*C.*], *quadridentatus* Eaton 1871 [*C.*], *scutellaris* Needham & Murphy 1924 [*C.*], *segnis* Needham & Murphy [*C.*], *striatus* Needham & Murphy 1924 [*C.*], *truncatus* Ulmer 1920 [*C.*], *ulmeri* Traver 1950 [*C.*], *violaceus* Needham & Murphy 1924 [*C.*], *wappaei* Weyenbergh 1883 [*Ephemera*], *zikani* Navás 1934 [*C.*].

Material examined: spp. (Bolivia, Brazil, French Guiana): ♉, ⊕'.

***

2.2;2,1-3/4.2;2,2-2. Tortopus/g(1)

(Fossoriae Cryptoprosternata Polymitarcys/f1=Ephoron/g2 Campsurus/fg1 .../fg2 *Tortopus*)

(Figs 78:F; 80)

Nomen hierarchicum: **Tortopus/g(1)** [g: *Tortopus* Needham & Murphy 1924: 23, typus *T. igaranus* Needham & Murphy 1924 (design. orig.)].
In circumscription matches:
— gen. *Tortopus* Needham & Murphy 1924: 23.

References. Needham & Murphy 1924: ♁; – Ulmer 1942: ⊕; – Traver 1950: ⊕; – Burks 1953: ⊕; – Scott & Berner & Hirsch 1959: ♉; – Koss 1968: ⊙; – McCafferty 1975: ♉; – Edmunds & Jensen & Berner 1976: ♉* ⊕*; – Domínguez 1987: ⊕; – McCafferty & Bloodgood 1989: ⊕; – Lugo-Ortiz & McCafferty 1996b: ⊕.

Autapomorphies of *Tortopus*.

(1) Pedestal of gonostylus [actively movable – see Campsurus/fg2 (2)] with a pointed lateral process – **parastylus** (FIG.80:C–D). Female has **receptors of parastyli** – a pair of impressions on abdominal sternum VIII (McCafferty & Bloodgood 1989:Figs 1–9). Unique apomorphy.

(2) Mandibular tusk [see Campsurus/fg1 (1)] on its median margin with a subapical denticle, sometimes with one more denticles proximad of it.

(3) Vestige of tergalius I [initially bilamellate – see Cryptoprosternata (6)] is unilamellate. Non-unique apomorphy (see Index of characters [1.3.33]).

Plesiomorphy of *Tortopus*. In contrast to Campsurus/fg3, middle and hind legs of male and female imago retain vestigial non-functional femur, tibia and tarsus.

Size. Fore wing length 9–20 mm.

Distribution. America; dominate in Neotropical Region.

Species composition of Tortopus/g(1). About 10 species are formally described, most of them are poorly known – *bellus* Lugo-Ortiz & McCafferty 1996 [*Tortopus*], *bruchianus* Navás 1926 [*Campsurus*], *circumfluus* Ulmer 1942 [*Tortopus*], *harrisi* Traver 1950 [*Tortopus*], *igaranus* Needham & Murphy 1924 [*Tortopus*], *obscuripennis* Domínguez 1985 [*Tortopus*], *parishi* Banks 1918 [*Campsurus*], *primus* McDunnough 1924 [*Campsurus*] (syn. subj.: *manitobensis* Ide 1941 [*Campsurus*]), *puella* Pictet 1843 [*Palingenia*] (syn. subj.: *incertus* Traver 1935 [*Campsurus*]), *sarae* Domínguez 1985 [*Tortopus*], *unguiculatus* Ulmer 1920 [*Campsurus*], *zottai* Navás 1919 [*Campsurus*].

Material examined: *puella* [*P.*]: ♉, ⊕', ♁/⊙.

*** *** *** *** ***

Fossoriae INCERTAE SEDIS

Here are included fossil Late Jurassic and Early Cretaceous burrowing larvae, whose preservation is not enough to determine systematic position more exactly.

Fossoriae INCERTAE SEDIS 1. † Mesopalingea/g [g: *Mesopalingea* Whalley & Jarzembowski 1985: 384, typus *M. lerida* Whalley & Jarzembowski 1985].

Reference. Whalley & Jarzembowski 1985: ♉.

Age. Late Jurassic / Early Cretaceous (Spain).

Material examined: –.

Fossoriae INCERTAE SEDIS 2. † Torephemera/fg [f: Torephemeridae Sinitshenkova 1989: 39; g: *Torephemera* Sinitshenkova 1989: 40, typus *T. longipes* Sinitshenkova 1989 (design. orig.)].

Reference. Sinitshenkova 1989: ♉.

Age. Early Cretaceous (Mongolia).

Material examined (Paleontol. Inst.): *longipes* [*T.*]: ♉.

Comment. In contrast to majority of Fossoriae, mandibular tusks are absent or, at least, invisible from above; the same in *Siphangarus*. Larval fore leg is burrowing, directed forward, its tibia is thickened, with incised distal margin [see Fossoriae (1)].

Fossoriae INCERTAE SEDIS 3. † Siphangarus/g [g: *Siphangarus* Sinitshenkova 2000a: 68, typus *S. rotundus* Sinitshenkova 2000 (design. orig.)].

Reference. Sinitshenkova 2000a: ♉.

Age. Late Jurassic (Siberia).

Material examined (Paleontol. Inst.): *rotundus* [*S.*]: ♉.

Comment. In contrast to majority of Fossoriae, mandibular tusks are absent or, at least, invisible from above (the same in *Torephemera*). As in other Fimbriatotergaliae, larval pronotum has a collar (well-visible on paratype 4626/254). Possibly, larval fore legs are directed forward and have shape similar to that of *Torephemera* (left fore leg is very poorly visible on holotype only). Larval hind tibia is slightly arched, with dense fine setae on outer margin and inner-apical angle (not shown in the original description). All larval abdominal segments I–IX with rounded lateral margins and without posterolateral spines [see Fimbriatotergaliae (10)]. Bases of tergalii VII are not transferred anteriorly (in contrast to Ephemera/fg9). Tergalii are not preserved on fossils. Larval caudalii (preserved in holotype) consist of numerous short segments with perpendicular margins, with dense secondary swimming setae on lateral and median sides of cerci and lateral sides of paracercus [see Furcatergaliae (6)].

*** ***

2.2;2,1-4. Caenotergaliae, or Caenis/fl=Brachycercus/g1

(Anteritorna Bidentiseta Furcatergaliae Fimbriatotergaliae Caenotergaliae)

(Figs 81–88)

Nomen hierarchicum: **Caenis/fl=Brachycercus/gl** (incl. *Neoephemera*) [f: Caenidae (orig. Coenidae) Newman 1853: 187); g: *Brachycercus* Curtis 1834: 122, typus *B. harrisella* Curtis 1834 (design. Lestage 1924c: 61); syn. obj.: ***Eurycaenis*** Bengtsson 1917: 186].

Nomen circumscribens: **Caenotergaliae** Kluge 2000: 252.

In circumscription matches:

— superfam. Caenoidea: Edmunds & Traver 1954a: 239.

— Caenotergaliae, or Caenis/f1=Brachycercus/g1: Kluge 2000: 252.

References. Edmunds & Allen & Peters 1963: ♉*; – Koss & Edmunds 1974: ⊙; – Wang & McCafferty & Bae 1997: ⊕*; – Kluge 1997b: ♉*.

Autapomorphies of Caenotergaliae.

(1) Tergalii beginning from II [initially bilamellate, with marginal processes – see Fimbriatotergaliae (6)] have unique structure (FIGS 82:D–E; 88:A) [about tergalius I – see (7)].

Tergalius II has a form of gill operculum: able to cover tergalii III–VI and lost ability for rhythmical respiratory movement (non-unique apomorphy, the same in Posteritorna and selected Pantricorythi – see Index of characters [1.3.32]). Shape of the gill operculum is roundish-quadrangular, with prominent anal-proximal angle reaching middle of posterior margin of abdominal tergum II. In this place posterior margin of tergum II usually has a median projection (which is absent in Brachycercus/fl=g3 only). Gill operculum [lacking primary marginal ribs – see Furcatergaliae (5)] is usually strengthened by secondary ribs: a **proximal rib** stretches obliquely from tergalial base (i. e. from its anterior-lateral angle) toward middle; a **median rib** stretches longitudinally from initial anal-proximal angle (i. e. anterior-middle angle) to initial distal margin (i. e. posterior margin). These two ribs can be independent (FIG.82:A, D) [see Neoephemera/fgl (1) below], or connected in the middle of tergalius, forming an integral **Y-shaped rib** (FIG.88:A, J) [see Caenoptera (18) below].

Tergalii III–VI have unique form: semicircular, with marginal processes pectinatedly branching (only in *Madecocercus* processes of tergalii IV–VI are non-branching); any ribs are absent. Bases of tergalii VI are strongly transferred anteriorly, that allows these tergalii to be hidden under the gill opercula (in contrast to other operculate-gilled mayflies – see [1.3.57]). Ventral lamella of tergalii II–VI is either diminished (in Neoephemera/fg1 – FIG.82: D–E), or lost (in Caenoptera).

Tergalii VII are lost (non-unique apomorphy – see Index of characters [1.3.21]).

Tergalii II–VI lie dorsally on abdomen and are directed by their apices posteriorly, overlapping tergalii located posteriad of them; lateral margins of abdominal segments are often expanded laterad of tergalii (FIG.88:A).

(2) Larval fore wing buds are fused with notum not only by their basitornal margins (as in majority of mayflies), but at least partly by their tornoapical margins as well. At the same time, in contrast to Coloburiscus/fg1, Discoglossata and Ephemerella/ /fg1, left and right fore wing buds in Caenotergaliae are not fused one with another, but each of them is fused with notum only; because of this, in Neoephemera/fg1 whose scutellum is non-enlarged, wing buds remain to be free in most part (FIG.82:F); in Caenoptera whose scutellum is strongly elongated, wing buds appear to be fused with it nearly all along their length (FIG.88:B).

(3) Larval penis bud is lost, thus subimaginal penis develops inside larval sternum IX, and in mature larva subimaginal penis and styliger are covered by a common cuticular sheath of larval subanal plate (FIGS 81:D; 82:H; 84:B); separate larval gonostyli buds can be either retained, or fused [see Caenis/f3=g1 (2) below]. In other mayflies (except for Turbanoculata) larva of last instar has distinct penis bud covered by larval cuticle and projected posteriad from articulatory membrane between subanal plate and paraprocts (FIG.33:F); such penis bud is retained even if gonostyli buds are completely fused with subanal plate (FIG.98:C).

(4) Imaginal and subimaginal furcasternal protuberances are separated, metathoracic nerve ganglion is inserted between them (FIGS 83:D; 87:F). Non-unique apomorphy (see Index of characters [2.2.23]); among Fimbriatotergaliae the same independently evolved in Campsurus/fg1 only.

Apomorphies of Caenotergaliae common with Potamanthus/fg1 and Ephemerella/fg1 (see Classifications of Furcatergaliae I).

(5) In larva [which has a collar on pronotum – see Fimbriatotergaliae (1)] mesonotum also has a **collar** – a concave band at anterior margin, separated from the rest part of mesonotum by a transverse crest (FIGS 82:F; 88:B). The same in Potamanthus/ /fg1 and Euthyplocia/fg1, and the same mesonotal collar in Ephemerella/fg1.

(6) Dorsal (anterior) surface of larval fore femur can bear a transverse row of setae (Provonsha 1990: Fig.82); sometimes this row is irregular, indistinct or lost. The same in Potamanthus/fg1 and Ephemerella/fg1.

(7) Tergalius I [different from others – see Fimbriatotergaliae (6)] is stick-like, setose, attached on a prominent cylindrical pedestal arisen from anterior part of abdominal segment I close to metathorax (the same in Potamanthus/fg1 and Ephemerella/fg1 only); pedestal is sclerotized; tergalius itself is either integral (Fig.88:A), or subdivided to 2 segments (FIG.82:B) (the same in Potamanthus/fg1 – FIG.67:A).

(8) Egg can have one or two polar caps; usually such cap at dormant stage looks as integral (Provonsha 1990:Figs 112–120), but in a few species of Caenis/f4=g2 cups are formed by very long spirally coiled threads. One cap is present in Brachycercus/f1=g3, one or two caps – in selected species of Caenis/f3=g1; in Neoephemera/fg1 and many Caenis/f3=g1 caps are absent.

(9) Cubital field of fore wing [see Anteritorna (1)] with bifurcate vein arising from CuA (x_1 and x_2 in FIGS 81:A; 86:F); in selected specimens these two branches independently arise from CuA; other veins can be absent [see Caenoptera (10) below], or 1 or 2 simple or branched veins arise from CuA more distally (1 simple vein is shown in FIG.81:A). The same, besides Potamanthus/fg1 and Ephemerella/ /fg1, in some other mayflies (see Index of characters [2.2.51]).

Characters of Caenotergaliae of unclear phylogenetic status: possible reductions.

(10) Mandibular tusks are absent [see Fimbriatotergaliae (8)].

(11) Dense setae on inner side of hind femur and tibia are absent [see Fimbriatotergaliae (9)].

(12) In larva, subimago and imago abdominal tergite X has lateral-ventral margins longitudinal, as long as the segment, reaching bases of cerci; lateroposterior angle of tergite forms a ventral condylus for cercal base and separates paraproct from cercotractor; cercotractor is transformed to a narrow semicircular sclerite exposed caudally and surrounding lateral half of cercal base (FIG.12:G), can be fused with cercal base. The same in Fossoriae and Geminovenata (but not in other mayflies whose lateral-ventral margins of tergite X are oblique, not articulating with cerci, and cercotractors are triangular, exposed laterally and widely connected with paraprocts – FIG.12:F).

Plesiomorphies of Caenotergaliae. Maxilla

[without apical-ventral row of setae – see Fimbriatotergaliae (10)] always has 3 canines, 2 dentisetae [see Bidentiseta (1)] and no apical field of setae adjacent to canines (in contrast to Potamanthus/fg1, Euthyplocia/fg1 and Ephemerella/fg1); usually maxilla is biting, only in Clypeocaenis/g1 specialized as filtering. Larval paracercus is always well-developed, subequal to cerci (while imaginal paracercus can be either well-developed, or vestigial). Imaginal and subimaginal claws are ephemeropteroid [except for fore legs of male – see below, *Leucorhoenanthus* (2), *Potamanthellus* (3) and Caenoptera (16)].

Size. Fore wing length 2–17 mm.

Age and distribution. Eocene (see Neoephemera/fg1 INCERTAE SEDIS) — Recent; world-wide.

Status and systematic position of Caenotergaliae. About possibility to unite Caenotergaliae with Potamanthus/fg1 and Ephemerella/fg1 – see above, Classifications of Furcatergaliae I and II.

Caenotergaliae are divided into Neoephemera/fg1 and Caenoptera (p.277).

2.2;2,1-4/1. Neoephemera/fg1

(Bidentiseta Furcatergaliae Fimbriatotergaliae Caenotergaliae Neoephemera/fg1)

(Figs 81–83)

Nomen hierarchicum: **Neoephemera/fg1** (incl. *Ochernova, Leucorhoenanthus, Potamanthellus*) [f: Neoephemerinae Traver (in Needham & Traver & Hsu) 1935: 288; g: *Neoephemera* McDunnough 1925: 168, typus *N. bicolor* McDunnough 1925 (design. orig.)].

In circumscription matches:

— gen. *Oreianthus* Traver 1931: 104;

— fam. Neoephemeridae: Burks 1953: 42;

— superfam. Neoephemeridea: Tshernova 1962b: 57;

— Neoephemera/fg1: Kluge 2000: 252.

References. Tshernova 1960: ♉* ⊕*; – Edmunds & Allen & Peters 1963: ♉*; – Tshernova 1970: ♉* ⊕*; – Bae & McCafferty 1998: ♉* ⊕*;– McCafferty & Wang 2000: ♉* ⊕*.

Autapomorphies of Neoephemera/fg1.

(1) Gill operculum [i. e. tergalius II – see Caenotergaliae (1)] has unique structure (FIG.82:A, D). Its margin directed medially (initial anal margin of tergalius) is bent ventrally; along its line of bend goes a sclerotized crest which bears dense setae directed medially or obliquely; the bent portion is membranous and directed ventrally, perpendicular to the rest portion of gill operculum. Left and right opercula are connected mesially by their lines of bend and firmly coupled by setae.

(2) Larval abdominal terga VI–VIII with more or less expressed unpaired projection on posterior margin [besides projection on tergum II – see Caenotergaliae (1)]. Non-unique apomorphy (see Index of characters [1.3.3]).

Plesiomorphies of Neoephemera/fg1. In larva: Maxillary palp is 3-segmented (in contrast to Brachycercus/f1=g3 and Clypeocaenis/g2). Labial palp is 3-segmented (in contrast to Brachycercus/f1=g3). Tergalii II–V [see Caenotergaliae (1)] retain vestigial ventral lamella with marginal processes (FIG.82:D–E) (in contrast to Caenoptera).

In imago and subimago (in contrast to Caenoptera): Eyes of male are larger than in female and retain division to dorsal and ventral portions. Thorax has no modifications peculiar for Caenoptera (FIG.83:A–F). Mesonotal suture is well-expressed both in subimago and imago, being strongly stretched posteriorly by sides of median line [see Fimbriatotergaliae (2)], and has a characteristic uniform structure in all representatives: its median components represent distinct grooves going longitudinally between median longitudinal suture and medioparapsidal sutures (FIG.83:E). Subimago is not shortly-moulting; subimaginal cuticle retains characteristic colour pattern [see Fimbriatotergaliae (2)–(4)] (FIG.83:C, F). Fore wing has curved MP_2 and CuA [see Fimbriatotergaliae (5)], large number of cross veins and connected or free marginal intercalaries; cubital field with variable number (1–3) of bifurcate and/or simple veins arising from CuA to wing margin [see Caenotergaliae (9)]; hind wing is well-developed, as long as 0.3–0.4 of fore wing length (FIG.81:A–B). Tarsus retains structure initial for Furcatergaliae: first segment is short, usually fused with tibia (FIG.83:G); only in *Ochernova* first tarsal segment is secondarily separated from tibia (FIG.81:C). Gonostylus is usually 4-segmented: with 2 short distal segments and usually (except for Potamanthellus/g1) with elongate 1st segment separated by a ledge from 2nd segment (FIGS 81:D, 82:H).

Size. Fore wing length 6–17 mm.

Age and distribution. Eocene (see Neoephemera/fg1 INCERTAE SEDIS) — Recent; Holarctic and Oriental Regions.

Neoephemera/fg1 is divided into Neoephemera/fg2, *Ochernova*, *Leucorhoenanthus* and *Potamanthellus*; some species have uncertain position (p.276).

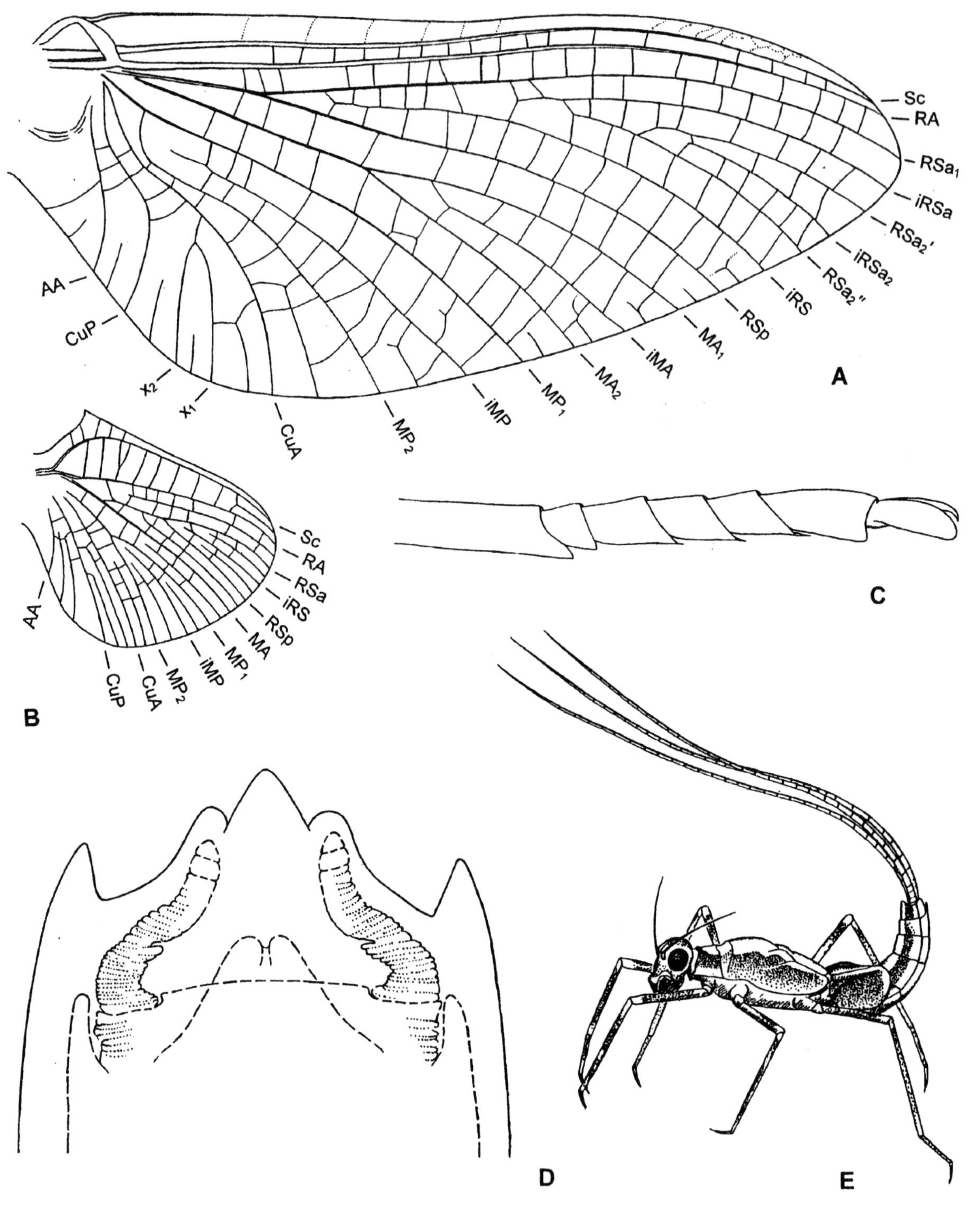

Figure 81. Caenotergaliae-Neoephemera/fg1-Ochernova/g* *tshernovae* [*Neoephemera*].
A–B – fore and hind wings; **C** – tarsus of subimaginal left middle leg (extracted from larva), anterior view; **D** – hind part of abdominal sternum IX of mature male larva, ventral view (internal subimaginal parts shown by interrupted lines); **E** – mature larva in most common pose.

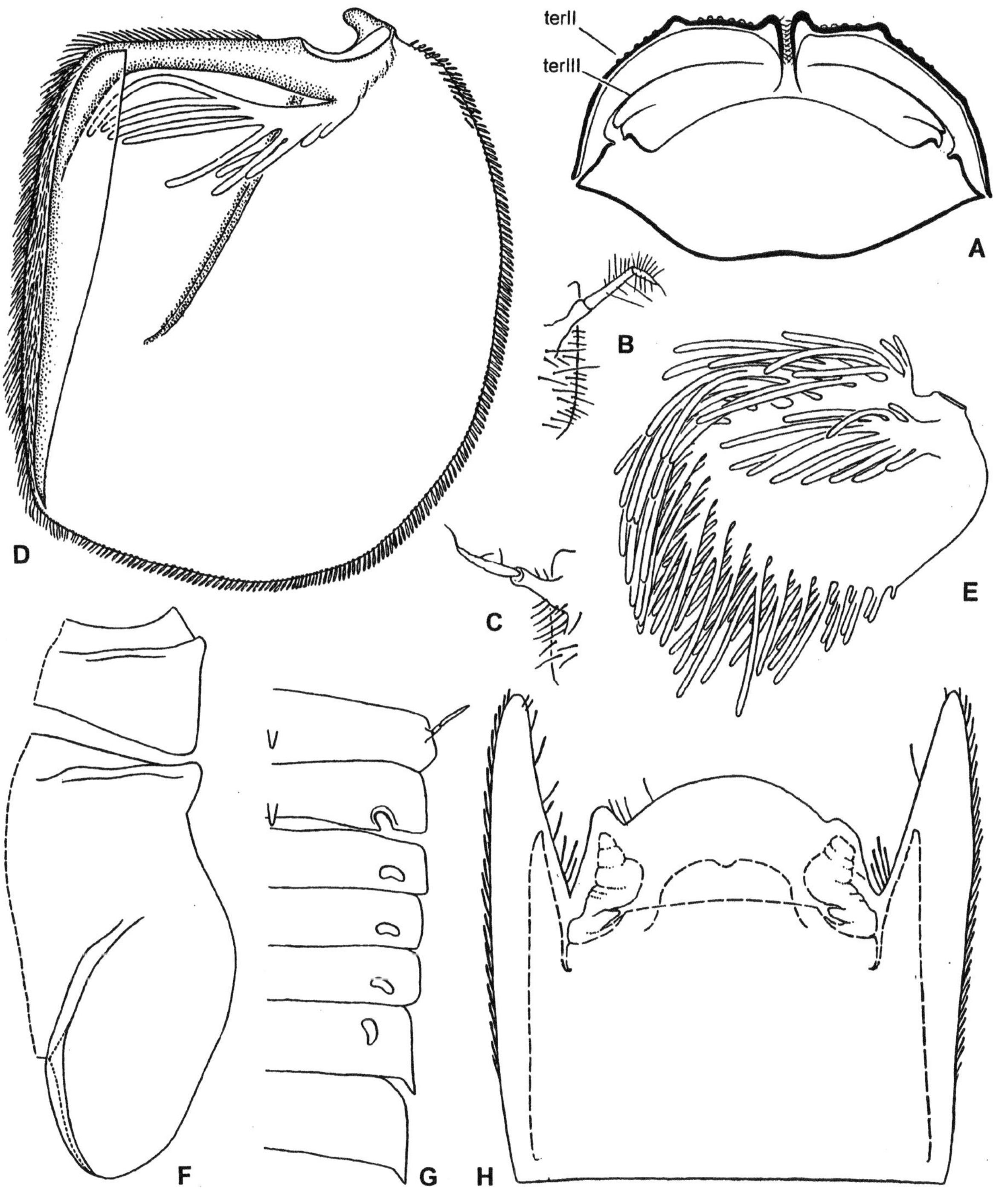

Figure 82. Neoephemera/fg1-Leucorhoenanthus/g* *maxima* [*Caenis*], larva.
A – transverse section on level of abdominal segment III. **B–E** – tergalii I–III on the same scale: **B–C** – right and left tergalii I of one specimen, dorsal view; **D** – left tergalius II (gill operculum), ventral view; **E** – left tergalius III, ventral view. **F** – larval exuviae of right half of pronotum and mesonotum. **G** – right half of larval abdominal terga I-VII (tergalii II-VI removed); **H** – abdominal sternum IX of mature male larva, ventral view (internal subimaginal parts shown by interrupted lines). (F, G – from Kluge 1997b).

Abbreviations: **terII** – gill operculum (tergalius II); **terIII** – tergalius III.

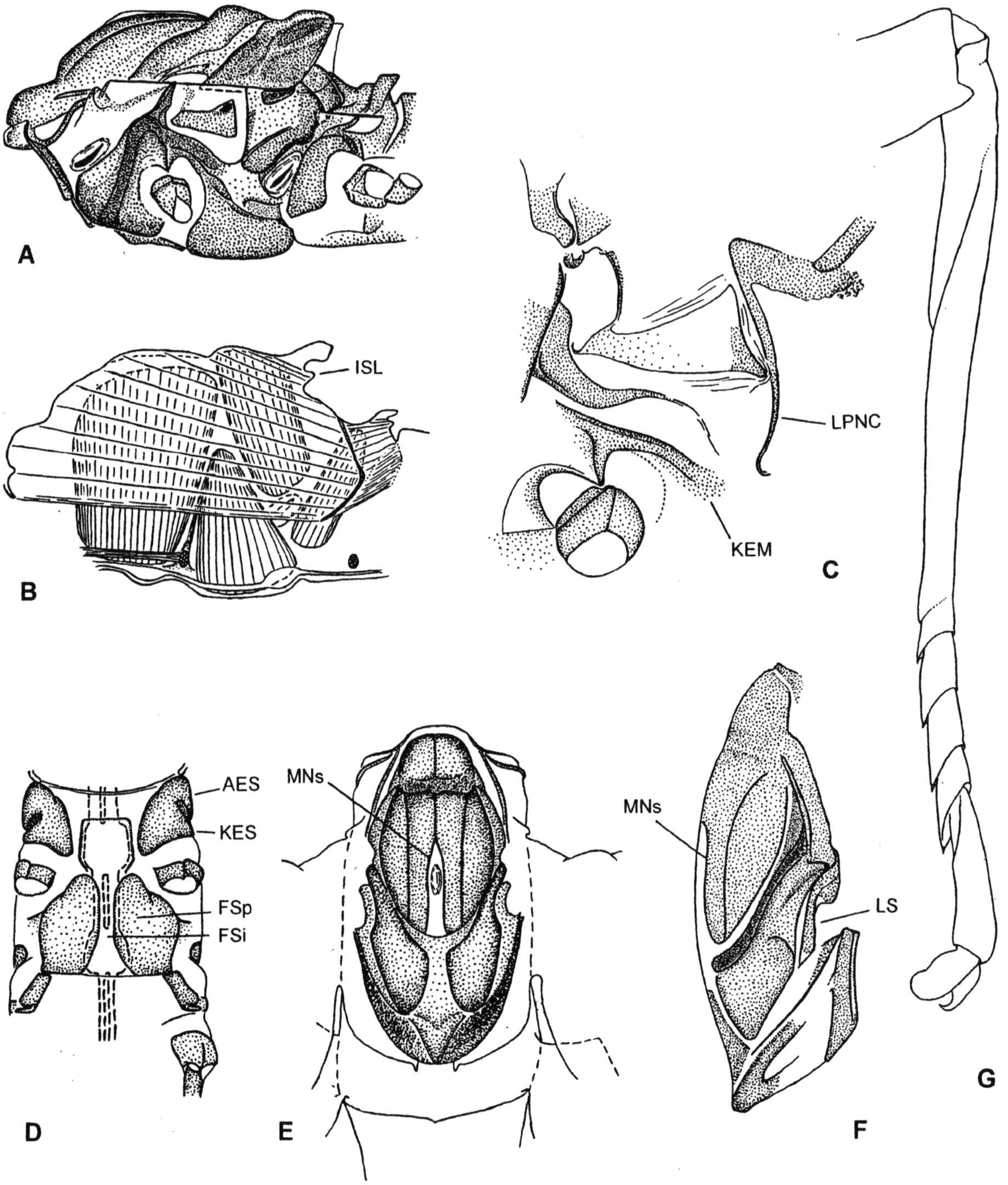

Figure 83. Leucorhoenanthus/g* *maxima* [*Caenis*], thorax of imago and subimago.
A – imaginal pterothorax, lateral view; **B** – imaginal pterothorax, median section; **C** – subimaginal exuviae of left mesopleuron with middle coxa (compare with Figs 5:D and 71:D); **D** – imaginal pterothorax, ventral view (nerve chain shown by interrupted line); **E** – imaginal mesonotum, dorsal view; **F** – subimaginal exuviae of right half of mesonotum; **G** – tibia and tarsus of imaginal left middle leg, anterior view. (A, B, F – from Kluge 1997b; E – from Kluge 1998).

Abbreviations: **AES** – anepisternum; **FSp** – furcasternal protuberance; **FSi** – furcasternal median impression; **ISL** – infrascutellum; **KEM** – katepimeron; **KES** – katepisternum; **LPNC** – lateropostnotal crest; **LS** – lateroscutum; **MNs** – mesonotal suture.

2.2;2,1-4/1.1. Neoephemera/fg2

(Furcatergaliae Fimbriatotergaliae Caenotergaliae Neoephemera/fg1 .../fg2)

Nomen hierarchicum: **Neoephemera/fg2** (sine *Ochernova*, *Leucorhoenanthus*, *Potamanthellus*; incl. *Oreianthus*).
In circumscription matches:
— gen. *Neoephemera* McDunnough 1925: 168;
— subfam. Neoephemerinae: Needham & Traver & Hsu 1935: 288.

References. Traver 1931 (*Oreianthus*): ♉' ♁'; – Needham & Traver & Hsu 1935: ♉* ⊕*; – Berner 1956: ♉* ⊕*; – Edmunds & Jensen & Berner 1976: ♉* ⊕*.

Autapomorphies of Neoephemera/fg2.

(1) Larval thoracic terga with more or less expressed protuberances; these are: a pair of submedian protuberances on anterior part of pronotum on the crest bordering collar [see Fimbriatotergaliae (1)]; a pair of submedian protuberances on anterior part of mesonotum on the crest bordering collar [see Caenotergaliae (5)]; and an unpaired protuberance near posterior margin of mesonotum. Non-unique apomorphy; dorsal protuberances occur in some other mayflies, but not in other Neoephemera/fg1.

Plesiomorphies of Neoephemera/fg2. In imago and subimago all claws are ephemeropteroid, including male fore legs. Gonostyli are not diminished, median incision of penis is small (in contrast to Potamanthellus/g1). Imaginal paracercus is well-developed (in contrast to *Leucorhoenanthus* and Potamanthellus/g1).

Size. Fore wing length 8–17 mm.

Distribution. Eastern Nearctic.

Nominal taxon included. Neoephemera/fg2 includes **Oreianthus/g** [g: *Oreianthus* Traver 1931: 104, typus *O. purpureus* Traver 1931 (design. orig.)] which is regarded either as a monotypic subgenus in the genus *Neoephemera* (Demoulin 1961: 67), or as a generic synonym of *Neoephemera* (Burks 1953: 42).

Species composition of Neoephemera/fg2. 4 species – *bicolor* McDunnough 1925 [*Neoephemera*], *compressa* Berner 1956 [*Neoephemera*], *purpureus* Traver 1931 [*Oreianthus*], *youngi* Berner 1953 [*Neoephemera*].

Material examined: *youngi* [*N.*]: ♂–♂, ♁.

***

2.2;2,1-4/1.2. Ochernova/g(1)

(Furcatergaliae Fimbriatotergaliae Caenotergaliae Neoephemera/fg1 *Ochernova*)

(Fig. 81)

Nomen hierarchicum: **Ochernova/g(1)** [g: *Ochernova* Bae & McCafferty 1998, typus *Neoephemera tshernovae* Kazlauskas 1963 (design. orig.)].
In circumscription matches:
— gen. *Ochernova* Bae & McCafferty 1998: 59.
Monospecific taxon.

Reference. Kazlauskas 1963: ♉.

Autapomorphies of *Ochernova*.

(1) Patella-tibial suture (initially present on middle and hind legs) is lost on all legs of larva, subimago and imago. Non-unique apomorphy (see Index of characters [1.2.18]).

(2) Larval legs have very long slender femora and tibiae [see (1)] (FIG.81:E). Larvae of the single species are adopted for inhabitancy inside accumulations of dead plant remainders in rivers running through sandy deserts.

Plesiomorphies of *Ochernova*. In larva: thoracic terga without protuberances (FIG.82:E) (in contrast to Neoephemera/fg2). In imago and subimago: gonostyli are not diminished; median incision of penis is small (FIG.81:D) (in contrast to Potamanthellus/g1); paracercus is well-developed (in contrast to Leucorhoenanthus and Potamanthellus/g1). Claw structure of male imago is unknown.

Size. Fore wing length 9 mm.

Distribution. Middle Asia: plain desert rivers of Turkmenistan, Kazakhstan and Tajikistan.

Species composition of Ochernova/g(1). 1 species – *tshernovae* Kazlauskas 1963 [*Neoephemera*].

Material examined: *tshernovae* [*N.*]: ♉, ♂/♂, ♀/♁.

***

2.2;2,1-4/1.3. Leucorhoenanthus/g(1)

(Furcatergaliae Fimbriatotergaliae Caenotergaliae Neoephemera/fg1 *Leucorhoenanthus*)

(Figs 82–83)

Nomen hierarchicum: **Leucorhoenanthus/g(1)** [g: *Leucorhoenanthus* Lestage 1931b: 134, typus *Rhoenanthus macedonicus* Ulmer 1920 (design. orig.); syn. subj.: *Neoephemera* subgen. ***Caenomera*** Demoulin 1961: 66, typus *Caenis maxima* Joly 1871 (design. orig.) (synn. subj.: *maxima* [*C.*] = *macedonicus* [*Rh.*])].
In circumscription matches:
— subgen. *Caenomera* Demoulin 1961: 66;
— gen. *Leucorhoenanthus* Lestage 1931b: 134.
Monospecific taxon.

References. Joly 1871 (*Caenis maxima*): ♉; – Eaton 1883–1888 (*Tricorythus* sp.): ♉; – Ulmer 1920a (*Rhoenanthus macedonicus*): ⊕; – Kazlauskas 1959: ♉ ⊕; – Ikonomov 1962: ♉ ⊕; – Jazdzewska 1975: ♉ ⊕.

Characters of unclear phylogenetic status.

(1) Imaginal and subimaginal paracercus is vestigial, consists of several segments. Non-unique apomorphy (see Index of characters [2.3.22]); among

Neoephemera/fg1 the same in Potamanthellus/g1; possibly synapomorphy.

(2) On fore leg of male imago both claws are blunt (in subimago ephemeropteroid). Non-unique apomorphy (see Index of characters [2.2.77]); among Neoephemera/fg1 the same in Potamanthellus/g1; possibly synapomorphy.

Plesiomorphies of *Leucorhoenanthus*. In larva: thorax without protuberances (in contrast to Neoephemera/fg2). In imago: gonostyli are not diminished; median incision of penis is small (FIG.82:H). (in contrast to Potamanthellus/g1). Larval and adult patella-tibial suture is developed on middle and hind legs (FIG.83:G) (in contrast to *Ochernova*).

Size. Fore wing length 8–11 mm.

Distribution. Europe.

Species composition of Leucorhoenanthus/g1. 1 species – *maxima* Joly 1871 [*Caenis*] (syn.subj.: *macedonicus* Ulmer 1920 [*Rhoenanthus*]).

Material examined: *maxima* [*C.*]: ☋–◎–⊕/⊙.

***

2.2;2,1-4/1.4. **Potamanthellus/g1**

(Furcatergaliae Fimbriatotergaliae Caenotergaliae Neoephemera/fg1 *Potamanthellus*)

Nomen hierarchicum: **Potamanthellus/g1** (incl. *Neoephemeropsis*) [g: *Potamanthellus* Lestage 1931b: 120, typus *P. horai* Lestage 1931 (design. orig.); syn. subj.: ***Rhoenanthodes*** Lestage 1931b: 136, typus *Rhoenanthus amabilis* Eaton 1892 (design. orig.) (synn. subj.: *amabilis* [*R.*] = *horai* [*P.*])].
In circumscription matches:
— gen. *Potamanthellus*: Bae & McCafferty 1998: 39.

References. Eaton 1892b (*Rhoenanthus amabilis*): ⊕; – Lestage 1931b: ⊕; – Hsu 1936: ⊕; – Ulmer 1939: ☋ ⊕; – Dang 1967: ☋; – Tiunova & Levanidova 1989: ☋; – Tiunova 1991: ⊕; – Bae & McCafferty 1998: ☋* ⊕*.

Autapomorphies of *Potamanthellus*.

(1) Gonostyli and penis are vestigial: Gonostylus is strongly diminished, 1st segment is not expressed, thus gonostylus is 3-segmented (in selected specimens 1- or 2-segmented if apical segments are fused). In most species (except for *caenoides* [*Neoephemeropsis*]) median part of penis is reduced, and penis represents a pair of widely separated narrow sticks curved medially.

Characters of unclear phylogenetic status.

(2) Imaginal and subimaginal paracercus is vestigial, consists of several segments. Non-unique apomorphy (see Index of characters [2.3.22]); among Neoephemera/fg1 the same in *Leucorhoenanthus*; possibly synapomorphy.

(3) On fore leg of male imago both claws are blunt (in subimago ephemeropteroid). Non-unique apomorphy (see Index of characters [2.2.77]); among Neoephemera/fg1 the same in *Leucorhoenanthus*; possibly synapomorphy.

(4) Larval caudalii [which have no primary swimming setae – see Furcatergaliae (6)] have dense secondary swimming setae: a row of setae is present on lateral and median side of each cercus and on each lateral side of paracercus. In other Neoephemera/fg2 swimming setae are absent; but they are present in many other Furcatergaliae, including selected species of Caenoptera.

Plesiomorphies of *Potamanthellus*. Larval thoracic terga without protuberances (in contrast to Neoephemera/fg2). Larval and adult patella-tibial suture is developed on middle and hind legs (as in FIG.83:G) (in contrast to *Ochernova*).

Size. Fore wing length 6–10 mm.

Distribution: Eastern Asia.

Nominal taxon included. Potamanthellus/g1 includes **Neoephemeropsis/g** [g: *Neoephemeropsis* Ulmer 1939: 483, typus *N. caenoides* Ulmer 1939 (design. orig.)], regarded as a generic synonyms of *Potamanthellus* (Hsu 1937: 137; Bae & McCafferty 1998: 40).

Species composition of Potamanthellus/g1. 5 species (Bae & McCafferty 1998) – *amabilis* Eaton 1892 [*Rhoenanthus*] (synn. subj.: *horai* Lestage 1930 [*Potamanthellus*], *cuaraoensis* Dang 1967 [*Neoephemeropsis*]), *caenoides* Ulmer 1939 [*Neoephemeropsis*], *chinensis* Hsu 1936 [*Potamanthellus*] (syn. subj.: *rarus* Tiunova & Levanidova 1989 [*Neoephemeropsis*]), *edmundsi* Bae & McCafferty 1998 [*Potamanthellus*], *ganges* Bae & McCafferty 1998 [*Potamanthellus*].

Material examined: *chinensis* [*P.*]: ☋–◎–⊕/⊙; spp. (Nepal, Thailand, Vietnam): ♂, ♀.

***

Neoephemera/fg1 INCERTAE SEDIS

The single fossil species of Neoephemera/fg1 is *antiqua* Sinitshenkova 1999 [*Neoephemera*] known as larva from Eocene of North America; it has coupled gill opercula, thus surely belongs to Neoephemera/fg1 (the holotype was re-examined by me), but its more exact systematic position is unclear. Another fossil larva which was attributed to Neoephemeridae – *rubiens* Lewis 1977 [*Potamanthellus*] – can not belong to Caenotergaliae, as it has triangular gill opercula.

A Recent species *projecta* Zhou & Zheng 2000 [*Neoephemera*] from China is described as larvae only; it has no dorsal thoracic tubercles (in contrast to Neoephemera/fg2), and no swimming setae on caudalii (in contrast to Potamanthellus/g1).

*** ***

2.2;2,1-4/2. **Caenoptera, or Caenis/f2=Brachycercus/g2**

(Bidentiseta Furcatergaliae Fimbriatotergaliae Caenotergaliae Caenoptera)

(Figs 84–88)

Nomen hierarchicum: **Caenis/f2=Brachycercus/g2** (sine *Neoephemera*).
Nomen circumscribens: **Caenoptera** Kluge 2000: 252.
In circumscription matches:
— gen. *Brachycercus* Curtis 1834: 122;
— gen. *Caenis* Stephens 1835: 60;
— subfam. Caeninae: Edmunds & Traver 1954a: 239;
— fam. Brachycercidae Lestage 1924c: 62;
— fam. Caenidae: Lestage 1942: 16.
— Caenoptera, or Caenis/f2=Brachycercus/g2: Kluge 2000: 252.

References. Lestage 1942: ⊕*; – Edmunds & Allen & Peters 1963: ⌚*; – Tshernova 1970: ⌚* ⊕*; – Edmunds & Jensen & Berner 1976: ⌚* ⊕*; – Kluge 1992c: ⊕*; – Malzacher 1997: ⊕*.

Unique autapomorphies of Caenoptera.

(1) In imago and subimago pterothorax is extremely specialized (FIG.87): mesothorax is greatly enlarged, has only cross wing musculature and modified sclerites; metathorax is extremely reduced. Mesothorax constitutes about 1/2 of total body length: length of mesonotum is subequal to 0.4–0.5 of fore wing length (which is subequal to body length, as in other mayflies); this is larger than in all other mayflies: in Neoephemera/fg1, Tricorythodes/fg1 and Prosopistoma/f1=g2 mesonotum is subequal to 0.3–0.4 of fore wing length, in majority of other mayflies – to 0.2–0.3 of fore wing length (TABLE 2). Some features of pterothorax structure are common with some Tricoryptera [see (8)], and the following features are unique.

Anterior transverse impression of mesonotum is completely lost (FIG.87:A).

Mesonotal suture (initially curved backward by sides of median line as in FIG.83:F) is transformed to a pair of arched sutures convergenting anteriorly and posteriorly by acute angles and bordering a small closed **medionotal membrane** (FIG.87:G), [non-unique apomorphy – see Fimbriatotergaliae (2)].

Subalar-sternal muscle (SA.Sm) is completely lost (FIG.87:B) (at the same time there are retained diminished and widely separated furcasternal protuberances which initially serve as places of attachment of the paired subalar-sternal muscles – FIG.87:F). In other Ephemeroptera subalar-sternal muscles are always present, while in some Leptohyphes/fg1 (particularly, in Tricorythodes/fg1) can be diminished.

In male (but not in female) left and right halves of phragma between mesothorax and metathorax are fused by mean of two newly appeared walls – anterior and posterior ones; thus, phragma of male is unpaired and has three cavities – a pair of opened cavities in the left and the right halves of the phragma (as well as in other mayflies), and a median unpaired cavity, closed from all sides (FIG.87:C–E).

(2) Fore wing [of single pair – see (9)] has unique features (FIG.86:F): iMP and MP_2 arise from extreme base of wing independently one from another and from neighbouring veins – thus they seem not as intercalaries or branches, but as two independent non-branched veins. Shape of wing strongly differs from that of all other Fimbriatotergaliae and is uniform in all Caenoptera – fan-like widened proximally, without curvation of MP_2 and CuA [see Fimbriatotergaliae (5)]; it has much in common with Tricorythodes/fg1 and some other Tricoryptera [see (10)–(12)].

(3) Ventral side of gill operculum [tergalius II – see Caenotergaliae (1)] near its lateral (i. e. costal) and posterior (i. e. apical) margins bears microtrichia – small immobile outgrowths of cuticle not homologous to setae. In Brachycercus/f1=g3 these microtrichia are spine-like (FIG.84:E) (as microtrichia on some other parts of body), while in Caenis/f3=g1 microtrichia have peculiar structure (FIGS 85:A; 88:D, G) (see below).

(4) Each tergalius lacks any vestige of ventral lamella [see Caenotergaliae (1)] (FIG.88:A, C).

(5) Genitals have following unique structure (FIG.84:A–D). Base of styliger is strongly transferred anteriorly and W-shapely speaks into sternite IX; boundary between styliger and sternite can be lost, but is indicated by muscle bases (Malzacher 1997); thus styliger is immobile connected with sternite, being unable to turn down. Dorsal wall of styliger, which is invariably pressed to penis, has sclerite with a pair of projections directed proximally (FIG. 84:D); ventral wall of styliger is membranous and translucent, thus the sclerite of dorsal wall is visible from ventral side (FIG.84:C).

Gonostylus is non-segmented – i.e. without distal segments and without traces of suture between 1st and 2nd segments (non-unique apomorphy – see Index of characters [2.3.12] and [2.3.10]).

Sensory globular papillae (which in Neoephemera/fg1 and other mayflies are present of inner surface of gonostylus – see Chapter II) are lost. Possibly, initially for Caenoptera, imaginal gonostylus has smooth sclerotized surface and a deep longitudinal groove opened medially, while subimaginal gonostylus is covered by microtrichia (as in other mayfly subimagoes) and retains normal roundish cross section; such structure is present in all Brachycercus/f1=g3 (where the groove can be so deep, that gonostylus has a form of leaf rolled in a tube – FIG.84:A–D), in *Madecocercus* and *Tillyardocaenis* (FIG.86:C–E); in Caenis/f5=g3 imaginal gonostyli resemble subimaginal ones: they are roundish in cross section and usually covered by microtrichia.

Larval gonostyli buds are extremely shortened or lost, but developing inside them gonostyli of adults are never crumpled, as their bases are shifted proximally (FIGS 84:B; 86:C) (in contrast to many other mayflies, particularly Neoephemera/fg1, where larval gonostyli buds are also very short, and subimaginal gonostyli are crumpled inside them – FIGS 81:D; 82:H).

In connection with immobility of styliger, penial arms are functionless, often weak, desclerotized, or non-expressed.

Penis [developing without larval penis bud – see Caenotergaliae (3)] is integral, poorly sclerotized or non-sclerotized.

(6) Imaginal and subimaginal caudalii [cerci and paracercus – see Plesiomorphy] are modified in a unique manner, differently in male and female. In male subimago caudalii are short (as in larva), segments are short and thick, with whorls of long setae on joinings only; in male imago caudalii are very long, segments are strongly elongated and slender, completely lack setae (in contrast to majority of mayflies whose caudalii of subimago and imago bear irregularly situated setae). In female caudalii are short and slender (as in larva), with usual irregularly situated setae; in contrast to majority of mayflies, caudalii do not moult when subimago moults to imago (while all other parts of body do moult, in contrast to females of some short-living mayflies). About sexual dimorphism of larval caudalii – see (14).

Apomorphies of Caenoptera common with Tricoryptera (see Classifications of Furcatergaliae II).

(7) Adults are shortly-moulting: in mature larva ready to moult to subimago, imaginal cuticle is already developed; shortly (in a few minutes) after the moult from larva to subimago, the next moult from subimago to imago follows. At the same time, imagoes are not short-living, and have functional legs. Subimaginal cuticle is colourless (FIG.87:H), without pigmented and sclerotized areas characteristic for Neoephemera/fg1 and many other mayflies [see Fimbriatotergaliae (2)–(4) and FIG.83:C, F]. The same in Tricoryptera (at least in Tricorygnatha and Leptohyphes/fg1).

(8) Imaginal and subimaginal mesothorax, besides unique apomorphies [see (1)] has following modifications common with all or some Tricoryptera (FIG.87):

On sublateroscutum transverse interscutal suture is present (FIG.87:A, G, H) (the same in Tricorygnatha and Leptohyphes/fg1 – see Classifications of Furcatergaliae IIb).

Anterior paracoxal suture is transferred posteriorly, close to anterior margin of coxal conjunctiva (FIG.87:A, F) (the same in Tricorygnatha and Leptohyphes/fg1).

Posterior scutal protuberances are enlarged and convergent posteriorly (FIG.87:A, G) (the same in Leptohyphes/fg1-Tricorythodes/fg1 – FIG.104).

Lateropostnotum is enlarged and stretches ventrally nearly up to furcasternum, reaches coxal conjunctiva, and ventrally is fused with epimeron (FIG.87:A) (the same in Tricorythodes/fg1 – FIG. 104). Fusion of lateropostnotum with epimeron (but not enlargement of lateropostnotum) takes place also in Fossoriae (FIG.71:D) and some others, but not in Neoephemera/fg1 (FIG.83:C).

Scutellum is large, its lateral impressions are enlarged; infrascutellum is interrupted medially (FIG.87:A–C) (the same in all Tricoryptera and Teloganodes/fg1 – FIG.89:B).

(9) Hind wings are always completely lost (non-unique apomorphy – see Index of characters [2.2.59]); metathorax is vestigial; at least in female wing muscles of metathorax, including scuto-episternal muscle (S.ESm) and median tergal muscle (MTm) are lost; in male at least S.ESm can be retained (FIG.87:B), but lost in *meridies* [*Caenopsella*]. The same in Tricorythodes/fg1; in other mayflies, even if hind wings are completely lost, wing muscles S.ESm and MTm are retained in both sexes.

(10) On fore wing [see (2)] CuP is not connected with CuA and looks as arising from AA; cubital field has only a bifurcate vein [see Caenotergaliae (9)] which is usually connected basally both with CuP and CuA (FIG.86:F: x_1, x_2). The same in Tricoryptera and some other mayflies (selected Leptophlebia/fg1, Tetramerotarsata and others).

(11) On fore wing [see (2)] marginal intercalaries are absent (non-unique character – see Index of characters [2.2.55]); number of cross veins is strongly diminished (FIG.86:F).

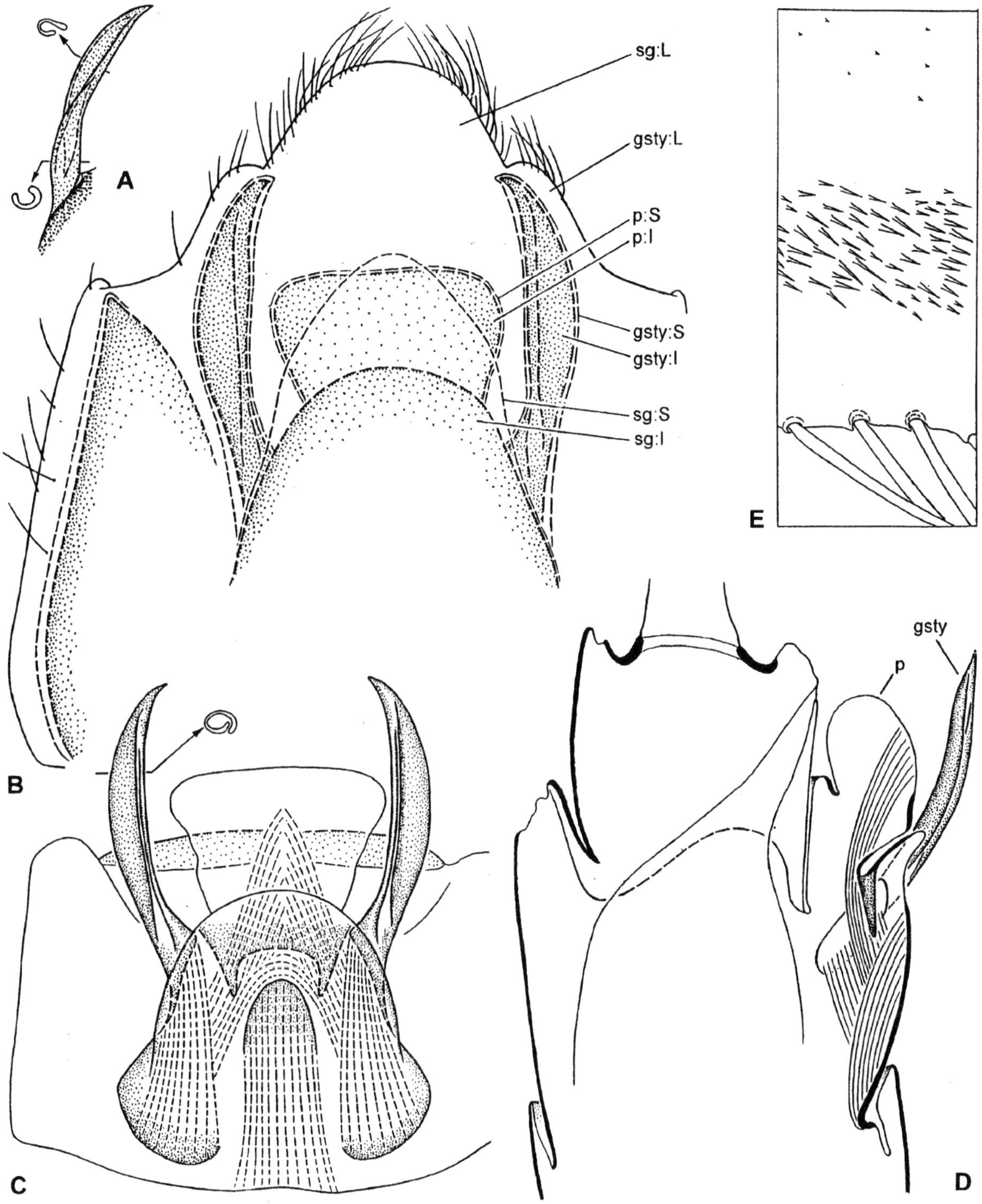

Figure 84. Caenoptera-Brachycercus/f1=g3.
A – *minutus* [*Brachycercus*] – left imaginal gonostylus. ventral view (arrows show cross sections). **B–E** – *harrisella* [*Brachycercus*]: **B** – abdominal sternum IX of mature male larva, ventral view (internal structures shown by interrupted lines: subimaginal cuticle without dots, imaginal parts with dots); **C** – genitals of male imago, ventral view (muscles of styliger, gonostyli and penis shown by interrupted lines); **D** – genitals of male imago, median section with muscles of styliger, gonostyli and penis; E – ventral side and lateral margin of gill operculum (tergalius II). (A – from Kluge 1991a).

Abbreviations: **gsty** – gonostylus; **p** – penis; **sg** – styliger. **I** – imaginal part; **L** – larval bud; **S** – subimaginal cuticle.

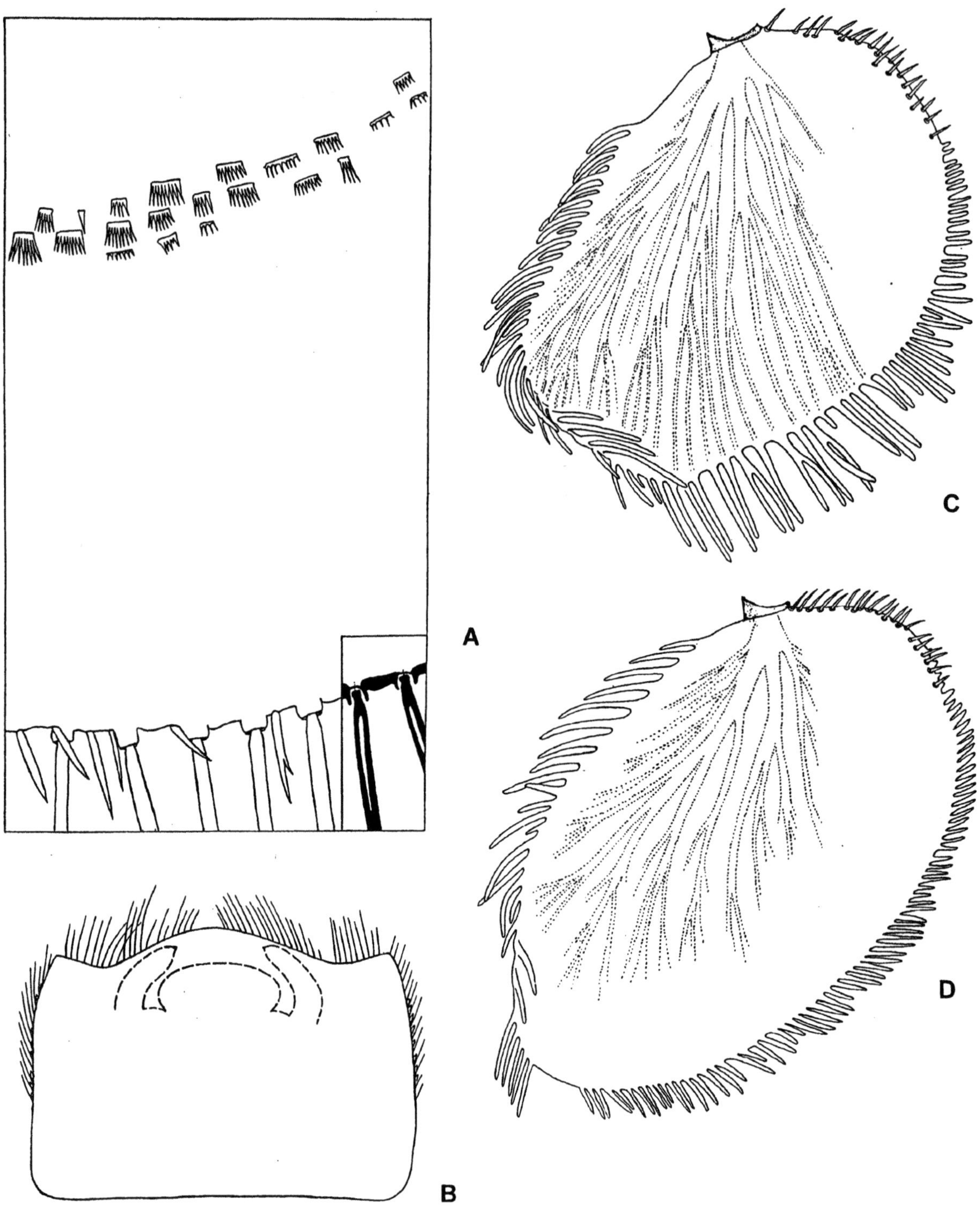

Figure 85. Caenoptera-Caenis/f3=g1-Madecocercus/g* *tauroides* [*Madecocercus*], larva.
A – ventral side and lateral margin of gill operculum (tergalius II); **B** – abdominal sternum IX of immature male larva, ventral view (buds of subimaginal gonostyli and penis shown by interrupted lines); **C** – tergalius III; **D** – tergalius IV (tracheae chown by dotted lines).

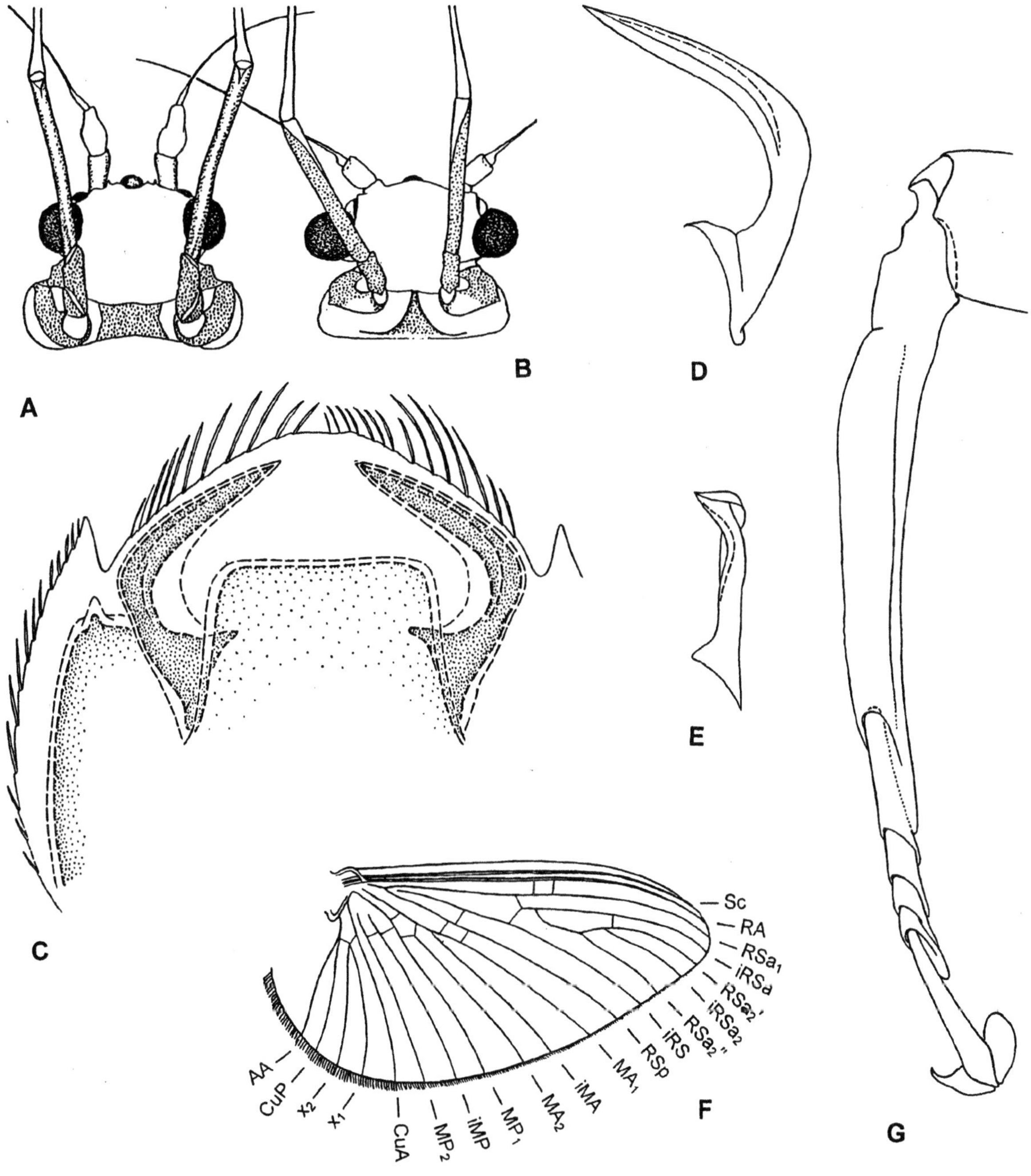

Figure 86. Caenoptera.
A – Brachycercus/f1=g3 *minutus* [*Brachycercus*], prothorax and head of male imago, ventral view; **B–G** – Caenis/f4=g2: **B** – Caenis/f6=g4 *horaria* [*Ephemera*], the same; **C–E** – Tillyardocaenis/g* *tillyardi* [*Caenis*]: **C** – abdominal sternum IX of mature male larva, ventral view (internal structures shown by interrupted lines: subimaginal cuticle without dots, imaginal parts with dots; penis not shown); **D** – right gonostylus of male imago, ventral view; **E** – the same, median view. **F–G** – Caenis/fg* *macrura* [*Caenis*]: **F** – fore wing; **G** – tibia and tarsus of imaginal right hind leg, anterior view. (A, B – from Tshernova & Kluge & Sinitshenkova & Belov 1986)

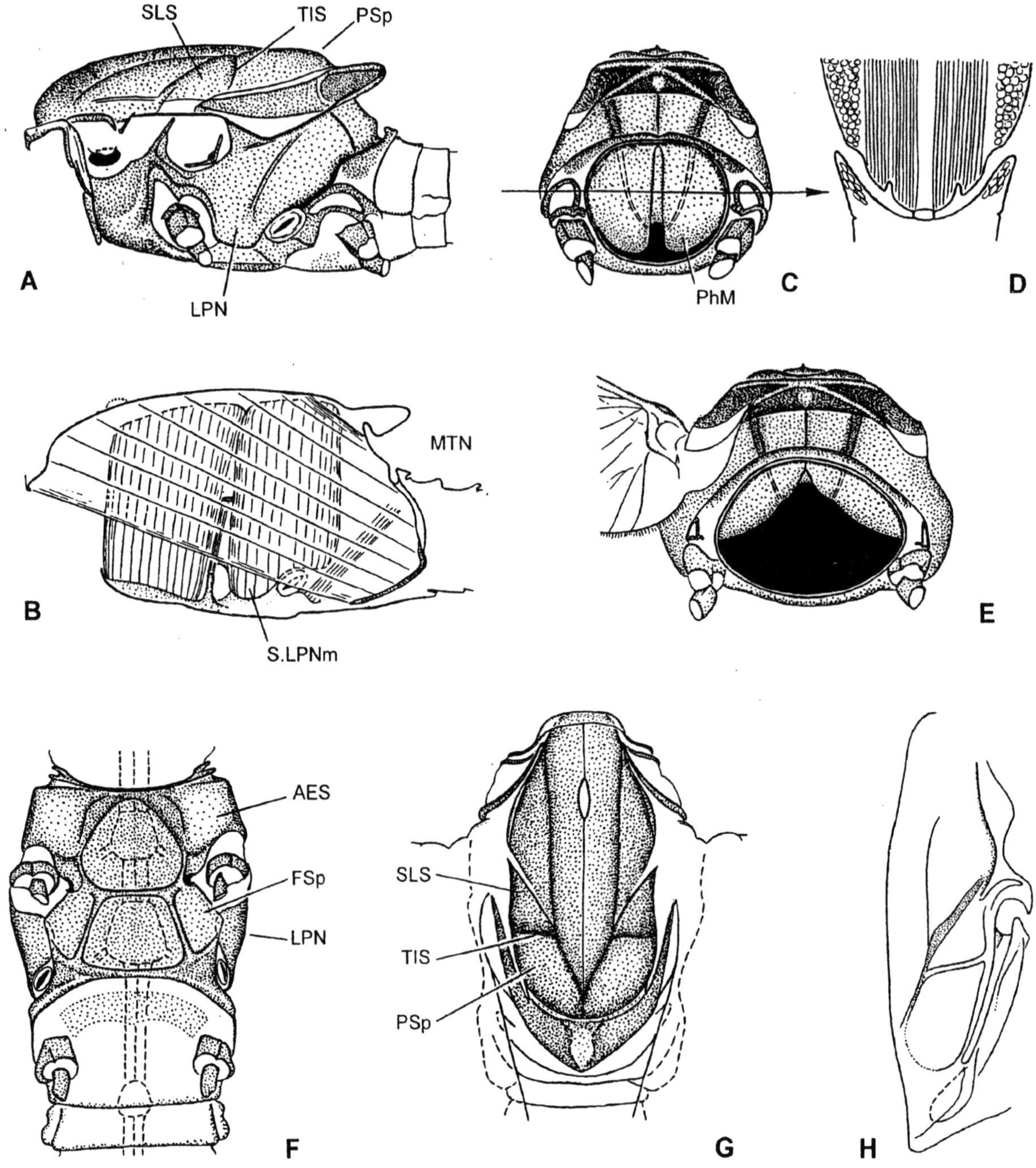

Figure 87. Caenoptera-Caenis/fg* *macrura* [*Caenis*], pterothorax structure.
A – male imaginal pterothorax, lateral view; **B** – male imaginal pterothorax, median section; **C** – male imaginal pterothorax, posterior view (arrow shows section in Fig.D); **D** – longitudinal section of hind part of male pterothorax (by line shown in Fig.C) (compare with Fig.104:C–D); **E** – female imaginal pterothorax, posterior view; **F** – male imaginal pterothorax, ventral view (nerve chain shown by interrupted line); **G** – male imaginal pterothorax, dorsal view; **H** – subimaginal exuviae of right half of mesonotum.

Abbreviations: **AES** – anepisternum; **FSp** – furcasternal protuberance; **LPN** – lateropostnotum; **MTN** – metanotum; **PhM** – middle phragma; **PSp** – posterior scutal protuberance; **S.LPNm** – scuto-lateropostnotal muscle; **SLS** – sublateroscutum; **TIS** - transverse interscutal suture.

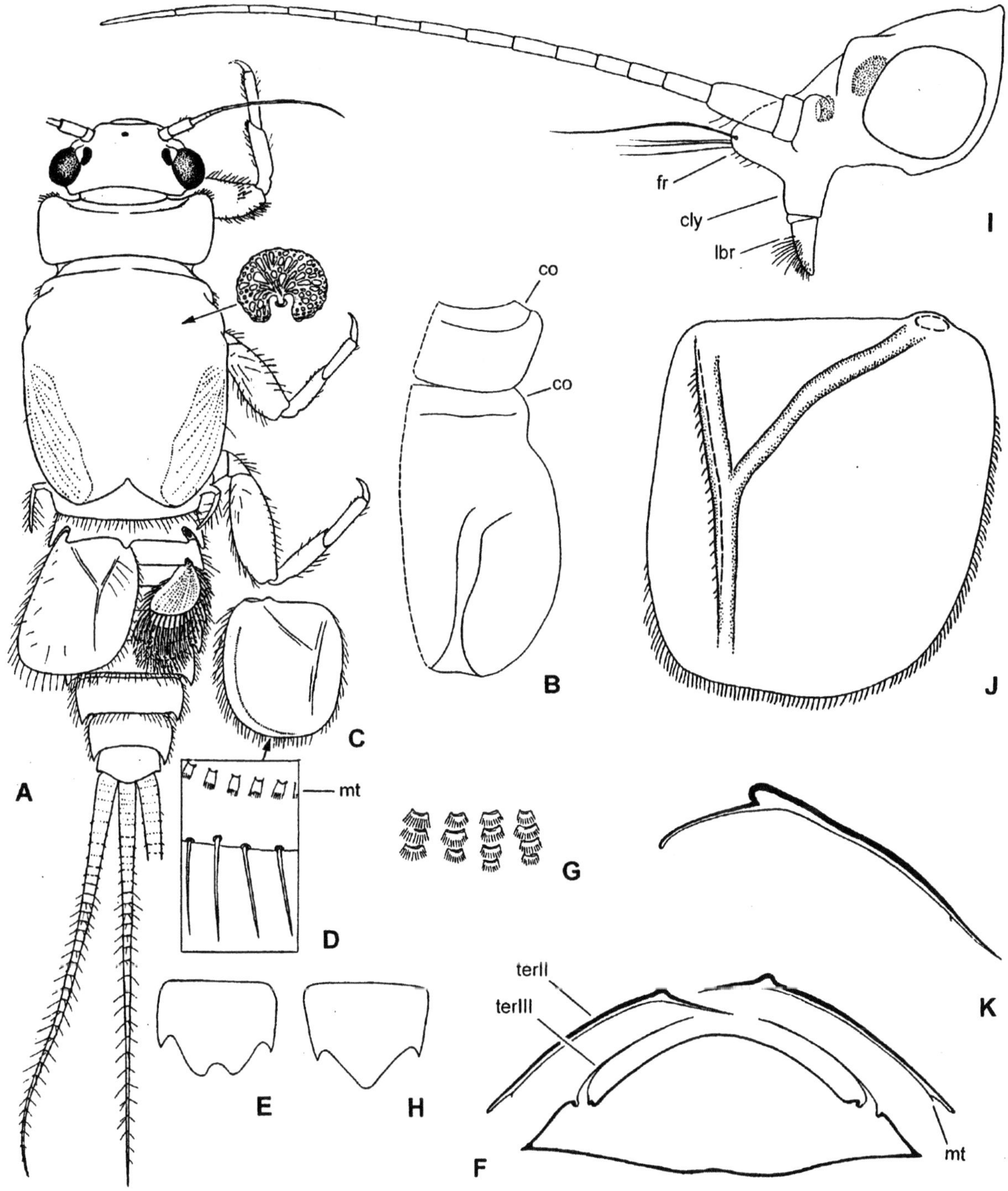

Figure 88. Caenis/f5=g3, larvae.
A–H – Caenis/f6=g4: **A–F** – *macrura* [*Caenis*]: **A** – mature male larva, dorsal view (right gill operculum removed); **B** – larval exuviae of right half of pronotum and mesonotum; **C** – right gill operculum (tergalius II), ventral view; **D** – ventral side and distal margin of gill operculum; **E** – abdominal sternum IX of male larva, ventral view; **F** – transverse section on level of abdominal segment III; **G** – *horaria* [*Ephemera*], microtrichiae on ventral side of gill operculum; **H** – *lactea* [*Oxycypha*], abdominal sternum IX of male larva, ventral view. **I–K** – Clypeocaenis/fg1: **I** – Clypeocaenis/g3 *oligosetosa* [*Clypeocaenis*], larval head, lateral view (mouthparts removed); **J–K** – Amercaenis/g* *ridens* [*Caenis*], left gill operculum (tergalius II): **G** – dorsal view; **H** – cross section. (A, C–E, G–H – from Kluge 1997d; B – from Kluge 1997b).

Abbreviations: **cly** – clypeus; **co** – collar; **fr** – frons; **lbr** – labrum; **mt** – microtrichia; **terII**, **terIII** – tergalii II and III.

(12) In imago amphitornal margin of wing [see (2)] bears setae (FIG.86:F) (as in subimago of all mayflies). Non-unique apomorphy (see Index of characters [2.2.27]); particularly, the same in Tricorygnatha and Leptohyphes/fg1.

(13) In imago and subimago of both sexes tarsi of all legs are distinctly 5-segmented, because 1st tarsal segment of middle and hind legs [initially fused with tibia and shortened – see Furcatergaliae (2)] on its outer side is secondarily distinctly separated from tibia; joining of tibia and tarsus is strongly inclined, thus outer side of 1st tarsal segment is elongate (FIG.86:G). The same in Tricorythodes/fg1 (FIG.105: A); in Tricorygnatha 1st tarsal segment is separated from tibia in the same manner, but is fused with 2nd (see Index of characters [2.2.84]).

(14) In male larva caudalii [cerci and paracercus – see Plesiomorphy] are thickened in proximal part (FIG.88:A), in contrast to caudalii of female, which have usual form. The same in Tricorygnatha and Leptohyphes/fg1 (FIG.103:E–F). About modification of adult caudalii – see (6).

Non-unique apomorphies of Caenoptera.

(15) Eyes of males lost division into dorsal and ventral portions, spherical, usually not large, have the same size and shape as in female. Non-unique apomorphy (see Index of characters [2.1.3]); particularly, the same in many Tricoryptera, which have other characters in common with Caenoptera (see above). Among Caenoptera eyes of male are larger than in female only in Caenopsella/g1; in this case they have the same spherical shape as in other Caenoptera, without division into dorsal and ventral portions; most probably, this is a secondarily enlargement.

(16) On fore leg of male imago and subimago both claws are blunt. Non-unique apomorphy (see Index of characters [2.2.77]).

(17) Patella-tibial suture (initially present on middle and hind legs) is lost on adult legs (FIG.86:G) and at least on larval middle leg; it can be retained on larval hind leg only (in many Caenis/f3=g1), or lost on all legs (in Brachycercus/f1=g3 and selected species and specimens of Caenis/f3=g1). Non-unique apomorphy (see Index of characters [1.2.18]).

Character of Caenoptera of unclear polarity.

(18) Gill operculum (tergalius II) usually has an integral Y-shaped rib convex on dorsal side: three apices of this rib are directed, correspondingly, to base of the gill operculum, to its anal-proximal angle (situated medially-anteriorly), and to its distal margin [see Caenotergaliae (1)] (FIG.88:A). In contrast to Neoephemera/fg1, median margins of gill opercula are non-modified, overlap one another (FIG.88:F). Among Caenoptera the Y-shaped rib is modified in Clypeocaenis/g1 (FIG.88:J–K) and absent in Caenopsella/g2 (see below).

Plesiomorphy of Caenoptera. In all stages paracercus is always well-developed, subequal to cerci (FIG.88:A) [see (6) and (14)].

Size. Small, fore wing length 2–8 mm; comparable with Tricoryptera (see TABLE 2).

Age and distribution. Miocene — Recent; world-wide. Fossil larvae are described only from Miocene of Japan as "*Caenis* sp." (Fujiyama & Nomura 1986). Probably Caenoptera are not older than Neogene: being well-recognizable, they had never been reported from Baltic amber and other fossil resins with rich entomofaunas; larvae of many species inhabit stagnant waters, that promote fossilization in lake deposits, but their impressions are not found in pre-Miocene formations. Caenoptera (and particularly the large plesiomorphon Caenis/f6=g4), being uniform in structure and very young, are distributed on all continents; thus, their Recent distribution can be connected not with continental drift or change of sea level, but with high settling ability.

Caenoptera are divided into Brachycercus/f1=g3 and Caenis/f3=g1; the taxon *Afrocercus* has uncertain systematic position (p.292).

2.2;2,1-4/2.1. Brachycercus/f1=g3

(Furcatergaliae Fimbriatotergaliae Caenotergaliae Caenoptera Brachycercus/f1=g3)

(Fig. 84; 86:A)

Nomen hierarchicum: **Brachycercus/f1=g3** [f: Brachycercidae Lestage 1924c: 62].

In circumscription matches:

— gen. *Eurycaenis* Bengtsson 1917: 186;

— gen. *Brachycercus*: Lestage 1924c: 62;

— subfam. Brachycercinae: Malzacher 1997: 550.

References. Needham & Traver & Hsu 1935: Ծ* ⊕*; – Koss & Edmunds 1974: ⊙*; – Edmunds & Jensen & Berner 1976: Ծ* ⊕*; – Soldán 1986: Ծ' ⊕'; – Kluge 1991a: Ծ' ⊕'; – Malzacher 1997: Ծ* ⊕*; – McCafferty & Wang 2000: Ծ* ⊕*.

Autapomorphies of Brachycercus/f1=g3.

(1) Maxillary palp is 2-segmented. Non-unique apomorphy (see Index of characters [1.1.42]); among Caenotergaliae the same in Clypeocaenis/g2 only.

(2) Labial palp is 2-segmented: its initial 2nd segment is not separated by any means from 3rd segment and lost muscle (in contrast to all other Caenotergaliae). Non-unique apomorphy (see Index of characters [1.1.55]).

(3) Larval head has 3 ocellar protuberances; in different species these protuberances can have form from small tubercles to long corners (Kluge 1991a: Fig.1). Non-unique apomorphy: among Caenoptera ocellar tubercles are present also in some Caenis/f4= =g2; similar ocellar protuberances are present in *Vietnamella* and some species of Drunella/fg1 (in Ephemerella/fg1).

(4) Larval body is more or less flattened dorsoventrally, sternites are relatively wider than in other Caenotergaliae; abdominal sternites are wider than tergites; abdominal tergite II lacks median projection [see Caenotergaliae (1)] (other projections of abdominal tergites are also absent); several middle abdominal segments (at least segments III–VI) have large lateral lobes which are curved dorsally and overlap by sides gill opercula [see Caenotergaliae (1)] (Kluge 1991a:Fig.2).

(5) Larval legs [lacking patella-tibial suture – see Caenoptera (17)] are slender; femur has uniform thickness, somewhat arched (in contrast to other Caenotergaliae, whose femur in its middle part is thicker than basally and apically); claws are long, slender, slightly curved, without denticles (in contrast to many Caenis/f3=g1); middle and hind legs are similar and long, fore legs are much smaller (in contrast to other Caenotergaliae whose legs are equal or gradually increase from fore to hind ones).

(6) Egg has longitudinal crests (which can be straight or somewhat spirally rolled, small and numerous or large and sparse) and with one polar cap [see Caenotergaliae (8)] (Kluge 1991a:Fig.4:12–18).

Plesiomorphies of Brachycercus/f1=g3. Prosternum is wide and rectangular (FIG.86:A); metanotum without transverse crest, flat, with scutellum distinctly projected posteriorly (the same in *Madecocercus*, in contrast to Caenis/f4=g2). Microtrichia near lateral and apical margins of ventral side of gill operculum [see Caenoptera (3)] have a form of simple spine-like processes (in contrast to Caenis/ /f3=g1); they do not form a regular row, being situated irregularly as a wide stripe; this stripe can be located at a distance from margin (in *harrisella* [*B.*] – FIG.84:E), or close to margin (in *tubulatus* [*B.*], *europaeus* [*B.*] and *corniger* [*B.*]), or reduced to a few microtrichia close to margin (in *minutus* [*B.*]). Gonostylus always with a longitudinal groove [see Caenoptera (5)] (FIG.84:A–D). Larval gonostyli buds are retained [being very small – see Caenoptera (5)] (FIG.84:B).

Size. Fore wing length 3–8 mm (see Caenoptera).

Distribution: Holarctic, Oriental and Neotropical Regions.

Nominal taxa included. Brachycercus/f1=g3 includes: (1) **Cercobrachys/g** [g: *Cercobrachys* Soldán 1986: 336, typus *C. etowah* Soldán 1986 (design. orig.)], regarded as a generic synonym of *Brachycercus* (Kluge 1991a: 14); (2) **Insulibrachys/g** [g: *Insulibrachys* Soldán 1986: 333, typus *I. needhami* Soldán 1986 (design. orig.)], known only as larva and female imago of a single Cuban species, and differing from other Brachycercus/f1=g3 only by an autapomorphy in shape of ocellar tubercles.

Species composition of Brachycercus/f1=g3 (here [*B.*]=[*Brachycercus*]). About 30 species. **Eurasian and north African species:** *corniger* Kluge 1991 [*B.*], *europaeus* Kluge 1991 [*B.*], *gilliesi* Soldán & Landa 1991 [*B.*], *harrisella* Curtis 1834 [*B.*] (synn. subj.: *pennata* Stephens 1836 [*Caenis (B.)*], *pallidus* Tshernova 1928 [*B.*], *magnus* Tshernova 1952 [*B.*]), *japonica* Gose 1985 [*B.*], *kabyliensis* Soldán 1986 [*B.*], *minutus* Tshernova 1952 [*B.*], *parviforceps* Zhou & Gui & Su 2000 [*B.*], *petersorum* Soldán 1986 [*Cercobrachys*], *tubulatus* Tshernova 1952 [*B.*]. **American species:** *arcticus* Soldán 1986 [*B.*], *berneri* Soldán 1986 [*B.*], *columbianus* Soldán 1986 [*Cercobrachys*], *cree* Sun & Webb & McCafferty 2002 [*Cercobrachys*], *edmundsi* Soldán 1986 [*B.*], *etowah* Soldán 1986 [*Cercobrachys*], *flavus* Traver 1935 [*B.*], *floridicola* Soldán 1986 [*B.*], *lacustris* Needham 1918 [*Caenis*] (syn. subj.: *pallida* Ide 1930 [*Eurycaenis*] = *idei* Lestage 1931 [*B.*]), *maculatus* Berner 1946 [*B.*], *nasutus* Soldán 1986 [*B.*], *needhami* Soldán 1986 [*Insulibrachys*], *nitida* Traver 1932 [*Eurycaenis*], *peruanicus* Soldán 1986 [*Cercobrachys*], *pini* Soldán 1986 [*B.*], *prudens* McDunnough 1931 [*Eurycaenis*], *serpentis* Soldán 1986 [*Cercobrachys*], *tuberculatus* Soldán 1986 [*B.*].

Material examined: *corniger* [*B.*]: ʘ, ♂/♂/⊕♂, ♀/♀/⊕♀/⊙; *europaeus* [*B.*]: ʘ, ʘ–♂/⊕♂, ♀–♀/⊙; *harrisella* [*B.*]: ʘ–◎–⊕/⊙; *minutus* [*B.*]: ʘ–◎–⊕/⊙; *tubulatus* [*B.*]: ʘ; sp. (Thailand): ⊕♂.

***

2.2;2,1-4/2.2. **Caenis/f3=g1**

(Furcatergaliae Fimbriatotergaliae Caenotergaliae Caenoptera Caenis/f3=g1)

(Figs 85; 86:B–G; 87–88)

Nomen hierarchicum: **Caenis/f3=g1** (sine *Brachycercus*; incl. *Madecocercus*) [g: *Caenis* Stephens 1835: 60, typus *C. macrura* Stephens 1835 (design. Westwood 1840: Synopsis, p. 47); synn. obj.: ***Coenis*** auct., ***Ordella*** Campion 1923: 518].

Autapomorphies of Caenis/f3=g1.

(1) Each microtrichion on ventral side of gill operculum [see Caenoptera (3)] has a form of thin wide plate with distal margin consisting of numerous pointed spine-like processes; all microtrichia form a row parallel to margin of the operculum; usually each microtrichion is directed toward the nearest margin of the operculum. In most species the row of microtrichia is regular (FIG.88:D; Provonsha 1990: Figs 85–87); in a few species it is irregular (in *tauroides* [*Madecocercus*] – FIG.85:A, *rivulorum* [*Caenis*], *deani* [*Irpacaenis*], *oligosetosa* [*Clypeocaenis*] and others). In *oligosetosa* [*C.*] each microtrichion consists of 2–3 spine-like processes only, and is directed along the operculum margin (not perpendicular to it). In some Caenis/f5=g3 instead of each microtrichion, a short row of the same microtrichia is present, which is directed perpendicular to the operculum margin (FIG.88:G).

(2) Larval gonostyli buds [see Caenoptera (5)] are completely fused with abdominal sternite IX, forming an integral preanal plate which completely lost its initial sexual dimorphism; shape of this plate is species-specific (FIG.88:E, H). When subimaginal gonostyli [non-crumpled – see Caenoptera (5)] develop under larval cuticle, they lie freely inside the preanal plate (FIGS 85:B; 86:C). As larval cuticle has no special penial sheath [see Caenotergaliae (3)], larval exuviae of male and female abdominal segment IX are non-distinguishable [sexual dimorphism is present in caudalial structure only – see Caenoptera (14)]. Unique apomorphy: in all other Ephemeroptera male larval preanal plate has more or less projected external gonostyli buds, thus differing from female preanal plate (only in Turbanoculata gonostyli buds and preanal plate of both sexes can be lost); in Brachycercus/f1=g3 gonostyli buds are retained, being extremely diminished (FIG.84:B).

Plesiomorphy of Caenis/f3=g1. In contrast to Brachycercus/f1=g3, labial palp is always 3-segmented.

Size. Fore wing length 2–8 mm (see Caenoptera).

Distribution. World-wide (as in Caenoptera).

Caenis/f3=g1 is divided into *Madecocercus* and Caenis/f4=g2.

2.2;2,1–4/2.2;1. Madecocercus/g(1)

(Fimbriatotergaliae Caenotergaliae Caenoptera Caenis/f3=g1 *Madecocercus*)

(Fig. 85)

Nomen hierarchicum: **Madecocercus/g(1)** [g: *Madecocercus* Malzacher 1995: 2, typus *M. tauroides* Malzacher 1995 (design. orig.); syn. subj.: ***Provonshaka*** McCafferty & Wang 1995: 179, typus *P. thomasorum* McCafferty & Wang 1995 (design. orig.) (synn. subj.: *tauroides* [*M.*] = *thomasorum* [*P.*])].
In circumscription matches:
— gen. *Madecocercus* Malzacher 1995: 2;
— gen. *Provonshaka* McCafferty & Wang 1995: 179.
Monospecific taxon.

References. Malzacher 1995: ⊕♂; – McCafferty & Wang 1995: ♉; – Elouard & Sartori 2001: ⊙ ♉ ♀.

Autapomorphies of *Madecocercus*.

(1) Maxillary palp is lost. Non-unique apomorphy, the same in some Ephemerella/fg1 (see Index of characters [1.1.41]).

(2) Marginal processes of tergalii IV–VI are non-branched, thus only tergalius III retains branched processes [see Caenotergaliae (1) and Caenoptera (4)]. On all tergalii III–VI on proximal-anal margin processes are substituted by setae of similar size (FIG.85:C–D).

(3) Larval body has peculiar shape: flattened and widened, femora strongly widened, with regular rows of long setae on outer and inner margins; dorsal surface of body and legs is warty, on fore femur warts form a transverse row (which originally was confused with a setal row in Pantricorythi).

Plesiomorphies of *Madecocercus*. Prosternum is wide and rectangular; metanotum without transverse crest, flat, with scutellum distinctly projected posteriorly (Malzacher 1995:Fig.1c,d) (the same in Brachycercus/f1=g3, in contrast to Caenis/f4=g2). Gonostylus with a longitudinal groove [see Caenoptera (5)].

Size. Larval body length 5–6 mm (see Caenoptera).

Distribution. Madagascar.

Species composition of Madecocercus/g(1). 1 species – *tauroides* Malzacher 1995 [*Madecocercus*] (syn.subj.: *thomasorum* McCafferty & Wang 1995 [*Provonshaka*]).

Material examined: *tauroides* [*M.*]: ♉♂.

***

2.2;2,1-4/2.2;2. Caenis/f4=g2

(Fimbriatotergaliae Caenotergaliae Caenoptera Caenis/f3=g1 .../f4=g2)

(Figs 86:B–G; 87–88)

Nomen hierarchicum: **Caenis/f4=g2** (sine *Madecocercus*; incl. *Tillyardocaenis*).
In circumscription matches:
— subfam. Caeninae: Malzacher 1997: 551.

References. Needham & Traver & Hsu 1935: ♉* ⊕*; – Edmunds & Jensen & Berner 1976: ♉* ⊕*; – Malzacher 1997: ♉* ⊕*; – McCafferty & Wang 2000: ♉* ⊕*.

Autapomorphies of Caenis/f4=g2.

(1) Imaginal, subimaginal and larval prosternum

is narrow, bases of fore legs are brought together; in imago and subimago prosternum has more or less triangular or trapezoid shape, strongly narrowing anteriorly (FIG.86:B; Malzacher 1995:Figs 2–6). Non-unique character, but differs from Brachycercus/f1=g3 and *Madecocercus*, where prosternum is wider and rectangular.

(2) Imaginal metanotum [vestigial – see Caenoptera (9)] has posterior border of scutellum bent up, thus area behind it is exposed dorsally, and this border looks as a transverse ridge on dorsal side of metanotum (FIG.87:A–B). At the same time, larval and subimaginal metanotum retains primitive structure – flat, with prominent scutellum projected posteriorly.

Character of unclear phylogenetic status.

(3) Larval claws usually with 1 row of denticles on inner margin (in some species these denticles are very small or absent). Non-unique apomorphy (see Index of characters [1.2.21]); other Fimbriatotergaliae have no denticles on claws.

Size. Fore wing length 2–8 mm (see Caenoptera).

Distribution. World-wide (as in Caenoptera).

Caenis/f4=g2 is divided into *Tillyardocaenis* and Caenis/f5=g3; some taxa known as larvae only have uncertain position (p.291).

2.2;2,1-4/2.2;2,1.
Plesiomorphon **Tillyardocaenis/g(1)**

(Fimbriatotergaliae Caenotergaliae Caenoptera Caenis/f3=g1 .../f4=g2 *Tillyardocaenis*)

(Fig. 86:C–E)

Nomen hierarchicum: **Tillyardocaenis/g(1)** [g: *Tillyardocaenis* Kluge, SUBGEN. NOV., typus *Caenis tillyardi* Lestage 1938].

References (as *Tasmanocoenis*, partim). Malzacher 1987: ⊕♂; – 1991: ⊕♂; – 1997: ⊕♂; – Alba-Tercedor & Suter 1990: ↺ ⊕.

Plesiomorphy of *Tillyardocaenis*. In contrast to Caenis/f5=g3, gonostylus retains a longitudinal groove [see Caenoptera (5)]; gonostyli are strong, sclerotized and sharply bowed (FIG.86:A–C).

Size. Fore wing length 2–4 mm (see Caenoptera).

Distribution. Australia.

Species composition of *Tillyardocaenis*/g(1). 3 species – *arcuata* Alba-Tercedor & Suter 1990 [*Tasmanocoenis*], *jillongi* Harker 1957 [*Tasmanocoenis*], *tillyardi* Lestage 1938 [*Caenis*] (syn. obj.: *scotti* Tillyard 1935 [*Caenis*] nom. praeocc.).

Material examined: *tillyardi* [*C.*]: ♂/♂/⊕♂, ⊕♂, ♀/♀/⊕♀/⊙.

***

2.2;2,1-4/2.2;2,2. **Caenis/f5=g3**

(Fimbriatotergaliae Caenotergaliae Caenoptera Caenis/f3=g1 .../f5=g3)

(Figs 86:B, F–G; 87–88)

Nomen hierarchicum: **Caenis/f5=g3** (sine *Tillyardocaenis*; incl. *Brasiliocaenis*, *Clypeocaenis*, *Caenopsella*).

Autapomorphy of Caenis/f5=g3.

(1) Grooves of gonostyli [see Caenoptera (5)] are lost. In a few species gonostyli retain bowed shape and strong musculature (Australian species *tonnoiri* [*Tasmanocoenis*] and *dostini* [*Wundacaenis*], sp.cf. *tonnoiri* from Thailand, Palaearctic species *rivulorum* [*Caenis*], and some others). In majority of Caenis/f5=g3 (including the type species of *Caenis*) gonostyli are more straight, less sclerotized, with weak muscles, apically often with a thin projection or a bunch of spines; it some species gonostyli are vestigial, non-functional, shorter than penis.

Size. Fore wing length 2–8 mm (see Caenoptera).

Distribution. World-wide (as in Caenoptera).

Caenis/f5=g3 is divided into Caenis/f6=g4, *Brasiliocaenis*, Clypeocaenis/g1 and Caenopsella/g1.

2.2;2,1-4/2.2;2,2-1.
Plesiomorphon **Caenis/f6=g4**

(Fimbriatotergaliae Caenotergaliae Caenoptera Caenis/f3=g1 .../f6=g4)

(Figs 86:B, F–G; 87; 88:A–H)

Nomen hierarchicum: **Caenis/f6=g4** (sine *Brasiliocaenis*, *Clypeocaenis*, *Caenopsella*).

Plesiomorphies of Caenis/f6=g4. Maxillary palp is 3-segmented (in contrast to Clypeocaenis/g2). Fore legs of larva without filtering setae (in contrast to Clypeocaenis/g1). Genitals have no such modification as in *Brasiliocaenis*. Male eyes are small (in contrast to Caenopsella/g1) [see Caenoptera (15)].

Size. Fore wing length 2–8 mm (see Caenoptera).

Distribution. World-wide (as in Caenoptera).

Nominal taxa included. Caenis/f6=g4 includes:

(1) **Oxycypha/g** [g: *Oxycypha* Burmeister 1839: 796, typus *O. luctuosa* Burmeister 1839 (design. Jacob 1974b: 96)], regarded as a generic synonym of *Caenis* (Jacob 1974b: 96).

(2) **Caenodes/g** [g: *Caenodes* Ulmer 1924b: 7, typus *Caenodes ulmeri* Kimmins 1949 (design. Kimmins 1949: 831) (= *Caenis cibaria*: Ulmer 1924b, non *C. cibaria* Eaton 1879; syn. obj. *Caenis douglasi* Malzacher 1993)], regarded as a generic synonym of *Caenis* (Kluge 1985: 1588; Malzacher 1993: 414).

(3) **Tasmanocoenis/g** [g: *Tasmanocoenis* Lestage 1931a: 53, typus *T. tonnoiri* Lestage 1931 (design. orig.)],

regarded as a generic synonym of *Caenis* (Kluge 1985: 1588). Traditionally (Malzacher 1987, 1997, et al.) the "genus *Tasmanocoenis*" is accepted as limited by Australian Region and including, besides the type species, all three species placed here to *Tillyardocaenis*; in this case it is characterized by bowed gonostyli only, and its diagnosis agrees with characters of *rivulorum* [*Caenis*] and some other non-Australian species.

(4) **Austrocaenis/g** [g: *Austrocaenis* Barnard 1932: 227, typus *capensis* Barnard 1932 (design.orig.)], regarded as a generic synonym of *Caenis* (Malzacher 1993: 414).

(5) **Caenomedea/g** [g: *Caenomedea* Thew 1960: 199, typus *Caenis kivuensis* Demoulin 1956 (design. orig.)], regarded as a generic synonym of *Caenis* (Kluge 1985: 1588; Malzacher 1993: 414).

(6) **Wundacaenis/g** [g: *Wundacaenis* Suter 1993: 788, typus *W. dostini* Suter 1993 (design. orig.)]. The genus *Wundacaenis* was established for three Australian species, one of which has bowed gonostyli as in type species of *Tasmanocoenis* (see above), and for two other species imagoes are unknown (see below, Caenis/f4=g2 INCERTAE SEDIS); this taxon was characterized by presence of a pair of lateral mesonotal projections in larva (the same in *cornuta* [*Ordella*], *horaria* [*Ephemera*] and other non-related species of Caenis/f4=g2) and a peculiar shape of posterior margin of larval abdominal sternum IX.

(7) **Irpacaenis/g** [g: *Irpacaenis* Suter 1999: 159, typus *I. deani* Suter 1999 (design. orig.)]. The genus *Irpacaenis* was established for three Australian species, one of which has soft straight gonostyli with thin apical projection (as in type species of *Caenis*), and for two other species imagoes are unknown (see below, Caenis/f4=g2 INCERTAE SEDIS).

Species composition of Caenis/f6=g4 (here [*C.*]= =[*Caenis*]). More than 100 species are described; probably here belong most species placed to Caenis/f4=g2 INCERTAE SEDIS (see below). **Eurasian species:** *amurensis* Kluge 1986 [*C.*], *antoniae* Malzacher 1992 [*C.*], *aspera* Tong & Dudgeon 2002 [*C.*], *bicornis* Tong & Dudgeon 2002 [*C.*], *cornuta* Tshernova 1952 [*Ordella*], *gilbonensis* Malzacher 1992 [*C.*], *hissari* Kluge 1985 [*C.*], *horaria* Linnaeus 1758 [*Ephemera*] (synn. subj.: *dimidiata* Stephens 1835 [*C.*], *lactella* Eaton 1884 [*C.*]), *insularis* Demoulin 1969 [*Caenomedea*], *jungi* Braasch 1980 [*C.*], *kopetdagi* Kluge 1985 [*C.*], *lactea* Burmeister 1939 [*Oxycypha*] (synn.subj.: *tumida* Bengtsson 1912 [*C.*], *nocturna* Bengtsson 1917 [*C.*], *undosa* Tiensuu 1939 [*C.*]), *lubrica* Tong & Dudgeon 2002 [*C.*], *luctuosa* Burmeister 1839 [*Oxycypha*] (synn. subj.: *moesta* Bengtsson 1917 [*C.*], *felsinea* Grandi 1951 [*C.*]), *macronyx* Kluge 1986 [*C.*], *macrura* Stephens 1835 [*C.*] (syn. subj.: *interrupta* Stephens 1835 [*C.*], *grisea* Pictet 1843 [*C.*], *oophora* Pictet 1843 [*C.*]), *maculata* Tshernova 1952 [*Ordella*], *martae* Belfiore 1984 [*C.*], *miliaria* Tshernova 1952 [*Ordella*], *nachoi* Alba-Tercedor & Zamora-Munoz 1993 [*C.*], *nigroforceps* Zhou & Gui & Su 1997 [*C.*], *nigropunctata* Klapálek 1905 [*C.*], *nishinoae* Malzacher 1996 [*C.*], *parabrevipes* Malzacher 1992 [*C.*], *perpusilla* Walker 1853 [*C.*], *philippinensis* Ulmer 1924 [*Caenodes*], *picea* Kimmins 1947 [*C.*], *piscina* Kimmins 1947 [*C.*], *pseudorivulorum* Keffermüller 1960 [*C.*] (synn. subj. et subspp.: *beskidensis* Sowa 1973 [*C.*], *belfiorei* Malzacher 1986 [*C.*]), *pusilla* Navás 1913 [*C.*] (syn. subj.: *rhenicola* Malzacher 1976 [*C.*]), *pustula* Tiunova 1999 [*C.*], *pygmaea* Costa 1888 [*C.*], *rivulorum* Eaton 1884 [*C.*] (syn. subj.: *nivea* Bengtsson 1917 [*C.*]), *robusta* Eaton 1884 [*C.*] (syn. subj.: *incus* Bengtsson 1912 [*C.*], *ulmeri* Brodsky 1930 [*C.*]); *srinagari* Traver 1939 [*C.*], *strugaensis* Ikonomov 1961 [*C.*], *valentinae* Grandi 1951 [*C.*]. **African species:** *alicae* Malzacher 1990 [*C.*], *antelucana* Malzacher 1990 [*C.*], *basuto* Demoulin 1970 [*C.*], *berneri* Kimmins 1956 [*C.*], *brevipes* Kimmins 1956 [*C.*], *capensis* Barnard 1932 [*Austrocaenis*], *cibaria* Eaton 1879 [*C.*], *cincta* Demoulin 1956 [*C.*], *corbeti* Malzacher 1990 [*C.*], *duodecima* Malzacher 1990 [*C.*], *edwardsi* Kimmins 1939 [*C.*], *elouardi* Malzacher 1990 [*C.*], *ghibana* Malzacher 1990 [*C.*], *gilliesi* Malzacher 1990 [*C.*], *hoggariensis* Grandi 1951 [*C.*], *inflexa* Kopelke 1981 [*Caenomedea*], *jinjana* Kimmins 1956 [*Caenodes*], *johannae* Malzacher 1996 [*C.*], *kivuensis* Demoulin 1956 [*C.*], *knowlesi* Gillies & Knowles 1990 [*C.*], *kungu* Eaton 1879 [*C.*], *liebenauae* Malzacher 1990 [*C.*], *magnipilosa* Kopelke 1981 [*Caenomedea*], *margherita* Malzacher 1990 [*C.*], *namorona* Malzacher 1995 [*C.*], *nervulosa* Malzacher 1990 [*C.*], *noctivaga* Malzacher 1990 [*C.*], *occulta* Malzacher 1990 [*C.*], *pallida* Malzacher 1990 [*C.*], *rugosa* Malzacher 1995 [*C.*], *rutila* Malzacher 1995 [*C.*], *scotti* Ulmer 1930 [*C.*], *spinosa* Malzacher 1995 [*C.*]. **Australian species:** *deani* Suter 1999 [*Irpacaenis*], *dostini* Suter 1993 [*Wundacaenis*], *tonnoiri* Lestage 1931 [*Tasmanocoenis*]. **North and South American species:** *amica* Hagen 1861 [*C.*] (syn. subj.: *simulans* McDunnough 1931 [*C.*]), *anceps* Traver 1935 [*C.*], *argentina* Navás 1915 [*C.*], *bajaensis* Allen & Murvosh 1983 [*C.*], *burmeisteri* Malzacher 1990 [*C.*], *candelata* Malzacher 1986 [*C.*], *candida* Harper & Harper 1981 [*C.*], *chamie* Alba-Tercedor & Mosquera 1999 [*C.*], *cigana* Pereira & Da-Silva 1990 [*C.*], *diminuta* Walker 1953 [*C.*], *dominguezi* Malzacher 2001 [*C.*], *fittkaui* Malzacher 1986 [*C.*], *gonseri* Malzacher 2001 [*C.*], *grimi* Malzacher 2001 [*C.*], *hilaris* Say 1839 [*Ephemera*], *latipennis* Banks 1907 [*C.*] (synn. subj.: *forcipata* McDunnough 1931 [*C.*], *jocosa* McDunnough 1931 [*C.*], *delicata* Traver 1935 [*C.*], *gigas* Burks 1953 [*C.*]), *ludiera* Navás 1920 [*C.*] (syn. obj.: *ludicra* Navás 1924 [*C.*]), *macafferti* Provonsha 1990 [*C.*], *panamensis* Malzacher 2001 [*C.*], *pflugfelderi* Malzacher 1990 [*C.*], *plaumanni* Malzacher 2001 [*C.*], *pseudamica* Malzacher 1990 [*C.*], *punctata* McDunnough 1931 [*C.*], *quatipuruica* Malzacher 1986 [*C.*], *reissi* Malzacher 1986 [*C.*], *sigillata* Malzacher 1985 [*C.*], *tardata* McDunnough 1931 [*C.*], *youngi* Roemhild 1984 [*C.*].

Material examined: *amurensis* [*C.*]: Ծ–◎–⊕/⊙; *cornuta* [*Ordella*]: ♂/♂/⊕♂, ♀/♀/⊕♀/⊙; *cubensis* [*Caenis*] (in litt.): Ծ–◎–⊕/⊙; *deani* [*Irpacaenis*]: ♂/♂, ♀/♀/⊙; *desertus* [*Caenis*] (in litt.): Ծ–◎–⊕/⊙; *gilliesi* [*C.*]: ⊕; *hissari* [*C.*]: Ծ–◎–⊕/⊙; *horaria* [*Ephemera*]: Ծ–◎–⊕/⊙; *kopetdagi* [*C.*]:

℧-o-⊕/⊙; *lactea* [*Oxycypha*]: ℧, ⊕̄, ♀-♀-⊕/⊙; *luctuosa* [*Oxycypha*]: ⊕̄, ⊕/⊙; *macronyx* [*C.*]: ℧, ℧/⊘/⊕̄; *macrura* [*C.*]: ℧-o-⊕/⊙; *maculata* [*Ordella*]: ℧, ⊕̄, ♀-♀-⊕/⊙; *martae* [*C.*]: ℧, ⊕; *miliaria* [*Ordella*]: ℧-o-⊕/⊙; *pseudorivulorum* [*C.*]: ℧-o-⊕/⊙; *punctata* [*C.*]: ℧; *pusilla* [*C.*]: ℧, ⊕; *pustula* [*C.*]: ℧; *rivulorum* [*C.*]: ℧-o-⊕/⊙; *robusta* [*C.*]: ℧-o-⊕/⊙; *tardata* [*C.*]: ℧; *tonnoiri* [*Tasmanocoenis*]: ℧/⊘/⊕̄; *youngi* [*C.*]: ℧; sp. (French Guyana): ⊕/⊙; sp.cf. *tonnoiri* [*T.*] (Thailand): ⊕̄.

***

2.2;2,1-4/2.2;2,2-2. **Brasiliocaenis/g(1)**

(Fimbriatotergaliae Caenotergaliae Caenoptera Caenis/f3=g1 .../f5=g3 *Brasiliocaenis*)

Nomen hierarchicum: **Brasiliocaenis/g(1)** [g: *Brasiliocaenis* Puthz 1975: 411, typus *B. irmleri* Puthz 1975].
In circumscription matches:
— gen. *Brasiliocaenis* Puthz 1975: 411.

References. Puthz 1975: ℧* ⊕*; – Malzacher 1986b: ℧* ⊕*; – 1991: ⊕*.

Autapomorphy of *Brasiliocaenis*.

(1) Styliger between bases of gonostyli is strongly elongate, and gonostyli [1-segmented – see Caenoptera (5)], arising by sides of this elongated styliger, are contiguous or fused by their median margins with lateral margins of styliger. Unique apomorphy.

Plesiomorphy of *Brasiliocaenis*. Maxillary palp is 3-segmented (in contrast to Clypeocaenis/g2).

Size. Fore wing length 2–3 mm (see Caenoptera).

Distribution. Amazon Region.

Species composition of Brasiliocaenis/g(1) (here [*B.*] =[*Brasiliocaenis*]). 5 or 6 species – *irmleri* Puthz 1975 [*B.*], *puthzi* Malzacher 1986 [*B.*], *renata* Malzacher 1986 [*B.*], *intermedia* Malzacher 1986 [*B.*] (known as larva); *septentrionalis* Malzacher 1990 [*B.*] (known as imago); *mendesi* Malzacher 1998 [*B.*] (known as imago).

Material examined: –.

***

2.2;2,1-4/2.2;2,2-3. **Clypeocaenis/g1**

(Fimbriatotergaliae Caenotergaliae Caenoptera Caenis/f3=g1 ...f5=g3 Clypeocaenis/g1)

(Fig. 88:I–K)

Nomen hierarchicum: **Clypeocaenis/g1** (incl. *Amercaenis*) [g: *Clypeocaenis* Soldán 1978a: 119, typus *C. bisetosa* Soldán 1978 (design. orig.)].

Reference. Provonsha & McCafferty 1985: ℧*

Autapomorphies of Clypeocaenis/g1.

(1) Larval fore legs and mouthparts are specialized as filtering (Soldán 1978:Figs 2–8; Provonsha & McCafferty 1985:Figs 1–6; 1995:Figs 3–5): Tibia and tarsus with long setae on inner and sometimes on dorsal side (non-unique apomorphy – see Index of characters [1.2.32]). Maxilla with apical field of long dense setae; canines [initially three – see Plesiomorphies of Caenotergaliae] are slender and more or less fused together; dentisetae [two – see Bidentiseta (1)] are weak and indistinguishable among setae. 3rd segment of maxillary palp has numerous long setae [and can be fused with 2nd segment – see Clypeocaenis/g2 (1) below]. Labial palp has 2nd segment shortened and 3rd segment elongate, with numerous long setae (in *Amercaenis* and Clypeocaenis/g3 the 2nd segment is several times shorter than 3rd, only in *Barnardara* they are subequal).

(2) Y-shaped rib of gill operculum is modified: median rib [which stretches longitudinally from antero-median angle to posterior margin – see Caenotergaliae (1) and Caenoptera (18)] is more or less expanded in median direction; it can be slightly expanded (FIG.88:J–K), or forms a wide plate reaching median margin (Soldán 1978a:Fig.13).

(3) Plumose microtrichia on ventral side of gill operculum [see Caenis/f3=g1 (1)] are more or less reduced and directed not strictly perpendicular to margin, but obliquely (somewhat similar to Brachycercus/f1=g3); in *ridens* [*Caenis*] a regular row of multipointed microtrichia is retained, while *oligosetosa* [*Clypeocaenis*] has irregular stripe of very small, mainly bipointed microtrichia (other species were not examined).

Size. Fore wing length 2–5 mm (see Caenoptera).

Distribution. Possibly, Arctogea: Holarctic, Oriental and Ethiopian Regions.

Clypeocaenis/g1 is divided into *Amercaenis* and Clypeocaenis/g2.

2.2;2,1-4/2.2;2,2-3/1. Plesiomorphon **Amercaenis/g(1)**

(Caenoptera Caenis/f3=g1 .../f5=g3 Clypeocaenis/g1 *Amercaenis*)

(Fig. 88:J–K)

Nomen hierarchicum: **Amercaenis/g(1)** [g: *Amercaenis* Provonsha & McCafferty 1985: 2, typus *Caenis ridens* McDunnough 1931 (design. orig.)].
In circumscription matches:
— gen. *Amercaenis* Provonsha & McCafferty 1985: 2.
Monospecific taxon.

Reference. Provonsha & McCafferty 1985: ℧ ⊕.

Plesiomorphies of *Amercaenis*. Maxillary palp is 3-segmented (in contrast to Clypeocaenis/g2).

Filtering setae on fore tibia and tarsus [see Clypeocaenis/g1 (1)] are situated only on inner side (in contrast to Clypeocaenis/g2), not forming rows (in contrast to Clypeocaenis/g3).

Size. Fore wing length 2–4 mm (see Caenoptera).

Distribution. Nearctic.

Species composition of Amercaenis/g(1). 1 species – *ridens* McDunnough 1931 [*Caenis*].

Material examined: *ridens* [*C.*]: ℧, ⊕̄, ⊕.

***

2.2;2,1-4/2.2;2,2-3/2. Clypeocaenis/g2

(Caenoptera Caenis/f3=g1 .../f5=g3 Clypeocaenis/g1 .../g2)

(Fig. 88:I)

Nomen hierarchicum: **Clypeocaenis/g2** (sine *Amercaenis*; incl. *Barnardara*).

Autapomorphies of Clypeocaenis/g2.

(1) Maxillary palp [filtering – see Clypeocaenis/g1 (1)] is 2-segmented, as its initial 2nd and 3rd segments are completely fused. Non-unique apomorphy (see Index of characters [1.1.42]); among Caenoptera the same in Brachycercus/f1=g3.

(2) Filtering setae on fore tibia and tarsus [see Clypeocaenis/g1 (1)] are situated not only on inner, but also on dorsal (anterior) side.

Size. Fore wing length 2–5 mm (see Caenoptera).

Distribution. Asia and Africa.

Clypeocaenis/g2 is divided into *Barnardara* and Clypeocaenis/g3.

2.2;2,1-4/2.2;2,2-3/2.1. Plesiomorphon Barnardara/g(1)

(Caenoptera Caenis/f3=g1 .../f5=g3 Clypeocaenis/g1 .../g2 *Barnardara*)

Nomen hierarchicum: **Barnardara/g(1)** [g: *Barnardara* McCafferty & Provonsha (in Provonsha & McCafferty) 1995: 242, typus *B. demoori* McCafferty & Provonsha 1995 (design. orig.)].

In circumscription matches:

— gen. *Barnardara*: Provonsha & McCafferty 1995: 242.

Monospecific taxon.

Reference. Provonsha & McCafferty 1995: ℧.

Plesiomorphies of *Barnardara* (in contrast to Clypeocaenis/g3). Larval frons without long setae. Long filtering setae on fore tibia and tarsus [see Clypeocaenis/g2 (2)] are situated irregularly.

Size. Larval body length 2–3 mm (see Caenoptera).

Distribution. South Africa.

Species composition of Barnardara/g(1). 1 species – *demoori* McCafferty & Provonsha 1995 [*Barnardara*].

Material examined: –.

***

2.2;2,1-4/2.2;2,2-3/2.2. Clypeocaenis/g3

(Caenoptera Caenis/f3=g1 .../f5=g3 Clypeocaenis/g1 .../g3)

(Fig. 88:I)

Nomen hierarchicum: **Clypeocaenis/g3** (sine *Barnardara*).

In circumscription matches:

— gen. *Clypeocaenis* Soldán 1978a: 119.

References. Soldán 1978a: ℧* ⊕*; – 1983: ℧' ⊕'; – Soldán & Landa 1991: ℧'; – Malzacher 1993: ℧* ⊕*; – Provonsha & McCafferty 1995: ℧'.

Autapomorphies of Clypeocaenis/fg3.

(1) Larval frons between antennae bases is projected anteriorly and bears 2 or more very long setae directed anteriorly (FIG.88:I). This frontal projection corresponds to anterior ocellar protuberance occurring in some other taxa of Caenoptera [see Index of characters [1.1.6]).

(2) Long filtering setae on inner and dorsal side of fore tibia and tarsus [see Clypeocaenis/fg2 (2)] form regular rows: two transverse rows on tibia and one oblique row on tarsus. Unique apomorphy.

(3) Each mandible on its outer side with 2 regular transverse rows of long setae (instead of irregular setae present in other Caenoptera).

Size. Fore wing length 2–5 mm (see Caenoptera).

Distribution. Asia and Africa (Iran, India, Vietnam, Sri Lanka, Upper Volta, South Africa).

Species composition of Clypeocaenis/g3 (here [*H.*]= =[*Clypeocaenis*]). 6 species – *afrosetosa* Soldán 1983 [*C.*], *bisetosa* Soldán 1978 [*C.*], *femorisetosa* Soldán & Landa 1991 [*C.*], *multisetosa* Soldán 1978 [*C.*], *oligosetosa* Soldán 1983 [*C.*], *umgeni* Provonsha & McCafferty 1995 [*C.*].

Material examined: *oligosetosa* [*C.*]: ℧.

*** *** ***

2.2;2,1-4/2.2;2,2-4. Caenopsella/g1

(Fimbriatotergaliae Caenotergaliae Caenoptera Caenis/f3=g1 .../f5=g3 Caenopsella/g1)

Nomen hierarchicum: **Caenopsella/g1** (incl. *Afrocaenis*) [g: *Caenopsella* Gillies 1977: 451, typus *C. meridies* Gillies 1977 (design. orig.)].

In circumscription matches:

— gen. *Caenopsella*: Gillies 1977: 451.

References. Gillies 1977: ℧ ⊕; – Malzacher 1991: ⊕; – 1993: ℧ ⊕.

Autapomorphy of Caenopsella/g1.

(1) Eyes of male are much larger than in female;

eye width is greater than half of distance between eyes, eye length exceeds head length, eyes are projected posteriad of posterior head margin (Malzacher 1993:Fig.1). At the same time, eyes have the same shape as in female, and the same as in males of other Caenoptera – spherical, without division into dorsal and ventral portion; this allows to conclude that male eyes of Caenopsella/g1 are enlarged secondarily, but not retain the sexual dimorphism initial for Ephemeroptera [see Caenoptera (15)].

Plesiomorphies of Caenopsella/g1. Maxillary palp is 3-segmented. (in contrast to Clypeocaenis/ /g2). Gonostyli are nearly straight, with one or two thin apical processes (similar to some Caenis/f5=g3 which have several processes of such kind).

Size. Fore wing length 2–8 mm (see Caenoptera).

Distribution. Ethiopian Region.

Caenopsella/g1 is divided into *Afrocaenis* and Caenopsella/g2.

2.2;2,1-4/2.2;2,2-4/1. Plesiomorphon **Afrocaenis/g(1)**

(Caenoptera Caenis/f3=g1 .../f5=g3 Caenopsella/g1 *Afrocaenis*)

Nomen hierarchicum: **Afrocaenis/g(1)** [g: *Afrocaenis* Gillies 1982: 15, typus *Caenopsella major* Gillies 1977 (design. orig.)].

In circumscription matches:

— gen. *Afrocaenis* Gillies 1982: 15.

References. Gillies 1977 (*Caenopsella major*): ⊕'; – 1982: ♉* ⊕*; – Malzacher 1991: ⊕'; – 1993: ♉* ⊕*.

Plesiomorphy of *Afrocaenis*. Gill operculum retains Y-shaped rib, in contrast to Caenopsella/g2.

Size. Fore wing length 5–8 mm (see Caenoptera).

Distribution. Ethiopian Region.

Species composition of Afrocaenis/g(1). 2 species – *major* Gillies 1977 [*Caenopsella*] and *browni* Gillies 1982 [*Afrocaenis*].

Material examined: –.

***

2.2;2,1-4/2.2;2,2-4/2. **Caenopsella/g2**

(Caenoptera Caenis/f3=g1 .../f5=g3 Caenopsella/g1 .../g2)

Nomen hierarchicum: **Caenopsella/g2** [g: *Caenopsella* Gillies 1977: 451, typus *C. meridies* Gillies 1977 (design. orig.)].

In circumscription matches:

— gen. *Caenopsella*: Gillies 1982: 15, 16.

Monospecific taxon.

References. Gillies 1977: ♉ ⊕; – Malzacher 1991: ⊕; – Malzacher 1993: ♉ ⊕.

Autapomorphy of Caenopsella/g2.

(1) Gill operculum [see Caenoptera (18)] has no Y-shaped rib, in contrast to other Caenoptera. As in other respects Caenopsella/g2 is related to *Afrocaenis* and other Caenis/f4=g2 which have developed Y-shaped rib, we can assume that this rib was secondarily lost in Caenopsella/g2.

Size. Fore wing length 2 mm (see Caenoptera).

Distribution. Ethiopian Region.

Species composition of Caenopsella/g2. 1 species – *meridies* Gillies 1977 [*Caenopsella*].

Material examined: *meridies* [*C.*]: ⊕̄.

*** ***

Caenis/f4=g2 INCERTAE SEDIS

Caenis/f4=g2 INCERTAE SEDIS 1. Pseudocaenis/g [g: *Pseudocaenis* Soldán 1978a: 124, typus *P. queenslandica* Soldán 1978 (design. orig.)].

Reference. Soldán 1978a: ♉.

Size. Larval body length 4–8 mm (see Caenoptera).

Distribution. Australia.

Material examined: –.

Comment. *Pseudocaenis* was regarded to be a generic synonym of Tasmanocoenis s. l. (inc. *Tillyardocaenis*) (Suter 1984: 105).

Caenis/f4=g2 INCERTAE SEDIS 2. Caenoculis/g [g: *Caenoculis* Soldán 1986: 347, typus *C. bishopi* Soldán 1986 (design. orig.)].

Reference. Soldán 1986: ♉.

Size. Larval body length 4 mm (see Caenoptera).

Distribution. Oriental Region.

Material examined: –.

Comment. Originally the genus *Caenoculis* was characterized by presence of three or at least two small ocellar tubercles on larval head; this character was regarded to be a synapomorphy with Brachycercus/f1=g3 (Soldán 1986: 356). However, ocellar tubercles evolve independently in various groups (see Index of characters [1.1.6]).

Other species of Caenis/f4=g2 INCERTAE SEDIS (here [*C.*]=[*Caenis*]. Besides type species of *Pseudocaenis* and *Caenoculis*, for following species male imagoes remain to be unknown: **Australian species:** *angulata* Suter 1993 [*Wundacaenis*], *coolooli* Suter 1999 [*Irpacaenis*], *flabellum* Suter 1993 [*Wundacaenis*], *kaapi* Suter 1999 [*Irpacaenis*], *rieki* Soldán 1978 [*Pseudocaenis*]. **Asian species:** *dangi* Soldán 1986 [*Caenoculis*], *argillosa* Kang & Yang 1994 [*C.*], *bella* Kang & Yang 1994 [*C.*], *carloi* Marie & Dia & Thomas 2000 [*C.*], *corana* Thomas & Sartori 1989 [*C.*], *cornigera* Kang & Yang 1994 [*C.*], *corpulenta* Kang & Yang 1994 [*C.*], *demoulini* Bruggen 1954 [*C.*], *granifera* Kang & Yang 1994 [*C.*], *kimminsis* Ali 1967 [*C.*], *moe* Hwang & Bae 1999 [*C.*], *montana* Kang & Yang 1994 [*C.*], *nhahoensis* Soldán 1986 [*Caenoculis*], *nitida* Kang & Yang 1994 [*C.*], *novaegiuneae* Bruggen 1957 [*Tasmanocoenis*] (presumably, as I did not see the description), *tuba* Hwang & Bae 1999 [*C.*], *yangi* Kang & Yang 1996 [*C.*]. **American species:** *catherinae* Hofmann & Sartori & Thomas 1999 [*C.*], *cuniana* Froechlich 1969 [*C.*], *femina* Hofmann & Sartori 1999 [*C.*].

*** *** ***

Caenoptera INCERTAE SEDIS

Caenoptera INCERTAE SEDIS. **Afrocercus/g** [g: *Afrocercus* Malzacher 1987: 2, typus *A. forcipatus* Malzacher 1987 (design. orig.)].

Reference. Malzacher 1987: ⊕♂.

Size. Fore wing length 2.5 mm (see Caenoptera).

Distribution. Uganda.

Material examined: –.

Comment. Only symplesiomorphies with Brachycercus/f1=g3 and *Madecocercus* are known: prosternum is wide, scutellum of imaginal metathorax is projected posteriorly, gonostyli are grooved.

*** *** ***

Fimbriatotergaliae INCERTAE SEDIS

Fimbriatotergaliae INCERTAE SEDIS 1.
† **Australiphemera/fg** [f: Australiphemeridae McCafferty 1991: 355; g: *Australiphemera* McCafferty 1990: 37, typus *A. relevata* McCafferty 1990 (design. orig.)].

Reference. McCafferty 1990: ⊕.

Age. Early Cretaceous (Brazil).

Material examined: –.

Comment. *Australiphemera* is placed to Fimbriatotergaliae presumably, basing on wing venation only. Mesothoracic structure is unknown, thus relationship with *Baba* can not be excluded (see below, Furcatergaliae INCERTAE SEDIS).

Fimbriatotergaliae INCERTAE SEDIS 2.
† **Microphemera/g** [g: *Microphemera* McCafferty 1990: 39, typus *M. neotropica* McCafferty 1990 (design. orig.)].

Reference. McCafferty 1990: ⊕.

Age. Early Cretaceous (Brazil).

Material examined: –.

Comment. *Microphemera* is placed to Fimbriatotergaliae presumably, basing on wing venation only. Mesothoracic structure is unknown, thus relationship with *Baba* can not be excluded (see below, Furcatergaliae INCERTAE SEDIS).

Fimbriatotergaliae INCERTAE SEDIS 3.
† **Pristiplocia/g** [g: *Pristiplocia* McCafferty 1990: 40, typus *P. rupestris* McCafferty 1990 (design. orig.)].

Reference. McCafferty 1990: ⊕.

Age. Early Cretaceous (Brazil).

Material examined: –.

Fimbriatotergaliae INCERTAE SEDIS 4.
† **Palaeoanthus/fg** [f: Palaeoanthidae Kluge 1993b: 44; g: *Palaeoanthus* Kluge 1993b: 44, typus *P. orthostylus* Kluge 1993 (design. orig.)].

Reference. Kluge 1993b: ◎* ⊕*.

Character of unclear phylogenetic status.

(1) In imago and subimago 1st segment of gonostylus is not expressed, completely fused with 2nd (so gonostylus is 3-segmented). Non-unique apomorphy (see Index of characters [2.3.10]); among Fimbriatotergaliae the same in Potamanthus/fg1, Euthyplocia/fg1, Behningia/fg2.

Plesiomorphies. Furcasternal protuberances are contiguous (in contrast to Campsurus/fg1). Wings are non-modified (in contrast to Euthyplocia/fg1, Behningia/fg1, Palingenia/f3=g2, Polymitarcys/f1=Ephoron/g2); on fore wing AA without bifurcation (in contrast to Potamanthus/fg1) and without veins going to basitornal margin (in contrast to Ephemera/fg9). Legs functional, all claws are ephemeropteroid. Bases of tergalii VII are not transferred anteriorly (in contrast to Ephemera/fg9). Gonostylus with 2 distal segments. Paracercus is well-developed in both sexes. Moult from subimago to imago takes place in both sexes.

Size. Fore wing length 4–9 mm.

Age. Late Cretaceous (Taimyr amber).

Species composition. 2 species – *minutus* Kluge 1993 [*Palaeoanthus*] and *orthostylus* Kluge 1993 [*Palaeoanthus*].

Material examined (Paleontol.Inst.): *minutus* [*P.*]: ♀, ⊕♀; *orthostylus* [*P.*]: ◎, ⊕♂.

Comment. Both species have wing venation characteristic for Fimbriatotergaliae and *Baba* [see Fimbriatotergaliae (5)]. The type species of *Palaeoanthus* – *orthostylus* [*P.*] – is known as incomplete imagoes and subimagoes, no one of which has mesonotum preserved. Thus, we do not know for certain, if this species really had mesonotal suture characteristic for Furcatergaliae, or its mesonotal suture was plesiomorphic as in *Baba* (see below, Furcatergaliae INCERTAE SEDIS). Another species attributed to *Palaeoanthus* – *minutus* [*P.*] – has well-preserved mesonotal suture of the structure typical for true Furcatergaliae (Kluge 1993:Fig.5m). But we do not know reliable synapomorphies which would prove that these two species constitute a holophyletic taxon. On hind wing of the both species common stem of RS+MA is relatively long, and non-branched part of RS is relatively short (as long as 0.25–0.5 triad RS), in contrast to *Baba* and many other mayflies.

Fimbriatotergaliae INCERTAE SEDIS 5. † Borephemera/g [g: *Borephemera* Sinitshenkova 2000c: 115, typus *B. goldmani* Sinitshenkova 2000 (design. orig.)].

Reference. Sinitshenkova 2000c: ⊕̃.

Age. Late Cretaceous (New Jersey amber).

Material examined: –.

Comment. *Borephemera* is placed to Fimbriatotergaliae presumably, basing on wing venation only. Mesothoracic structure is unknown, thus relationship with *Baba* can not be excluded (see below, Furcatergaliae INCERTAE SEDIS).

*** ***

2.2;2,2. Ephemerella/fg1

(Euplectoptera Anteritorna Bidentiseta Furcatergaliae Ephemerella/fg1)

(Figs 89–105)

Nomen hierarchicum: **Ephemerella/fg1** (incl. *Tricorythus*) [f: Ephemerellidae Klapálek 1909: 13; g: *Ephemerella* Walsh 1862: 377, typus *E. excrucians* Walsh 1862 (design. Eaton 1868b: 87)].

In circumscription matches:
— superfam. Ephemerelloidea: Koss 1973: 74;
— Ephemerella/fg1: Kluge 2000: 252.

References. Koss & Edmunds 1974: ⊙*; – Kluge 1997b: ♉*; – McCafferty & Wang 2000: ♉* ⊕*.

Autapomorphies of Ephemerella/fg1.

(1) Maxillary palp lost muscles (while in other mayflies it has muscles attached to its base and muscles in 1st segment, which move 2nd segment). In most representatives maxillary palp is diminished, but in Tricorygnatha and *Machadorythus* it is secondarily elongate; usually all 3 segments are retained, but when the palp is strongly diminished, its segments can be fused together.

In connection with loss of musculature, maxillary palp disappeared many times independently in various taxa of Ephemerella/fg1: It is absent in Eurylophella/fg1, but retained in other Timpanoga/ /fg1; absent in selected species of Cincticostella/g3, but retained in other Ephemerella/fg4; absent in selected species of Torleya/g2 and present in others; absent in *Teloganopsis*, *Uracanthella* and *Amurella*; among Hyrtanella/f1=Crinitella/g1 – absent in sp. HC1, but retained as non-segmented vestige in two other species; present in *Vietnamella*; absent or vestigial in Melanemerella/fg1 and Teloganodes/fg1; among Afrotricorythi – absent in *Ephemerythus*, but retained in other taxa; among Leptohyphes/fg1 – absent in *popayanicus* [*Tricorythodes*], but retained in other Tricorythodes/fg1, as well as in Leptohyphes/fg2 and *Allenhyphes*.

Among other mayflies loss of maxillary palp took place only once, in *Madecocercus*, and loss of musculature – only in Acanthametropus/fg2 which has strongly specialized mouth apparatus (see Index of characters [1.1.41]). In Ephemerella/fg1 maxillae have no initial specialization, and are variably specialized in selected representatives [see (16)].

(2) Paraglossae are fused with mentum: at least suture separating paraglossae from mentum is always lost (at the same time glossae can be distinctly separated from mentum) (FIG.101:B). Non-unique apomorphy, the same in Posteritorna.

(3) In labial palp 2nd segment lacks muscle-adductor of 3rd (terminal) segment; non-unique apomorphy; 3rd segment is usually diminished, but nearly always retained (FIG.101:B).

(4) Larval pronotum with more or less expressed V-shaped impression which goes from anterior-lateral angles medially-posteriorly (FIG.92:C–D). Possibly, this structure is homologous to the pronotal collum [see Fimbriatotergaliae (1) above].

(5) Larval fore wing buds are fused one with another by mean of a mesial plate which has a median incision on posterior margin; fore wing buds convergent by their apices (FIG.91:A); sometimes (particularly in Afrotricorythi) they are fused by their apices, in this case the apex of the mesial plate between them is not expressed (FIG.97: E); only in Timpanoga/fg4 wing buds somewhat diverge (that is connected with strong widening of body). When adult fore wings develop inside wing buds, a pair of membranous processes can develop inside the mesial plate (compare FIGS 101:A and 102:A–B); thus adult can have a pair of short or long membranous processes (**plumidia**) on posterior margin of mesonotal scutellum (FIG.89:A–B). Non-unique apomorphy (see Index of characters [1.2.6] and [2.2.13]).

(6) On fore wing CuP in its proximal part is connected with CuA by a constant crossvein *cua-cup*, and with AA – by a constant crossvein *cup-aa*. Presence of these constant crossveins is not connected with course of tracheae: veins *cua-cup*, *cup-aa*, AA, CuP and proximal portion of CuA lack tracheae, and distal portion of CuA is supplied by trachea from MP (as in majority of mayflies, in contrast to Campsurus/fg1); the common trachea of MP and CuA can penetrate into the wing either with other tracheae, passing in front of basal wing plate, or rounds the basal wing plate from behind (among Ephemerella/fg1, tracheation was examined in *thani* [*Vietnamella*], *aculea* [*E.*] and *commodema* [*E.*]).

All Ephemerella/fg1 except for Tricoryptera, have uniform structure of wing base (FIG.90:A): *cua-cup* is located more distally than *cup-aa*; CuP arises from CuA under acute angle (plesiomorphic condition) and is sharply bent at the place of connection with *cua-cup*. In Tricoryptera this structure is modified, portion of CuP proximad of connection with *cup-aa* is lost (FIG.89:K–M) [see below, Tricoryptera (1)].

(7) Each tergalius II–VI [bilobed and lacking marginal ribs – see Furcatergaliae (5)] initially consists of dorsal and ventral lobes; the **dorsal lobe** represents initial tergalial lamella and can retain costal and anal ribs (see below, Plesiomorphies of Pantricorythi); the **ventral lobe** bifurcates to two branches – **costal branch** (located most laterally, under costal margin of dorsal lobe) and **anal branch** (located most medially, under anal margin of dorsal lobe); each branch bears on its outer side a row of processes (FIGS 89:C–G; 91:D–F; 97:F–G). Unique apomorphy. Tergalius VII is diminished; it also consists of the dorsal and ventral lobes, but its ventral lobe is never bifurcate.

Usually tergalii II–VII lie dorsally on abdomen and are directed by their apices posteriorly, overlapping tergalii located posteriad of them; lateral margins of abdominal segments are often expanded laterad of tergalii (FIG.91:A, I). In connection with this, in selected groups of Ephemerella/fg1 tergalii of the anteriormost pair are transformed to gill opercula; these can be tergalii II, III or IV (see TABLE 8 and Index of characters [1.3.32]).

In various groups of Ephemerella/fg1 tergalii of this or that pairs can be simplified in structure or lost (see TABLE 8 and Index of characters [1.3.19–21] and [1.3.54–59]). Only in *Dicercomyzon* tergalii are modified so strongly, that completely lost the structure initial for Ephemerella/fg1.

About tergalius I – see (13).

(8) Imaginal and subimaginal gonostylus with one distal segment only (instead of two initial ones). Non-unique apomorphy (see Index of characters [2.3.12]). In some Ephemerella/fg1 (Afrotricorythi and some others) the distal segment is lost.

Characters of Ephemerella/fg1 of unclear phylogenetic status: unique apomorphies, but present not in all representatives.

(9) Imaginal and subimaginal mesonotal suture is transverse; in the point of its crossing with median longitudinal suture its margins can be somewhat separated, between them a small membranous area can be expressed (FIG.90:D); such distinctly expressed mesonotal suture is present in Ephemerella/fg2, *Vietnamella* and Melanemerella/fg1, but in Teloganodes/fg1 and Tricoryptera it can be poorly expressed, up to disappearance (FIG.104). In all cases mesonotal suture is never transferred backward (in contrast to Fimbriatotergaliae, Leptophlebia/fg1 and Tetramerotarsata), only at crossings with medioparapsidal sutures it can be narrowly stretched backward (FIG.90:D).

(10) Lateroparapsidal suture is curved by its posterior end laterally, and subimaginal lateral sclerotized area is medially bordered by a **relief line** which passes mediad of lateroparapsidal suture and is arched by its convexity medially; correspondingly, the lateral sclerotized area occupies the lateroparapsidal suture, whole sublateroscutum and most part of lateroscutum (FIG.90:D).

Such structure of the lateral sclerotized area is present in all Ephemerella/fg2, and among Pantricorythi distinctly expressed in *Vietnamella* and examined representative of Melanemerella/fg1. In other taxa of Pantricorythi structure of mesonotum is quite various: lateroparapsidal suture can be either curved laterally (FIG.99:D), or convergent with medioparapsidal suture (FIG.98:D); the relief line can be ether present or absent; lateral sclerotized area can have different shape (FIG.96:A) or absent (FIG.104:I) (see Index of characters [2.2.9] and [2.2.14]).

Apomorphies of Ephemerella/fg1 common with Potamanthus/fg1 and Caenotergaliae (see Classifications of Furcatergaliae I).

(11) Larval mesonotum has a **collar** – a concave band at anterior margin, separated from the rest part of mesonotun by a transverse crest (FIG.92:C–D). The same in Potamanthus/fg1, Euthyplocia/fg1 and Caenotergaliae.

(12) Dorsal (anterior) surface of larval fore femur can bear a transverse row of setae (FIG.89: H–J); sometimes this row is irregular, indistinct or lost. The same in Caenotergaliae and Potamanthus/fg1. About peculiar shape of this row in Pantricorythi – see below.

(13) Tergalius I is stick-like, setose, attached on a prominent cylindrical pedestal which can arise either from anterior part of the abdominal segment I (Fig.95:D–E) (the same in Potamanthus/fg1 and Caenotergaliae), or from posterior part of the segment (Fig.91:G–H); apical part of pedestal can be membranous and retractile.

In many Ephemerella/fg1 tergalii I are completely lost; among Ephemerella/fg2 they are retained only in most Timpanoga/fg1, and among

Tricorythus/fg1 – in *Vietnamella* and selected Melanemerella/fg1 (see Index of characters [1.3.33]) (in contrast to Potamanthus/fg1 and Caenotergaliae, where they are retained in all representatives).

(14) Egg usually with a large cap on one pole; only in selected cases the polar cap is absent (in Eurylophella/fg2 and selected species of Tricoryptera) or caps are present on both poles (in *simplex* [*Ephemerella*] and *Tricorythopsis*). Sometimes several anchors are present; each anchor consists of a terminal knob and a skein of threads which surround the knob in a form of regular ring (Studemann & Landolt 1997a:Figs 1–54). Similar polar caps and anchors are present in Potamanthus/fg1, and caps – also in some Caenotergaliae.

(15) In cubital field of fore wing [see Anteritorna (1)] the most constant element is a bifurcate vein initially arising from CuA – 1st cubital fork (x_1 and x_2 in FIGS 90:A, 95:A, 99:A and 100:A). More distally from CuA can arise two other bifurcate veins – 2nd and 3rd cubital forks. All 3 cubital forks occur in *Vietnamella* only (FIG.95:A: x, y, z). Ephemerella/fg2, *Austremerella* and Teloganodes/fg1 have only the 1st and 2nd cubital forks or their vestiges; both forks in form of integral bifurcate veins arising from CuA occur in a few largest representatives of Ephemerella/fg2 (Allen & Edmunds 1962a:Fig.11); in most species one or both branches of the 2nd cubital fork lost connection with CuA and become intercalaries (FIG.90:A: y_2); sometimes the same happens with the 1st cubital fork also. Tricoryptera have only the 1st cubital fork, which sometimes lost connection with CuA (FIGS 99:A and 100:A: x_1, x_2).

The same cubital forks, besides Potamanthus/fg1 and Caenotergaliae, occur in some other mayflies (see Index of characters [2.2.51]).

Non-unique apomorphies of Ephemerella/fg1.

(16) Apical-ventral side of maxilla [see (1)] bears a field of long setae, usually situated densely and regularly; this field is close to base of maxillary canines, being located on a protuberance, from which canines arise (FIG.92:B). In representatives with biting maxillae this setal field can be strongly diminished (FIG.95:C); in representatives with filtering maxillae (Cincticostella/g2, *Uracanthella* and others) this field is enlarged and maxilla has a truncate shape (FIGS 92:H and 93:B); scraping maxillae of *Dicercomyzon* have this field extremely enlarged (see Index of characters [1.1.29]). The same field of setae is present in Potamanthus/fg1 and Euthyplocia/fg1, a similar field – in Leptophlebia/fg1. Apical-ventral row of setae is absent (in contrast to many Leptophlebia/fg1, Radulapalpata and Tridentiseta – see Index of characters [1.1.31]).

(17) Imaginal and subimaginal furcasternal protuberances of mesothorax are separated (FIG.90:E). Non-unique apomorphy (see Index of characters [2.2.23]).

(18) Larval claw usually with one row of denticles on inner side (non-unique apomorphy – see Index of characters [1.2.21]); sometimes denticles are lost [see Timpanoga/fg3 (1) below]; sometimes, apart of this row, there are also subapical denticles on anterior and/or posterior side (FIG.98:A–B). In an exceptional case, two regular rows of denticles are present (in *crassi* [*Ephemerellina*] and *brincki* [*E.*]).

(19) Larval abdominal terga often have paired submedian projections at posterior margin (FIG.91:A); such projections are present in many Ephemerella/fg2 and most primitive Pantricorythi – *Vietnamella*, *Austremerella* and some Melanemerella/fg1, that allows to conclude that this structure is initial for Ephemerella/fg1. In many species paired abdominal projections are poorly expressed or absent. Only in a few taxa of Ephemerella/fg1 abdominal terga have unpaired median projections instead of paired ones – these are Ephemerella/fg3-*-Amurella* and some Pantricorythi (see Index of characters [1.3.3]).

Plesiomorphies of Ephemerella/fg1. Imaginal and subimaginal claws are ephemeropteroid (only on fore legs of male can be blunt – see Index of characters [2.2.77]). Larval and adult paracercus is nearly always well-developed, subequal to cerci; only in Teloganodes/fg1 and *Dicercomyzon* paracercus is reduced both in larva and adults.

Variable characters of Ephemerella/fg1. Larval and adult patella-tibial suture is usually developed on middle and hind legs (as in majority of mayflies), but in selected taxa this suture is lost on all legs: it is retained in all the species examined of Timpanoga/fg1, Torleya/g2 and Ephemerella/fg5; among Drunella/g1 – lost in *lepnevae* [*Ephemerella*], *doddsi* [*E.*], *coloradensis* [*E.*] and *spinifera* [*E.*], but retained in other species examined; among Cincticostella/g1 – lost in larva of *tshernovae* [*Ephemerella*], but retained in its adults and in other species examined; lost in larvae of the both Caudatella/g(1) species examined; among the examined taxa of Pantricorythi, this suture is lost in larvae and adults of Teloganodes/fg1, Tricorygnatha and *Dicercomyzon*, but retained in *Vietnamella*, *Ephemerythus*, *Machadorythus* and Leptohyphes/fg1 (FIG.105:A).

Size. Fore wing length 2–19 mm.

Age and distribution. Palaeogene – Recent; world-wide. A single known extinct species is *visca-ta* [*Ephemerella (Timpanoga)*] from Baltic amber (see below, Ephemerella/fg1 INCERTAE SEDIS).

Systematic position of Ephemerella/fg1. About possibility to unite Ephemerella/fg1 with Potamanthus/fg1 and Caenotergaliae – see above, Classifications of Furcatergaliae I.

Ephemerella/fg1 is divided into Ephemerella/fg2 and Pantricorythi (p.316). Some insufficiently described taxa have uncertain systematic position (p.346).

2.2;2,2-1. Ephemerella/fg2

(Anteritorna Bidentiseta Furcatergaliae Ephemerella/fg1 .../fg2)

(Figs 90–94)

Nomen hierarchicum: **Ephemerella/fg2** (sine *Tricorythus*; incl. *Timpanoga*).

In circumscription matches:
— gen. *Ephemerella*: Eaton 1883–1888: 124;
— subfam. Ephemerellinae: Allen 1965: 264;
— fam. Ephemerellidae Klapálek 1909: 13;
— Ephemerella/fg2: Kluge 2000: 252.

References. Eaton 1883–1888: ♉* ⊕*; – Needham & Traver & Hsu 1935: ♉* ⊕*; – Edmunds & Jensen & Berner 1976: ♉* ⊕*; – Studemann & Landolt 1997a: ⊙'; – McCafferty & Wang 2000: ♉* ⊕*.

Autapomorphies of Ephemerella/fg2.

(1) Tergalii II are lost. Thus, bilobed tergalii [see Ephemerella/fg1 (7)] are present either beginning from abdominal segment III (in Ephemerella/fg3 – FIG.91:A) or from abdominal segment IV (in Timpanoga/fg1 – FIG.91:I) (TABLE 8).

(2) Tergalial ribs [non-marginal – see Furcatergaliae (5)] are lost (in contrast to Pantricorythi). Probably initial for Ephemerella/fg2 shape of dorsal lobe of tergalii III–VI [see Ephemerella/fg1 (7)] is truncate, with oblique or incised distal margin, with anal margin (directed medially) longer than costal margin (directed laterally); often dark colour pattern in a form of clover leaf is developed: "graft" arises from base of tergalius, and three "lives" are sided to costal margin (directed laterally), distal margin (directed posteriorly) and anal margin (directed medially) (FIG.91:A). These shape and colour pattern occur in many (but not all) Ephemerella/fg2, being never found in other mayflies.

Character of unclear phylogenetic status.

(3) Fore wing with free marginal intercalaries (FIG.90:A). The same in many other Ephemerella/ /fg1, except for *Vietnamella* which has connected intercalaries, and some taxa which lost intercalaries (see Index of characters [2.2.55]).

Plesiomorphies of Ephemerella/fg2. In contrast to Pantricorythi: Transverse row of setae on larval fore femur [see Ephemerella/fg1 (12)] is usually irregular or lost; setae on outer margin of femur can be irregular, or form a regular longitudinal row not continued with the transverse row. 1st segment of gonostylus is not elongate, shorter than 1/2 of 2nd segment, sometimes completely fused with 2nd segment (FIG.93:C).

In contrast to selected groups of Pantricorythi, imago and subimago is non-specialized: Eyes of male are always large, divided into two portions. Mesonotum with distinct mesonotal suture, curved lateroparapsidal suture and subimaginal lateral pigmented area bordered by relief line [see Ephemerella/fg1 (9) and (10)] (FIG.90:D). Cubital field of fore wing has up to 2 bifurcate veins arising from CuA (in largest species only); often the distal of these veins is vestigial (FIG.90:A: y). Hind wing is always well-developed, as long as 0.2–0.3 of fore wing length; at least RS with bifurcation; usually MP also has bifurcation; usually hind wing is oval, without prominent costal projection and with Sc terminating near apex (FIG.90:B); only in *Teloganopsis* and Hyrtanella/f1=Crinitella/g1 hind wing is modified (see below). Imaginal wings without setae on hind margin.

Maxilla always with 2 dentisetae [see Bidentiseta (1)] and usually with 3 canines (FIG.92:B) (only in *Uracanthella* and Cincticostella/g1 canines are modified or lost).

Size. Fore wing length 5–19 mm.

Distribution. Holarctic and Oriental Region.

Ephemerella/fg2 is divided into Timpanoga/fg1 and Ephemerella/fg3 (p.303).

Table 8. Structure of tergalii in various taxa of Ephemerella/fg1

Taxon	Pairs of tergalii												
	I	II		III		IV		V		VI		VI	
Ephemerella/fg1 (7), (13)													
Ephemerella/fg2 (1)													
Timpanoga/fg1 (1), (P)													
Attenella/g1 (P)	+	–	–	–	–	I	Ʉ	I	Ʉ	I	Ʉ	I	I
–"–	–	–	–	–	–	I	Ʉ	I	Ʉ	I	Ʉ	I	I
Timpanoga/fg2 (1) (FIG.91:G–I)	+	–	–	–	–	O	Ʉ	I	Ʉ	I	Ʉ	I	I
Ephemerella/fg3 (1)													
Ephemerella/fg4 (1) (FIG.91:C–F)	–	–	–	I	Ʉ	I	Ʉ	I	Ʉ	I	ǂ	I	I
Torleya/fg1 (P) (FIG.91:A–B)	–	–	–	I	Ʉ	I	Ʉ	I	Ʉ	I	Ʉ	I	I
–"– (FIG.94:A–E)	–	–	–	O	Ʉ	I	Ʉ	I	Ʉ	I	Ʉ	I	I
Pantricorythi (P)													
Vietnamella (P) (FIG.95:D–E)	+	I	Ʉ	I	Ʉ	I	Ʉ	I	Ʉ	I	Ʉ	I	I
Austremerella (P) (FIG.95:F–H)	–	I	Ʉ	I	Ʉ	I	Ʉ	I	Ʉ	I	ǂ		
Melanemerella/fg1													
harrisoni [*Lithogloea*]	+	O	Ʉ	I	Ʉ	I	Ʉ	I	Ʉ	I	–	–	–
brasiliana [*Melanemerella*]	–	O	Ʉ	I	Ʉ	I	Ʉ	I	Ʉ	I	–	–	–
barnardi [*Ephemerellina*]	–	O	Ʉ	I	Ʉ	I	Ʉ	I	Ʉ	I	–	–	–
brincki [*Ephemerellina*]	–	O	Ʉ	I	Ʉ	I	Ʉ	I	Ʉ	–	–	–	–
crassi [*Ephemerellina*]	+	O	Ʉ	I	Ʉ	I	Ʉ	I	Ʉ	–	–	–	–
penicillata [*Lithogloea*]	+	O	Ʉ	I	Ʉ	I	–	–	–	–	–	–	–
Teloganodes/fg1 (1), (8)													
insignis [*Macafertiella*] (FIG.96:E)	–	O	Ʉ	U	Ʉ	U	Ʉ	U	Ʉ	I	–	–	–
sp. T4 (FIG.96:F)	–	O	Ʉ	U	Ʉ	U	Ʉ	U	–	–	–	–	–
sp.T1, sp. T2 (FIG.96:G)	–	O	Ʉ	U	Ʉ	U	Ʉ	I	–	–	–	–	–
sp. T3 (FIG.96:H)	–	O	Ʉ	U	Ʉ	I	–	–	–	–	–	–	–
Tricoryptera (5), (6)													
Afrotricorythi													
Tricorygnatha (P):													
jacobsoni [*Tricorythus*]	–	I	Ʉ	I	Ʉ	I	Ʉ	I	Ʉ	I	Ʉ	I	–
other species (FIG.97:F–G)	–	I	Ʉ	I	Ʉ	I	Ʉ	I	Ʉ	I	Ʉ	–	–
Ephemerythus (2), (VAR)	–	O	Ʉ	I	Ʉ	I	Ʉ	I	–	I	–	–	–
–"–	–	O	Ʉ	I	Ʉ	I	Ʉ	I	–	–	–	–	–
Dicercomyzon (4)	–	ǂ		ǂ		ǂ		ǂ		ǂ		–	–
Machadorythus (5)	–	O	I	I	I	I	–	I	–	I	–	–	–
Leptohyphes/fg1 (4), (10)													
Leptohyphes/fg2 (P) (FIG.101)	–	O	U	I	Ʉ	I	Ʉ	I	Ʉ	I	I	–	–
Allenhyphes (3)	–	O	I	I	U	I	U	I	U	I	I	–	–
Tricorythodes (6)	–	O	U	I	U	I	U	I	U	I	I	–	–
–"–	–	O	I	I	U	I	U	I	U	I	I	–	–
–"– (FIG.103:A–C)	–	O	U	I	U	I	U	I	U	I	–	–	–
Tricorythopsis (8)	–	O	Ʉ	I	Ʉ	I	Ʉ	I	I	I	–	–	–
Coryphorus/g1 (3)													
aquilis [*Coryphorus*]	–	O	Ʉ	I	Ʉ	I	Ʉ	I	Ʉ	–	–	–	–
sp. Cor1 (FIG.105:G)	–	O	Ʉ	I	Ʉ	I	Ʉ	I	–	–	–	–	–
Teloganella	–	O	Ʉ	I	Ʉ	I	Ʉ	I	Ʉ	I	–	–	–

Tergalius I [stick-like – see Ephemerella/fg1 (13)]: "+" – present, "–" – absent.
Tergalii of other pairs [see Ephemerella/fg1 (7)]: first symbol – dorsal lobe, second symbol – ventral lobe: "O" – operculate, "Ʉ" – bifurcate with marginal processes, "U" – bifurcate without processes, "ǂ" – non-bifurcate with marginal processes, "I" – non-bifurcate without processes, "–" – absent.

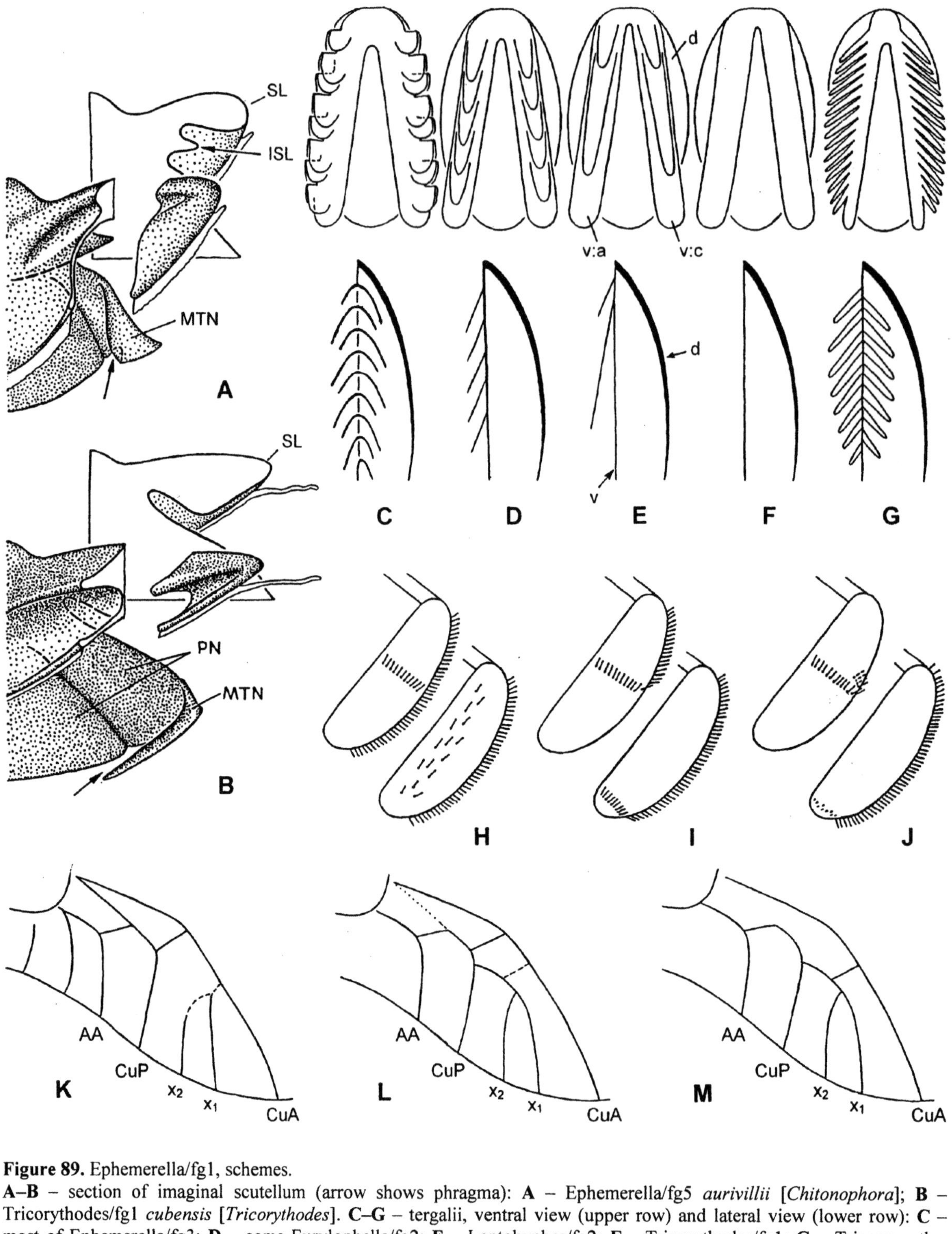

Figure 89. Ephemerella/fg1, schemes.
A–B – section of imaginal scutellum (arrow shows phragma): **A** – Ephemerella/fg5 *aurivillii* [*Chitonophora*]; **B** – Tricorythodes/fg1 *cubensis* [*Tricorythodes*]. **C–G** – tergalii, ventral view (upper row) and lateral view (lower row): **C** – most of Ephemerella/fg3; **D** – some Eurylophella/fg2; **E** – Leptohyphes/fg2; **F** – Tricorythodes/fg1; **G** – Tricorygnatha, *Ephemerythus*, *Teloganodes*. **H–J** – position of setae on larval fore femur (above) and middle or hind femur (below): **H** – Ephemerella/fg2 (not all); **I** – Pantricorythi (not all); **J** – Tricorygnatha. **K–M** – transformation of cubital field venation from initial for Ephemerella/fg1 (**K**) to that of Tricoryptera (**M**). (A–J – from Kluge 1992c; K–L– from Kluge 1997b).

Abbreviations: **d** – dorsal lobe of tergalius; **ISL** – infrascutellum; **MPN** – mediopostnotum; **MTN** – metanotum; **SL** – scutellum; **v** – ventral lobe of tergalius; **v:a** – anal branch of ventral lobe; **v:c** – costal branch of ventral lobe.

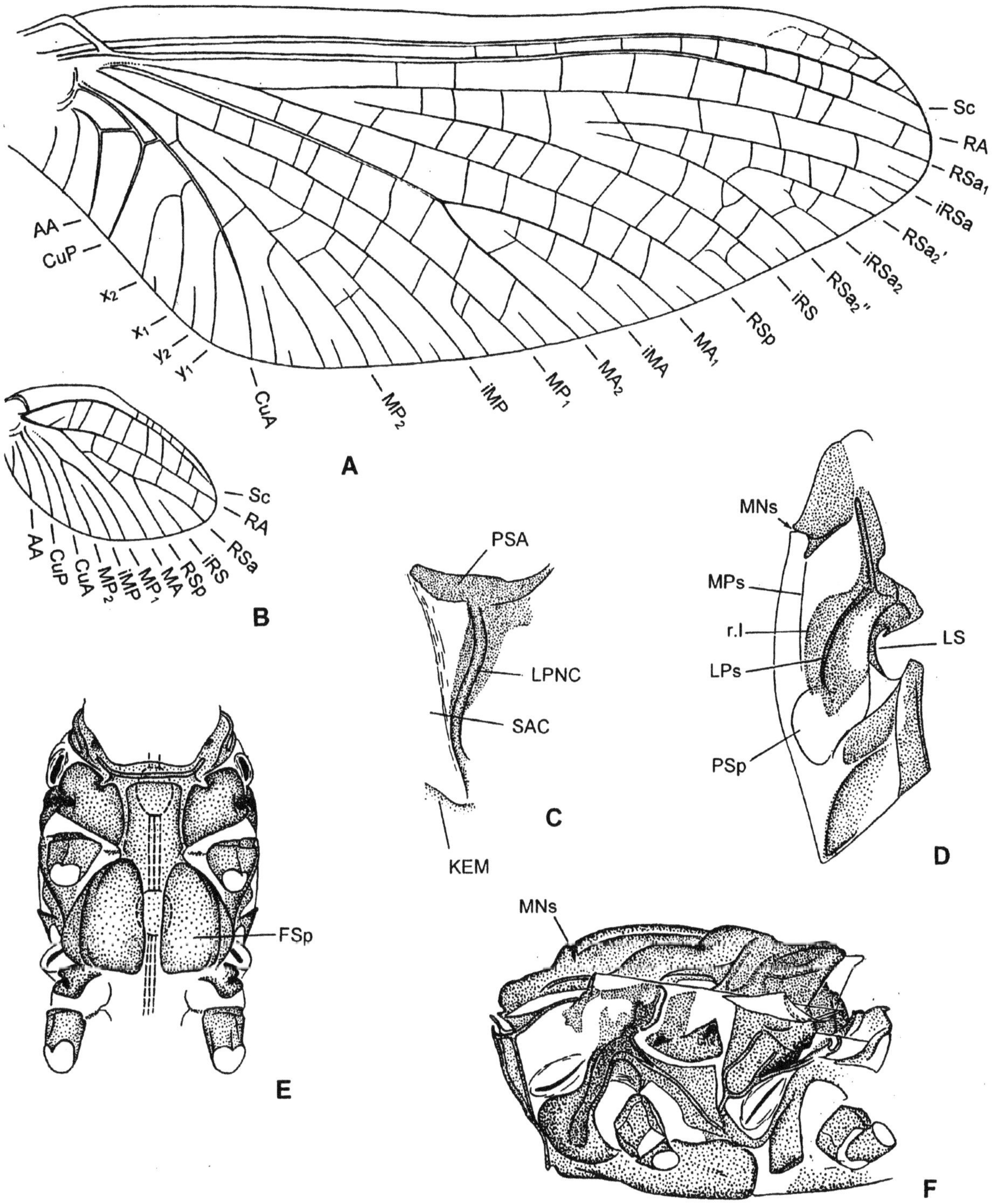

Figure 90. Ephemerella/fg3, thorax of imago and subimago.
A–B – Torleya/g2 *ignita* [*Ephemera*], fore and hind wings. **C–D** – Ephemerella/fg5 *aurivillii* [*Chitonophora*]: **C** – lateropostnotal sclerotization of subimaginal exuviae (compare with Fig.5:D); **D** – subimaginal exuviae of right half of mesonotum. **E–F** – Drunella/g1 *aculea* [*Ephemerella*], imaginal pterothorax: **E** – ventral view (nerve chan shown by interrupted line); **F** – lateral view. (C–D – from Kluge 1997b).

Abbreviations: **FSp** – furcasternal protuberance; **KEM** – katepimeron; **LPNC** – lateropostnotal crest; **LPs** – lateroparapsidal suture; **LS** – lateroscutum; **MNs** – mesonotal suture; **MPs** – medioparapsidal suture; **PSA** – postsubalar sclerite of lateropostnotum; **PSp** – posterior scutal protuberance; **r.l** – relief line; **SAC** – subalar conjunctiva.

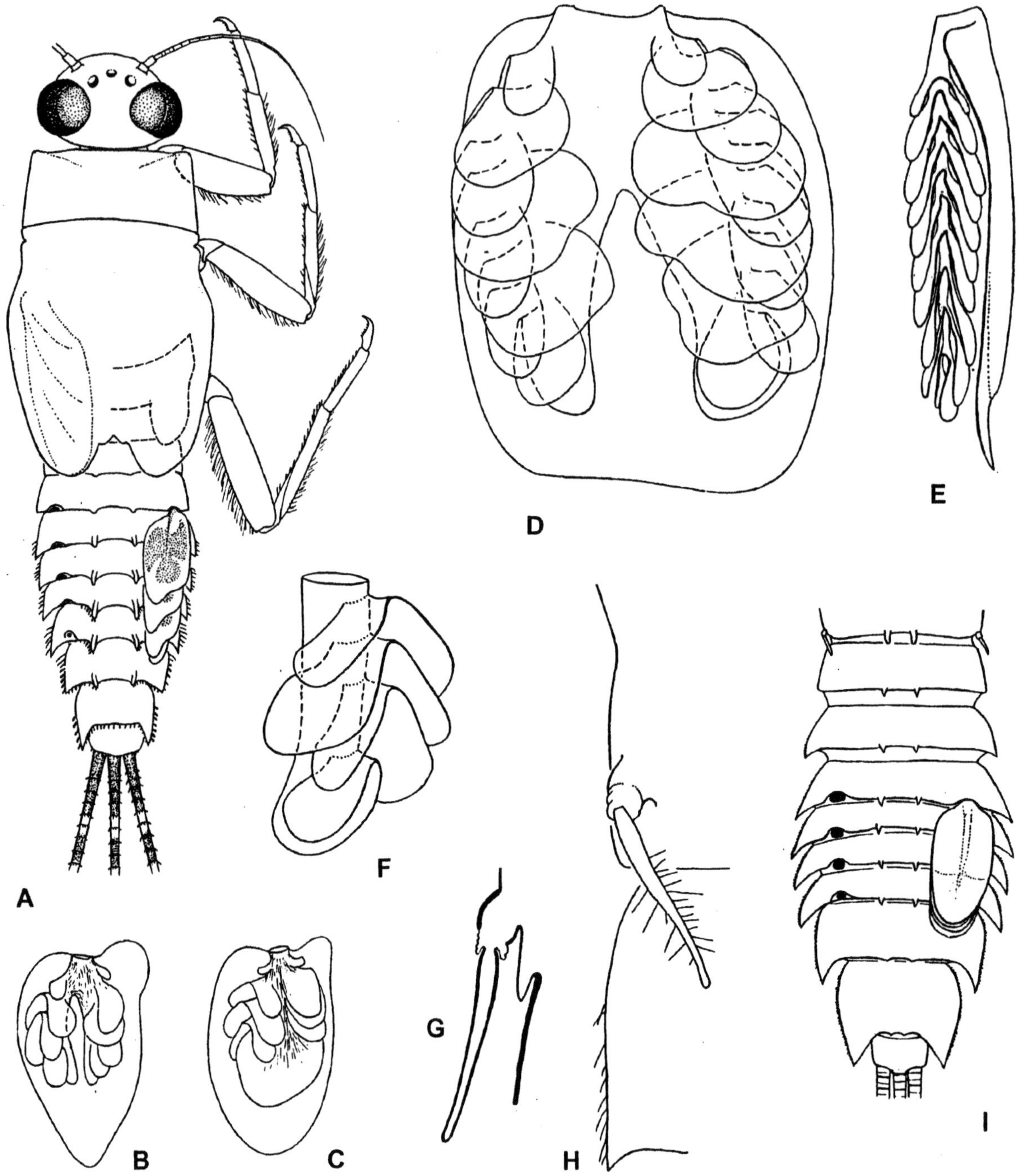

Figure 91. Ephemerella/fg2, larvae.
A–F – Ephemerella/fg3: **A–B** – Torleya/g2 *ignita* [*Ephemera*]: **A** – dorsal view (left legs and tergalii removed, right half of metanotum and anterior abdominal terga shown by interrupted line); **B** – left tergalius VI, ventral view. **C** – Ephemerella/fg5 *aurivillii* [*Chitonophora*], left tergalius VI, ventral view. **D–F** – Drunella/g1 *aculea* [*Ephemerella*], tergalius III: **D** – ventral view; **E** – lateral view; **F** – apex of one of branches of ventral lobe, schematic. **G–I** – Timpanoga/fg2 Eurylophella/fg2 *karelica* [*Eurylophella*]: **G** – optic section of tergalius I; **H** – left margin of abdominal segment I (fused with metathorax) and segment II, dorsal view (tergalius I in natural position); **I** – larval abdomen, dorsal view (left tergalii IV–VII removed). (A–C, I – from Kluge 1997d).

2.2;2,2-1/1. Timpanoga/fg1

(Anteritorna Bidentiseta Furcatergaliae Ephemerella/fg1 .../fg2 Timpanoga/fg1)

(Fig. 91:G–I)

Nomen hierarchicum: **Timpanoga/fg1** (incl. *Attenella*) [f: Timpanogae Allen 1984: 246; g: *Ephemerella* subgen. *Timpanoga* Needham 1927: 115, typus *Ephemerella hecuba* Eaton 1884 (design. orig.)].

In circumscription matches:
— complex genn. *Timpanoga*: McCafferty & Wang 1994a: 571;
— subtribus Timpanogae Allen 1984: 246;
— tribus Timpanoginae: McCafferty & Wang 2000: 41.

References. McCafferty & Wang 1994a: ♉*; – 2000: ♉*.

Autapomorphy of Timpanoga/fg1.

(1) Tergalii III are lost, thus bilamellate tergalii of structure typical for Ephemerella/fg1 are present on abdominal segments IV–VII only (FIG.91:I) [see Ephemerella/fg1 (7)]. Unique apomorphy (TABLE 8).

Plesiomorphies and variable characters of Timpanoga/fg1. In contrast to Ephemerella/fg3, tergalii I [stick-like – see Ephemerella/fg1 (13)] are usually retained, being attached to posterior part of the segment (FIG.91:G–H) (while in other mayflies with stick-like tergalii I – *Vietnamella*, Melanemerella/fg1, Caenotergaliae and Potamanthus/fg1 – they are attached anteriorly – Fig.95:D); in Timpanoga/fg4 tergalii I are transferred to extreme posterior margin of the segment; in Attenella/g1 *delantala* [*Ephemerella*]) tergalii I are lost. Paired tergal projections [see Ephemerella/fg1 (19)] are nearly always retained at least on some segments, rarely lost (in *Danella*). In contrast to *Teloganopsis* and Hyrtanella/f1=Crinitella/g1, hind wing without prominent costal projection and with Sc terminating near apex (as in FIG.90:B).

Size. Fore wing length 6–14 mm.

Distribution. Mainly Nearctic, also in Europe.

Timpanoga/fg1 is divided into *Attenella* and Timpanoga/fg2.

2.2;2,2-1/1.1. Attenella/fg(1)

(Bidentiseta Furcatergaliae Ephemerella/fg1 .../fg2 Timpanoga/fg1 *Attenella*)

Nomen hierarchicum: **Attenella/fg(1)** [f: Attenellini McCafferty 2000: 158; g: *Ephemerella* subgen. *Attenella* Edmunds 1971: 152, typus *Ephemerella attenuata* McDunnough 1925 (design. orig.); syn. obj.: *Ephemerella* subgen. ***Attenuatella*** Edmunds 1959: 546 (non *Attenuatella* Stehli 1954)].

In circumscription matches:
— subgen. *Attenuatella* Edmunds 1959: 546;
— gen. *Attenella*: Allen 1980: 84;
— tribus Attenellini McCafferty 2000: 158.

References. Edmunds 1959: ♉ ⊕; – Allen & Edmunds 1961: ♉ ⊕; – Edmunds & Jensen & Berner 1976: ♉* ⊕*; – McCafferty & Wang 1994a (gen.): ♉* ⊕*.

Autapomorphy of *Attenella*.

(1) In imago distal segment of gonostylus [see Ephemerella/fg1 (8)] is strongly elongate, its length 5–6 times exceeds width.

Plesiomorphies of *Attenella*. In contrast to Timpanoga/fg2, larva has no tendency to development of gill opercula: tergalius IV is not larger or only slightly larger than next one; abdominal segments bearing tergalii are not shortened and not widened if compare with other segments.

Size. Fore wing length 6 mm.

Distribution. Nearctic.

Species composition of Attenella/fg(1). 4 species – *attenuata* McDunnough 1925 [*Ephemerella*] (syn. subj.: *hirsuta* Berner 1945 [*Ephemerella*]), *margarita* Needham 1927 [*Ephemerella*], *delantala* Mayo 1952 [*Ephemerella*], *soquele* Day 1954 [*Ephemerella*] (for last two species imagoes are unknown).

Material examined: *attenuata* [*E.*]: ♉; *margarita* [*E.*]: ♉; *soquele* [*E.*]: ♉.

***

2.2;2,2-1/1.2. Timpanoga/fg2

(Bidentiseta Furcatergaliae Ephemerella/fg1 .../fg2 Timpanoga/fg1 .../fg2)

(Fig. 91:G–I)

Nomen hierarchicum: **Timpanoga/fg2** (sine *Attenella*; incl. *Eurylophella*).

Reference. McCafferty & Wang 1994a ("*Eurylophella – Dentatella – Timpanoga – Danella*" lineage): ♉*.

Autapomorphy of Timpanoga/fg2.

(1) Tergalii are more or less operculate: tergalius IV [the anteriormost among lamellate tergalii – see Timpanoga/fg1 (1)] is enlarged, covers not less than 1/2 of tergalius V. Abdominal segments bearing tergalii have enlarged lateral lobes: at least on each of abdominal segments V–VII posterolateral process is longer than segment in its middle part (FIG.91:I).

Plesiomorphy of Timpanoga/fg2. In contrast to *Attenella*, distal segment of gonostylus is not elongate.

Size. Fore wing length 6–14 mm.

Distribution. Mainly Nearctic, also in Europe.

Timpanoga/fg2 is divided into Eurylophella/fg1 and Timpanoga/fg3.

2.2;2,2-1/1.2;1. **Eurylophella/fg1**

(Bidentiseta Furcatergaliae Ephemerella/fg1 .../fg2 Timpanoga/fg1 .../fg2 Eurylophella/fg1)

(Fig. 91:G–I)

Nomen hierarchicum: **Eurylophella/g1** (incl. *Dentatella*) [f: Eurylophellini McCafferty 2000: 158; g: *Eurylophella* Tiensuu 1935: 20, typus *E. karelica* Tiensuu 1935 (monotypy); syn.subj.: ***Melanameletus*** Tiensuu 1935: 15, typus *M. brunnescens* Tiensuu 1935 (monotypy) (synn. subj.: *karelica* [*E.*] = *brunnescens* [*M.*])].
In circumscription matches:
— subgen. *Eurylophella*: McCafferty 1978: 138;
— gen. *Eurylophella*: McCafferty & Wang 1994a: 574;
— tribus Eurylophellini McCafferty 2000: 158.

Reference. McCafferty & Wang 1994a (genus): Ծ*.

Autapomorphies of Eurylophella/fg1.

(1) Tergalius IV [gill operculum – see Timpanoga/fg2 (1)] is oval, with distal margin not truncate and not incised [see Ephemerella/fg2 (2)]; these gill opercula completely or nearly completely cover the rest of tergalii (FIG.91:I). Non-unique apomorphy.

Plesiomorphy of Eurylophella/fg1. In contrast to Timpanoga/fg3, larval claws with a row of denticles on inner margin [see Ephemerella/fg1 (18)].

Size. Fore wing length 6–11 mm.

Distribution. Mainly Nearctic, also in Europe.

Eurylophella/fg1 is divided into *Dentatella* and Eurylophella/fg2.

2.2;2,2-1/1.2;1,1. Plesiomorphon **Dentatella/g(1)**

(Ephemerella/fg1 .../fg2 Timpanoga/fg1 .../fg2 Eurylophella/g1 *Dentatella*)

Nomen hierarchicum: **Dentatella/g(1)** [g: *Danella* subgen. *Dentatella* Allen 1980: 88, typus *Ephemerella (Danella) bartoni* Allen 1977 (design. orig.)].
In circumscription matches:
— subgen. *Dentatella* Allen 1980: 88;
— gen. *Dentatella*: McCafferty 2000: 158.

References. Allen 1980: Ծ; – McCafferty 1978 (*Ephemerella bartoni*): Ծ; – McCafferty & Wang 1994a: Ծ.

Plesiomorphy of *Dentatella*. In contrast to Eurylophella/g2, abdominal segment IX is not elongate, of the same length as segment VIII.

Size. Larval body length 7–8 mm.

Distribution. Canada (Ontario).

Species composition of Dentatella/g(1). 2 species – *bartoni* Allen 1977 [*Ephemerella (Danella)*] and *danutae* McCafferty 2000 [*Dentatella*].

Material examined: –.

***

2.2;2,2-1/1.2;1,2. **Eurylophella/fg2**

(Ephemerella/fg1 .../fg2 Timpanoga/fg1 .../fg2 Eurylophella/g1 .../g2)

(Fig. 91:G–I)

Nomen hierarchicum: **Eurylophella/fg2** (sine *Dentatella*).
In circumscription matches:
— subgen. *Eurylophella*: Edmunds 1959: 546;
— gen. *Eurylophella* Tiensuu 1935: 20;
— gen. *Melanameletus* Tiensuu 1935: 15.

References. Edmunds 1959: Ծ* ⊕*; – Allen & Edmunds 1963c: Ծ* ⊕*; – Edmunds & Jensen & Berner 1976: Ծ* ⊕*; – McCafferty & Wang 1994a (subgenus): Ծ*; – Studemann & Landolt 1997a: ⊙*.

Autapomorphy of Eurylophella/fg2.

(1) Larval abdominal segment IX is elongate: medially it is approximately 1.4 times longer than segment VIII (FIG.91:I). Non-unique apomorphy, the same in many Ephemerella/fg3 and some other Ephemeroptera; in other Timpanoga/fg1 abdominal segment IX is subequal or only slightly longer than segment VIII.

Character of unclear phylogenetic status.

(2) Egg without polar cap (in contrast to majority of other Ephemerella/fg1 which have one polar cap), with polygonal relief, without anchors [see Ephemerella/fg1 (14)]. It is unclear if this is an autapomorphy of Eurylophella/g2 or Eurylophella/g1, as egg structure of *Dentatella* is unknown.

Size. Fore wing length 6–11 mm.

Distribution. Mainly Nearctic, also in Europe.

Species composition of Eurylophella/fg2 (here [*Eph.*]=[*Ephemerella*]). 17 species. Two **European species:** *iberica* Keffermüller & Da-Terra 1978 [*Eph. (Eurylophella)*], *karelica* Tiensuu 1935 [*Eurylophella*] (synn. subj.: *brunnescens* Tiensuu 1935 [*Melanameletus*], *lithuanica* Kazlauskas 1959 [*Eph.*]). **North American species:** *aestiva* McDunnough 1931 [*Eph.*], *bicolor* Clemens 1913 [*Eph.*], *bicoloroides* McDunnough 1938 [*Eph.*], *coxalis* McDunnough 1926 [*Eph.*], *doris* Traver 1934 [*Eph.*] (syn. subj.: *trilineata* Berner 1945 [*Eph.*]), *enoensis* Funk 1994 [*Eurylophella*], *funeralis* McDunnough 1925 [*Eph.*], *lodi* Mayo 1952 [*Eph.*], *lutulenta* Clemens 1913 [*Eph.*], *macdunnoughi* Funk 1994 [*Eurylophella*], *minimella* McDunnough 1931 [*Eph.*], *poconoensis* Funk 1994 [*Eurylophella*], *prudentalis* McDunnough 1931 [*Eph.*], *temporalis* McDunnough 1924 [*Eph.*] (synn. subj.: *lineata* Clemens 1913 [*Eph.*], *doris* Traver 1934 [*Eph.*]), *verisimilis* McDunnough 1930 [*Eph.*].

Material examined: *karelica* [*Eu.*]: Ծ, ♀, ⊕♂, ⊕♀; *prudentalis* [*Eph.*]: Ծ, ⊕; *temporalis* [*Eph.*]: Ծ, ♂, ⊕♂; *verisimilis* [*Eph.*]: Ծ.

*** ***

2.2;2,2-1/1.2;2. **Timpanoga/fg3**
(Bidentiseta Furcatergaliae Ephemerella/fg1 .../fg2 Timpanoga/fg1 .../fg3)

Nomen hierarchicum: **Timpanoga/fg3** (sine *Eurylophella*; incl. *Danella*).
In circumscription matches:
— gen. *Timpanoga*: McCafferty & Wang 1994a: 574;
— tribus Timpanogini: McCafferty 2000: 158.

Reference. McCafferty & Wang 1994a (genus): Ꙩ*.

Autapomorphy of Timpanoga/fg3.

(1) Larval claws without denticles [see Ephemerella/fg1 (18)]. Non-unique apomorphy.

Plesiomorphies of Timpanoga/fg3. In contrast to Eurylophella/g1, tergalius IV with distal margin obliquely truncate or incised [see Ephemerella/fg2 (2)], only partly covers tergalii behind it.

Size. Fore wing length 6–14 mm.

Distribution. Nearctic.

Timpanoga/fg3 is divided into *Danella* and Timpanoga/fg4.

2.2;2,2-1/1.2;2,1. Plesiomorphon **Danella/g(1)**
(Bidentiseta Furcatergaliae Ephemerella/fg1 .../fg2 Timpanoga/fg1 .../fg3 *Danella*)

Nomen hierarchicum: **Danella/g(1)** [g: *Ephemerella* subgen. *Danella* Edmunds 1959: 546, typus *Ephemerella simplex* McDunnough 1925 (design. orig.)].
In circumscription matches:
— subgen. *Danella* Edmunds 1959: 546;
— gen. *Danella*: Allen 1980: 85.

References. Edmunds 1959: Ꙩ ⊕; – Allen & Edmunds 1962b: Ꙩ ⊕; – Edmunds & Jensen & Berner 1976: Ꙩ* ⊕*; – McCafferty & Wang 1994a: Ꙩ*.

Plesiomorphy of *Danella*. In contrast to Timpanoga/fg4, larva has no head shield and no spine-like projections on femora.

Size. Fore wing length 6 mm.

Distribution. Eastern Nearctic.

Species composition of Danella/g(1). 3 species – *lita* Burks 1949 [*Ephemerella*], *provonshai* McCafferty 1977 [*Ephemerella (Danella)*], *simplex* McDunnough 1925 [*Ephemerella*].

Material examined: *lita* [*E.*]: Ꙩ; *simplex* [*E.*]: Ꙩ.

***

2.2;2,2-1/1.2;2,2. **Timpanoga/fg4**
(Bidentiseta Furcatergaliae Ephemerella/fg1 .../fg2 Timpanoga/fg1 .../fg4)

Nomen hierarchicum: **Timpanoga/fg4** (sine *Danella*).
In circumscription matches:
— subgen. *Timpanoga* Needham 1927: 115;
— gen. *Timpanoga*: Demoulin 1954e: 324.
Monospecific taxon.

References. Edmunds 1959: Ꙩ ⊕; – Allen & Edmunds 1959: Ꙩ ⊕; – Edmunds & Jensen & Berner 1976: Ꙩ ⊕; – McCafferty & Wang 1994a (subgenus): Ꙩ.

Autapomorphies of Timpanoga/fg4.

(1) Larval frons anteriorly forms a shield which is widened anteriorly and covers from above clypeus and mouthparts.

(2) In larva each femur apically with a spine-like projection.

Size. Fore wing length 12–14 mm.

Distribution. Western Nearctic.

Species composition of Timpanoga/fg4. 1 species – *hecuba* Eaton 1884 [*Ephemerella*].

Material examined: *hecuba* [*E.*]: Ꙩ.

*** *** *** ***

2.2;2,2-1/2. **Ephemerella/fg3**
(Anteritorna Bidentiseta Furcatergaliae Ephemerella/fg1 .../fg3)
(Figs 90; 91:A–F; 92–94)

Nomen hierarchicum: **Ephemerella/fg3** (sine *Timpanoga*; incl. *Torleya*).
In circumscription matches:
— Ephemerellinae: McCafferty & Wang 2000: 39;
— Ephemerella/fg3: Kluge [Internet publication 1998].

Reference: McCafferty & Wang 2000: Ꙩ* ⊕*.

Autapomorphy of Ephemerella/fg3.

(1) Tergalii I [initially stick-like – see Ephemerella/fg1 (13)] are lost (non-unique apomorphy – see Index of characters [1.3.33]); thus only tergalii III–VII are present [see Ephemerella/fg2 (1)] (FIG. 91:A).

Plesiomorphies of Ephemerella/fg3. In contrast to Timpanoga/fg1, tergalii III are retained, having structure typical for Ephemerella/fg1 [see Ephemerella/fg1 (7)]; usually tergalii III have approximately the same size as tergalii IV–VI (FIG.91:A), but sometimes tergalii III are enlarged, transformed to gill opercula and cover the others (TABLE 8).

Size. Fore wing length 5–19 mm.

Distribution. Holarctic and Oriental Region.

Ephemerella/fg3 is divided into Ephemerella/fg4 and Torleya/g1 (p.312). Some insufficiently described species have uncertain systematic position (p.315).

2.2;2,2-1/2.1. **Ephemerella/fg4**

(Anteritorna Bidentiseta Furcatergaliae Ephemerella/fg1 .../fg4)

(Figs 90:C–F; 91:C–F; 92)

Nomen hierarchicum: **Ephemerella/fg4** (sine *Torleya*; incl. *Drunella, Caudatella, Cincticostella*).

Reference: Kluge 1997 (subgenn. *Drunella, Ephacerella, Cincticostella, Ephemerella*): ⍥*.

Autapomorphy of Ephemerella/fg4.

(1) Tergalius VI has both branches of ventral lobe [see Ephemerella/fg1 (7)] fused together, so its ventral lobe is integral, not bifurcate and bears processes on both sides (FIG.91:C) (at the same time, this tergalius is not diminished if compare with tergalii III–V, in contrast to vestigial tergalius VII). The same in *Austremerella* only; in all other Ephemerella/fg1 tergalius VI either has bifurcate ventral lobe (FIG.91:B), or is vestigial or lost (in some Pantricorythi – see TABLE 8).

Plesiomorphies of Ephemerella/fg4. In contrast to *Teloganopsis* and Hyrtanella/f1=Crinitella/g1, hind wing always without prominent costal projection and with Sc terminating near apex (as in FIG.90:B). In contrast to *Amurella*, larval abdominal terga either with paired projections [see Ephemerella/fg1 (19)] or without projections, but never have unpaired projections.

Size. Fore wing length 6–19 mm.

Distribution. Holarctic and Oriental Region.

Ephemerella/fg4 is divided into Ephemerella/fg5, Drunella/g1, *Caudatella* and Cincticostella/g1.

2.2;2,2-1/2.1;1. Plesiomorphon **Ephemerella/fg5**

(Anteritorna Bidentiseta Furcatergaliae Ephemerella/fg1 .../fg5)

(Figs 90:C–D; 91:C; 92:B)

Nomen hierarchicum: **Ephemerella/fg5** (sine *Drunella, Caudatella, Cincticostella*).

In circumscription matches:

— subgen. *Ephemerella*: Edmunds 1959: 544;

— gen. *Ephemerella*: Allen 1980: 74.

References. Edmunds 1959 (subgenus): ⍥* ⊕*; – Allen & Edmunds 1965 (subgenus): ⍥* ⊕*; – Edmunds & Jensen & Berner 1976 (subgenus): ⍥* ⊕*; – Kluge 1997d (subgenus): ⍥*.

Plesiomorphies of Ephemerella/fg5. Maxilla is biting [see Ephemerella/fg1 (16)] (FIG.92:B), vestige of maxillary palp is always present, 3-segmented [see Ephemerella/fg1 (1)]. Larval thorax without projections characteristic for Cincticostella/g1. Larval fore legs are non-grasping (in contrast to Drunella/g1). Larval paracercus has the same length as cerci (in contrast to *Caudatella*); secondary swimming setae are present on lateral and medial sides of cerci and on lateral sides of paracercus [see Furcatergaliae (6)] (among Ephemerella/fg3 the same in some Drunella/fg1 and some Torleya/fg1).

Size. Fore wing length 6–13 mm.

Distribution. Holarctic and Oriental Region.

Nominal taxa included. Ephemerella/fg5 includes: (1) **Chitonophora/g** [g: *Chitonophora* Bengtsson 1908: 243, typus *Ch. aurivillii* Bengtsson 1908 (monotypy)], regarded as a generic or subgeneric synonym of *Ephemerella* (Allen 1984: 246); (2) **Caurinella/g** [g: *Caurinella* Allen 1984: 245, typus *C. idahoensis* Allen 1984 (design. orig.)].

Species composition of Ephemerella/fg5 (here [*E.*]= =[*Ephemerella*]). More than 30 described species can be presumably attributed here; for many of them tergalial structure, which allows to separate Ephemerella/fg4 from Torleya/g1, is unknown (see below, Ephemerella/fg3 INCERTAE SEDIS – p.315). Following species are attributed here for certain. One **Trans-Holarctic species:** *aurivillii* Bengtsson 1908 [*Chitonophora*] (synn. subj.: *aronii* Eaton 1908 [*Ephemerella*], *norda* McDunnough 1924 [*E.*], *concinnata* Traver 1934 [*E.*], *taeniata* Tshernova 1952 [*E.*], *maxima* Allen 1971 [*E.*], *ezoensis* Gose 1985 [*E. (Cincticostella)*]). **Eurasian species:** *atagosana* Imanishi 1937 [*E.*] (synn. subj.: *dentata* Bajkova 1967 [*E.*], *denticula* Allen 1971 [*E.*], *keijoensis* Allen 1971 [*E.*]), *kozhovi* Bajkova 1967 [*E.*] (syn. subj.: *notofascia* Yoon & Bae 1988 [*E.*]), *mucronata* Bengtsson 1909 [*Chitonophora*] (synn. subj.: *krieghoffi* Ulmer 1920 [*Chitonophora*], *unicolorata* Ikonomov 1961 [*Chitonophora*]), *notata* Eaton 1887 [*E.*]. **North American species:** *excrucians* Walsh 1862 [*E.*] (syn. subj.: *semiflava* McDunnough 1926 [*E.*]), *idahoensis* Allen 1984 [*Caurinella*], *inermis* Eaton 1887 [*E.*], *infrequens* McDunnough 1924 [*E.*].

Material examined: *aurivillii* [*Ch.*]: ⍥–O–⊕/⊙; *dentata* Bajkova 1967 [*E.*]: ⍥; *excrucians* [*E.*]: ⍥; *inermis* [*E.*]: ⍥; *infrequens* [*E.*]: ⍥; *kozhovi* [*E.*]: ⍥–O–⊕/⊙; *mucronata* [*Ch.*]: ⍥, ⊕♂, ⊕♀/⊙; *notata* [*E.*]: ⍥.

***

2.2;2,2-1/2.1;2. **Drunella/g1**

(Anteritorna Bidentiseta Furcatergaliae Ephemerella/fg1 .../fg4 *Drunella*)

(Figs 90:E–F; 92:C–D)

Nomen hierarchicum: **Drunella/g1** [g: *Drunella* Needham 1905: 42, typus *Ephemerella grandis* Eaton 1884 (monotypy)].

In circumscription matches:

— subgen. *Drunella*: McDunnough 1931b: 210;

— gen. *Drunella*: Allen 1980: 79.

References. Edmunds 1959: ♉* ⊕*; – Allen & Edmunds 1962a: ♉* ⊕*; – Edmunds & Jensen & Berner 1976: ♉* ⊕*.

Autapomorphies of *Drunella*.

(1) Larva is carnivorous, fore legs are specialized as grasping: fore femur is usually thickened, usually widest in its proximal half; transverse row of setae [see Ephemerella/fg1 (12)] is always absent; inner (directed anteriorly) margin of fore femur is usually serrate, that allows to keep a prey, clutching it between the femur and tibia; fore tibia on its apex on inner side has a projection directed distally, along inner side of tarsus (Kluge 1997d:Fig.18:4–8). In selected representatives of Drunella/fg1 some of these characters are not expressed or weakly expressed: fore femur can be nearly parallel-sided, and its denticles can be the same as on middle and hind femora or absent. Among the characters named here, the most constant is the projection on tibia. Besides Drunella/g1, similar specialization of fore femur (but not tibia) independently appeared in Tricorythus/ /fg1-*Vietnamella*.

(2) Larval clypeus is crossed transversely by a distinct stripe of hairs projected forward (Allen 1973:Fig.8). Unique apomorphy: some other Ephemerella/fg1 have similar transverse row of setae on frons, but not on clypeus.

Plesiomorphies of *Drunella*. Maxilla is biting [see Ephemerella/fg1 (16)] (as in FIG.92:B), vestige of maxillary palp is always present, 3-segmented [see Ephemerella/fg1 (1)]. Larval thorax without projections characteristic for Cincticostella/g1. Larval paracercus has the same length as cerci (in contrast to *Caudatella*).

Size. Fore wing length 6–19 mm.

Distribution. Amphipacific: Nearctic and Eastern Palaearctic; the most western representatives inhabit Tien-Shan–Himalayan Region.

Nominal taxa included. Drunella/g1 includes: (1) **Eatonella/g** [g: *Ephemerella* subgen. *Eatonella* Needham 1927: 108, typus *Ephemerella (Eatonella) doddsi* Needham 1927 (monotypy)]; (2) **Myllonella/g** [g: *Drunella* subgen. *Myllonella* Allen 1980: 80, typus *Ephemerella coloradensis* Dodds 1923 (design. orig.)]; (3) **Tribrochella/g** [g: *Drunella* subgen. *Tribrochella* Allen 1980: 80, typus *Ephemerella trispina* Ueno 1928 (design. orig.)]; (4) **Unirachella/g** [g: *Drunella* subgen. *Unirachella* Allen 1980: 80, typus *Ephemerella tuberculata* Morgan 1911 (design. orig.)]. These taxa were regarded as subgenera of the genus *Drunella*; they are partly natural ones, but have no clear diagnoses.

Species composition of Drunella/g1 (here [*E.*]= =[*Ephemerella*], [*D.*]=[*Drunella*]). About 30 species are described. **Asian species:** *aculea* Allen 1971 [*E.(D.)*], *basalis* Imanishi 1937 [*E.*], *cryptomeria* Imanishi 1937 [*E.*] (syn. subj.: *bicornis* Gose 1980 [*E.*]), *gilliesi* Allen & Edmunds 1963 [*E.(D.)*], *latipes* Tshernova 1952 [*E.*], *lepnevae* Tshernova 1949 [*E.*] (synn. subj.: *longipes* Tshernova 1952 [*E.*], *fusongensis* Su & Gui 1995 [*D.*]), *kohnoi* Allen 1971 [*E.(D.)*], *sachalinensis* Matsumura 1931 [*E.*] (syn. subj.: *bifurcata* Allen 1971 [*E.(D.)*]), *solida* Bajkova 1980 [*E.*], *submontana* Brodsky 1930 [*E.*] (synn. subj.: *traverae* Allen & Edmunds 1963 [*E.(D.)*], *nasiri* Ali 1971 [*E.*], *borakensis* Allen 1971 [*E.(D.)*], *kabulensis* Allen 1973 [*E.(D.)*]), *triacantha* Tshernova 1949 [*E.*] (synn.subj.: *tenax* Tshernova 1952 [*E.*], *ezoensis* Gose 1980 [*Ephemerella trispina*] (nom.praeocc.), *trispina* Ueno 1928 [*E.*], *yoshinoensis* Gose 1963 [*E.*]. **North American species:** *alleghensis* Traver 1934 [*E.*], *coloradensis* Dodds 1923 [*E.*] (syn. subj.: *wilsoni* Mayo 1952 [*E.*]), *cornuta* Morgan 1911 [*E.*] (syn.subj.: *depressa* Ide 1930 [*E.*]), *cornutella* McDunnough 1931 [*E.*], *doddsi* Needham 1927 [*E.*], *flavilinea* McDunnough 1926 [*E.*] (syn. subj.: *lapidula* McDunnough 1935 [*E.*]), *grandis* Eaton 1884 [*E.*] (synn. subj.: *flavitincta* McDunnough 1934 [*E.*], *glacialis* Traver 1930 [*E.*], *ingens* McDunnough 1934 [*E.*], *proserpina* Traver 1934 [*E.*], *yosemite* Traver 1934 [*E.*]), *lata* Morgan 1911 [*E.*] (syn. subj.: *inflata* McDunnough 1926 [*E.*]), *longicornis* Traver 1932 [*E.*], *pelosa* Mayo 1951 [*E.*], *spinifera* Needham 1927 [*E.*] (synn. subj.: *autumnalis* McDunnough 1934 [*E.*], *sierra* Mayo 1952 [*E.*]), *tuberculata* Morgan 1911 [*E.*] (synn. subj.: *cherokee* Traver 1937 [*E.*], *wayah* Traver 1932 [*E.*], *conestee* Traver 1932 [*E.*]), *walkeri* Eaton 1885 [*E.*] (syn. subj.: *bispina* Needham 1905 [*E.*]).

Material examined: *aculea* [*E.*]: ♉–○–⊕/⊙; *basalis* [*E.*]: ♉; *coloradensis* [*E.*]: ♉, ♀–⊕; *cornutella* [*E.*]: ♉; *cryptomeria* [*E.*]: ♉–○–⊕/⊙; *doddsi* [*E.*]: ♉, ⊕; *grandis* [*E.*]: ♉, ⊕; *lepnevae* [*E.*]: ♉–○–⊕/⊙; *solida* [*E.*]: ♉–○–⊕/⊙; *spinifera* [*E.*]: ♉; *submontana* [*E.*]: ♉–○–⊕/⊙; *triacantha* [*E.*]: ♉–○–⊕/⊙; *tuberculata* [*E.*]: ♉.

***

2.2;2,2-1/2.1;3. Caudatella/g(1)

(Anteritorna Bidentiseta Furcatergaliae Ephemerella/fg1 .../fg4 *Caudatella*)

Nomen hierarchicum: **Caudatella/g(1)** [g: *Ephemerella* subgen. *Caudatella* Edmunds 1959: 546, typus *Ephemerella heterocaudata* McDunnough 1929 (design. orig.)].

In circumscription matches:

— subgen. *Caudatella* Edmunds 1959: 546;

— gen. *Caudatella*: Allen 1980: 77.

References. Edmunds 1959: ♉* ⊕*; – Allen & Edmunds 1960: ♉* ⊕*; – Edmunds & Jensen & Berner 1976: ♉* ⊕*.

Autapomorphy of *Caudatella*.

(1) Larval, subimaginal and imaginal paracercus

is much (1.3–4 times) longer than cerci (in contrast to other Ephemerella/fg2, whose paracercus is subequal to cerci).

Plesiomorphies of *Caudatella*. Maxilla is biting [see Ephemerella/fg1 (16)] (as in FIG.92:B). Larval thorax without projections characteristic for Cincticostella/g1. Larval fore legs are non-grasping (in contrast to Drunella/g1).

Size. Fore wing length 6–12 mm.

Distribution. Western Nearctic.

Species composition of Caudatella/g(1) (here [*E.*]= =[*Ephemerella*]). 6 species – *cascadia* Allen & Edmunds 1961 [*E. (Caudatella)*], *edmundsi* Allen 1959 [*E.*], *heterocaudata* McDunnough 1929 [*E.*] (syn. subj.: *columbiella* McDunnough 1935 [*E.*]), *hystrix* Traver 1934 [*E.*], *jacobi* McDunnough 1939 [*E.*], *orestes* Allen & Edmunds 1961 [*E. (Caudatella)*].

Material examined: *cascadia* [*E.*]: ℧; *hystrix* [*E.*]: ℧.

***

2.2;2,2-1/2.1;4. Cincticostella/g1

(Anteritorna Bidentiseta Furcatergaliae Ephemerella/fg1 .../fg4 Cincticostella/g1)

(Fig. 92: A, E–H)

Nomen hierarchicum: **Cincticostella/g1** (incl. *Notacanthella*) [g: *Ephemerella* subgen. *Cincticostella* Allen 1971: 513, typus *Ephemerella nigra* Ueno 1928 (design. orig.); syn.obj.: ***Asiatella*** Tshernova 1972:611].

Autapomorphies of Cincticostella/g1.

(1) On maxilla apical field of setae [see Ephemerella/fg1 (16)] is enlarged, its setae are elongated; medially this field is limited by a more or less sharp impression which separates it from bases of maxillary canines (FIG.92:F, H). This impression can be rather shallow (in *commodema* [*Ephemerella*]) or very sharp and deep (in *quadrata* [*Ephemerella*] and Cincticostella/g2. Other Ephemerella/fg3 can have only a shallow impression at this place (FIG.92:B). In Cincticostella/g1-*Notacacanthella* maxilla retains biting structure (FIG.92:G), while in Cincticostella/g2 it becomes filtering, with weak canines and greatly enlarged apical setal field (see below).

(2) Larval thorax has following peculiarities: pronotum has anterior-lateral angles more or less expanded forward; mesonotum has paired lateral projections near anterior margin. Shape of mesonotal projections is variable: in *Notacanthella* and *Ephacerella* they are sharp and narrow (FIG.92:E), while in Cincticostella/g3 they have peculiar lamellate form (FIG.92:A). As *Ephacerella* and Cincticostella/g3 have a synapomorphy in maxillary structure [see Cincticostella/g2 (1) below], sharp narrow projections are most probably plesiomorphic within Cincticostella/g1.

Plesiomorphies of Cincticostella/g1. Larval fore legs are non-grasping (in contrast to Drunella/ /g1). Larval paracercus has the same length as cerci (in contrast to *Caudatella*).

Size. Fore wing length 6–16 mm.

Distribution. Eastern Asia.

Cincticostella/g1 is divided into *Notacanthella* and Cincticostella/g2.

2.2;2,2-1/2.1;4,1. Plesiomorphon Notacanthella/g(1)

(Bidentiseta Furcatergaliae Ephemerella/fg1 .../fg4 Cincticostella/g1 *Notacanthella*)

(Fig. 92:E–G)

Nomen hierarchicum: **Notacanthella/g(1)** [g: *Notacanthella* Kluge, SUBGEN. NOV., typus *Ephemerella quadrata* Kluge & Zhou & Jacobus & McCafferty (in press)].

References. Allen 1971 (*E. commodema, E. perculta*): ℧'; – Kang & Yang 1995 (*Acerella*): ℧*.

Plesiomorphy of *Notacanthella* (in contrast to Cincticostella/g2). Maxilla [see Cincticostella/g1 (1)] retains well-developed stout biting canines projecting medially (FIG.92:F–G).

Size. Length of mature larva 10–16 mm.

Distribution. Oriental Region.

Species composition of Notacanthella/g(1). 5 species – *commodema* Allen 1971 [*Ephemerella (Acerella)*], *glebosa* Kang & Yang 1995 [*Acerella*], *montana* Kang & Yang 1995 [*Acerella*], *perculta* Allen 1971 [*Ephemerella (Acerella)*] and *quadrata* Kluge & Zhou & Jacobus & McCafferty [*Ephemerella*] (in press).

Material examined: *commodema* [*E.*]: ℧; *quadrata* [*E.*]: ℧.

***

2.2;2,2-1/2.1;4,2. Cincticostella/g2

(Bidentiseta Furcatergaliae Ephemerella/fg1 .../fg4 Cincticostella/g1 .../g2)

(Fig. 92: A, H)

Nomen hierarchicum: **Cincticostella/g2** (sine *Notacanthella*; incl. *Ephacerella*).

Autapomorphies of Cincticostella/g2.

(1) The most ventral maxillary canine [basally separated by a deep impression – see Cincticostella/g1 (1)] is not projected medially, but widened and flattened in such a manner that projects apically;

its ventral margin is denticulate (sometimes translucent and poorly visible on background of apical setae); two other maxillary canines are diminished (FIG.92:H) or lost. Unique apomorphy. Generally, maxilla is specialized as filtering: apical margin is more or less widened, the apical setal field is enlarged, with setae long (Kang & Yang 1995:Figs 3C, 4E). Similar filtering specialization of maxillae independently appeared in Torleya/g1-*Uracanthella*, but without such widening of ventral canine and without impression at its base (FIG.93:B,D).

Size. Fore wing length 6–16 mm.

Distribution. Eastern Asia.

Cincticostella/g2 is divided into *Ephacerella* and Cincticostella/g3.

2.2;2,2-1/2.1;4,2-1.
Plesiomorphon **Ephacerella/g(1)**

(Furcatergaliae Ephemerella/fg1 .../fg4 Cincticostella/g1 .../g2 *Ephacerella*)

(Fig. 92:H)

Nomen hierarchicum: **Ephacerella/g(1)** [g: *Ephacerella* Paclt 1994: 283, typus *Ephemerella longicaudata* Ueno 1928 (design. orig.); syn. obj.: *Ephemerella* subgen. ***Acerella*** Allen 1971: 517 (non *Acerentulus* subgen. *Acerella* Berlese 1909)].

References. Ueno 1928 (*E. longicaudata*): Ꝺ'; – Allen 1971: Ꝺ' ⊕'; – Yoon & Bae 1988: Ꝺ' ⊕'; – Kang & Yang 1995 (*Cincticostella*): Ꝺ*.

Plesiomorphy of *Ephacerella* (in contrast to Cincticostella/g3). In larva lateral projections of mesothorax are narrow and sharply projected [see Cincticostella/g1 (2)] (Ueno 1928:Fig.8; Allen 1971: Figs 23, 27, 28; Kang & Yang 1995:Figs 3B, 4F).

Size. Fore wing length 13–16 mm.

Distribution. Eastern Asia.

Species composition of Ephacerella/g(1). 3 species – *colossa* Kang & Yang 1995 [*Cincticostella*], *fusca* Kang & Yang 1995 [*Cincticostella*] and *longicaudata* Ueno 1928 [*Ephemerella*].

Material examined: *longicaudata* [*E.*]: Ꝺ.

***

2.2;2,2-1/2.1;4,2-2. **Cincticostella/g3**

(Furcatergaliae Ephemerella/fg1 .../fg4 Cincticostella/g1 .../g3)

(Fig. 92:A)

Nomen hierarchicum: **Cincticostella/g3** (sine *Ephacerella*; incl. *Rhionella*).
In circumscription matches:
— subgen. *Cincticostella* Allen 1971: 513;
— gen. *Asiatella* Tshernova 1972: 611;
— gen. *Cincticostella*: Allen 1980: 82.

References. Allen 1971: Ꝺ* ⊕*; – Tshernova 1972 (*Asiatella*): Ꝺ*; – Allen 1975: Ꝺ*; – Yoon & Bae 1988 (genus): Ꝺ* ⊕*.

Autapomorphy of Cincticostella/g3.

(1) In larva lateral projection at anterior part of mesonotum [see Cincticostella/g1 (2)] is flat; lateral margin of this projection anteriorly arises from anterior-lateral angle of mesonotum, posteriorly it passes dorsad of lateral margin of mesonotum and terminates by a ledge; pronotum is wide, its posterior margin has the same width as anterior margin of mesonotum together with its lateral projections (FIG.92). Unique apomorphy.

Size. Fore wing length 6–14 mm.

Distribution. Eastern Asia.

Cincticostella/g3 is divided into Cincticostella/g4 and *Rhionella*.

2.2;2,2-1/2.1;4,2-2/1.
Plesiomorphon **Cincticostella/g4**

(Furcatergaliae Ephemerella/fg1 .../fg4 Cincticostella/g1 .../g4)

(Fig. 92:A)

Nomen hierarchicum: **Cincticostella/g4** (sine *Rhionella*).
In circumscription matches:
— subgen. *Cincticostella*: Allen 1980: 83.

References. Tshernova 1952: Ꝺ; – Allen 1971: Ꝺ ⊕; – 1975: Ꝺ; – Yoon & Bae 1988: Ꝺ ⊕.

Plesiomorphy of Cincticostella/g4. In contrast to *Rhionella*, larval femora are not so strongly widened, frons without tubercles.

Size. Fore wing length 8–14 mm.

Distribution: Eastern Asia.

Species composition of Cincticostella/g4 (here [*E.*]= =[*Ephemerella*], [*C.*]=[*Cincticostella*]). 5 species are described – *gosei* Allen 1975 [*E.(C.)*] (syn.obj.: *thailandensis* Allen 1980 [*Serratella*]), *levanidovae* Tshernova 1952 [*E.*] (synn. subj.: *orientalis* Tshernova 1952 [*E.*], *castanea* Allen 1971 [*E.(C.)*], *delicata* Allen 1971 [*E.(C.)*]), *nigra* Ueno 1928 [*E.*], *okumai* Gose 1980 [*E.(C.)*], *tshernovae* Bajkova 1962 [*E.*] (syn. subj.: *imanishii* Allen 1971 [*E.*]).

Material examined: *levanidovae* [*E*]: Ꝺ, Ꝺ♂–◎♂–⊕♂; *nigra* [*E.*]: Ꝺ; *tshernovae* [*E.*]: Ꝺ, Ꝺ♂–◎♂–⊕♂; sp.Cin1 (China): Ꝺ; sp.Cin2 (Thailand): Ꝺ.

***

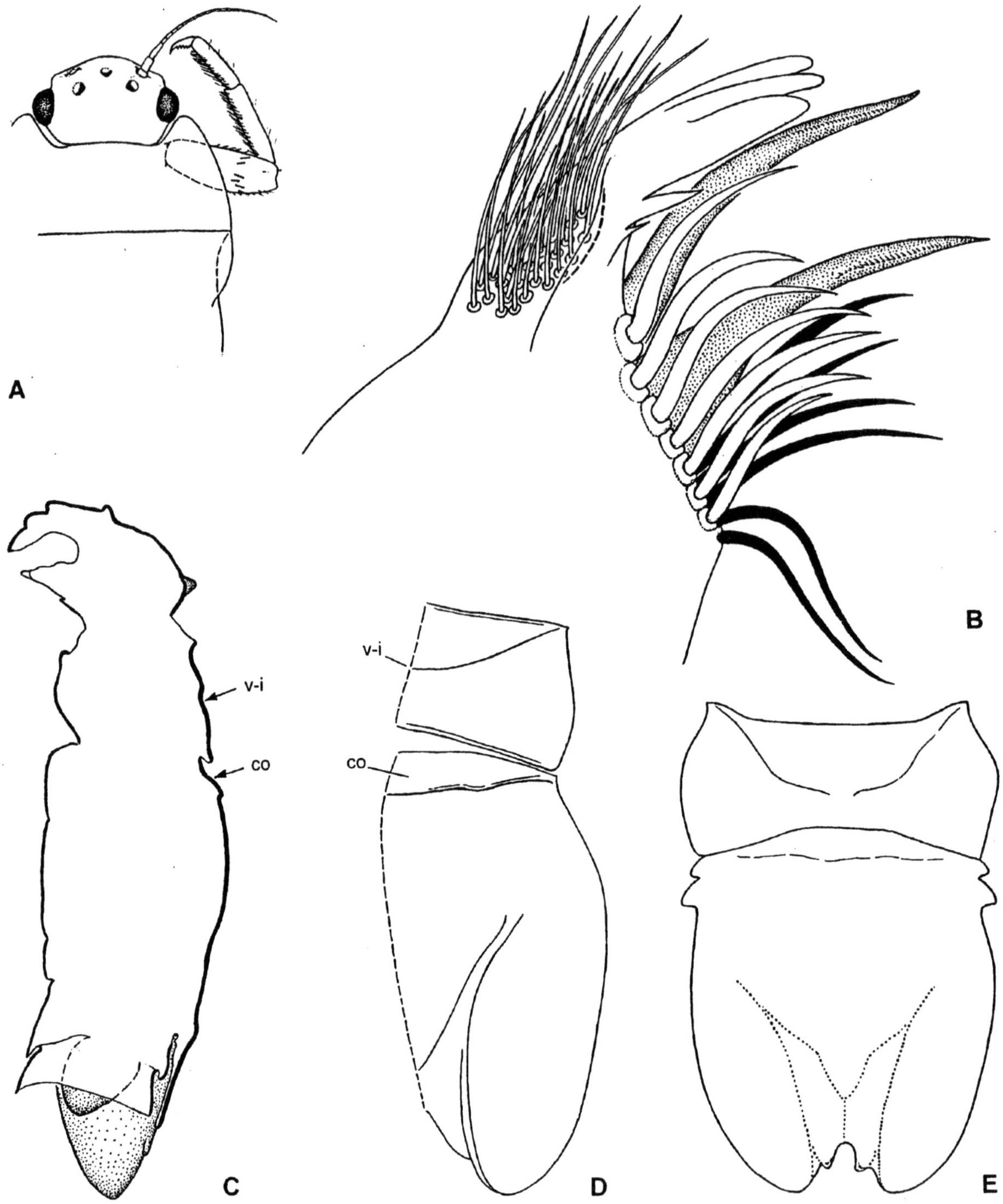

Figure 92. Ephemerella/fg4, larvae.
A – Cincticostella/g4 *levanidovae* [*Ephemerella*], anterior part of larva, dorsal view. **B** – Ephemerella/fg* *excrucians* [*Ephemerella*], apex of right maxilla, ventral view (dentiseta shown by dots, other setae of median-dorsal row shown by black). **C–D** – Drunella/g1 *submontana* [*Ephemerella*]: **C** – head and thorax, median section; **D** – larval exuviae of right half of pronotum and mesonotum. **E–G** – Notacanthella/g* *quadrata* [*Ephemerella*]: **E** – larval pronotum and mesonotum; **F** – apex of right maxilla, ventral view (dentiseta shown by dots, other setae of median-dorsal row shown by black); **G** – right maxilla, ventral view. **H** – Ephacerella/g* *longicaudata* [*Ephemerella*], apex of right maxilla, ventral view (dentiseta shown by dots, other setae of median-dorsal row shown by black, instead of some apical setae shown their bases and area occupied). Abbreviations: **co** – collar; **v-i** – V-shape impression.

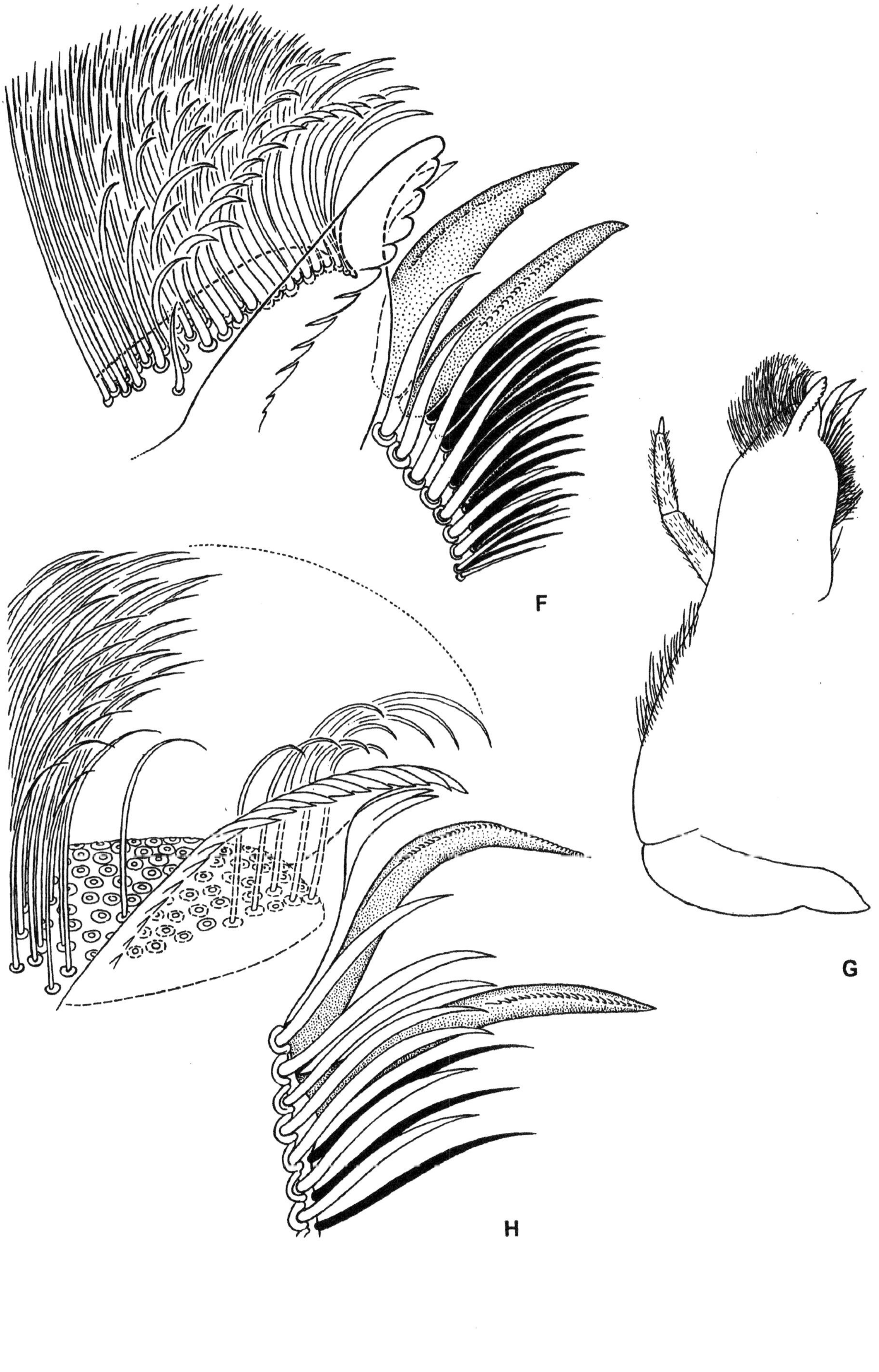
F
G
H

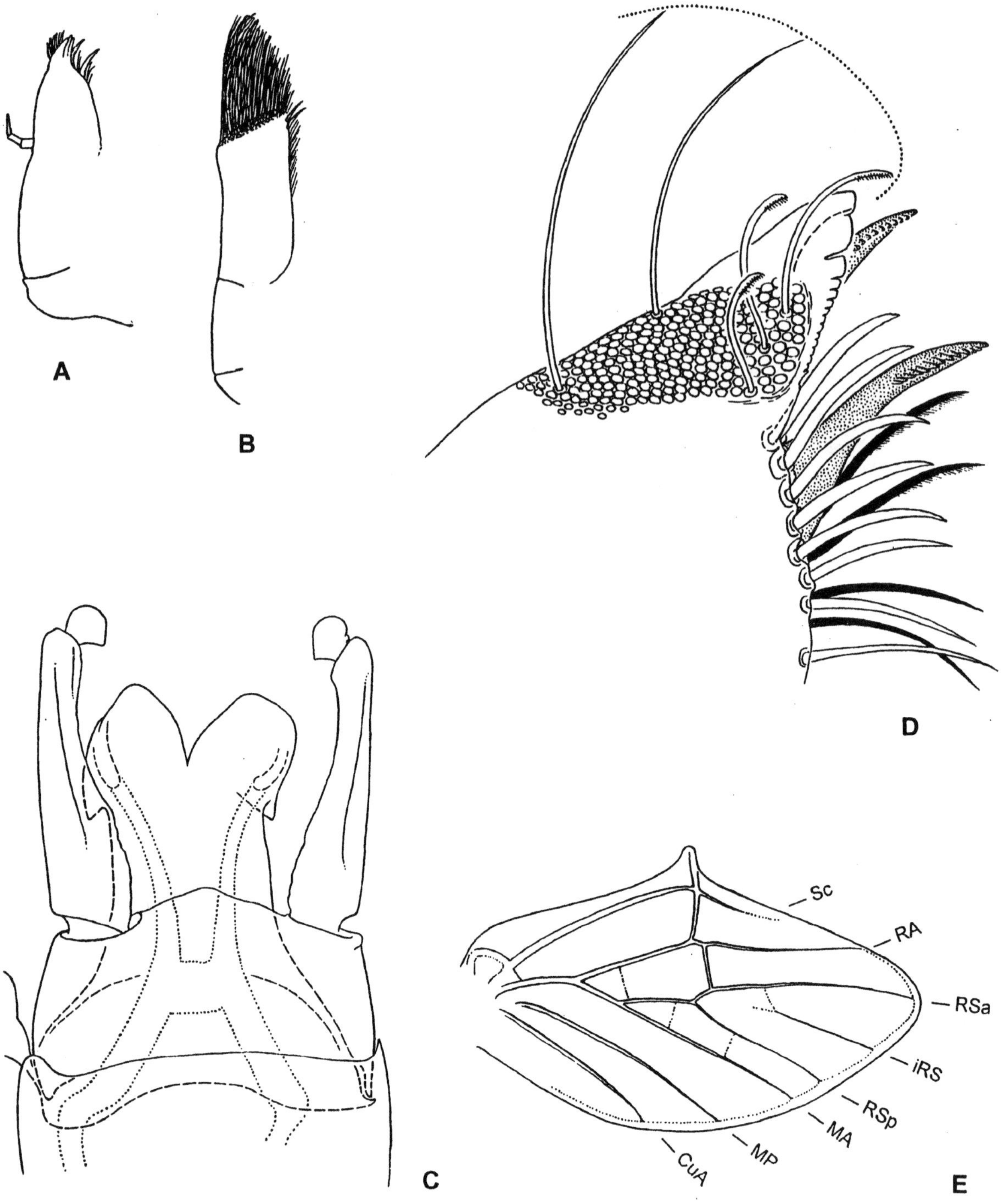

Figure 93. Ephemerella/fg3-Torleya/g1.
A – Torleya/g2 *ignita* [*Ephemera*], maxilla. **B–D** – Uracanthella/g1: **B–C** – *lenoki* [*Ephemerella*]: **B** – maxilla; **C** – male imaginal genitals, ventral view (gonoducts shown by dotted lines). **D** – *hispanica* [*Ephemerella*], apex of right maxilla, ventral view (dentiseta shown by dots, other setae of median-dorsal row shown by black, instead of apical setae shown their bases and area occupied). **E** – Teloganopsis/g1 sp., hind wing (drawn from subimago, but marginal setae not shown).

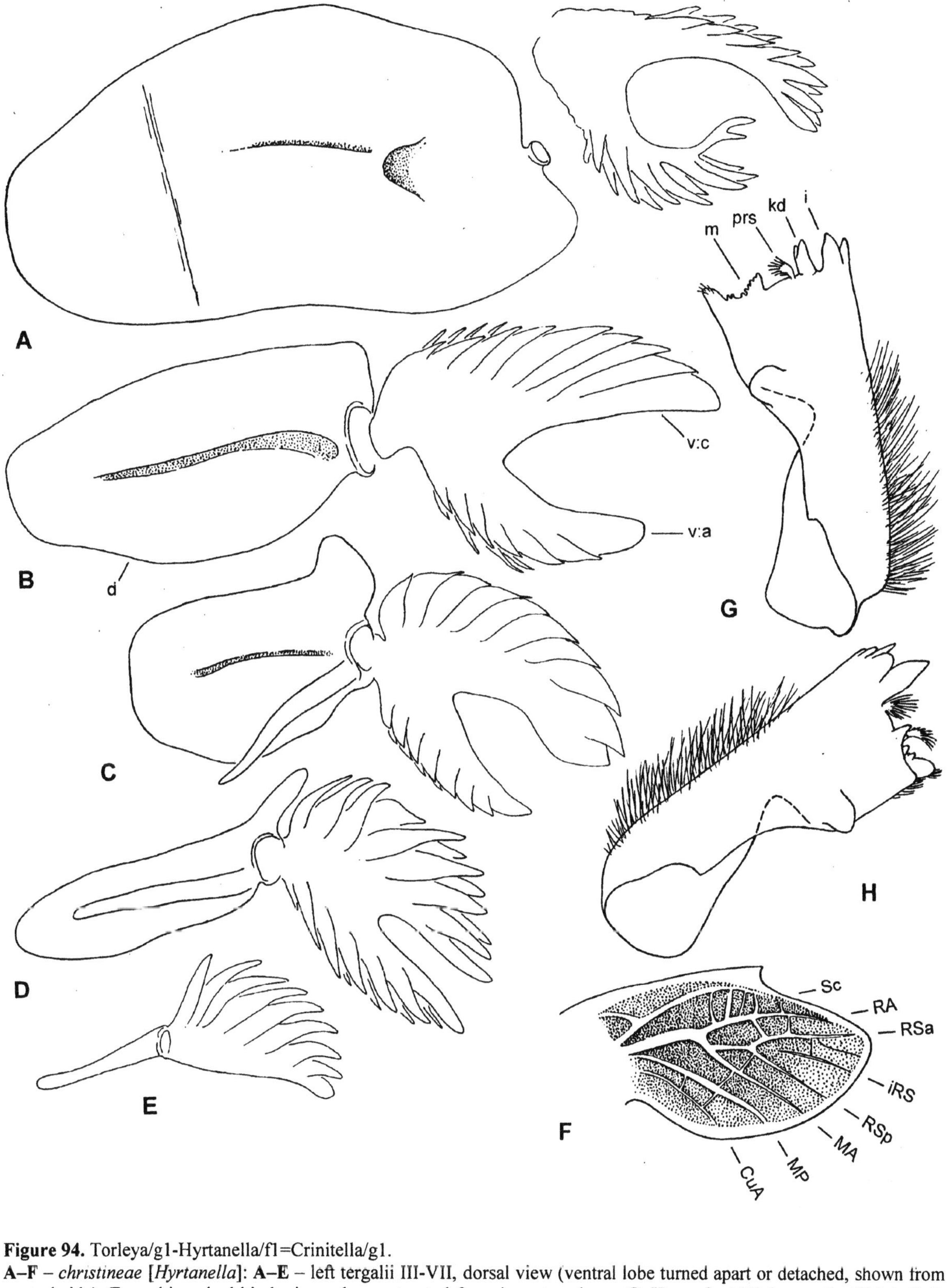

Figure 94. Torleya/g1-Hyrtanella/f1=Crinitella/g1.
A–F – *christineae* [*Hyrtanella*]: **A–E** – left tergalii III-VII, dorsal view (ventral lobe turned apart or detached, shown from ventral side); **F** – subimaginal hind wing anlage extracted from immature larva. **G–H** – *coheri* [*Ephemerella (Crinitella)*], left and right mandibles, dorsal view.

Abbreviations: **d** – dorsal lobe of tergalius; **v:a** – anal branch of ventral lobe; **v:c** – costal branch of ventral lobe.

2.2;2,2-1/2.1;4,2-2/2. **Rhionella/g(1)**

(Furcatergaliae Ephemerella/fg1 .../fg4 Cincticostella/g1 .../g3 *Rhionella*)

Nomen hierarchicum: **Rhionella/g(1)** [g: *Cincticostella* subgen. *Rhionella* Allen 1980: 83, typus *Ephemerella (Cincticostella) insolta* Allen 1971 (design. orig.)].
In circumscription matches:
— subgen. *Rhionella* Allen 1980: 83.

References. Gose 1969 (*Ephemerella* TEA): ♉'; – Allen 1971 (*Ephemerella insolta*): ♉'; – Tshernova 1972 (*Asiatella femorata*): ♉'; – Allen 1975 (*E. boja, E. femorata, E. insolta*): ♉'.

Autapomorphies of *Rhionella*.

(1) Larval middle and hind femora are very strongly widened, with serrate margins. Non-unique apomorphy.

(2) Larval head with a pair of preoccipital tubercles. Non-unique apomorphy.

Size. Larval body length 6–12 mm.

Distribution. Indo-China.

Species composition of Rhionella/g(1). 3 species – *boja* Allen 1975 [*Ephemerella (Cincticostella)*], *femorata* Tshernova 1972 [*Asiatella*] and *insolta* Allen 1971 [*Ephemerella (Cincticostella)*].

Material examined: *femorata* [*E.*]: ♉.

*** *** *** *** ***

2.2;2,2-1/2.2. **Torleya/g1**

(Anteritorna Bidentiseta Furcatergaliae Ephemerella/fg1 .../fg3 Torleya/g1)

(Figs 90:A–B; 91:A–B; 93; 94)

Nomen hierarchicum: **Torleya/g1** (incl. *Teloganopsis, Crinitella, Uracanthella, Amurella*) [g: *Torleya* Lestage 1917: 366, typus *T. belgica* Lestage 1917 (monotypy; syn. subj. *Ephemerella major* Klapálek 1905)].

Autapomorphies of Torleya/g1. Distinct autapomorphies are unknown. The features listed here, are characteristic for Torleya/g1, while can not serve as diagnostic to separate Torleya/fg1 from Ephemerella/fg4.

(1) Maxillary palp is usually smaller than in majority of Ephemerella/fg3, often lost [see Ephemerella/fg1 (1)].

(2) Larval caudalii often (but not always) with whorls of heavy spine-like setae on posterior margins of segments, and often (but not always) without swimming setae.

Plesiomorphy of Torleya/g1. In contrast to Ephemerella/fg4, ventral lobe of tergalius VI is bifurcate (FIG.91:B), as that of tergalii III–V [see Ephemerella/fg3 (1)].

Size. Fore wing length 4–10 mm.

Distribution. Holarctic and Oriental Region.

Torleya/g1 is divided into Torleya/g2, *Teloganopsis*, Hyrtanella/f1=Crinitella/g1, *Uracanthella* and *Amurella.*

2.2;2,2-1/2.2;1. Plesiomorphon **Torleya/g2**

(Bidentiseta Furcatergaliae Ephemerella/fg1 .../fg3 Torleya/g1 .../g2)

(Figs 90:A–B; 91:A–B; 93:A)

Nomen hierarchicum: **Torleya/g2** (sine *Teloganopsis, Crinitella, Uracanthella, Amurella*; incl. *Serratella*).

References. Allen & Edmunds 1963b (*Serratella*): ♉', ⊕'; – Kluge 1997: ♉*.

Plesiomorphies of Torleya/g2. In contrast to *Teloganopsis* and Hyrtanella/f1=Crinitella/g1, hind wing without prominent costal projection and with Sc terminating near apex (FIG.90:B) (as in all other Ephemerella/fg2). In contrast to *Uracanthella*, maxilla retains biting structure (FIG.93:A) and has well-developed canines (as in FIG.92:B). In contrast to *Amurella*, larval abdominal terga either with paired projections (FIG.91:A) [see Ephemerella/fg1 (19)], or without projections, but never have unpaired projections.

Variable characters of Torleya/g2. Tergalii III [anteriormost – see Ephemerella/fg3 (1)] can have normal size (FIG.91:A), or more or less strongly enlarged and form gill opercula covering the rest pairs of tergalii; the same variability in Hyrtanella/ /f1=Crinitella/g1.

Size. Fore wing length 5–10 mm.

Distribution. Holarctic and Oriental Region.

Nominal taxon included. Torleya/g2 includes **Serratella/g** [g: *Ephemerella* subgen. *Serratella* Edmunds 1959: 544, typus *Ephemerella serrata* Morgan 1911 (design. orig.)]. The type species of *Serratella* has well-developed paired abdominal projections, and tergalii non-operculate; the type species of *Torleya* has abdominal terga without projections, and tergalii operculate. It is used to separate genera or subgenera *Torleya* and *Serratella* basing on one or both of these characters, but such taxa appear to be artificial.

Species composition of Torleya/g2 (here [*E.*]= =[*Ephemerella*]). More than 30 described species can be presumably attributed here; for many of them tergalial structure, which allows to separate Torleya/g1 from Ephemerella/fg4, is unknown (see below, Ephemerella/fg3 INCERTAE SEDIS – p.315). Following species are attributed here for certain. One **Trans-Palaearctic species:** *ignita* Poda 1761 [*Ephemera*] (synn. subj.: *erytrophthalma* Schrank 1798 [*Ephemera*], *apicalis* Stephens 1935 [*Ephemera*], *diluta* Stephens 1935 [*Ephemera*], *obscura*

Stephens 1935 [*Baetis*], *rosea* Stephens 1935 [*Ephemera*], *rufescens* Stephens 1935 [*Ephemera*], *aeneus* Pictet 1843 [*Potamanthus*], *gibbus* Pictet 1843 [*Potamanthus*], *dilectus* Pictet 1843 [*Potamanthus*], *lactata* Bengtsson 1909 [*E.*], *torrentium* Bengtsson 1917 [*E.*], *sibirica* Tshernova 1952 [*E.*], *karasuensis* Kustareva 1976 [*Drunella*], *antuensis* Su & You 1989 [*E. (E.)*]). **Other Eurasian species:** *maculocaudata* Ikonomov 1961 [*E.*], *major* Klapálek 1905 [*E.*] (syn.subj.: *belgica* Lestage 1917 [*Torleya*]), *mikhaili* Tiunova 1995 [*Torleya*], *nazarita* Alba-Tercedor & Derka 2002 [*Torleya*], *nuda* Tshernova 1949 [*E.*] (synn. subj.: *thymalli* Tshernova 1952 [*E.*], *verrucosa* Kluge 1980 [*E.*]), *padunica* Kazlauskas 1963 [*E. (Torleya)*], *paradinasi* Gonzalez-del-Tanago & Garcia-de-Jalon 1981 [*Drunella*], *setigera* Bajkova 1965 [*E.*], *uenoi* Allen & Edmunds 1963 [*Ephemerella (Drunella)*] (syn. obj.: *undatella* Allen 1971 [*Ephemerella (Acerella)*], *zapekinae* Bajkova 1967 [*E.*]. **North American species:** *molita* McDunnough 1930 [*Ephemerella*] (syn. subj.: *serratoides* McDunnough 1931 [*E.*]), *serrata* Morgan 1911 [*E.*], *tibialis* McDunnough 1924 [*E.*] (synn. subj.: *angusta* Traver 1934 [*Ephemerella*], *angustata* Berner 1950 [*Ephemerella*]).

Material examined: *chantauense* [*Ephemerella*] (in litt.): Ծ, ⊕♂, Ծ♀–◎♀–⊕♀/⊙; *ignita* [*E.*]: Ծ–◎–⊕/⊙; *kogistana* [*Ephemerella*] (in litt.): Ծ–◎–⊕/⊙; *maculocaudata* [*E.*]: Ծ–◎–⊕/⊙; *major* [*E.*]: Ծ, ◎, ⊕; *mikhaili* [*T.*]: Ծ, ⊕; *molita* [*E.*]: Ծ, Ծ♂–◎♂; *nuda* [*E.*]: Ծ–◎–⊕/⊙; *padunica* [*E.*]: Ծ, Ծ♂–◎♂; *paradinasi* [*D.*]: Ծ♂–◎♂; *serrata* [*E.*]: Ծ; *setigera* [*E.*]: Ծ–◎–⊕/⊙; *tibialis* [*E.*]: Ծ; *zapekinae* [*E.*]: Ծ–◎–⊕/⊙; spp. (Japan, Thailand): Ծ.

***

2.2;2,2-1/2.2;2. Teloganopsis/g(1)

(Bidentiseta Furcatergaliae Ephemerella/fg1 .../fg3 Torleya/g1 *Teloganopsis*)

(Fig. 93:E)

Nomen hierarchicum: **Teloganopsis/g(1)** [g: *Teloganopsis* Ulmer 1939: 513, typus *T. media* Ulmer 1939 (design. orig.)].

In circumscription matches:
— subgen. *Teloganopsis*: Allen 1965: 265;
— gen. *Teloganopsis* Ulmer 1939: 513.

Reference. Ulmer 1939: Ծ ⊕.

Autapomorphy of *Teloganopsis*.

(1) Hind wing with a prominent costal projection at midlength of costal margin, and with a strong constant cross-vein connecting Sc and RA close to costal projection; no other crossveins between Sc and RA; Sc terminates at this cross-vein or close to it (FIG.93:E). Non-unique apomorphy (see Index of characters [2.2.63]). Hyrtanella/f1=Crinitella/g1 can have similar shape of hind wing, but without constant cross-vein.

Plesiomorphy of *Teloganopsis*. In contrast to *Uracanthella*, maxilla retains biting structure with well-developed canines.

Size. Fore wing length 4–7 mm.

Distribution. Oriental Region.

Species composition of Teloganopsis/g(1). 2 species are described – *jinghongensis* Xu & You & Hsu 1984 [*Ephemerella (Serratella)*] and *media* Ulmer 1939 [*Teloganopsis*].

Material examined: *media* [*T.*]: Ծ (from Zool. Mus. Hamburg, examined in Entomol. Inst.); sp. (Thailand): Ծ (deposited in Entomol. Inst.); sp. (Malaysia, Sabah): ◎♂ (FIG.93:E).

***

2.2;2,2-1/2.2;3. Hyrtanella/f1=Crinitella/g1

(Bidentiseta Furcatergaliae Ephemerella/fg1 .../fg3 Torleya/g1 Hyrtanella/f1=Crinitella/g1)

(Fig. 94)

Nomen hierarchicum: **Hyrtanella/f1=Crinitella/g1** [f: Hyrtanellini Allen 1980: 88; g: *Ephemerella* subgen. *Crinitella* Allen & Edmunds 1963a: 17, typus *Ephemerella (Crinitella) coheri* Allen & Edmunds 1963 (design. orig.)].

References. Allen & Edmunds 1963a (*Crinitella*): Ծ'; – Allen & Edmunds 1976b (*Hyrtanella*): Ծ' ⊕'.

Autapomorphies of Hyrtanella/f1=Crinitella/g1.

(1) Tergalius [consisting of dorsal integral lamella and ventral bifurcate pectinate lamella – see Ephemerella/fg1 (7) and Plesiomorphy of Torleya/g1] can bear an additional narrow lamella dorsad of the dorsal lobe; this additional lamella represents an outgrowth of anal margin of the dorsal lobe arising close to its base. In *christineae* [*Hyrtanella*] and sp.HC1 additional lamella is present on tergalii IV–VI or V–VI (FIG.94:A–E); in *coheri* [*Ephemerella (Crinitella)*] – on tergalius VI only.

In other respects tergalial structure is plesiomorphic or varies among species:

Tergalii III [anteriormost – see Ephemerella/fg3 (1)] can have normal size (in *coheri* [*E.*]) or are strongly enlarged (FIG.94:A) and form gill opercula, which reach crests of tergum VIII [see (2)] and cover the rest pairs of tergalii (in *christineae* [*H.*] and sp.HC1); the same variability in plesiomorphon Torleya/fg2.

Tergalius VI has ventral lobe bifurcate, as that of tergalii III–V (FIG.94:D) (in contrast to Ephemerella/fg4).

Tergalius VII [with ventral lobe non-bifurcate – see Ephemerella/fg1 (7)] retains both lamellate lobes (FIG.94:E) [originally the genus *Hyrtanella* was wrongly described as having "gills on segments 3–6" only (Allen & Edmunds 1976)].

(2) Abdominal terga IV–VIII form a pair of cavities, into which tergalii [lying dorsally – see Ephemerella/fg1 (7)] can be inserted: these cavities are deeper posteriorly and distinctly outlined from behind by a pair of semicircular transverse crests on tergum VIII; in *coheri* [*E.*] and sp.HC1 each such crest bears a row of stout setae, in *christineae* [*H.*] these crests are very prominent and lack setae. In some other Ephemerella/fg1 similar cavities occur, but without crests on segment VIII.

Character of unclear phylogenetic status.

(3) Hind wing can have a prominent costal projection at midlength of costal margin, and Sc terminating just distad of its base (FIG.94:F). Non-unique apomorphy (see Index of characters [2.2.62]); among Ephemerella/fg2 the same in *Teloganopsis* only; in contrast to *Teloganopsis*, subcostal field retains indeterminate number of non-specialized cross-veins. These features of hind wing structure are found in larval wing buds of selected specimens of *coheri* [*E.*] and *christineae* [*H.*]; possibly, they vary individually.

Variable characters of Hyrtanella/f1=Crinitella/g1. Larval head either has normal shape (in *christineae* [*H.*] and sp.HC1), or is flattened dorsoventrally, with mandibles and other mouthparts strongly shortened (in *coheri* [*E.*] – FIG.94:G–H) (in the original description of *E. coheri* left and right mandibles were drawn in different positions that makes an illusion of asymmetry). Abdominal terga either with paired projections [see Ephemerella/fg1 (19)] on segments IV–IX (in sp.HC1), or with paired projections on segments IV–VII and unpaired on modified segments VIII–IX (in *christineae* [*H.*]), or without projections (in *coheri* [*C.*]).

Size. Larval body length 5–7 mm.

Distribution. Oriental Region.

Nominal taxon included. Hyrtanella/f1=Crinitella/g1 includes **Hyrtanella/fg** [g: *Hyrtanella* Allen & Edmunds 1976b: 133, typus *H. christineae* Allen & Edmunds 1976].

Species composition of Hyrtanella/f1=Crinitella/g1. 2 species are described – *christineae* Allen & Edmunds 1976 [*Hyrtanella*] and *coheri* Allen & Edmunds 1963 [*Ephemerella (Crinitella)*] (syn. subj. *permkami* Wang & Sites 1999 [*Crinitella*]).

Material examined: *coheri* [*E.*]: ♉/♂; *christineae* [*H.*]: ♉/♂; sp.HC1 (Nepal): ♉.

***

2.2;2,2-1/2.2;4. Uracanthella/g1

(Bidentiseta Furcatergaliae Ephemerella/fg1 .../fg3 Torleya/g1 *Uracanthella*)

(Fig. 93: B–D)

Nomen hierarchicum: **Uracanthella/g1** [g: *Uracanthella* Belov 1979: 575, typus *Ephemerella lenoki* Tshernova 1952 (design. orig.)].

In circumscription matches:

— subgen. *Uracanthella*: Tshernova & Kluge & Sinitshenkova & Belov 1986: 138;

— gen. *Uracanthella* Belov 1979: 575.

References. Tshernova 1952 (*Ephemerella lenoki*): ♉; – Belov 1979: ♉; – Kang & Yang 1995 (*Eburella*): ♉.

Autapomorphies of *Uracanthella.*

(1) Maxilla is specialized as filtering: field of apical setae [see Ephemerella/fg1 (16)] is enlarged and expanded up to tip; maxillary canines are either reduced to a row of short denticles bordering median margin of this field (FIG.93:D), or lost at all (in *lenoki* [*E.*] and *brocha* [*E.*]). Maxillary palp is always lost [see Ephemerella (1)]. Such filtering maxillary specialization is similar to that of Cincticostella/g2 (FIG.92:H), but ventral canine is not flattened and widened.

In other respects structure of mouthparts is rather variable. The type species of *Uracanthella* and *Eburella* are closely related and have highly specialized mouth apparatus: maxillae are flattened, parallel-sided, with wide apical margin lacking canines (FIG.93:B); mandibles are elongated with incisor, kinetodontium and prostheca located on a stretched pedestal; labium is enlarged, and labial palp is slender (Tshernova 1952:Figs 83–85; Belov 1979:Figs a–f; Kang & Yang 1995:Fig.5). In other species maxillae vary from narrowed apically, as in majority of Ephemerella/fg1 (FIG.93:D; Studemann & Tomka 1987:Fig.14b) to widened apically (in sp.U1); mandibles can be non-specialized; labium can be either of a usual form, or strongly widened (in sp.U1).

Character of unclear phylogenetic status.

(2) Abdominal terga always without projections [see Ephemerella/fg1 (19)]. The same in some species of Torleya/g2; possibly synapomorphy.

Plesiomorphy of *Uracanthella.* In contrast to *Teloganopsis* and Hyrtanella/f1=Crinitella/g1, hind wing without prominent costal projection and with Sc terminating near apex (as in all other Ephemerella/fg2 – FIG.90:B).

Size. Fore wing length 5–10 mm.

Distribution. Holarctic and Oriental Region.

Nominal taxon included. Uracanthella/g1 includes **Eburella/g** [g: *Eburella* Kang & Yang 1995: 104, typus *E. brocha* Kang & Yang 1995] [see (1)].

Species composition of Uracanthella/g1. 6 described species can be attributed here for certain – *albai* Gonzalez-del-Tanago & Garcia-de-Jalon 1984 [*Serratella*], *albostriata* Tong & Dudgeon 2000 [*Serratella*], *brocha* Kang & Yang 1995 [*Eburella*], *deficiens* Morgan 1911 [*Ephemerella*] (syn. subj.: *atrescens* McDunnough 1925 [*Ephemerella*]), *hispanica* Eaton 1888 [*Ephemerella*] and *lenoki* Tshernova 1952 [*Ephemerella*] (syn. subj.: *markevitshi* Belov 1979 [*Uracanthella*]).

Probably here also belong *changbaishanensis* Su & You 1988 [*Ephemerella*], *punctisetae* Matsumura 1931 [*Drunella*] (syn. subj.: *rufa* Imanishi 1937 [*Ephemerella*]), *yixingensis* Wu & Gui 1993 [*Ephemerella*] and some other species described as imagoes.

Material examined: *albai* [*S.*]: ℧, ◎♂–⊕♂, ⊕♀/⊙; *deficiens* [*E.*]: ℧♂–◎♂; *hispanica* [*E.*]: ℧; *lenoki* [*E.*]: ℧–◎–⊕/⊙; sp.U1 (China): ℧; sp.U2 (Japan): ℧; sp.U3 (Russian Far East): ℧, ⊕♂.

***

2.2;2,2-1/2.2;5. Amurella/g(1)

(Bidentiseta Furcatergaliae Ephemerella/fg1 .../fg3 Torleya/g1 *Amurella*)

Nomen hierarchicum: **Amurella/g** [g: *Ephemerella* subgen. *Amurella* Kluge 1997a: 235, typus *Ephemerella gracilis* Tshernova 1952 (design. orig.)].
In circumscription matches:
— subgen. *Amurella* Kluge 1997a: 235.

References. Tshernova 1952 (*Ephemerella gracilis*): ℧; – Kluge 1997a: ℧.

Autapomorphy of *Amurella*.

(1) Larval abdominal terga have unpaired median projections on posterior margin (Tshernova 1952: Fig.82). In other Ephemerella/fg2 abdominal terga either have paired submedian projections, or lack projections [see Ephemerella/fg1 (19)]; unpaired projections are present also in some Tricorythus/fg1 and some other Ephemeroptera (see Index of characters [1.3.3]).

Plesiomorphies of *Amurella*. In contrast to *Teloganopsis* and Hyrtanella/f1=Crinitella/g1, hind wing without prominent costal projection and with Sc terminating near apex (as in all other Ephemerella/fg2 – FIG.90:B). In contrast to *Uracanthella*, maxilla retains biting structure with well-developed canines.

Size. Fore wing length 5–10 mm.

Distribution. East Palaearctic and Nearctic.

Species composition of Amurella/g(1). 2 species – *gracilis* Tshernova 1952 [*Ephemerella*] (Russian Far East) and *septentrionalis* McDunnough 1925 [*Ephemerella*] (Nearctic).

Material examined: *gracilis* [*Ephemerella*]: ℧, ℧/◎.

***

Torleya/g1 INCERTAE SEDIS

Torleya/g1 *wahensis* Ali 1971 [*Ephemerella*] is described from Pakistan as larvae only; its hind wing structure is unknown, thus it can belong either to Torleya/g2, or to Teloganopsis/g(1).

*** ***

Ephemerella/fg3 INCERTAE SEDIS

Here [*E.*]=[*Ephemerella*]. For the following species tergalial structure [see Ephemerella/fg4 (1)] is not described, so their systematic position is not quite clear; most of them belong to plesiomorphons Ephemerella/fg5 and Torleya/ /g2. **Eurasian species:** *arenosa* Tong & Dudgeon 2000 [*Torleya*], *bauernfeindi* Marie & Dia & Thomas 1999 [*Serratella*], *chinoi* Gose 1980 [*E. (Serratella)*], *cornutus* Gose 1980 [*E.(E.)*] (nom. praeocc.), *corpulenta* Braasch 1981 [*E. (Drunella)*], *elongatula* McLachlan 1875 [*Leptophlebia*], *euphratica* Kazanci 1987 [*Drunella*], *fusongensis* Su & You 1988 [*E.*], *glareosa* Kang & Yang 1995 [*Torleya*], *hainanensis* She & Gui 1995 [*Serratella*], *ikonomovi* Puthz 1971 [*E.*], (syn. obj.: *spinosa* Ikonomov 1961 [*E.*] nom. praeocc., syn. subj.: *andaluciaca* Kazanci 1990 [*Drunella*]), *imanishii* Gose 1980 [*E.(E.)*] (nom. praeocc.), *indica* Kapur & Kripalani 1961 [*E.*], *ishiwatai* Gose 985 [*E.(E.)*], *ishiyamana* Matsumura 1932 [*Drunella*], *japonica* Gose 1980 [*E. (Torleya)*], *karia* Kazanci 1990 [*Drunella*], *longipennis* Zhou & Gui & Su 1997 [*Serratella*], *lutosa* Kang & Yang 1995 [*Torleya*], *mesoleucus* Brauer 1857 [*Potamanthus*], *nepalica* Allen & Edmunds 1963 [*E. (Torleya)*], *nigromaculata* Xu & You & Su & Hsu 1980 [*E. (Serratella)*], *serrata* Braasch 1981 [*E. (Drunella)*] (nom. praeocc.), *soldani* Allen 1986 [*Drunella (Drunella)*], *subsolana* Allen 1973 [*E. (Serratella)*], *svenhedini* Ulmer 1936 [*E.*], *swatensis* Ali 1971 [*E.*], *tianmushanensis* Xu & You & Su & Hsu 1980 [*E. (Serratella)*], *tumiforceps* Zhou & Su 1997 [*Serratella*], *xiaosimaensis* You 1987 [*Ephemerellina*]. **North American species:** *alleni* Jensen & Edmunds 1966 [*E.*], *altana* Allen 1968 [*E.*], *apopsis* McCafferty 1992 [*E.*], *argo* Burks 1949 [*E.*], *berneri* Allen & Edmunds 1958 [*E.*], *carolina* Berner & Allen 1961 [*E.*], *catawba* Traver 1932 [*E.*], *choctawhatchee* Berner 1945 [*E.*], *consimilis* Walsh 1862 [*E.*], *crenula* Allen & Edmunds 1965 [*E.*], *dorothea* Needham 1908 [*E.*], *floripara* McCafferty 1985 [*E.*], *fratercula* McDunnough 1925 [*E.*], *frisoni* McDunnough 1927 [*E.*], *hispida* Allen & Edmunds 1965 [*E.*], *inconstans* Traver 1932 [*E.*], *invaria* Walker 1853 [*Baetis*], *lacustris* Allen & Edmunds 1965 [*E.*], *levis* Day 1954 [*E.*], *maculata*

Traver 1934 [*E.*] (syn. subj.: *euterpe* Traver 1934 [*E.*]), *micheneri* Traver 1934 [*E.*], *moffatae* Allen 1977 [*E.*], *mollitia* Seemann 1927 [*E.*], *needhami* McDunnough 1925 [*E.*], *ora* Burks 1949 [*E.*], *rama* Allen 1968 [*E.*], *rossi* Allen & Edmunds 1965 [*E.*], *rotunda* Morgan 1911 [*E.*] (synn. subj.: *vernalis* Banks 1914 [*E.*], *feminina* Needham 1924 [*E.*]), *sequoia* Allen & Cohen 1968 [*E.*], *simila* Allen & Edmunds 1965 [*E.*], *sordida* McDunnough 1925 [*E.*], *spiculosa* Berner & Allen 1961 [*E.*], *spinosa* Morgan 1911 [*E.*], *subvaria* McDunnough 1931 [*E.*], *teresa* Traver 1934 [*E.*] (syn. subj.: *cognata* Traver 1934 [*E.*]), *unicornis* Needham 1905 [*E.*], *velmae* Allen & Edmunds 1963 [*E.*].

*** *** ***

2.2;2,2-2. **Pantricorythi**, or **Tricorythus/fg1**

(Anteritorna Bidentiseta Furcatergaliae Ephemerella/fg1 Pantricorythi)

(Figs 95–105)

Nomen hierarchicum: **Tricorythus/fg1** (incl. *Vietnamella*, *Austremerella*, *Melanemerella*, *Teloganodes*) [f: Tricorythidae Lestage 1942:15; g. *Tricorythus* Eaton 1868b: 82, typus *Caenis varicauda* Pictet 1843 (design. orig.)].
Nomen circumscribens: **Pantricorythi** Kluge, TAXON NOV.
In circumscription matches:
— Tricorythus/fg1: Kluge 2000: 252.

Autapomorphies of Pantricorythi.

(1) Larval fore femur bears a regular row of setae, which crosses anterior (dorsal) side of femur near its middle and continues distally as a longitudinal row on outer margin of femur; proximad of this row on outer margin of femur such setae are absent; on middle and hind legs, in difference to the fore leg, row of setae borders entire outer margin and usually crosses dorsal side of femur close to its base (FIGS 89:I–J; 100:E–F). In some species femoral setal rows are irregular or lost, so this character can not be used as diagnostic for every species, but it occurs at least in some representatives of each major group of Pantricorythi (*Vietnamella*, *Austremerella*, many Melanemerella/fg1, *Teloganodes*, Tricoryptera).

Unique apomorphy; in Ephemerella/fg2, Caenotergaliae and Potamanthus/fg1 transverse row of setae also can be developed on dorsal side of fore femur [see Ephemerella/fg1 (12)], but in this case setae on outer margin of fore femur are arranged in the same manner as on middle and hind femora (FIG. 89:H). In Caenotergaliae-*Madecocercus* (which was initially wrongly attributed to the family Tricorythidae), femora bear wart-like tubercles, which arrangement somewhat resembles that of setae in Pantricorythi; these tubercles were taken for "spinelike setae" (McCafferty & Wang 1995: p.3 and Fig.1).

(2) Imaginal gonostyli, being quite variable, have some features in common. 1st segment is usually (but not always) elongate: its length usually exceeds width, sometimes being subequal to length of 2nd segment (while in Ephemerella/fg2 its length usually is less than width). 1st and 2nd segments (immobile connected in majority of mayflies) can be movably articulated; 2nd segment either retains initial caudal direction, or is inclined laterally (in *Machadorythus*, Leptohyphes/fg2, *Allenhyphes*, *Tricorythopsis*). 3rd segment [the last – see Ephemerella/fg1 (8)] is short, its length usually is subequal to width (the same in some Ephemerella/fg2); sometimes 3rd segment is lost (see Index of characters [2.3.12]).

Plesiomorphies of Pantricorythi. In contrast to Ephemerella/fg2, bilobed tergalii [see Ephemerella/ /fg1 (7)] are present beginning from abdominal segment II (tergalii V, VI and VII can be present or lost – see TABLE 8 and Index of characters [1.3.21]). Costal and anal ribs [located at a distance from tergalial margins – see Furcatergaliae (5)] can be more or less retained (FIGS 95:F–G; 96:E–H; 101:A; 103:C) (but sometimes they are indistinct or lost, especially on posteriormost tergalii).

Size. Fore wing length 2–16 mm.

Distribution. Ethiopian, Oriental, Australian, Neotropical and Nearctic Regions.

Pantricorythi are divided into *Vietnamella*, *Austremerella*, Melanemerella/fg1 (p.319), *Teloganodes* (p.320) and Tricoryptera (p.326).

2.2;2,2-2/1. **Vietnamella/fg(1)**

(Bidentiseta Furcatergaliae Ephemerella/fg1 Pantricorythi *Vietnamella*)

(Fig. 95:A–E)

Nomen hierarchicum: **Vietnamella/fg(1)** [f: Vietnamellae Allen 1984: 247; g: *Vietnamella* Tshernova 1972: 609, typus *V. thani* Tshernova 1972 (design. orig.)].
In circumscription matches:
— gen. *Vietnamella*: Wang & McCafferty 1995a: 193;
— subtribus Vietnamellae Allen 1984: 247;
— tribus Vietnamellini: Edmunds & Murvosh 1995: 159.

References. Tshernova 1972 (*Vietnamella thani*): ☉; – ibid. (*Ephemerellina ornata*): ♂; – You & Su 1987: ☉ ⊕; – Wang & McCafferty 1995a; – McCafferty & Wang 1997: ☉ ⊕.

Autapomorphies of *Vietnamella*.

(1) Larval head bears pointed projections (spines) directed anteriorly; these are: a pair of very large spines between eyes and antennal bases, a pair of small spines ventrad of antennal bases and a pair of small spines on lateral ocelli (Tshernova 1972: Fig.4).

(2) Asymmetry of mandibles is lost (but in contrast to other mayflies with symmetrical mandibles, molae are developed). Each mandible has enlarged incisor and two prostheca-like structures, among which the distal one is probably a vestige of kinetodontium (Tshernova 1972:Fig.4).

(3) In larva fore femur is widened in its proximal part, with denticulate inner margin (Tshernova 1972: Fig.4); its shape resembles that of many *Drunella*, and probably is also used for grasping prey.

(4) Hind wing has characteristic roundish shape with strongly arched fore margin (FIG.95:B). In other respects it retains plesiomorphic features: Sc reaches wing apex, RS forms a triad (in contrast to *Teloganodes* and Tricoryptera). Behind MA [simple – see Furcatergaliae (4)], venation is variable: in *dabieshanensis* [*V.*] it consists of simple longitudinal veins and intercalaries; in *thani* [*V.*] and *ornata* [*E.*] MP is simple, but CuA and CuP form secondary triads not found in other mayflies.

Character of unclear phylogenetic status.

(5) On maxilla all three canines are completely fused forming a single elongate canine (FIG.95:C). Non-unique apomorphy; the same in *Teloganodes* (FIG.96:C); possibly synapomorphy. Maxilla is biting, with field of apical setae small [see Ephemerella/fg1 (16)] and both dentisetae retained [see Bidentiseta (1)].

(6) Fore wing has marginal intercalaries connected with longitudinal veins (FIG.95:A). Possibly this is a plesiomorphy, as the same in majority of Ephemeroptera apart of Ephemerella/fg1; but such condition never occurs in other Ephemerella/fg1, which have either free, or non marginal intercalaries (see Index of characters [2.2.55]).

Plesiomorphies of *Vietnamella*. All tergalii I–VII are present (unique plesiomorphy among Ephemerella/fg1 – see TABLE 8): tergalius I is stick-like [see Ephemerella/fg1 (13)] (FIG. 95:D–E); tergalii II–VI with bifurcate ventral lobes bearing numerous marginal processes; tergalius VII with non-bifurcate ventral lobe [see Ephemerella/fg1 (7)]; in contrast to *Teloganodes* and some others, tergalii II are not operculate. Larval abdominal terga with paired submedian projections [see Ephemerella/fg1 (19)] (Tshernova 1972:Fig.3a).

In imago and subimago: Eyes of male are large (in contrast to some Tricoryptera and others). Imaginal and subimaginal mesonotum has structure primitive for Ephemerella/fg1 [see Ephemerella/fg1 (9) and (10)]: mesonotal suture is distinct, lateroparapsidal suture is curved laterally; subimaginal lateral pigmented area of mesonotum is bordered by relief line (as in FIG.90:D); scutellum is not modified, infrascutellum is present [as in Ephemerella/fg2 (FIG.89:A) and Melanemerella/fg1, in contrast to *Teloganodes* and Tricoryptera]. Cubital field of fore wing has up to 3 bifurcate veins arising from CuA (FIG.95:A: x, y, z) [other Ephemerella/fg1 have no more than two such veins – see Ephemerella/fg1 (15)].

Size. Fore wing length 15–16 mm.

Distribution. Oriental Region.

Species composition of Vietnamella/fg(1). 6 species are formally described – *dabieshanensis* You & Su 1987 [*Vietnamella*], *guadunensis* Zhou & Su 1995 [*Vietnamella*], *ornata* Tshernova 1972 [*Ephemerellina*], *qingyuanensis* Zhou & Su 1995 [*Vietnamella*], *sinensis* Hsu 1936 [*Ephemerella*], *thani* Tshernova 1972 [*Vietnamella*].

Material examined: *thani* [*V.*]: Ꝺ; *ornata* [*E.*]: ♂.

***

2.2;2,2-2/2. Austremerella/fg(1)

(Bidentiseta Furcatergaliae Ephemerella/fg1 Pantricorythi *Austremerella*)

(Fig. 95:F–H)

Nomen hierarchicum: **Austremerella/fg(1)** [f: Austremerellinae McCafferty & Wang 1997: 392; g: *Austremerella* Riek 1963: 50, typus *A. picta* Riek 1963 (design. orig.)].

In circumscription matches:

— subgen. *Austremerella*: Allen 1965: 264;

— gen. *Austremerella* Riek 1963: 50;

— subfam. Austremerellinae: McCafferty & Wang 1997:392.

Monospecific taxon.

References. Riek 1963: Ꝺ ♀; – Allen 1965: Ꝺ ♀; – Riek 1970: Ꝺ ♀; – Peters & Campbell 1991: Ꝺ; – McCafferty & Wang 1997: Ꝺ ⊕.

Characters of unclear phylogenetic status.

(1) Tergalii I [initially stick-like – see Ephemerella/fg1 (13)] are lost. Non-unique apomorphy (see Index of characters [1.3.33]).

(2) Tergalius VI has ventral lobe non-bifurcate, thus only tergalii II–V have bifurcate ventral lobes [see Ephemerella/fg1 (7)] (FIG.95:F–H). The same structure of tergalius VI occurs in Ephemerella/fg4 only. Possibly, synapomorphy with *Teloganodes* and/or Tricoryptera, where ventral lobe of tergalius VI is vestigial or lost (see TABLE 8 and Index of characters [1.3.57]).

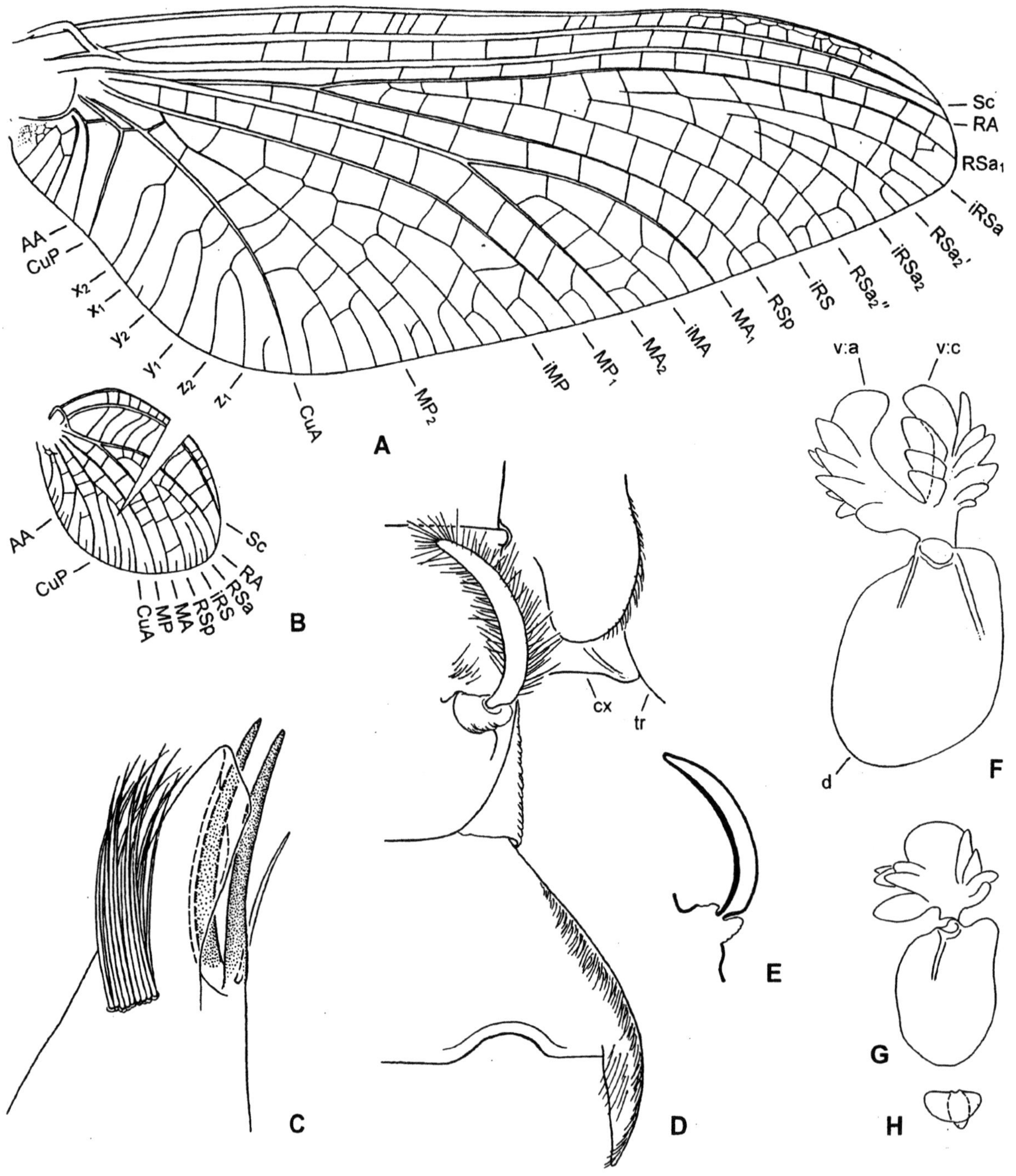

Figure 95. Pantricorythi: *Vietnamella* and *Austremerella*.

A–E – Vietnamella/fg1: **A–B** – *ornata* [*Ephemerellina*], holotype, fore and hind wings (drawn from subimago, but marginal setae not shown); **C–E** – *thani* [*Vietnamella*]: **C** – apex of right maxilla, ventral view (dentisetae shown by dots); **D–E** – holotype, right tergalius I: **D** – right margin of abdominal segment I (fused with metathorax) and segment II, dorsal view (tergalius I in natural position, tergalius II removed); **E** – optic section of tergalius I. **F–G** – Austremerella/fg* *picta* [*Austremerella*], right tergalii II, VI and VII, dorsal view (on tergalii II and VI ventral lobe turned apart and shown from ventral side).

Abbreviations: **d** – dorsal lobe of tergalius; **cx**, **tr** – posterior margin of hind coxa and trochanter; **v:a** – anal branch of ventral lobe; **v:c** – costal branch of ventral lobe.

(3) Tergalii VII [small, with non-bifurcate ventral lobes – see Ephemerella/fg1 (7), FIG.95:H] are strongly shifted anteriorly, being located at mid-length of tergite (in contrast to *Vietnamella* and Ephemerella/fg2, where tergalii VII retain position close to posterior margin of tergite). In Tricoryptera and *Teloganodes* these tergalii are lost, thus their initial position is unknown.

(4) Fore wing with free marginal intercalaries. (as in many other Ephemerella/fg1 – see Index of characters [2.2.55]).

Plesiomorphies of *Austremerella*. Larval head, mandibles and fore legs without modifications peculiar for *Vietnamella*. Larval abdominal terga with paired submedian projections [see Ephemerella/fg1 (19)] (Riek 1963:Fig.1). In contrast to Melanemerella/fg1, *Teloganodes* and some others, tergalii II are not operculate. Cubital field of fore wing has up to 2 bifurcate veins arising from CuA [see Ephemerella/fg1 (15)]; hind wing is well-developed, as long as 0.2 of fore wing length, with Sc reaching wing apex, RS branching, and veins behind MP_1 present (Riek 1963:Fig.10) (in contrast to *Teloganodes* and Tricoryptera).

Size. Fore wing length 12 mm.

Distribution. Australia (this is the only representative of Ephemerella/fg1 in Australian Region).

Species composition of Austremerella/g(1). 1 species – *picta* Riek 1963 [*Austremerella*].

Material examined (Entomol. Inst.): *picta* [*A.*]: ℧.

***

2.2;2,2-2/3. Melanemerella/fg1

(Bidentiseta Furcatergaliae Ephemerella/fg1 Pantricorythi Melanemerella/fg1)

Nomen hierarchicum: **Melanemerella/fg1** (incl. *Ephemerellina, Lithogloea, Lestagella, Nadinetella*) [f: Melanemerellinae Demoulin 1955d: 216; g: *Melanemerella* Ulmer 1920a: 43, typus *M. brasiliana* Ulmer 1920 (design. orig.)].

References. Lestage 1924a (gen. *Ephemerellina*): ⊕'; – Allen & Edmunds 1963a (*Ephemerellina crassi*): ℧' ⊕'; – Barnard 1932 (genn. *Ephemerellina, Lithogloea*): ℧' ⊕'; – 1940 (*E. barnardi*): ℧'; – Demoulin 1970e (genn. *Ephemerellina, Lestagella*): ℧'; – Wang & McCafferty 1996 (*Melanemerella*): ♀'; – McCafferty & Wang 1997 (genn. *Ephemerellina, Nadinetella, Lithogloea, Lestagella*): ℧' ⊕'; – Molineri & Domínguez 2003 (*Melanemerella*): ℧'.

Characters of unclear phylogenetic status.

(1) Tergalii [see Ephemerella/fg1 (7)] are specialized: tergalii II are larger than others and serve as gill opercula, completely or partly covering other tergalii; tergalii VII are always lost; tergalii VI if present, are diminished and lack ventral lobe. Non-unique apomorphy, the same in *Teloganodes* (possibly, synapomorphy) and some Tricoryptera – *Ephemerythus, Machadorythus* and Leptohyphes/fg1 (see Index of characters [1.3.39], [1.3.59]).

In various species number of tergalii is different (TABLE 8): Tergalii I [stick-like – see Ephemerella/fg1 (13)] are either present (in *crassi* [*E.*], *penicillata* [*L.*], *harrisoni* [*L.*], *animosa* [*M.*], *sphyxia* [*M.*]), or lost (in *brasiliana* [*M.*], *barnardi* [*E.*] and *brincki* [*E.*]); lamellate tergalii can be developed either up to pair VI (in *brasiliana* [*M.*], *harrisoni* [*L.*] and *barnardi* [*E.*]), or up to pair V (in *brincki* [*E.*] and *crassi* [*E.*]), or up to pair IV only (in *penicillata* [*L.*]).

(2) Fore wing with free marginal intercalaries (as in many other Ephemerella/fg1 – see Index of characters [2.2.55]).

Plesiomorphies of Melanemerella/fg1. In imago and subimago: Eyes of male are large (in contrast to some Tricoryptera and others). Imaginal and subimaginal mesonotum (examined for *harrisoni* [*L.*] only) has structure primitive for Ephemerella/fg1 [see Ephemerella/fg1 (9) and (10)]: mesonotal suture is distinct, lateroparapsidal suture is curved laterally; subimaginal lateral pigmented area of mesonotum is bordered by relief line (as in FIG.90:D); scutellum is not modified, infrascutellum is present (as in Ephemerella/fg2 and *Vietnamella*, in contrast to *Teloganodes* and Tricoryptera). Cubital field of fore wing with vestiges of 2 bifurcate veins [see Ephemerella/ /fg1 (15)] variably developed, sometimes reduced to 1–4 intercalaries. Hind wing is well-developed, as long as 0.2–0.25 of fore wing length, with Sc reaching wing apex, RS branching, and veins behind MP_1 developed (in contrast to *Teloganodes* and Tricoryptera).

Variable characters of Melanemerella/fg1. Larval head either has normal shape, or (in *penicillata* [*L.*]) is flattened with frons forming a wide flat shield which is margined by a row of long setae and overlaps clypeus, labrum and mandibles; in this case mandibles are strongly shortened (Demoulin 1970e: Figs 15a–d) (similar to *Dicercomyzon*, some *Teloganodes* and some others – see Index of characters [1.1.4]).

Each larval claws can bear 1 row of denticles [see Ephemerella/fg1 (18)], or 2 equal rows (in *crassi* [*E.*] and *brincki* [*E.*], united in *Nadinetella*) (the same in some other taxa – see Index of characters [1.2.21]).

Projections of posterior margin of larval abdominal terga [see Ephemerella/fg1 (19)] can either be paired (in *brasiliana* [*M.*]), or vary individually from paired to unpaired (in *crassi* [*E.*] and *brincki* [*E.*]), or are constantly unpaired (in *barnardi* [*E.*], *harrisoni* [*L.*], *animosa* [*M.*], *sphyxia* [*M.*]), or absent (in *penicillata* [*L.*]) (Demoulin 1970e:Figs 12–15; McCafferty & Wang 1997:Figs 4–7, 74–76; Molineri & Domínguez 2003:Figs 1–2).

Size. Fore wing length 5–12 mm.

Distribution. Ethiopian and Neotropical Regions.

Nominal taxa included. Melanemerella/fg1 includes:

(1) **Ephemerellina/g** [g: *Ephemerellina* Lestage 1924a: 346, typus *E. barnardi* Lestage 1924 (design. orig.)];

(2) **Lithogloea/g** [g: *Lithogloea* Barnard 1932: 252, typus *Lithogloea harrisoni* Barnard 1932 (monotypy)];

(3) **Lestagella/g** [g: *Lestagella* Demoulin 1970e: 130, typus *Lithogloea penicillata* Barnard 1940 (design. orig.)];

(4) **Nadinetella/g** [g: *Nadinetella* McCafferty & Wang 1998: 344, typus *Ephemerellina crassi* Allen & Edmunds 1963 (design. orig.); syn. obj.: ***Nadinella*** McCafferty & Wang 1997: 399 (non *Nadinella* Penard 1899)].

(5) Possibly here belong **Manohyphella/g**, whose characters are unknown for certain (see below, Ephemerella/fg1 INCERTAE SEDIS).

Species composition of Melanemerella/fg1. One **South American species** – *brasiliana* Ulmer 1920 [*Melanemerella*]. Five **African species** are described – *barnardi* Lestage 1924 [*Ephemerellina*], *brincki* Demoulin 1970 [*Ephemerellina (Ephemerellina)*], *crassi* Allen & Edmunds 1963 [*Ephemerellina*], *harrisoni* Barnard 1932 [*Lithogloea*], *penicillata* Barnard 1940 [*Lithogloea*]. Presumably here belong also two **Madagascar species** described as larvae only – *animosa* McCafferty & Benstead 2002 [*Manohyphella*] and *sphyxia* McCafferty & Benstead 2002 [*Manohyphella*].

Material examined: *crassi* [*E.*]: ℧ (Entomol. Inst.); *harrisoni* [*L.*]: ♀ exuviae (Purdue Univ.), ℧.; *penicillata* [*L.*]: ℧; sp.n.: ℧ (Albany Mus.).

Status of Melanemerella/fg1. Diagnosis of this taxon does not contain any unique autapomorphies, thus its paraphyly is not excluded. Some authors regard this group to be a plesiomorphon ancestral for Teloganodes/fg1 or Tricoryptera (McCafferty & Wang 1997; Molineri & Domínguez 2003). However, an argued conclusion about status of Melanemerella/fg1 and, possibly, its splitting could be made only after re-examination of all stages of development of all its representatives.

***

2.2;2,2-2/4. Teloganodes/fg1

(Bidentiseta Furcatergaliae Ephemerella/fg1 Pantricorythi *Teloganodes*)

(Fig. 96)

Nomen hierarchicum: **Teloganodes/fg1** (incl. *Macafertiella*) [f: Teloganodinae Allen 1965: 263; g: *Teloganodes* Eaton 1882: 208, typus *Cloe tristis* Hagen 1858 (design. orig.)].

In circumscription matches:

— gen. *Teloganodes* Eaton 1882: 208.

References. Eaton 1883–1888: ⊕̧*; – Ulmer 1924c: ⊕'; – 1939: ℧' ⊕' – Wang & McCafferty 1996a (*Macafertiella*): ℧'.

Autapomorphies of *Teloganodes*.

(1) Tergalii beginning from pair III, have dorsal lobe cleft into two branches – costal and anal ones; costal rib is located on the costal branch, and anal rib – on the anal branch [ventral lobe is bifurcate with processes, like that of pair II – see Ephemerella/fg1 (7)] (FIG.96:E–H).

Besides this unique character, tergalii have non-unique operculate specialization: tergalii II are larger than others and serve as gill opercula, completely or partly covering other tergalii; tergalii VII are lost; tergalii VI if present, are diminished and lack ventral lobe. The same in Melanemerella/fg1 and some Tricoryptera – *Ephemerythus*, *Machadorythus* and Leptohyphes/fg1 (see Index of characters [1.3.39], [1.3.59]).

Various species of Teloganodes/fg1 have different number of tergalii (TABLE 8): in *insignis* [*Macafertiella*] tergalii III–V have dorsal lobe cleft and ventral lobe bifurcate with processes, tergalius VI has simple dorsal lobe and no ventral lobe (FIG.96:E); in some species tergalii III–IV have dorsal lobe cleft and ventral lobe bifurcate with processes, tergalius V has dorsal lobe either cleft (in sp.T1 and sp.T4 – FIG.96:F), or simple (in sp.T2 – FIG.96:G), and no ventral lamella, tergalius VI being lost; in sp.T3 only tergalius III has dorsal lobe cleft and ventral lobe bifurcate with processes, tergalius IV has integral dorsal lobe and no ventral lobe, tergalii V–VI being lost (FIG.96:H).

(2) Larval, subimaginal and imaginal paracercus is vestigial. Non-unique apomorphy (see Index of characters [1.3.64]); among Ephemerella/fg1 the same in *Dicercomyzon* only.

Character of *Teloganodes* of unclear phylogenetic status: possible synapomorphy with *Vietnamella*.

(3) Maxilla is somewhat modified (FIG.96:C): canines (which are initially three) are fused in a single elongate canine with a denticle on inner-ventral margin; at least proximal dentiseta [among two dentisetae – see Bidentiseta (1)] is pectinate; biting edge of maxilla is shortened, bears only 2 setae of inner-dorsal row proximad of dentisetae, and only 3 setae of inner-ventral row (as larva uses its maxillae for biting, canines and dentisetae can be ground off, thus larva shortly after moult should be examined). Non-unique apomorphies; maxillary canines of *Vietnamella* are similarly fused and elongate (FIG.95:C).

Characters of *Teloganodes* of unclear phylogenetic status: possible synapomorphies with Tricoryptera.

(4) On mesonotum [see Ephemerella/fg1 (10)] lateroparapsidal suture is not curved laterally, but convergent with medioparapsidal suture; relief line of submedioscutum is lost, and subimaginal lateral pigmented area occupies the whole sublateroscutum, whole submedioscutum and a lateral part of medioscutum, leaving a non-pigmented median area which narrows posteriorly (FIG.96:A). Microtrichia cover most part of scutum (both pigmented and non-pigmented areas), being absent on lateroscutum and posterior scutal protuberances. This is different from Ephemerella/fg2, *Vietnamella* and Melanemerella/ /fg1, where pigmented area is smaller [see Ephemerella/fg1 (10) and FIG.90:D] and from Tricoryptera, where pigmented area is larger or non-expressed (FIGS 98:D and 104:I); possibly, synapomorphy with Tricoryptera.

(5) Infrascutellum is interrupted medially, scutellum with enlarged lateral impressions (FIG.96:A). Among Pantricorythi the same in Tricoryptera (FIG.89:B) (possibly synapomorphy); the same in Caenoptera (see Classifications of Furcatergaliae II).

(6) Hind wing is diminished (as long as 0.13–0.14 of fore wing length) and modified (FIG.96:D): costal projection is well-expressed, situated at the middle of costal margin; Sc far not reaches wing apex, terminates at costal projection; RS [initially forming a single triad – see Euplectoptera (1)] is non-branched or lost; MP is non-branched; there are no veins behind MP. Non-unique character, the same in some Ephemerella/fg2 (*Teloganopsis*, Hyrtanella/f1=Crinitella/g1) and some Leptophlebia/fg1. In Tricoryptera hind wing is under greater reduction, and in other Pantricorythi Sc goes up to wing apex. Possibly, synapomorphy with Tricoryptera.

(7) Larval abdominal terga always lack paired projections [see Ephemerella/fg1 (19)]: they can either bear unpaired median projection on hind margin, or have no projections (non-unique character – see Index of characters [1.3.3]).

(8) Tergalii I [initially stick-like – see Ephemerella/fg1 (13)] are lost. Non-unique apomorphy (see Index of characters [1.3.33]).

Other character of unclear phylogenetic status.

(9) Fore wing with free marginal intercalaries (as in many other Ephemerella/fg1 – see Index of characters [2.2.55]).

Plesiomorphies of *Teloganodes*. Eyes of male are large, divided into two portions. Cubital field of fore wing with vestiges of 2 bifurcate veins [see Ephemerella/fg1 (15)] variably developed, sometimes reduced to 1–4 intercalaries.

Variable characters of *Teloganodes*. Larval head either has normal shape, or is strongly flattened, with frons margined by a row of long setae and overlapping clypeus, labrum and mandibles (in sp.T3); in this case mandibles are strongly shortened (McCafferty & Wang 1997:Figs 10, 28, 37) (similar to some Melanemerella/fg1 and others – see Index of characters [1.1.4]).

Size. Fore wing length 4–12 mm.

Distribution: Oriental Region.

Nominal taxon included. Teloganodes/fg1 includes **Macafertiella/g(1)** [g: *Macafertiella* Wang (in Wang & McCafferty 1996a): 15, typus *M. insignis* Wang & McCafferty 1996 (design. orig.)]; monospecific taxon characterized by the most complete number of tergalii among Teloganodes/fg1 [see (1)].

Species composition of Teloganodes/fg1. 5 species are formally described – *dentata* Navás 1932 [*Teloganodes*], *insignis* Wang & McCafferty 1996 [*Macafertiella*], *lugens* Navás 1933 [*Teloganodes*], *major* Eaton 1885 [*Teloganodes*], *tristis* Hagen 1858 [*Cloe*].

Material examined: *insignis* [*M.*]: ⊘, ♂/♂; sp.T1 (Malaysia; FIG.96:A–B): ♂/♂, ♂, ♀/♀/⊙; sp.T2 (Sri Lanka; FIG.96:G): ♂/♂, ♀/♀/⊙; sp.T3 (East Malaysia; FIG.96:H; McCafferty & Wang 1997:Fig.10): ♂/♂, ♀/♀/⊙; sp.T4 (Thailand; FIG.96:F): ⊘.

***

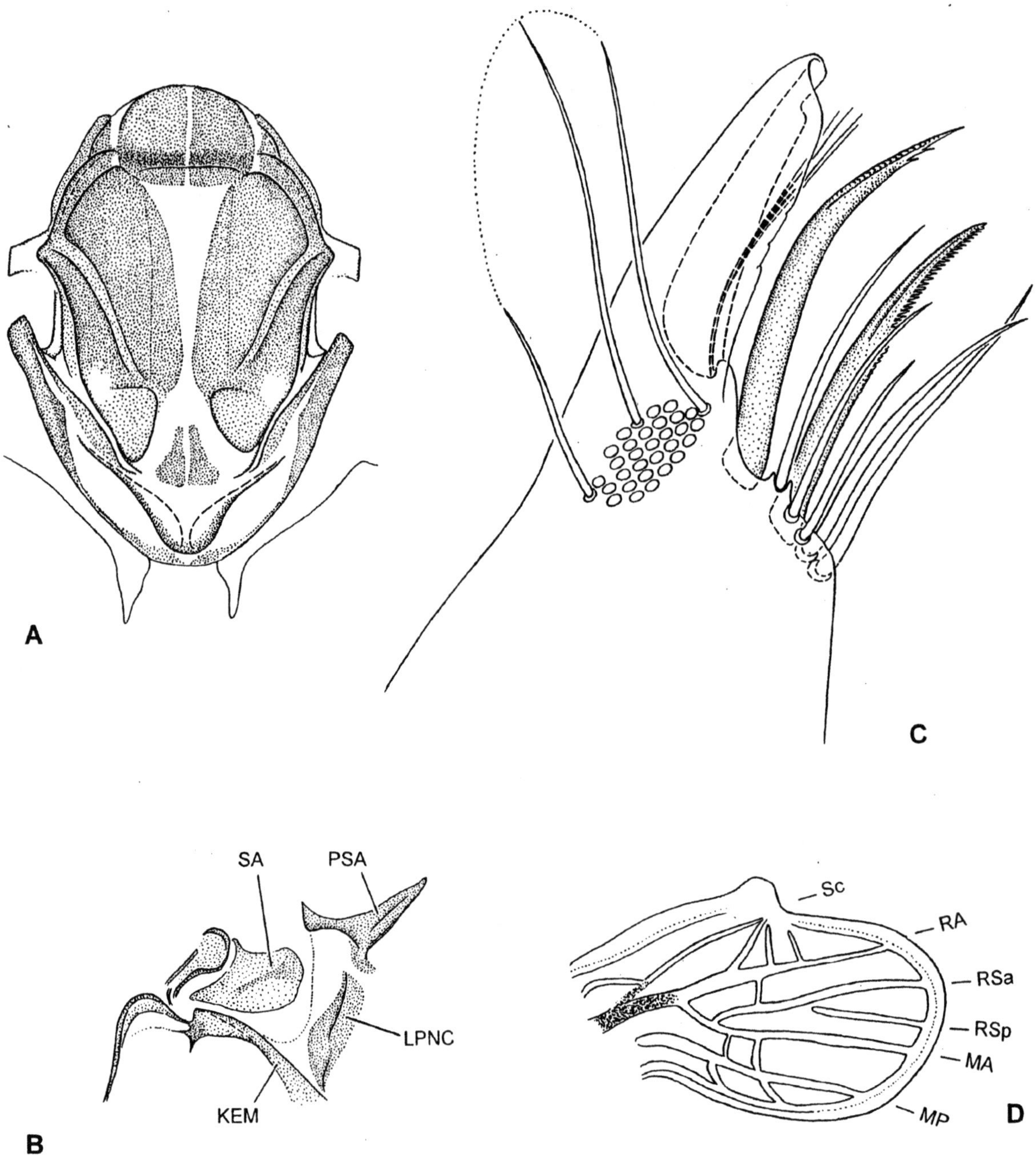

Figure 96. Pantricorythi-Teloganodes/g1.

A–B – sp.T1, female subimago: **A** – mesonotum; **B** – left mesopleuron (compare with Fig.5:D). C–E – *insignis* [*Macafertiella*]: **C** – apex of right maxilla, ventral view (dentisetae shown by dots); **D** – subimaginal hind wing anlage extracted from immature larva; **E** – right tergalii, dorsal view (ventral lobe turned apart and shown from ventral side). **F** – sp.T4, the same; **G** – sp.T2, the same; **H** – sp.T3, the same.

Abbreviations: **II, III, IV, V, VI** – tergalii of II–VI pairs; **d** – dorsal lobe of tergalius; **KEM** – katepimeron; **LPNC** – lateropostnotal crest; **PSA** – postsubalar sclerite of lateropostnotum; **SA** – subalare; **v:a** – anal branch of ventral lobe; **v:c** – costal branch of ventral lobe.

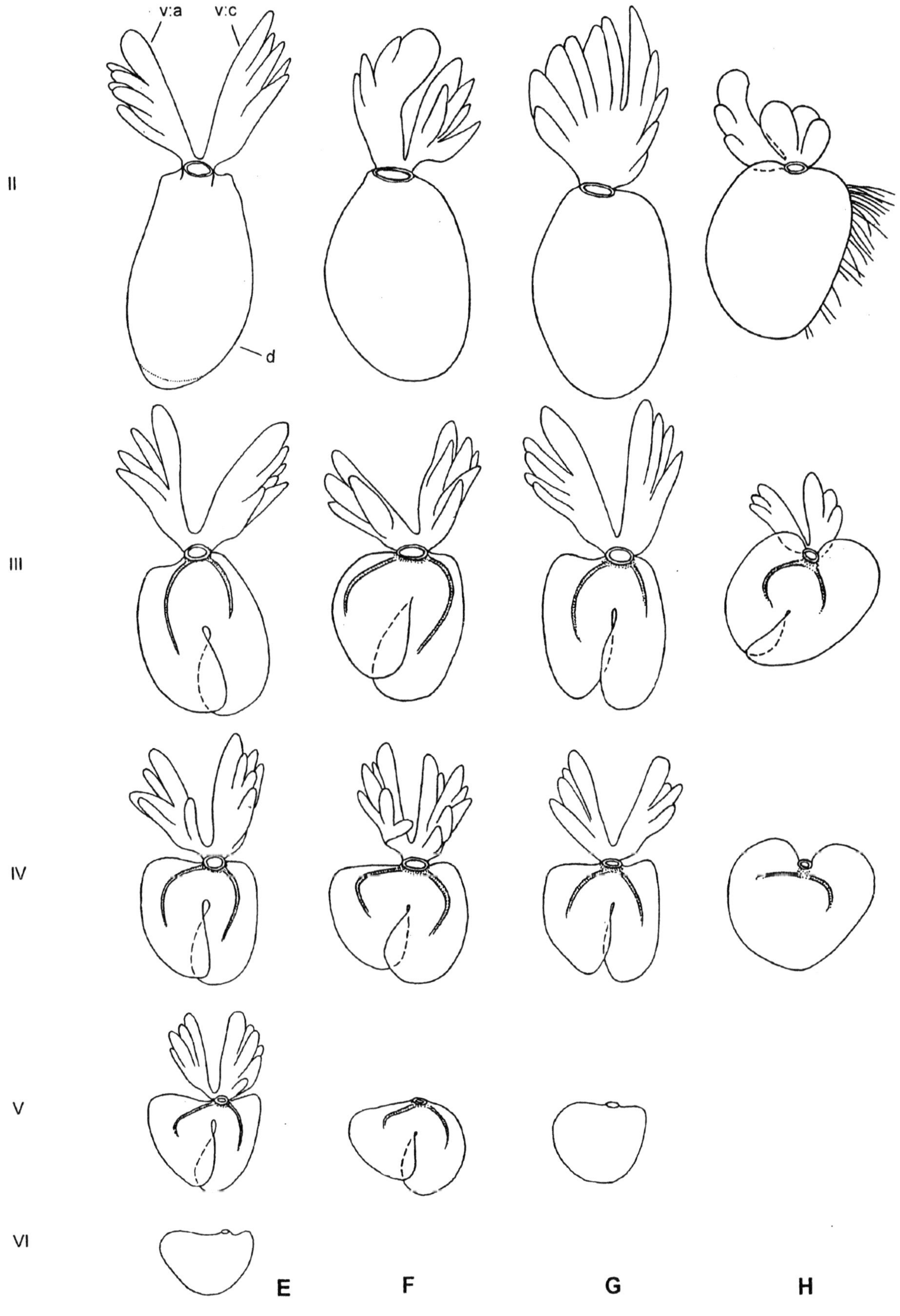
v:a
v:c
d
II
III
IV
V
VI
E
F
G
H

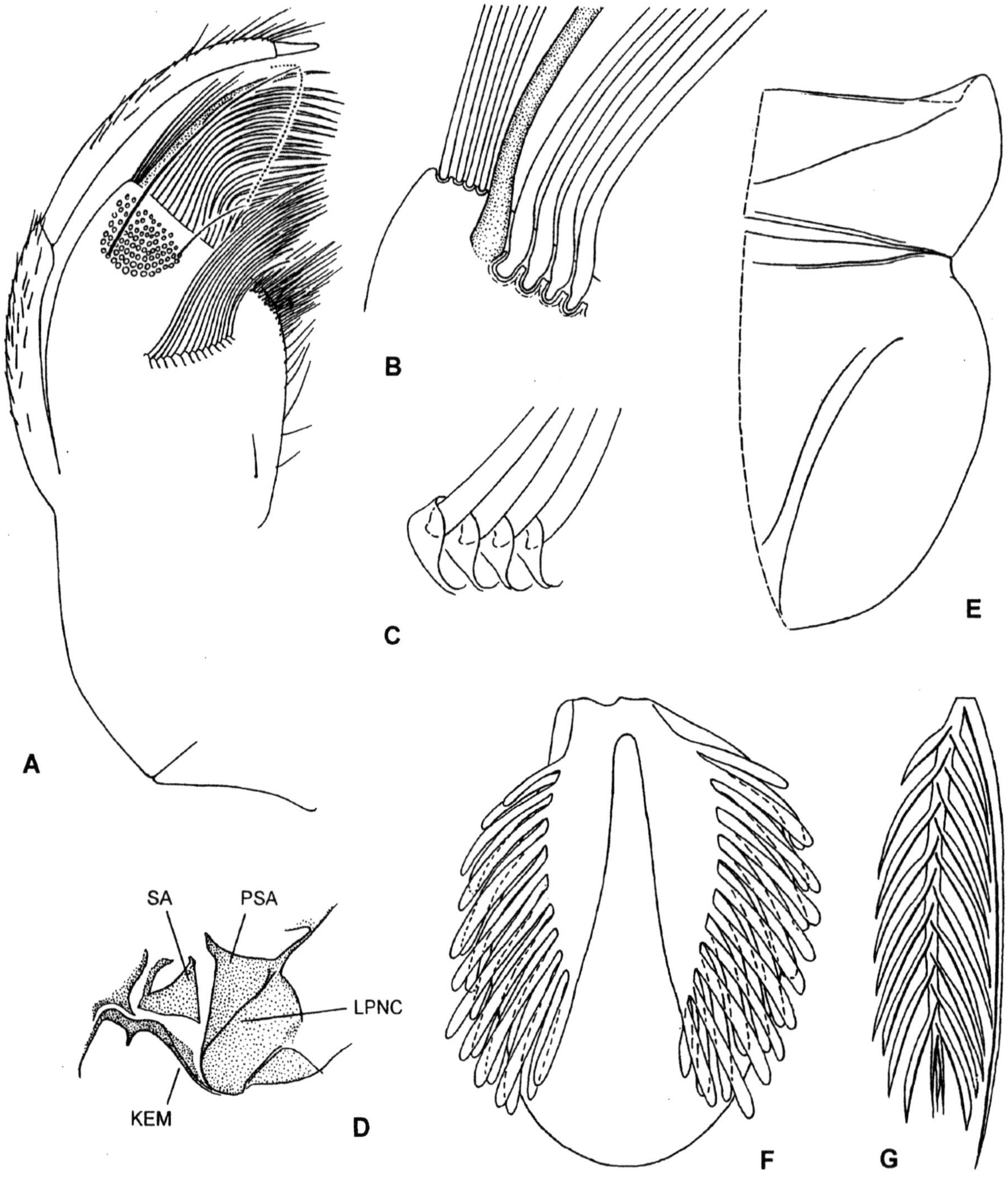

Figure 97. Pantricorythi-Tricoryptera-Afrotricorythi-Tricorygnatha.
A–D – Tricorythus/fg5 *discolor* [*Oxycypha*]: **A** – right maxilla, ventral view (among setae of subapical ventral field, only two setae are shown; dentiseta shown by dots); **B** – apex of maxilla with bases of setae, dorsal view (dentiseta shown by dots); **C** – bases of setae of ventral row on maxilla, ventral view; **D** – subimaginal exuviae of left mesopleuron (compare with Figs 96:B and 5:D). **E–G** – Tricorythus/fg5 *tinctus* [*Tricorythus*]: **E** – larval exuviae of right half of pronotum and mesonotum; **B–G** – tergalius II: **B** – ventral view; **G** – lateral view.

Abbreviations: **KEM** – katepimeron; **LPNC** – lateropostnotal crest; **PSA** – postsubalar sclerite of lateropostnotum; **SA** – subalare.

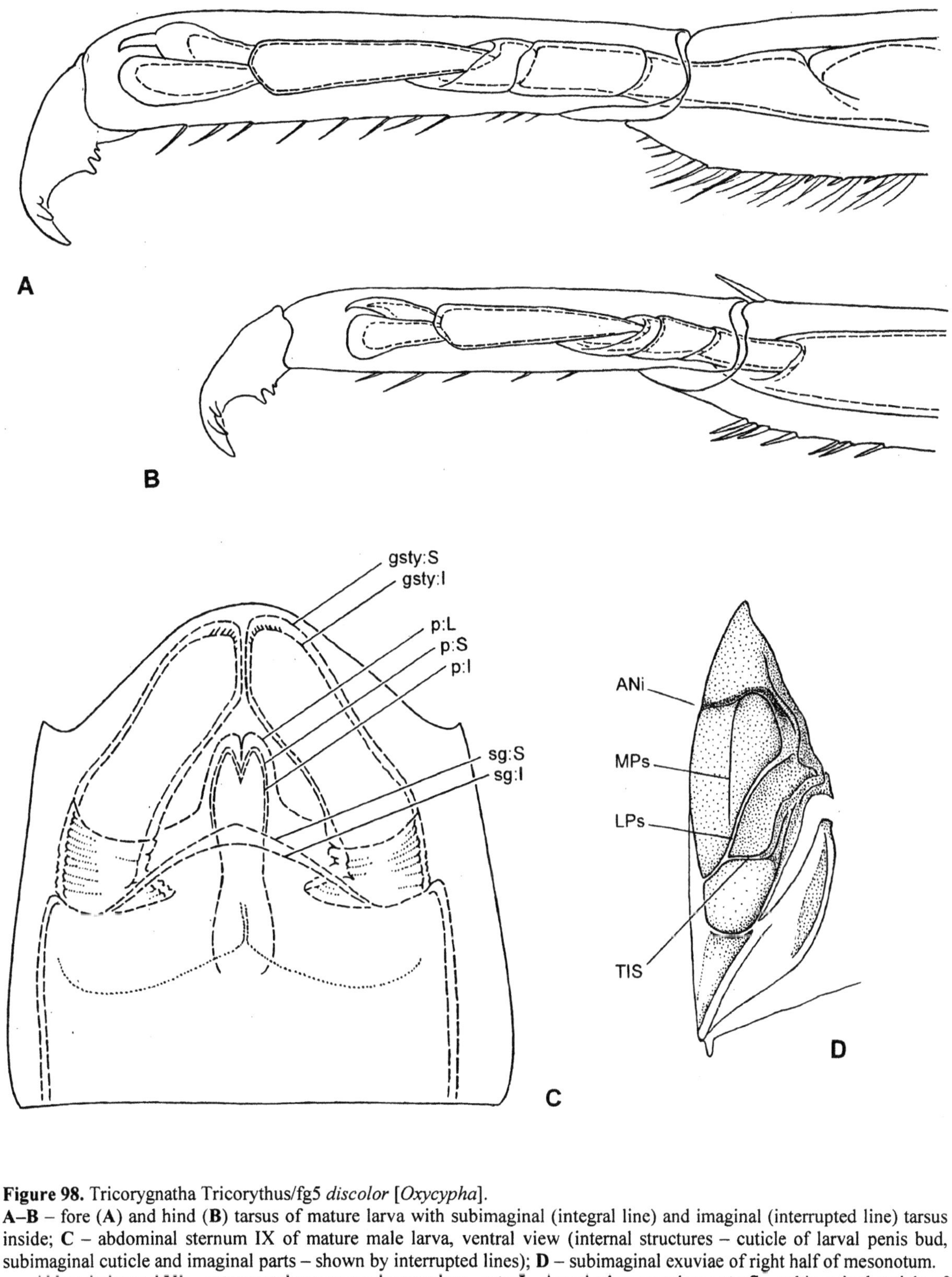

Figure 98. Tricorygnatha Tricorythus/fg5 *discolor* [*Oxycypha*].
A–B – fore (**A**) and hind (**B**) tarsus of mature larva with subimaginal (integral line) and imaginal (interrupted line) tarsus inside; **C** – abdominal sternum IX of mature male larva, ventral view (internal structures – cuticle of larval penis bud, subimaginal cuticle and imaginal parts – shown by interrupted lines); **D** – subimaginal exuviae of right half of mesonotum.

Abbreviations: **ANi** – anteronotal transverse impression; **gsty:I** – imaginal gonostylus; **gsty:S** – subimaginal cuticle of gonostylus; **LPs** – lateroparapsidal suture; **MPs** – medioparapsidal suture; **p:I** – imaginal penis; **p:L** – larval penis bud; **p:S** – subimaginal cuticle of penis; **sg:I** – imaginal styliger; **sg:S** – subimaginal cuticle of styliger; TIS – transverse intersutal suture.

2.2;2,2-2/5. **Tricoryptera**, or **Tricorythus/fg2**

(Bidentiseta Furcatergaliae Ephemerella/fg1 Pantricorythi Tricoryptera)

(Figs 97–104; 105:A–F)

Nomen hierarchicum: **Tricorythus/fg2** (sine *Vietnamella*, *Austremerella*, *Melanemerella*, *Teloganodes*; incl. *Leptohyphes*).
Nomen circumscribens: **Tricoryptera** Kluge, TAXON NOV.
In circumscription matches:
— fam. Tricorythidae Lestage 1942: 15.

References. Lestage 1942: ⊕*; – Edmunds & Allen & Peters 1963: ♉*; – Tshernova 1970: ♉* ⊕*.

Autapomorphies of Tricoryptera. Characters listed here are evidently apomorphic in comparison with other Ephemerella/fg1; however, the characters (1)–(3) and (5)–(6) are common with Caenoptera, that has no explanation (see Classifications of Furcatergaliae II). About other characters common with Caenoptera and occurring in selected taxa of Tricoryptera – see below, Tricorygnatha (2)–(7), Leptohyphes/fg1 (3)–(8) and *Tricorythodes* (2)–(5).

(1) On fore wing tornus is lost [in connection with reduction of hind wing – see (2)], and venation is modified: portion of CuP proximad of the point where it connects with *cup-aa* [see Ephemerella/fg1 (6)] is partly or completely lost; because of this CuP looks as arising not from CuA, but from AA; cubital field has only one bifurcate vein – 1st cubital fork (x_1, x_2) [see Ephemerella/fg1 (15)] which usually (but not always) arises from CuP (FIGS 89:M, 99:A, 100:A). This is similar to Caenoptera [see Classifications of Furcatergaliae II, Caenoptera (10) and FIG.86].

Similar structure, when CuP arises close to base of AA, occurs in Tetramerotarsata and some Leptophlebia/fg1, but origin of such veins position is different: in Tetramerotarsata and Leptophlebia/fg1 CuP at its base is curved and set apart from CuA (FIG.106:F), while in Ephemerella/fg1 the base of CuP is never curved or set apart from CuA, but can disappear.

(2) Hind wings are small (as long as 0.1–0.2 of fore wing length) or lost; if hind wing is present, it has a stretched pointed costal projection, and Sc terminates close to base of costal projection, i. e. far from wing apex (non-unique apomorphy – see Index of characters [2.2.62]). Small hind wings are present in selected specimens of *Ephemerythus* (FIG.99:C), highly modified hind wings are present in selected taxa of Leptohyphes/fg1 (FIG.100:B); in other Tricoryptera hind wings are completely lost. Non-unique apomorphy (see Index of characters [2.2.59]).

Characters of unclear phylogenetic status.

(3) Infrascutellum is interrupted medially, scutellum is more or less enlarged, with enlarged lateral impressions (FIG.89:B). The same in *Teloganodes* (possibly synapomorphy) and Caenoptera (see Classifications of Furcatergaliae II).

(4) Larval abdominal terga without paired dorsal projections [see Ephemerella/fg1 (19)]; usually terga have no dorsal projections, rarely have unpaired projections; the same in some other taxa (see Index of characters [1.3.3]), particularly in *Teloganodes*.

(5) Tergalii I [initially stick-like – see Ephemerella/fg1 (13)] are lost. Non-unique apomorphy (see TABLE 8 and Index of characters [1.3.33]); the same in *Teloganodes* (possibly synapomorphy) and Ephemerella/fg3.

(6) Tergalii VII are reduced; usually they are completely absent, only for *jacobsoni* [*Tricorythus*] vestiges of tergalii VII are described (Ulmer 1939: Figs 339–342). Non-unique apomorphy (see Index of characters [1.3.59]); particularly, tergalii VII are lost in *Teloganodes* (possibly synapomorphy) and Caenoptera.

(7) Adults are shortly-moulting: in mature larva ready to moult to subimago, imaginal cuticle is already developed; shortly (in a few minutes) after the moult from larva to subimago, the next moult from subimago to imago follows. At the same time, imagoes are not short-living, and have functional legs. Subimaginal cuticle has no distinct pigmented and sclerotized areas characteristic for many other mayflies [see Ephemerella/fg1 (10)]. This is true at least for Tricorygnatha and Leptohyphes/fg1. Other Ephemerella/fg1 (at least Ephemerella/fg2, *Vietnamella*, Melanemerella/fg1 and *Teloganodes*) are not shortly-moulting. Besides Tricoryptera, adults are shortly-moulting non-short-living in Caenoptera.

Size. Small, fore wing length 2–10 mm; comparable with Caenoptera (see TABLE 2).

Distribution. Ethiopian, Oriental, Neotropical and Nearctic Regions.

Tricoryptera are divided into Afrotricorythi and Leptohyphes/fg1 (p.333). Several taxa placed here in Ephemerella/fg1 INCERTAE SEDIS (p.346), probably also belong to Pantricorythi.

2.2;2,2-2/5.1. **Afrotricorythi**, or **Tricorythus/fg3**

(Furcatergaliae Ephemerella/fg1 Pantricorythi Tricoryptera Afrotricorythi)

(Figs 97–99)

Nomen hierarchicum: **Tricorythus/fg3** (sine *Leptohyphes*; incl. *Ephemerythus, Dicercomyzon, Machadorythus*).
Nomen circumscribens: **Afrotricorythi** Kluge, TAXON NOV.

Autapomorphies of Afrotricorythi.

(1) The 1st cubital fork [a single vein in cubital field of fore wing – see Tricoryptera (1)] arises from CuP forming with it a symmetrical wide Y-shape figure; 1st cubital fork can retain own bifurcation, or its posterior branch (x_2) is transformed to intercalary, thus x_1, x_2 and CuP look as a triad (FIG.99:A).

(2) Gonostylus lacks distal segment [see Ephemerella/fg1 (8) and Pantricorythi (2)] (FIG.98:C). Non-unique apomorphy (see Index of characters [2.3.12]); among Pantricorythi the same in some Leptohyphes/fg1.

Size. Fore wing length 4–10 mm (see Tricoryptera).

Distribution: Tropical areas of Old World: Ethiopian and Oriental Regions.

Afrotricorythi are divided into Tricorygnatha, *Ephemerythus* (p.330), *Dicercomyzon* (p.332) and *Machadorythus* (p.332).

2.2;2,2-2/5.1;1. **Tricorygnatha**, or **Tricorythus/fg4**

(Ephemerella/fg1 Pantricorythi Tricoryptera Afrotricorythi Tricorygnatha)

(Figs 97–98)

Nomen hierarchicum: **Tricorythus/fg4** (sine *Ephemerythus, Dicercomyzon, Machadorythus*; incl. *Ranorythus, Madecassorythus*).
Nomen circumscribens: **Tricorygnatha** Kluge, TAXON NOV.
In circumscription matches:
— gen. *Tricorythus*: Barnard 1932: 229;
— subfam. Tricorythinae: Edmunds & Traver 1954a: 238.

Reference. Edmunds & Allen & Peters 1963: ♉*.

Autapomorphies of Tricorygnatha.

(1) Mouth apparatus is strongly modified:

Mandible is enlarged, with regular row of long setae on outer margin [however, Kimmins (1949) mentioned a mature larva of "? *Tricorythus* sp." without setal row on mandible].

Maxilla has unique structure (FIG.97:A–C): flat, truncate in such a manner that its apical margin is formed by expansion of initial inner (median) margin, and its apical-lateral angle corresponds to initial apical-median angle. Apical canines are lost. A single dentiseta [instead of two – see Bidentiseta (1)] is long and slender, situated near apical-lateral angle. Setae of initial inner-dorsal row are as long as dentiseta, and form a regular row on apical margin; each seta of this row has such basal articulation which allows it to turn dorsally or distally only. Initial inner-ventral row of setae is lost. Ventral side besides the subapical field of dense ordered setae [see Ephemerella/fg1 (16)] bears a regular transverse row of long setae; each seta of this row has such basal articulation which allows it to turn ventrally or distally only. Maxillary palp is long, with slender long arched 1st and 2nd segments and small 3rd segment [while muscles in the maxillary palp are lost – see Ephemerella/fg1 (1)].

Labium has unique structure: Submentum is strongly shortened; mentum, glossae and paraglossae [see Ephemerella/fg1 (2)] are fused together forming an integral semicircular plate with dorsal concavity, into which hypopharynx is inserted; shape of this concavity repeats shape of hypopharynx. Labial palp has 2nd segment long, arched, with a regular row of long filtering setae on outer margin, and 3rd segment short (Barnard 1932:Fig.20g; Ulmer 1939:Fig.334; Crass 1947a:Fig.31; Oliarinony & al. 2000:Fig.9).

Apomorphies of Tricorygnatha common with Leptohyphes/fg1 and Caenoptera (see Classifications of Furcatergaliae II).

(2) Imaginal and subimaginal mesothorax [see Tricoryptera (3)] has following modifications: On mesonotum [see Ephemerella/fg1 (10)] lateroparapsidal suture is not curved laterally, but convergent with medioparapsidal suture; sublateroscutum has a transverse interscutal suture; relief line of submedioscutum is absent, and in subimago [shortly-moulting – see Tricoryptera (7)] cuticular pigmentation is dispersed (FIG.98:D). Anterior paracoxal suture is transferred posteriorly, closer to anterior margin of coxal conjunctiva. The same in Leptohyphes/fg1 and Caenoptera.

(3) On fore wing marginal intercalaries are absent. Among Ephemerella/fg1 the same in *Dicercomyzon* and Leptohyphes/fg1 (see Index of characters [2.2.55]); possibly synapomorphy; the same in Caenoptera.

(4) In imago amphitornal margin of wing with setae (as in subimago of all mayflies). Non-unique apomorphy (see Index of characters [2.2.27]); particularly, the same in Leptohyphes/fg1 and Caenoptera.

(5) Hind wings [initially vestigial – see Tricoryptera (2)] are lost. Non-unique apomorphy (see Index of characters [2.2.59]).

(6) In male larva caudalii (cerci and paracercus) are thickened in proximal part, in contrast to caudalii of female which have usual form. In male imago (but not in subimago) caudalii are very long. The same in Leptohyphes/fg1 (FIG.103:E–F) and Caenoptera (male larvae of other Tricoryptera – *Ephemerythus*, *Dicercomyzon* and *Machadorythus* – were not examined).

(7) On imaginal and subimaginal tarsi [see (8)] of middle and hind legs 1st segment [initially fused with tibia and shortened – see Furcatergaliae (2)] on its outer side is secondarily distinctly separated from tibia, somewhat resembling that of larva (FIG.98:B). The same in *Tricorythodes*, *Tricorythopsis* and Caenoptera (see Index of characters [2.2.84]).

Other character of unclear phylogenetic status.

(8) In imago and subimago 1st tarsal segment of all legs [separated from tibia – see (7)] is completely fused with 2nd segment, without suture between them; thus, all tarsi of winged instars are 4-segmented (FIG.98:A–B). Fore leg of male imago (initially elongate) has length subequal with middle and hind legs (the same in some other Afrotricorythi – see Index of characters [2.2.74]). Besides Tricorygnatha, fusion of 1st and 2nd tarsal segments takes place in *Machadorythus* (where these segments are non-fused only on fore legs of male subimago – see Index of characters [2.2.76] and [2.2.84], FIG.99: E–F).

Plesiomorphies of Tricorygnatha. Tergalii II–VI with all lobes completely developed [see Ephemerella/fg1 (7)] (FIG.97:F–G) (in contrast to *Dicercomyzon*); tergalii II are non-operculate (in contrast to *Ephemerythus*, *Machadorythus*, Leptohyphes/fg1 and some others – see TABLE 8).

Size. Fore wing length 4–10 mm (see Tricoryptera).

Distribution. Ethiopian and Oriental Regions.

Tricorygnatha are divided into Tricorythus/fg5, *Ranorythus* and Madecassorythus/fg1.

2.2;2,2-2/5.1;1,1. Tricorythus/fg5

(Pantricorythi Tricoryptera Afrotricorythi Tricorygnatha Tricorythus/fg5)

(Figs 97–98)

Nomen hierarchicum: **Tricorythus/fg5** (sine *Ranorythus*, *Madecassorythus*).

In circumscription matches:

— gen. *Tricorythus*: Oliarinony & Sartori & Elouard 2000:377.

References. Barnard 1932: ☊ ⊕; – Ulmer 1939: ☊ ⊕; – Lestage 1942: ☊ ⊕; – Crass 1947a: ☊ ⊕; – Demoulin 1954d (*Tricorythus* + *Neurocaenis*): ⊕; Demoulin 1970e (subfam.): ⊕.

Character of unclear phylogenetic status.

(1) In male imago on fore leg both claws are blunt (in subimago ephemeropteroid) (FIG.98:A). Non-unique apomorphy (see Index of characters [2.2.77]); among Tricorygnatha the same in Madecassorythus/fg1-*Spinirythus*.

Plesiomorphy of Tricorythus/fg5. In contrast to Madecassorythus/fg1, penis without two pairs of processes (FIG.98:C).

Variable characters of Tricorythus/fg5. Eyes of male are either large and distinctly divided into two portions, or small as in female.

Size. Fore wing length 4–7 mm (see Tricoryptera).

Distribution. Ethiopian and Oriental Regions.

Nominal taxa included. Tricorythus/fg5 includes: (1) **Neurocaenis/g** [g: *Neurocaenis* Navás 1936: 365, typus *N. fuscata* Navás 1936 (design. orig.)], regarded as a generic synonym of *Tricorythus* (Oliarinony & Elouard & Raberiaka 1998b: 74); (2) **Tricorythurus/g** [g: *Tricorythus* subgen. *Tricorythurus* Lestage 1942: 13, typus *Tricorythus latus* Ulmer 1916 (monotypy)], regarded as a generic synonym of *Tricorythus* (Demoulin 1954d: 265). True meanings of the names *Tricorythus*, *Neurocaenis* and *Tricorythurus* are unclear, because larvae of their type species are unknown.

Species composition of Tricorythus/fg5 (here [*T.*]= =[*Tricorythus*]). About 20 species are formally described (some of them are poorly known and can actually belong to other taxa in Afrotricorythi) – *abyssinicus* Ulmer 1930 [*T.*], *ambinintsoae* Oliarinony & Elouard 1998 [*T.*], *discolor* Burmeister 1839 [*Oxycypha*], *fyae* Oliarinony & Raberiaka 1998 [*T.*], *fuscata* Navás 1936 [*Neurocaenis*], *goodmani* Elouard & Oliarinony 1998 [*T.*], *jacobsoni* Ulmer 1913 [*T.*], *jeannae* Oliarinony & Elouard 1998 [*T.*], *lanceolatus* Kimmins 1960 [*T.*], *latus* Ulmer 1916 [*T.*], *longus* Ulmer 1916 [*T.*] (synn. subj.: *collarti* Navás 1933 [*Caenis*], *regia* Navás 1932 [*Caenis*]), *pierrei* Elouard & Oliarinony 1998 [*T.*], *poincinsi* Navás 1926 [*T.*], *reticulatus* Barnard 1932 [*T.*], *rolandi* Oliarinony & Raberiaka 1998 [*T.*], *sylvestris* Oliarinony & Elouard 1998 [*T.*], *tinctus* Kimmins 1956 [*T.*], *variabilis* Oliarinony & Raberiaka 1998 [*T.*], *varicauda* Pictet 1843 [*Caenis*], *vulgaris* Raberiaka & Oliarinony 1998 [*T.*].

Material examined: *discolor* [*O.*]: ☊, ♂/♂/⊕♂, ♀/♀/⊕♀; sp.cf. *tinctus* [*T.*]:☊; spp. (Mali, Cameroon, Rwanda): ☊,⊕.

***

2.2;2,2-2/5.1;1,2. **Ranorythus/fg(1)**

(Pantricorythi Tricoryptera Afrotricorythi Tricorygnatha *Ranorythus*)

Nomen hierarchicum: **Ranorythus/fg(1)** [f: Ranorythinae Oliarinony & Elouard 1997: 441; g: *Ranorythus* Oliarinony & Elouard 1997: 439, typus *R. violettae* Oliarinony & Elouard 1997].
In circumscription matches:
— gen. *Ranorythus* Oliarinony & Elouard 1997: 439;
— subfam. Ranorythinae Oliarinony & Elouard 1997: 441.

Reference. Oliarinony & Elouard 1997: ⊕.

Autapomorphy of *Ranorythus*.

(1) Paracercus of male (at least in imago) is shortened, several times shorter than cerci; at the same time in female paracercus is equal to cerci. Non-unique apomorphy (see Index of characters [2.3.22]); among Tricorygnatha paracercus is shortened also in *Spinirythus*.

Plesiomorphies of *Ranorythus*. Eyes of male are larger than in female (in contrast to some Tricorythus/fg5). In imago all claws, including claws of fore legs of male, are ephemeropteroid (in contrast to Tricorythus/fg5 and *Spinirythus*). Penis without processes (in contrast to Madecassorythus/fg1).

Size. Fore wing length 4–9 mm (see Tricoryptera).

Distribution. Madagascar.

Species composition of Ranorythus/fg(1). 2 species – *violettae* Oliarinony & Elouard 1997 [*Ranorythus*] and *longrandi* Oliarinony & Elouard 1997 [*Ranorythus*].

Material examined: –.

***

2.2;2,2-2/5.1;1,3. **Madecassorythus/fg1**

(Pantricorythi Tricoryptera Afrotricorythi Tricorygnatha Madecassorythus/fg1)

Nomen hierarchicum: **Madecassorythus/fg1** (incl. *Spinirythus*) [f: Madecassorythinae Elouard & Oliarinony 1997: 225; g: *Madecassorythus* Elouard & Oliarinony 1997: 225, typus *M. hertui* Elouard & Oliarinony 1997 (design. orig.)].
In circumscription matches:
— subfam. Madecassorythinae Elouard & Oliarinony 1997: 225.

References. Elouard & Oliarinony 1997: ⊕*; – Oliarinony & Sartori & Elouard 2000: ⊙ ⊅ ⊕.

Autapomorphy of Madecassorythus/fg1.

(1) Penis with 2 pairs of long processes directed posteriorly – lateroventral and mediodorsal ones; mediodorsal processes are especially long, curved laterally.

Plesiomorphies of Madecassorythus/fg1. Eyes of male are larger than in female (in contrast to some Tricorythus/fg5).

Size. Fore wing length 5–10 mm (see Tricoryptera).

Distribution. Madagascar.

Madecassorythus/fg1 is divided into Madecassorythus/fg2 and *Spinirythus*.

2.2;2,2-2/5.1;1,3-1. **Madecassorythus/fg2**

(Tricoryptera Afrotricorythi Tricorygnatha Madecassorythus/fg1 .../fg2)

Nomen hierarchicum: **Madecassorythus/fg2** (sine *Spinirythus*).
In circumscription matches:
— gen. *Madecassorythus* Elouard & Oliarinony 1997: 225.

References. Elouard & Oliarinony 1997: ⊕*; – Oliarinony & Sartori & Elouard 2000: ⊙ ⊅ ⊕.

Autapomorphy of Madecassorythus/fg2.

(1) Mediodorsal processes of penis [see Madecassorythus/fg1 (1)] are longer than gonostyli [2-segmented – see Afrotricorythi (2)].

Plesiomorphies of Madecassorythus/fg2. In imago all claws, including claws of fore legs in male, are ephemeropteroid (in contrast to *Spinirythus* and Tricorythus/fg5). Paracercus is subequal to cerci (in contrast to *Spinirythus* and *Ranorythus*).

Size. Fore wing length 5–7 mm (see Tricoryptera).

Distribution. Madagascar.

Species composition of Madecassorythus/fg2 (here [*M.*]=[*Madecassorythus*]). 4 species – *hertui* Elouard & Oliarinony 1997 [*M.*], *linae* Elouard & Oliarinony 1997 [*M.*], *ramanankasinae* Elouard & Oliarinony 1997 [*M.*] and *raphaeli* Elouard & Oliarinony 1997 [*M.*].

Material examined: –.

***

2.2;2,2-2/5.1;1,3-2. **Spinirythus/g(1)**

(Tricoryptera Afrotricorythi Tricorygnatha Madecassorythus/fg1 *Spinirythus*)

Nomen hierarchicum: **Spinirythus/g(1)** [g: *Spinirythus* Oliarinony & Elouard (in Oliarinony & Elouard & Raberiaka) 1998a: 237, typus *S. martini* Oliarinony & Elouard 1998].
In circumscription matches:
— gen. *Spinirythus* Oliarinony & al. 1998a: 237.

Reference. Oliarinony & al. 1998a: ⊕*.

Non-unique apomorphies of *Spinirythus*.

(1) Imaginal paracercus is vestigial, many times shorter than cerci; larval paracercus is as long as cerci, but much thinner. Among Tricorygnatha the same in *Ranorythus* (see Index of characters [2.3.22]).

(2) In male imago on fore leg both claws are blunt (subimagoes unknown). Among Tricorygnatha the same in Tricorythus/fg5 (see Index of characters [2.2.77]).

Size. Fore wing length 5–10 mm (see Tricoryptera).

Distribution. Madagascar.

Species composition of Spinirythus/g(1). 3 species (Oliarinony & Elouard & Raberiaka 1998a) – *colasi* Elouard & Oliarinony 1998 [*Spinirythus*], *martini* Oliarinony & Elouard 1998 [*Spinirythus*] and *rosae* Oliarinony & Raberiaka 1998 [*Spinirythus*].

Material examined: *martini* [*S.*]: Ŏ♂, ⊕♂.

*** *** ***

2.2;2,2-2/5.1;2. **Ephemerythus/fg1**
(Ephemerella/fg1 Pantricorythi Tricoryptera Afrotricorythi *Ephemerythus*)
(Fig. 99:A–B)

Nomen hierarchicum: **Ephemerythus/fg1** (incl. *Tricomerella*) [f: Ephemerythinae Gillies 1960: 35; g: *Ephemerythus* Gillies 1960: 35, typus *E. niger* Gillies 1960 (design. orig.)].

In circumscription matches:
— gen. *Ephemerythus* Gillies 1960: 35;
— subfam. Ephemerythinae Gillies 1960: 35;
— fam. Ephemerythidae: McCafferty & Wang 2000: 49.

References. Kimmins 1955 (*Ephemerellidae*, genus uncertain): Ŏ; – Gillies 1960: Ŏ* ⊕*; – Edmunds & Allen & Peters 1963: Ŏ*; Demoulin 1964b: Ŏ'; – 1965c: Ŏ' – McCafferty & Wang 2000: Ŏ* ⊕*.

Characters of unclear phylogenetic status.

(1) Tergalii II are enlarged and serve as gill opercula, covering the rest tergalii (Demoulin 1964b: Fig.1A). Non-unique apomorphy (see Index of characters [1.3.39] and TABLE 8).

In the specimens examined by me, only tergalii II–IV have bifurcate ventral lobe with processes [see Ephemerella/fg1 (7)], tergalii V have no ventral lobe, and tergalii VI are lost; according to Gillies (1960), some larvae have vestigial single-lamellate tergalii VI [tergalii I and VII are always lost – see Tricoryptera (5) and (6)].

(2) On fore wing free marginal intercalaries are present, mainly 2 in each space (FIG.99:A). These intercalaries are free also in most other Ephemerella/fg1 – see Index of characters [2.2.55]).

(3) Hind wing [small – see Tricoryptera (2)] has a long costal projection directed anteriorly-distally; Sc terminates at wing margin just distad of costal projection, its apex is connected with RA by a constant oblique crossvein (FIG.99:C). Non-unique character (see Index of characters [2.2.62] and [2.2.63]); it can not be compared with other Afrotricorythi, as their hind wings are always lost. In selected specimens of *Ephemerythus* hind wings are also lost.

Plesiomorphies of *Ephemerythus*. In contrast to Tricorygnatha and *Dicercomyzon*, mouth apparatus is non-specialized, maxilla is biting [see Ephemerella/fg1 (16)], with 3 canines (ventral of them can be serrate) and 2 dentisetae [see Bidentiseta (1)].

In imago: Eyes of male are diminished, but slightly larger than in female (see Index of characters [2.1.3]). Mesonotum [see Ephemerella/ /fg1 (10)] retains curvation of lateroparapsidal suture and has no transverse interscutal suture (in contrast to Tricorygnatha some others – see Index of characters [2.2.9] and [2.2.10]). Imaginal wing without setae on amphitornal margin (in contrast to Tricorygnatha and Leptohyphes/fg1). All claws, including male fore leg, are ephemeropteroid (at least in *kiboensis* [*E.*] and *pictus* [*E.*]).

Size. Fore wing length 5–6 mm (see Tricoryptera).

Distribution. Ethiopian Region.

Nominal taxon included. Ephemerythus/fg1 includes **Tricomerella/g** [g: *Ephemerythus* subgen. *Tricomerella* Demoulin 1964b: 17, typus *E. (T.) straeleni* Demoulin 1964 (monotypy)] which was described as a subgenus differing by apomorphies: mandible is more robust, maxilla has proximal dentiseta massive and opposed to canines (while in subgenus *Ephemerythus* s.str. mouth apparatus is non-specialized, and larval body is more robust). The type species of *Ephemerythus* is known only as imago, and the type species of *Tricomerella* only as larvae, thus separate diagnoses of these taxa are nonreliable.

Species composition of Ephemerythus/fg1. 5 species – *dissimillimus* Kopelke 1980 [*Ephemerythus*], *kiboensis* Gillies 1960 [*Ephemerythus*], *niger* Gillies 1960 [*Ephemerythus*], *pictus* Gillies 1960 [*Ephemerythus*], *straeleni* Demoulin 1964 [*Ephemerythus (Tricomerella)*].

Material examined: *kiboensis* [*E.*]: ⊕♂; *straeleni* [*E.*]: Ŏ; sp.: Ŏ.

***

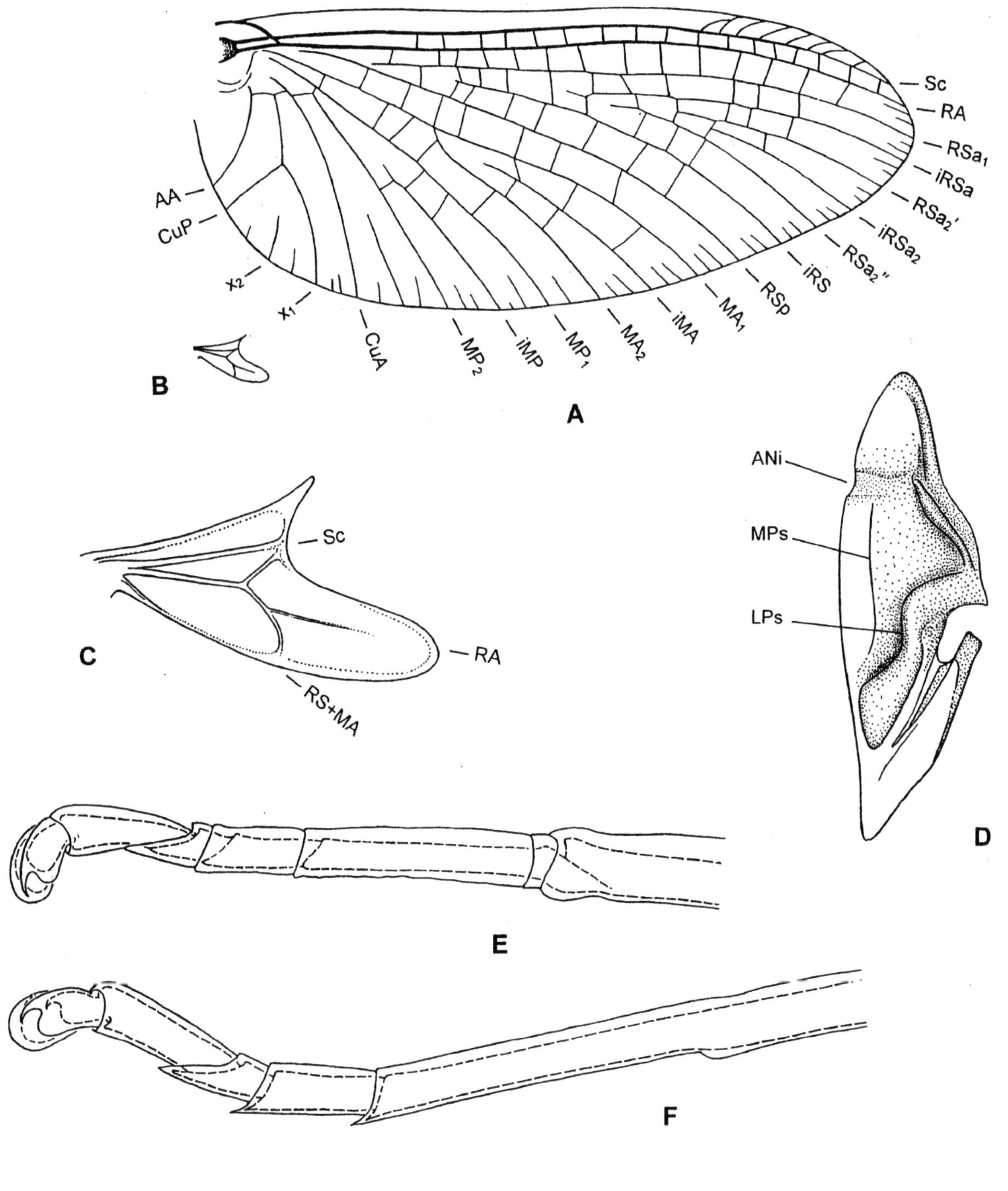

Figure 99. Afrotricorythi: *Ephemerythus*, *Dicercomyzon* and *Machadorythus*.
A–C – Ephemerythus/fg1 *kiboensis* [*Ephemerythus*]: **A–B** – fore and hind wings of male imago; **C** – its hind wing, enlarged. **D** Dicercomyzon/fg1 sp., subimaginal exuviae of right half of mesonotum. **E–F** – Machadorythus/fg1 sp.M1, male subimaginal fore (**E**) and hind (**F**) tarsus with imaginal tarsus inside (shown by interrupted line).

Abbreviations: **ANi** – anteronotal transverse impression; **LPs** – lateroparapsidal suture; **MPs** – medioparapsidal suture.

2.2;2,2-2/5.1;3. Dicercomyzon/fg(1)

(Ephemerella/fg1 Pantricorythi Tricoryptera Afrotricorythi *Dicercomyzon*)

(Fig. 99:D)

Nomen hierarchicum: **Dicercomyzon/fg(1)** [f: Dicercomyzinae Edmunds & Traver 1954a: 238; g: *Dicercomyzon* Demoulin 1954a: 1, typus *D. femorale* Demoulin 1954 (design. orig.)].
In circumscription matches:
— gen. *Dicercomyzon* Demoulin 1954a: 1;
— subfam. Dicercomyzinae Edmunds & Traver 1954a: 238.

References. Demoulin 1954a: ♉; – 1954d: ♉ ♀; – Kimmins 1957: ♉ ⊕; – Edmunds & Allen & Peters 1963: ♉; – Demoulin 1964b: ♉ ◎.

Autapomorphies of *Dicercomyzon*.

(1) Larval head is prognathous, frons forms a shield covering mouthparts from above; clypeus is strongly reduced, so base of labrum is hidden under the head shield (Demoulin 1964b:Fig.4A). Non-unique apomorphy (see index of characters [1.1.4]).

(2) Mouth apparatus is strongly modified (Demoulin 1964b:Figs 4–5): Mandibles are shortened perpendicular to axis of articulation. Superlinguae are stretched laterally. Maxilla [with vestigial palp – see Ephemerella/fg1 (1)] has a unique structure, being strongly specialized as scraping: short and stout; field of closely situated setae on apical-ventral side [see Ephemerella/fg1 (16)] is greatly enlarged and occupies largest part of maxilla; maxillary canines are lost; only 1 dentiseta is retained [instead of two – see Bidentiseta (1)] (Demoulin 1964b:Fig. 5D). Labium is strongly widened, with widened paraglossae.

(3) Larval sterna form two roundish disks, the anterior of which overlaps by its posterior margin the posterior one; margins of these disks are fringed by dense setae; the anterior disc is formed by sternites of prothorax and mesothorax, and the posterior one – by sternites of metathorax, abdominal segment I and partly segment II (Demoulin 1964b:Fig.3A). Probably these disks are used as adhesive suckers. Similar disk on sterna is present in Ephemerella/fg4-Drunella/g1 *doddsi* [*E.*], but there it is formed by abdominal sterna II–IX.

(4) Each tergalius [II–VI – see Tricoryptera (5) and (6)] consists of a single lamella with numerous long processes on margins (Demoulin 1964b:Figs 3B-F). Such tergalius sharply differs from the initial one [see Ephemerella/fg1 (7)]; probably it is formed by complete fusion of initial two branches of ventral lobe, while dorsal lobe is completely lost. In some other Ephemerella/fg1 branches of ventral lobe can be fused, but dorsal lobe never has tendency to reduction or loss.

(5) Larval, subimaginal and imaginal paracercus is vestigial, non-segmented. Non-unique apomorphy (see Index of characters [1.3.64]); among Ephemerella/fg1 the same in *Teloganodes* only.

Characters of unclear phylogenetic status.

(6) In male imago fore leg is not elongate. The same in Tricorygnatha and some others (see Index of characters [2.2.74]). On all legs tarsal segments 1–4 are extremely shortened.

(7) On fore wing marginal intercalaries are absent. Among Ephemerella/fg1 the same in Tricorygnatha and Leptohyphes/fg1 (see Index of characters [2.2.55]); possibly synapomorphy.

(8) Hind wings [initially vestigial – see Tricoryptera (2)] are lost. Non-unique apomorphy (see Index of characters [2.2.59]).

Plesiomorphies of *Dicercomyzon*. Eyes of male are large. Imaginal and subimaginal mesonotum [without outlined lateral pigmented area – see Tricoryptera (7)] retains curved lateroparapsidal suture [see Ephemerella/fg1 (10)] and has no transverse interscutal suture (FIG.99:D) (in contrast to Tricorygnatha and some others – see Index of characters [2.2.9] and [2.2.10]). Imaginal wing without setae on amphitornal margin (in contrast to Tricorygnatha and Leptohyphes/fg1). Claws on fore leg of male imago are ephemeropteroid (as well as other claws of imago and subimago – see Plesiomorphies of Ephemerella/fg1).

Size. Fore wing length 4–6 mm (see Tricoryptera).

Distribution. Ethiopian Region.

Species composition of Dicercomyzon/fg(1). 4 species – *costale* Kimmins 1957 [*Dicercomyzon*], *femorale* Demoulin 1954 [*Dicercomyzon*], *sjostedti* Ulmer 1909 [*Caenis*] (syn. subj.: *marginatum* Kimmins 1957 [*Dicercomyzon*]), *verrierae* Demoulin 1954 [*Dicercomyzon*].

Material examined: *costale* [*D.*]: ⊕♂ (Entomol. Inst.); *sjostedti* [*C.*]: ♀/⊙; sp.: ◎♂

***

2.2;2,2-2/5.1;4. Machadorythus/fg(1)

(Ephemerella/fg1 Pantricorythi Tricoryptera Afrotricorythi *Machadorythus*)

(Fig. 99:E–F)

Nomen hierarchicum: **Machadorythus/fg(1)** [f: Machadorythinae Edmunds & Allen & Peters 1963: 17; g: *Machadorythus* Demoulin 1959: 251, typus *M. palanguim* Demoulin 1959 (monotypy; syn. subj. *Tricorythus maculatus* Kimmins 1949)].
In circumscription matches:
— gen. *Machadorythus* Demoulin 1959: 251;
— subfam. Machadorythinae Edmunds & Allen & Peters 1963:17;

— fam. Machadorythidae: McCafferty & Wang 2000: 50.

References. Demoulin 1959: Ꝺ; – Edmunds & Allen & Peters 1963: Ꝺ; – Peters 1981: Ꝺ; – Elouard & Gillies 1989: Ꝺ ⊕; – McCafferty & Wang 2000: Ꝺ* ⊕*.

Autapomorphy of *Machadorythus*.

(1) Abdominal tergum VII [not bearing tergalii – see Tricoryptera (6)] in larva, subimago and imago is enlarged, with elevated middle of posterior margin and with a pair of ridges which stretch from this convexity to anterior-lateral corners of the tergum (Demoulin 1959:Fig.1; Peters 1981:Figs 27–28; Elouard & Gillies 1989:Figs 1, 3). Unique character; probably this is connected with an operculate-gilled specialization of larva [see (5)].

Characters of unclear phylogenetic status: possible synapomorphies with Tricorygnatha.

(2) Imaginal 1st tarsal segment [shortened and fused with tibia – see Furcatergaliae (2)] is completely fused with 2nd tarsal segment, without any trace of suture between them; the same on subimaginal middle and hind legs; only on fore leg of male subimago suture between 1st and 2nd segments is retained (FIG.99:E–F). Because of fusion of 1st+2nd segments, tarsi can be interpreted as 4-segmented (if take into account long 1st+2nd segment immobile fused with tibia), or as 3-segmented (if count only mobile 3rd–5th segments). Thus, tarsi of middle and hind legs are indistinguishable from that of Tetramerotarsata (FIG.27:E). Besides *Machadorythus*, fusion of 1st and 2nd tarsal segments takes place in Tricorygnatha (where these segments are fused on all legs of imago and subimago – see FIG.98:A–B, Index of characters [2.2.76] and [2.2.84]). In contrast to Tricorygnatha, in *Machadorythus* fusion of 1st tarsal segment with tibia is retained.

(3) Maxillary palp [initially vestigial – see Ephemerella/fg1 (1)] has 1st and 2nd segments secondarily elongated, 3rd segment is lost. Similar structure of maxillary palp in Tricorygnatha; possibly, synapomorphy.

(4) Glossae, paraglossae and mentum are completely fused together. The same in Tricorygnatha and *Coryphorus* (but shape of labium in all three taxa is different).

Other characters of unclear phylogenetic status.

(5) Tergalii II are larger than others and serve as gill opercula. Non-unique apomorphy (see Index of characters [1.3.39] and TABLE 8).

(6) On fore wing free marginal intercalarics are present at least in female; in male they are either present, or absent (personal communication by J.G. Peters). Non-unique character (see Index of characters [2.2.55]).

(7) Hind wings [initially vestigial – see Tricoryptera (2)] are lost. Non-unique apomorphy (see Index of characters [2.2.59]).

(8) On fore leg of male imago both claws are blunt (in subimago ephemeropteroid) (FIG.99:E). Non-unique apomorphy (see Index of characters [2.2.77]); the same in many other Tricoryptera.

Plesiomorphies of *Machadorythus*. In contrast to Tricorygnatha and *Dicercomyzon*, maxilla is biting, non-specialized, with well-developed canines and 2 dentisetae [see Bidentiseta (1)].

In imago and subimago: Eyes of male are large. Mesonotum [without outlined lateral pigmented area – see Tricoryptera (7)] retains curved lateroparapsidal suture [see Ephemerella/fg1 (10)] and has no transverse interscutal suture (in contrast to Tricorygnatha and some others – see Index of characters [2.2.9] and [2.2.10]). Imaginal wing without setae on amphitornal margin (in contrast to Tricorygnatha and Leptohyphes/fg1).

Size. Fore wing length 4–7 mm (see Tricoryptera).

Distribution. Ethiopian Region.

Species composition of Machadorythus/fg(1). 1 species is described – *maculatus* Kimmins 1949 [*Tricorythus*] (syn. subj.: *palanguim* Demoulin 1959 [*Machadorythus*]).

Material examined: *maculatus* [*T.*]: ⊕♂; sp.M1 (Uganda): ♂, ⊕♂.

*** ***

2.2;2,2-2/5.2. Leptohyphes/fg1

(Furcatergaliae Ephemerella/fg1 Pantricorythi Tricoryptera Leptohyphes/fg1)

(Figs 100–104; 105:A–F)

Nomen hierarchicum: **Leptohyphes/fg1** (incl. *Allenhyphes*, *Tricorythodes*, *Tricorythopsis*) [f: Leptohyphinae Edmunds & Traver 1954a: 238; g: *Leptohyphes* Eaton 1882: 208, typus *L. eximius* Eaton 1882 (monotypy); syn. subj.: ***Bruchella*** Navás 1920: 56, typus *B. nigra* Navás 1920 (design. orig.) (synn. subj.: *eximius* [*L.*] = *nigra* [*B.*]).

In circumscription matches:

— subfam. Leptohyphinae Edmunds & Traver 1954a: 238;

— fam. Leptohyphidae: Landa 1973: 156.

References. Traver 1958: ⊕; – Edmunds & Allen & Peters 1963: Ꝺ*; – Edmunds & Jensen & Berner 1976: Ꝺ*, ⊕*; – Kluge 1992c: Ꝺ* ◎* ⊕*;– McCafferty & Wang 2000: Ꝺ* ⊕*; – Molineri & Peters & Zuñiga-de-Cardoso 2001: ⊕♂*.

Autapomorphy of Leptohyphes/fg1.

(1) Tergalii are modified in following manner:

Ventral lobe [initially bifurcate and bearing processes – see Ephemerella/fg1 (7)] lost initial numerous marginal processes and bifurcates from its base, thus two its branches (costal and anal ones) represent two separate lamellae; these branches can either lack processes at all (in *Allenhyphes* and *Tricorythodes* – FIG.103:C), or have a few lamellate processes arising basally from ventral side (tergalii III–V of Leptohyphes/fg2 and *Tricorythopsis* – FIG.101:D–I), or have by one marginal process on each branch (tergalius II of *Tricorythopsis* and FIG.105:H). Probably initially for Leptohyphes/fg1, costal branch is widened and projects from under costal margin of dorsal lobe (FIG.101:F–H) (while in other Ephemerella/fg1 both branches are usually completely covered by dorsal lobe); this is connected with a peculiar construction of operculate tergalius II.

Tergalius II has dorsal lobe transformed to a gill operculum which covers the rest tergalii [III–VI – see Tricoryptera (5) and (6)] (non-unique apomorphy – see Index of characters [1.3.32]). Its costal branch of ventral lobe is bent longitudinally so that its portion projected from under dorsal lobe, orientates perpendicular to the flatness of the rest tergalius (FIG.103:A–B); this portion of costal branch can be called **costal flap**. Dorsal lobe and costal flap together form a gill chamber enclosing respiratory tergalii III–VI of one side: dorsal lobe limits this gill chamber dorsally, and costal flap limits it laterally. In the group *sacculobranchis* costal flap is elongate, apically curved and closes gill chamber not only laterally, but posteriorly as well (FIG.103:D). Tergalii-bearing abdominal segments III–VI have especially wide and flat paranota laterad of tergalii bases; when tergalius II is pressed to the body, costal margin of its dorsal lobe touches lateral margins of paranota, and costal flap appears to be hidden (FIG.103:B). Such opercula structure is present in *Allenhyphes* and *Tricorythodes*; in Leptohyphes/fg2 costal flap is reduced to a spine-like basal vestige (FIG.101:D); only in Tricorythopsis costal flap is absent [see Tricorythopsis (8) below].

Tergalii III–V have the following peculiarity. Costal rib is lost, thus dorsal lobe has only anal rib in a form of thin sclerotized arched line [at a distance from anal margin – see Fimbriatotergaliae (6)] (in contrast to tergalius II which can have both costal and anal ribs developed). The projected portion of costal branch lies in the same flatness as the whole tergalius and sometimes can bear a thin additional rib resembling lost costal rib of dorsal lobe (FIG.103:C). Tergalius VI (the last one) has simplified structure.

In some species of Leptohyphes/fg1 certain tergalii lost anal branch or entire ventral lobe.

(2) Hind wing, if present, has unique structure (FIG.100:B): costal projection [pointed – see Tricoryptera (2)] is extremely shifted to wing base and arises perpendicular to the main wing axis; this projection is very long, slender and often arched at proximal direction. Sc is short, curved anteriorly and terminates close to base of the costal projection; sometimes Sc is lost. The rest part of wing is narrow, with 1–3 non-branched veins. In spite of the peculiar form of hind wing in imago and subimago, larval hind wing bud retains a usual form, being wide, without prominent costal projection (FIG.100: C–D), like in Ephemerella/fg2 (FIG.91:A) and other mayflies with normal hind wings (hind wing buds are examined only for *eximius* [*L.*], *zalope* [*L.*] and *flinti* [*L.*], where they are present in males only).

Hind wings are present in both sexes of some Leptohyphes/fg2 and *Haplohyphes*; more often (in some Leptohyphes/fg2, *Allenhyphes*, *Vacupernius*, *Yaurina*, *Traverhyphes*) hind wings are present in males only. In *Tricorythodes*, *Tricorythopsis* and some others hind wings are lost in both sexes.

It is not quite clear if the peculiar hind wing structure is an autapomorphy of Leptohyphes/fg1, because in many other Tricoryptera hind wings are lost (see Index of characters [2.2.59]), and their initial structure in unknown.

Characters of Leptohyphes/fg1 of unclear phylogenetic status (including apomorphies common with Tricorygnatha and Caenoptera – see Classifications of Furcatergaliae II).

(3) Eyes of male are usually as small as in female (non-unique apomorphy – see Index of characters [2.1.3]); only in Tricorythodes/fg1-*Homoleptohyphes* male eyes are enlarged, probably secondarily.

(4) Imaginal and subimaginal mesothorax [see Tricoryptera (3)] has the following modifications common with Tricorygnatha and Caenoptera: On mesonotum [see Ephemerella/fg1 (10)] relief line of submedioscutum is lost, and in subimago [shortly-moulting – see Tricoryptera (6)] cuticular pigmentation is lost (FIG.104:I). Sublateroscutum has a transverse intercutal suture. Lateroparapsidal suture is not so strongly curved laterally as is initial for Ephemerella/fg1; some curvation is retained in Leptohyphes/fg2 and *Allenhyphes*, where lateroparapsidal suture merges with transverse interscutal suture laterad of medioparapsidal suture (FIG.102:A); in *Tricorythodes* lateroparapsidal suture is more straightened and merges with transverse interscutal suture posteriad of its merging with medioparapsidal suture (FIG.104:A). Anterior paracoxal suture is transferred posteriorly, nearer to anterior margin of coxal conjunctiva (FIGS 102:B; 104:B).

(5) On fore wing marginal intercalaries are absent (FIG.100:A). Among Ephemerella/fg1 the same in *Dicercomyzon* and Tricorygnatha (see Index of characters [2.2.55]); the same in Caenoptera.

(6) In imago amphitornal margin of wing with setae (as in subimago of all mayflies) (FIGS 66:C; 100:A–B). Non-unique apomorphy (see Index of characters [2.2.27]); particularly, the same in Tricorygnatha and Caenoptera; probably synapomorphy.

(7) On fore leg of male imago both claws are blunt, in subimago ephemeropteroid (FIG.105:B–D). Non-unique apomorphy (see Index of characters [2.2.77]). Other claws are ephemeropteroid (see Plesiomorphies of Ephemerella/fg1). Fore some species ephemeropteroid imaginal fore claws were reported; in these cases subimagoes were taken for imagoes (Molineri & Peters & Zuñiga-de-Cardoso 2001: 118).

(8) In male larva caudalii [cerci and paracercus – see Plesiomorphies of Ephemerella/fg1] are thickened at proximal part, in contrast to caudalii of female which have usual form (FIG.103:E–F). In male imago (but not in subimago) caudalii are very long. The same in Tricorygnatha and Caenoptera (FIG.88:E). In female imago and subimago caudalii can be either normally developed, or diminished [see *Tricorythodes* (6) below].

Plesiomorphy of Leptohyphes/fg1. On fore wing CuP is more or less strongly arched by its convexity anteriorly and by apex posteriorly, thus the bifurcate vein iCu, which arises from it anteriorly [see Tricoryptera (1)], does not form symmetrical fork with CuP (FIG.100:A) (in contrast to Afrotricorythi).

Size. Fore wing length 2–9 mm (see Tricoryptera).

Distribution. America; dominate in Neotropical Region.

Leptohyphes/fg1 is divided into Leptohyphes/fg2, *Allenhyphes* (p.336), *Tricorythodes* (p.343) and *Tricorythopsis* (p.344); a number of insufficiently described taxa have uncertain systematic position (p.345).

2.2;2,2-2/5.2;1. Leptohyphes/fg2

(Ephemerella/fg1 Pantricorythi Tricoryptera Leptohyphes/fg1 ...fg2)

(Figs 100–102)

Nomen hierarchicum: **Leptohyphes/fg2** (sine *Allenhyphes*, *Tricorythodes*, *Tricorythopsis*).
In circumscription matches:
— gen. *Leptohyphes*: Kluge 1992c: 6.

References. Kluge 1992c: Ꝺ* ♀*; – Hofmann & Sartori & Tomas 1999: Ꝺ ⊕; – Baumgardner & McCafferty 2000 (*L. zalope*): Ꝺ; – Wiersema & McCafferty 2000: Ꝺ ⊕; – Molineri 2003a: Ꝺ* ⊕* ⊙*.

Autapomorphy of Leptohyphes/fg2.

(1) Costal flap of ventral lobe of gill operculum [see Leptohyphes/fg1 (1)] is reduced to a spine-like process reaching base of the next tergalius only (FIG.101:D–E).

(2) In imago and subimago [shortly-moulting – see Tricoryptera (7)] lateropostnotal crest is either lost, or sharply curved toward anterior margin of lateropostnotum (FIG.102:B). In other Ephemerella/ /fg1 lateropostnotal crest is more straight and looks as a continuation of infrascutellar-lateropostnotal arm (FIGS 97:D; 104:B).

Plesiomorphies of Leptohyphes/fg2. In contrast to *Allenhyphes* and *Tricorythodes*, ventral lobe of tergalii III–V retains processes in a form of one or several lamellae arising ventrally near base [see Leptohyphes/fg1 (1)] (FIG.101:F–H). In contrast to *Haplohyphes*, mouth apparatus is non-specialized, maxilla [see Ephemerella/fg1 (16)] with 3 canines and 2 dentisetae [see Bidentiseta (1)] (FIG.101:C).

In imago and subimago: Scuto-lateropostnotal muscle of metathorax is not so large as in *Tricorythodes* [while transverse interscutal suture is present – see Leptohyphes/fg1 (4)]; correspondingly, posterior scutal protuberances do not convergent posteriorly (can be divergent or parallel – FIG.102:A), lateropostnotum [see (2)] is not fused with epimeron, and is separated by epimeron from coxal conjunctiva (FIG.102:B). Lateroparapsidal suture somewhat retains curvation initial for Ephemerella/ /fg1 [see Leptohyphes/fg1 (4)], merging with transverse interscutal suture laterad of medioparapsidal suture (FIG.102:A). Hind wings [of peculiar structure – see Leptohyphes/fg1 (2) and FIG.100:B–D] are present at least in male; in female they are either present, or lost; metathoracic wing muscles S.ESm and MTm are retained even in females which have no hind wings (FIG.102:C–D). 1st tarsal segment is short and fused with tibia (as in FIG.83:G). Gonostylus is 3-segmented, i. e. with distal segment [short –

see Pantricorythi (2)] distinguishable at least in imago.

Size. Fore wing length 4–8 mm (see Tricoryptera).

Distribution. Neotropical and south of Nearctic Regions.

Species composition of Leptohyphes/fg2 (here [*L.*]= =[*Leptohyphes*]). Several dozens species were placed to the "genus *Leptohyphes*", but for many of them true systematic position is still unclear (see below, Leptohyphes/fg1 INCERTAE SEDIS). The following species can be attributed here for certain (Baumgardner & McCafferty 2000; Molineri 2003) – *carinus* Allen 1973 [*L.*], *cornutus* Allen 1967 [*L.*], *ecuador* Mayo 1968 [*L.*], *eximius* Eaton 1882 [*L.*] (synn. subj.: *bruchi* Navás 1913 [*L.*], *nigra* Navás 1920 [*Bruchella*]), *guadeloupensis* Hofmann & Sartori 1999 [*L.*], *illiesi* Allen 1967 [*L.*], *invictus* Allen 1973 [*L.*], *jodiannae* Allen 1967 [*L.*], *liniti* Wang & Sites & McCafferty 1998 [*L.*], *maculatus* Allen 1967 [*L.*] (synn. subj.: *comatus* Allen 1967 [*L.*], *hirsutus* Allen & Roback 1969 [*L.*], *myllonotus* Allen & Roback 1969 [*L.*]), *mollipes* Needham & Murphy 1924 [*L.*], *nigripunctum* Traver 1943 [*L.*], *peterseni* Ulmer 1919 [*L.*], *petersi* Allen 1967 [*L.*], *plaumanni* Allen 1967 [*L.*] (syn. subj.: *pereirae* Da-Silva 1993 [*L.*]), *populus* Allen 1973 [*L.*], *setosus* Allen 1967 [*L.*] (syn. subj.: *echinatus* Allen & Roback 1969 [*L.*]), *tacajalo* Mayo 1968 [*L.*] (syn. subj.: *albus* Mayo 1968 [*L.*]), *tuberculatus* Allen 1967 [*L.*], *zalope* Traver 1958 [*L.*] (synn. subj.: *apache* Allen 1967 [*L.*], *ferruginus* Allen & Brusca 1973 [*L.*], *hispidus* Allen & Brusca 1973 [*L.*], *lumas* Allen & Brusca 1973 [*L.*], *piraticus* Allen 1978 [*L.*], *succinus* Allen 1978 [*L.*], *zelus* Allen 1978 [*L.*]).

Material examined: *eximius* [*L.*]: ♁/⊙ (holotypus, deposited in Harvard Univ.); Ơ/♂, ♀/♀/⊙; *zalope* [*L.*]: Ơ, ⊕♂, ♁; spp. (Argentina and Ecuador): ⊕♂, ♁; sp. (Brazil, deposited in Entomol. Inst.): Ơ, ○.

***

2.2;2,2-2/5.2;2. Allenhyphes/g(1)

(Ephemerella/fg1 Pantricorythi Tricoryptera Leptohyphes/fg1 *Allenhyphes*)

(Fig. 105:F)

Nomen hierarchicum: **Allenhyphes/g(1)** [g: *Allenhyphes* Hofmann & Sartori (in Hofmann & Sartori & Thomas) 1999: 67, typus *Leptohyphes flinti* Allen 1973 (design. orig.)].

In circumscription matches:

— gen. *Allenhyphes*: Hofmann & al. 1999: 67.

References. Henry 1986: (*Leptohyphes vescus*): ⊕♂; – Hofmann & Sartori & Thomas 1999: ⊙ Ơ ⊕; – Wiersema & McCafferty 2000: Ơ ⊕; – Molineri & Flowers 2001: ⊕♂.

Autapomorphy of *Allenhyphes*.

(1) In contrast to all other mayflies, male imaginal paracercus is specialized and involved into genital apparatus: apex of its 2nd segment bears a stout spine-like seta directed ventrally, perpendicular to paracercus (FIG.105:F; Molineri & Flowers 2001: Fig.2). Unique character.

Characters of unclear phylogenetic status.

(2) Mouth apparatus is somewhat specialized (Hofmann & al. 1999:Figs 147–152): maxilla is filtering, with enlarged tuft of apical setae [see Ephemerella/fg1 (16)] (but it retains blunt canines, well-developed two dentisetae and vestigial 3-segmented maxillary palp); labial submentum with enlarged lamellate margins overlapping maxillae. Non-unique apomorphies.

(3) Tergalii [see Leptohyphes/fg1 (1)] have no processes on ventral bifurcate lobe, thus each tergalius II–V is 3-lamellate; tergalius VI is 1-lamellate, without ventral lobe. The same in *Tricorythodes*.

Plesiomorphies of *Allenhyphes*. In contrast to *Tricorythodes*: Scuto-lateropostnotal muscle of metathorax is not so large, posterior scutal protuberances are not convergent posteriorly, lateropostnotum is not fused with epimeron (the same in Leptohyphes/fg2). Lateroparapsidal suture somewhat retains curvation initial for Ephemerella/fg1 [see Leptohyphes/fg1 (4)], merging with transverse interscutal suture laterad of medioparapsidal suture (as in FIG.102:A). Hind wings [of peculiar structure – see Leptohyphes/fg1 (2) and FIG.100:B] are present in male, but lost in female; however, metathoracic wing muscles S.ESm and MTm are retained not only in male, but in female as well (as in FIG.102:D). 1st tarsal segment is short and fused with tibia (as in FIG.83:G). In contrast to Leptohyphes/fg2: Tergalius II (gill operculum) has well-developed costal flap of ventral lobe [see Leptohyphes/fg1 (1)] (as in FIG.103:A–B). Lateropostnotum retains straight lateropostnotal crest forming one line with infrascutellar–lateropostnotal arm (as in FIG.97:D). In contrast to *Haplohyphes*, gonostylus is 3-segmented, i. e. with developed distal segment [short – see Pantricorythi (2)].

Size. Fore wing length 3 mm (see Tricoryptera).

Distribution. Neotropical Region.

Species composition of Allenhyphes/g(1). 2 species are known as larvae and adults associated for certain – *flinti* Allen 1973 [*Leptohyphes*] and *vescus* Allen 1978 [*Leptohyphes*]. Besides this, here were presumably attributed some other species of Leptohyphes/fg1, known as larvae only (Wiersema & McCafferty 2000).

Material examined: *flinti* [*L.*]: Ơ, ♀/♀/⊙, ⊕♂, ♁; *vescus* [*L.*]: Ơ.

***

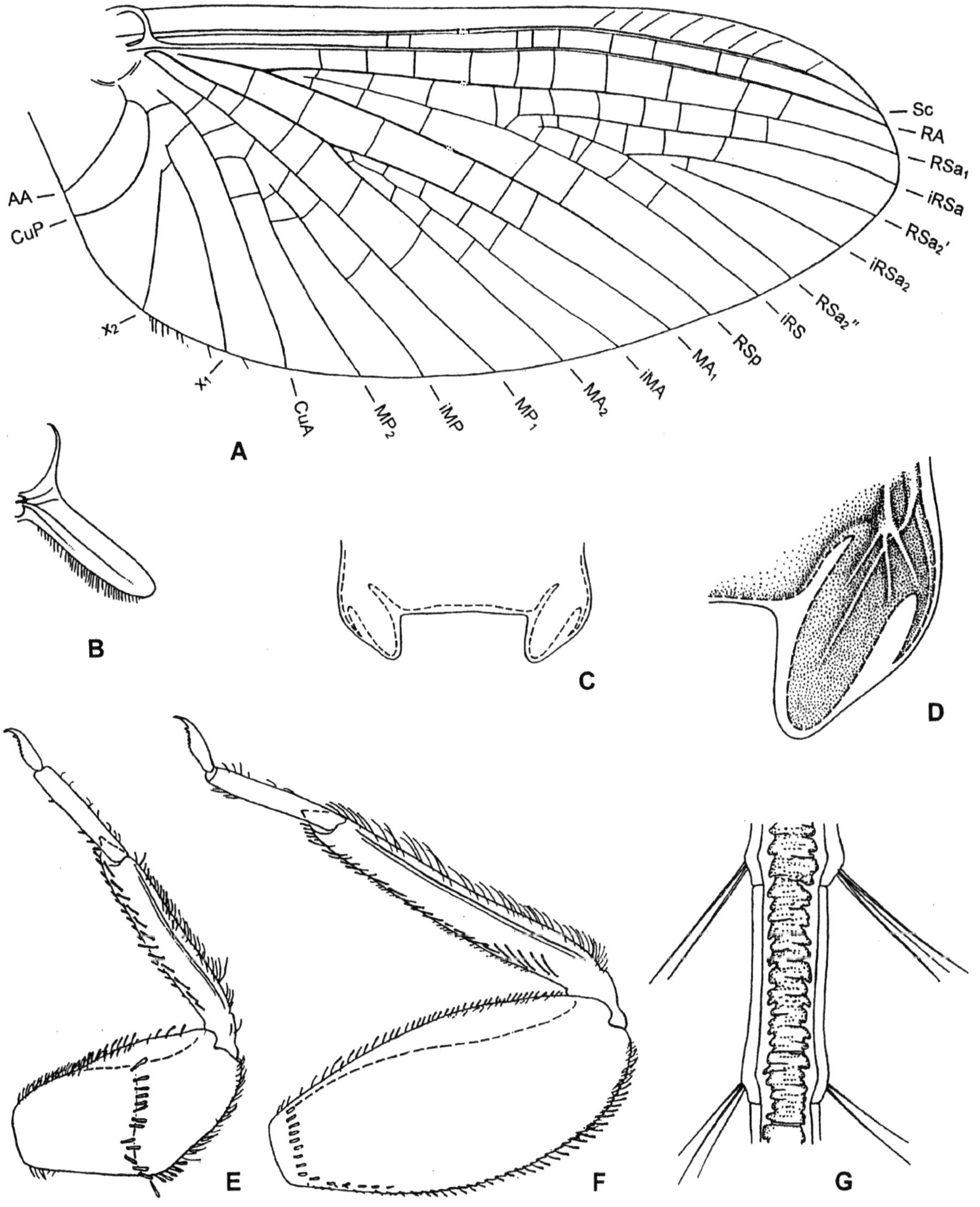

Figure 100. Leptohyphes/fg2.
A–D – *zalope* [*Leptohyphes*]: **A–B** – fore and hind wings of male imago; **C** – immature male larva of last instar, metanotum with hind wing buds (scale as in Figs A–B); **D** – left wing bud enlarged (subimaginal hypodermal parts inside shown by interrupted line and dots, venation shown by light strips). **E–G** – *eximius* [*Leptohyphes*], larva: **E** – right fore leg, dorsal view; **F** – right hind leg, dorsal view; **G** – optic section of median part of cercus of mature female lava, with crumpled subimaginal paracercus inside (compare with Fig.105:E). (E–G – from Kluge 1992c).

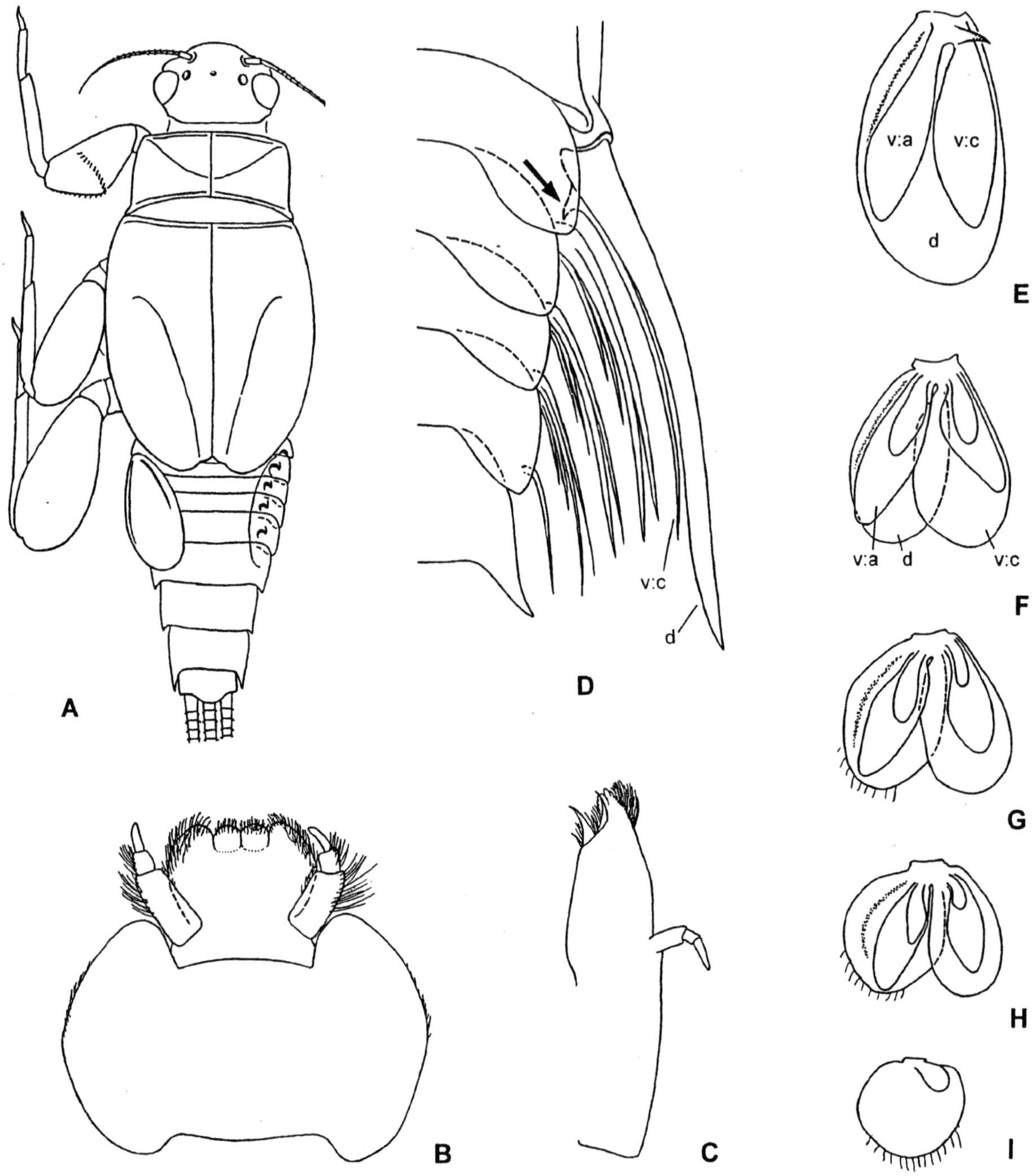

Figure 101. Leptohyphes/fg* *eximius* [*Leptohyphes*], larva.
A – female larva of last instar (right legs and tergalii removed); **B** – labium, ventral view; **C** – right maxilla, dorsal view; **D** – right margin of abdominal segments II-VII with right tergalii, ventro-lateral view, tergalii not pressed to the body (arrow shows spine-like vestige of costal flap) (compare with Fig.103:A–B); **E–I** – left tergalii II-VI (ventral view). (A–C, E–I – from Kluge 1992c).

Abbreviations: **d** – dorsal lobe of tergalius; **v:a** – anal branch of ventral lobe; **v:c** – costal branch of ventral lobe.

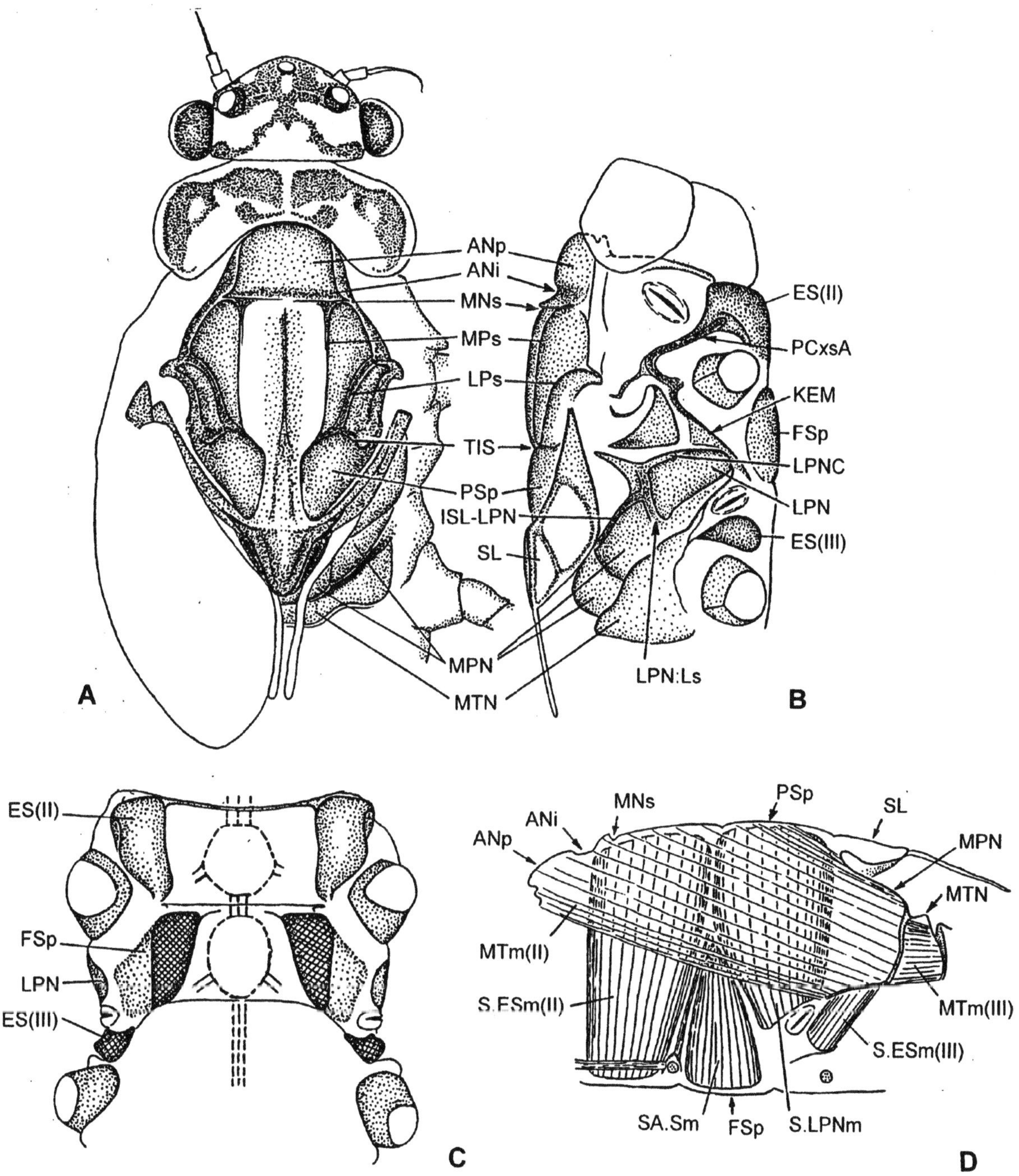

Figure 102. Leptohyphes/fg* *eximius* [*Leptohyphes*], female subimago dissected from mature larva.
A – subimaginal head and thorax, dissected from larva (left fore wing folded as in larval wing bud, right wing removed); **B** – thorax, lateral view; **C** – pterothorax, ventral view (nerve chain shown by interrupted line, muscle bases shown by checks); **D** – pterothorax, median section. (From Kluge 1992c)

Abbreviations: **ANi** – anteronotal transverse impression; **ANp** – anteronotal protuberance; **ES(II)** – episternum of mesothorax; **ES(III)** – episternum of metathorax; **FSp** – furcasternal protuberance; **ISL-LPN** – infrascutellar-lateropostnotal arm; **KEM** – katepimeron; **LPN** – lateropostnotum; **LPNC** – lateropostnotal crest; **LPs** – lateroparapsidal suture; **LPN:Ls** – longitudinal suture of lateropostnotum; **MNs** – mesonotal suture; **MPN** – mediopostnotum; **MPs** – medioparapsidal suture; **MTm(II)** – median tergal muscle of mesothorax; **MTm(III)** – median tergal muscle of metathorax; **MTN** – metanotum; **PCxsA** – anterior paracoxal suture; **PSp** – posterior scutal protuberance; **SA.Sm** – subalar-sternal muscle; **S.ESm(II)** – scuto-episternal muscle of mesothorax; **S.ESm(III)** – scuto-episternal muscle of metathorax; **SL** – scutellum; **S.LPNm** – scuto-lateropostnotal muscle; **TIS** – transverse interscutal suture.

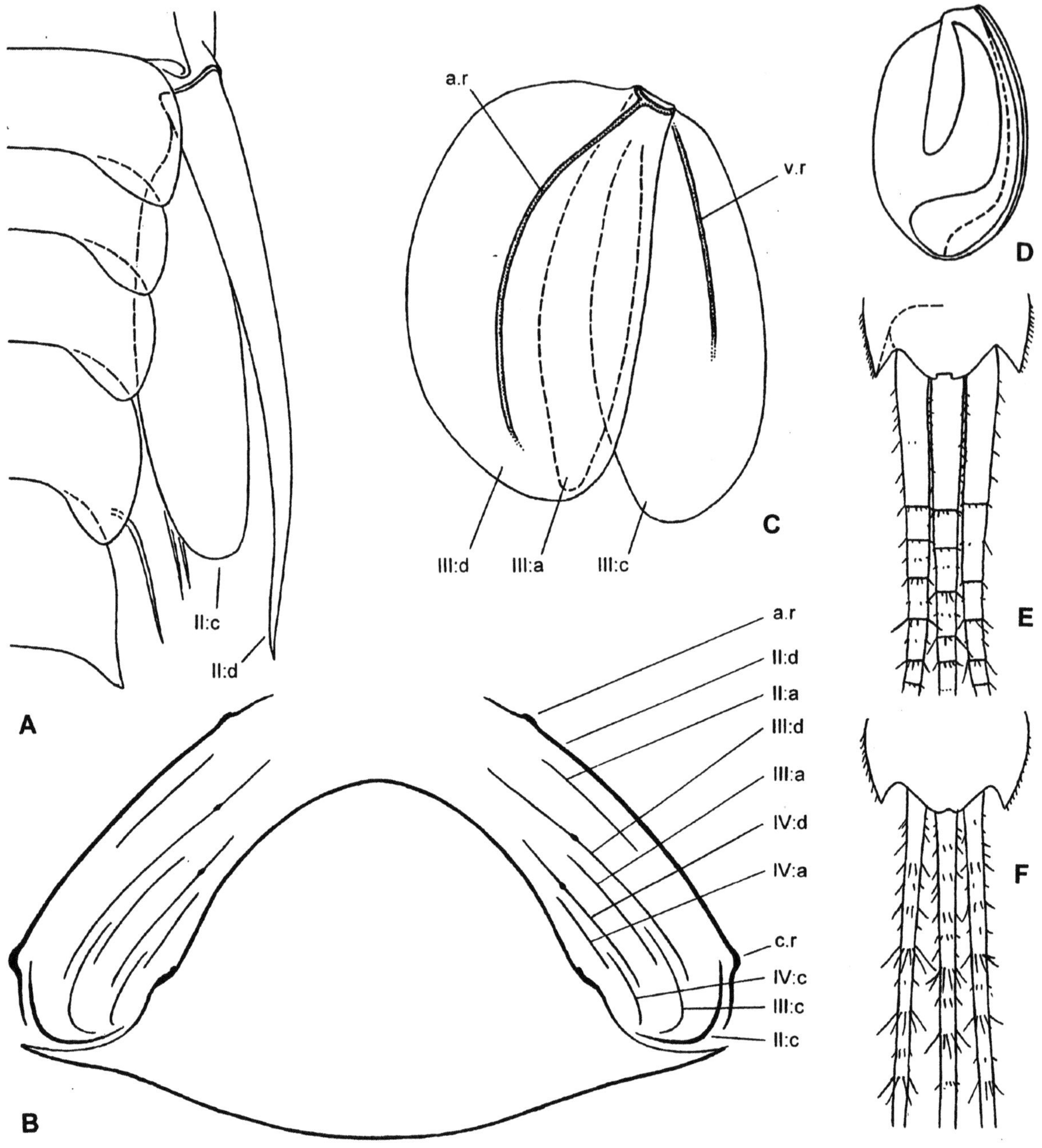

Figure 103. Leptohyphes/fg1-Tricorythodes/fg1, larva.
A–B – *minutus* [*Tricorythodes*]: **A** – right margin of abdominal segments II-VII with right tergalii, ventro-lateral view, tergalii not pressed to the body (compare with Fig.101:D); **B** – transverse section on level of end of abdominal segment V, tergalii pressed to the body. **C** – *cubensis* [*Tricorythodes*], right tergalius III, dorsal view. **D–F** – *sacculobranchis* [*Tricorythodes*]: **D** – left tergalius II (gill operculum), ventral view; **E** – abdominal sternum IX and caudalii of male larva; **F** – the same of female larva. (D–F – from Kluge & Naranjo 1990).

Abbreviations: **II:a** – anal branch of ventral lobe of tergalius II; **II:c** – costal branch of ventral lobe of tergalius II; **II:d** – dorsal lobe of tergalius II (gill operculum); **III:a** – anal branch of ventral lobe of tergalius III; **III:c** – costal branch of ventral lobe of tergalius III; **III:d** – dorsal lobe of tergalius III; **IV:a** – anal branch of ventral lobe of tergalius IV; **IV:c** – costal branch of ventral lobe of tergalius IV; **IV:d** – dorsal lobe of tergalius IV; **a.r** – anal rib; **c.r** – costal rib; **v.r** – additional rib on costal branch of ventral lobe

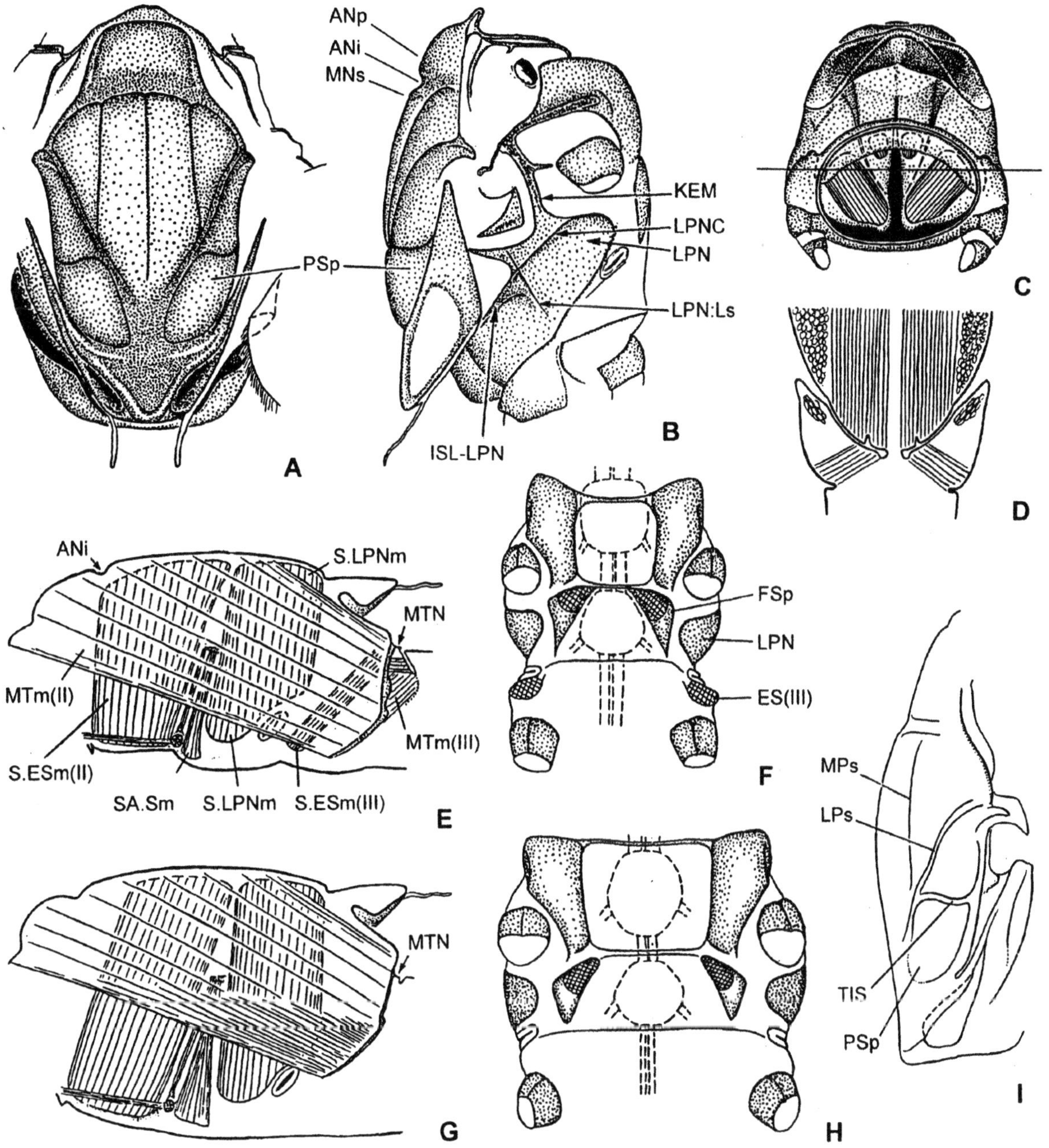

Figure 104. Tricorythodes/fg1 *cubensis* [*Tricorythodes*], pterothorax.
A – female imaginal pterothorax, dorsal view; **B** – female imaginal pterothorax, lateral view; **C** – male imaginal pterothorax, posterior view (line shows section in Fig.D); **D** – longitudinal section of hind part of male pterothorax (by line shown in Fig.C) (compare with Fig.87:C–D); **E** – male imaginal pterothorax, median section; **F** – male imaginal pterothorax, ventral view (nerve chain shown by interrupted line, muscle bases shown by checks); **G** – female imaginal pterothorax, median section; **H** – female imaginal pterothorax, ventral view (nerve chain shown by interrupted line, muscle bases shown by checks); **I** – subimaginal exuviae of right half of mesonotum.

Abbreviations: **ANi** – anteronotal transverse impression; **ANp** – anteronotal protuberance; **ES(III)** – episternum of metathorax; **FSp** – furcasternal protuberance; **ISL-LPN** – infrascutellar-lateropostnotal arm; **KEM** – katepimeron; **LPN** – lateropostnotum; **LPNC** – lateropostnotal crest; **LPN:Ls** – longitudinal suture of lateropostnotum; **LPs** – lateroparapsidal suture; **MNs** – vestige of mesonotal suture; **MPs** – medioparapsidal suture; **MTm(II)** – median tergal muscle of mesothorax; **MTm(III)** – the same of metathorax; **MTN** – metanotum; **PSp** – posterior scutal protuberance; **SA.Sm** – subalar-sternal muscle; **S.ESm(II)** – scuto-episternal muscle of mesothorax; **S.ESm(III)** – the same of metathorax; **SL** – scutellum; **S.LPNm** – scuto-lateropostnotal muscle; **TIS** – transverse interscutal suture.

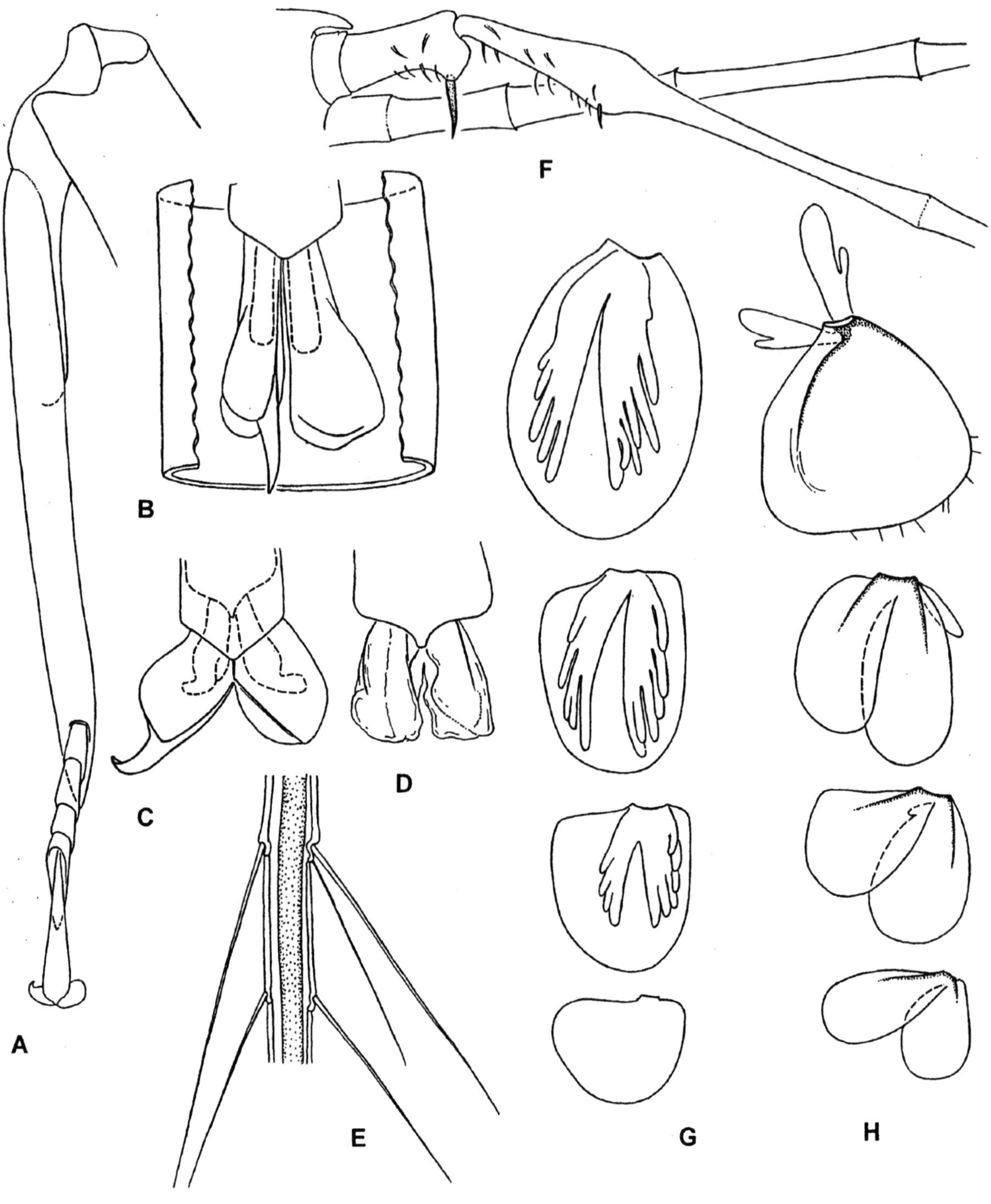

Figure 105. Tricorythodes/fg1, *Allenhyphes* and Ephemerella/fg1 INCERTAE SEDIS.
A–E – Tricorythodes/fg1 *cubensis* [*Tricorythodes*]: **A** – tibia and tarsus of imaginal right hind leg, anterior view; **B–D** – claws of male left fore leg: **B** – subimaginal claws inside tarsus of mature larva (imaginal cuticle shown by interrupted line); **C** – subimaginal claws (imaginal cuticle shown by interrupted line); **D** – imaginal claws; **E** – optic section of median part of cercus of mature female lava, with subimaginal paracercus inside (compare with Fig.100:G). **F** – Allenhyphes/g* *flinti* [*Leptohyphes*], base of paracercus and right cercus of male imago, lateral view (left cercus removed). **G** – Coryphorus/fg1 sp.Cor1, left tergalii II–V, ventral view. **H** – ?Leptohyphodes/g sp. (Brazil), right tergalii II–V, dorsal view.

2.2;2,2-2/5.2;3. Tricorythodes/fg1

(Ephemerella/fg1 Pantricorythi Tricoryptera Leptohyphes/fg1 *Tricorythodes*)
(Figs 103–104; 105:A–E)

Nomen hierarchicum: **Tricorythodes/fg1** incl. *Tricorythafer, Tricoryhyphes, Homoleptohyphes, Asioplax*) [f: Tricorythodinae Wiersema & McCafferty 2000: 346: *Tricorythodes* Ulmer 1920a: 51, typus *Tricorythus explicatus* Eaton 1892 (design. orig.)].
In circumscription matches:
— gen. *Tricorythodes* Ulmer 1920a: 51.

References. Ulmer 1920a: ⊕; – Needham & Traver & Hsu 1935: ♉* ⊕*; – Kluge 1992c: ♉* ◎* ⊕*; – Molineri 2002: ♉* ⊕*.

Autapomorphy of *Tricorythodes*.

(1) 2nd segment of gonostylus is basally strongly swollen, bulbous (Kluge & Naranjo 1990:Figs 26–29, 51–56, 70–71). In other respects gonostylus retains plesiomorphic features: 1st segment [elongate – see Pantricorythi (2)] is well-developed, 3rd (distal) segment is present (in contrast to *Tricorythopsis* and some others – see Index of characters [2.3.12]).

Apomorphies of *Tricorythodes* common with Caenoptera (see Classifications of Furcatergaliae II).

(2) Posterior scutal protuberances [bearing dorsal bases of S.LPNm – see (3)] are strongly enlarged and convergent posteriorly; each of them has a common boundary with medioscutum all over its length, because lateroparapsidal suture lost its curvation and merges with medioparapsidal suture anteriad of transverse interscutal suture [see Leptohyphes/fg1 (4)] (FIG.104:A). Similar in Tricorygnatha.

(3) Lateropostnotum [bearing ventral base of S.LPNm – see (2)] is strongly enlarged, stretches up to sternum, reaches coxal conjunctiva and is fused with epimeron (FIG.104:B). The same in *Tricorythopsis*, in contrast to all other Ephemerella/fg1 (see Index of characters [2.2.16]).

(4) Metathorax is strongly reduced (FIG.104:E–G): hind wings [see Leptohyphes/fg1 (2)] are lost in both sexes; in female metathoracic scuto-episternal muscles (S.ESm) are lost (while in male non-functional indirect wing muscles S.ESm and MTm of metathorax are retained). The same in Caenoptera; in other mayflies, even if hind wings are completely lost, metathoracic indirect wing muscles are retained in both sexes.

(5) In imago and subimago 1st tarsal segment of middle and hind legs [initially fused with tibia and shortened – see Furcatergaliae (2)] on its outer side is secondarily distinctly separated from tibia; joining of tibia and tarsus is strongly inclined resembling larval condition, thus outer side of 1st tarsal segment is strongly elongate. Tarsus retains 5 distinctly separated segments (FIG.105:A). The same in Caenoptera (FIG.86:G); in *Tricorythopsis* and Tricorygnatha tarsus is separated in the same manner, but has four segments only (see Index of characters [2.2.84]).

(6) In female imago and subimago caudalii [cerci and paracercus – see Plesiomorphies of Ephemerella/fg1] are strongly diminished, shorter than the body. When adult caudalius develops under larval cuticle, it is not crumpled (FIG.105:E), in contrast to Leptohyphes/fg2 (FIG.100:G), *Allenhyphes*, Tricorygnatha and other taxa. This character is not connected with sexual dimorphis of larval caudalii [see Leptohyphes/fg1 (8)].

Apomorphy of *Tricorythodes* common with Leptohyphes/fg2 and *Allenhyphes*.

(7) Tergalii completely lost processes of ventral bifurcate lobes, thus each tergalius consists of no more than 3 simple lamellae arising from base (FIG.103:B–C) [see Leptohyphes/fg1 (1)]. The same on tergalii II of Leptohyphes/fg2 and all tergalii of *Allenhyphes*.

In selected species (*bullus* [*T.*], *cubensis* [*T.*], *grallator* [*T.*], *santarita* [*T.*]) tergalius II (gill operculum) lost anal branch of ventral lobe, thus is 2-lamellate; in *santarita* [*T.*] costal branch of its ventral lobe is vestigial (Molineri 2002:Fig.136); in other species this costal branch is well-developed and has completely developed costal flap [see Leptohyphes/fg1 (1)] (FIG.103:A–B) (in contrast to Leptohyphes/fg2). Tergalii III–V are always 3-lamellate. In various species tergalius VI is 2-lamellate or 1-lamellate (Kluge & Naranjo 1990: Figs 18–65; Molineri 2002:Figs 14–172).

Plesiomorphies of *Tricorythodes*. Mouth apparatus [see Ephemerella/fg1 (1)–(3), (16)] is non-specialized, maxilla with 3 canines and 2 dentisetae [see Bidentiseta (1)]. In contrast to Leptohyphes/fg2, lateropostnotum [see (3)] retains well-developed straight lateropostnotal crest forming one line with infrascutellar-lateropostnotal arm (FIG. 104:B).

Variable characters of *Tricorythodes*. In the group *sacculobranchis* larval femoral setae [see Pantricorythi (1)] are short and blunt, and gill operculum is oval, with maximum width near apex (FIG.103:D) (the same in Leptohyphes/fg2 and *Allenhyphes*); in many other *Tricorythodes* femoral setae are pointed, and gill operculum is triangular, with maximum width near base.

Size. Fore wing length 3–9 mm (see Tricoryptera).

Distribution. America: Neotropical and Nearctic Regions.

Nominal taxa included. Tricorythodes/fg1 includes: (1) **Tricorythafer/g** [g: *Tricorythafer* Lestage 1942: 4, typus *Caenopsis fugitans* Needham 1920 (monotypy); synn. obj.: ***Caenopsis*** Needham 1920: 39 (non *Caenopsis* Bach 1854), ***Needhamocoenis*** Lestage 1945: 85] (probably erroneously described from Africa as male imago of a single species); (2) **Tricoryhyphes/g** [g: *Tricorythodes* subgen. *Tricoryhyphes* Allen & Murvosh 1987: 38, typus *Tricorythodes condylus* Allen 1987 (design. orig.)]; (3) **Homoleptohyphes/g** [g: *Tricorythodes* subgen. *Homoleptohyphes* Allen & Murvosh 1987: 39, typus *Tricorythodes dimorphus* Allen 1987 (design. orig.)]; (4) **Asioplax/g** [g: *Asioplax* Wiersema & McCafferty 2000: 347, typus *Tricorythodes edmundsi* Allen 1964 (design. orig.)], regarded as a generic synonym of *Tricorythodes* (Molineri 2002).

Species composition of Tricorythodes/fg1 (here [*T.*]=[*Tricorythodes*]). About 40 species are described (many of them are poorly known and actually can belong to other taxa in Leptohyphes/fg1) – *albilineatus* Berner 1946 [*T.*], *allecta* Needham 1905 [*Caenis*] (syn. subj.: *atrata* McDunnough 1923 [*Tricorythus*]), *angulatus* Traver 1959 [*T.*], *arequita* Traver 1959 [*T.*], *barbus* Allen 1967 [*T.*], *bullus* Allen 1967 [*T.*], *cobbi* Alba-Tercedor & Flannagan 1995 [*T.*], *comus* Traver 1959 [*T.*], *condylus* Allen 1987 [*T.*], *corpulentus* Kilgore & Allen 1973 [*T.*], *cristatus* Allen 1967 [*T.*], *cubensis* Kluge & Naranjo 1990 [*T.*], *curiosus* Lugo-Ortiz & McCafferty 1995 [*Leptohyphes*], *curvatulus* Allen & Cohen 1977 [*T.*], *dimorphus* Allen 1987 [*T.*], *edmundsi* Allen 1964 [*T.*], *explicatus* Eaton 1892 [*Tricorythus*], *fictus* Traver 1935 [*T.*], *fallacina* McDunnough 1939 [*T.*], *fugitans* Needham 1920 [*Caenopsis*], *grallator* Kluge & Naranjo 1990 [*T.*], *griseus* Hofmann & Sartori 1999 [*T.*], *hiemalis* Molineri 2001 [*T.*], *lichyi* Traver 1943 [*T.*], *minutus* Traver 1935 [*T.*], (syn. subj.: *fallax* Traver 1935 [*T.*]), *mirca* Molineri 2002 [*T.*], *mirus* Allen 1967 [*Leptohyphes*] (syn. subj.: *baumanni* Kilgore & Allen 1973 [*Leptohyphes*]), *molequs* Alba-Tercedor & Flannagan 1995 [*T.*], *montanus* Kluge & Naranjo 1990 [*T.*], *mulaiki* Traver 1959 [*T.*], *nicholsae* Wang & Sites & McCafferty 1997 [*Leptohyphes*], *notatus* Allen & Brusca 1973 [*T.*], *numinuh* Wiersema & McCafferty & Baumgardner 2001 [*Asioplax*], *ocellus* Allen & Roback [*T.*], *peridius* Burks 1953 [*T.*], *popayanicus* Domínguez 1982 [*T.*], *quizeri* Molineri 2002 [*T.*], *sacculobranchis* Kluge & Naranjo 1990 [*T.*], *santarita* Traver 1959 [*T.*], *sierramaestrae* Kluge & Naranjo 1990 [*T.*], *sordidus* Allen 1967 [*T.*], *stygiatus* McDunnough 1931 [*T.*], *texanus* Traver 1935 [*T.*], *ulmeri* Allen & Brusca 1973 [*T.*], *undatus* Lugo-Ortiz & McCafferty 1995 [*T.*], *yura* Molineri 2002 [*T.*], *zunigae* Molineri 2002 [*T.*].

Material examined: **Group *explicatus*:** *allecta* [*C.*]: ♉♀–◎♀–⊕♀/⊙, ♂/⊕♂, ⊕♂, ⊕♀; *fictus* [*T.*]: ♉, ⊕♂, ⊕♀; *minutus* [*T.*]: ♉, ♉♂/♂/⊕♂, ♉♀/♀/⊕♀/⊙, ⊕♂. **Group *cubensis*:** *cubensis* [*T.*]: ♉–◎–⊕/⊙; *montanus* [*T.*]: ♉–◎–⊕/⊙. **Group *dimorphus* (Homoleptohyphes/g):** *dimorphus* [*T.*]: ♉; *mirus* [*L.*]: ♉. **Group *grallator*:** *grallator* [*T.*]: ♉–◎–⊕/⊙; *sordidus* [*T.*]: ♉; sp. (Jamaica): ♂/⊕♂, ⊕♂, ♀/⊕♀/⊙. **Group *edmundsi* (Asioplax/g):** *numinuh* [*A.*]: ♉–⊕♂, ♉–⊕♀. **Group *sacculobranchis*:** *sacculobranchis* [*T.*]: ♉–◎–⊕/⊙; *sierramaestrae* [*T.*]: ♉–◎–⊕/⊙.

***

2.2;2,2-2/5.2;4. Tricorythopsis/g(1)

(Ephemerella/fg1 Pantricorythi Tricoryptera Leptohyphes/fg1 *Tricorythopsis*)

Nomen hierarchicum: **Tricorythopsis/g(1)** [g: *Tricorythopsis* Traver 1958: 491, typus *T. artigas* Traver 1958 (design. orig.)].
In circumscription matches:
— gen. *Tricorythopsis* Traver 1958: 491.

References. Traver 1958: ⊕♂; – Molineri 1999b: ♉' ⊕*; – Wiersema & McCafferty 2000: ♉ ⊕*; – Molineri 2001c: ♉* ⊕*.

Autapomorphies of *Tricorythopsis.*

(1) On fore wing proximal end of iMP is adjacent or connected with CuA, while MP_2 originates more distally; thus iMP, MP_2 and CuA form a secondary triad (Molineri 1999:Figs 19–37; 2001c: Figs 51–107). In other Pantricorythi MP_2 is usually also detached from MP-MP_1, but is not shorter than iMP (FIG.100:A).

(2) On middle and hind legs of imago and subimago tarsus is 4-segmented: it has only 2 articulations between tibio-tarsal articulation [see (7)] and the articulation of claw-bearing segment (initial segment 5) (Molineri 2001c:Fig.20). The same in Tricorygnatha (FIG.98:B) and some others (see Index of characters [2.2.83]).

(3) Gonostyli are non-functional and have peculiar structure: distal segment is lost (non-unique character – see Index of characters [2.3.12]); the last (2nd) segment is directed laterally [see Pantricorythi (2)]; styliger forms a pair of more or less prominent narrow gonostyli pedestals lacking muscles (Molineri 1999b:Figs 18–35; 2001c:Figs 1–102).

(4) Egg [initially having a polar cap – see Ephemerella (14)] has 2 polar caps (Molineri 2001c: Figs 122–131). Non-unique character (see Index of characters [3.5]).

Characters of *Tricorythopsis* common with *Tricorythodes* and Caenoptera.

(5) Lateropostnotum is large, stretches up to sternum and fused with epimeron.

(6) Hind wings [see Leptohyphes/fg1 (2)] are lost in both sexes (see Index of characters [2.2.59]).

(7) On middle and hind legs of imago and subimago [see (2)] 1st tarsal segment [initially fused with tibia and shortened – see Furcatergaliae (2)] on outer side is secondarily distinctly separated from tibia; joining of tibia and tarsus is strongly inclined resembling larval condition, thus outer side of 1st tarsal segment is strongly elongate (Molineri 2001c: Fig.20). The same is *Tricorythodes* (possibly synapomorphy) and Caenoptera.

Characters of unclear phylogenetic status.

(8) Tergalial ventral lobes [see Leptohyphes/fg1 (1)] have peculiar structure uniform in all species: on tergalius II (operculate) each of two branches has one process (the same in *Leptohyphodes* only), costal flap is probably lost; on each tergalius III–IV costal branch (widely projected, as in other Leptohyphes/fg1) has one process arising from base, anal branch is narrow and lacks processes; on tergalius V costal branch is also wide, but lacks processes, anal branch is lost; tergalius VI has no ventral lobe (Molineri 2001c:Figs 6–121).

Size. Fore wing length 2–3 mm (see Tricoryptera).

Distribution. Neotropical Region.

Species composition of Tricorythopsis/g(1) (here [*T.*]=[*Tricorythopsis*]). 9 species – *artigas* Traver 1958 [*T.*], *chiriguano* Molineri 2001 [*T.*], *gibbus* Allen 1967 [*Leptohyphes*], *fictus* Molineri 1999 [*T.*], *minimus* Allen 1973 [*Leptohyphes*] (syn. subj.: *viriosus* Allen 1973 [*Leptohyphes*]), *sigillatus* Molineri 1999 [*T.*], *undulatus* Allen 1967 [*Leptohyphes*] (syn. subj.: *petersorum* Molineri 1999 [*T.*]), *volsellus* Molineri 1999 [*T.*], *yacutinga* Molineri 2001 [*T.*].

Material examined: *artigas* [*T.*]: ⊕.

***

Leptohyphes/fg1 INCERTAE SEDIS

Leptohyphes/fg1 INCERTAE SEDIS 1. Haplohyphes/g(1) [g: *Haplohyphes* Allen 1966: 566, typus *H. huallaga* Allen 1966 (design. orig.)].

References. Allen 1966: ⊕; – Domínguez 1984: ℧; – Molineri 1999a: ⊕; – Wiersema & McCafferty 2000: ℧ ⊕; – Molineri 2003b: ⊙ ℧ ⊕.

Autapomorphies.

(1) Larval mouthparts are modified, maxilla is filtering, widened apically, with long apical setae.

(2) Distal segment of gonostylus [initially single – see Ephemerella/fg1 (8)] is lost (non-unique apomorphy – see Index of characters [2.3.12]).

Plesiomorphy. Hind wings are present in both sexes [see Leptohyphes/fg1 (2)].

Size. Fore wing length 4–6.5 mm (see Tricoryptera).

Distribution. Neotropical Region.

Species composition: 6 species – *aquilonius* Lugo-Ortiz & McCafferty 1995 [*Haplohyphes*], *baritu* Domínguez 1984 [*Haplohyphes*] (syn. subj. *furtiva* Domínguez 1984 [*Haplohyphes*]), *dominguezi* Molineri 1999 [*Haplohyphes*], *huallaga* Allen 1966 [*Haplohyphes*], *mithras* Traver 1958 [*Leptohyphes*], *yanahuicsa* Molineri 2003 [*Haplohyphes*].

Material examined: –.

Leptohyphes/fg1 INCERTAE SEDIS 2. Cotopaxi/g(1) [g: *Cotopaxi* Mayo 1968: 301, typus *C. macuchae* Mayo 1968 (design. orig.)].

References. Mayo 1968: ⊕ (larvae unknown); – Molineri 2003: ⊕.

Material examined: –.

Comment. Described as a single adult specimen from Equador, which was regarded to be a male with unusually modified genitals, lacking gonostyli and imaginal caudalii (Mayo 1968). Molineri (2003) regards this specimen to be a gynandromorph belonging to Leptohyphes/fg2.

Leptohyphes/fg1 INCERTAE SEDIS 3. Vacupernius/g(1) [g: *Vacupernius* Wiersema & McCafferty 2000: 345, typus *Leptohyphes packeri* Allen 1967 (design. orig.)].

References. Henry 1986 (*Leptohyphes packeri*): ⊕; – Wiersema & McCafferty 2000: ℧ ⊕.

Plesiomorphy. Hind wings are present in males only [see Leptohyphes/fg1 (2)].

Size. Fore wing length 3–3.5 mm (see Tricoryptera).

Distribution. Neotropical Region and south of Nearctic Region.

Species composition: 1 species is attributed here for certain – *packeri* Allen 1967 [*Leptohyphes*] (syn. subj.: *phalarobranchus* Kilgore & Allen 1973 [*Leptohyphes*]); 2 other species of Leptohyphes/fg1 were placed here presumably (Wiersema & McCafferty 2000).

Material examined: –.

Leptohyphes/fg1 INCERTAE SEDIS 4. Epiphrades/g(1) [g: *Epiphrades* Wiersema & McCafferty 2000: 349, typus *Tricorythodes undatus* Lugo-Ortiz & McCafferty 1995 (design. orig.)].

Reference. Wiersema & McCafferty 2000: ℧.

Character of unclear phylogenetic status.

(1) Hind wings [see Leptohyphes/fg1 (2)] are lost in both sexes (see Index of characters [2.2.59]).

Distribution. Neotropical Region.

Species composition: 1 species – *undatus* Lugo-Ortiz & McCafferty 1995 [*Tricorythodes*].

Material examined: –.

Leptohyphes/fg1 INCERTAE SEDIS 5. Yaurina/g(1) [g: *Yaurina* Molineri 2001a: 338, typus *Y. yuta* Molineri 2001 (design. orig.)].

Reference. Molineri 2001a: ℧* ⊕*.

Autapomorphy.

(1) Penis with a pair of long processes arising from ventral side of base and directed distally (Molineri 2001a:Figs 1–9).

Plesiomorphy. Hind wings are present in males only [see Leptohyphes/fg1 (2)].

Size. Fore wing length 3 mm (see Tricoryptera).

Distribution. Neotropical Region.

Species composition: 3 species – *mota* Molineri 2001 [*Yaurina*], *yapa* Molineri 2001 [*Yaurina*] and *yuta* Molineri 2001 [*Yaurina*].

Material examined: –.

Leptohyphes/fg1 INCERTAE SEDIS 6. Traverhyphes/g(1) [g: *Traverhyphes* Molineri 2001b: 130, typus *Leptohyphes indicator* Needham & Murphy 1924 (design. orig.)].

Reference. Molineri 2001b: ♉' ⊕*.

Autapomorphy.

(1) Genitals have peculiar structure, penis with a pair of processes arising on ventral side and directed proximally (Molineri 2001b:Figs 8–15, 38–41).

Plesiomorphy. Hind wings are present in males only [see Leptohyphes/fg1 (2)].

Size. Fore wing length 3–4 mm (see Tricoryptera).

Distribution. Neotropical Region.

Species composition: 2 species – *indicator* Needham & Murphy 1924 [*Leptohyphes*] and *pirai* Molineri 2001 [*Traverhyphes*].

Material examined: –.

Other species of Leptohyphes/fg1 INCERTAE SEDIS (here [*L.*]=[*Leptohyphes*]. Following species have insufficient descriptions, which do not allow to determine their systematic position more exactly – *alleni* Brusca 1971 [*L.*], *asperulus* Allen 1967 [*L.*], *australis* Banks 1913 [*Tricorythus*], *berneri* Traver 1958 [*L.*], *brevissimus* Eaton 1892 [*L.*], *brunneus* Allen & Brusca 1973 [*L.*], *castaneus* Allen 1967 [*L.*], *costaricanus* Ulmer 1920 [*L.*], *dicinctus* Allen & Brusca 1973 [*L.*], *dolani* Allen 1967 [*L.*], *edmundsi* Allen 1973 [*L.*], *jamaicanus* Allen 1973 [*L.*], *lestes* Allen & Brusca 1973 [*L.*], *melanobranchus* Allen & Brusca 1973 [*L.*], *michaeli* Allen 1978 [*L.*], *murdocki* Allen 1967 [*L.*], *musseri* Allen 1967 [*L.*], *nanus* Allen 1967 [*L.*], *paraguttatus* Allen 1978 [*L.*], *pilosus* Allen & Brusca 1973 [*L.*], *priapus* Traver 1958 [*L.*], *quercus* Kilgore & Allen 1973 [*L.*], *rallus* Allen 1967 [*L.*], *robacki* Allen 1967 [*L.*], *rolstoni* Allen 1973 [*L.*], *sabinas* Traver 1958 [*L.*] (syn. subj.: *consortis* Allen & Brusca 1973 [*L.*]), *spinosus* Allen & Roback 1969 [*L.*], *tarsos* Allen & Murvosh 1987 [*L.*], *tinctus* Allen 1973 [*L.*], *vulturnus* Allen 1978 [*L.*].

*** *** *** ***

Ephemerella/fg1 INCERTAE SEDIS 1. Coryphorus/fg(1)

(Fig. 105:G)

Nomen hierarchicum: **Coryphorus/fg(1)** [f: Coryphoridae Molineri & Peters & Zuñiga-de-Cardoso 2001: 118; g: *Coryphorus* Peters 1981: 209, typus *C. aquilis* Peters 1981 (design. orig.)].

In circumscription matches:

— gen. *Coryphorus* Peters 1981: 209;

— fam. Coryphoridae Molineri & Peters & Zuñiga-de-Cardoso 2001: 118.

References. Peters 1981: ♉''; – Molineri & al. 2001: ⊙' ◎' ⊕'.

Autapomorphy of *Coryphorus*.

(1) On fore wings CuA is shortened, thus cubital field [see (4)] is so strongly diminished that lacks intercalaries or other veins (Peters 1981:Fig.34; Molineri & al. 2001:Fig.8). Apomorphy unique among Recent mayflies (see Index of characters [2.2.51]).

Characters of unclear phylogenetic status.

(2) In imago amphitornal margin of wing with setae (as in subimago of all mayflies). Non-unique apomorphy (see Index of characters [2.2.27]).

(3) Hind wings are lost. Non-unique apomorphy (see Index of characters [2.2.59]); among Ephemerella/fg1 the same in many Tricoryptera.

(4) Tergalii are operculate, 4 pairs only: tergalii I [initially stick-like – see Ephemerella/fg1 (13)] are lost; tergalii II form gill opercula; tergalii VI and VII are lost (FIG.105:G). Non-unique apomorphies, the same in some Pantricorythi (see TABLE 8 and Index of characters [1.3.33], [1.3.39], [1.3.58]).

Plesiomorphies of *Coryphorus*. In contrast to Tricoryptera, veins CuA and CuP [shortened – see (1)] retain portions proximad of *cua-cup* [see Ephemerella/fg1 (6)] (Peters 1981:Fig.34; Molineri & al. 2001:Fig.8). All claws of imago and subimago, including male imaginal fore claws, are ephemeropteroid (see Index of characters [2.2.77]). In contrast to Leptohyphes/fg1, tergalii [see (3)] retain numerous marginal processes of ventral bifurcate lobe [see Ephemerella/fg1 (7)]; in *aquilis* [*C.*] such ventral lobes are retained on tergalii II–V (Peters 1981:Figs 22–23), in sp.Cor1 – on tergalii II–IV only (FIG.105: G). Male larval caudalii are not thickened (Peters 1981:Fig.1) (in contrast to Leptohyphes/fg1 and some others – see Index of characters [1.3.63]).

Variable characters of *Coryphorus*. Mouthparts have variable structure; particularly, glossae and paraglossae can be either developed (in sp.Cor1) or completely lost (in *aquilis* [*C.*]). Abdominal terga can be either simple (in sp.Cor1) or sharply modified, with lateral and dorsal projections forming gill chamber (in *aquilis* [*C.*]).

Size. Larval body length 3–6 mm.

Distribution. Neotropical Region.

Species composition of Coryphorus/fg(1). 1 species is described from Brazil – *aquilis* Peters 1981 [*Coryphorus*].

Material examined: sp. Cor1: 1 ☿ with doubtful label "Argentina, Tucuman, Rio Augostura, Tafi del Valle, 28 I 1969, leg. W.L. Peters" (FIG.105:G).

***

Ephemerella/fg1 INCERTAE SEDIS 2. **Teloganella/fg(1)**

Nomen hierarchicum: **Teloganella/fg(1)** [f: Teloganellidae McCafferty & Wang 2000: 47; g: *Teloganella* Ulmer 1939: 516, typus *T. umbrata* Ulmer 1939 (design. orig.)].

In circumscription matches:
— gen. *Teloganella* Ulmer 1939: 516;
— fam. Teloganellidae McCafferty & Wang 2000: 47.

References. Ulmer 1939: ♀; – Wang & McCafferty & Edmunds 1995: ♉ ♂; – McCafferty & Wang 2000: ♉ ⊕.

Distribution. Oriental Region.

Comment. Originally a single species was described as female subimago from Sumatra (Ulmer 1939). Larva and male imago from Malaysia attributed to this species (Wang & McCafferty & Edmunds 1995) are described not enough completely and non-reliably. I had an opportunity to examine three young larvae of another related species from southern India (Boukal, Palace, Tadiy, Dalace, 21–23 XII 1998) deposited in the Entomological Institute in České Budějovice (Czech Republic).

These larvae have following features. Tergalius I [see Ephemerella/fg1 (13)] is absent; tergalii are present beginning from pair II (in contrast to Ephemerella/fg2); tergalii II–V with ventral lamella bifurcate and bearing processes [see Ephemerella/ /fg1 (7)], dorsal lamella without costae (in contrast to many Pantricorythi) and non-cleft (in contrast to *Teloganodes*); tergalius II forms an opercula; tergalius VI is attached to anterior part of tergite (in contrast to other Ephemerella/fg1) and lacks ventral lamella; tergalius VII is absent. On fore femur setae are irregular, transverse row is poorly expressed [that does nor allow to use the character of Pantricorythi (1)]. Hind wing buds are present. Paracercus is well-developed (in contrast to *Teloganodes*).

***

Other Ephemerella/fg1 INCERTAE SEDIS

Ephemerella/fg1 INCERTAE SEDIS 3. Leptohyphodes/g [g: *Leptohyphodes* Ulmer 1920a: 50, typus *Potamanthus inanis* Pictet 1843 (design. orig.)].

References. Pictet 1843–1845 (*Potamanthus inanis*): ♂; – Banks 1913 (*Tricorythus australis*): ♂; – Ulmer 1920a: ♂; – Traver 1944: ♉ (associated presumably).

Characters of unclear phylogenetic status.

(1) Eyes of male are large (in contrast to majority of Leptohyphes/fg1).

(2) Hind wings are lost in both sexes (see Index of characters [2.2.59]).

Distribution. Neotropical Region.

Material examined: sp. (Brazil): ♉, ☿ (associated presumably; FIG.105:H).

Ephemerella/fg1 INCERTAE SEDIS 4. Manohyphella/g [g: *Manohyphella* Allen 1973: 160, typus *M. keiseri* Allen 1973 (design. orig.)].

Reference. Allen 1973: ♀.

Distribution. Madagascar.

Material examined: –.

Comment. Fore wing retains structure initial for Ephemerella/fg1 [see Ephemerella/fg1 (6)] (in contrast to Tricoryptera), but hind wing is strongly diminished, roundish, with prominent pointed costal projection near apex (Allen 1973:Figs 2–3). McCafferty & Benstead (2002) attributed *Manohyphella* to the family Teloganodidae (= Melanemerella/fg1 + Teloganodes/fg1) basing on larvae which were placed by them to *Manohyphella* without enough ground (here they are attributed to Melanemerella/fg1).

Other species of Ephemerella/fg1 INCERTAE SEDIS. A single extinct species of Ephemerella/fg1 – *viscata* Demoulin 1968 [*Ephemerella (Timpanoga)*] – was described as a male subimago from Baltic amber; it was attributed to Timpanoga/fg1 basing on presence of tergalial insertions on abdominal segments IV–VII (Demoulin 1968a); but it is unclear if such insertions are really absent on preceding abdominal segments.

*** ***

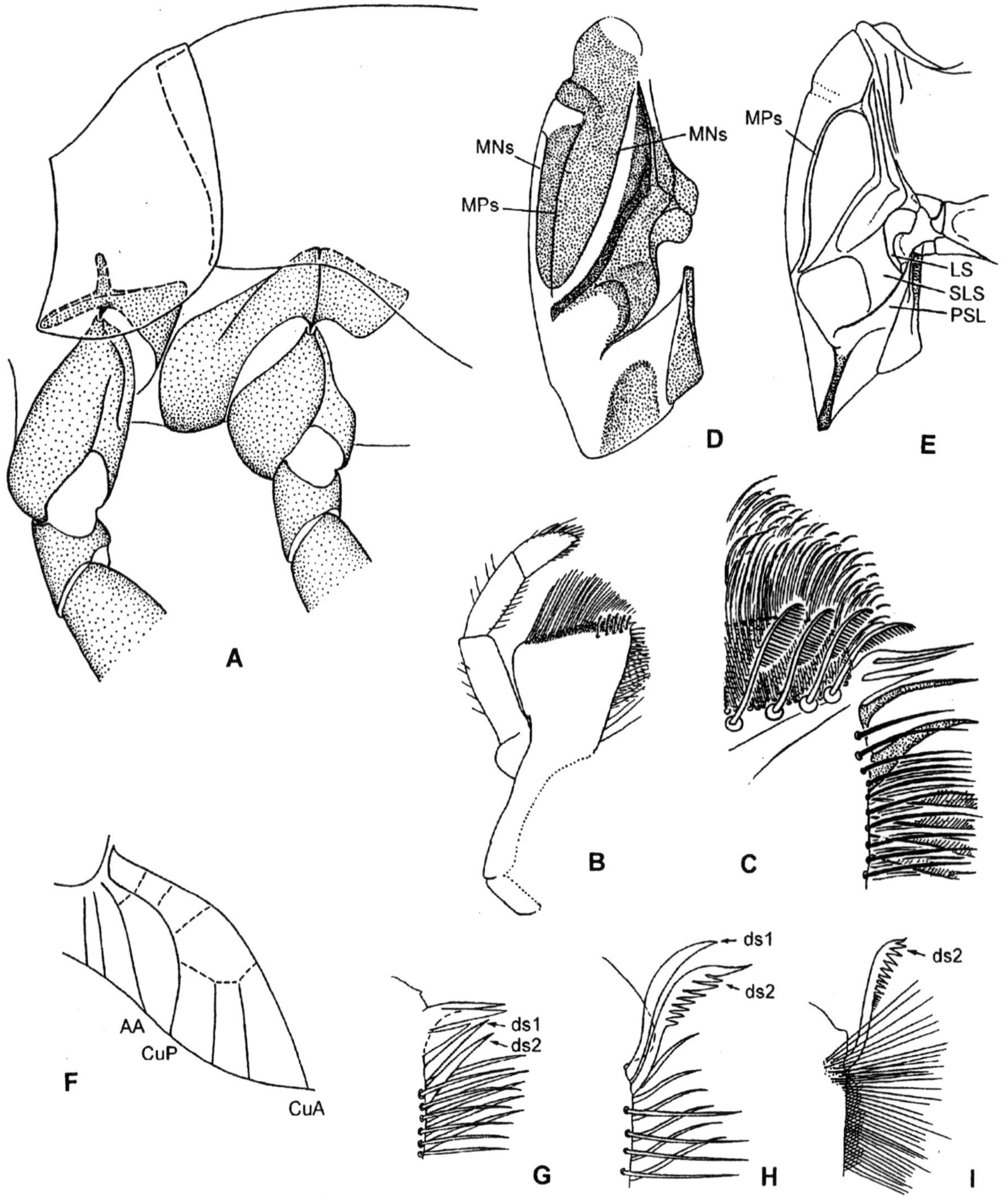

Figure 106. Leptophlebia/fg1.
A–C – Leptophlebia/fg4 *marginata* [*Ephemera*]: **A** – larval exuviae, left side of prothorax and mesothorax (pleurites and leg bases shown by dots); **B** – right maxilla, ventral view; **C** – apex of right maxilla, ventral view (dentisetae shown by dots). **D** – Neoleptophlebia/g* *japonicus* Matsumura 1931 [*Baetis*] (= *chocolata* Imanishi 1937 [*Paraleptophlebia*]), subimaginal exuviae of right half of mesonotum. **E** – Farrodes/g1 *bimaculatus* Peters & Alayo 1971 [*Farrodes*], subimaginal exuviae of right half of mesonotum. **F** – scheme of venation of cubital field of fore wing. **G–I** – schemes of maxillary apex: **G** – Leptophlebia/fg2 (corresponds to Fig.C); **H** – Habrophlebia/fg1; **I** – Atalophlebia/fg1. (B – from Kluge 1997d; C – from Kluge 1998; D – from Kluge 1997b; E – from Kluge 1994b; G–I – from Kluge 1994c).

Abbreviations: **ds1** – distal dentiseta; **ds2** – proximal dentiseta; **LS** – lateroscutum; **MNs** – mesonotal suture; **MPs** – medioparapsidal suture; **PSL** – parascutellum; **SLS** – sublateroscutum.

2.2;2,3. Leptophlebia/fg1

(Euplectoptera Anteritorna Bidentiseta Furcatergaliae Leptophlebia/fg1)

(Fig. 106)

Nomen hierarchicum: **Leptophlebia/fg1** (incl. *Calliarcys*, *Habrophlebia*, *Atalophlebia*) [f: Leptophlebini Banks 1900: 246; g: *Leptophlebia* Westwood 1840: 31; typus *Ephemera vespertina* Linnaeus 1767 (monotypy); syn. subj.: ***Euphyurus*** Bengtsson 1909: 4, typus *E. albitarsus* Bengtsson 1909 (monotypy) (synn. subj.: *vespertina* [*E.*] = *albitarsus* [*E.*])].

In circumscription matches:

— "section 5 of genera": Eaton 1883–1888: 82;

— subfam. Leptophlebiinae: Lameere 1917: 66; Lestage 1917: 323.

— fam. Leptophlebiidae: Edmunds & Traver 1954a: 238;

— superfam. Leptophlebioidea: McCafferty & Edmunds 1979: 6;

— Leptophlebia/fg1: Kluge 2000: 253.

References. Eaton 1883–1888: ♉* ⊕*; – Needham & Traver & Hsu 1935: ♉* ⊕*; – Edmunds & Allen & Peters 1963: ♉*; – Peters & Edmunds 1970: ♉* ⊕*; – Tshernova 1970: ♉* ⊕* – Edmunds & Jensen & Berner 1976: ♉* ⊕*; – Tomka & Elpers 1991: ♉*; – Kluge 1993a: ⊕*.

Autapomorphy of Leptophlebia/fg1.

(1) Maxilla is specialized as filtering (FIG.106: B–C): its apical margin is widened and truncate, with a large field of very densely and regularly situated long filtering setae; ventrad of this field, an apical-ventral row is developed – a short row of small elongate pectinate setae; this row reaches inner margin of maxilla but fare not reaches its outer margin. These elements (the field of filtering setae and the apical-ventral row of pectinate setae) occur also in some other mayfly groups, but not in such combination (see Index of characters [1.1.31] and [1.1.32]).

In connection with the filtering specialization, canines and dentisetae undergo reduction: if all 3 canines and 2 dentisetae are retained, they are diminished and brought together (FIG.106:C, G); in *Calliarcys*, Habrophlebia/fg1 and Atalophlebia/fg1 canines are lost (FIG.106:H–I); in Atalophlebia/fg1 distal dentiseta is also lost, and proximal dentiseta is pectinate and involved into apical setal crown (FIG.106:I). In a few Atalophlebia/fg1 (Hermanella/g1 and *Choroterpides*) this pectinate dentiseta is lost, the apical-ventral row of pectinate setae also can be vestigial or lost. In some independent taxa of Atalophlebia/fg1 (e. g. Hermanella/g1, Hagenulus/ /fg*, Choroterpes/g1-*Choroterpides* and others) median-apical angle of maxilla is produced forming a tusk-like projection which functionally substitutes the lost canines. In various Leptophlebia/fg1 mouth apparatus apart of maxillae either retains primitive biting structure, or gets more or less deep filtering specialization.

Non-unique apomorphies of Leptophlebia/fg1.

(2) In imago and subimago left and right halves of mesonotal suture are curved posteriorly and stretch backward so strongly that medially they go along median longitudinal suture, and laterally – along lateroparapsidal sutures. In subimago mesonotal suture separates pigmented area from a light median stripe and a pair of light lateral stripes; lateral pigmented area of subimaginal mesonotal occupies lateroparapsidal suture up to its end and all area laterad of it – sublateroscutum and lateroscutum; thus subimaginal mesonotal pigmentation has a characteristic pattern (FIG.106:D). Non-unique apomorphy: the same in Fimbriatotergaliae and Tetramerotarsata. In some Leptophlebia/fg1 (particularly in *Farrodes*) mesonotal suture completely disappeared, being fused with median longitudinal and lateroparapsidal sutures, and subimaginal mesonotum lost pigmentation (FIG.106:E).

(3) Imaginal and subimaginal furcasternal protuberances are separated, between them metathoracic nerve ganglion is situated. Non-unique apomorphy (see Index of characters [2.2.23]).

(4) On fore wing CuP in its extreme base strongly diverges from CuA and close to base is strongly curved, thus outstanding from CuA (FIG.106:F). The transverse portion of CuP base can disappear, in this case base of CuP appears to be widely separated from the base of CuA. Non-unique apomorphy; the same in Tetramerotarsata only.

(5) Cubital field of fore wing [see Anteritorna (1)] usually with 2 intercalaries (FIG.106:F) or one bifurcate vein arising from CuA and/or CuP. Non-unique apomorphy (see Index of characters [2.2.51]).

(6) Larval claw usually with one row of denticles on inner side (non-unique apomorphy – see Index of characters [1.2.21]); rarely denticles are secondarily lost (in *Aprionyx*, *Atalomicria*), or form two rows (in *Dactylophlebia*), or in addition to the main row several subapical denticles are present.

(7) Tergalii lack costal and anal ribs or their vestiges [see Furcatergaliae (5)]. Probably the initial shape of tergalius is either bifurcate, or bilamellate; each of two branches (or lamellae) are slender, with stretched pointed apex (costal branch of bifurcate tergalius corresponds to dorsal lamella of bilamellate tergalius, and anal branch corresponds to ventral lamella). Tergalii of such two types and types intermediates between them are present in majority of Leptophlebia/fg1. Besides this, tergalii of various other forms are found: tergalii can be unilamellate (when anal branch, or ventral lamella, is lost); both or one lamella can have margins smooth or with numerous processes; several times independently evolved such bilamellate tergalii where each lamella has three apical processes, middle of which represents the initial pointed apex. Ability for rhythmical respiratory movements can be retained or lost. Some taxa (Australian *Kirrara*, New Zealand *Deleatidium* and New Caledonian *Lepeorus* and *Lepegenia*) have strongly modified lamellate tergalii which form a sucking disk similar to that of Rhithrogena/fg3 and Iron/g1; such tergalii can have secondary ribs reminding primary ribs of other mayflies. *Adenophlebiodes* has tergalii I transformed as gill opercula. Usually all tergalii I–VII are present, but in some taxa tergalius I or VII are lost.

Plesiomorphies of Leptophlebia/fg1. In larva: Maxillary palp is always 3-segmented. Labial palp is always 3-segmented. Paracercus is always multisegmented, usually subequal to cerci, rarely vestigial.

In imago and subimago: Gonostylus with 2 distal segments (the only exception is *braueri* Ulmer 1920 [*Hagenulodes*] which has gonostylus with single distal segment). Paracercus is nearly always well-developed, usually subequal to cerci, rarely vestigial.

Variable characters of Leptophlebia/fg1. In larva: Patella-tibial suture can be normally developed on middle and hind legs (in Leptophlebia/fg2 and some Atalophlebia/fg1), or lost on middle legs (in Habrophlebia/fg1 and some Atalophlebia/fg1), or lost on all legs (in some Atalophlebia/fg1). Abdominal terga are usually smooth, rarely with median projections. Usually all tergalii [see (7)] have normal dorsal attachment on posterior margins of segments, but in *Isca* tergalial bases are transferred ventrally.

In imago and subimago: In most representatives moult from subimago to imago takes place in both sexes, but in selected taxa females do not moult. Male eyes have usual for mayflies division to dorsal and ventral portions; in selected taxa the dorsal portion is transformed into turban eye (like in Turbanoculata); in Fulleta/g1 (incl. *Fulletomimus*) division to dorsal and ventral portions is lost. Imaginal wings usually have no setae, but sometimes marginal setae are present, like in subimago. Hind wing can be either well-developed (as long as 0.25–0.35 of fore wing length), or diminished, or lost; if developed, Sc can reach wing apex or terminates much more proximally; costal margin can be smooth or with prominent projection just proximad of Sc apex; triads of RS and MP [see Furcatergaliae (4)] can be either developed, or MP is non-branched. Claws can be either ephemeropteroid, or both pointed; claws of male fore leg can have the same structure as others (ephemeropteroid or pointed), or both blunt.

Size. Fore wing length 2–15 mm.

Age and distribution. Late Cretaceous — Recent; world-wide. The oldest known representative is *Aureophlebia sinitshenkovae* Peters & Peters 2000, described as a female subimago from Upper Cretaceous New Jersey amber (Peters & Peters 2000). Other fossil representatives are known from Baltic amber (Kluge 1993b). Some authors attributed to Leptophlebia/fg1 also Jurassic and Cretaceous taxa *Mesobaetis*, *Mesoneta*, *Cretoneta*, *Leptoneta* and *Conovirilus* (Demoulin 1954e, Tshernova 1962b 1971, Sinitshenkova 1989, McCafferty 1997a). Placing here *Cretoneta* (known as well-preserved adults in Taimyr amber) was surely wrong (Kluge 1993b). *Mesobaetis* and *Mesoneta* are knows as larvae; judging by presence of primary swimming setae and oval tergalii with marginal ribs, they also can not belong to Leptophlebia/fg1 (Kluge 1989a). *Leptoneta* is known as larvae which have no details allowing to determine their systematic position. *Conovirilus* is described as male imago from Lower Cretaceous Lebanese amber, but its description does not contain characters which could prove its belonging to Leptophlebia/fg1. In the present book all these taxa are listed in the divisions Anteritorna INCERTAE SEDIS and Euplectoptera INCERTAE SEDIS (see below).

Classification of Leptophlebia/fg1. Here belong about 600 species. W. Peters (1980) divided the family Leptophlebiidae into 2 subfamilies – Leptophlebiinae and Atalophlebiinae (nomen hierarchicum: **Atalophlebia/fg1** [f: Atalophlebiinae Peters 1980: 38; g: *Atalophlebia* Eaton 1881: 193, typus *Ephemera australis* Walker 1853 (design. orig.)]). Atalophlebia/fg1 is a large holophyletic taxon characterized by a number of autapomorphies (Peters 1980, Kluge 1994b, Peters & Gillies 1995), while the taxon Leptophlebiinae SENSU Peters is probably paraphyletic.

Because of this, it was suggested (Kluge 1994c) to divide the family Leptophlebiidae into 3 subfamilies, separating Leptophlebiinae SENSU Peters to two subfamilies – a plesiomorphon Leptophlebiinae s. str. and a holophyletic subfamily

Habrophlebiinae (nomen hierarchicum: **Habrophlebia/fg1** (incl. *Habroleptoides*) [f: Habrophlebiinae Kluge 1994c: 37; g: *Habrophlebia* Eaton 1881: 195, typus *Ephemera fusca* Curtis 1834 (design. orig.)]). Habrophlebia/fg1 and Atalophlebia/ /fg1 have some characters in common which can be synapomorphies (Kluge 1994c).

A monospecific taxon **Calliarcys/g(1)** [g: *Calliarcys* Eaton 1881: 21, typus *C. humilis* Eaton 1881 (design. orig.)] does not fall neither to Atalophlebia/fg1, nor to Habrophlebia/fg1, and was presumably attributed to Leptophlebiinae SENSU Kluge 1994c. However, its larva has no maxillary canines; this can be a synapomorphy with Habrophlebia/fg1 and Atalophlebia/fg1.

Detailed revision of supraspecies taxa of Leptophlebia/fg1 is recently far from its finish. It is necessary to re-examine representatives of many genera and subgenera described in Atalophlebia/fg1, particularly their subimaginal sclerotization which is now described for a few taxa only (Kluge 1994a 1994b, 1994c). It is planned to publish supraspecies revision of Leptophlebia/fg1 as a 3rd volume of this book (after the 2nd volume about Turbanoculata).

***

Furcatergaliae INCERTAE SEDIS 1. † **Baba/fg(1)**

Nomen hierarchicum: Baba/fg(1) [f: Babidae Kluge & Godunko & Krzeminski (in press); g: *Baba* Kluge & Godunko & Krzeminski (in press)].

Reference. Kluge & Godunko & Krzeminski, in press: ⊕.

Characters of unclear phylogenetic status.

(1) Mesonotal suture is transverse (in contrast to Fimbriatotergaliae and Leptophlebia/fg1 – see Index of characters [2.2.8]).

(2) In proximal part of fore wing MP_2 and CuA are sharply curved and divergent from MP_1. The same in most Fimbriatotergaliae [see above, Fimbriatotergaliae (5)].

Plesiomorphies of *Baba*. Furcasternal protuberances are contiguous. On fore wing [see (2)] cubital field with several branched and simple veins arising from CuA; AA without bifurcation (in contrast to Potamanthus/fg1) and without veins going to basitornal margin (in contrast to Ephemera/fg9). Hind wing [see Furcatergaliae (4)] is well-developed. Legs [see Furcatergaliae (2)] with ephemeropteroid claws. Paracercus is equal to cerci at least in female.

Size. Fore wing length 7 mm.

Age. Palaeogene (Baltic amber).

Material examined (Mus.Nat.Hist. of Inst.Syst.Evol. Anim. in Krakow): ⊕.

Comment. It can be assumed that *Baba* represents a sister taxon for Fimbriatotergaliae: character (2) can be a synapomorphy of *Baba* and Fimbriatotergaliae, and character (1) – a plesiomorphy. Another possible assumption is that *Baba* is an aberrant representative of Fimbriatotergaliae with reversed mesonotal structure.

It is not excluded that here belong some earlier described extinct taxa whose thorax structure is unknown (see above, Fimbriatotergaliae INCERTAE SEDIS).

***

Other Furcatergaliae INCERTAE SEDIS

Furcatergaliae INCERTAE SEDIS 2. † Blasturophlebia/g [g: *Blasturophlebia* Demoulin 1968a: 268, typus *B. hirsuta* Demoulin 1968 (design. orig.)].

Reference. Demoulin 1968a: ◎ (exuviae).

Age. Eocene (Baltic amber).

Material examined: –.

Comment. Demoulin (1968a) placed *Blasturophlebia* to Leptophlebiidae (i. e. Leptophlebia/fg1); in order to confirm or disprove this, thorax structure of the holotype (deposited in Bayer. Staatssammlung) should be examined.

Other species of Furcatergaliae INCERTAE SEDIS (here [*C.*]=[*Caenis*]). Some species attributed to *Caenis* are insufficiently described and can belong either to Caenoptera, or to Tricorythus/fg2 (see Classifications of Furcatergaliae II); it is not excluded that some of them belong not to Furcatergaliae, but to Prosopistoma/fl=g2. **South Asian species:** *annulata* Navás 1923 [*C.*], *nemoralis* Navás 1922 [*C.*], *nigrostriata* Navás 1931 [*C.*], *pumila* Navás 1923 [*C.*]. **African species:** *aethiopica* Navás 1935 [*C.*], *fasciata* Navás 1927 [*C.*], *haywardi* Navás 1926 [*C.*]. **South American species:** *albata* Navás 1920 [*C.*], *axillata* Navás 1931 [*C.*], *nigella* Navás 1932 [*C.*], *tenella* Navás 1933 [*Eurycaenis*].

*** ***

Appendix

Mayflies of uncertain systematic position

Most fossil (mainly Mesozoic and Palaeogenous) and some insufficiently described Recent mayflies have an uncertain position among Ephemeroptera. Here generic and species names belonging to such taxa are accounted in the following method:

Anteritorna INCERTAE SEDIS – names given for fossil supraspecies taxa and selected Recent species of mayflies, whose anteritornous wing structure is known for certain.

Euplectoptera INCERTAE SEDIS (p.356) – names given for fossil mayflies known as larvae or adults with poorly preserved wings. Probably, most or all of these mayflies belong to Anteritorna, but this is not proven because wing venation and number of dentisetae are unknown.

Euephemeroptera INCERTAE SEDIS (p.359) – Mesozoic taxa, one of which (Litophlebia/fg1) has unusual wing structure and others which were established for indeterminable fragments.

Pterygota INCERTAE SEDIS (p.360) – Palaeozoic taxon Syntonoptera/fg1 possibly related to Ephemeroptera s.l., and some taxa erroneously attributed to Ephemeroptera.

Euarthropoda and Animalia INCERTAE SEDIS (p.361) – fossils, which some authors regard to belong to mayfly larvae.

Nomina dubia (p.361) – old forgotten species names, possibly belonging to well-known species.

Anteritorna INCERTAE SEDIS 1
† Hexagenites/fg1

Nomen hierarchicum: **Hexagenites/fg1** [f: Hexagenitinae Lameere 1917: 74; g: *Hexagenites* Scudder 1880: 6, typus *H. weyenberghii* Scudder 1880 (design. orig.)]; synn. subj.: **Paedephemera/fg** [f: Paedephemeridae Lameere 1917: 49; g: *Paedephemera* Handlirsch 1906: 601, typus *Ephemera multinervosa* Oppenheim 1888 (design. Tshernova 1961: 859)]; **Stenodicranum/fg** [f: Stenodicranidae Demoulin 1954f: 553; g: *Stenodicranum* Demoulin 1954f: 571, typus *Ephemera cellulosa* Hagen 1862 (design. orig.)] [synn. subj.: *cellulosa* [*E.*] = *weyenberghii* [*H.*] = *multinervosa* [*E.*] (Demoulin 1970c: 7)].

In circumscription matches:
— subfam. Ephemeropsinae Cockerell 1924: 136;
— fam. Hexagenitidae: Tshernova 1961: 589.

References. Cockerell 1924: ♉ ⊕; – Tshernova 1961: ⊕*; Meshkova 1961: ♉; – Tshernova 1970: ♉ ⊕ – Tshernova & Sinitshenkova 1974: ⊕'; – McCafferty 1990: ♉'; – Martins-Neto 1996: ♉'.

Autapomorphies of Hexagenites/fg1.

(1) On fore wing CuA bifurcates to CuA_1 and "CuA_2" (non-unique character – see Index of characters [2.2.50]), from the bifurcation a vein "iCu" arises, which forms several (3–5) triads following one another: anterior branch of each triad forms next triad; each of these triads has anterior branch arched by its convexity anteriorly; all branches of these triads go to basitornal margin of wing. Unique apomorphy.

(2) Fore and hind wings with more or less expressed vein gemination: at least there are pairs of veins brought together: $RSp+MA_1$, $iMA+MA_2$, MP_2+CuA; MP_1 and iMP are more or less approximated to $iMA+MA_2$. Non-unique apomorphy (see Index of characters [2.2.28]).

(3) Tergalius VII is enlarged, distinctly longer than others, while tergalii I–VI have subequal length (unique apomorphy). On all tergalii at least costal rib is thickened and anal rib is situated at a distance from anal margin (as in some other mayflies – see Index of characters [1.3.28]).

Plesiomorphies of Hexagenites/fg1. Larva has siphlonuroid appearance, with dense primary swimming setae on caudalii.

Size. Fore wing length 16–42 mm.

Age. Late Jurassic – Late Cretaceous (Europe, Asia, South America).

Nominal taxa included. Hexagenites/fg1 includes: (1) **Ephemeropsis/fg** [f: Ephemeropsinae Cockerell 1924: 136; g: *Ephemeropsis* Eichwald 1864: 21, typus *E. trisetalis* Eichwald 1864 (monotypy)]; (2) **Phacelobranchus/g** [g: *Phacelobranchus* Handlirsch 1906: 604, typus *Ph. braueri* Handlirsch 1906 (monotypy)], regarded as a generic synonym of *Ephemeropsis* (Tshernova 1961: 861); (3) **Protoligoneuria/g** [g: *Protoligoneuria* Demoulin 1955e: 271, typus *P. limai* Demoulin 1955 (design. orig.); (4) **Hexameropsis/g** [g: *Hexameropsis* Tshernova & Sinitshenkova 1974: 132, typus *H. selini* Tshernova & Sinitshenkova 1974 (design. orig.)]; (5) **Mongologenites/g** [g: *Mongologenites* Sinitshenkova 1986: 45, typus *M. lagueatus* Sinitshenkova 1986 (design. orig.)]; (6) **Palaeobaetodes/g** [g: *Palaeobaetodes* Brito 1987: 594, typus *P. costalimai* Brito 1987]; (7) **Cratogenites/g** [g: *Cratogenites* Martins-Neto 1996: 180, typus *C. corratiniae* Martins-Neto 1996 (design. orig.)]; (8) **Cratogenitoides/g** [g: *Cratogenitoides* Martins-Neto 1996: 181, typus *C. delclosi* Martins-Neto 1996 (design. orig.)]; (9) **Caenoephemera/g** [g: *Caenoephemera* Lin & Huang 2001: 748, typus *C. shanguanensis* Lin & Huang 2001 (design. orig.)]. Possibly, here also belongs *Costalimella* (see below, Euplectoptera INCERTAE SEDIS).

Species composition of Hexagenites/fg1. Following formally described species belong here for certain: From **Mesozoic of Europe:** *cellulosa* Hagen 1862 [*Ephemera*] (synn. subj.: *mortua* Hagen 1862 [*Ephemera*], *weyenberghii* Scudder 1880 [*Hexagenites*], *multinervosa* Oppenheim 1888 [*Ephemera*], *oppenheimi* Handlirsch 1906 [*Paedephemera*]). From **Mesozoic of Siberia, Mongolia and China:** *lagueatus* Sinitshenkova 1986 [*Mongologenites*], *martynovae* Tshernova 1961 [*Ephemeropsis*], *selini* Tshernova & Sinitshenkova 1974 [*Hexameropsis*], *shanguanensis* Lin & Huang 2001 [*Caenoephemera*], *trisetalis* Eichwald 1864 [*Ephemeropsis*] (synn. subj.: *rossica* Eichwald 1866 [*Lethaea*], *orientalis* Eichwald 1868 [*Estheria*], *braueri* Handlirsch 1906 [*Phacelobranchus*], *middendorfi* Handlirsch 1906 [*Ephemeropsis*], *melanurus* Cockerell 1924 [*Ephemeropsis*], *reisi* Handlirsch 1939 [*Ephemeropsis*]). From **Lower Cretaceous of Brazil:** *britoi* Martins-Neto 1996 [*Palaeobaetodes*], *costalimai* Brito 1987 [*Palaeobaetodes*], *corratiniae* Martins-Neto 1996 [*Cratogenites*], *delclosi* Martins-Neto 1996 [*Cratogenitoides*], *limai* Demoulin 1955 [*Protoligoneuria*].

Material examined (Paleontol. Inst.): *limai* [*P.*]: ♉; *lagueatus* [*M.*]: ⊕, ♉ (presumably associated); *martynovae* [*E.*]: ⊕; *selini* [*H.*]: ⊕; *trisetalis* [*E.*]: ♉, ⊕.

* * *

Other Anteritorna INCERTAE SEDIS

Anteritorna INCERTAE SEDIS 2. † Cronicus/g [g: *Cronicus* Eaton 1871:133, typus *Baetis anomala* Pictet & Hagen 1856 (monotypy; syn.subj.: *Baetis longipes* Pictet & Hagen 1856)].

Reference. Demoulin 1968a: ◎ ⊕.

Characters of unclear phylogenetic status.

(1) On hind wing bifurcation of MP_2 is transferred distally or lost (see Index of characters [2.2.69]).

(2) Imaginal paracercus is vestigial (see Index of characters [2.3.20]).

(3) Gonostylus with 3 distal segments (see Index of characters [2.3.12]).

Plesiomorphies. Fore wing with several (5–6) simple or branched veins going from CuA to basitornal and tornoapical margins. Imaginal and subimaginal claws are ephemeropteroid.

Age. Eocene (Baltic amber).

Material examined: –.

Comment. Demoulin (1968a) attributed *Cronicus* to Coloburiscinae (i. e. Coloburiscus/fg1) basing on plesiomorphies only. In order to clarify its systematic position, thorax structure of types (deposited in Humboldt University) should be examined.

Anteritorna INCERTAE SEDIS 3. † Siphlurites/g [g: *Siphlurites* Cockerell 1923: 170, typus *S. explanatus* Cockerell 1923 (design. orig.)].

Reference. Demoulin 1970d: fore wing.

Character of unclear phylogenetic status.

(1) Cubital field of fore wing with 2 veins going from CuA to basitornal margin, 1 intercalary going to tornoapical margin, and several (9–10) simple or branched veins going from this intercalary to basitornal margin (see Index of characters [2.2.51]). In other respects wing venation is plesiomorphic.

Age. Miocene (USA).

Material examined: –.

Anteritorna INCERTAE SEDIS 4. † Electrogenia/g [g: *Electrogenia* Demoulin 1956a: 95, typus *E. dewalschei* Demoulin 1956 (design. orig.)].

Reference. Demoulin 1956a: ⊕'.

Characters of unclear phylogenetic status.

(1) Cubital field of fore wing with 2 pairs (4 ones) intercalaries, among which hind pair is longer. The same in Heptagennota and some others (see Index of characters [2.2.51–52]).

(2) Imaginal paracercus is vestigial (see Index of characters [2.3.20]).

Plesiomorphies. 1st tarsal segment is fused with tibia (in contrast to Heptagennota-Pentamerotarsata).

Imaginal claws of all legs are ephemeropteroid. Gonostylus with 2 distal segments.

Age. Eocene (Baltic amber).

Material examined: –.

Comment. Demoulin (1956a) attributed *Electrogenia* to Arthropleinae (i. e. Arthroplea/fg1); this is wrong, as *Electrogenia* has no articulation of 1st tarsal segment with tibia [see Pentamerotarsata (1)]. In order to clarify its systematic position, thorax structure should be examined.

Anteritorna INCERTAE SEDIS **5. † Stackelbergisca/g** [g: *Stackelbergisca* Tshernova 1967: 323, typus *S. sibirica* Tshernova 1967 (design. orig.)].

Reference. Tshernova 1967: fore wing, ℧ (presumably associated).

Plesiomorphy. Cubital field of fore wing with many (about 15) veins going from CuA to basitornal margin.

Age. Jurassic (Siberia).

Material examined (Paleontol. Inst.): *sibirica* [*S.*]: ⊕, ℧ (presumably associated).

Comment. Larva attributed to *Stackelbergisca sibirica* without argumentation, has plesiomorphic siphlonuroid appearance, with primary swimming setation on caudalii.

Anteritorna INCERTAE SEDIS **6. † Balticophlebia/g** [g: *Balticophlebia* Demoulin 1968a: 237, typus *B. hennigi* Demoulin 1968 (design. orig.)].

Reference. Demoulin 1968a: ⊕.

Character of unclear phylogenetic status.

(1) Imaginal paracercus is vestigial (see Index of characters [2.3.20]).

Plesiomorphies. Fore wing with 5 simple veins going from CuA to basitornal and tornoapical margins. Imaginal claws are ephemeropteroid.

Age. Eocene (Baltic amber).

Material examined: –.

Comment. Demoulin (1968a) attributed *Balticophlebia* to Ameletopsinae (i. e. Ameletopsis/fg1) basing on plesiomorphies only. In order to clarify its systematic position, thorax structure of the holotype (deposited in Staatl. Mus. Naturk. Stuttgart) should be examined.

Anteritorna INCERTAE SEDIS **7. † Baltameletus/g** [g: *Baltameletus* Demoulin 1968a: 238, typus *B. oligocaenicus* Demoulin 1968 (design. orig.)].

Reference. Demoulin 1968a: ♂.

Character of unclear phylogenetic status.

(1) Imaginal paracercus is vestigial (see Index of characters [2.3.20]).

Plesiomorphies. Fore wing with several (3–4) simple veins going from CuA to basitornal margin. Subimaginal claws are ephemeropteroid.

Age. Eocene (Baltic amber).

Material examined: –.

Comment. Demoulin (1968a) attributed *Baltameletus* to Siphlonurinae (i. e. Tridentiseta non-Tetramerotarsata) basing on plesiomorphies only. In order to clarify its systematic position, thorax structure of the holotype (deposited in Staatl. Mus. Naturk. Stuttgart) should be examined.

Anteritorna INCERTAE SEDIS **8. † Xenophlebia/g** [g: *Xenophlebia* Demoulin 1968a: 267, typus *X. aenigmatica* Demoulin 1967 (design. orig.)].

Reference. Demoulin 1968a: ⊕.

Character of unclear phylogenetic status.

(1) Hind wing is diminished; cubital field of fore wing is short, with 2 intercalaries (see Index of characters [2.2.51–52]).

Plesiomorphies. Imaginal claws are ephemeropteroid. Gonostylus with 2 distal segments.

Age. Eocene (Baltic amber).

Material examined: –.

Comment. Demoulin (1968a) attributed *Xenophlebia* to Leptophlebiidae (i. e. Leptophlebia/fg1). In order to confirm or disprove this, thorax structure of the holotype (deposited in Bayer. Staatssamml.) should be re-examined.

Anteritorna INCERTAE SEDIS **9. † Olgisca/g** [g: *Olgisca* Demoulin 1970c: 6, typus *Paedephemera schwertschlageri* Handlirsch 1906 (design. orig.)].

Reference. Demoulin 1970c: fore wing.

Plesiomorphy. Fore wing with several (about 8) veins going from CuA to basitornal margin.

Age. Jurassic (Germany).

Material examined: –.

Anteritorna INCERTAE SEDIS **10. † Cretoneta/g** [g: *Cretoneta* Tshernova 1971: 614, typus *C. zherichini* Tshernova 1971 (design. orig.)].

References. Tshernova 1971: ⊕♂; – Kluge 1993b: ◎* ⊕*.

Characters of unclear phylogenetic status.

(1) On hind wing MA and MP are non-forked (see Index of characters [2.2.67] and [2.2.69]). In both known species hind wing is somewhat diminished; basitornal margin of fore wing is shortened, tornus is poorly expressed, number of veins in cubital field is diminished (non-unique apomorphy).

(2) Imaginal and subimaginal paracercus is vestigial (see Index of characters [2.3.20]).

Plesiomorphies. Mesonotal suture is well expressed, nearly transverse, stretched backward medially (as in Siphlonurus/fg4 and Ameletus/fg1). Anterior paracoxal suture is complete, anepisternum and katepisternum are subequal. Furcasternal protuberances are contiguous. On fore wing CuA and CuP basally diverge under acute angle from one point (in contrast to Tetramerotarsata, Leptophlebia/fg1 and some others); in cubital field several (1–4) veins go from CuA to basitornal margin [see (1)]. Tarsi are 5-segmented, 1st segment is non-shortened (in contrast to Furcatergaliae), fused with tibia (except for fore tarsus of male). Claws (at least on middle and hind legs of male and all legs of female) are ephemeropteroid. Gonostylus with 2 distal segments.

Size. Fore wing length 3–5 mm.

Age. Late Cretaceous (Taimyr amber).

Species composition. 2 species – *zherichini* Tshernova 1971 [*Cretoneta*], *acmoptera* Kluge 1993 [*Cretoneta*].

Material examined (Paleontol. Inst.): *zherichini* [*C.*]: ◎, ⊕; *acmoptera* [*C.*]: ⊕.

Anteritorna INCERTAE SEDIS 11. † Proameletus/g [g: *Proameletus* Sinitshenkova 1976a: 86, typus *P. caudatus* Sinitshenkova 1976 (design. orig.)].

Reference. Sinitshenkova 1976a: ⊕, ⥁ (presumably associated).

Character of unclear phylogenetic status.

(1) Imaginal and subimaginal paracercus is strongly shortened (see Index of characters [2.3.20]).

Plesiomorphy. Fore wing with several (about 10) veins going from CuA to basitornal margin.

Age. Early Cretaceous (Siberia).

Material examined (Paleontol. Inst.): *caudatus* [*P.*]: ⊕, ⥁.

Comment. Larva attributed to *Proameletus caudatus* basing only on abundance in the same deposits, has plesiomorphic siphlonuroid appearance, with primary swimming setation on caudalii.

Anteritorna INCERTAE SEDIS 12. † Philolimnias/g [g: *Philolimnias* Hong 1979: 336, typus *Ph. sinica* Hong 1979 (design. orig.)].

Reference. Hong 1979: ⊕♂.

Age. Eocene (China, Liaoning amber).

Material examined: –.

Comment. Originally *Philolimnias* was attributed to Ephemerellidae; however, original description probably contains errors and does not allow to understand the true systematic position.

Anteritorna INCERTAE SEDIS 13. † Conovirilus/g [g: *Conovirilus* McCafferty 1997: 78, typus *C. poinari* McCafferty 1997 (design. orig.)].

Reference. McCafferty 1997: ⊕.

Age. Early Cretaceous (Lebanese amber).

Material examined: –.

Comment. Originally *Conovirilus* was attributed to Leptophlebiidae Atalophlebiinae; however, characters which would allow to place it into Leptophlebia/fg1, are not described, and are invisible on the photos in the original description.

Other extinct species of Anteritorna INCERTAE SEDIS. Besides the type species of generic names listed above, following species from Baltic amber have uncertain position – *henningseni* Demoulin 1965 [*Metretopus*], *jaegeri* Demoulin 1968 [*Siphloplecton*], *major* Demoulin 1968 [*Cronicus*], *trinervis* Demoulin 1968 [*Metretopus*] (see also below, Euplectoptera INCERTAE SEDIS).

Recent species of Anteritorna INCERTAE SEDIS. Descriptions of following Recent species do not allow to determine their systematic position. **African species:** *nimia* Navás 1915 [*Ephemera*]. **Asian species:** *davidica* Navás 1934 [*Heptagenia*], *rubescens* You 1987 [*Cinygma*], *towadensis* Matsumura 1931 [*Ameletus*], *validus* Navás 1920 [*Chirotonetes*]. **South American species:** *atrostoma* Weber 1871 [*Palingenia*], *bullocki* Navás 1930 [*Euthyplocia*], *hollermayeri* Navás 1936 [*Metamonius*], *guttata* Pictet 1843 [*Baetis*].

* * *

Euplectoptera INCERTAE SEDIS 1
† Mesogenesia/g1

Nomen hierarchicum: **Mesogenesia/g1** (incl. *Archaeobehningia*) [g: *Mesogenesia* Tshernova 1977: 92, typus *M. petersae* Tshernova 1977 (design. orig.)].

Reference. Tshernova 1977 (*Mesogenesia*, *Archaeobehningia*): ⥁.

Autapomorphy of Mesogenesia/g1.

(1) Larval fore leg is specialized as grasping (but not burrowing, as it was originally described): it is enlarged (at least larger than middle one), directed forward with knee directed dorsally and claw directed ventrally; its coxa is elongate; femur is widened, with dense setation on inner side. In contrast to burrowing legs of Fossoriae, tibia is not widened (in the original description of *petersae* [*M.*] femur was taken for tibia, and boundary between tibia and tarsus was overlooked). In *petersae* [*M.*] fore leg looks as having two claws, possibly because of a pointed spine-like inner-apical angle of tarsus which is similar to the true claw, but a little shorter.

Character of unclear phylogenetic status.

(2) Larval middle leg is the smallest, and hind leg the largest; in the both specimens examined (fossils pressed laterally) middle legs are directed forward like fore legs, and hind legs are directed backward. The same in Furcatergaliae-Fossoriae.

Plesiomorphy of Mesogenesia/g1. In contrast to Recent Furcatergaliae, caudalii retain primary swimming setation on median sides of cerci and lateral sides of paracercus, and have no secondary swimming setae on lateral sides of cerci (only basal parts of cerci and paracercus are preserved in holotype of *petersae* [*M.*], while caudalii of *edmundsi* [*A.*] are unknown).

Age. Jurassic (Siberia).

Nominal taxon included. Mesogenesia/g1 includes **Archaeobehningia/g** [g: *Archaeobehningia* Tshernova 1977: 94, typus *A. edmundsi* Tshernova 1977 (design. orig.)].

Material examined (Paleontol. Inst.): *petersae* [*Mesogenesia*]: Ꝺ; *edmundsi* Tshernova 1977 [*Archaeobehningia*]: Ꝺ.

* * *

Other Euplectoptera INCERTAE SEDIS

Euplectoptera INCERTAE SEDIS 2. † Mesoneta/fg [f: Mesonetidae Tshernova 1969: 158; *Mesoneta* Brauer & Redtenbacher & Ganglbauer 1889: 4, typus *M. antiqua* Brauer & Redtenbacher & Ganglbauer 1889 (monotypy)].

References. Demoulin 1968c; – Tshernova 1969: Ꝺ.

Age. Jurassic (Siberia). Here were also attributed some Triassic species (Sinitshenkova 2000d), that is not grounded (see below, Euephemeroptera INCERTAE SEDIS).

Material examined (Paleontol. Inst.): *antiqua* [*M.*]: Ꝺ. Other species attributed here are listed among "Other extinct species of Euplectoptera INCERTAE SEDIS".

Euplectoptera INCERTAE SEDIS 3. † Mesobaetis/g [g: *Mesobaetis* Brauer & Redtenbacher & Ganglbauer 1889: 5, typus *M. sibirica* Brauer & Redtenbacher & Ganglbauer 1889 (monotypy)].

References. Brauer & al. 1889: Ꝺ; – Demoulin 1968c; – Sinitshenkova 1985: Ꝺ, ⊕ (presumably associated).

Age. Jurassic (Siberia). Here was also attributed Triassic species (Sinitshenkova 2000d), that is not grounded (see below, Euephemeroptera INCERTAE SEDIS).

Material examined (Paleontol. Inst.): *sibirica* [*M.*]: Ꝺ. Other species attributed here are listed among "Other extinct species of Euplectoptera INCERTAE SEDIS".

Euplectoptera INCERTAE SEDIS 4. † Aphelophlebodes/fg [f: Aphelophlebodidae Pierce 1945: 3; g: *Aphelophlebodes* Pierce 1945: 3, typus *A. stocki* Pierce 1945 (design. orig.)].

Reference. Demoulin 1962: part of wing.

Age. Later Miocene (USA).

Material examined: –.

Euplectoptera INCERTAE SEDIS 5. † Turfanerella/g [g: *Turfanerella* Demoulin 1954e: 324, typus *Ephemeropsis tingi* Ping 1935 (design. orig.)].

Reference. Demoulin 1954e: Ꝺ (abdomen only).

Age. Jurassic (China).

Material examined: –.

Euplectoptera INCERTAE SEDIS 6. † Parabaetis/g [g: *Parabaetis* Haupt 1956: 32, typus *P. eocaenicus* Haupt 1956 (design. orig.)].

Reference. Demoulin 1957b: part of hind wing.

Age. Eocene (Germany).

Material examined: –.

Comment. Demoulin (1968a) presumably placed it to Ephemerldae.

Euplectoptera INCERTAE SEDIS 7. † Lepismophlebia/g [g: *Lepismophlebia* Demoulin 1968c: 7, typus *Lepisma platymera* Scudder 1890 (design. orig.)].

References. Scudder 1890: Ꝺ; – Demoulin 1956d; – Demoulin 1968c.

Age. Miocene (USA).

Material examined: –.

Comment. Demoulin (1968c) presumably placed it to Leptophlebiidae (i.e. Leptophlebia/fg1).

Euplectoptera INCERTAE SEDIS 8. † Brevitibia/g [g: *Brevitibia* Demoulin 1968a: 245, typus *B. intricans* Demoulin 1968 (design. orig.)].

Reference. Demoulin 1968a: ♂ (wing venation unknown).

Age. Eocene (Baltic amber).

Material examined: –.

Comment. Demoulin (1968a) placed it to Ametropodidae (i. e. Ametropus/fg1); in order to confirm or disprove this, thorax structure of the holotype and paratypes (deposited in Staatl. Mus. Naturk. Stuttgart) should be examined.

Euplectoptera INCERTAE SEDIS 9. † Epeoromimus/fg [f: Epeoromimidae Tshernova 1969: 154; g: *Epeoromimus* Tshernova 1969: 156, typus *E. kazlauskasi* Tshernova 1969 (design. orig.)].

Reference. Tshernova 1969 (*E. kazlauskasi*): Ꝺ.

Character of unclear phylogenetic status.

(1) Tergalii of all pairs I–VII are very long, with strong costal ribs, spread laterally. In other respect larvae have plesiomorphic siphlonuroid appearance, with primary swimming setation on caudalii.

Age. Jurassic (Siberia).

Material examined (Paleontol. Inst.): *kazlauskasi* [*E.*]: Ꝺ.

Euplectoptera INCERTAE SEDIS 10. † Mogzonurus/g [g: *Mogzonurus* Sinitshenkova 1985: 15, typus *M. elevatus* Sinitshenkova 1985 (design. orig.)].

Reference. Sinitshenkova 1985: genitals of ⊕̸.

Plesiomorphy. Gonostylus with 2 distal segments.

Age. Jurassic (Siberia).

Material examined (Paleontol. Inst.): *elevatus* [*M.*]: ⊕.

Euplectoptera INCERTAE SEDIS 11. † Mogzonurella/g [g: *Mogzonurella* Sinitshenkova 1985: 16, typus *M. dissimilis* Sinitshenkova 1985 (design. orig.)].

Reference. Sinitshenkova 1985: Ꝺ.

Plesiomorphy. Larva has siphlonuroid appearance, with primary swimming setation on caudalii.

Age. Jurassic (Siberia).

Material examined (Paleontol. Inst.): *dissimilis* [*M.*]: Ꝺ.

Euplectoptera INCERTAE SEDIS 12. † Siberiogenites/g [g: *Siberiogenites* Sinitshenkova 1985: 20, typus *S. angustatus* Sinitshenkova 1985 (design. orig.)].

Reference. Sinitshenkova 1985: Ɒ.

Character of unclear phylogenetic status.

(1) Each tergalius has a prominent anal rib at a distance from anal margin. The same in Hexagenites/fg1 and some Recent taxa (see Index of characters [1.3.28]).

Plesiomorphy. Larva has siphlonuroid appearance, with primary swimming setation on caudalii.

Age. Jurassic (Siberia).

Species composition: *angustatus* Sinitshenkova 1985 [*Siberiogenites*], *medius* Sinitshenkova 1989 [*Siberiogenites*], *recticostalis* Sinitshenkova 2000 [*Siberiogenites*], *rotundatus* Sinitshenkova 1985 [*Siberiogenites*].

Material examined (Paleontol. Inst.): *angustatus* [*S.*]: Ɒ, *medius* [*S.*]: Ɒ, *recticostalis* [*S.*]: Ɒ, *rotundatus* [*S.*]: Ɒ.

Euplectoptera INCERTAE SEDIS 13. † Foliomimus/g [g: *Foliomimus* Sinitshenkova 1985: 21, typus *Epeoromimus tertius* Tshernova 1969 (design. orig.)].

Reference. Sinitshenkova 1985: Ɒ.

Plesiomorphy. Larva has siphlonuroid appearance.

Age. Jurassic (Siberia).

Material examined (Paleontol. Inst.): *tertius* [*E.*]: Ɒ.

Euplectoptera INCERTAE SEDIS 14. † Australurus/g [g: *Australurus* Jell & Duncan 1986: 120, typus *A. plexus* Jell & Dunkan 1986 (design. orig.)].

Reference. Jell & Duncan 1986: Ɒ.

Plesiomorphy. Larva has siphlonuroid appearance, with primary swimming setation on caudalii.

Age. Early Cretaceous (Australia).

Material examined: –.

Euplectoptera INCERTAE SEDIS 15. † Dulcimanna/g [g: *Dulcimanna* Jell & Duncan 1986: 124, typus *D. sculptor* Jell & Duncan 1986 (design. orig.)].

Reference. Jell & Duncan 1986: Ɒ.

Plesiomorphy. Larva has siphlonuroid appearance, with primary swimming setation on caudalii.

Age. Early Cretaceous (Australia).

Material examined: –.

Euplectoptera INCERTAE SEDIS 16. † Leptoneta/g [g: *Leptoneta* Sinitshenkova 1989: 37, typus *L. calyptrata* Sinitshenkova 1989 (design. orig.)].

Reference. Sinitshenkova 1989: Ɒ.

Age. Early Cretaceous (Mongolia).

Material examined (Paleontol. Inst.): *calyptrata* [*L.*]: Ɒ.

Euplectoptera INCERTAE SEDIS 17. † Albisca/g [g: *Albisca* Sinitshenkova 1989: 39, typus *A. tracheata* Sinitshenkova 1989 (design. orig.)].

Reference. Sinitshenkova 1989: Ɒ.

Plesiomorphy. Larva has siphlonuroid appearance, with primary swimming setation on caudalii.

Age. Early Cretaceous (Mongolia).

Material examined (Paleontol. Inst.): *tracheata* [*A.*]: Ɒ.

Euplectoptera INCERTAE SEDIS 18. † Siphgondwanus/g [g: *Siphgondwanus* McCafferty 1990: 28, typus *S. occidentalis* McCafferty 1990 (design. orig.)].

Reference. McCafferty 1990: Ɒ.

Autapomorphy.

(1) Each tergalius I–VII with a row of long spine-like setae on anal margin (McCafferty 1990:Fig.9).

Plesiomorphy. Larva has siphlonuroid appearance, with primary swimming setation on caudalii.

Age. Early Cretaceous (Brazil).

Material examined: –.

Euplectoptera INCERTAE SEDIS 19. † Bolbonyx/g [g: *Bolbonyx* Sinitshenkova 1990: 17, typus *B. ludibriosus* Sinitshenkova 1990].

Reference. Sinitshenkova 1990: Ɒ.

Age. Late Jurassic / Lower Cretaceous (Siberia).

Material examined (Paleontol. Inst.): *ludibriosus* [*B.*]: Ɒ.

Euplectoptera INCERTAE SEDIS 20. † Furvoneta/g [g: *Furvoneta* Sinitshenkova 1990: 18, typus *Mesoneta lata* Sinitshenkova 1976].

Reference. Sinitshenkova 1976a: Ɒ; – 1990: Ɒ.

Age. Late Jurassic / Lower Cretaceous (Siberia).

Material examined (Paleontol. Inst.): *lata* [*M.*]: Ɒ.

Euplectoptera INCERTAE SEDIS 21. † Costalimella/g [g: *Costalimella* Martins-Neto 1996: 184, typus *C. nordestina* Martins-Neto 1996 (design. orig.)].

Reference. Martins-Neto 1996: ⊕.

Age. Early Cretaceous (Brazil).

Material examined: –

Comment. Venation of cubital field of fore wing is unknown; judging by distal position of MA bifurcation, it can belong to Hexagenites/fg1.

Euplectoptera INCERTAE SEDIS 22. † Cretomitarcys/fg [f: Cretomitarcyinae Sinitshenkova 2000c: 113; g: *Cretomitarcys* Sinitshenkova 2000c: 114, typus *C. luzzii* Sinitshenkova 2000 (design. orig.)].

Reference. Sinitshenkova 2000c: ⊕.

Age. Late Cretaceous (New Jersey amber).

Material examined: –

Comment. The original description does not allow to conclude if this specimen belongs to Posteritorna (with AA falling to tornoapical margin), or to Anteritorna (with 2 additional longitudinal veins in some space between MA and CuP).

Euplectoptera INCERTAE SEDIS 23. † Palaeometropus/g [g: *Palaeometropus* Sinitshenkova 2000c: 117, typus *P. cassus* Sinitshenkova 2000 (design. orig.)].

Reference. Sinitshenkova 2000c: ♂.

Age. Late Cretaceous (New Jersey amber).
Material examined: –

Other extinct species of Euplectoptera INCERTAE SEDIS. Besides the type species of generic names listed above, following extinct species have uncertain systematic position: Species poorly described as adults from **Baltic amber** – *gigantea* Pictet & Hagen 1856 [*Baetis*], *grossa* Pictet & Hagen 1856 [*Baetis*], *macrops* Pictet & Hagen 1856 [*Palingenia*]. Species described as adults with unclear wing venation from **Tertiary of North America** – *howarthi* Cockerell 1908 [*Ephemera*]. Species described as larvae from **Jurassic, Cretaceous and Coenozoic of Eurasia, Africa, Australia and North America** – *africana* Sinitshenkova 1975 [*Hexameropsis*] (examined), *africanus* Haughton 1924 [*Phtharthus*], *alderensis* Lewis 1977 [*Isonychia*], *allata* Sinitshenkova 1985 [*Mesobaetis*] (examined), *amplectus* Sinitshenkova 2000 [*Mesobaetis*] (examined), *beipiaoensis* Wang 1980 [*Mesoneta*], *beybienkoi* Tshernova 1969 [*Epeoromimus*] (examined), *citima* Sinitshenkova 2000 [*Clavineta*] (examined), *clara* Sinitshenkova 2000 [*Stackelbergisca*] (examined), *colorata* Sinitshenkova 1990 [*Mogzonurella*] (examined), *cretaceous* Sinitshenkova 1976 [*Epeoromimus*] (examined), *deusta* Sinitshenkova 2000 [*Mesoneta*] (examined), *emmavillensis* Riek 1954 [*Cloeon*], *imitans* Sinitshenkova 1985 [*Foliomimus*] (examined), *infractus* Sinitshenkova 1989 [*Epeoromimus*] (examined), *longa* Sinitshenkova 1985 [*Mogzonurella*] (examined), *maculata* Hong & Liang & Hu 1995 [*Mesobaetis*], *magna* Sinitshenkova 1985 [*Mesoneta*] (examined), *mandalensis* Sinitshenkova 1989 [*Mesobaetis*] (examined), *mongolica* Sinitshenkova 1989 [*Mesoneta*] (examined), *rubiens* Lewis 1977 [*Potamanthellus*] (for comment – see above, Neoeophemera/fg1 INCERTAE SEDIS), *sanjianfangensis* Hong & Liang & Hu 1995 [*Mesobaetis*], *shanwangensis* Hong 1983 [*Ephemerella*], *tushilgae* Sinitshenkova 1989 [*Mesoneta*] (examined), *undina* Sinitshenkova 1976 [*Mesoneta*] (examined), *utriculata* Sinitshenkova 1985 [*Mesoneta*] (examined), *zolensis* Sinitshenkova 1990 [*Mesoneta*] (examined).

* * *

Euephemeroptera INCERTAE SEDIS 1
† Litophlebia/fg(1)

Nomen hierarchicum: **Litophlebia/fg(1)** [f: Litophlebiidae Hubbard & Riek 1978: 260; g: *Litophlebia* Hubbard & Riek 1978: 260, typus *Xenophlebia optata* Riek 1976 (design. orig.); syn. obj.: ***Xenophlebia*** Riek 1976: 150, nom. praeocc.].
In circumscription matches:
— superfam. Xenophlebioidea Riek 1976: 149;
— fam. Xenophlebiidae Riek 1976: 150;
— gen. *Xenophlebia* Riek 1976: 150;
— superfam. Litophlebioidea Hubbard & Riek 1978: 260;
— fam. Litophlebiidae Hubbard & Riek 1978: 260;
— gen. *Litophlebia* Hubbard & Riek 1978: 260.

Reference. Riek 1976: fore wing.

Autapomorphy of *Litophlebia*.

(1) Cubital and anal fields of fore wing are greatly reduced, CuA is extremely shortened, cubital field has no intercalaries or other veins going to wing margin. At the same time MP is not shortened, terminates far from CuA and partly goes parallel to wing margin (like CuA in Anteritorna or AA in Posteritorna).

Age. Triassic (South Africa).
Material examined: –.

* * *

Other Euephemeroptera INCERTAE SEDIS

Euephemeroptera INCERTAE SEDIS 2.
† Thuringopteryx/g [g: *Thuringopteryx* Kuhn 1937: 191, typus *T. gimmi* Kuhn 1937 (monotypy)].

References. Kuhn 1937: distal part of wing; – Müller 1965: distal part of wing; – Kluge & Sinitshenkova 2002: belongs to Ephemeroptera.

Age. Triassic (Germany).
Material examined: –.

Euephemeroptera INCERTAE SEDIS 3.
† Triassomachilis/fg [f: Triassomachilidae Sharov 1948; g: *Triassomachilis* Sharov 1948, typus *T. uralensis* Sharov 1948].

References. Rasnitsyn 1980; – Sinitshenkova 2000d: ☉.

Age. Triassic (Urals).

Material examined (Paleontol. Inst.): *uralensis* [*T.*]: ☉.

Comment. Originally (Sharov 1948) these fossils were described as Triplura with segmented abdominal styli (that was used to argue a wrong statement that insect styli were initially segmented and originated from legs); Subsequently (Sinitshenkova 2000d), they were described as larvae of the euplectopteran genus *Mesoneta*, which have mesothoracic wing buds larger than metathoracic ones. Actually on these fossils styli are absent, and wing buds are not preserved.

Euephemeroptera INCERTAE SEDIS 4.
† Aenigmephemera/fg [f: Aenigmephemeridae Tshernova 1968: 23; g: *Aenigmephemera* Tshernova 1968: 23, typus *A. demoulini* Tshernova 1968 (design. orig.)].

References. Tshernova 1968: distal part of fore wing; – 1970.

Age. Jurassic (Kazakhstan).

Material examined (Paleontol. Inst.): *demoulini* [*A.*]: distal part of fore wing.

Euephemeroptera INCERTAE SEDIS 5.
† Palinephemera/g [g: *Palinephemera* Lin 1986: 26,

typus *Palinephemera densivena* Lin 1986 (design. orig.)].
Reference. Lin 1986: wing.
Age. Early Jurassic (China).
Material examined: –.

Euephemeroptera INCERTAE SEDIS **6.**
† **Clephemera/g** [g: *Clephemera* Lin 1986: 27, typus *C. clava* Lin 1986 (design. orig.)].
Reference. Lin 1986: ♉.
Age. Early Jurassic (China).
Material examined: –.

Other species of Euephemeroptera INCERTAE SEDIS. Besides *uralensis* [*Triassomachilis*] (see above), following species are described as larvae from Triassic of Ukraine, Urals and West Siberia – *minuta* Sinitshenkova 2000 [*Mesoneta*], *triassica* Sinitshenkova 2000 [*Mesoneta*], *picta* Sinitshenkova 2000 [*Mesoneta*], *ornata* Sinitshenkova 2000 [*Mesobaetis*]. They were attributed to artificial Mesozoic euplectopteran genera *Mesoneta* and *Mesobaetis*; in the original descriptions (Sinitshenkova 2000d) there are figured large mesonotal and small metanotal wing buds typical for Euplectoptera; however, all specimens (deposited in Palaeontololgical Institute in Moscow and re-examined by me) have thoraces very poorly preserved, with wing buds indistinct. As no Triassic Euplectoptera are known for certain, there is no reason to regard these fossils as belonging to Euplectoptera.

* * *

Pterygota INCERTAE SEDIS 1
† Syntonoptera/fg1

Nomen hierarchicum: **Syntonoptera/fg1** (incl. *Lithoneura, Bojophlebia*) [f: Syntonopteridae Handlirsch 1911: 299; g: *Syntonoptera* Handlirsch 1911: 299, typus *S. schucherti* Handlirsch 1911 (monotypy)].
In circumscription matches:
— superfam. Syntonopteroidea: Kukalova-Peck 1985:939.

References. Kukalova-Peck 1985: adult; – Carpenter 1987: adult.

Characters of unclear phylogenetic status.

(1) Sc and RA go nearly up to wing apex; RS begins as common stem, concave, independent from RA; MA convex, MP concave, CuA convex, CuP concave. The same in Panephemeroptera, Protorrhynchota and Odonatoptera.

(2) MA begins independently from RS, but in its base is curved anteriorly and at a short distance approximated or fused with RS (in contrast to Euephemeroptera, where MA is basally completely fused with RS). Non-unique character.

(3) RS, MA, MP and CuA are branched, in some cases forming triads. More constant triads are peculiar for Panephemeroptera and Odonatoptera.

Size. Fore wing length presumably 100–200 mm.

Age. Late Carboniferous (USA).

Nominal taxa included. Syntonoptera/fg1 includes: (1) **Lithoneura/g** [g: *Lithoneura* Carpenter 1938: 446, typus *L. lameerei* Carpenter 1938 (design. orig.)]; (2) **Bojophlebia/fg** [f: Bojophlebiidae Kukalova-Peck 1985: 934; g: *Bojophlebia* Kukalova-Peck 1985: 936, typus *B. procopi* Kukalova-Peck 1985 (design. orig.)].

Species composition of Syntonoptera/fg1. 5 species – *carpenteri* Richardson 1956 [*Lithoneura*], *lameerei* Carpenter 1938 [*Lithoneura*], *mirifica* Carpenter 1944 [*Lithoneura*], *procopi* Kukalova-Peck 1985 [*Bojophlebia*], *schucherti* Handlirsch 1911 [*Syntonoptera*].

Material examined: –.

* * *

Other Pterygota INCERTAE SEDIS

Pterygota INCERTAE SEDIS **2.** † **Geisfeldiella/g** [g: *Geisfeldiella* Kuhn 1951: 61, typus *G. benkerti* Kuhn 1951 (monotypy)].
Reference. Kuhn 1951: wing.
Age. Liass (Germany).
Material examined: –.
Comment. Originally *Geisfeldiella* was described as a member of Lepidoptera; later it was assumed to belong to Ephemeroptera (Kluge & Sinitshenkova 2002: 97). Actually, judging by the photograph (Kuhn 1951: Abb.1) RS seems to form no more than 2 successive triads (instead of 3 triads characteristic for Ephemeroptera); possibly this insect belongs to Odonata.

Pterygota INCERTAE SEDIS **3.** † **Montralia/g** [g: *Montralia* Via & Calsada 1987: 278, typus *M. muelleri* Via & Calsada 1987 (design. orig.)].
Reference. Via&Calsada 1987: wing.
Age. Triassic (Spain).
Material examined: –.
Comment. Originally *Montralia* was described as a member of Ephemeroptera. Actually, veins do not form triads (Via & Calsada 1987:Fig.1); probably this insect belongs to Neoptera.

Pterygota INCERTAE SEDIS **4.** † **Tintorina/fg** [f: Tintorinidae Krzeminski & Lombardo 2001: 72; g: *Tintorina* Krzeminski & Lombardo 2001: 72, typus *T. meridensis* Krzeminski & Lombardo 2001 (design. orig.)].
Reference. Krzeminski&Lombardo 2001: adults.
Age. Middle Triassic (Switzerland).
Material examined: –.
Comment. Originally described as a member of Ephemeroptera. Venation is similar to that of Ephemeroptera (while some veins are not preserved); in contrast to all Euephemeroptera, costal brace is absent, and Sc falls to costal margin.

Euarthropoda INCERTAE SEDIS

A single fossil specimen from Middle Triassic of West Siberia was described as mayfly larva *mogutshevae* Sinitshenkova 2000 [*Archaeobehningia*]. This fossil (deposited in Paleontol. Inst. and re-examined by me) has distinct trunk tergites, but other details described and figured (Sinitshenkova 2000d: Fig.3b) – antenna, legs and fore wing bud – are indistinct and possibly absent.

* * *

Animalia INCERTAE SEDIS

Animalia INCERTAE SEDIS. † Asthenopodichnium/g [g: *Asthenopodichnium* Thenius 1979: 185, typus *A. xylobiontum* Thenius 1979 (design. orig.)].

Age. Miocene (Austria).

Material examined: –.

Comment. Two species – *xylobiontum* Thenius 1979 [*A.*] and *ossibiontum* Thenius 1988 [*A.*] – are described only as traces, and attributed to Ephemeroptera-Fossoriae. Actually, such traces can belong to any group of animals.

* * *

Nomina dubia

The following old species names are recently out of use; possibly they belong to well-known species and should be regarded as NOMINA OBLITA. **European species:** *angustipennis* Rambur 1842 [*Ephemera*], *argentata* Pictet 1843 [*Caenis*], *brevicauda* Fabricius 1793 [*Ephemera*], *chironomiformis* Curtis 1834 [*Brachycercus*], *fusca* Schneider 1845 [*Cloe*], *fuscula* Schrank 1798 [*Ephemera*], *gemmata* Scopoli 1763 [*Ephemera*], *halterata* Fabricius 1777 [*Ephemera*], *minima* (Linnaeus 1747) Müller 1776 [*Ephemera*], *montana* Pictet 1843 [*Baetis*], *nervosa* Villers 1789 [*Ephemera*], *plumosa* Müller 1776 [*Ephemera*], *rufa* Rambur 1842 [*Ephemera*], *speciosa* Poda 1761 [*Ephemera*], *stigma* Gmelin (in Linnaeus) 1790 [*Ephemera*], *vitrea* Zetterstedt 1840 [*Ephemera*]. **North American species:** *noveboracana* Lichtenstein 1796 [*Ephemera*], *pudica* Hagen 1861 [*Ephemera*].

* * *

INDEX OF CHARACTERS

In detail about these characters see in the characteristics of the taxa: number in brackets corresponds to a number of an autapomorphy or a character of unclear phylogenetic status, "(P)" means that the character is placed in a list of plesiomorphies without a number, "VAR" means that the character is placed in a list of variable characters without a number. Words "... (not all)" mean that this is possibly initial (but not unique) condition of the character in this taxon; words "part of ..." mean that this is secondary (and not unique) condition of the character in this taxon; more exact explanations of each case is given in the taxon characteristics. Taxon name in brackets: {...} – taxon (or taxa) subordinate to the previous one.

Characters are given in following order:

1. LARVA

Head and its appendages [1.1], thorax and its appendages [1.2], abdomen and its appendages [1.3].

1.1. LARVAL HEAD AND ITS APPENDAGES

Head capsule [1.1.1–5], eyes [1.1.6–7], antennae [1.1.8–11], mouth apparatus [1.1.12–59]

— **on ventral side arched row:** Cryptoprosternata (3)
— **on ventral side regular transverse arched row:** Polymitarcys/f1=Ephoron/g2 (1A)
— **on tusk** – see [1.1.20]
— **on median margin** – see [1.1.23]

[1.1.20] **Tusk**
— **present:** Fimbriatotergaliae (not all) (8)
— **absent:** part of Fimbriatotergaliae (8) {Behningia/fg1 (13), Caenotergaliae (10)}
— **curved medially:** Potamanthus/fg1 (7), Euthyplocia/fg1 (2), *Ichthybotus* (1), Polymitarcys/f1=Ephoron/g2 (8)
— **curved laterally:** Ephemera/fg9 (2), Palingenia/f2=g1 (8)
— **long:** *Anthopotamus* (1), part of Rhoenanthus/g1 (VAR)
— **short:** Potamanthus/fg3 (1), part of Rhoenanthus/g1 (VAR), Afromera/g1 (1)
— **round in section:** Ephemera/fg9 (2)
— **with lateral serrate ridge:** Palingenia/f2=g1 (2)
— **with spine-like seta on each denticle:** Palingenia/f2==g1 (not all) (2) {*Pentagenia* (P), *Chankagenesia* (P), Anagenesia/g2 (P), *Cheirogenesia* (P)}
— **without spine-like seta on denticle:** part of Palingenia/f2=g1 (2) {Palingenia/f4=g3 (1), *Mortogenesia* (1), *Plethogenesia* (2)}
— **with tubercles dorsally and laterally:** *Ichthybotus* (1), Polymitarcys/f2=Ephoron/g3 (7)
— **with denticles on inner side:** Campsurus/fg1 (1)
— **with largest denticle in proximal part of inner side:** Campsurus/fg3 (2)
— **with largest denticle in distal part of inner side:** *Tortopus* (2)
— **specialized as biting:** Asthenopus/fg1 (1)
— **with long setae on inner side:** Euthyplocia/fg1 (2)

[1.1.21] **Incisor (apical canine of mandible)**
— **diminished:** Discoglossata (2), Campsurus/fg1 (1)
— **enlarged:** Nesameletus/f1=Metamonius/g1 (1), *Vietnamella* (2)
— **with combs:** *Rallidens* (1)
— **with 3 apical denticles:** Heptagenia/f6=g5 (not all) (1)
— **stout:** Epeorus/fg3 (1)
— **of right mandible with large denticle near apex:** Ecdyonurus/fg1 (2)
— **of right mandible with process at base:** Polymitarcys/f2=Ephoron/g3 (1)

[1.1.22] **Kinetodontium (subapical canine of mandible)**
— **vestigial:** Nesameletus/f1=Metamonius/g1 (1), *Vietnamella* (2)
— **diminished:** Discoglossata (2), Campsurus/fg1 (1)
— **jointed:** *Siphlaenigma* (P)
— **not jointed:** Turbanoculata (5)
— **elongate:** Ameletopsis/fg1 (1)
— **stout:** Epeorus/fg3 (1)

[1.1.23] **Median margin (between kinetodontium and mola)**
— **bears prostheca** – see [1.1.24–25]
— **with setae:** Turbanoculata (not all) (6), Heptagennota (not all) (7) {Heptagenia/f6=g5 (not all) (1), Ecdyonurus/fg1 (2), Epeorus/fg3 (1)}

[1.1.24] **Prostheca of left mandible**
— **brush-like:** *Siphlaenigma* (P)
— **stick-like:** Turbanoculata (6)
— **bristle-like:** Ameletus/fg1 (1), Heptagennota (not all) (7) {Heptagenia/f6=g5 (1), Epeorus/fg3 (1)}
— **absent:** Acanthametropus/fg1 (1), Ameletopsis/fg1 (1), part of Heptagennota (7) {*Pseudiron* (1)}

[1.1.25] **Prostheca of right mandible**
— **brush-like:** *Siphlaenigma* (P)
— **stick-like:** Turbanoculata (not all) (6)
— **bifurcate:** part of Turbanoculata (6)
— **bristle-like:** part of Turbanoculata (6)
— **absent:** Acanthametropus/fg1 (1), Ameletus/fg1 (1), Ameletopsis/fg1 (1), Heptagennota (7) {*Pseudiron* (1)}

[1.1.26] **Mola**
— **absent:** Prosopistoma/f1=g2 (1)
— **without grater:** Acanthametropus/fg1 (1), Ameletopsis/fg1 (1), *Pseudiron* (1), *Raptoheptagenia* (1), *Acanthomola* (1), *Anepeorus* (1), Behningia/fg1 (2)

SUPERLINGUAE
[1.1.27]

[1.1.27] **Superlinguae**
— **vestigial or absent:** Prosopistoma/f1=g2 (1), Acanthametropus/fg1 (1), Ameletopsis/fg1 (1), *Pseudiron* (1), *Raptoheptagenia* (1), *Anepeorus* (1)
— **curved laterally:** *Arthroplea* (7), Heptagenia/f5=g4 (1), *Raptoheptagenia* (3), *Dicercomyzon* (2)
— **strongly curved laterally:** Stenonema/g1 (1)
— **laterally concave, apically rounded:** *Kageronia* (1)
— **laterally concave, apically angulate:** *Dacnogenia* (3)
— **laterally with subapical incision:** Heptagenia/f8=g7 (1)
— **laterally with subbasal concavity:** Ecdyonurus/fg1 (2)
— **widened apically:** *Cinygma* (4)
— **with divergent margins:** *Bleptus* (4), *Ironodes* (3)
— **with maximum width in middle and convergent toward apex:** Rhithrogena/fg2 (5), Epeorus/fg3 (3)
— **parallel-sided:** *Paegniodes* (P)

HYPOPHARYNX
[1.1.28]

[1.1.28] **Hypopharynx**
— **truncate:** Metretopus/fg1 (2)
— **incised apically:** *Ametropus* (2)
— **shortened:** *Dacnogenia* (1)

MAXILLAE
Maxilla [1.1.29], ventral setae [1.1.30], apical setae [1.1.31–32], canines [1.1.33–34], setae of median margin [1.1.35–40], maxillary palp [1.1.41–46]

[1.1.29] **Maxilla**
— **filtering** [1.1.31–32]: Ameletus/fg1 (1), Polymitarcys/f1==Ephoron/g2 (2), part of Ephemerella/fg1 (16) {Cincticostella/g2 (1), *Uracanthella* (1), *Allenhyphes* (2), *Haplohyphes* (1)}, Leptophlebia/fg1 (1)
— **scraping:** part of Ephemerella/fg1 (16) {*Dicercomyzon* (2)}
— **diminished:** Behningia/fg2 (2)
— **widened:** *Rallidens* (2)
— **elongate-rectangle:** Radulapalpata (not all) (3) {*Bleptus* (P), *Ironodes* (P), *Paegniodes* (P)}
— **narrowed toward apex:** Discoglossata (2)

— **truncate (with apical margin formed by expansion of outer margin):** Radulapalpata (not all) (3), Polymitarcys/f1=Ephoron/g2 (2), part of Ephemerella/ /fg1 (16), Leptophlebia/fg1 (1)
— **truncate (with apical margin formed by expansion of inner margin):** Tricorygnatha (1)
— **with acute median-apical angle:** Isonychia/fg1 (10), Discoglossata (2), Rhithrogena/fg2 (1), Epeorus/fg3 (1)
— **with tusk-like process on median-apical angle:** part of Leptophlebia/fg1 (1)
— **with 5 movable appendages:** Ameletopsis/fg1 (1)
— **with gill at joining of cardo and stipes:** Nesameletus/ /f1=Metamonius/g1 (7), *Rallidens* (4)
— **with gill at joining with head:** Eusetisura (3)

Ventral setae of maxilla
[1.1.30-31]

[1.1.30] **Setae on ventral side of maxilla**
— **form transverse row:** Tricorygnatha (1)
— **form longitudinal row:** Branchitergaliae (1)
— **longitudinal row straight:** *Cinygma* (5), *Paegniodes* (P), Epeorus/fg1 (6) {Epeorus/fg3 (1)}
— **longitudinal row slightly curved laterally:** Heptagenia/f6=g5 (2)
— **longitudinal row strongly curved laterally:** Rhithrogena/fg2 (1)
— **irregular field:** Ecdyonurus/fg1 (2)

Apical setae of maxilla
[1.1.31-32]

[1.1.31] **Setae of apical-ventral row** (among all Ephemeroptera)
— **absent:** Posteritorna (7), *Siphlonisca* (1), Acanthametropus/fg1 (1), *Ametropus* (2), part of Nesameletus/f1= =Metamonius/g1 (2), Ameletopsis/fg1 (1), *Rallidens* (2), Eusetisura (4), *Pseudiron* (1), *Arthroplea* (1), Raptoheptagenia (1), *Cinygma* (2), Epeorus/fg3 (1), *Acanthomola* (1), *Anepeorus* (1), Fimbriatotergaliae (11), Ephemerella/fg1 (16), part of Leptophlebia/fg1 (1)
— **vestigial:** *Dipteromimus* (1), Nesameletus/f1=Metamonius/g1 (not all) (2), Stenonema/g2 (VAR), Epeorus/ /fg1 (not all) (5) {*Bleptus* (P), *Ironodes* (P)}, part of Leptophlebia/fg1 (1)
— **non-pectinate, transferred ventrally:** Metretopus/fg1 (2), Tetramerotarsata (7)
— **non-pectinate or pectinate:** Siphlonurus/fg1 (not all) (P), Vetulata (P)
— **pectinate elongate:** Ameletus/fg1 (1), Leptophlebia/fg1 (not all) (1)
— **pectinate oblique brush-shaped:** Radulapalpata (not all) (3)

[1.1.32] **Setae on apical side of maxilla** (apart of apical-ventral row – see [1.1.31])
— **absent:** Nesameletus/f1=Metamonius/g1 (2), Ameletopsis/fg1 (1), *Raptoheptagenia* (1), Epeorus/fg3 (1)
— **long, irregular:** *Arthroplea* (1), *Cinygma* (2), *Bleptus* (P), *Ironodes* (P)
— **long, form regular field not adjacent to canines:** Polymitarcys/f1=Ephoron/g2 (2), Clypeocaenis/g1 (1)
— **long, form regular field adjacent to canines:** Potamanthus/fg1 (8), Euthyplocia/fg1 (6), Ephemerella/fg1 (16) {Cincticostella/g1 (1) {Cincticostella/g2 (1)}, *Uracanthella* (1)}, Leptophlebia/fg1 (1)
— **short, scraping:** *Dicercomyzon* (2)

Apical canines of maxilla
[1.1.33–34]

[1.1.33] **Number of maxillary canines** (among all Ephemeroptera)
— **0:** Ameletus/fg1 (1), part of *Uracanthella* (1), Tricorygnatha (1), *Dicercomyzon* (2), part of Leptophlebia/ /fg1 (1)
— **1:** Prosopistoma/f1=g2 (1), Discoglossata (2), *Arthroplea* (1), *Cinygma* (2), Behningia/fg2 (2), Palingenia/ /f2=g1 (3), *Vietnamella* (5), *Teloganodes* (3)
— **2:** Metretopus/fg1 (2), Acanthametropus/fg1 (1), *Ametropus* (2), Isonychia/fg1 (10), *Pseudiron* (1), Hexagenia/g1 (2), *Protobehningia* (P), part of Polymitarcys/f1=Ephoron/g2 (2)
— **3:** Baetisca/f2=g1 (P), Siphlonurus/fg1 (P), *Dipteromimus* (P), Tetramerotarsata (P), Vetulata (P), Nesameletus/f1=Metamonius/g1 (2), *Rallidens* (2), Coloburiscus/fg1 (not all) (8), Heptagenia/f5=g4 (P), Rhithrogena/fg1 (P), Potamanthus/fg1 (P), Euthyplocia/fg1 (P), *Ichthybotus* (P), Ephemera/fg10 (P), Polymitarcys/f1=Ephoron/g2 (not all) (2), Caenotergaliae (P), Ephemerella/fg2 (not all) (P), *Ephemerythus* (P), Tricorythodes/fg1 (P), Leptohyphes/fg2 (P), *Allenhyphes* (2), Leptophlebia/fg1 (not all) (1)

[1.1.34] **Form of maxillary canines**
— **in one plane with dentisetae:** Acanthametropus/fg1 (1)
— **stout:** Nesameletus/f1=Metamonius/g1 (2)
— **flat and wide:** *Rallidens* (2)
— **strongly enlarged:** Epeorus/fg3 (1)
— **diminished:** *Uracanthella* (1)
— **completely fused together:** *Vietnamella* (5), *Teloganodes* (3)
— **separated by impression:** Cincticostella/g1 (1)
— **ventral canine flattened and denticulate:** Cincticostella/g2 (1)

Setae of inner (median) margin of maxilla
[1.1.35–40]

[1.1.35] **Setae of inner-ventral row**
— **absent:** Tricorygnatha (1)
— **3:** *Teloganodes* (3)
— **distal seta canine-like:** Discoglossata (2)

[1.1.36] **Setae of inner-dorsal row**
— **2:** *Teloganodes* (3)
— **on distal margin:** Tricorygnatha (1)
— **dentisetae** – see [1.1.37–40]

[1.1.37] **Number of dentisetae** (among all Ephemeroptera)
— **0:** part of Bidentiseta (1) {Protobehningia (3), part of Polymitarcys/f1=Ephoron/g2 (2), part of Leptophlebia/ fg1 (1)
— **1:** Ameletus/fg1 (1), part of Bidentiseta (1) {part of Isonychia/fg1 (2), Palingenia/f2=g1 (3), Tricorygnatha (1), *Dicercomyzon* (2), part of Leptophlebia/fg1 (1)}

— **2:** part of Acanthametropus/fg1 (1), Bidentiseta (not all) (1) {Coloburiscus/fg1 (not all) (8), Isonychia/fg1 (not all) (2), Heptagennota (P), Potamanthus/fg1 (P), Euthyplocia/fg1 (P), *Ichthybotus* (P), Ephemera/fg9 (P), Behningia/fg2 (P), Polymitarcys/f1=Ephoron/g2 (not all) (2), Caenotergaliae (P), Ephemerella/fg2 (P), *Ephemerythus* (P), *Machadorythus* (P), Tricorythodes/ /fg1 (P), Leptohyphes/fg2 (P), *Allenhyphes* (2), *Vietnamella* (5), *Teloganodes* (3), Leptophlebia/fg1 (not all) (1)
— **3:** Prosopistoma/f1=g2 (1), Anteritorna (not all) (2) {Tridentiseta (not all) (P) {Siphlonurus/fg1 (P), *Dipteromimus* (P), Metretopus/fg1 (P), Acanthametropus/ /fg1 (not all) (1), *Ametropus* (2), Tetramerotarsata (P), Nesameletus/f1=Metamonius/g (3), Vetulata (P), *Rallidens* (2)}, part of Coloburiscus/fg1 (8)
— **more than 3:** Posteritorna (not all) (P) {Baetisca/f2=g1 (P)}
— **indeterminate:** Posteritorna (not all) (P) {Baetisca/f2= =g1 (P)}, part of Anteritorna (2) {part of Bidentiseta (1) {Coloburiscus/fg1 (8)}}

[1.1.38] **1st dentiseta (distal)**
— **absent:** part of Bidentiseta (1) {part of Isonychia/fg1 (2), Palingenia/f2=g1 (3), Tricorygnatha (1), *Dicercomyzon* (2), part of Leptophlebia/fg1 (1)
— **vestigial:** Isonychia/fg1 (not all) (2), Euthyplocia/fg1 (P), Hexagenia/g1 (2)
— **canine-like:** Acanthametropus/fg1 (1)
— **massive, pressed to canines:** part of Turbanoculata (VAR), Nesameletus/f2=Metamonius/g2 (2)
— **branched from base:** Atopopus/g1 (1)
— **not branched:** Ecdyonurus/fg2 (P), Afghanurus/g1 (P)

[1.1.39] **2nd dentiseta (proximal in Bidentiseta)**
— **canine-like:** Acanthametropus/fg1 (1)
— **on common base with 3rd:** Nesameletus/f1=Metamonius/g1 (3)
— **on common plate with 3rd:** Nesameletus/f2=Metamonius/g2 (3)
— **straight:** Isonychia/fg1 (2)
— **widened and dentate:** *Pseudiron* (1)
— **bifurcate:** Ecdyonurus/fg1 (2)
— **pectinate:** *Teloganodes* (3)

[1.1.40] **3rd dentiseta (proximal in Tridentiseta)**
— **absent:** part of Acanthametropus/fg1 (1), Bidentiseta (1)
— **on common base with 2nd:** Nesameletus/f1=Metamonius/g1 (3)
— **on common plate with 2nd:** Nesameletus/f2=Metamonius/g2 (3)

Maxillary palp
[1.1.41–46]

[1.1.41] **Maxillary palp**
— **absent** (among all Ephemeroptera): *Madecocercus* (1), part of Ephemerella/fg1 (1) {part of Torleya/g1 (1)}
— **vestigial, without muscles:** Acanthametropus/fg2 (2), Ephemerella/fg1 (not all) (1)
— **secondarily elongate:** Tricorygnatha (1), *Machadorythus* (3)
— **strongly elongate:** *Arthroplea* (1)
— **strongly enlarged:** Behningia/fg2 (2)
— **specialized as scraping:** Rhithrogena/fg1 (1)

[1.1.42] **Number of segments in maxillary palp** (among all Ephemeroptera)
— **1:** Acanthametropus/fg2 (2), part of Ephemerella/fg1 (1)
— **2:** part of Turbanoculata (VAR), part of Eusetisura (2) {Isonychia (2), Discoglossata (2)}, *Arthroplea* (1), Rhithrogena/fg1 (1), *Anepeorus* (2), part of Ephemera/fg9 (P), Cryptoprosternata (4), Brachycercus/f1= =g3 (1), Clypeocaenis/g2 (1)
— **3:** Posteritorna (P), Siphlonurus/fg1 (P), *Dipteromimus* (P), Ameletus/fg1 (P), Metretopus/fg1 (2), *Analetris* (P), *Ametropus* (P), *Siphlaenigma* (2), Turbanoculata (not all) (VAR), Vetulata (P), Nesameletus/f1= =Metamonius/g1 (4), *Rallidens* (P), Eusetisura (not all) (2) {Coloburiscus/fg1 (P)}, *Pseudiron* (P), Heptagenia/f5=g4 (P), Potamanthus/fg1 (P), Euthyplocia/fg1 (P), *Ichthybotus* (P), Ephemera/fg9 (not all) (P), Behningia/fg1 (P), Neoephemera/fg1 (P), Caenis/f5= =g3 (P), *Amercaenis* (P), Caenopsella/g1 (P), *Brasiliocaenis* (P), *Caenoculis* (P), Ephemerella/fg1 (not all) (1), Leptophlebia/fg1 (P)
— **many secondary segments:** Ameletopsis/fg1 (1)

[1.1.43] **1st segment of maxillary palp**
— **shortened:** Cryptoprosternata (4) {Discoglossata (2)}
— **elongate:** *Arthroplea* (1)
— **with fan-form flexor attached to outer margin:** Rhithrogena/fg1 (1)
— **with fan-form flexor attached to inner margin:** Behningia/fg2 (2)

[1.1.44] **Distal (2nd+3rd) segment of 2-segmented maxillary palp**
— **divided into secondary segments:**Ameletopsis/fg1 (1)
— **thickened:** Eusetisura (2), Cryptoprosternata (4)
— **long, with 2 rows of filtering setae:** *Arthroplea* (1)
— **with scraping setae:** Rhithrogena/fg1 (1)
— **with scraping setae occupying nearly whole length of segment:** Rhithrogena/fg3 (1)
— **with pigmented ridge bordering field of scraping setae:** Rhithrogena/fg2 (2)

[1.1.45] **2nd segment of 3-segmented maxillary palp**
— **shortened:** Metretopus/fg1 (2), Clypeocaenis/g1 (1)
— **curved:** Prosopistoma/f1=g2 (1)

[1.1.46] **3rd segment of 3-segmented maxillary palp**
— **shortened:** Nesameletus/f1=Metamonius/g1 (4), *Siphlaenigma* (2), Tricorygnatha (1)
— **vestigial:** Heptagenia/f5=g4 (P)
— **vestigial small:** Heptagenia/f6=g5 (3)
— **vestigial triangular:** Ecdyonurus/fg1 (2)

LABIUM

Labium [1.1.47], submentum [1.1.48], mentum [1.1.49], glossae [1.1.50–51], paraglossae [1.1.52–53], labial palps [1.1.54–59]

[1.1.47] **Labium**
— **with gills:** *Siphluriscus* (1), part of Eusetisura (3)
— **scraping:** Radulapalpata (not all) (1)
— **raptorial:** *Raptoheptagenia* (1)
— **diminished:** Behningia/fg2 (2)
— **wide:** *Dicercomyzon* (2)

[1.1.48] **Submentum**
— **enlarged, cover labial palps:** Prosopistoma/fl=g2 (1)
— **with pair of flat lateral lobes:** Ameletopsis/fg1 (1)
— **shortened:** Tricorygnatha (1)
— **membranous without muscles:** Pentamerotarsata (1)

[1.1.49] **Mentum**
— **enlarged, with palpi adductors transverse:** Radulapalpata (1)

Glossae
[1.1.50–51]

[1.1.50] **Glossae**
— **fused together forming disc:** Discoglossata (2)
— **fused with mentum and paraglossae:** Prosopistoma//fl=g2 (1), Tricorygnatha (1) *Machadorythus* (4), part of *Coryphorus* (VAR)
— **curved ventrally:** Metretopus/fg1 (2)
— **petiolate:** *Dipteromimus* (1)
— **basally widened ventrad of paraglossa:** Turbanoculata (7), Discoglossata (2)
— **projected laterad of glossae:** Discoglossata (2)
— **truncate, with apical row of flat setae:** Ameletus/fg2 (1)
— **thickened and truncate:** *Chiloporter* (1)
— **elongate and narrowed, with stout setae on apex:** Nesameletus/fl=Metamonius/g1 (5)
— **crescent:** *Arthroplea* (1)
— **semicircular:** *Kageronia* (2)
— **curved medially:** Heptagenia/f8=g7 (2), Stenonema/g1 (4)
— **pyramidal:** Radulapalpata (not all) (1) {*Dacnogenia* (P), *Paegniodes* (P), Rhithrogena/fg2 (P), Epeorus/fg2 (P)}
— **rhomboid:** Ecdyonurus/fg1 (2)
— **widening distally, roundish-triangular:** *Cinygma* (3)
— **roundish:** *Bleptus* (1)
— **shortened:** Behningia/fg2 (2)

[1.1.51] **Setation of glossa**
— **stout apical setae:** Nesameletus/fl=Metamonius/g1 (5)

Paraglossae
[1.1.52–53]

[1.1.52] **Paraglossae**
— **fused with mentum and glossae:** Prosopistoma/fl=g2 (1), Tricorygnatha (1), *Machadorythus* (4), part of *Coryphorus* (VAR)
— **not separated from mentum:** Posteritorna (7), Ephemerella/fg1 (2)
— **curved ventrally:** Metretopus/fg1 (2)
— **elongate and narrowed:** Nesameletus/fl=Metamonius/g1 (5)
— **petiolate:** *Dipteromimus* (1)
— **hidden in cavities on dorsal side of glossal disc:** Discoglossata (2)
— **crescent:** *Arthroplea* (1), Campsurus/fg1 (2)
— **widened:** Radulapalpata (1), *Dicercomyzon* (2)
— **extended proximally:** Hexagenia/fg1 (1)
— **shortened:** Behningia/fg2 (2)

[1.1.53] **Setation of paraglossa**
— **longitudinal row of setae:** Tetramerotarsata (8)

Labial palps
[1.1.54–59]

[1.1.54] **Labial palp**
— **specialized as scraping:** Radulapalpata (1)
— **strongly enlarged:** Behningia/fg2 (2)

[1.1.55] **Number of segments of labial palp** (among all Ephemeroptera)
— **2:** Metretopus/fg1 (2), part of Turbanoculata (VAR), part of Eusetisura (2) {Isonychia/fg1 (2), Discoglossata (2)}, Heptagennota (8) {*Pseudiron* (1), *Arthroplea* (1), Radulapalpata (1)}, part of Behningia/fg1 (2) Cryptoprosternata (4), Brachycercus/fl=g3 (2)
— **3:** Posteritorna (P), Siphlonurus/fg1 (P), *Dipteromimus* (P), Ameletus/fg1 (P), Acanthametropus/fg1 (P), Vetulata (P), Nesameletus/fl=Metamonius/g1 (P), *Rallidens* (P), *Siphlaenigma* (3), Turbanoculata (not all) (VAR), *Ametropus* (P), Eusetisura (not all) (2) {Coloburiscus/fg1 (P)}, Potamanthus/fg1 (P), *Euthyplocia* (P), *Ichthybotus* (P), Ephemera/fg9 (P), Behningia/fg1 (not all) (2), Neoephemera/fg1 (P), Caenis//f3=g1 (P), Ephemerella/fg1 (not all) (3), Leptophlebia/fg1 (P)
— **many secondary segments:** Ameletopsis/fg1 (1)

[1.1.56] **1st segment of labial palp**
— **shortened:** Cryptoprosternata (4)
— **widened, but with weak muscles:** *Pseudiron* (1)
— **widened, with fan-form flexor attached to outer margin:** *Ametropus* (2), Radulapalpata (1)
— **widened, with fan-form flexor attached to inner margin:** Behningia/fg2 (2)

[1.1.57] **2nd+3rd segment of 2-segmented labial palp**
— **divided into secondary segments:**Ameletopsis/fg1 (1)
— **thickened:** Metretopus/fg1 (2), Eusetisura (2), Cryptoprosternata (4)
— **widest near middle:** Metretopus/fg2 (3)
— **widest near apex:** *Siphloplecton* (3), *Metreplecton* (3)
— **with filtering setae:** *Arthroplea* (1), Isonychia/fg1 (2)
— **scraping:** Radulapalpata (not all) (1)
— **with field of scraping setae terminating at distance from sclerotized ridge:** *Kageronia* (3), Ecdyonurus//fg1 (5), Rhithrogena/fg1 (4)
— **with field of scraping setae terminating close to sclerotized ridge:** Heptagenia/f7=g6 (1), *Cinygma* (6)
— **in form of sclerotized hook:** *Raptoheptagenia* (1)
— **with bifurcate dorsal setae:** Rhithrogena/fg1 (2)

[1.1.58] **2nd segment of 3-segmented labial palp**
— **widened or with inner-apical projection:** Baetisca/f3==g2 (3), Parameletus/fg2 (1), *Ametropus* (2), Acanthametropus/fg1 (1), *Siphlaenigma* (3)
— **with row of long setae on outer side:** Tricorygnatha (1)
— **without muscle** (among all Ephemeroptera, including 2-segmented palps – see [1.1.55]): Metretopus/fg1 (2), part of Turbanoculata (VAR), Heptagennota (8), part of Ephemera/fg9 (P), Cryptoprosternata (4), Behningia/fg1 (2), Brachycercus/fl=g3 (2), Ephemerella/fg1 (3)
— **diminished:** Tricorygnatha (1)

1.2. LARVAL THORAX AND ITS APPENDAGES

Thorax [1.2.1], prothorax [1.2.2–3], mesothorax and fore wing buds [1.2.4–8], metathorax and hind wing buds [1.2.9–10], legs [1.2.11–53]

[1.2] **Thorax**
— **forms notal shield** - see [1.2.1]

[1.2.1] **Notal shield**
— **present:** Posteritorna (3)
— **oval:** Prosopistoma/f1=g2 (2)
— **with pair of lateral projections:** Baetisca/f3=g2 (4)
— **compressed dorsoventrally:** *Fascioculus* (2)
— **strongly convex:** Baetisca/f4=g3 (2)

PROTHORAX OF LARVA

[1.2.2–4]

[1.2.2] **Pronotum**
— **fused with mesonotum:** Posteritorna (3), Coloburiscus/fg1 (2)
— **fused with propleura:** Campsurus/fg1 (3)
— **with projected anterior angles:** Cincticostella/g1 (2)
— **with antero-lateral denticles:** part of Fimbriatotergaliae (1)
— **with anterior setose projections:** Behningia/fg1 (1)
— **with lateral projections:** *Ametropus* (1),
— **with postero-lateral projections:** *Pseudiron* (1), Ecdyonurus/fg2 (1), Behningia/fg2 (1)
— **with pair of cavities:** *Pseudiron* (1)
— **with collar:** Fimbriatotergaliae (1)
— **with V-shape impression:** Ephemerella/fg1 (4)
— **with submedian tubercles:** Neoephemera/fg2 (1)

[1.2.3] **Propleuron**
— **transformed to apodeme:** Furcatergaliae (1)
— **fused with pronotum:** part of Furcatergaliae (1) {Campsurus/fg1 (3)}

[1.2.4] **Prosternum**
— **with unpaired projection:** Acanthametropus/fg1 (4), *Ametropus* (3)
— **with gill:** part of Eusetisura (3)
— **narrowed, with furcate apodeme:** Cryptoprosternata (1)

MESOTHORAX AND FORE WING BUDS OF LARVA

[1.2.5–8]

[1.2.5] **Mesonotum**
— **together with pronotum forms notal shield** - see [1.2.1]
— **has paranota anteriad of wing buds:** *Ametropus* (1)
— **with collar:** [Classifications of Furcatergaliae I] (3), Potamanthus/fg1 (2), Euthyplocia/fg1 (7), Caenotergaliae (5), Ephemerella/fg1 (11)
— **with lateral projections:** Cincticostella/g1 (2)
— **with sharp lateral projections:** Cincticostella/g1 (not all) (2) {*Ephacerella* (P)}
— **with flat lateral projections:** part of Cincticostella/g1 (2) {Cincticostella/g3 (1)}
— **with dorsal tubercles:** Neoephemera/fg2 (1)

[1.2.6] **Fore wing buds**
— **spread:** *Ametropus* (1), *Pseudiron* (1)
— **with divergent costal margins:** *Ametropus* (1)
— **divergent:** Discoglossata (3)
— **convergent:** Coloburiscus/fg1 (2), Ephemerella/fg1 (not all) (5)
— **fused with notum up to end of scutellum:** Turbanoculata (8), Eusetisura (5), Pentamerotarsata (2), part of Caenotergaliae (2)
— **fused by mesal plate behind notum:** Posteritorna (3), part of Turbanoculata (8), part of Eusetisura (5) {Coloburiscus/fg1 (2), Discoglossata (3)}, Ephemerella/fg1 (5)
— **enclosed into notal shield:** Posteritorna (3)

[1.2.7] **Mesopleuron**
— **with projection and setae:** Behningia/fg1 (1)

[1.2.8] **Mesosternum**
— **with pair of longitudinal ridges:** Posteritorna (3)
— **with unpaired projection:** Acanthametropus/fg1 (4)
— **with bifurcate projection:** Isonychia/fg1 (4)
— **with gill:** part of Eusetisura (3)

METATHORAX AND HIND WING BUDS OF LARVA

[1.2.9–10]

[1.2.9] **Hind wing buds**
— **differ from hind wings:** Leptohyphes/fg1 (not all) (2)
— **absent** - see [2.2.59]

[1.2.10] **Metasternum**
— **with pair of projections:** Isonychia/fg1 (4)
— **with gill:** part of Eusetisura (3)

LEGS OF LARVA

Leg [1.2.11], coxa [1.2.12], trochanter [1.1.13], femur [1.2.14–16], tibia [1.2.17–18], tarsus [1.2.19], claw [1.2.20–24]; fore leg [1.2.25–37], middle leg [1.2.38–43], hind leg [1.2.33–49], common features of certain leg pairs [1.2.50–53]
(about each pair - see [1.2.25–49])

[1.2.11] **Larval legs of all 3 pairs**
— **able to stretch posteriorly:** Siphlonurus/fg1 (P), *Dipteromimus* (P), Ameletus/fg1 (P), Metretopus/fg1 (P), Acanthametropus/fg1 (3), *Siphlaenigma* (P), Turbanoculata (not all) (VAR), Nesameletus/f1=Metamonius/ /g1 (P), *Rallidens* (P), Isonychia/fg1 (P)
— **unable to stretch posteriorly:** part of Turbanoculata (VAR), Heptagennota (1), Furcatergaliae (6)
— **used for swimming:** part of Heptagennota (1)
— **folded and pressed by sides to body when swimming:** part of Heptagennota (1) {Rhithrogena/fg2 (3)}

[1.2.12] **Coxa** (see also [1.2.26, 1.2.39, 1.2.45])
— **with gill basally:** part of Turbanoculata (VAR), *Siphluriscus* (1), part of Eusetisura (3) {Isonychia/fg1 (12)}

[1.2.13] **Trochanter** - see fore leg [1.2.27]

Femora of larva

[1.2.14–16]
(about femora of certain pairs - see also [1.2.28–29], [1.2.40], [1.2.46])

[1.2.14] **Femur**
— **inserted into concavity:** Prosopistoma/f1=g2 (2)
— **flattened:** Heptagennota (1)
— **narrow, arched:** Brachycercus/f1=g3 (5)
— **with curved apical projection:** Turbanoculata (11)

[1.2.15] **Femur setation** (see also [1.2.16])
— **spiny setae in 2 transverse rows:** Coloburiscus/fg1 (4)
— **on outer margin** - see [1.2.16]

[1.2.16] **Outer margin of femur** (see also [1.2.15])
— **without regular row of setae:** Heptagenia/f6=g5 (P)
— **with long hairs and long stout irregular setae:** Heptagenia/f7=g6 (2), *Raptoheptagenia* (4)
— **without long setae:** *Kageronia* (4)
— **with regular row of setae:** part of Turbanoculata (VAR), Ecdyonurus/fg1 (4), *Cinygma* (7), Rhithrogena/fg1 (5), *Madecocercus* (3), Ephemerella/fg2 (not all) (P), Pantricorythi (not all) (1)

Tibiae of larva
[1.2.17–18]
(about tibiae of certain pairs – see also [1.2.30–35], [1.2.41], [1.2.47])

[1.2.17] **Tibia**
— **shortened:** Acanthametropus/fg1 (3), *Ametropus* (1)

[1.2.18] **Patella-tibial suture** (among all Ephemeroptera)
— **absent on fore leg, present on middle and hind leg:** Siphlonurus/fg1 (P), *Dipteromimus* (P), Ameletus/fg1 (P), Metretopus/fg1 (P), Acanthametropus/fg1 (P), *Ametropus* (P), *Siphlaenigma* (P), Afroptilum/f1= =Centroptiloides/g1 (see Classification of Turbanoculata), Nesameletus/f1=Metamonius/g1 (P), Vetulata (P), Ameletopsis/fg1 (P), *Rallidens* (P), Coloburiscus/ /fg1 (P), Isonychia/fg1 (P), *Arthroplea* (P), Heptagenia/f5=g4 (P), *Cinygma* (P), *Paegniodes* (P), *Cinygmula* (P), part of Epeorus/fg1 (8) {Belovius/g2 (2)}, Potamanthus/fg1 (P), *Ichthybotus* (P), Ephemera/fg9 (P), *Dolania* (P), Palingenia/f2=g1 (P), Polymitarcys/ /f2=Ephoron/g3 (P), *Leucorhoenanthus* (P), *Potamanthellus* (P), Ephemerella/fg1 (not all) (VAR), Leptophlebia/fg1 (not all) (VAR)
— **present on all legs:** Anteropatellata (not all) (see Classification of Turbanoculata), Rhithrogena/fg3 (2)
— **absent on all legs:** Posteritorna (11), Discoglossata (8), *Pseudiron* (2), Epeorus/fg1 (not all) (8), part of Euthyplocia/fg1 (VAR), *Protobehningia* (2), Campsurus/fg1 (7), *Ochernova* (1), part of Caenoptera (17), part of Ephemerella/fg1 (VAR), part of Leptophlebia/fg1 (VAR)
— **absent on fore and middle legs, present on hind leg:** Caenoptera (not all) (17), part of Leptophlebia/fg1 (VAR)
— **absent on fore and hind legs, present on middle leg:** part of Euthyplocia/fg1 (VAR), Behningia/fg3 (2)

Tarsus of larva
[1.2.19]
(about tarsi of certain pairs – see also [1.2.36], [1.2.42], [1.2.48])

[1.2.19] **Tarsus**
— **segmented:** Permoplectoptera (P), Ameletopsis/fg1 (6)
— **arched:** *Pseudiron* (1)

Claw of larva
[1.2.20–24]
(about claws of certain pairs – see also [1.2.37], [1.2.43], [1.2.49])

[1.2.20] **Claw**
— **absent:** Behningia/fg2 (1)
— **elongate:** Acanthametropus/fg1 (6), *Ametropus* (1), *Pseudiron* (1), Brachycercus/f1=g3 (5)
— **swollen in male:** Palingenia/f3=g2 (4)

[1.2.21] **Denticles on inner side of claw** (not subapical denticles on anterior side – see [1.2.22])
— **absent or very small and irregular:** Posteritorna (P), Siphlonurus/fg1 (P), *Dipteromimus* (P), Ameletus/fg2 (P), Metretopus/fg1 (1), Acanthametropus/fg1 (P), *Ametropus* (P), part of Tetramerotarsata (11), *Siphluriscus* (2), Vetulata (P), Ameletopsis/fg1 (P), *Rallidens* (P), part of Homoeoneuria/fg1 (2), *Pseudiron* (1), Heptagenia/f7=g6 (3), Fimbriatotergaliae (not all) (P) {Brachycercus/f1=g3 (5)}, part of Ephemerella/fg1 (18) {Timpanoga/fg3 (1)}, part of Leptophlebia/fg1 (6)
— **in 1 row:** *Metreletus* (2), part of Tetramerotarsata (11), Eusetisura (not all) (6), Pentamerotarsata (not all) (4) {*Kageronia* (5)}, Caenis/f4=g2 (not all) (3), Ephemerella/fg1 (not all) (18), Leptophlebia/fg1 (not all) (6)
— **in 2 rows:** Tetramerotarsata (not all) (11), Nesameletus/f2=Metamonius/g2 (4), part of Homoeoneuria/fg1 (2) {*Oligoneurisca* (1)}, part of Ephemerella/fg1 (18) {part of Melanemerella/fg1 (VAR)}, part of Leptophlebia/fg1 (6)

[1.2.22] **Subapical denticles on anterior (dorsal) side of claw**
— **present:** Pentamerotarsata (not all) (4), part of Ephemerella/fg1 (5), part of Leptophlebia/fg1 (6)

[1.2.23] **Proximal projection**
— **flexible:** *Siphluriscus* (2)
— **present:** *Chiloporter* (2)
— **absent:** Ameletopsis/fg2 (P), *Mirawara* (P), *Chaquihua* (P)

[1.2.24] **Subapical setae of claw**

FORE LEG OF LARVA
Fore leg [1.2.25], coxa [1.2.26], trochanter [1.2.27], femur [1.2.28–29], tibia [1.2.30–35], tarsus [1.2.36], claw [1.2.37]

[1.2.25] **Fore leg (in contrast to middle and/or hind)**
— **diminished:** *Ametropus* (3), Behningia/fg2 (1), Brachycercus/f1=g3 (5)
— **palp-like:** *Ametropus* (3), Behningia/fg2 (1)
— **filtering:** Eusetisura (1), Clypeocaenis/g1 (1)
— **grasping:** Drunella/g1 (1), *Vietnamella* (3), Mesogenesia/g1 (1)
— **burrowing:** Exeuthyplocia/fg1 (3), Fossoriae (not all) (1) {*Ichthybotus* (2), Ephemera/fg10 (3), Hexagenia/fg1 (4), *Protobehningia* (1), Cryptoprosternata (2) {Palingenia/f2=g1 (4), (5), (9), Polymitarcys/f2= Ephoron/g3 (2), Campsurus/fg1 (6)}, "*Mesopalingea*", "*Torephemera*", "*Siphangarus*"}

[1.2.26] **Coxa of fore leg** (see also [1.2.12])
— **with condylus close to median line of body:** Cryptoprosternata (1)
— **elongate:** Mesogenesia/g1 (1)

— **with projection:** *Ametropus* (3)
— **with gill:** part of Turbanoculata (VAR), *Siphluriscus* (1), part of Eusetisura (3) {Isonychia/fg1 (12)}

[1.2.27] **Trochanter of fore leg**
— **wedged into anterior side of femur:** Cryptoprosternata (2, 3B)

Femur of larval fore leg
[1.2.28–29]

[1.2.28] **Femur of fore leg** (see also [1.2.14–16])
— **widened (as middle leg):** Acanthametropus/fg1 (2)
— **with concave anterior side (as middle):** Fossoriae (1)
— **with inner-proximal convexity bearing long setae:** Cryptoprosternata (2, 3B)
— **with protuberances on proximal convexity:** Polymitarcys/f2=Ephoron/g3 (2)
— **widened, with denticulate inner margin:** Drunella/g1 (not all) (1), *Vietnamella* (3)
— **widened, with setose inner margin:** Mesogenesia/g1 (1)

[1.2.29] **Setation of fore femur** (see also [1.2.15–16])
— **on inner side 2 rows of long setae:** Eusetisura (1)
— **on proximal convexity long setae:** Cryptoprosternata (3)
— **on proximal convexity regular U-shape row of long setae:** part of Cryptoprosternata (3B) {Polymitarcys/ /f1=Ephoron/g2 (1B)}
— **on dorsal side transverse row of setae:** [Classifications of Furcatergaliae I] (2), Potamanthus/fg1 (not all) (3) {Rhoenanthus/g1 (not all) (VAR)}, Caenotergaliae (not all) (6), Ephemerella/fg1 (not all) (12)
— **on dorsal side transverse row continued distally on outer margin:** Pantricorythi (not all) (1)

Tibia of larval fore leg
[1.2.30–35]

[1.2.30] **Tibia of fore leg** (see also [1.2.17–18])
— **without patella-tibial suture** – see [1.2.18]

[1.2.31] **Outer margin of fore tibia**
— **with spine-like setae:** Palingenia/f2=g1 (not all) (5) {*Pentagenia* (P), *Chankagenesia* (P), Anagenesia/g2 (P), *Plethogenesia* (P), *Cheirogenesia* (P)}
— **with denticles:** part of Palingenia/f2=g1 (5) {Palingenia/f4=g3 (2), Mortogenesia (2)}

[1.2.32] **Inner margin of fore tibia**
— **proximally convex:** Eusetisura (1)
— **with long setae:** Eusetisura (1), Cryptoprosternata (3D), Clypeocaenis/g1 (1)
— **with 2 longitudinal rows of long setae:** Eusetisura (1), part of Cryptoprosternata (3D) {Polymitarcys/f1= =Ephoron/g2 (1D)}
— **with 2 transverse rows of long setae:** Clypeocaenis/g3 (2)

[1.2.33] **Anterior (dorsal) side of fore tibia**
— **with long setae:** Clypeocaenis/g2 (2)
— **with long setae in oblique row:** Cryptoprosternata (3C)
— **with long setae in oblique row continuing as 2 longitudinal rows:** Polymitarcys/f1=Ephoron/g2 (1C)
— **with long setae in 2 oblique rows:** Palingenia/f2=g1 (4)

[1.2.34] **Posterior (ventral) side of fore tibia**
— **with tubercles:** Polymitarcys/f2=Ephoron/g3 (2)

[1.2.35] **Apex of fore tibia**
— **with appendage:** Isonychia/fg1 (5)
— **with projection:** Euthyplocia/fg1 (3), Drunella/g1 (1)
— **with incision:** Fossoriae (not all) (1) {*Ichthybotus* (2), Hexagenia/fg1 (4), Palingenia/f2=g1 (9), "*Torephemera*"
— **without incision:** Ephemera/fg10 (3)

Tarsus of larval fore leg
[1.2.36]

[1.2.36] **Tarsus of fore leg** (see also [1.2.19])
— **articulated to anterior side of tibia:** *Ametropus* (3)
— **on inner side with 2 rows of setae:** Isonychia/fg1 (5)
— **with oblique row of long setae:** Clypeocaenis/g3 (2)
— **with apical projection:** Euthyplocia/fg2 (2), *Polyplocia* (4)
— **vestigial:** Homoeoneuria/g1 (1)
— **fused with tibia:** *Protobehningia* (1), Campsurus/fg1 (6)

Claw of larval fore leg
[1.2.37]

[1.2.37] **Claw of fore leg** (see also [1.2.20–24])
— **bifurcate:** Metretopus/fg1 (1)
— **with setae:** *Ametropus* (3)
— **with denticles:** Metretopus/fg1 (1), Asthenopus/fg1 (2)
— **vestigial:** Homoeoneuria/g1 (not all) (1) {*Oligoneurisca* (P)}, *Eatonigenia* (1)
— **absent:** part of Homoeoneuria/g1 (1) {Homoeoneuria/ /g2 (1)}

MIDDLE LEG OF LARVA
[1.2.38–43]

[1.2.38] **Middle leg** (see also [1.2.11–24])
— **directed forward (as fore leg):** Fossoriae (1), Mesogenesia/g1 (2)
— **specialized for moving in sand:** Behningia/fg2 (1)

[1.2.39] **Coxa of middle leg** (see also [1.2.12])
— **elongate (as hind leg):** Homoeoneuria/g1 (2)
— **with gill (as fore leg):** *Siphluriscus* (1)

[1.2.40] **Femur of middle leg** (see also [1.2.14–16])
— **widened (as fore leg):** Acanthametropus/fg1 (2)
— **widened (as hind leg):** *Rhionella* (1)
— **with filtering setae (as fore leg):** Coloburiscus/fg1 (3)
— **with concave anterior side (as fore leg):** Exeuthyplocia/fg1 (3), Fossoriae (1)
— **shortened:** Behningia/fg2 (1)

[1.2.41] **Tibia of middle leg** (see also [1.2.17–18])
— **shortened (as hind leg):** *Oligoneurisca* (1)
— **wide, with inner-apical projection:** Behningia/fg2 (1)
— **without patella-tibial suture (in contrast to hind leg):** Caenoptera (not all) (17), part of Leptophlebia/ /fg1 (VAR)
— **with patella-tibial suture (in contrast to other legs):** part of Euthyplocia/fg1 (VAR), Behningia/fg3 (2)
— **with patella-tibial suture (as hind leg)** – see [1.2.18]

[1.2.42] **Tarsus of middle leg** (see also [1.2.19])
— **with reversed setation (as hind leg):** *Oligoneurisca* (1)

— **with paired submedian projections:** part of Turbanoculata (VAR), part of Epeorus/fg1 (VAR), Ephemerella/fg1 (not all) (19) {Timpanoga/fg1 (not all) (P), Ephemerella/fg4 (not all) (P), Torleya/fg1 (not all) (P), Hyrtanella/f1=Crinitella/g1 (not all) (VAR), *Vietnamella* (P), *Austremerella* (P), part of Melanemerella/fg1 (VAR)}
— **without median projections:** part of Ephemerella/fg1 (19) {part of Timpanoga/fg1 (P), part of Ephemerella//fg4 (P), part of Torleya/g1 (P), part of Hyrtanella//f1=Crinitella/g1 (VAR), *Uracanthella* (2), part of Melanemerella/fg1 (VAR), *Teloganodes* (not all) (7), Tricoryptera (not all) (4)}
— **with pair of longitudinal ridges:** Behningia/fg1 (1)
— **with lateral expansions:** Behningia/fg2 (1)
— **I with elevated hind margin:** Vetulata (1)
— **II with unpaired projection:** Caenotergaliae (not all) (1)
— **VI enlarged and elevated medially:** Posteritorna (5)
— **VII enlarged and elevated medially:** *Machadorythus* (1)

[1.3.4] **Setation of abdominal terga**
— **no stout setae:** Dipteromimus (6), Metretopus/fg2 (1), *Metreplecton* (1)
— **spine-like setae:** Siphlonurus/fg1 (6), *Siphloplecton* (1)
— **stout scales in narrow nests:** Acanthametropus/fg1 (7)
— **scales in wide nests:** Turbanoculata (not all) (12)

[1.3.5] **Posterior margins of abdominal terga**
— **without denticles:** Parameletus/fg1 (3), *Dipteromimus* (6), Acanthametropus/fg1 (7)
— **denticles fused to integral plate:** Coloburiscus/fg1 (5)

Abdominal sterna of larva
[1.3.6–8]

[1.3.6] **Abdominal sterna (shape)**
— **with median projections:** Acanthametropus/fg2 (3)
— **widened:** Brachycercus/f1=g3 (4)
— **I with finger-like process:** Homoeoneuria/g2 (2)
— **I with pair of projected plates:** Rhithrogena/fg3 (3)
— **II–III with setose convexity:** Homoeoneuria/g1 (3)

[1.3.7] **Setation of abdominal sterna**
— **no stout setae:** *Dipteromimus* (6), Metretopus/fg2 (1), *Metreplecton* (1)
— **spine-like setae:** Siphlonurus/fg1 (6), *Siphloplecton* (1)
— **stout scales:** Acanthametropus/fg1 (7)

[1.3.8] **Posterior margins of abdominal sterna**
— **denticles fused to integral plate:** Coloburiscus/fg1 (5)

Selected abdominal segments of larva
[1.3.9–18]

[1.3.9] **Segment I** (see also [1.3.1–8])
— **dorsally with elevated hind margin:** Vetulata (1)
— **ventrally not expressed:** Posteritorna (3)
— **ventrally with unpaired process:** Homoeoneuria/g2 (2)
— **ventrally with projected plates:** Rhithrogena/fg3 (3)

[1.3.10] **Segment II** (see also [1.3.1–8])
— **fused with I–VI:** Prosopistoma/f1=g2 (2)
— **dorsally with projection:** Caenotergaliae (not all) (1)
— **dorsally without projection:** Brachycercus/f1=g3 (4)
— **ventrally setose (as III):** Homoeoneuria/g1 (3)

[1.3.11] **Segment III** (see also [1.3.1–8])
— **fused with I–VI:** Prosopistoma/f1=g2 (2)
— **with lateral processes (as IV–VII):** Palingenia/f2=g1 (6)
— **ventrally setose (as II):** Homoeoneuria/g1 (3)

[1.3.12] **Segment IV** (see also [1.3.1–8])
— **fused with I–VI:** Prosopistoma/f1=g2 (2)
— **with lateral processes (as III–VII):** Palingenia/f2=g1 (6)

[1.3.13] **Segment V** (see also [1.3.1–8])
— **fused with I–VI:** Prosopistoma/f1=g2 (2)
— **with pair of lateral incisions:** Prosopistoma/f1=g2 (2)
— **with lateral processes (as III–VII):** Palingenia/f2=g1 (6)

[1.3.14] **Segment VI** (see also [1.3.1–8])
— **fused with I–V:** Prosopistoma/f1=g2 (2)
— **with lateral processes (as III–VII):** Palingenia/f2=g1 (6)
— **enlarged and elevated medially:** Posteritorna (5)

[1.3.15] **Segment VII** (see also [1.3.1–8])
— **with lateral processes (as III–VI):** Palingenia/f2=g1 (6)
— **enlarged, elevated medially:** *Machadorythus* (1)

[1.3.16] **Segment VIII** (see also [1.3.1–8])
— **with pair of crests:** Hyrtanella/f1=Crinitella/g1 (2)

[1.3.17] **Segment IX** (see also [1.3.1–8])
— **without posterolateral spines:** *Chiloporter* (3), Radulapalpata (2)
— **with posterolateral spines:** *Arthroplea* (P)
— **elongate:** Eurylophella/fg2 (1)
— **bears gonostyli buds** – see [1.3.60]

[1.3.18] **Segment X** – see [1.3.61–62]

TERGALII
Position of tergalii [1.3.19–22], shape of tergalii [1.3.23–25], margins of tergalii [1.3.26], ribs of tergalii [1.3.27–29], function of tergalii [1.3.30–32], tergalii of selected pairs [1.3.33–59]

Position of tergalii
[1.3.19–22]

[1.3.19] **Tergalii present on abdominal segments** (among all Ephemeroptera)
— **I–IX:** Euephemeroptera (not all) (6)
— **I–VII:** Euplectoptera (not all) (2) {Siphlonurus/fg1 (7), *Dipteromimus* (5), Ameletus/fg1 (VAR), Metretopus//fg1 (3), Acanthametropus/fg1 (5), *Ametropus* (5), *Siphlaenigma* (P), Turbanoculata (not all) (VAR), Nesameletus/f1=Metamonius/g1 (9), Oniscigaster/fg2 (P), Ameletopsis/fg1 (4), *Rallidens* (P), *Coloburiscoides* (P), Coloburiscus/fg3 (P), Isonychia/fg1 (P), Discoglossata (P), Heptagennota (not all) (P), Fimbriatotergaliae (not all) (6) {Potamanthus/fg1 (P), Euthyplocia/fg1 (P), Fossoriae (3)}, Ephemerella/fg1 (not all) (7) {*Vietnamella* (P)}, Leptophlebia/fg1 (not all) (7)}
— **I–VI:** Posteritorna (4), part of Turbanoculata (VAR), part of *Macdunnoa* (1), Caenotergaliae (1), Melanemerella/fg1 (not all) (1), *Teloganodes* (not all) (1), (8), part of Leptophlebia/fg1 (7)
— **I–V:** part of Turbanoculata (VAR), part of Melanemerella/fg1 (1)
— **I–IV:** Tasmanophlebia/g1 (2), part of Melanemerella//fg1 (1)
— **I and IV–VII:** Timpanoga/fg1 (1)

— **II–VII:** part of Turbanoculata (VAR), *Austremerella* (1)–(3), part of Leptophlebia/fg1 (7)
— **II–VI:** part of Melanemerella/fg1 (1), *Teloganodes* (not all) (1), (8), Tricoryptera (not all) (5), (6) {*Ephemerythus* (not all) (1)}, "*Teloganella*"
— **II–V:** part of *Ephemerythus* (1), part of *Teloganodes* (1), (8), *Coryphorus* (4)
— **II–IV:** part of *Teloganodes* (1), (8)
— **III–VII:** Ephemerella/fg3 (1)

[1.3.20] **Absent tergalii of anterior pairs** (among all Ephemeroptera)
— **I:** part of Turbanoculata (VAR), Ephemerella/fg3 (1), *Austremerella* (1), part of Melanemerella/fg1 (1), *Teloganodes* (8), Tricoryptera (5), *Coryphorus* (4), "*Teloganella*", part of Leptophlebia/fg1 (7)
— **I–II:** Ephemerella/fg3 (1)
— **I–VII:** *Murphyella* (1)
— **II:** Ephemerella/fg2 (1)
— **II–III:** Timpanoga/fg1 (1)

[1.3.21] **Absent tergalii of posterior pairs** (among all Ephemeroptera)
— **VII:** Posteritorna (4), part of Turbanoculata (VAR), part of *Macdunnoa* (1), Caenotergaliae (1), Melanemerella/fg1 (not all) (1), *Teloganodes* (not all) (1), Tricoryptera (not all) (6), "*Teloganella*", part of Leptophlebia/fg1 (7)
— **VI–VII:** part of Turbanoculata (VAR), part of Melanemerella/fg1 (1), part of *Teloganodes* (1), part of *Ephemerythus* (1), *Coryphorus* (4)
— **V–VII:** Tasmanophlebia/g1 (2), part of Melanemerella/ /fg1 (1), part of *Teloganodes* (1)
— **I–VII:** *Murphyella* (1)

[1.3.22] **Places of tergalii attachment** transferred
— **ventrally:** part of Turbanoculata (VAR), *Raptoheptagenia* (2), Behningia/fg2 (1), part of Leptophlebia/ /fg1 (VAR)
— **I ventrally:** Geminovenata (1)
— **I anteriorly:** Potamanthus/fg1 (4), Caenotergaliae (7), Ephemerella/fg1 (not all) (13)
— **II–IV anteriorly:** Tasmanophlebia/g1 (2)
— **VI anteriorly:** Posteritorna (5), Caenotergaliae (1)
— **VII anteriorly:** Ephemera/fg9 (1), *Austremerella* (3)

Shape of tergalii
[1.3.23–25]

[1.3.23] **Tergalius of each pair** (about specialization of selected pairs - see [1.3.33–59])
— **with second lamella or bent lobe** - see [1.3.24]
— **with fibrillose ventral lobe** - see [1.3.25]
— **round:** *Ametropus* (5)
— **cup-shaped:** Discoglossata (not all) (10)
— **with mid-ventral process:** *Pseudiron* (1)
— **with dorsal process:** Hyrtanella/f1=Crinitella/g1 (1)
— **with anterior-proximal lobe:** *Arthroplea* (3)
— **secondarily unilamellate:** part of Fimbriatotergaliae (6) {Caenoptera (4)}, part of Ephemerella/fg1 (7) {*Dicercomyzon* (4)}, part of Leptophlebia/fg1 (7)
— **unilamellate with processes:** *Dicercomyzon* (4), part of Leptophlebia/fg1 (7)
— **pointed apically:** *Kageronia* (7), *Stenacron* (2)
— **enlarged, spread laterally:** "*Epeoromimus*" (1)
— **bidentate:** Coloburiscus/fg1 (not all) (6)
— **bifurcate:** Furcatergaliae (not all) (4)
— **bilamellate** - see [1.2.24]

[1.3.24] **Tergalius with additional lamella formed as**
— **costal lobe bent ventrally:** part of Metretopus/fg1 (3)
— **anal lobe bent ventrally** (see [1.3.25]): I–II in Siphlonurus/fg2 (1), Acanthametropus/fg1 (5), part of Turbanoculata (VAR)
— **anal lobe bent dorsally:** III–VII in *Siphlurella* (1), II–IV in Tasmanophlebia/g1 (2), part of Turbanoculata (VAR), Hyrtanella/f1=Crinitella/g1 (1)
— **outgrowth of ventral surface** (see [1.3.25]): part of Ameletopsis/fg1 (4) {*Mirawara* (2), *Chiloporter* (4), *Promirara* (2)}, *Rallidens* (5), Bidentiseta (not all) (2) {Branchitergaliae (not all) (3) {*Coloburiscoides* (P), Isonychia/fg1 (P)}, Furcatergaliae (not all) (5) {Fimbriatotergaliae (6) {Neoephemera/fg1 (P)}}, Ephemerella/fg1 (7)}}

[1.3.25] **Ventral lobe** (see [1.2.24])
— **with marginal processes or fibrillose:** part of Ameletopsis/fg1 (4) {*Mirawara* (2), *Chiloporter* (4), *Promirara* (2)}, *Rallidens* (5), Branchitergaliae (not all) (3) {*Coloburiscoides* (P), Isonychia/fg1 (P)}, II–VI in Neoephemera/fg1 (P)
— **bifurcate with processes:** Ephemerella/fg1 (not all) (7) {Leptohyphes/fg2 (P)}
— **bifurcate without processes:** part of Ephemerella/fg1 (7) {*Allenhyphes* (3), *Tricorythodes* (7)}
— **II–III bifurcate:** part of Ephemerella/fg1 (7) {part of Melanemerella/fg1 (1), part of *Teloganodes* (1)}
— **II–IV bifurcate:** part of Ephemerella/fg1 (7) {part of *Teloganodes* (1), *Tricorythopsis* (8)}
— **II–V bifurcate:** part of Ephemerella/fg1 (7) {*Austremerella* (2), *Teloganodes* (not all) (1), *Tricorythodes* (not all) (7), "*Teloganella*"}
— **II–VI bifurcate:** Ephemerella/fg1 (not all) (7)
— **III–V bifurcate:** part of Ephemerella/fg1 (7) {part of *Tricorythodes* (7)}
— **III–VI bifurcate:** part of Ephemerella/fg1 (7) {Torleya/ /g1 (P)}
— **III–V bifurcate:** part of Ephemerella/fg1 (7) {Ephemerella/fg4 (1)}
— **IV–VI bifurcate:** part of Ephemerella/fg1 (7) {Timpanoga/fg1 (1)}
— **folded as 2 halves:** Acanthametropus/fg1 (5)
— **lost on all tergalii:** Ameletopsis/fg2 (1), Chaquihua (1), Coloburiscus/fg3 (P), *Arthroplea* (3), *Cinygmula* (not all) (2), Caenoptera (4)
— **diminished or lost on tergalius VII** - see [1.3.38] and [1.3.59]

Margin of tergalius
[1.3.26]

[1.3.26] **Margin of tergalius** (about ventral lobe - see [1.3.25])
— **with branched processes:** Acanthametropus/fg2 (4), Caenotergaliae (1)
— **with non-branched processes:** Fimbriatotergaliae (not all) (6), *Dicercomyzon* (4), part of Leptophlebia/fg1 (7)

— **with semicircular festoons:** part of Rhithrogena/fg3 (3)
— **with denticles:** Ameletus/fg1 (VAR), Metretopus/fg2 (2), *Metreplecton* (2), Nesameletus/f1=Metamonius/g1 (9), *Rallidens* (5), Isonychia/fg1 (8)
— **with peculiar setation:** *Ametropus* (5), Coloburiscus/ /fg1 (not all) (6), Homoeoneuria/fg1 (6), *Fittkauneuria* (3), Epeorus/fg2 (2) {Epeorus/fg3 (2)}, *Siphgondwanus* (1)

Ribs of tergalii
[1.3.27–29]

[1.3.27] **Costal rib of tergalius**
— **absent:** I–II in part of Ameletus/fg1 (VAR), I–III in Metretopus/fg1 (3), part of Turbanoculata (VAR), *Arthroplea* (3), part of Heptagenia/f8=g7 (VAR), Stenonema/fg1 (2), part of Furcatergaliae (5) {part of Fimbriatotergaliae (6) {Caenotergaliae (1)}, Ephemerella/fg2 (2), part of Pantricorythi (P) {III–VI in Leptohyphes/fg1 (1)}, Leptophlebia/fg1 (7)}
— **at distance from costal margin:** part of Turbanoculata (VAR), I in part of Nesameletus/f1=Metamonius/g1 (9), Rhithrogena/fg3 (3), I in Epeorus/fg2 (1), Furcatergaliae (not all) (5) {Fimbriatotergaliae (not all) (6), Pantricorythi (not all) (P)}
— **sharply curved:** *Epeiron* (1)
— **strongly thickened:** Coloburiscus/fg1 (not all) (6), Hexagenites/fg1 (3)
— **with projection:** *Caucasiron* (1)
— **with denticles:** Ameletus/fg1 (VAR), Metretopus/fg2 (2), *Metreplecton* (2), Nesameletus/f1=Metamonius/g1 (9), *Rallidens* (5), Isonychia/fg1 (8)
— **without denticles:** *Siphloplecton* (2)
— **with spine-like setae:** Coloburiscus/fg1 (not all) (6)
— **with field of curved spine-like setae:** Epeorus/fg2 (2)
— **with hairs in apical part:** Epeorus/fg3 (2)
— **secondary:** part of Leptophlebia/fg1 (7)

[1.3.28] **Anal rib of tergalius**
— **absent:** Metretopus/fg1 (3), Acanthametropus/fg1 (5), Vetulata (2), part of Turbanoculata (VAR), part of Ameletopsis/fg1 (4) {Ameletopsis/fg2 (1), *Chiloporter* (4)}, *Arthroplea* (3), part of Heptagenia/f8=g7 (VAR), part of Furcatergaliae (5) {part of Fimbriatotergaliae (6) {Caenotergaliae (1)}, Ephemerella/fg2 (2), part of Pantricorythi (P), Leptophlebia/fg1 (7)}
— **at distance from anal margin:** Siphlonurus/fg1 (7) {Siphlonurus/fg2 (1)}, *Dipteromimus* (5), Ameletus/ /fg1 (not all) (VAR), Nesameletus/f1=Metamonius/g1 (9), part of Ameletopsis/fg1 (4) {Mirawara (2)}, *Rallidens* (5), Isonychia/fg1 (8), Stenonema/fg1 (2), Belovius/g1 (1), Furcatergaliae (not all) (5) {Fimbriatotergaliae (not all) (6), Pantricorythi (not all) (P)}, Hexagenites/fg1 (3), "*Siberiogenites*" (1)
— **on anal margin:** Ameletus/fg1 (not all) (VAR), *Ametropus* (5), Tetramerotarsata (not all) (12) {Turbanoculata (not all) (VAR)}, part of Ameletopsis/fg1 (4) {*Chaquihua* (1)}, Heptagennota (not all) (9) {*Kageronia* (P), Heptagenia/f8=g7 (not all) (VAR), Iron/g1 (P)}
— **strongly thickened:** Coloburiscus/fg1 (not all) (6)
— **arched:** Belovius/g2 (1)
— **with denticles:** Isonychia/fg1 (8)
— **with spine-like setae:** Coloburiscus/fg1 (not all) (6)

[1.3.29] **Additional rib**
— **behind costal margin:** *Promirara* (1)
— **on anal margin:** part of Nesameletus/f1=Metamonius/ /g1 (9) {Nesameletus/f2=Metamonius/g2 (5)}, *Rallidens* (5), Isonychia/fg1 (8)
— **on costal branch of ventral lobe:** Leptohyphes/fg1 (1)

Function of tergalii
[1.3.30–32]

[1.3.30] **Ability of tergalii for respiratory movements**
— **present:** Siphlonurus/fg1 (P), Metretopus/fg1 (P), part of Turbanoculata (VAR), Heptagenia/f5=g4 (P), part of Leptophlebia/fg1 (7)
— **lost:** Ameletus/fg1 (4), part of Turbanoculata (VAR), Eusetisura (7) {Isonychia/fg1 (8)}, *Cinygma* (8), Rhithrogena/fg1 (6), part of Leptophlebia/fg1 (7)
— **lost by gill opercula** – see [1.3.32]

[1.3.31] **Specialization of tergalii**
— **form sucking disc:** Rhithrogena/fg3 (3), Iron/g1 (1), part of Leptophlebia/fg1 (7)
— **form gill opercula** – see [1.3.32]

[1.3.32] **Tergalii forming gill opercula** (among all Ephemeroptera)
— I (see [1.3.33]): Vetulata (1) {Tasmanophlebia/g1 (2)}, part of Leptophlebia/fg1 (7)
— **II** (see [1.3.39]): Posteritorna (4), Caenotergaliae (1), Melanemerella/fg1 (1), *Teloganodes* (1), *Ephemerythus* (1), *Machadorythus* (5), Leptohyphes/fg1 (1), *Coryphorus* (4), "*Teloganella*"
— **III** (see [1.3.45]): part of Ephemerella/fg3 (P) {part of Torleya/g1 (VAR), part of Hyrtanella/f1=Crinitella/g1 (1)}
— **IV** (see [1.3.50]): Timpanoga/fg2 (1) {Eurylophella/ /fg1 (1)}

Tergalii of certain pairs
[1.3.33–59]

[1.3.33] **Tergalius I (in contrast to others)**
— **absent:** part of Turbanoculata (VAR), part of Ephemerella/fg1 (13) {Ephemerella/fg3 (1), Tricoryptera (5), *Austremerella* (1), part of Melanemerella/fg1 (1), *Teloganodes* (8), *Coryphorus* (4), "*Teloganella*"}, part of Leptophlebia/fg1 (7)
— **attached ventrally:** Geminovenata (1)
— **consists of fibrillose lobe only:** Homoeoneuria/g2 (3), *Fittkauneuria* (2)
— **specialized as gill operculum:** Vetulata (1) {Tasmanophlebia/g1 (2)}, part of Leptophlebia/fg1 (7)
— **specialized similar to gill operculum:** Vetulata (1) {Oniscigaster/fg2 (1)}
— **stick-like:** [Classifications of Furcatergaliae I] (1), part of Fimbriatotergaliae (6) {Potamanthus/fg1 (4), Caenotergaliae (7)}, Ephemerella/fg1 (not all) (13) {Timpanoga/fg1 (not all) (P), *Vietnamella* (P), Melanemerella/fg1 (not all) (1)}
— **unilamellate:** part of Fimbriatotergaliae (6) {Behningia/fg1 (15)}

— **with costal rib:** Metretopus/fg1 (3)
— **with additional dorsal lamella:** part of Hyrtanella/f1= =Crinitella/g1 (1)
— **hidden under gill opercula:** part of Ephemerella/fg3 (P) {part of Torleya/g1 (VAR), part of Hyrtanella/f1= =Crinitella/g1 (1)}

[1.3.54] **Tergalius V (in contrast to others)**
— **without ventral lobe:** part of Ephemerella/fg1 (7) {part of *Ephemerythus* (1), part of *Teloganodes* (1)}

[1.3.55] **Tergalii V–VI (in contrast to others)**
— **without ventral lobe:** part of Ephemerella/fg1 (7) {part of *Ephemerythus* (1)}

[1.3.56] **Tergalii V–VII (in contrast to others)**
— **absent:** Tasmanophlebia/g1 (2), part Melanemerella/ /fg1 (1), part of *Teloganodes* (1)
— **hidden under gill opercula:** Timpanoga/fg2 (1)
— **with additional dorsal lamella:** part of Hyrtanella/f1= =Crinitella/g1 (1)

[1.3.57] **Tergalius VI (in contrast to others)**
— **attached anteriorly:** Posteritorna (5), Caenotergaliae (1), "*Teloganella*"
— **blade-like:** Posteritorna (4)
— **diminished:** *Macdunnoa* (1)
— **with incised posterior-distal margin:** *Epeiron* (1)
— **without ventral lobe:** part of Ephemerella/fg1 (7) {Melanemerella/fg1 (not all) (1), *Teloganodes* (not all) (1), part of Leptohyphes/fg1 (1), "*Teloganella*"}
— **with non-bifurcate ventral lobe:** part of Ephemerella/ /fg1 (7) {Ephemerella/fg4 (1), *Austremerella* (2)}

[1.3.58] **Tergalii VI–VII (in contrast to others)**
— **absent:** part of Turbanoculata (VAR), part of part of *Teloganodes* (1), part of Melanemerella/fg1 (1), *Ephemerythus* (1), *Coryphorus* (4)
— **with additional dorsal lamella:** part of Hyrtanella/f1= =Crinitella/g1 (1)

[1.3.59] **Tergalius VII (in contrast to others)**
— **absent:** Posteritorna (4), part of Turbanoculata (VAR), part of *Macdunnoa* (1), Caenotergaliae (1), part of Ephemerella/fg1 (7) {Melanemerella/fg1 (1), *Teloganodes* (1), Tricoryptera (not all) (6), "*Teloganella*"}, part of Leptophlebia/fg1 (7)
— **attached anteriorly:** Ephemera/fg9 (1), *Austremerella* (3)
— **enlarged:** *Promirara* (1), Hexagenites/fg1 (3)
— **vestigial:** *Macdunnoa* (not all) (1)
— **without fibrillose ventral lobe:** part of *Coloburiscoides* (P), *Kageronia* (6), part of Heptagenia/f8=g7 (VAR), Stenonema/fg1 (3), Ecdyonurus/fg1 (not all) (6), *Ironodes* (2)
— **with non-bifurcate ventral lobe:** Ephemerella/fg1 (not all) (7)
— **with longitudinal fold:** Rhithrogena/fg3 (3), Iron/g1 (1)

GENITAL BUDS OF LARVA
[1.3.60]

[1.3.60] **Genital buds of mature male larva**
— **reduced:** Turbanoculata (13) (see [2.3.9])
— **without gonostyli buds:** part of Turbanoculata (13), Caenis/f3=g1 (2)
— **without penis buds:** Liberevenata (6), Caenotergaliae (3)

ABDOMINAL SEGMENT X OF LARVA
[1.3.61–62]

[1.3.61] **Abdominal tergite X**
— **with apodemes:** Prosopistoma/f1=g2 (3)
— **with angles reaching cerci ventro-laterally:** Geminovenata (8), Fossoriae (5), Caenotergaliae (12)

[1.3.62] **Paraproct**
— **elongate:** Prosopistoma/f1=g2 (3)
— **with spine:** Siphlonurus/fg2 (2), Metretopus/fg1 (7), *Siphluriscus* (3), Vetulata (7), *Rallidens* (8)
— **separated from cercotractor:** Geminovenata (8), Fossoriae (5), Caenotergaliae (12)

CAUDALII OF LARVA
[1.3.63–67]

[1.3.63] **Caudalii of larva** (see also [1.3.64–65])
— **retractile:** Prosopistoma/f1=g2 (3)
— **thickened in male:** [Classifications of Furcatergaliae IIb] (7), Caenoptera (14), Tricorygnatha (6), Leptohyphes/fg1 (8)
— **with fused denticles:** Coloburiscus/fg1 (5)
— **with whorls of stout setae:** Heptagenia/f5=g4 (2)
— **without whorls of stout setae:** *Raptoheptagenia* (5), *Cinygma* (9), Rhithrogena/fg1 (7)

[1.3.64] **Paracercus of larva** (see also [1.3.53])
— **absent:** part of Oligoneuria/f4=g5 (6)
— **non-segmented:** part of Turbanoculata (VAR), *Murphyella* (2), part of Oligoneuriella/g1 (VAR), part of Oligoneuria/f4=g5 (6) {*Lachlania* (3)}, part of Pentamerotarsata (6) {Epeorus/fg1 (2), *Acanthomola* (2), *Anepeorus* (3)}, part of Ephemerella/fg1 (P) {*Teloganodes* (2), *Dicercomyzon* (5)}
— **short multisegmented:** part of Turbanoculata (VAR), Coloburiscus/fg2 (2) (not all) {Coloburiscus/fg3 (P)}, *Chromarcys* (1), part of Oligoneuriella/g1 (VAR), *Fittkauneuria* (P), part of Leptophlebia/fg1 (P)
— **long multisegmented:** Siphlonurus/fg1 (P), *Dipteromimus* (P), Ameletus/fg1 (P), Metretopus/fg1 (P), Acanthametropus/fg1 (P), *Ametropus* (P), *Siphlaenigma* (P), Turbanoculata (not all) (VAR), Nesameletus/f1= =Metamonius/g1 (P), Vetulata (P), Ameletopsis/fg1 (P), *Rallidens* (P), *Coloburiscoides* (P), Isonychia/fg1 (P), Oligoneuriella/g1 (not all) (VAR), Homoeoneuria/ /g1 (P), Elassoneuria/g1 (P), Oligoneuria/f5=g6 (3), *Pseudiron* (P), Pentamerotarsata (6) (not all) {*Paegniodes* (P), Rhithrogena/fg2 (P)}, Fimbriatotergaliae (P), Ephemerella/fg1 (not all) (P), Leptophlebia/fg1 (not all) (P)
— **longer than cerci:** *Caudatella* (1)

[1.3.65] **Cerci of larva** (see also [1.3.53])
— **contiguous basally:** part of Oligoneuria/f4=g5 (6) {*Lachlania* (3)}

[1.3.66] **Primary swimming setae**
— **lost:** part of Turbanoculata (VAR), Coloburiscus/fg2 (2), part of Pentamerotarsata (6) {part of *Kageronia* (VAR), Leucrocuta/g1 (1), Atopopus/g1 (2), *Cinygma* (10), *Cinygmula* (1), Epeorus/fg1 (2)}, part of Furcatergaliae (6) {part of Fossoriae (P) {Campsurus/fg1 (12)}}

— **vestigial:** part of Turbanoculata (VAR), Pentamerotarsata (not all) (6) {*Arthroplea* (P), part of *Kageronia* (VAR), Heptagenia/f7=g6 (P), Ecdyonurus/fg2 (P), Afghanurus/g2 (P), *Paegniodes* (P), Rhithrogena/fg3 (P)}
— **dense:** Euephemeroptera (not all) (5) {Posteritorna (P), Siphlonurus/fg1 (P), *Dipteromimus* (P), Ameletus/fg1 (P), Metretopus/fg1 (P), Acanthametropus/fg1 (P), *Ametropus* (P), *Siphlaenigma* (P), Turbanoculata (not all) (VAR), Nesameletus/f1=Metamonius/g1 (P), Vetulata (P), Ameletopsis/fg1 (P), *Rallidens* (P), Eusetisura (not all) (P) {*Coloburiscoides* (P)}, *Pseudiron* (P), Furcatergaliae (not all) (6) {Fossoriae (not all) (P)}, Hexagenites/fg1 (P), "*Stackelbergisca*" (P), "*Proameletus*" (P), Mesogenesia/g1 (P), "*Epeoromimus*" (P), "*Mogzonurella*" (P), "*Siberiogenites*" (P), "*Australurus*" (P), "*Dulcimanna*" (P), "*Albisca*" (P), "*Siphgondwanus*" (P)}

[1.3.67] **Secondary swimming setae**
— **present:** Posteritorna (13), *Ametropus* (6), part of Turbanoculata (VAR), part of Vetulata (P), Ameletopsis/fg1 (not all) (5) {Ameletopsis/fg2 (2), *Chiloporter* (5), *Chaquihua* (2)}, part of Pentamerotarsata (6), Furcatergaliae (not all) (6) {Potamanthus/fg1 (P), Euthyplocia/fg1 (P), Fossoriae (P) {Campsurus/fg1 (12), "*Siphangarus*"}, Potamanthellus/g1 (4), Ephemerella/fg5 (P)}
— **absent:** Siphlonurus/fg1 (P), *Dipteromimus* (P), Ameletus/fg1 (P), Metretopus/fg1 (P), Acanthametropus/ /fg1 (P), Turbanoculata (not all) (VAR), Nesameletus/ /f1=Metamonius/g1 (P), Vetulata (not all) (P), part of Ameletopsis/fg1 (5) {*Mirawara* (P)}, *Rallidens* (P), *Siphlaenigma* (P), Eusetisura (P), Pentamerotarsata (not all) (6), *Pseudiron* (P), part of Furcatergaliae (6)

2. SUBIMAGO AND IMAGO

Head and its appendages [2.1], thorax and its appendages [2.2], abdomen and its appendages [2.3]

[2] **Adults**
— **short-living:** Prosopistoma/f1=g2 (4), Geminovenata (2), Exeuthyplocia/fg1 (1)}, Behningia/fg1 (3), Palingenia/ /f3=g2 (1), Polymitarcys/f1=Ephoron/g2 (3)
— **shortly-moulting, but non-short-living:** [Classifications of Furcatergaliae IIb] (8), Caenoptera (7), Tricoryptera (7)
— **have imaginal moult in male only:** Prosopistoma/f1= =g2 (4), Geminovenata (2), Behningia/fg1 (3), Palingenia/f3=g2 (1), part of Polymitarcys/f1=Ephoron/g2 (3) {Polymitarcys/f2=Ephoron/g3 (6)}, part of Leptophlebia/fg1 (VAR)
— **have no imaginal moult in both sexes:** part of Anagenesia/g1 (2)
— **non-flying:** part of Heptagennota (5), part of Anagenesia/g1 (1)

2.1. HEAD OF SUBIMAGO AND IMAGO

Head [2.1], antennae [2.1.1], eyes [2.1.2]

[2.1] **Head**
— **with lateral incisions:** Oligoneuria/f4=g5 (1)
— **with anterior projection curved backward:** part of Oligoneuria/f4=g5 (4) {*Spaniophlebia* (1), Oligoneuria/f5=g6 (1)}
— **with face fold projecting anteriorly:** part of Heptagennota (1)
— **with pair of projections between antennae:** Acanthametropus/fg2 (1)
— **with unpaired projection on vertex:** Nesameletus/f1= =Metamonius/g1 (not all) (6)

Antennae of subimago and imago
[2.1.1]

[2.1.1] **Antennae**
— **located on projections:** Behningia/fg1 (1)

Eyes of subimago and imago
[2.1.2–3]

[2.1.2] **Eyes**
— **with vertical bands:** *Fascioculus* (1)

[2.1.3] **Eyes of male**
— **as small as in female:** Prosopistoma/f1=g2 (8), part of Turbanoculata (1), part of Potamanthus/fg1 (VAR), Caenoptera (not all) (15), Tricorythus/fg5 (not all) (VAR), Leptohyphes/fg1 (not all) (3), part of Leptophlebia/fg1 (VAR)
— **large, but not turban:** Baetisca/f2=g1 (P), *Siphlaenigma* (P), *Palaeocloeon* (P), Potamanthus/fg1 (not all) (VAR), Neoephemera/fg1 (P), part of Caenoptera (15) {Caenopsella/g1 (1)}, Ephemerella/fg2 (P), *Vietnamella* (P), *Teloganodes* (P), Tricorythus/fg5 (not all) (VAR), Madecassorythus/fg1 (P), *Ephemerythus* (P), *Dicercomyzon* (P), *Machadorythus* (P), *Ranorythus* (P), part of Leptohyphes/fg1 (3), Leptophlebia/fg1 (not all) (VAR)
— **turban:** Turbanoculata (not all) (1), part of Leptophlebia/fg1 (VAR)

2.2. THORAX OF SUBIMAGO AND IMAGO AND ITS APPENDAGES

Prothorax [2.2.1–2], mesothorax [2.2.3–25], metathorax [2.2.26], wings [2.2.27–72], legs [2.2.73–87]

PROTHORAX OF SUBIMAGO AND IMAGO
[2.2.1–2]

[2.2.1] **Pronotum**

[2.2.2] **Prosternum**
— **with unpaired projection:** Acanthametropus/fg1 (4), *Ametropus* (3)
— **with bispinate projection:** Baetisca/f2=g1 (1)
— **with transverse ridge:** Heptagenia/f6=g5 (4)
— **narrowed, with furca fork-like:** Cryptoprosternata (1)
— **narrowing anteriorly:** Caenis/f4=g2 (1)
— **wide:** Brachycercus/f1=g3 (P), *Madecocercus* (P), "*Afrocercus*" (P)

MESOTHORAX OF SUBIMAGO AND IMAGO

Mesothorax [2.2.3], prealar bridge [2.2.4], tergite [2.2.5–18], pleurite [2.2.19–21], sternite [2.2.22–25]

[2.2.3] **Mesothorax**
— **strongly enlarged and modified:** [Classification of Furcatergaliae II], Caenoptera (1), Tricorygnatha (2) {Leptohyphes/fg1 (4) {Tricorythodes/fg1 (2), (3)}

PREALAR BRIDGE

[2.2.4]

[2.2.4] **Posterior arm of prealar bridge**
— **reduced:** Branchitergaliae (2)

TERGITE OF MESOTHORAX OF SUBIMAGO AND IMAGO

Mesonotum [2.2.5–13], subimaginal mesonotal pigmentation [2.2.14–15], postnotum and phragma [2.2.16–18]

Mesonotum of subimago and imago

[2.2.5–13]

[2.2.5] **Anteronotal protuberance**

[2.2.6] **Anteronotal transverse impression**
— **absent:** Caenoptera (1)

[2.2.7] **Median longitudinal suture**
— **with medionotal membrane:** part of Fimbriatotergaliae (2) {Caenoptera (1)}

[2.2.8] **Mesonotal suture** (among all Ephemeroptera)
— **stretched backward medially or transverse:** Baetisca/f2=g1 (P), Siphlonurus/fg1 (P), *Dipteromimus* (P), Ameletus/fg1 (P), Metretopus/fg1 (P), Acanthametropus/fg1 (P), *Ametropus* (P), Nesameletus/f1=Metamonius/g1 (P), Vetulata (P), Ameletopsis/fg1 (P), *Rallidens* (P), Branchitergaliae (not all) (P), Ephemerella/fg1 (not all) (9), *Baba* (1), "*Cretoneta*" (P)
— **curved backward as a whole:** part of Branchitergaliae (P) {Geminovenata (4)}
— **curved backward at each side:** Prosopistoma/f1=g2 (7), Tetramerotarsata (4), Fimbriatotergaliae (not all) (2) {Potamanthus/fg1 (P), *Ichthybotus* (P), *Sinephemera* (P), Ephemera/fg11 (P), Hexagenia/fg1 (P), *Pentagenia* (P), Neoephemera/fg1 (P)}, Leptophlebia/fg1 (2)
— **close to medioparapsidal suture:** Afromera/g1 (2)
— **lost being integrated with lateroparapsidal suture:** Tetramerotarsata (4), [Classifications of Furcatergaliae IIc] (11), part of Fimbriatotergaliae (2) {Palingenia/f3=g2 (1), Polymitarcys/f1=Ephoron/g2 (3), Caenoptera (1)}, part of Leptophlebia/fg1 (2)
— **lost being smoothed:** part of Branchitergaliae (P) {Epeorus/fg1 (1)}, part of Ephemerella/fg1 (9)}

[2.2.9] **Lateroparapsidal suture**
— **curved laterally:** Coloburiscus/fg2 (1), Rhithrogena/ /fg1 (3), Ephemerella/fg1 (not all) (10) {Ephemerella/fg2 (P), *Vietnamella* (P), Melanemerella/fg1 (P), *Ephemerythus* (P), *Dicercomyzon* (P), *Machadorythus* (P)}
— **connected with medioparapsidal suture:** Heptagenia/ /f5=g4 (P), *Cinygma* (P), part of Ephemerella/fg1 (10) {*Teloganodes* (4), Tricorygnatha (2), Leptohyphes/fg1 (4)}

[2.2.10] **Sublateroscutum of mesonotum**
— **with sublateroscutal suture:** Tetramerotarsata (2)
— **with transverse interscutal suture:** [Classifications of Furcatergaliae IIb] (3), Caenoptera (8), Tricorygnatha (2), Leptohyphes/fg1 (4)

[2.2.11] **Posterior scutal protuberances of mesonotum**
— **convergent:** Caenoptera (8), *Tricorythodes* (2)
— **not convergent:** Leptohyphes/fg2 (P), *Allenhyphes* (P)

[2.2.12] **Lateroscutal suture of mesonotum**
— **straight:** Fimbriatotergaliae (not all) (3)

[2.2.13] **Scutellum of mesonotum**
— **elongate:** Posteritorna (6)
— **with lateral impressions enlarged:** Caenoptera (8), *Teloganodes* (5), Tricoryptera (3)
— **with unpaired plumidium:** part of Posteritorna (3)
— **with pair of plumidia:** part of Eusetisura (5) {part of Discoglossata (3)}, part of Ephemerella/fg1 (5)
— **with infrascutellum** – see [2.2.17]

Pigmentation of subimaginal mesonotum

[2.2.14–15]

[2.2.14] **Lateral pigmented area of subimaginal mesonotum**
— **bifurcates backward into lateroparapsidal stripe** (see [2.2.15]) **and lateral portion, occupying smaller part of sublateroscutum:** Siphlonurus/fg1 (P), *Ametropus* (P), Acanthametropus/fg1 (P), Vetulata (P), *Coloburiscoides* (P), Heptagennota (2)
— **occupies most part of sublateroscutum:** *Dipteromimus* (3), Metretopus/fg1 (4), Nesameletus/f2=Metamonius/g2 (6), Ameletopsis/fg1 (3), Coloburiscus/fg2 (1), Fimbriatotergaliae (not all) (2) {Potamanthus/fg1 (P), *Ichthybotus* (P), Ephemera/fg9 (P), *Pentagenia* (P), Neoephemera (P)}, Leptophlebia/fg1 (not all) (2)
— **medially limited by curved relief line mediad of lateroparapsidal suture:** Ephemerella/fg1 (not all) (10) {Ephemerella/fg2 (P), *Vietnamella* (P), Melanemerella/fg1 (P)}
— **includes whole submedioscutum up to medioparapsidal suture:** Ameletus/fg1 (2), Isonychia/fg1 (3)
— **includes whole submedioscutum and part of medioscutum:** part of Ephemerella/fg1 (10) {*Teloganodes* (4)}
— **dispersed or non-expressed:** Prosopistoma/f1=g2 (4), part of Fimbriatotergaliae (2) {Palingenia/f3=g2 (1), Polymitarcys/f1=Ephoron/g2 (3), Caenoptera (7)}, part of Ephemerella/fg1 (10) {Tricoryptera (7) {Tricorygnatha (2), *Dicercomyzon* (P), *Machadorythus* (P), Leptohyphes/fg1 (4)}}, part of Leptophlebia/fg1 (2)

[2.2.15] **Lateroparapsidal stripe of pigmented area of mesonotum**
— **short:** Siphlonurus/fg1 (4), Vetulata (7)
— **long, laterally limited by relief line:** Heptagennota (2)
— **posteriorly widened:** Heptagennota (not all) (2) {Heptagenia/f5=g4 (P), *Paegniodes* (P)}
— **posteriorly curved laterally:** part of Rhithrogena/fg1 (3) {Rhithrogena/fg2 (6), Epeorus/fg1 (7)}
— **with portion behind apex of lateroparapsidal suture:** Rhithrogena/fg3 (4)

Postnotum and phragma of mesothorax of subimago and imago
[2.2.16–18]

[2.2.16] **Lateropostnotum of mesothorax**
— **fused with epimeron:** [Classifications of Furcatergaliae IIc] (9), Euthyplocia/fg1 (8), Fossoriae (4), Caenoptera (8), *Tricorythodes* (3), *Tricorythopsis* (5)
— **strongly enlarged:** [Classifications of Furcatergaliae IIc] (9), Caenoptera (8), *Tricorythodes* (3), *Tricorythopsis* (5)
— **with crest elongate:** Turbanoculata (9)
— **with crest on anterior margin:** Fimbriatotergaliae (not all) (4), Leptohyphes/fg2 (2)

[2.2.17] **Infrascutellum**
— **interrupted medially:** [Classifications of Furcatergaliae IIa] (1), Caenoptera (8), *Teloganodes* (5), Tricoryptera (3)

[2.2.18] **Phragma between meso- and metathorax**
— **in male unpaired:** Caenoptera (1)

PLEURON OF MESOTHORAX OF SUBIMAGO AND IMAGO
[2.2.19–21]

[2.2.19] **Anterior paracoxal suture of mesothorax**
— **complete (reaching sternum)** (among all Ephemeroptera): Posteritorna (P), Siphlonurus/fg1 (P), *Dipteromimus* (P), Ameletus/fg1 (P), Acanthametropus/fg1 (P), *Ametropus* (P), Metretopus/fg1 (P), Tetramerotarsata (3), Nesameletus/f1=Metamonius/g1 (P), Vetulata (P), Ameletopsis/fg1 (P), Eusetisura (P), "*Cretoneta*" (P)
— **incomplete (not reaching sternum)** (among all Ephemeroptera): *Rallidens* (6), Heptagennota (3), Furcatergaliae (3)
— **transferred anteriorly:** Posteritorna (8), Tetramerotarsata (3)
— **transferred posteriorly:** Heptagennota (3), [Classifications of Furcatergaliae IIa] (4), Caenoptera (8), Tricorygnatha (2), Leptohyphes/fg1 (4)
— **curved backward and reaching furcasternum:** Geminovenata (5)

[2.2.20] **Epimeron of mesothorax**
— **with membranous area:** Ameletus/fg1 (3)
— **in subimago not elongate:** *Siphlaenigma* (P)
— **in subimago elongate:** Turbanoculata (9)
— **fused with lateropostnotum:** Euthyplocia/fg1 (8), Fossoriae (4), Caenoptera (8), *Tricorythodes* (3), *Tricorythopsis* (5)

[2.2.21] **Lateropostnotum** – see [2.2.16]

STERNUM OF MESOTHORAX OF SUBIMAGO AND IMAGO
[2.2.22–25]

[2.2.22] **Basisternum of mesothorax**
— **contains synganglion:** Posteritorna (2)
— **with bifurcate projection:** Isonychia/fg1 (4)

[2.2.23] **Furcasternal protuberances of mesothorax** (among all Ephemeroptera)
— **contiguous (i.e. without median impression):** Posteritorna (2), Siphlonurus/fg1 (P), *Dipteromimus* (P), Ameletus/fg1 (P), Acanthametropus/fg1 (P), *Ametropus* (P), Metretopus/fg1 (P), *Siphlaenigma* (P), *Palaeocloeon* (P), Eusetisura (not all) (P), Potamanthus/fg1 (P), Euthyplocia/fg1 (P), *Ichthybotus* (P), Ephemera/fg9 (P), Behningia/fg1 (P), Palingenia/f2= =g1 (P), Polymitarcys/f2=Ephoron/g3 (P)}, *Baba* (P), "*Cretoneta*" (P)}
— **separated at least at posterior part (i.e. with median impression** – see [2.2.24]**):** Turbanoculata (2), part of Nesameletus/f1=Metamonius/g1 (P), Vetulata (6), Ameletopsis/fg1 (2), *Rallidens* (7), Heptagennota (4), Campsurus/fg1 (4), Caenotergaliae (4), Ephemerella/ /fg1 (17), Leptophlebia/fg1 (3)

[2.2.24] **Median impression of furcasternum**
— **present only in posterior part:** *Pseudiron* (P)
— **narrowed anteriorly:** part of Nesameletus/f2=Metamonius/g2 (P), Ameletopsis/fg1 (2), Heptagennota (not all) (4) {Heptagenia/f6=g5 (P)}
— **not narrowed anteriorly:** Vetulata (6), Ecdyonurus/ /fg1 (1)

[2.2.25] **Subalar-sternal muscle**
— **absent:** Caenoptera (1)

METATHORAX OF SUBIMAGO AND IMAGO
[2.2.26]

[2.2.26] **Metathorax**
— **vestigial:** [Classifications of Furcatergaliae IIc] (10), Caenoptera (9), *Tricorythodes* (4)
— **in imago with transverse ridge:** Caenis/f4=g2 (2)
— **without transverse ridge:** Brachycercus/f1=g3 (P), *Madecocercus* (P), "*Afrocercus*" (P)

WINGS
Wings [2.2.27], fore wing [2.2.28–58], hind wing [2.2.59–72]

[2.2.27] **Wings**
— **unable to fold at rest:** Panephemeroptera (1)
— **unavailable for flight:** part of Heptagennota (5), part of Anagenesia/g1 (1)
— **in imago with marginal setae as in subimago:** Prosopistoma/f1=g2 (6), [Classifications of Furcatergaliae IIb] (6), Exeuthyplocia/fg1 (2), Behningia/fg1 (7), Polymitarcys/f1=Ephoron/g2 (VAR), Caenoptera (12), Tricorygnatha (4), Leptohyphes/fg1 (6), *Coryphorus* (2), part of Leptophlebia/fg1 (VAR)
— **in imago without marginal setae:** *Ephemerythus* (P), *Dicercomyzon* (P), *Machadorythus* (P)

FORE WING
Fore wing [2.2.28], costal brace [2.2.29], vein Sc [2.2.30], vein RA [2.2.31], vein RS [2.2.32–41], vein MA [2.2.42–44], vein MP [2.2.45–49], cubital veins [2.2.50–53], vein AA [2.2.54], additional intercalaries [2.2.55–56], crossveins [2.2.57], coloration [2.2.58]

[2.2.28] **Fore wing**
— **with tornus behind CuP:** Posteritorna (1)
— **with tornus between CuA and CuP:** Anteritorna (1)
— **strongly narrowed:** *Dipteromimus* (2)
— **fan-like widened proximally:** Caenoptera (2)
— **with undulate tornoapical margin:** Behningia/fg3 (1), Anagenesia/g1 (1)

— **with this or that type of gemination:** *Palingeniopsis* (in Permoplectoptera), Discoglossata (5) {Geminovenata (3)}, part of Euthyplocia/fg1 (11), Behningia/fg3 (1), Palingenia/f3=g2 (2) {Anagenesia/g1 (1) {*Cheirogenesia* (1)}}, Hexagenites/fg1 (2)
— **with veins connected in pairs:** Geminovenata (3)
— **without gemination:** *Pentagenia* (P)
— **with additional intercalaries** – see [2.2.56]

[2.2.29] **Costal brace of fore wing**
— **present:** Euephemeroptera (1)
— **approximated to Sc:** Discoglossata (4)

[2.2.30] **Sc of fore wing**
— **terminates near wing apex:** Panephemeroptera (2), Syntonoptera/fg1 (1)

[2.2.31] **RA of fore wing**
— **terminates near wing apex:** Panephemeroptera (2), Syntonoptera/fg1 (1)

Vein RS of fore wing
[2.2.32–41]

[2.2.32] **Common stem of RS+MA of fore wing**
— **present:** Euephemeroptera (2)
— **reduced:** Posteritorna (9)

[2.2.33] **RS of fore wing**
— **begins as common stem:** Panephemeroptera (2), Syntonoptera/fg1 (1)
— **arises from RA:** Protephemeroidea (1)
— **arises from MA:** Euephemeroptera (2)
— **approximated with MA:** Syntonoptera/fg1 (2)
— **concave:** Panephemeroptera (3), Syntonoptera/fg1 (2)
— **forms triad RSa–iRS–RSp:** Panephemeroptera (4), Syntonoptera/fg1 (3)

[2.2.34] **RSa of fore wing**
— **concave:** Panephemeroptera (4)
— **forms triad RSa_1–iRSa–RSa_2:** Panephemeroptera (not all) (4)
— **without branches and intercalaries, approximated with iRS:** Geminovenata (not all) (3)
— **absent:** Homoeoneuria/g2 (4), Elassoneuria/g1 (2)
— **proximally approximated to RA:** part of Oligoneuria/f4=g5 (5) {*Lachlania* (2), Oligoneuria/f5=g6 (2)}
— **not connected with RSp, non-branched:** Prosopistoma/f1=g2 (5)

[2.2.35] **RSa_1 of fore wing:**
— **concave:** Panephemeroptera (4)

[2.2.36] **iRSa of fore wing**
— **convex:** Panephemeroptera (4)

[2.2.37] **RSa_2 of fore wing**
— **concave:** Panephemeroptera (4)
— **non-branched:** Protephemeroidea (2), part of Euephemeroptera (3) {Geminovenata (3), Behningia/fg1 (4), Anagenesia/g1 (1), Campsurus/fg1 (not all) (5)}
— **forms triad RSa_2'–iRSa_2–RSa_2":** Euephemeroptera (not all) (3)
— **connected with RSa:** Baetisca/f2=g1 (P), Radulapalpata (P)
— **not connected with RSa:** *Arthroplea* (4)
— **approximated with iRS:** Behningia/fg3 (1), Palingenia/f3=g2 (2)

[2.2.38] **RSa_2' of fore wing**
— **concave:** Euephemeroptera (not all) (3)
— **absent:** Geminovenata (3), Behningia/fg1 (4), Anagenesia/g1 (1), Campsurus/fg1 (not all) (5)

[2.2.39] **iRSa_2 of fore wing**
— **convex:** Euephemeroptera (not all) (3)
— **absent:** Geminovenata (3), Behningia/fg1 (4), Anagenesia/g1 (1), Campsurus/fg1 (5)

[2.2.40] **iRS of fore wing**
— **convex:** Panephemeroptera (4)
— **arise from RSa:** Discoglossata (not all) (6)
— **approximated with RSa:** Geminovenata (not all) (3)
— **approximated with RSa_2:** Behningia/fg3 (1), Palingenia/f3=g2 (2)
— **absent:** Homoeoneuria/g2 (4), Elassoneuria/g1 (2)
— **proximally approximated to RA:** part of Oligoneuria/f4=g5 (5) {*Lachlania* (2), Oligoneuria/f5=g6 (2)}

[2.2.41] **RSp of fore wing**
— **concave:** Panephemeroptera (4)
— **approximated with MA_1:** *Palingeniopsis* (in Permoplectoptera), Discoglossata (5), Behningia/fg3 (1), Palingenia/f3=g2 (2), Hexagenites/fg1 (2)
— **closely approximated with MA_1:** Geminovenata (3)

Vein MA of fore wing
[2.2.42–44]

[2.2.42] **MA of fore wing**
— **convex:** Panephemeroptera (3), Syntonoptera/fg1 (1)

[2.2.43] **MA bifurcation of fore wing**
— **present:** Euephemeroptera (not all) (4)
— **near middle:** Euephemeroptera (not all) (4) {Baetisca/ /f2=g1 (P), *Siphlaenigma* (P), Elassoneuria/g1 (1), Palingenia/f4=g3 (P)}
— **transferred proximally:** Euthyplocia/fg1 (4), Behningia/fg1 (5), Anagenesia/g1 (1), Polymitarcys/f1=Ephoron/g2 (4)
— **strongly transferred proximally:** Geminovenata (not all) (3) {Oligoneuriella/g1 (P), Homoeoneuria/g1 (P), Oligoneuria/f4=g5 (P), *Fittkauneuria* (P)}, Behningia/fg1 (5), Campsurus/fg1 (5)
— **absent, MA_2 free:** Liberevenata (1)
— **absent, MA_2 absent:** Protephemeroidea (3), Prosopistoma/f1=g2 (5)

[2.2.44] **iMA of fore wing:**
— **absent:** Geminovenata (3)
— **approximated with MA_2:** *Palingeniopsis* (in Permoplectoptera), Behningia/fg3 (1), Palingenia/f3=g2 (2), Hexagenites/fg1 (2)

Vein MP of fore wing
[2.2.45–49]

[2.2.45] **MP of fore wing**
— **concave:** Panephemeroptera (3), Syntonoptera/fg1 (1)

[2.2.46] **MP bifurcation of fore wing**

[2.2.47] **MP_1 of fore wing**
— **approximated with iMP:** Behningia/fg3 (1), *Cheirogenesia* (1)
— **approximated with MA_2:** Discoglossata (5)
— **closely approximated with MA_2:** Geminovenata (3)

[2.2.48] **iMP of fore wing**
— **approximated with MP_1:** Behningia/fg3 (1), *Cheirogenesia* (1)
— **beginning from wing base:** Caenoptera (2)
— **arises from CuA forming secondary triad:** *Tricorythopsis* (1)

[2.2.49] **MP_2 of fore wing**
— **not connected with MP_1, begins from wing base:** Caenoptera (2)
— **not connected with MP_1, begins near wing base:** Posteritorna (9)
— **not connected with MP_1, begins far from wing base:** Tetramerotarsata (9), *Tricorythopsis* (1)
— **in proximal part curved and divergent from MP_1:** Fimbriatotergaliae (not all) (5), *Baba* (2)
— **approximated with CuA:** Discoglossata (5), Behningia/fg3 (1), Hexagenites/fg1 (2)
— **closely approximated with CuA_1:** Geminovenata (3)

Cubital veins of fore wing
[2.2.50–53]

[2.2.50] **CuA of fore wing**
— **convex:** Panephemeroptera (3), Syntonoptera/fg1 (1)
— **approximated with MP_2:** Discoglossata (5), Behningia/fg3 (1)
— **closely approximated with MP_2:** Geminovenata (3)
— **in proximal part curved and divergent from MP_1:** Fimbriatotergaliae (not all) (5), *Baba* (2)
— **short:** *Coryphorus* (1), *Litophlebia* (1)
— **supplied by trachea from MP:** Ephemerella/fg1 (6)
— **supplied by trachea from behind:** Campsurus/fg1 (5)
— **with bifurcation or posterior branches** – see [2.2.51]

[2.2.51] **Cubital field of fore wing**
— **without veins going to wing margin:** Misthodotes/fg (in Permoplectoptera), Posteritorna (1), part of Anteritorna (1) {*Coryphorus* (1), *Litophlebia* (1)}
— **bordered by basitornal margin:** Anteritorna (1)
— **with several veins going from CuA to basitornal margin:** Anteritorna (not all) (1) {Siphlonurus/fg1 (P), *Dipteromimus* (2), Ameletus/fg2, Acanthametropus/fg1 (P), Vetulata (P), Nesameletus/f1=Metamonius/g1 (P), Ameletopsis/fg1 (P), *Rallidens* (P), Coloburiscus/fg1 (P), Isonychia/fg1 (P), Potamanthus/fg1 (6), Euthyplocia/fg2 (P), *Proboscidoplocia* (P), Exeuthyplocia/fg1 (P), *Ichthybotus* (P), Ephemera/fg9 (P), *Pentagenia* (P), Caenotergaliae (not all) (9) {Neoephemera/fg1 (not all) (P)}, Ephemerella/fg1 (not all) (15), *Baba* (P), "*Cronicus*" (P), "*Stackelbergisca*" (P), "*Balticophlebia*" (P), "*Baltameletus*" (P), "*Olgisca*" (P), "*Proameletus*" (P)}
— **with triad or strong vein "CuA_2":** Protereisma/fg (in Permoplectoptera), *Palingeniopsis* (ibid.), part of Anteritorna (1) {Geminovenata (3), Behningia/fg2 (3), Anagenesia/g2 (1), Plethogenesia/g1 (1), *Cheirogenesia* (2), Hexagenites/fg1 (1)}
— **with intercalaries** – see also [2.2.52]
— **with intercalaries going to tornoapical margin and veins going from them to basitornal margin:** *Campylocia* (1), *Polyplocia* (1), Polymitarcys/f1=Ephoron/g2 (not all) (5) {*Eopolymitarcys* (1)}, "*Siphlurites*" (1)
— **with first vein arising from CuP bifurcate (x_1, x_2):** [Classifications of Furcatergaliae I] (5), Potamanthus/fg1 (6), Caenotergaliae (9), Ephemerella/fg1 (15)
— **with 3 bifurcate veins (x, y, z) arising from CuP:** Ephemerella/fg1 (not all) (15) {*Vietnamella* (P)}
— **with 2 bifurcate veins (x, y) arising from CuP:** Ephemerella/fg1 (not all) (15) {Ephemerella/fg2 (not all) (P), *Austremerella* (P)}
— **with 1 bifurcate vein (x) arising from CuP:** [Classifications of Furcatergaliae IIa] (2), part of Caenotergaliae (9) {part of Neoephemera/fg1 (P), Caenoptera (10)}, part of Ephemerella/fg1 (15) {Tricoryptera (1)}
— **with vein (x) symmetrically arising from CuP:** Afrotricorythi (1)
— **with 2 intercalaries going to tornoapical margin:** *Polyplocia* (1), Asthenopus/fg1 (P)
— **with 2 intercalaries going to basitornal margin:** Campsurus/fg2 (1)
— **with vein forming several triads:** Hexagenites/fg1 (1)

[2.2.52] **Number of intercalaries in cubital field**
— **variable:** part of *Metreletus* (1), Metretopus/fg1 (5), part of Acanthametropus/fg1 (P), Behningia/fg1 (6) Polymitarcys/f1=Ephoron/g2 (5)
— **1:** part of Metretopus/fg1 (5) {*Metreplecton* (4)}, part of Tetramerotarsata (10), part of Behningia/fg1 (6), *Campylocia* (1), part of Polymitarcys/f1=Ephoron/g2 (5) {*Eopolymitarcys* (1)}, "*Siphlurites*" (1)
— **2 (1 pair) or 1 bifurcate vein:** part of Anteritorna (1) {part of Metretopus/fg1 (5) {Metretopus/fg2 (4)}, Tetramerotarsata (not all) (10), part of Heptagennota (5), part of Behningia/fg1 (6), *Polyplocia* (1), part of Polymitarcys/f1=Ephoron/g2 (5) {Campsurus/fg1 (5) {Asthenopus/fg1 (P), Campsurus/fg2 (1)}}, part of Caenotergaliae (9) {part of Neoephemera/fg1 (P), Caenoptera (10)}, part of Ephemerella/fg1 (15) {Tricoryptera (1)}, Leptophlebia/fg1 (5), "*Xenophlebia*" (1)}
— **4 (2 pairs), posterior pair longest:** part of Anteritorna (1) {part of Metretopus/fg1 (5) {*Siphloplecton* (4)}, Heptagennota (not all) (5), "*Electrogenia*" (1)}
— **more than 2, anterior pair longest:** *Ametropus* (8), part of Behningia/fg1 (6), part of Polymitarcys/f1==Ephoron/g2 (5) {Polymitarcys/f3=Ephoron/g4 (1)}

[2.2.53] **CuP of fore wing**
— **concave:** Panephemeroptera (3), Syntonoptera/fg1 (1)
— **sharply divergent or separated from CuA:** Tetramerotarsata (5), Leptophlebia/fg1 (4)
— **independent from CuA:** Posteritorna (9), Tetramerotarsata (5), Caenoptera (10), part of Ephemerella/fg1 (6) {Tricoryptera (1)}, Leptophlebia/fg1 (4)
— **connected with CuA and AA by constant crossveins:** Ephemerella/fg1 (6) {Tricoryptera (1)}
— **seems arising from AA:** Caenoptera (10), part of Ephemerella/fg1 (6) {Tricoryptera (1)}
— **strongly arched:** Leptohyphes/fg1 (P)
— **terminates distad of tornus:** Posteritorna (1)
— **terminates close to tornus:** *Ametropus* (7)
— **terminates proximad of tornus:** Anteritorna (1)

Vein AA of fore wing
[2.2.54]

[2.2.54] **AA of fore wing**
— **on common stem with AP:** Epeorus/fg1 (3)
— **with vein arising from base:** Campsurus/fg1 (5)
— **with veins arising posteriorly:** Ephemera/fg9 (6)
— **with bifurcation:** Potamanthus/fg1 (1)
— **terminates distad of tornus:** Posteritorna (1)
— **terminates close to tornus:** *Ametropus* (7)

Additional intercalaries of fore wing
[2.2.55–56]

[2.2.55] **Marginal intercalaries**
— **absent:** *Siphlaenigma* (P), [Classifications of Furcatergaliae IIa] (5), Caenoptera (11), Tricorygnatha (3), *Dicercomyzon* (7), part of *Machadorythus* (6), Leptohyphes/fg1 (5)
— **connected:** *Vietnamella* (6)
— **free:** Liberevenata (2), Ephemerella/fg2 (3), *Austremerella* (4), Melanemerella/fg1 (2), *Teloganodes* (9), *Ephemerythus* (2), *Machadorythus* (not all) (6)
— **1 in each space:** Liberevenata (not all) (2) {*Palaeocloeon* (P)}
— **2 in each space:** part of Liberevenata (2), *Ephemerythus* (2)
[2.2.56] **Other additional intercalaries**
— **border longitudinal veins:** ♂ Prosopistoma/f1=g2 (5)

Crossveins of fore wing
[2.2.58]

[2.2.57] **Crossveins**
— **absent:** Prosopistoma/f1=g2 (5), Homoeoneuria/g2 (4)
— **constant *cua-cup* and *cup-aa*:** Ephemerella/fg1 (6)

Coloration of fore wing
[2.2.58]

[2.2.58] **Coloration of fore wing**
— **at anterior and posterior margins:** Atopopus/g2 (3)

HIND WING
Hind wing [2.2.59–62], veins [2.2.63–72]

[2.2.59] **Hind wing**
— **not exceeds 1/2 of fore wing:** Euplectoptera (1)
— **normally developed:** Posteritorna (P), Siphlonurus/fg1 (P), Ameletus/fg1 (P), Metretopus/fg1 (P), Acanthametropus/fg1 (P), *Ametropus* (9), *Siphlaenigma* (P), Nesameletus/f1=Metamonius/g1 (P), Vetulata (P), Ameletopsis/fg1 (P), *Rallidens* (3), Branchitergaliae (P), Potamanthus/fg1 (P), Euthyplocia/fg1 (not all) (P), Fossoriae (P), Neoephemera/fg1 (P), Ephemerella/fg2 (P), *Vietnamella* (4), *Austremerella* (P), Melanemerella/fg1 (P), Leptophlebia/fg1 (not all) (VAR)
— **without intercalaries:** Geminovenata (3)
— **diminished (veins behind MP_1 absent):** Turbanoculata (not all) (10), *Teloganodes* (6), Tricoryptera (not all) (2) {*Ephemerythus* (not all) (3), Leptohyphes/fg1 (not all) (2)}
— **diminished (veins RS, MA, MP without furcations):** *Dipteromimus* (2)
— **diminished (2 longitudinal veins behind triad MP):** *Bleptus* (2)
— **present in both sexes:** *Ephemerythus* (not all) (3), Leptohyphes/fg1 (not all) (2) {Leptohyphes/fg2 (not all) (P), *Haplohyphes* (P)}
— **present in male, absent in female** (among all Ephemeroptera): part of Turbanoculata (10), part of Leptohyphes/fg1 (2) {part of Leptohyphes/fg2 (P), *Allenhyphes* (P), *Vacupernius* (P), *Yaurina* (P), *Traverhyphes* (P)}
— **absent in both sexes** (among all Ephemeroptera): part of Turbanoculata (10), Caenoptera (9), part of Tricoryptera (2) {Tricorygnatha (5), part of *Ephemerythus* (3), *Dicercomyzon* (8), *Machadorythus* (7), part of Leptohyphes/fg1 (2) {*Tricorythodes* (4), *Tricorythopsis* (6), *Epiphrades* (1)}, *Coryphorus* (3), part of Leptophlebia/fg1 (VAR)
[2.2.60] **Shape of hind wing**
— **roundish:** Baetisca/f2=g1 (2), *Vietnamella* (4)
— **narrow:** Prosopistoma/f1=g2 (5), Turbanoculata (not all) (10), Leptohyphes/fg1 (not all) (2)
[2.2.61] **Costal field of hind wing** (length – see [2.2.63])
[2.2.62] **Costal projection of hind wing**
— **proximally:** Prosopistoma/f1=g2 (5), Leptohyphes/fg1 (not all) (2)
— **close to apex of Sc:** *Teloganopsis* (1), Hyrtanella/f1==Crinitella/g1 (3), *Teloganodes* (6), *Ephemerythus* (not all) (3), "*Manohyphella*", part of Leptophlebia/fg1 (VAR)
— **curved distally:** Turbanoculata (not all) (10)
— **curved proximally:** Leptohyphes/fg1 (not all) (2)
— **with thickened crossvein:** Campsurus/fg1 (5)

Veins of hind wing
[2.2.63–72]

[2.2.63] **Sc of hind wing**
— **terminates at apex of fore margin:** Ephemerella/fg2 (not all) (P) {Timpanoga/fg1 (P), Ephemerella/fg4 (P), Torleya/g1 (P), *Uracanthella* (P), *Amurella* (P)}, *Vietnamella* (4), *Austremerella* (P), Melanemerella/fg1 (P), part of Leptophlebia/fg1 (VAR)
— **terminates at middle of fore margin:** *Teloganopsis* (1), Hyrtanella/f1=Crinitella/g1 (3), *Teloganodes* (6), Tricoryptera (not all) (2) {*Ephemerythus* (not all) (3)}, part of Leptophlebia/fg1 (VAR)
— **terminates at proximal part of fore margin:** Tricoryptera (not all) (2) {Leptohyphes/fg1 (not all) (2)}
— **apically connected with constant cross-vein:** *Teloganopsis* (1), *Ephemerythus* (not all) (3)
— **absent:** part of Leptohyphes/fg1 (2)
[2.2.64] **RA of hind wing**
— **curved proximally:** part of Rhoenanthus/g1 (VAR)
[2.2.65] **Between RA and RS of hind wing**
— **additional vein:** Polymitarcys/f2=Ephoron/g3 (4)
[2.2.66] **RS of hind wing**
— **on common stem with RA:** Protephemeroidea (1), Euplectoptera (1)

— **on common stem with MA:** Euephemeroptera (2) {Euplectoptera (1)}
— **concave:** Panephemeroptera (3)
— **with 1 bifurcation:** Euplectoptera (not all) (1) {*Vietnamella* (4), *Austremerella* (P), Melanemerella/ fg1 (P)}
— **without bifurcation:** *Dipteromimus* (2), Turbanoculata (10), *Teloganodes* (6)

[2.2.67] **MA of hind wing**
— **convex:** Panephemeroptera (3)
— **on common stem with RA:** Euplectoptera (1)
— **with bifurcation:** Euephemeroptera (not all) (4) {*Chromarcys* (VAR), Branchitergaliae (not all) (P)}
— **without bifurcation:** Protephemeroidea (3), Posteritorna (10), Turbanoculata (10), *Ametropus* (9), part of *Chromarcys* (not all) (VAR), Geminovenata (3), *Arthroplea* (5), part of *Cinygma* (VAR), Furcatergaliae (4), "*Cretoneta*" (1)
— **approximated with RSp and MP_1:** Geminovenata (3)

[2.2.68] **MP of hind wing**
— **concave:** Panephemeroptera (3)
— **bifurcate** – see [2.2.69] and [2.2.70]

[2.2.69] **Bifurcation of MP of hind wing**
— **not transferred distally:** Baetisca/f2=g1 (P), *Siphlonisca* (P), Oniscigaster/fg2 (P), Coloburiscus/fg2 (P)
— **transferred distally:** Tasmanophlebia/g1 (1), *Rallidens* (3), *Coloburiscoides* (1), Isonychia/fg1 (7), "*Cronicus*" (not all) (1)
— **absent:** Parameletus/fg2 (2), Turbanoculata (10), part of Campsurus/fg1 (5), *Vietnamella* (4), *Teloganodes* (6), part of Leptophlebia/fg1 (VAR), "*Cronicus*" (not all) (1), "*Cretoneta*" (1)

[2.2.70] **MP_2 of hind wing**
— **arises from CuA:** part of Fimbriatotergaliae (5)
— **intercalary:** *Ametropus* (9), Campsurus/fg1 (not all) (5)
— **absent** – see [2.2.69]

[2.2.71] **CuA of hind wing**
— **convex:** Panephemeroptera (3)
— **forms triad:** part of *Vietnamella* (4)
— **absent:** Turbanoculata (10), *Teloganodes* (6)

[2.2.72] **Crossveins of hind wing**
— **absent:** Prosopistoma/f1=g2 (5)
— **constant *sc-ra*:** *Teloganopsis* (1)

LEGS OF SUBIMAGO AND IMAGO

Fore leg [2.2.73–79], both middle and hind legs [2.2.80–85], middle leg [2.2.86], hind leg [2.2.87]

FORE LEGS OF SUBIMAGO AND IMAGO

In both sexes [2.2.73], in male [2.2.74–77], in female [2.2.78–79]

[2.2.73] **Fore leg**
— **with gill:** *Siphluriscus* (1), Isonychia/fg1 (12)
— **shortened:** Discoglossata (7)

Fore legs of male subimago and imago

[2.2.74–77]

[2.2.74] **Fore leg of male**
— **shortened:** Prosopistoma/f1=g2 (4), Discoglossata (7), Palingenia/f2=g1 (7) {Palingenia/f3=g2 (3) {Anagenesia/g1 (2)}}, Tricorygnatha (8), *Dicercomyzon* (6)

[2.2.75] **Tarsus of male fore leg**
— **5-segmented:** Tetramerotarsata (1)
— **4-segmented:** Tricorygnatha (7)–(8), ⊕ *Machadorythus* (2)
— **non-segmented:** Behningia/fg1 (8)
— **conic in subimago:** Palingenia/f3=g2 (3)
— **conic in subimago and imago:** Anagenesia/g1 (2)
— **with oblique segments:** *Dicercomyzon* (6)

[2.2.76] **1st tarsal segment of male fore leg**
— **non-shortened:** Baetisca/f2=g1 (P), Siphlonurus/fg1 (P) Ameletus/fg1 (P), Metretopus/fg1 (P), Acanthametropus/fg1 (3), *Ametropus* (P), *Siphlaenigma* (P), Isonychia/fg1 (P), part of *Kageronia* (VAR), part of Stenonema/g1 (VAR), part of Ecdyonurus/fg2 (VAR), *Epeiron* (2), *Himalogena* (1), Epeorus/fg1 (4)
— **strongly shortened:** Liberevenata (3), Coloburiscus/fg1 (7), part of *Kageronia* (VAR), Heptagenia/f8=g7 (3), *Dacnogenia* (4), part of Stenonema/g1 (VAR), part of Ecdyonurus/fg2 (VAR), Afghanurus/g1 (2), *Raptoheptagenia* (6), Rhithrogena/fg4 (1), *Sibirigena* (1), Furcatergaliae (not all) (2)
— **fused with 2nd:** Tricorygnatha (8), *Machadorythus* (2)

[2.2.77] **Claws of male fore leg** (among all Ephemeroptera)
— **ephemeropteroid in imago and subimago:** Ameletus/fg1 (P), *Analetris* (P), *Ametropus* (P), Metretopus/fg1 (P), Vetulata (P), Ameletopsis/fg1 (P), *Rallidens* (P), Tetramerotarsata (1), Coloburiscus/fg1 (P), Heptagennota (not all) (VAR), Potamanthus/fg1 (not all) (VAR), *Ichthybotus* (P), Neoephemera/fg2 (P), "*Palaeoanthus*" (P), Ephemerella/fg1 (not all) (P) {Madecassorythus/fg2 (P), *Ranorythus* (P), *Ephemerythus* (P), *Dicercomyzon* (P), *Coryphorus* (P)}, Leptophlebia/fg1 (not all) (VAR), "*Electrogenia*" (1)
— **pointed in imago and subimago:** Siphlonurus/fg1 (4), *Dipteromimus* (4), Nesameletus/f1=Metamonius/g1 (8), part of Heptagennota (VAR), part of Leptophlebia/fg1 (VAR)
— **blunt in imago, ephemeropteroid in subimago:** part of Heptagennota (VAR), part of Potamanthus/fg1 (VAR), Euthyplocia/fg1 (not all) (10), Ephemera/fg9 (5), Polymitarcys/f1=Ephoron/g2 (not all) (6) {Polymitarcys/f2=Ephoron/g3 (P)}, *Leucorhoenanthus* (2), *Potamanthellus* (3), part of Ephemerella/fg1 (P) {Tricorythus/fg5 (1), *Spinirythus* (2), *Machadorythus* (8), Leptohyphes/fg1 (7)}
— **blunt in imago, pointed in subimago:** Isonychia/fg1 (11), part of Euthyplocia/fg1 (10) {*Polyplocia* (2)}
— **blunt in imago and subimago:** Baetisca/f3=g2 (5), Acanthametropus/fg2 (6), *Ametropus* (4), Discoglossata (9), Palingenia/f2=g1 (7), Campsurus/fg1 (8), Caenoptera (16), part of Leptophlebia/fg1 (VAR)
— **single blunt:** Prosopistoma/f1=g2(4), Behningia/fg1(8)

Fore legs of female subimago and imago

[2.2.78–79]

[2.2.78] **Tarsus of female fore leg**
— **5-segmented (in contrast to middle and hind):** Liberevenata (3)
— **4-segmented (as middle and hind):** *Siphlaenigma* (1), Tricorygnatha (8), *Machadorythus* (2)

— **non-segmented:** Anagenesia/g1 (2)

[2.2.79] **Claws of fore leg of female (in contrast to middle and hind leg)**

— **blunt in imago and subimago:** *Ametropus* (4), Palingenia/f3=g2 (not all) (4)

— **blunt single:** Palingenia/f3=g2 (not all) (4)

BOTH MIDDLE AND HIND LEGS OF SUBIMAGO AND IMAGO [2.2.80–85]

[2.2.80] **Middle and hind legs** (see also [2.2.86–87])

— **non-functional at least in imago** (among all Ephemeroptera): Prosopistoma/f1=g2 (4), Geminovenata (2), Exeuthyplocia/fg1 (1), Behningia/fg1 (3), Palingenia/ /f3=g2 (1), Polymitarcys/f1=Ephoron/g2 (3)

— **consist of coxa and trochanter only:** Campsurus/fg3 (1)

[2.2.81] **Coxa of middle and hind leg:**

— **elongate, with narrow base:** Homoeoneuria/g1 (2)

[2.2.82] **Tibia of middle and hind leg**

— **shorter than tarsomere 1 :** Acanthametropus/fg1 (3)

— **with vestige of patella-tibial suture:** Posteritorna (not all) (11), Siphlonurus/fg1 (P), *Dipteromimus* (P), Ameletus/fg1 (P), Metretopus/fg1 (P), Acanthametropus/fg1 (P), *Ametropus* (P), *Siphlaenigma* (P), *Palaeocloeon* (P), Afroptilum/f1=Centroptiloides/g1 & Anteropatellata (see Classification of Turbanoculata), Nesameletus/f1=Metamonius/g1 (P), Vetulata (P), Ameletopsis/fg1 (P), *Rallidens* (P), Coloburiscus/fg1 (P), Isonychia/fg1 (P), *Arthroplea* (P), Heptagenia/f5= =g4 (P), *Cinygma* (P), *Paegniodes* (P), *Cinygmula* (P), Epeorus/fg1 (not all) (8), Potamanthus/fg1 (P), *Ichthybotus* (P), Ephemera/fg9 (P), Palingenia/f2=g1 (P), Polymitarcys/f2=Ephoron/g3 (P), *Leucorhoenanthus* (P), *Potamanthellus* (P), Ephemerella/fg1 (not all) (VAR), Leptophlebia/fg1 (not all) (VAR), Leptophlebia/fg1 (not all) (VAR)

— **without vestige of patella-tibial suture:** Discoglossata (8), *Pseudiron* (2), part of Epeorus/fg1 (8), part of Euthyplocia/fg1 (VAR), *Ochernova* (1), Caenoptera (17), part of Ephemerella/fg1 (VAR), part of Leptophlebia/fg1 (VAR)

[2.2.83] **Tarsus of middle and hind leg: number of distinguishable segments less than 5** (among all Ephemeroptera)

— **segmentation non-expressed:** ♀ Geminovenata (2), part of ♂ Homoeoneuria/g1 (4), part of Furcatergaliae (2) {Behningia/fg1 (3), Campsurus/fg1 (9), ♀ Anagenesia/g1 (2)}

— **2:** Prosopistoma/f1=g2 (4), ♂ Homoeoneuria/g1 (not all) (4), part of Furcatergaliae (2) {♂ Behningia/fg1 (3)}

— **3:** ♂ Oligoneuriella/g1 (2), ♂ Elassoneuria/g2 (1)

— **4:** Tetramerotarsata (1), ♂ *Madeconeuria* (P), ♂ Oligoneuria/f4=g5 (P), ♂ *Fittkauneuria* (P), part of Furcatergaliae (2) {Tricorygnatha (7), (8), *Machadorythus* (2), *Tricorythopsis* (2)}

[2.2.84] **1st tarsal segment of middle and hind leg** (among all Ephemeroptera)

— **fused with 2nd tarsal segment:** part of Furcatergaliae (2) {Tricorygnatha (8), *Machadorythus* (2)}

— **fused with tibia and non-shortened:** Baetisca/f2=g1 (P), Siphlonurus/fg1 (P), *Dipteromimus* (P), Ameletus/fg1 (P), Metretopus/fg1 (P), Acanthametropus/fg1 (P), *Ametropus* (P), Nesameletus/f1=Metamonius/g1 (P), Vetulata (P), Ameletopsis/fg1 (6), *Rallidens* (P), Tetramerotarsata (1), Eusetisura (P), *Pseudiron* (P), *Machadorythus* (2), "*Cretoneta*" (P)

— **fused with tibia and shortened:** Furcatergaliae (not all) (2) {Euthyplocia/fg1 (not all) (VAR), Neoephemera/fg1 (not all) (P), Leptohyphes/fg2 (P), *Allenhyphes* (P)}

— **secondarily elongate:** part of Furcatergaliae (2) {part of Euthyplocia/fg1 (VAR), Caenoptera (13)}

— **secondarily separated from tibia:** Pentamerotarsata (3), part of Furcatergaliae (2) {part of Euthyplocia/fg1 (VAR), part of Neoephemera/fg1 (P), Caenoptera (13), Tricorygnatha (7), *Tricorythodes* (5), *Tricorythopsis* (7)}

[2.2.85] **Claws (except for fore legs of male)** (among all Ephemeroptera)

— **ephemeropteroid:** Baetisca/f2=g1 (P), Ameletus/fg1 (P), Acanthametropus/fg1 (P), Metretopus/fg1 (P), Vetulata (P), Ameletopsis/fg1 (P), *Rallidens* (P), Tetramerotarsata (1), *Ametropus* (P), Coloburiscus/fg1 (P), *Chromarcys* (P), Heptagennota (not all) (VAR), Potamanthus/fg1 (P), *Proboscidoplocia* (P), *Afroplocia* (P), *Ichthybotus* (P), Ephemera/fg9 (P), *Pentagenia* (P), Palingenia/f4=g3 (P), Polymitarcys/f2=Ephoron/g3 (P), "*Palaeoanthus*" (P), Caenotergaliae (P), Ephemerella/fg1 (P), Leptophlebia/fg1 (not all) (VAR), *Baba* (P), "*Cronicus*" (P), "*Electrogenia*" (P), "*Balticophlebia*" (P), "*Baltameletus*" (P), "*Xenophlebia*" (P), "*Cretoneta*" (P)

— **pointed:** Siphlonurus/fg1 (4), *Dipteromimus* (4), Nesameletus/f1=Metamonius/g1 (8), Isonychia/fg1 (6), part of Heptagennota (VAR) {part of *Cinygmula* (VAR), part of *Himalogena* (VAR), part of *Caucasiron* (VAR)}, Euthyplocia/fg2 (1), *Campylocia* (2), *Polyplocia* (2), part of Leptophlebia/fg1 (VAR)

— **blunt:** Geminovenata (2), Exeuthyplocia/fg2 (2), Anagenesia/g1 (2)

— **blunt single:** Prosopistoma/f1=g2 (4), ♂ Behningia/fg1 (3), ♀ Anagenesia/g1 (2), *Chankagenesia* (1), Anagenesia/g2 (2)

— **vestigial or absent:** ♀ Geminovenata (2), Behningia/ /fg1 (3), Campsurus/fg1 (9)

Middle legs of subimago and imago [2.2.86] (see also [2.2.80–85])

[2.2.86] **Middle leg** (see also [2.2.80–85])

— **elongate:** Discoglossata (7)

— **with remainder of gill:** *Siphluriscus* (1)

Hind legs of subimago and imago [2.2.87] (see also [2.2.80–85])

[2.2.87] **Hind leg** (see also [2.2.80–85])

— **vestigial:** Exeuthyplocia/fg2 (1)

— **in male specialized:** Homoeoneuria/g2 (5)

2.3. ABDOMEN OF SUBIMAGO AND IMAGO AND ITS APPENDAGES

Segments I–VIII [2.3.1-5], sternum IX [2.3.6–7], gonostyli [2.3.8–14], penis [2.3.15–17], segment X [2.3.18], caudalii [2.3.19–22]

Abdominal segments I–VIII
[2.3.1-5]

[2.3.1] **Traces of larval tergalial attachment**
— **VI transferred anteriorly:** Posteritorna (5), Caenotergaliae (1)
— **VII transferred anteriorly:** Ephemera/fg9 (1)

[2.3.2] **Lateral lobes**
— **on segments V–IX:** *Siphlonisca* (2), part of Oniscigaster/fg2 (VAR)

[2.3.3] **Abdominal segment VI**
— **enlarged:** Posteritorna (5)
— **with transverse dorsal elevation:** Baetisca/f3=g2 (1)

[2.3.4] **Abdominal segment VII**
— **enlarged, with mid-dorsal elevation:** *Machadorythus* (1)
— **in female with copulatory pouch:** Siphlonurus/fg1 (1)
— **in female with copulatory pouch with basal lobes:** Parameletus/fg1 (1)

[2.3.5] **Abdominal sternum VIII**
— **in female with pair of impressions:** *Tortopus* (1)

Abdominal sternum IX of subimago and imago
[2.3.6–7]

[2.3.6] **Postgenital (subanal) plate of female**
— **normally developed:** *Siphlaenigma* (P)
— **reduced:** Liberevenata (4)
— **bipointed:** Baetisca/f3=g2 (2)

[2.3.7] **Styliger of male**
— **normally developed:** *Siphlaenigma* (P)
— **with reduced median part and separated gonostyli pedestals:** Liberevenata (4), Isonychia/fg2 (1), Polymitarcys/f1=Ephoron/g2 (not all) (7) {Campsurus/fg2 (2), *Asthenopus* (P)}
— **reduced, without gonostyli pedestals:** Geminovenata (6)
— **medially elongate and contiguous with gonostyli:** *Brasiliocaenis* (1)
— **dorsally with membranous area:** Ameletus/fg1 (5)
— **W-shapely speaks into sternite IX:** Caenoptera (5)
— **with projection on gonostylus pedestal:** *Tortopus* (1)

Gonostyli
[2.3.8–14]

[2.3.8] **Gonostylus**
— **absent** (among all Ephemeroptera): part of Geminovenata (6) {Homoeoneuria/g2 (6), *Fittkauneuria* (1)}, *Cotopaxi*
— **vestigial:** *Potamanthellus* (1), part of Caenis/f5=g3 (1)
— **without muscle:** Campsurus/fg2 (2)
— **grooved:** Caenoptera (not all) (5) {Brachycercus/f1=g3 (P), *Madecocercus* (P), *Tillyardocaenis* (P), "*Afrocercus*" (P)}
— **non-grooved:** part of Caenoptera (5) {Caenis/f5=g3 (1)}
— **contiguous with styliger:** *Brasiliocaenis* (1)

[2.3.9] **Anlage of subimaginal gonostylus in larva has**
— **2nd segment directed laterally:** Turbanoculata (not all) (13)
— **2nd segment directed posteriorly or medially:** part of Turbanoculata (13)

[2.3.10] **1st segment of gonostylus**
— **movably articulated with 2nd:** part of Pantricorythi (2)
— **distinct from 2nd:** *Chromarcys* (P), *Ichthybotus* (P), Ephemera/fg9 (P), *Pentagenia* (P), *Protobehningia* (P)
— **not distinct from 2nd:** Geminovenata (6), Potamanthus/fg1 (9), Euthyplocia/fg1 (5), Behningia/fg2 (4), Palingenia/f3=g2 (5), Polymitarcys/f1=Ephoron/g2 (7), "*Palaeoanthus*" (1), Potamanthellus/g1 (1), Caenoptera (5)
— **elongate:** Pantricorythi (2)
— **not elongate:** Ephemerella/fg2 (P)

[2.3.11] 2nd segment of gonostylus
— **with bulbous base:** *Tricorythodes* (1)
— **directed laterally:** part of Pantricorythi (2) {*Tricorythopsis* (3)}

[2.3.12] **Number of distal segments of gonostylus** (among all Ephemeroptera)
— **more than 2:** *Mirawara* (1), part of Discoglossata (12), *Arthroplea* (6), part of Palingenia/f3=g2 (5), "*Cronicus*" (3)
— **2:** Siphlonurus/fg1 (P), *Dipteromimus* (P), Ameletus/fg1 (P), Metretopus/fg1 (P), Acanthametropus/fg2 (P), Vetulata (P), Nesameletus/f1=Metamonius/g1 (P), Ameletopsis/fg2 (P), *Chiloporter* (P), *Chaquihua* (P), *Rallidens* (P), *Siphlaenigma* (P), *Ametropus* (P), Coloburiscus/fg1 (P), Isonychia/fg1 (P), *Oligoneurisca* (P), Oligoneuriella/g1 (not all) (P), Elassoneuria/g1 (not all) (P), Heptagennota (not all) (P), Radulapalpata (P), Neoephemera/fg1 (P), Potamanthus/fg1 (9), *Ichthybotus* (P), Ephemera/fg10 (P), Hexagenia/fg3 (P), *Litobrancha* (P), *Pentagenia* (P), part of Palingenia/f3=g2 (5), Polymitarcys/f2=Ephoron/g3 (P), Neoephemera/fg1 (P), Leptophlebia/fg1 (not all) (P), "*Electrogenia*" (P), "*Xenophlebia*" (P), "*Cretoneta*" (P), "*Mogzonurus*" (P)
— **1:** Posteritorna (12), *Analetris* (1), Liberevenata (not all) (5), Oligoneuria/f4=g5 (2), Euthyplocia/fg1 (not all) (5) {Euthyplocia/fg2 (P), *Proboscidoplocia* (P)}, *Pseudeatonica* (1), Eatonica/fg2 (1), *Eatonigenia* (3), Ephemerella/fg1 (not all) (8), part of Leptophlebia/fg1 (P)
— **0:** part of Liberevenata (5), *Campylocia* (3), *Polyplocia* (3), Exeuthyplocia/fg1 (4), Behningia/fg1 (10), Campsurus/fg1 (10), *Caenoptera* (5), part of Pantricorythi (2) {Afrotricorythi (2), *Tricorythopsis* (3), *Haplohyphes* (1)}

[2.3.13] **Distal segments of gonostylus**
— **shortened:** Neoephemera/fg1 (P)

[2.3.14] **Single distal segment of gonostylus**
— **small:** Pantricorythi (not all) (2)
— **elongate:** *Attenella* (1)

Penis
[2.3.15–17]

[2.3.15] **Penial arms**
— **articulated with gonostyli pedestals:** Campsurus/fg2 (2)

— **lost articulation with tergite IX:** Campsurus/fg2 (2), part of Caenoptera (5)

[2.3.16] **Penis**

— **with titillators** – see [2.3.16]

— **vestigial:** Liberevenata (6)

— **unpaired:** *Eatonigenia* (2), Caenoptera (5)

— **longer than gonostyli:** Behningia/fg1 (9), Madecassorythus/fg2 (1)

— **with median dorsal condylus:** Elassoneuria/g1 (3) {*Madeconeuria* (P)}, Oligoneuria/f4=g5 (3)

— **with extensible lateral and median sclerites:** Elassoneuria/g2 (2)

— **with telescopic apical process:** Oligoneuriella/g1 (1), Homoeoneuria/g2 (7)

— **bipointed, with lateral widenings:** Potamanthus/fg2 (1)

— **divided into 2 pairs of lobes:** Siphlonurus/fg1 (3)

— **with latero-dorsal denticles:** *Prionoides* (1), *Borisonychia* (1), Pentamerotarsata (not all) (5) {Rhithrogena/fg4 (2), *Epeiron* (not all) (3), *Himalogena* (2), Iron/g2 (P)}

— **without latero-dorsal denticles:** Isonychia/fg2 (P), part of Pentamerotarsata (5) {part of *Epeiron* (3), *Sibirigena* (3), Ironopsis/g1 (2)}

— **with paired ventral processes or denticles of this or that peculiar type** (see also [2.3.17]): Parameletus/ /fg1 (2), Oligoneuriella/g1 (1) {Oligoneuriella/g2 (1), *Oligoneuriopsis* (1)}, Afghanurus/g1 (not all) (1), Madecassorythus/fg1 (1), *Traverhyphes* (1), *Yaurina* (1)

[2.3.17] **Titillators**

— **median present:** Pentamerotarsata (not all) (5) {*Arthroplea* (8), Heptagenia/f6=g5 (P), Ecdyonurus/fg2 (P), Afghanurus/fg1 (P), Notacanthurus/g1 (P), *Cinygmula* (P), *Paegniodes* (P), *Proepeorus* (P), Belovius/g1 (not all) (VAR), Iron/g2 (P), *Caucasiron* (P)}

— **median contiguous:** Ecdyonurus/fg2 (P), Afghanurus/ /g1 (P), Notacanthurus/g1 (P)

— **median separated:** Afronurus/g1 (not all) (1)

— **median and ventral present:** *Arthroplea* (8), Rhithrogena/fg2 (not all) (4)

— **ventral flat:** Rhithrogena/fg3 (not all) (5) {Rhithrogena/fg4 (2), *Himalogena* (2), *Sibirigena* (not all) (3)}, part of *Afromera* (VAR)

— **ventral rolled:** Ephemera/fg11 (1)

— **absent:** part of Afronurus/g1 (1), part of Rhithrogena/fg3 (5) {*Epeiron* (3), part of *Sibirigena* (3)} *Bleptus* (3), *Ironodes* (2), Epeorus/fg4 (1), part of Belovius/g1 (VAR), Ironopsis/g2 (1), part of *Afromera* (VAR)

Abdominal segment X of subimago and imago
[2.3.18]

[2.3.18] **Abdominal tergite X**

— **with angles reaching cerci ventro-laterally:** Fossoriae (5), Caenotergaliae (12), Geminovenata (8)

Caudalii of subimago and imago
[2.3.19–22]

[2.3.19] **Caudalii of subimago and imago**

— **in male imago shorter than abdomen:** Acanthametropus/fg2 (5)

— **in male subimago (but not imago) shorter than abdomen:** Prosopistoma/f1=g2 (9), Geminovenata (7), Caenoptera (6)

— **in female do not moult:** Caenoptera (6)

— **in female diminished:** Caenoptera (6), Tricorythodes/ /fg1 (6)

— **without distinct segmentation:** Prosopistoma/f1=g2 (9), Behningia/fg1 (11)

[2.3.20] **Setae on caudalii**

— **irregular:** *Chromarcys* (P)

— **in whorls:** ⊕♂ Geminovenata (not all) (7), ◎♂ Caenoptera (6)

— **absent:** ⊕♂ Caenoptera (6)

[2.3.21] **Cerci of subimago and imago**

— **contiguous basally:** part of Oligoneuria/f4=g5 (6) {*Lachlania* (3), *Spaniophlebia* (3)}

[2.3.22] **Paracercus of subimago and imago** (among all Ephemeroptera)

— **absent:** part of Oligoneuria/f4=g5 (6) {*Spaniophlebia* (4)}

— **vestigial or markedly diminished:** Baetisca/f2=g1 (3), Siphlonurus/fg1 (5), Ameletus/fg1 (6), Metretopus/fg1 (6), Liberevenata (7), Tasmanophlebia/g1 (3), part of Nesameletus/f1=Metamonius/g1 (P), Ameletopsis/fg2 (3), *Mirawara* (4), *Chaquihua* (3), *Rallidens* (9), Coloburiscus/fg1 (9), Isonychia/fg1 (9), *Chromarcys* (1), part of Oligoneuriella/g1 (VAR), part of Oligoneuria/f4=g5 (6) {*Lachlania* (3)}, Heptagennota (6), part of Fimbriatotergaliae (P) {part of Rhoenanthus/g1 (VAR), Hexagenia/fg1 (3), Palingenia/f3=g2 (6), Campsurus/fg1 (11), *Leucorhoenanthus* (1), Potamanthellus/g1 (2), *Teloganodes* (2), *Spinirythus* (1), *Dicercomyzon* (5)}, part of Leptophlebia/fg1 (P), "*Cronicus*" (2), "*Electrogenia*" (2), "*Balticophlebia*" (P), "*Baltameletus*" (1), "*Cretoneta*" (2), "*Proameletus*" (1)

— **developed:** Prosopistoma/f1=g2 (9), *Dipteromimus* (P), Acanthametropus/fg1 (P), Oniscigaster/fg2 (P), Nesameletus/f1=Metamonius/g1 (not all) (P), *Chiloporter* (P), *Ametropus* (P), *Siphlaenigma* (P), Discoglossata (not all) (P) {Oligoneuriella/g1 (not all) (VAR), Homoeoneuria/g1 (P), Elassoneuria/g1 (P), Oligoneuria/f5=g6 (4), *Fittkauneuria* (P)}, Fimbriatotergaliae (not all) (P) {Potamanthus/fg2 (P), Rhoenanthus/g1 (not all) (VAR), *Campylocia* (P), *Polyplocia* (P), *Proboscidoplocia* (P), Ephemera/fg10 (P), Neoephemera/fg2 (P), *Ochernova* (P), Caenoptera (P), "*Palaeoanthus*" (P)}, Ephemerella/fg1 (not all) (P), Leptophlebia/fg1 (not all) (P), *Baba* (P)

— **developed in female, vestigial in male:** Rhoenanthus/ /g1 (not all) (VAR), Euthyplocia/fg2 (VAR), Exeuthyplocia/fg1 (5), *Ichthybotus* (4), Behningia/fg1 (12), Cryptoprosternata (not all) (5) {*Pentagenia* (P), Polymitarcys/f2=Ephoron/g3 (P)}, *Ranorythus* (1)

— **longer than cerci:** *Caudatella* (1)

— **with spine-like ventral seta:** *Allenhyphes* (1)

3. **EGGS**
[3.1-6]

[3.1] **Viviparity**
— **present:** part of Turbanoculata (VAR)

[3.2] **Shape of egg:**
— **biconvex lens:** *Prionoides* (2), Palingenia/f3=g2 (7)
— **concave-convex:** Campsurus/fg2 (3)

[3.3] **Chorion sculpture**
— **longitudinal ridges:** Brachycercus/f1=g3 (6)
— **polygonal:** Eurylophella/fg2 (2)

[3.4] **Anchor or its vestige (except for polar structures)**
— **rough spot formed by apices of threads:** Siphlonurus/fg1 (2)
— **thread, coiled under knob:** Branchitergaliae (not all) (4)
— **mushroom-like:** Discoglossata (11)
— **skein of threads, surrounds knob:** Potamanthus/fg1 (5), Ephemerella/fg1 (not all) (14)
— **flat spiral:** Afronurus/g1 (not all) (2)
— **lamellate:** Palingenia/f3=g2 (not all) (7)

[3.5] **Number of polar caps or anchors** (see [3.6])
— **2:** *Stenacron* (1), [Classifications of Furcatergaliae I] (4), Potamanthus/fg1 (5), part of Euthyplocia/fg1 (VAR), *Eopolymitarcys* (2), part of Campsurus/fg1 (VAR) {part of Asthenopus/fg1 (P)}, Caenotergaliae (not all) (8), part of Ephemerella/fg1 (14) {*Tricorythopsis* (4)}
— **1:** *Ametropus* (10), [Classifications of Furcatergaliae I] (2), part of Euthyplocia/fg1 (VAR), Polymitarcys/f3= =Ephoron/g4 (2), part of Campsurus/fg1 (WAR) {part of Campsurus/fg2 (3)}, Caenotergaliae (not all) (8) {Brachycercus/f1=g3 (6)}, Ephemerella/fg1 (not all) (14)
— **absent:** part of Campsurus/fg2 (3), Caenotergaliae (not all) (8), Ephemerella/fg1 (not all) (14) {Eurylophella/fg2 (2)}

[3.6] **Structure of polar cap or anchor** (see [3.5])
— **long threads coiled on pole:** *Stenacron* (1), part of Euthyplocia/fg1 (VAR), part of Campsurus/fg1 (WAR) {part of Asthenopus/fg1 (P), part of Campsurus/fg2 (3)}, part of Caenotergaliae (8)
— **cap with integral external surface:** [Classifications of Furcatergaliae I] (4), Potamanthus/fg1 (5), part of Euthyplocia/fg1 (VAR), part of Campsurus/fg1 (WAR) {part of Campsurus/fg2 (3)}, Caenotergaliae (not all) (8), Ephemerella/fg1 (not all) (14)
— **bunch of folded threads with knobs:** *Ametropus* (10)
— **cap bearing threads with knobs:** Polymitarcys/f2= =Ephoron/g3 (5) {*Eopolymitarcys* (2), Polymitarcys/f3= =Ephoron/g4 (2)}

REFERENCES

Besides sources directly cited in the text, here are also included sources mentioned as authorship of genus-group and family-group names of Ephemeroptera (but not that of species-group names).

AGNEW, J.D. 1973. Two new species of Oligoneuriopsis Crass from the Republic of South Africa (Oligoneuriidae: Ephemeroptera). — In: W.L. Peters & J.G. Peters (eds.). *Proc. 1st Int. Conf. Ephemeroptera*, Florida A & M Univ. 17–20 August 1970. Leiden, Brill: 114–121.

ALBA-TERCEDOR, J. & SUTER, P.J. 1990. A new species of Caenidae from Australia: Tasmanocoenis arcuata sp.n. (Insecta, Ephemeroptera). — *Aquatic Insects* **12** (2): 85–94.

ALBARDA, H. 1878. Descriptions of three new Ephemeridae. — *Entomol. Month. Mag.* **15**: 128–130.

ALI, S.R. 1970. Certain mayflies (Order: Ephemeroptera) of West Pakistan. — *Pakistan Jour. Sci.* **22**: 119–124.

ALLEN, R.K. 1965. A review of the subfamilies of Ephemerellidae (Ephemeroptera). — *Journ. Kansas Entomol. Soc.* **38**: 262–266.

ALLEN, R.K. 1966. *Haplohyphes*, a new genus of Leptohyphinae (Ephemeroptera: Tricorythidae). — *Journ. Kansas Entomol. Soc.* **39**: 262–266.

ALLEN, R.K. 1971. New Asian *Ephemerella* with notes (Ephemeroptera: Ephemerellidae). — *Canad. Entomol.* **103** (4): 512–528.

ALLEN, R.K. 1973. New Ephemerellidae from Madagascar and Afghanistan (Ephemeroptera). — *Pan-Pacif. Entomol.* **49**: 160–164.

ALLEN, R.K. 1975. *Ephemerella* (*Cincticostella*): a revision of the nymphal stages (Ephemeroptera, Ephemerellidae). — *Pan-Pacif. Entomol.* **51** (1): 16–22.

ALLEN, R.K. 1980. Geographic distribution and reclassification of the subfamily Ephemerellinae (Ephemeroptera: Ephemerellidae). — In: J.F. Flannagan & K.E. Marshall (eds.). *Advances in Ephemeroptera Biology* (Proc. 3rd Int. Conf. Ephemeroptera, 4–10 July 1979, Winnipeg, Canada). Plenum Press, New York: 71–91.

ALLEN, R.K. 1984. A new classification of the subfamily Ephemerellinae and the description of a new genus. — *Pan-Pacif. Entomol.* **60**: 245–247.

ALLEN, R.K. & COHEN, S.D. 1977. Mayflies (Ephemeroptera) of Mexico and Central America: new species, descriptions, and records. — *Canad. Entomol.* **109**: 399–414.

ALLEN, R.K. & EDMUNDS, G.F., JR. 1959. Revision of the Genus *Ephemerella*. I. The subgenus *Timpanoga*. — *Canad. Entomol.* **91**: 51–59.

ALLEN, R.K. & EDMUNDS, G.F., JR. 1960. A Revision of the genus *Ephemerella* (Ephemeroptera: Ephemerellidae) II. The subgenus *Caudatella*. — *Ann. Entomol. Soc. Amer.* **54**: 603–612.

ALLEN, R.K. & EDMUNDS, G.F., JR. 1961. Revision of the genus *Ephemerella* (Ephemeroptera: Ephemerellidae) III. The subgenus *Attenuatella*. — *Journ. Kansas. Entomol., Soc.* **34** (4): 161–173.

ALLEN, R.K. & EDMUNDS, G.F., JR. 1962a. A revision of the genus *Ephemerella* (Ephemeroptera) V. The subgenus *Drunella* in North America. — *Misc. Publ. Entomol. Soc. Amer.* **3** (5): 147–179.

ALLEN, R.K. & EDMUNDS, G.F., JR. 1962b. Revision of the genus *Ephemerella* (Ephemeroptera: Ephemerellidae) IV. The subgenus *Dannella*. — *Journ. Kansas Entomol. Soc.* **35**: 333–338.

ALLEN, R.K. & EDMUNDS, G.F., JR. 1963a. New and little known Ephemerellidae from southern Asia, Africa and Madagascar. — *Pacific Insects* **5**: 11–22.

ALLEN, R.K. & EDMUNDS, G.F., JR. 1963b. Revision of the genus *Ephemerella* (Ephemeroptera: Ephemerellidae) VI. The subgenus *Serratella* in North America. — *Ann. Entomol. Soc. Amer.* **56**: 533–600.

ALLEN, R.K. & EDMUNDS, G.F., JR. 1963c. Revision of the genus *Ephemerella*. VII. The subgenus *Eurylophella*. — *Canad. Entomol.* **95**: 597–623.

ALLEN, R.K. & EDMUNDS, G.F., JR. 1965. A revision of the genus *Ephemerella* (Ephemeroptera, Ephemerellidae). VIII. The subgenus *Ephemerella* in North America. — *Misc. Publ. Entomol. Soc. Amer.* **4** (6): 242–282.

ALLEN, R.K. & EDMUNDS, G.F., JR. 1976a. A revision of the genus *Ametropus* in North America. — *Journ. Kansas Entomol. Soc.* **49** (4): 625–635.

ALLEN, R.K. & EDMUNDS, G.F., JR. 1976b. *Hyrtanella*: a new genus of Ephemerellidae from Malaysia (Ephemeroptera). — *Pan-Pacif. Entomol.* **52**: 133–137.

ALLEN, R.E. & MURVOSH, CH.M. 1987. Mayflies (Ephemeroptera, Tricorythidae) of the Southwestern United States and North Mexico. — *Ann. Entomol. Soc. Amer.* **80** (1): 35–40.

ARO, J.E. 1910. Piirteitä päiväkorennoisten (Ephemeridae) elämäntavoista ja kehityksetstä. — *Jlipainos Viipurin suom. realilyseon vuosikertomuksesta*: 1–32.

ASHLOCK, P.D. 1971. Monophyly and associated term. — *Systematic Zoology* **20** (1): 63-69.

AUDOUIN, V. 1824. Recherches anatomiques sur les Thorax des animaux articulés et celui des insectes hexapodes en particulier. — *Ann. Sci. Nat.* **1**: 97–135; Atlas des T. 1–3: 22–23, Pl. 8.

BAE, Y.J. & MCCAFFERTY, W.P. 1991. Phylogenetic systematics of the Potamanthidae (Ephemeroptera). — *Trans. Amer. Entomol. Soc.* **117** (3/4): 1–145.

BAE, Y.J. & MCCAFFERTY, W.P. 1998. Phylogenetic systematics and biogeography of the Neoephemeridae. — *Aquatic Insects* **20** (1): 35–68.

BAE, Y.J. & MCCAFFERTY, W.P. & EDMUNDS, J.F., JR. 1990. *Stygifloris*, a new genus of mayflies (Ephemeroptera: Potamanthidae) from southeast Asia. — *Ann. Entomol. Soc. Amer.* **83**: 887–891.

[BAJKOVA, O.JA.] БАЙКОВА О.Я. 1970. Новые и малоизвестные виды поденок (Ephemeroptera) из бассейна Амура [New and little-known species of may-flies (Ephemeroptera) from the basin of the Amur River]. — *Энтомол. Обозрение* [*Rev. Entomol. URSS*] **49** (1): 146–155.

[BAJKOVA, O.JA.] БАЙКОВА О.Я. 1975. Новый род поденок из Приморья (Ephemeroptera, Heptageniidae) [New genus of Ephemeroptera from the Primor'ye]. — *Изв. Сиб. Отдел. АН*

CCCP, сер. биол. наук ["*Izvestiya Sibir. Otdel. Akad. Nauk SSSR*", Ser. Biol.] №1: 54–57.

BALASUBRAMANIAN, C. & VENKATARAMAN, K. & SIVARAMAKRISHNAN, K.G. 1991. Life stages of a South Indian burrowing mayfly *Ephemera (Aethephemera) nadinae* McCafferty and Edmunds 1973 (Ephemeroptera: Ephemeridae). — *Aquatic Insects* **13** (4): 223–228.

BALTHASAR, V. 1937. Athropleidae, eine neue Familie der Ephemeropteren mit Beschreibung einer neuen Art der Gattung *Athroplea* Bengtss. — *Zool. Anz.* **120**: 204–230.

BANKS, N. 1900. New genera and species of Nearctic Neuropteroid insect. — *Tr. Amer. Entomol. Soc.* **26**: 239–259.

BANKS, N. 1913. Neuropteroid insects from Brazil. — *Psyche*, 20: 83–89.

BARNARD, K.H. 1932. South African May-flies (Ephemeroptera). — *Trans. Roy. Soc. S. Afr.* **20**: 201–259.

BARNARD, K.H. 1937. A new may-fly from Natal (Ephemeroptera). — *Ann. Natal Mus.*, London, **8**: 275–278.

BARNARD, K.H. 1940. Additional records, and descriptions of new species of South-African alder-flies (Megaloptera), May-flies (Ephemeroptera), caddies-flies (Trichoptera), stone-flies (Plecoptera), and dragon-flies (Odonata). — *Ann. S. Afr. Mus.* **32**: 609–661.

BAUMGARDNER, D.E. & MCCAFFERTY, W.P. 2000. *Leptohyphes zalope* (Ephemeroptera: Leptohyphidae): a polytypic North and Central American species. — *Entomol. News* **111** (1): 49–59.

[BECKER, E.G.] БЕККЕР Е.Г. 1954. К вопросу о происхождении и развитии крыла насекомых. Часть II. К строению, механике и происхождению летательного аппарата поденок (Ephemeroptera). — *Вестн. Моск. ун-та* ["*Vestnik Moskov. Univ.*"] № 5.

BEDNARIK, A.F. 1979. Subgeneric classification of *Stenonema* (Ephemeroptera: Heptageniidae). — *Journ. Gorgia Entomol. Soc.* **14**: 190–191.

BEDNARIK, A.F. & MCCAFFERTY, W.P. 1979. Biosystematic revision of the Genus *Stenonema* (Ephemeroptera: Heptageniidae). — *Canad. Bull. Fisheries and Aquatic Sci.* **201**: 1–73.

BELFIORE, C. & HAYBACH A. & KLONOWSKA-OLEJNIK, M. 1999. Taxonomy and phenetic relationships of *Electrogena affinis* (Eaton 1883) (Ephemeroptera: Heptageniidae). — *Annals Limnol.* **35** (4): 245–256.

BELFIORE, C. & KAZANCI, N. & TANATMIS, M. (in litt). [paper with description of a new species of *Electrogena* from Turkey]

[BELOV, V.V.] БЕЛОВ В.В. 1979. Новий рід одноденок (Ephemeroptera, Ephemerellidae) у фауні СРСР [A new may-fly genus (Ephemeroptera, Ephemerellidae) in the USSR fauna]. — *Доповіди АН Укр. РСР*, сер. Б. геол., хим., биол. ["*Dopovidi Akad. Nauk Ukrain. RSR*", ser B. geol., chim., biol.], № 7: 577–580.

BENGTSSON, S. 1904. Reseberättelse afgifven af Regnellske stipendiaten docenten Simon Bengtsson för en zoologisk sera til Umea Lappmark 1903. — *Kungliga Svenska Vetenskaps-Akademiens Arsbok*: 117–131. (Cit. ex: Hubbard 1990).

BENGTSSON, S. 1908. Berättelse öfver en resa i entomologiskt syfte till mellersta Sverige sommaren 1907. — *Kungl. Svenska*, Vetenskapakad. Arsbok, **6**: 237–246. (Cit. ex: Hubbard 1990).

BENGTSSON, S. 1909. Beiträge zur Kenntnis der westpaläarktischen Ephemeriden. — *Lunds Univ. Arsbok*, N.F., Afd. 2, **5** (4): 237–246.

BENGTSSON, S. 1912. Neue Ephemeriden aus Schweden. — *Entomol. Tidskr.* **33**: 107–117.

BENGTSSON, S. 1913. Undersökningar öfver äggen hos Ephemeriderna. — *Entomol. Tidskr.* **34**: 271–320.

BENGTSSON, S. 1917. Weitere Beiträge zur Kenntnis der nordischen Eintagsfliegen. — *Entomol. Tidskr.* **38**: 174–194.

BENGTSSON, S. 1930. Kritische Bemerkungen über einige nordische Ephemeropteren, nebst Beschreibung neuer Larven. — *Lunds Univ. Arss.*, N.F., Avd. 2, **26** (3): 1–27.

BERNER, L. 1946. New species of Florida mayflies (Ephemeroptera). *Florida Entomol.* **28** (4): 60–82.

BERNER, L. 1954. Phoretic association between a species of *Simulium* and a mayfly nymph, with a description of the nymph. — *Ann. Mag. Nat. Hist.* London (12) **7**: 116–121.

BERNER, L. 1956. The genus *Neoephemera* in North America. — *Ann. Entomol. Soc. Amer.* **49**: 33–42.

BERNER, L. 1975. The mayfly family Leptophlebiidae in the southern United States. — *The Florida Entomologist* **8** (3): 137–156.

BERNER, L. & PESCADOR, M.L. 1980. The mayfly family Baetiscidae (Ephemeropterta). Part I. — In: J.F. Flannagan & K.E. Marshall (eds.). *Advances in Ephemeroptera Biology* (Proc. 3rd Int. Conf. Ephemeroptera, 4–10 July 1979, Winnipeg, Canada). Plenum Press, New York: 511–524.

BERNER, L. & THEW, T.B. 1961. Comments on the mayfly genus *Campylocia*, with a description of a new species (Euthyplociidae: Euthyplociinae). — *Amer. Midland Naturalist* **66** (2): 329–336.

BERTHÉLEMY, C. & THOMAS, A. 1967. Note taxonomique sur *Epeorus torrentium* Eaton 1881 et *E. assimilis* Eaton 1885 [Ephemeroptera, Heptageniidae]. — *Annales Limnol.* **3** (1): 65–74.

BIANCHERI, E. 1953. Note sugli Efemerotteri italiani - III. Sulla variabuilita delle ali posteriori di *Habrophlebia fusca* (Curtis). — *Mem. Soc. Entomol. Ital., Genova* **32**: 54–59.

BILLBERG, G.J. 1820. *Enumeratio insectorum in museo Gust. Joh. Billberg*. Typus Gadelianus: 1–138.

BLAIR, K.G. 1929. Two new British Mayflies. — *Entomol. Mon. Mag.* **65**: 253–255.

BOGOESCU, C. 1947. Un génre nouveau d'Ephéméroptère trouve en Romanie. — *Bull. Sect. Sci. Acad. Roum.* **29**: 602–606.

BOGOESCU, C. & TABACARU, I. 1962. Beiträge zur Kenntnis der Untersuchungsmerkmale zwichen den Gattungen *Ecdyonurus* und *Heptagenia*. — *Beitr. Entomol.* **12** (3/4): 273–291.

BÖRNER, C. 1904. Zur Systematic der Hexapoden. — *Zool. Anz.* **27**: 511–533.

BÖRNER, C. 1909. Neue Homologien zwischen Crustaceen und Hexapoden. Die Beissmandibel der Insecten und ihre phylogenetische Bedeutung. Archi- und Metapterygota. — *Zool. Anz.* **34**: 100–125.

BOUDREAUX, H.B. 1979. *Arthropod phylogeny, with special reference to insects*. New Yourk, J. Willey & sons: 1–320.

BRAASCH, D. 1981. Beitrag zur Kenntnis der Heptageniidae des Himalaya (Ephemeroptera). — *Reichenbachia Staatl. Mus. Tierk. Dresden* **19** (23): 127–132.

BRAASCH, D. 1990. Neue Eintagsfliegen aus Thailand, nebst einigen Bemerkungen zu deren generischen Status (Insecta, Ephemeroptera, Heptageniidae). — *Reichenbachia Staatl. Mus. Tierk. Dresden* **28** (1): 7–14.

BRAASCH, D. & SOLDÁN, T. 1984. Beitrag zur Kenntnis der Gattung *Thalerosphyrus* Eaton 1881 im Hinblik auf die Gattung *Ecdyonuroides* Thanh 1967 (Ephemeroptera, Heptageniidae). — *Reichenbachia Staatl. Mus. Tierk. Dresden* **22** (27): 201–206.

BRAASCH, D. & SOLDÁN, T. 1986a. *Asionurus* n.gen., eine neue Gattung der Heptageniidae aus Vietnam (Insecta, Ephemeroptera, Heptageniidae). — *Reichenbachia Staatl. Mus. Tierk. Dresden* **23** (28): 155–159.

BRAASCH, D. & SOLDÁN, T. 1986b. Die Heptageniidae des Gombak River in Malaysia (Ephemeroptera). — *Reichenbachia Staatl. Mus. Tierk. Dresden* **24** (3): 41–52.

BRAASCH, D. & SOLDÁN, T. 1988. *Trichogenia* gen. n., eine neue Gattung der Eintagsfliegen aus Vietnam (Ephemeroptera). — *Reichenbachia Staatl. Mus. Tierk. Dresden* **25** (25): 119–124.

BRAUER, F. 1885. Systematisch-zoologische Studien. — *Sitzungber. der k. Akademie der Wiss. Mat.-Naturw. Classe*, Bd. **91**, Hf. 5, Abt. 1: 237–413.

BRAUER, F. & REDTENBACHER, J. & GANGLBAUER, L. 1889. Fossile insecten dem Juraformation Ost-Siberiens. — *Mem. Acad. Sci. St.-Petersbourg*, Ser. 7, **36** (15): 1–22, Taf. I, II.

BRINCK, P. 1957. Reproductive system and mating in Ephemeroptera. — *Opuscula Entomol.* **22** (1): 1–37.

BRITO, I.M. 1987. Nota preliminar sobre uma nova Efemera do Cretáceo do Ceara (Insecta Ephemeroptera). — *Anais do X Congr. Brasil. de Paleontol.*: 593–597. (Cit. ex: Hubbard 1990).

[BRODSKY, A.K.] БРОДСКИЙ А.К. 1973. Роевое поведение поденок (Ephemeroptera) [Swarming behaviour of may-flies (Ephemeroptera)]. — *Энтомол. Обозрение* [*Rev. Entomol. URSS*] **52**: 51–61.

[BRODSKY, A.K.] БРОДСКИЙ А.К. 1974. Эволюция крылового аппарата у Ephemeroptera [Evolution of the flight apparatus of Ephemeroptera]. — *Энтомол. Обозрение* [*Rev. Entomol. URSS*] **53** (2): 291–303.

BRONGNIART, C. 1893. Recherches pour servir à l'histoire des insectes fossiles des temps primaires. Études sur terrain houiller de Commentry. III. Faune entomologique. Saint-Etienne: 1–493.

[BULDOVSKY, A.T.] БУЛДОВСКИЙ А.Т. 1935a. О новых формах сем. Palingeniidae (Ephemeroptera) из Дальневосточного края СССР. — *Изв. АН СССР*, отд. мат. и естеств. наук [*Bull. Acad. Sci. URSS*]: 831–836.

[BULDOVSKY, A.T.] БУЛДОВСКИЙ А.Т. 1935b. Новые представители сем. Palingeniidae (Ephemeroptera) из Дальнего Востока СССР [New representatives of the family Plingeniidae (Ephemeroptera) from the Far East of the USSR]. — *Вестн. ДВФ АН СССР* [*"Vestnik Dalnevost. Filiala Akad. Nauk SSSR"*] № 14: 151–166.

BURIAN, S.K. & GIBBS, K.E. 1988. A redescription of *Siphlonisca aerodromia* Needham (Ephemeroptera: Siphlonuridae). — *Aquatic Insects* **10** (4): 237–248.

BURKS, B.D. 1953. The mayflies, or Ephemeroptera, of Illinois. — *Bull. Illinois Nat. Hist. Survey*, **26** (1): 1–216.

BURMEISTER, H. 1829. *Die insectorum systemate naturali.* Diss. inaug. quam consensu Facultatis medicae halensis. Hales Saxonum, typus Grunertorum Partis Filique (2). 1–45.

BURMEISTER, H. 1839. *Handbuch der Entomologie*, Band II, Abt. 2. Fam. Ephemerina: 788–804.

CAMPION, H. 1923. On the use of the generic name *Brachycercus*. — *Ann. Mag. Nat. Hist.* **11**: 515–518.

CARPENTER, F.M. 1933. The Lower Permian insects of Kansas. Part 6. Delopteridae, Protelytroptera, Plectoptera, and a new collection of Protodonata, Odonata, Megasecoptera, Homoptera and Psocoptera. — *Proc. Amer. Acad. Arts & Sci.* **68** (11): 411–503.

CARPENTER, F.M. 1938. Two Carboniferous insects from the vicinity of Mazon Creek, Illinois. — *Amer. Journ. Sci.* (5), **36**: 445–452.

CARPENTER, F.M. 1963. Studies on Carboniferous insects from Commentry, France: Part IV. The genus *Triplosoba*. — *Psyche* **70** (2): 120–128.

CARPENTER, F.M. 1979. Lower Permian insects from Oklahoma. Part 2. Orders Ephemeroptera and Palaeodictyoptera. — *Psyche* **86**: 261–290.

CARPENTER, F.M. 1987. Review of the extinct family Syntonopteridae (order uncertain). — *Psyche* **94** (3–4): 373–388.

CHEN, S.H. 1958. [Cit. ex: Chen 1962. Types of insect metamorphosis and a new system of insect classification. — *Acta Entomol. Sinica* **11** (1): 1–15 (Chinese, with English summary)].

CHOPRA, B. 1927. The Indian Ephemeroptera (mayflies). Part I. The suborder Ephemeroidea: families Palingeniidae and Polymitarcyidae. — *Rec. Indian Mus.* **29** (2): 91–138.

CLEMENS, W.A. 1915. Mayflies of the *Siphlonurus* group. — *Canad. Entomol.* **47**: 245–260, pl. 9–11.

COCKERELL, T.D.A. 1923. A new genus of mayflies from the Miocene of Florissant, Colorado. — *Psyche* **30**: 170–172.

COCKERELL, T.D.A. 1924. Fossils in the Ondai Sair Formation, Mongolia. — *Bull. Amer. Mus. Nat. Hist.* **51**: 129–144.

COMSTOCK, J.H. & NEEDHAM, J.G. 1898–1899. The wings of insects. — *Amer. Nat.* **32** (373): 43–48; (374): 81–89; (376): 231–257; et al.

CRAMPTON, G.G. 1914. The ground plan of a typical thorax segment in winged insects. — *Zool. Anz.* **44**: 56–67.

CRAMPTON, G.C. 1916. The lines of descent of the lower pterygotan insects. — *Entomol. News*, **27** (6): 244–258; (7): 297–307.

CRAMPTON, G.C. 1928. The grouping of the insect orders and their lines of descent. — *Entomologist* **61**: 85.

CRAMPTON, G.C. 1938. The interrelationships and lines of descent of living insects. — *Psyche* **45** (4): 165–181.

CRASS, R.S. 1947a. The May-flies (Ephemeroptera) of Natal and the Eastern Cape. — *Ann. Natal Mus.* **11**: 37–100.

CRASS, R.S. 1947b. Mayflies (Ephemeroptera) collected by J. Omer-Cooper in the Eastern Cape Provinz, South Africa, with a description of a new genus and species (*Notonurus cooperi*). — *Proc. Roy. Entomol. Soc. London* (B) **16**: 124–128.

CURTIS, J. 1834. Descriptions of some nondescribt British species of May-flies of Anglers. — *Lond. Edinb. Phil. Mag.* **4**: 120–122.

CUVEIR, G. 1798. *Tableau élementaire de l'histoire naturelle des animaux.* — Paris, Baudouin, 710 p., 28 pl.

DA-SILVA, E.R. 1992. Description of the nymph of *Homoeoneuria (Notochoa) fittkaui* Pescador & Peters 1980 from northeastern Brazil (Ephemeroptera, Oligoneuriidae, Oligoneuriinae). — *Rev. Brasl. Entomol.* **36** (3): 693–696.

DAKKI, M. & GIUDICELLI, J. 1980. Ephéméroptères d'Afrique du Nord. 2 – Description d' *Oligoneuriella skoura* n.sp. et d' *Oligoneuriopsis skountate* n.sp. avec notes sur leur ecologie (Ephem., Oligoneuriidae). — *Bull. Inst. Sci.*, Rabat, **4**: 13–28.

DANG, N.T. 1967. [New genera, new species of the invertebrate fauna of fresh and brackish waters of North Vietnam]. — *Sinh Vat-Dia Hoc* **6**: 155–165.

DAY, W.C. 1952. New species and noted on California mayflies (Ephemeroptera). — *Pan-Pacif. Entomol.* **28** (1): 17–39.

DAY, W.C. 1953. A new mayfly genus from California (Ephemeroptera). — *Pan-Pacif. Entomol.* **29**: 19–24.

DEMOULIN, G. 1951. A propos de *Metretopus goetghebueri* Lestage 1938, et des Metretopodidae (Insectes Ephéméroptères). — *Bull. Inst. Roy. Sci. Nat. Belg.* **27** (49): 1–11.

DEMOULIN, G. 1952a. Sur deux Palingeniidae (Insectes Ephéméroptères) mal connus. — *Bull. Inst. Roy. Sci. Nat. Belg.* **28** (13): 1–11.

DEMOULIN, G. 1952b. Les *Behningia* Lestage 1929, et leur position dans la classification des Ephéméres. — *Bull. Inst. Roy. Sci. Nat. Belg.* **28** (21): 1–15.

DEMOULIN, G. 1952c. Sur les affinities siphlonuridiennes du génre *Metreletus* Demoulin 1938 (Insectes Ephéméroptères). — *Bull. Inst. Roy. Sci. Nat. Belg.* **28** (31): 1–11.

DEMOULIN, G. 1952d. Essai de nouvelle clé pour le détermination des Oligoneuriidae (Insectes Ephéméroptères). — *Bull. Inst. Roy. Sci. Nat. Belg.* **28** (43): 1–22.

DEMOULIN, G. 1952e. Contibution à l'étude des Ephoronidae

Euthyplociinae (Insectes Ephéméroptères). — *Bull. Inst. Roy. Sci. Nat. Belg.* **28** (45): 1–22.

DEMOULIN, G. 1953a. Les Chromarcynae subfam. nov., Ephéméroptères Oligoneuriides orientaux. — *Bull. Inst. Roy. Sci. Nat. Belg.* **29** (17): 1–12.

DEMOULIN, G. 1953b. A propos des *Polyplocia* de Borneo (Insectes Ephéméroptères). — *Bull. Inst. Roy. Sci. Nat. Belg.* **29** (17): 1–12.

DEMOULIN, G. 1954a. Description préliminaire d'un type larvaire nouveau d'Ephéméroptères Tricorythidae du Congo Belge. — *Bull. Inst. Roy. Sci. Nat. Belg.* **30** (6): 1–4.

DEMOULIN, G. 1954b. Une question de nomenclature: *Prosopistoma foliaceum* (Fourcroy) ou *Binoculus pennigerus* (Müller)? —*Bull. & Ann. Soc. Roy. Entomol. Belg.* **90** (3–4): 99–103.

DEMOULIN, G. 1954d. Recherches critiques sur les Ephéméroptères Tricorythidae d'Afrique et d'Asie. — *Bull. & Ann. Soc. Entomol. Belg.* **90** (9–10): 264–277.

DEMOULIN, G. 1954e. Les Ephéméroptères jurassiques du Sinkiang. — *Bull. & Ann. Soc. Entomol Belg.* **90** (11–12): 322–326.

DEMOULIN, G. 1954f. Essai sur quelques Ephéméroptères fossils adultes. — *Volume Jubilaire Victor van Straelen* (Bruxelles) **1**: 549–574.

DEMOULIN, G. 1955b. Ephéméroptères nouveaux ou rares du Chili. — *Bull. Inst. Roy. Sci. Nat. Belg.* **31** (22): 1–15.

DEMOULIN, G. 1955c. Ephéméroptères nouveaux ou rares du Chili. II. — *Bull. Inst. Roy. Sci. Nat. Belg.* **31** (58): 1–16.

DEMOULIN, G. 1955d. *Melanemerella brasiliana* Ulmer, Ephémérellide ou Tricorythide? (Ephemeroptera). — *Bull. & Ann. Soc. Roy. Entomol. Belg.* **91** (7–8): 214–216.

DEMOULIN, G. 1955e. Sur une larve siphlonuridienne d'Ephemere fossil du Bresil. — *Bull. & Ann. Soc. Roy. Entomol. Belg.* **91** (11–12): 270.

DEMOULIN, G. 1955f. *Afromera* gen. nov., Ephemeridae de la faune éthiopienne (Ephemeroptera). — *Bull. & Ann. Soc. Roy. Entomol. Belg.* **91** (11–12): 291–295.

DEMOULIN, G. 1955g. Quelques remarques sur les composantes de la famille Ametropodidae (Ephemeroptera). — *Bull. & Ann. Soc. Roy. Entomol. Belg.* **91** (11–12): 342–346.

DEMOULIN, G. 1955h. Bréve note sur la position systématique de *Siphluriscus chinensis* Ulmer (1920) (Ephemeroptera). — *Bull. Inst. Roy. Sci. Nat. Belg.* **31** (20): 1–32.

DEMOULIN, G. 1955i. Contribution à l'étude morphologique, systematique et phylogenique des éphéméroptères jurassiques d'Europe central. — *Bull. Inst. Roy. Sci. Nat. Belg.* **31** (39): 1–14.

DEMOULIN, G. 1956a. *Electrogenia dewalschei* n. gen. n. sp., Ephéméroptère fossile de l'ambre. — *Bull. & Ann. Soc. Roy. Entomol. Belg.* **92** (3–4): 95–100.

DEMOULIN, G. 1956b. Nouvelles recherches sur *Triplosoba pulchella* (Brongniart) (Insectes Ephéméroptères). — *Bull. Inst. Roy. Sci. Nat. Belg.* **32** (14): 1–8.

DEMOULIN, G. 1956c. Ephemeroptera. — In: *Exploration hydrobiologioque du Tanganika (1946–1947). Resultats scientifiques* **3** (7): 1–24.

DEMOULIN, G. 1956d. Note sur un "Lepisme" du Miocene de Florissant. — *Bull. & Ann. Soc. Roy. Entomol. Belg.* **92** (9–10): 263–264.

DEMOULIN, G. 1957a. Remarques critiques sur la position systématique des *Ichthybotes* Eaton, Ephéméroptères de Nouvelle-Zélande. — *Bull. & Ann. Soc. Roy. Entomol. Belg.* **93** (11–12): 335–337.

DEMOULIN, G. 1957b. A propos de deux insectes éocénes. — *Bull. Inst. Roy. Sci. Nat. Belg.* **33** (45): 1–4.

DEMOULIN, G. 1958. Nouveau schema de classification des Archodonates et des Ephéméroptères. — *Bull. Inst. Roy. Sci. Nat. Belg.* **34** (27): 1–19.

DEMOULIN G. 1959. Une curieuse larve d'Ephéméroptères de l'Angola portugais. — *Bull. & Ann. Soc. Roy. Entomol. Belg.* **95** (7–8): 249–252.

DEMOULIN, G. 1961. A propos des donées récentes sur la "*Caenis" maxima* Joly (= *Neoephemera maxima*). — *Bull. & Ann. Soc. Roy. Entomol. Belg.* **97** (1–2): 63–68.

DEMOULIN, G. 1964a. Mission H.G. Amsel en Afghanistan (1956). Ephemeroptera). — *Bull.& Ann. Soc. Roy. Entomol. Belg.* **100** (28): 351–363.

DEMOULIN, G. 1964b. Ephemeroptera. — In: *Parc National de l'Upemba. Mission G.F. de Witte* **68** (2): 13–27.

DEMOULIN, G. 1965a. Contribution à la connaissance des Ephéméroptères de l'ambre oligocene de la Baltique. — *Entomologiske Meddelelser* **34**: 143–153.

DEMOULIN, G. 1965b. Contribucion à l'étude des Palingeniidae (Insecta, Ephemeroptera). *Nova Guinea*, N.S., Zoology **10** (33): 305–344.

DEMOULIN, G. 1965c. LXXXVIII. — Ephemeroptera. — *Ann. Mus. Roy. Afr. Centr.*, in-8, Zool. 138: 91–114.

DEMOULIN, G. 1966a. Contribution à l'tude des Ephéméroptères du Surinam. — *Bull. Inst. Roy. Sci. Nat. Belg.* **42** (37): 1–22.

DEMOULIN, G. 1966b. Fixation des types de *Plethogenesia papuana* (Eaton) (Insecta, Ephemeroptera). — *Bull. Inst. Roy. Sci. Nat. Belg.* **42** (40): 1–3.

DEMOULIN, G. 1966c. Contribution à l'étude des Euthyplociidae III (Insectes Ephéméroptères). — *Zool. Mededelingen* **41** (7): 137–141.

DEMOULIN, G. 1966d. Quelques Ephéméroptères nouveaux de Madagascar. — *Ann. Soc. Entomol. France* (N.S) **2** (3): 711–717.

DEMOULIN, G. 1966e. Contribution à l'étude des Euthyplociidae [Ephemeroptera]. IV. Un nouveau génre de Madagascar. — *Ann. Soc. Entomol. France* (N.S.) **2** (4): 941–949.

DEMOULIN, G. 1967a. Redescription de l'holotype ♀ imago des *Chromarcys magnifica* Navas et discussion des affinites phyletiques de génre *Chromarcys* Navas (Ephemeroptera, Chromarcyinae). — *Bull. Inst. Roy. Sci. Nat. Belg.* **43** (31): 1–10.

DEMOULIN, G. 1968a. Deuxieme contibution à la connaissance des Ephéméroptères de l'ambre oligocene de la Baltique. — *Deutsche Entomol. Zeitschr.* (N.F.) **15** (1–3): 233–276.

DEMOULIN, G. 1968b. Les larves des Ephemeridae (Ephemeroptera) d'Afrique. — *Bull. Inst. Roy. Sci. Nat. Belg.* **44** (10): 1–3.

DEMOULIN, G. 1968c. Remarques sur la position systématique de deux éphéméroptères du Jurassique inférieur de Siberie orientale. — *Bull. Inst. Roy. Sci. Nat. Belg.* **44** (18): 1–8.

DEMOULIN, G. 1968d. Quelques Ephéméroptères nouveaux de Madagascar. II. — *Bull. Inst. Roy. Sci. Nat. Belg.* **44** (32): 1–9.

DEMOULIN, G. 1969. Le génre *Eatonica* Navas (Ephemeroptera, Ephemeridae) existe-t-il à Madagascar? — *Bull. Inst. Roy. Sci. Nat. Belg.* **45** (6): 1–9.

DEMOULIN, G. 1970a. Troisième contribution à la connaissance des Ephéméroptères de l'ambre oligocène de la Baltique. — *Bull. Inst. Roy. Sci. Nat. Belg.* **46** (2): 1–11.

DEMOULIN, G. 1970b. Remarques critiques sur des larves "éphéméromorphes" du Permien. — *Bull. Inst. Roy. Sci. Nat. Belg.* **46** (3): 1–10.

DEMOULIN, G. 1970c. Contribution à l'étude morphologique systématique et phylogenique des Ephéméroptères jurassiques d'Europe centrale. V. Hexagenitidae = Paedephemeridae (syn. nov.). — *Bull. Inst. Roy. Sci. Nat. Belg.* **46** (4): 1–8.

DEMOULIN, G. 1970d. Contribution à la connaissance des Ephéméroptères du Miocène I. *Siphlurites explanatus* Cockerell. — *Bull. Inst. Roy. Sci. Nat. Belg.* **46** (5): 1–4.

DEMOULIN, G. 1970e. Ephemeroptera des faunes éthiopienne et malgache. — *South African Animal Life (Results of the Lunds Univ. Exped. 1950–1951)* **14**: 34–170.

DEMOULIN, G. 1973. Ephéméroptères de Madagascar. III. — *Bull. Inst. Roy. Sci. Nat. Belg.* **49** (7): 1–20.

DEMOULIN, G. 1974. Remarques critiques sur les Acanthametropodinae et sur certaines formes affines (Ephemeroptera Siphlonuridae). — *Bull. Inst. Roy. Sci. Nat. Belg.* **50** (2): 1–5.

DEMOULIN, G. 1975. Remarques sur la nervation alaire des Oligoneuriidae (Ephemeroptera). — *Bull. Inst. Roy. Sci. Nat. Belg.* **5** (1): 1–4.

DOMÍNGUEZ, E. 1987. Dos nuevos combinaciones de Polymitarcyidae (Ephemeroptera), descriptar par L. Navas. — *Physis (Buenos Aires)* (sect. B) **45** (108): 9–12.

DOMÍNGUEZ, E. 1988. *Asthenopus gilliesi* sp.n. y su importanicia la taxonomia de la subfamilia Asthenopodinae (Ephemeroptera: Polymitarcidae). — *An. Mus. Hist. Nat. Valpso* **19**: 21–26.

EATON, A.E. 1866. Notes on some species of the orthopterous genus *Cloeon*, Leach (as limited by M. Pictet). — *Ann. & Mag. Nat. Hist.* (3) **18**: 145–148.

EATON, A.E. 1868a. Remarks upon the homologies of the ovipositor. — *Trans. Entomol. Soc. London*: 141–144. (Cit. ex: Eaton 1883–1888; Hubbard 1990).

EATON, A.E. 1868b. An outline of a re-arrangement of the genera of Ephemeridae. — *Entomol. Month. Mag.* **5**: 82–91.

EATON, A.E. 1869. On *Centroptilum*, a new genus of the Ephemeridae. — *Entomol. Month. Mag.* **6**: 131–132.

EATON, A.E. 1871. A monograph on the Ephemeridae. — *Trans. Entomol. Soc. London*: 1–164.

EATON, A.E. 1881. An announcement of new genera of the Ephemeridae. — *Entomol. Month. Mag.* **17**: 191–197; **18**: 21–27 (Cit. ex: Eaton 1883–1888, Hubbard 1990).

EATON, A.E. 1882. An announcement of new genera of the Ephemeridae. — *Entomol. Month. Mag.* **18**: 207–208 (Cit. ex: Eaton 1883–1888, Hubbard 1990).

EATON, A.E. 1883–1888. A revisional monograph of recent Ephemeridae or mayflies. — *Trans. Linn. Soc. London* (2) **3**: 352p., 65 pl.

EATON, A.E. 1892a. Ephemeridae. — In: *Biologia Centrali-Americana*. Zoologia. Class Insecta. Order Neuroptera: 1–16.

EATON, A.E. 1892b. New species of Ephemeridae from the Tenasserim Valley. — *Trans. Entomol. Soc. London* **2**: 185–190, 1 Fig.

EATON, A.E. 1899. An annotated list of the Ephemeridae of New Zealand. — *Trans. Entomol. Soc. London*: 285–293, 1 pl.

EATON, A.E. 1901. Ephemeridae collected by Herr E. Strand in South and Arctic Norway. — *Entomol. Month. Mag.* **37**: 252–255.

EDMUNDS, G.F., JR. 1951. New species of Utah mayflies. I. Oligoneuriidae (Ephemeroptera). — *Proc. Entomol. Soc. Washington* **53** (6): 327–331.

EDMUNDS, G.F., JR. 1957. The systematic relationships of the Paleantarctic Siphlonuridae (including Isonychiidae) (Ephemeroptera). — *Proc. Entomol. Soc. Washington* **59**: 245–246.

EDMUNDS, G.F., JR. 1959. Subgeneric groups within the mayfly genus *Ephemerella* (Ephemeroptera: Ephemerellidae). — *Ann. Entomol. Soc. Amer.* **52**: 543–547.

EDMUNDS, G.F., JR. 1960. Two generic synonyms in the Siphlonuridae (Ephemeroptera). — *Bull. Broocklyn Entomol. Soc.* **55**: 24.

EDMUNDS, G.F., JR. 1961. A key to the genera of known nymphs of the Oligoneuriidae (Ephemeroptera). — *Proc. Entomol. Soc. Washigton* **63** (4): 255–256.

EDMUNDS, G.F., JR. 1962. The type localities of the Ephemeroptera of North America north of Mexico. — *Univ. Utah Biol. Ser.* **12** (5): 1–39.

EDMUNDS, G.F., JR. 1971. A new name for a subgeneric homonym in *Ephemerella* (Ephemeroptera: Ephemerellidae). — *Proc. Entomol. Soc. Washington* **73**: 152.

EDMUNDS, G.F., JR. 1975. Phylogenetic biogeography of mayflies. — *Ann. Mo. bot. Gdn.* **62**: 251–263.

EDMUNDS, G.F., JR. & ALLEN, R.K. 1957. A check list of the Ephemeroptera of North America north to Mexico. — *Ann. Entomol. Soc. Amer.* **50** (4): 317–324.

EDMUNDS, G.F., JR. & ALLEN R.K. & PETERS, W.L 1963. An annotated key to the nymphs of the families and subfamilies of mayflies (Ephemeroptera). — *Univ. Utah Biol. Ser.* **13** (1): 1–49

EDMUNDS, G.F., JR. & BERNER L. & TRAVER J.R. 1958. North American mayflies of the family Oligoneuriidae. — *Ann. Entomol. Soc. Amer.* **51**: 375–382.

EDMUNDS, G.F., JR. & JENSEN, S.L. 1974. A new genus and subfamily of North American Heptageniidae (Ephemeroptera). — *Proc. Entomol. Soc. Washington* **76**: 495–497.

EDMUNDS, G.F., JR. & JENSEN, S.L. & BERNER, L. 1976. *The mayflies of North and Central America*. — Univ. Minnesota Press., 330 pp.

EDMUNDS, G.F., JR. & KOSS, R.W. 1972. A review of the Acanthametropodinae with a description of a new genus (Ephemeroptera: Siphlonuridae). — *Pan-Pacif. Etomol.* **48**: 136–144.

EDMUNDS, G.F., JR. & MURVOSH, C.M. 1995. Systematic changes in certain Ephemeroptera studied by R.K. Allen. — *Pan-Pacific Entomologist* **71** (3): 157–160.

EDMUNDS, G.F., JR. & TRAVER, J.R. 1954a. An outline of a reclassification of the Ephemeroptera. — *Proc. Entomol. Soc. Washington* **56**: 236–240

EDMUNDS, G.F., JR. & TRAVER, J.R. 1954b. The flight mechanics and evolution of the wings of Ephemeroptera, with notes on the archetype insect wing. — *Journ. Washington Acad. Sci.* **44**: 390–400.

EDMUNDS, G.F., JR. & TRAVER, J.R. 1959. The classification of Ephemeroptera. I. Ephemeroidea: Behningiidae. — *Ann. Entomol. Soc. Amer.* **52**: 43–51.

D'EICHWALD, E. 1864. Sur un terrain jurassique a poissons et Insectes d'eau douce de la Siberie orietale. — *Bull. Soc. Geol. France* (2) **21**: 19–25.

ELOUARD, J.-M. 1986a. Ephemeres d'Afrique de l'Ouest. Le génre *Eatonica* (Ephemeridae). — *Rev. Hydrobiol. trop.* **19** (2): 87–92.

ELOUARD, J.-M. 1986b. Ephéméres d'Afrique de l'Ouest. Le génre *Afromera* (Ephemeridae). — *Rev. Hydrobiol. trop.* **19** (3–4): 169–176.

ELOUARD, J.-M. & GILLIES, M.T. 1989. West African Ephemeroptera. The Genus *Machadorythus* (Tricorythidae). — *Aquatic Insects* **11** (1): 1–10.

ELOUARD, J.-M. & OLIARINONY, R. 1997. Biodiversité aquatique de Madagascar. 6. — *Madacassorythus* un nouveau génre de Tricorythidae definissant la nouvelle sous-famille des Madecassorythinae (Ephemeroptera Pannota). — *Bull. Soc. Entomol. France* **102** (3): 225–232.

ELOUARD, J.-M. & OLIARINONY, R. & SARTORI, M. 1998. Biodiversité aquatique de Madagascar. 9. Le génre *Eatonica* Navás (Ephemeroptera, Ephemereidae). — *Mitt. Scweizer. Entomol. Ges.* **71**: 1–9.

ELOUARD, J.-M. & SARTORI, M. 1997. *Proboscidoplocia* a singular plural (Ephemeroptera: Polymitarcyidae: Euthypliciinae). — In: P. Landolt & M. Sartori (eds.). *Ephemeroptera & Plecoptera. Biology-Ecology-Systematics* (Proc. 8th Int. Conf. Ephemeroptera & 12th Int. Symp. Plecoptera 14–20 August 1995, Losanne, Switzerland). Mauron+Tinguely & Lacht SA, Fribourg / Switzerland: 439–448.

ELOUARD, J.-M. & SARTORI, M. 2001. A revision of the Malagasy genus *Madecocercus* Malzacher 1995 (Ephemeroptera, Caenidae). — *Bull. Soc. Vaud. Sci. Nat.* **87** (3): 229–235.

ELPERS, C. & TOMKA, I. 1992. Structur der Mundwerkzeuge und Nahrungsaufnahme bei den Larven von *Oligoneuriella rhenana* Imhoff (Ephemeroptera: Oligoneuriidae). — *Mitt. Schweiz. Entomol. Ges.* **65**: 119–139.

ELPERS, C. & TOMKA, I. 1994a. Structure of mouthparts and feedind habits of *Potamanthus luteus* (Linne) (Ephemeroptera: Potamanthidae). — *Arch. Hydrobiol.* Suppl. 99 (1/2): 73–96.

ELPERS, C. & TOMKA, I. 1994b. Mouthparts of the predaceous larvae of the Behningiidae (Insecta: Ephemeroptera). — *Arch. Hydrobiol.* Suppl. 99 (4): 381–413.

ENGBLOM, E. & LINGDELL, P.-E. & NILSSON, A.N. & SAVOLAINEN, E. 1993. The genus *Metretopus* (Ephemeroptera, Siphlonuridae) in Fennoscandia – identification, faunistics and natural history. — *Entomol. fenn.* **4** (4): 213–222.

ESBEN-PETERSEN, T. 1909. New Ephemeridae from Denmark, Arctic Norway and the Argentine Republik. — *Deutsche Entomol. Zeitschr.* 551–556.

ESBEN-PETERSEN, T. 1913. Ephemeridae from South Africa. — *Ann. S. Afr. Mus.* **10** (6): 177–187.

FIZAINE, G. 1931. Ephéméroptère nouveau appartenant au génre *Ameletus, Ameletus Hessei* n.sp. — *Bull. Soc. Zool. France* **56**: 25–29.

FLOWERS, R.W. 1980. Two new genera of nearctic Hepageniidae (Ephemeroptera). — *Florida Entomologist* **63** (3): 296–307.

FLOWERS, R.W. 1982. Review of the genus *Macdunnoa* (Ephemeroptera: Heptageniidae) with description of a new species from Florida. — *The Great Lakes Entomologist* **5** (1): 25–30.

FLOWERS, R.W. & PESCADOR, M.L. 1984. A new *Afronurus* (Ephemeroptera: Heptageniidae) from the Philippines. — *Int. Journ. Entomol.* **26** (4): 362–365.

FONTAINE, J. 1968. Contribution à l'étude des Ephéméroptères malgaches: la superfamille des Ephemeroidea. — *Bull. Mens. Soc. Linn. Lyon* **37** (6): 228–242.

FROECHLICH, C.G. 1964. The feeding apparatus of the nymph of *Athroplea congener* Bengtsson. — *Opuscula Entomol.* **29**: 188–208.

FUJIYAMA, I. & NOMURA, T. 1986. Early Miocene insect fauna of Hachiya Formation, Gifu Prefecture, Japan. — *Bull. Mizunami Fossil Mus.* No.13: 1–14, Pl.1–2.

GAINO, E. & REBORA, M. 2002. Sensory receptors on the forceps of the male imago and subimago of *Ecdyonyrus venosus* (Ephemeroptera: Heptageniidae). — *Ann. Entomol. Soc. Amer.* **95** (2): 244–249.

GILLIES, M.T. 1954. The adult stages of *Prosopistoma* Latreille (Ephemeroptera), with descriptions of two new species from Africa. — *Trans. Roy. Entomol. Soc. London* **105**: 355–372.

GILLIES, M.T. 1956. A supplementary note on *Prosopistoma* Latreille (Ephemeroptera). — *Proc. Roy. Entomol. Soc. London* **31** (10/12): 165–166.

GILLIES, M.T. 1960. A new genus of Tricorythidae (Ephemeroptera) from East Africa. — *Proc. Roy. Entomol. Soc. London* (B) **29**: 35–40.

GILLIES, M.T. 1974. Three new species of *Elassoneuria* from tropical Africa. — *Journ. Entomol.* B43 (1): 73–82.

GILLIES, M.T. 1977. A new genus of Caenidae (Ephemeroptera) from East Afrika. — *Journ. Nat. Hist.* **11**: 451–455.

GILLIES, M.T. 1980a. The African Euthyplociidae (Ephemeroptera) (Exeuthyplociinae subfam. n.). — *Aquatic Insects* **2**: 217–224.

GILLIES, M.T. 1982. A second large-eyed genus of Caenidae (Ephemeroptera) from Africa. — *Journ. Nat. Hist.* **16**: 15–22.

GILLIES, M.T. & ELOUARD, J.-M. 1990. The mayfly-mussel association, a new example from the river Niger basin. — In: J.C. Campbell (ed.). *Mayflies and Stoneflies* (Proc. 5th Int. Conf. Ephemeroptera & 9th Int. Symp. Plecoptera 18–24 February 1987, Marysville, Australia). Kluwer Acad. Publ.: 289–297.

GOSE, K. 1969. Mayflies (Ephemeroptera) from Thailand. — *Nature Life S.E. Asia* **6**: 125–138 (Cit. ex: Allen 1975).

GOTTANKA, J. & BÜNING, J. 1993. Mayflies (Ephemeroptera), the most "primitive" winged insects, have telotrophic meroistic ovaries. — *Roux's Arch. Dev. Biol.* **203** (1–2): 18–27.

GRANDI, M. 1942–1943. Contribuli allo studio degli Efemeroidei Italiani. V. Reperti su *Habrophlebia fusca* (Curtis). — *Boll. Inst. Entomol. R. Univ. Bologna.* **14**: 114–130.

GRANDI, M. 1953. Contributi allo studio degli Efemeroidei italiani. XVII. Ecdyonuridae. — *Boll. Inst. Entomol. Univ. Bologna* **19**: 307–386.

GRANDI, M. 1960. Ephemeroidea. — *Fauna d'Italia.* Bologna: 1–472.

GRAVELY, F.M. 1920. Notes on some Asiatic species of *Palingenia* (order Ephemeroptera). — *Rec. Indian Mus.* **18**: 137–143, pl. 18–20.

GROS, A.J. & LESTAGE, J.A. 1927. Contribution à l'étude des larves des Ephéméroptères. Serie IV. Le groupe Euthyplocien. — *Ann. Biol. Lacustre* **15**: 119–162, 1 pl.

HAECKEL, E. 1866. *Generale Morphologie der Organismen.* Berlin. Bd.2: 1–574.

HAGEN, H.A. 1855. Die Ephemeriden-Gattung *Oligoneuria.* — *Entomol. Zt. Stettin* 16: 262–270, tab. 1.

HAGEN, H.A. 1868.: On *Lachlania abnormis*, a new genus and species from Cuba belonging to the Ephemerina. — *Proc. Boston Soc. Nat. Hist.* **11**: 372–375. (Cit. ex: Hubbard 1990).

HAMILTON, K.G.A. 1971–1972. The insect wing. — *Journ. Kansas Entomol. Soc.* **44** (4): 421–433; **54** (1): 54–58; (2): 145–162.

HANDLIRSCH, A. 1903. Zur Pylogenie der Hexapoden. — *Sitzunsber. K. Akad. Wiss. Wien. Mathem.-Natur. Klasse* **112**, Abt. 1: 716–738.

HANDLIRSCH, A. 1904a. Über einige Insektenreste aus der Permformation Russlands. — *Mem. Acad. St.Petersburg* (8) **16** (5): 1–8 1 tab.

HANDLIRSCH, A. 1904b. Zur Systematik der Hexapoden. — *Zoologischer Anzeiger* **27** (23/24): 733-759.

HANDLIRSCH, A. 1906–1908. *Die fossile Insekten und die Phylogenie der rezenten Formen.* Ein Handbuch für Palaontologen und Zoologen. — Verlag Wilhelm Engelmann, Leipzig: 1–1430.

HANDLIRSCH, A. 1911. New Paleozoic insects from the vicinity of Mazon Creek, Illinois. — *Amer. Journ. Sci.* (4), **31**: 297–326, 353–377.

HANDLIRSCH, A. 1918. Fossile Ephemeridenlarven aus dem Buntsandstein der Vogesen. — *Verhdl. Kaiserl.-Konigl. Zool.-Bot. Ges. in Wien* **68**: 112–114.

HANDLIRSCH, A. 1919. Revision der Paläozoischem Insecten. *Denkschriften Akad. Wiss. Wien. Mat.-Natur. Kl.* **96**: 1–82 [511–592].

HANDLIRSCH, A. 1925. Geschichte, Literatur, Technik, Paläontologie, Phylogenie, Systematik. — In book: Schröder Ch. (ed.). *Handbuch der Entomologie*, Bd.3, Jena: 1201 S.

HARKER, J.E. 1954. The Ephemeroptera of Eastern Australia. — *Trans. Roy. Entomol. Soc. London* (B) **105** (12): 241–268.

HARKER, J.E. 1957. Some new Australian Ephemeroptera. — *Proc. Roy. Entomol. Soc. London* (B) **26**: 63–78.

HAUPT, H. 1956. Beitrag zur Kenntnis der eozanen Arthropodenfauna des Geiseltales. — *Nova Acta Leopoldina* N.F. **18** (128): 1–90.

HENRY, B.C., JR. 1986. Mayflies (Ephemeroptera) of the Concho River, Texas. — *The Southeastern Naturalist* **31** (1): 15–21.

HIDEUX, P. 1987. Ephéméres d'Afrique de l'Ouest. Le génre *Exeuthyplocia* (Exeuthyplociidae). — *Rev. Hydrobiol. trop.* **20** (2): 113–119.

HITCHINGS, T.R. & STANICZEK, A. 2003. Nesameletidae (Insecta: Ephemeroptera). — *Fauna of New Zealand* **46**: 1–72.

HOFMANN, C. & SARTORI, M. & TOMAS, A. 1999. Les Ephéméroptères (Ephemeroptera) de la Guadeloupe (petites Antilles francaises). — *Mem. Soc. Vaudoise Sci. Nat.* **20** (1): 1–96.

HONG, Y.-CH. 1979. On Eocene *Philolimnias* gen.nov. (Ephemeroptera, Insecta) in amber from Fushung Coalfield, Liaoning Province. — *Scientia Sinica* **22** (3): 331–339, Pl. I–II.

HSU, YIN-CHI 1932. Two new species of mayflies from China (order Ephemeroptera). — *Peking Nat. Hist. Bull.* **6** (2): 39–41, 1 pl.

HSU, YIN-CHI 1936. New Chines mayflies from Kiangsi Province (Ephemeroptera). — *Peking Nat. Hist. Bull.* **10** (4): 319–325.

HSU, YIN-CHI 1937. The mayflies of China. — *Peking Nat. Hist. Bull.* **11** (2): 129–148.

HUBBARD, M.D. 1979a. Designation of type species for 2 genera of Siphlonuridae (Ephemeroptera: Siphlonuridae). — *Florida Entomologist* **62** (4): 412.

HUBBARD, M.D. 1979b. Genera and subgenera of Recent Ephemeroptera. — *Eatonia*, Suppl. 2: 1–8.

HUBBARD, M.D. 1981. Type-species designation for the Jurassic Mayfly genus *Mesephemera* (Ephemeroptera: Mesephemeridae). — *Great Lakes Entomol.* **14**: 69.

HUBBARD, M.D. 1982. Catalog of the Ephemeroptera: Family-group taxa. — *Aquatic Insects* **4** (1): 49–53.

HUBBARD, M.D. 1984. A revision of the genus *Povilla* (Ephemeroptera: Polymitarcyidae). — *Aquatic Insects* **6** (1): 17–35.

HUBBARD, M.D. 1990. Mayflies of the world. A Catalog of the Family and Genus Group Taxa (Insecta: Ephemeroptera). — *Flora & Fauna Handbook* No.8. The Sandhill Crane Press, Inc. Gainesville Flo. 119 p.

HUBBARD, M.D. & DOMINGUEZ, E. 1988. Synonymy of the Neotropical mayfly genera *Asthenopus* and *Asthenopodes* (Ephemeroptera: Polymitarcyidae: Asthenopodinae). — *Florida Entomolopgist* **71** (2): 207–210.

HUBBARD, M.D. & KUKALOVA-PECK, J. 1980. Permian mayfly nymphs: new taxa and systematic characters. — In: J.F. Flannagan & K.E. Marshall (eds.). *Advances in Ephemeroptera Biology* (Proc. 3rd Int. Conf. Ephemeroptera, 4–10 July 1979, Winnipeg, Canada). Plenum Press, New York: 19–31.

HUBBARD, M.D. & PETERS, W.L. 1978. A catalogue of the Ephemeroptera of the Indian Subregion. — *Oriental Insects*, Suppl. 9: 1–43.

HUBBARD, M.D. & RIEK, E.F. 1978. New name for a triassic mayfly from South Africa (Ephemeroptera). — *Psyche* **83** (3–4): 260–261.

HYATT, A. & ARMS, J. M. 1891. A novel diagrammatic representation of the orders of insects. — *Psyche* **6** (177): 11–13 (reprinted from: *Guides for Science Teaching*, No. 8. Insecta. By A. Hyatt & J. M. Arms).

IDE, F.P. 1935. Life history notes on *Ephoron, Potamanthus, Leptophlebia* and *Blasturus* with descriptions (Ephemeroptera). — *Canad. Entomol.* **67**: 113–125.

[IKONOMOV, P.] ИКОНОМОВ П. 1961. Еднодниввите (Ephemeroptera) на Македонија. Сем. Ephemerellidae [Eintarsfliegen (Ephemeroptera) Mazedoniens. Fam. Ephemerellidae]. — *Изданија на Природонаучниот Музеј*, Скопје [*Musei Macedonici Scientiarum Naturalium*, Skopje] **8** (3/71): 53–74.

IKONOMOV, P. 1962. *Oreianthus macedonicus* (Ulmer) (= *Rhoenanthus* Ulmer). — *Fragm. balcan.* **4**: 1–10.

ILLIES, J. 1960. Penturoperlidae, eine neue Plecopteren-familie. — *Zool. Anz.* **164**: 26–41.

IMANISHI, K. 1940. [Ephemeroptera of Manchoukuo, Inner Mongolia, and Chosen] (in Japanese). — *Rep. Limnobiol. Survey. Kwant. and Manch.*: 169–263.

INTERNATIONAL CODE OF ZOOLOGICAL NOMENCLATURE / CODE I NTERNATIONAL DE NOMENCLATURE ZOOLOGIQUE. 4th edition. — *Int. Trust for Zool. Nomencl.*, London 1999.

ISHIWATA, SHIN-ICHI 1996. A study of the genus *Ephoron* from Japan (Ephemeroptera, Polymitarcyidae). — *Canad. Entomol.* **128**: 551–572.

JACOB, U. 1974a. Die bisher nachgewiesenen Ephemeropteren der Deutschen Demokratischen Republik. — *Entomol. Nachr. Dresden* **18** (1): 1–7.

JACOB, U. 1974b. Zur Kenntnis zweier *Oxycypha*-Arten Hermann Burmeisters (Ephemeroptera, Caenidae). — *Reichenbachia* 15: 93–97.

JACOB, U. & DORN, A. & HAYBACH, A. 1996. Systematik und Verbreitung der Gattung *Heptagenia* und nahestehender Taxa in Europa. — *Verhandlungen der Westdeutschen Entomologen Tagung* 1995: 93–105.

JACOB, U. & SARTORI, M. 1984. Die europäischen Arten der Gattung *Habrophlebia* Eaton (Ephemeroptera Leptophlebiidae). — *Entomol. Abh. Tierk. Dresden* **48** (1): 45–52.

[JACOBSON, G.G. & BIANCHI, V.L.] ЯКОБСОНЪ Г.Г., БІАНКИ В.Л. 1905. *Прямокрылыя и ложнос□тчатокрылыя Российской имперіи и сопредельныхъ странъ.* [Orthoptera et Pseudoneuroptera]. S.-Petersburg, Publ. A.F. Devrien: 952 pp.

JACOBUS, L.M. & MCCAFFERTY, W.P., 2001. Contribution to the systematics of Afrobaetodes Demoulin (Ephemeroptera: Baetidae). — *African Entomology* **9** (2): 97–103.

JAZDZEWSKA, T. 1975. *Neoephemera maxima* (Joly 1870) (Ephemeroptera, Neoephemeridae) in Poland. — *Pol. Pismo Entomol.* **45** (2): 227–237.

JELL, P.A. & DUNCAN, P.M. 1986. Invertebrates, mainly insects, from freshwater, Lower Cretaceous, Coonwarra Fossil Bed (Korumburra Group), South Gippsland, Victoria. — *Mem. Ass. Auastralas. Palaeontol.* **3**: 111–205.

JENSEN, S.L. 1974. A new genus of mayflies from North America (Ephemeroptera: Heptageniidae). — *Proc. Entomol. Soc. Washington* **76**: 225–228.

JOLY, E. 1871. Contributions pour servir a l'histoire naturelle des Ephemerines. — *Bull. Soc. Hist. Nat. Toulouse* **4**: 142–151.

JOLY, N. & JOLY, E. 1872. Études sur le pretendu crustace. — *Mem. Acad. Sci., Inscriptions et Belles-Lettres de Toulouse* (7) **4**: 437–438.

JOLY, M. & JOLY, E. 1877. Contributions à l'histoire naturelle et à l'anatomie des Ephemeres. — *Rev. Acad. Sci.* **5**: 305–330 (Cit. ex. Hubbard 1990).

KANG S.C. & YANG, C.T. 1994. Heptageniidae of Taiwan (Ephemeroptera). — *Journ. Taiwan Mus.* **47** (1): 5–36.

KANG, S.C. & YANG, C.T. 1995. Ephemerellidae of Taiwan (Insecta, Ephemeroptera). — *Bull. Nat. Mus. Natur. Sci.* No.5: 95–116.

[KAZLAUSKAS, R.S.] КАЗЛАУСКАС Р.С. 1959. Материалы по фауне поденок (Ephemeroptera) Литовской ССР с описанием нового вида *Eurylphella lithuanica* Kazlauskas sp. n. и имаго *Neoephemera maxima* (Joly) [Materials about the mayflies (Ephemeroptera) in Lithuanian SSR with description of new space *Eurylophella lithuanica* Kazlauskas and imago *Neoephemera maxima* (Joly)]. — *Vilniaus Univ. Mokslo Darbai*, Biol., Geogr., Geol. **23** (6): 157–174.

[KAZLAUSKAS, R.S.] КАЗЛАУСКАС Р.С. 1962. Новые данные по фауне поденок (Ephemeroptera) Прибалтики [Neue Angaben über die Eintagsfliegenfauna (Ephemeroptera) des Baltikums]. — *Гидробиол. исслед. Ин-т Зоол. Бот. Эст. ССР ["Hydrobiol. issled. Inst. Zool. & Bot. Eston.SSR"]* **3**: 147–150.

[KAZLAUSKAS, R.S.] КАЗЛАУСКАС Р.С. 1963. Новые и малоизвестные поденки (Ephemeroptera) фауны СССР [New and little-known may flies (Ephemeroptera) from the USSR]. — *Энтомол. Обозрение* [*Rev. Entomol. URSS*] **42** (3): 582–592.

KAZLAUSKAS, R.S. 1972. Neues über das System der Eintagsfliegen der Familie Baetidae (Ephemeroptera). — *Proc. XIII Int. Congr. Entomol.*, Moscow, 2–9 August 1968, **3**: 337–338.

KEFFERMÜLLER, M. 1957. Kilka rzedkich w Polsce gatunkow jętek (Ephemeroptera) [Some rare species of may-flies (Ephemeroptera) in Poland]. — *Fragm. Faun.* **7** (9): 253–292.

KEFFERMÜLLER, M. 1959. Nowe dane dotyczane jętek z rodzaju *Ametropus* i *Behningia*. — *Pozn. Towarz. Prz. Nauk.* **19** (5): 1–32.

KEFFERMÜLLER, M. 1960. Badnania nad fauna jętek (Ephemeroptera) Wielkopolski. — *Prace Kom. Biol. Poznan Tow. Przyi Nauk*, Wyd. Mat.-Prz. **19** (8): 411–467, pl. 1–11.

KEFFERMÜLLER, M. 1964. Uzupelnienie badan nad fauna jętek (Ephemeroptera) Wielkopolski. — *Badania Fizjograf. Polska Zachodnia* **14**: 69–86.

KEFFERMÜLLER, M. 1967. Badania nad fauna jętek (Ephemeroptera) Wielkopolski. III. — *Badania Fizjograf. Polska Zachodnia* **20**: 15–28.

KELTNER, J.W. & MCCAFFERTY, W.P. 1986. Functional morphology of burrowing in the mayfly *Hexagenia limbata* and *Pentagenia vittigera*. — *Zool. Journ. Linn. Soc.* **87**: 139–162.

KIMMINS, D.E. 1934. Notes on the Ephemeroptera of the Godman and Salvin colllection, with description of two new species. — *Ann. & Mag. Nat. Hist.* (10) **14**: 338–353.

KIMMINS, D.E. 1937. Some new Ephemeroptera. — *Ann. & Mag. Nat. Hist.* (10) **19**: 430–440.

KIMMINS, D.E. 1949. Ephemeroptera from Nyasaland, with descriptions of new species. — *Ann. & Mag. Nat. Hist.* (12) **1** (11): 825–836.

KIMMINS, D.E. 1955. Ephemeroptera from Nyasaland, with description of three new species and some interesting nymphal forms. — *Ann. & Mag. Nat. Hist.* (12) **8** (95): 859–880.

KIMMINS, D.E. 1956. New species of Ephemeroptera from Uganda. — *Bull. Brit. Mus. (Nat. Hist)*, Entomol. **4**: 71–87.

KIMMINS, D.E. 1957. New species of the genus *Dicercomyzon* Demoulin (Ephemeroptera, Fam. Tricorythidae). — *Bull. Brit. Mus. (Nat. Hist.)*, Entomol. **6** (5): 129–136.

KIMMINS, D.E. 1960. The Ephemeroptera types of species described by A.E. Eaton, R. McLachlan and F. Walker, with particular reference to those in the British Museum (Natural History). — *Bull. Brit. Mus. (Nat. Hist.)*, Entomology **9** (4): 269–318.

KLAPÁLEK, F. 1905. Plecopteren und Ephemeriden aus Java, gesammelt von Prof. K. Kraepelin 1904. — *Mitt. Naturhist. Mus. Hamburg* **22**: 103–107.

KLAPÁLEK, F. 1909. I. Ephemerida, Eintagsfliegen. — In: Brauer (ed.). *Die Susswasserfauna Deutschlands* **8**: 1–32.

[KLUGE, N.J.] КЛЮГЕ Н.Ю. 1980. К познанию поденок (Ephemeroptera) Таймырского национального округа [To the knowledge of mayflies (Ephemeroptera) of Taimyr National District]. — *Энтомол. Обозрение* [*Rev. Entomol. URSS*] **59** (3): 561–579.

[KLUGE, N.J.] КЛЮГЕ Н.Ю. 1982. Новые и малоизвестные поденки Дальнего Востока СССР. Семейства Leptophlebiidae и Siphlonuridae (Ephemeroptera) [New and little known mayflies of the USSR Far East. Families Leptophlebiidae and Siphlonuridae (Ephemeroptera)]. — *Вестник Ленинград. ун-та* [*"Vestnik Leningrad. Univ."*] № 9: 112–116.

[KLUGE, N.J.] КЛЮГЕ Н.Ю. 1983a. Новые и малоизвестные поденки Дальнего Востока СССР. Род *Ecdyonurus* (Ephemeroptera, Heptageniidae). [New and little known mayfies of Far East of the USSR. The genus *Ecdyonurus* (Ephemeroptera, Heptageniidae)]. — In: *Экология и систематика пресноводных организмов Дальнего Востока* [*Ecology and systematics of freshwater organisms of Far East*], Vladivostok: 27–36.

KLUGE, N.J. 1983b. A generic revision of the family Heptageniidae (Ephemeroptera). — *IVth Int. Conf. on Ephemeroptera. Abstracts.*

[KLUGE, N.J.] КЛЮГЕ Н.Ю. 1983c. *Rhithrogena eugeniae* sp.n. (Ephemeroptera, Heptageniidae) из Таджикистана [*Rhithrogena eugeniae* sp.n. (Ephemeroptera, Heptageniidae) from Tadjikistan]. — *Зоол. Журнал* [*Zool. Zhurnal*] **62** (7): 1116–1118.

[KLUGE, N.J.] КЛЮГЕ Н.Ю. 1985. О дальневосточных видах поденок группы *aestivalis* рода *Siphlonurus* Etn. (Ephemeroptera, Siphlonuridae) [On Far Eastern mayflies of the group *aestivalis* of the genus *Siphlonurus* Etn. (Ephemeroptera, Siphlonuridae)]. — *Вестник Ленинград. ун-та* [*"Vestnik Lenigrad. Univ."*] № 10: 12–20.

[KLUGE, N.J.] КЛЮГЕ Н.Ю. 1986. Современный вид поденок (Ephemeroptera, Heptageniidae) в балтийском янтаре [Recent mayfly species (Ephemeroptera, Heptageniidae) in Baltic amber]. — *Палеонтол. Журнал* [*Paleontol. Journal*] № 2: 111–112.

[KLUGE, N.J.] КЛЮГЕ Н.Ю. 1987a. Поденки рода *Heptagenia* Walsh (Ephemeroptera, Heptageniidae) фауны СССР. [Mayflies of the genus *Heptagenia* Walsh (Ephemeroptera, Heptageniidae) of the fauna of the USSR]. — *Энтомол. Обозрение* [*Rev. Entomol. URSS*] **66** (2): 302–320; English translation in: *Entomol. Review* 1988, **67** (1): 60–79.

[KLUGE, N.J.] КЛЮГЕ Н.Ю. 1987b. Новый вид поденок (Ephemeroptera, Heptageniidae) из Закавказья [A new species of mayflies (Ephemeroptera, Heptageniidae) from Transcaucasien]. — *Зоол. Журнал* [*Zool. Zhurnal*] **66** (5): 775–777.

[KLUGE, N.J.] КЛЮГЕ Н.Ю. 1988a. Ревизия родов сем. Heptageniidae (Ephemeroptera). 1. Диагнозы триб, родов и подродов подсем. Heptageniinae [Revision of genera of the family Heptageniidae (Ephemeroptera). I. Diagnoses of tribes, genera and subgenera of the subfamily Heptageniinae]. — *Энтомол. Обозрение* [*Rev. Emtomol. URSS*] **67** (2):291–313; English translation in: *Entomol. Review* 1989, **68** (3): 1–24.

[KLUGE, N.J.] КЛЮГЕ Н.Ю. 1988b. Новый вид поденок рода *Ephemera* (Ephemeroptera, Ephemeridae) из Армении [A new mayfly species of the genus *Ephemera* (Ephemeroptera, Ephemeridae) from Armenia]. — *Вестник зоологии* [*Vestnik Zoologii*] № 3: 76-78.

[KLUGE, N.J.] КЛЮГЕ Н.Ю. 1989a. Вопрос о гомологии трахейных жабр и паранотальных выростов личинок поденок и крыльев насекомых в связи с систематикой и филогенией отряда поденок (Ephemeroptera) [The problem of the homology of the tracheal gills and paranotal processi of the mayfly larvae and wings of the insects with reference to the taxonomy and phylogeny of the order Ephemeroptera]. — *Чтения памяти Холодковского: доклады на 41-м ежегодном чтении памяти Н.А. Холодковского 1 апреля 1988 г.*, Ленинград, "Наука". [*"Chteniya pamyati N.A. Kholodkovskogo"*, Leningrad, "Nauka"] **41**: 48–77.

[KLUGE, N.J.] КЛЮГЕ Н.Ю. 1989b. Поденки рода *Heptagenia* Walsh (Ephemeroptera, Heptageniidae) фауны СССР. 2. Дополнения и описание нового вида с Памира [Mayflies of the genus *Heptagenia* Walsh (Ephemeroptera, Heptageniidae) of the USSR fauna. 2. Additions with description of a new species from Pamirs]. — *Вестник. Ленинград. ун-та* [*"Vestnik Leningrad. Univ."*] ser.3, № 4 (24): 3–13.

[KLUGE, N.J.] КЛЮГЕ Н.Ю. 1991a. Поденки рода *Brachycercus*

(Ephemeroptera, Furcatergalia, Caenidae) фауны СССР [Mayflies of the genus *Brachycercus* (Ephemeroptera, Furcatergalia, Caenidae) of the fauna of the USSR]. — *Вестник Зоол.* [*"Vestnik Zool."*] № 2: 14–22.

[KLUGE, N.J.] КЛЮГЕ Н.Ю. 1991b. Кубинские поденки семейства Baetidae (Ephemeroptera). 1. Роды *Callibaetis*, *Cloeodes* и *Paracloeodes* [Cuban mayflies of the family Baetidae (Ephemeroptera). 1. Genera *Callibaetis*, *Cloeodes* and *Paracloeodes*]. — *Зоол. Журнал* [*Zool. Zhurnal*] **70** (12): 128–135.

[KLUGE, N.J.] КЛЮГЕ Н.Ю. 1992a. Кубинские поденки семейства Baetidae (Ephemeroptera). 2. Подроды *Caribaetis* subgen. n. и *Americabaetis* subgen. n. рода *Baetis* s. l. [Cuban mayflies of the family Baetidae (Ephemeroptera). 2. Subgenera *Caribaetis* subgen. n. and *Americabaetis* subgen. n. of the genus *Baetis* s.l.]. — *Зоол. Журнал* [*Zool. Zhurnal*] **71** (4): 13–20.

[KLUGE, N.J.] КЛЮГЕ Н.Ю. 1992b. Кубинские поденки семейства Baetidae (Ephemeroptera). 3. Подрод *Fallceon* рода *Baetis* s. l. [Cuban mayflies of the family Baetidae (Ephemeroptera). 3. Subgenus *Fallceon* of the genus *Baetis* s.l.]. — *Зоол. Журнал* [*Zool. Zhurnal*] **71** (5): 38–48.

KLUGE, N.J. 1992c. Redescription of *Leptohyphes eximius* Etn. and diagnoses of the genera Leptohyphes and Tricorythodes based on structure of pterothorax (Ephemeroptera: Tricorythidae, Leptohyphinae). — *Opuscula zool. fluminensia* **98**: 1–16.

KLUGE, N.J. 1992d. Phylogeny and higher classification of Ephemeroptera. — *VIIth Int. Conf. on Ephemeroptera, August 3–6 1992, Orono, Maine, USA. Program & Abstracts*: 10.

[KLUGE, N.J.] КЛЮГЕ Н.Ю. 1992e. Насекомые-поденки (отряд Ephemeroptera). I–II. С.-Петербургский ун-т. (text-book).

[KLUGE, N.J.] КЛЮГЕ Н.Ю. 1993a. Ревизия родов сем. Heptageniidae (Ephemeroptera). II. Филогения [Revision of genera of the family Heptageniidae (Ephemeroptera). II. Phylogeny]. — *Энтомол. Обозрение* [*Rev. Entomol. URSS*] **72** (1): 39–54.

KLUGE, N.J. 1993b. New data on mayflies (Ephemeroptera) from fossil Mesozoic and Cenozoic resins. — *Palaeontol. Journal* **27** (1A): 35–49.

KLUGE, N.J. 1994a. Pterothorax structure of mayflies (Ephemeroptera) and its use in systematics. — *Bull. Soc. entomol. France* **99** (1): 41–61.

KLUGE, N.J. 1994b. A revision of Leptophlebiidae of Cuba (Insecta, Ephemeroptera). — *Zoosyst. Rossica* **2** (2) 1993: 247–285.

KLUGE, N.J. 1994c. Habrophlebiinae subfam. n. with description of a new species of *Habroleptoides* from Caucasus (Ephemeroptera: Leptophlebiidae). — *Zoosyst. Rossica* **3** (1): 35–43.

[KLUGE, N.J.] КЛЮГЕ Н.Ю. 1995. *Каталог типовых экземпляров коллекции Зоологического института РАН.* Поденки (Ephemeroptera) [*A catalogue of the type-specimens in the collection of the Zoological Institute, Russian Academy of Sciences.* Insecta, Ephemeroptera]. — СПб, Зоологический институт РАН [St.Petersburg, Zoological Institute RAS]: 1–52.

KLUGE, N.J. 1996a. A new suborder of Thysanura for the Carboniferous insect initially described as larva of Bojophlebia, with comments on characters of the orders Thysanura and Ephemeroptera (Insecta). — *Zoosyst. Rossica* **4** (1) 1995: 71–75.

KLUGE, N.J. 1996b. The Palaearctic Metretopodidae, with description of a new genus and species from Siberia (Ephemeroptera). — *Zoosyst. Rossica* **4** (1) 1995: 76–80.

[KLUGE, N.J.] КЛЮГЕ Н.Ю. 1996c. Мифы в систематике насекомых и принципы зоологической номенклатуры [Myths in insect systematics and principles of zoological nomenclature]. — *Энтомол. Обозрене* [*Rev. Entomol. URSS*] **75** (4): 939–944.

KLUGE, N.J. 1997a. New subgenera of Holarctic mayflies (Ephemeroptera: Heptageniidae, Leptophlebiidae, Ephemerellidae). — *Zoosyst. Rossica*, **5** (2) 1996: 233–235.

KLUGE, N.J. 1997b. A paradoxical problem in the phylogeny of Furcatergalia (Ephemeroptera). — In: P. Landolt & M. Sartori (eds.). *Ephemeroptera & Plecoptera. Biology-Ecology-Systematics* (Proc. VIII Int. Conf. on Ephemeroptera and XII Int. Symposium on Plecoptera, August 1995, Losanne). Mauron+ Tinguely & Lacht SA, Fribourg/Switzerland: 520–526.

KLUGE, N.J. 1997c. Classification and phylogeny of the Baetidae (Ephemeroptera) with description of the new species from the Upper Cretaceous resins of Taimyr. — In: P. Landolt & M. Sartori (eds.). *Ephemeroptera & Plecoptera. Biology-Ecology-Systematics* (Proc. VIII Int. Conf. on Ephemeroptera and XII Int. Symposium on Plecoptera, August 1995, Losanne). Mauron+ Tinguely & Lacht SA, Fribourg/Switzerland: 527–535.

[KLUGE, N.J.] КЛЮГЕ Н.Ю. 1997d. Отряд поденки Ephemeroptera. — In: *Определитель пресноводных беспозвоночных России и сопредельных территорий.* Ред. С.Я. Цалолихин. С.-Петербург [S.J. Tsalolikhin (ed.). *Key to frashwater invertebrates of Russia and adjacent lands.* S-Petersburg] **3**: 176–220.

KLUGE, N.J. 1998. Phylogeny and higher classification of Ephemeroptera. — *Zoosyst. Rossica* **7** (2): 255–269.

[KLUGE, N.J.] КЛЮГЕ Н.Ю. 1999a. Система альтернативных номенклатур надвидовых таксонов [A system of alternative nomenclatures of the supraspecific taxa. Linnaean and post-Linnaean principles of systematics]. — *Энтомол. Обозрение* [*Rev. Entomol. URSS*] **78** (1): 224–243; English translation in: *Entomol. Review* **79** (2): 106–110

[KLUGE, N.J.] КЛЮГЕ Н.Ю. 1999b. Линнеевская и постлиннеевская систематика надвидовых таксонов и новые принципы номенклатуры. [Linnaean and post-Linnaean taxonomy of supraspecific taxa and new principles for nomenclature]. — *Русский орнитологический журнал* [*The Russian Journal of Ornitology*] Express-issue **79**: 3–21.

KLUGE, N.J. 1999c. Mitos en sistematica y principos de nomenclatura zoologica / Myths in systematics and principles of zoological nomenclature. — *Bol. Soc. Entomol. Aragonesa* **26**: 347–377.

[KLUGE, N.J.] КЛЮГЕ Н.Ю., 2000. *Современная систематика насекомых.* Часть I. Принципы систематики живых организмов и общая система насекомых с классификацией первичнобескрылых и древнекрылых. С.-Петербург, Лань [*Modern systematics of insects.* Part I. Principles of systematics of living organisms and general system of insects with classification of primary wingless and palepterous insects. S-Petersburg, Lan']: 336 pp.

[KLUGE, N.J. & NARANJO, J.C.] КЛЮГЕ Н.Ю., НАРАНХО Х.К. 1990. Поденки сем. Leptohyphidae (Ephemeroptera) Кубы [Mayflies of the family Leptohyphidae (Ephemeroptera) of Cuba]. — *Энтомол. Обозрение* [*Rev. Entomol. URSS*] **69** (3): 564–578; English translation in: *Entomol. Review* 1991, **70** (4): 1–16.

[KLUGE, N.J. & NARANJO, J.C.] КЛЮГЕ Н.Ю., НАРАНХО Х.К. 1994. Новый необычный вид поденок рода *Euthyplocia* (Ephemeroptera, надсем. Ephemeroidea) с Кубы [A peculiar new species of the mayfly genus *Euthyplocia* (Ephemeroptera, Ephemeroidea) from Cuba]. — *Энтомол. Обозрение* [*Rev. Entomol. URSS*] **73** (4): 777–782.

[KLUGE, N.J. & NOVIKOVA, E.A.] КЛЮГЕ Н.Ю., НОВИКОВА Е.А. 1992. Ревизия палеарктических родов и подродов подсемейства Cloeoninae (Ephemeroptera, Baetidae) с описанием новых видов из СССР. [Revision of the Palaearctic genera and subgenera of mayflies of the subfamily Cloeoninae (Ephemeroptera, Baetidae) with description of new species from the USSR]. — *Энтомол. Обозрение* [*Rev. Entomol. URSS*] **71** (1): 38–55; English translation in: *Entomol. Review* **71** (9): 29–54.

[KLUGE, N.J. & NOVIKOVA, E.A. & BRODSKY, A.K.] КЛЮГЕ Н.Ю., НОВИКОВА Е.А., БРОДСКИЙ А.К. 1984. Движения личинок поденок при плавании, дыхании и чистке [Movements of larvae of the Ephemeroptera during swimming, respiration and claening]. — *Зоол. Журнал* [*Zool. Zhurnal*] **63** (9): 1345–1354.

KLUGE, N.J. & SINITSHENKOVA, N.D. 2002. Order Ephemerida Latreille 1810. The true mayflies (= Ephemeroptera Hyatt et Arms 1892 (s.l.); = Euephemeroptera Kluge, 2000). — In: A.P. Rasnitsyn & D.L.J. Quicke (eds.). *History of insects*. Kluwer Acad. Publ., Dordrecht, Boston, London: 89–97.

KLUGE, N.J. & STUDEMANN, D. & LANDOLT, P. & GONSER, T. 1995. A reclassification of Siphlonuroidea (Ephemeroptera). — *Bull. Soc. Entomol. Suisse* **68**: 103–132.

KONDRATIEFF, B.C. & VOSHELL, J.R., JR. 1983. Subgeneric and species-group classification of the mayfly genus *Isonychia* in North America (Ephemeroptera: Oligoeuriidae). — *Proc. Entomol. Soc. Washington* **85**: 128–138.

KONDRATIEFF, B.C. & VOSHELL, J.R., JR. 1984. The North and Central American species of *Isonychia* (Ephemeroptera: Oligoneuriidae). — *Trans. Amer. Entomol. Soc.* **110**: 129–244.

KOPELKE, J.-P. 1980. Morphologische Studien an den Eiern der Eintagsfliegen (Ephemeroptera) aus der Emergenz des zentralafrikanischen Bergbaches Kalengo. — *Mitt. Schweiz. Entomol. Ges.* **53** (2/3): 287–311.

[KOSOVA, A.A.] КОСОВА А.А. 1967. Экология поденки *Palingenia sublongicauda* Tshern. в дельте Волги. [A contribution to the ecology of the may-fly *Palingenia sublongicauda* Tshern. in the Volga Delta]. — *Зоол. Журнал* [*Zool. Zhurnal*] **46** (12): 1856–1859.

KOSS, R.W. 1968. Morphology and taxonomic use of Ephemeroptera eggs. — *Ann. Entomol. Soc. Amer.* **61**: 696–721.

KOSS, R.W. 1973. The significance of egg stage to taxonomic and phylogenetic studies of the Ephemeroptera. — In: W.L. Peters & J.G. Peters (eds.). *Proc. 1st Int. Conf. Ephemeroptera*, Florida A&M Univ. 17–20 August 1970. Leiden, Brill: 73–78.

KOSS, R.W. & EDMUNDS, G.F., JR. 1970. A new species of *Lachlania* from New Mexico with notes on the genus (Ephemeroptera: Oligoneuriidae). — *Proc. Entomol. Soc. Washington* **72** (1): 55–65.

KOSS, R.W. & EDMUNDS, G.F., JR. 1974. Ephemeroptera eggs and their contribution to phylogenetic studies of the order. — *Zool. Lourn. Linn. Soc.* **55** (4): 267–349, Pl. 1–24.

KRAUSSE, A. & WOLFF, M. 1919. Eine Üebersicht über die bisher aufgestellten fossilen und rezenten Insectenordungen. — *Arch. Naturg.* (A) **85** (3): 151–171.

KRZEMINSKI, W. & LOMBARDO, C. 2001. New fossil Ephemeroptera and Coleoptera from the Ladinian (Middle Triassic) of Canton Ticino (Switzerland). — *Rev. Ital. Paleontol. Stratigr.* **107** (1): 69–78.

KUHN, O. 1937. Insecten aus dem Buntsandstein von Thüringen. — *Beitr. Geol. Thuringen* **4** (5): 190–193.

KUHN, O. 1951. Ein vermutlicher Schmetterling, *Geisfeldiella benkerti* n.g. n.sp. aus dem Lias ε1 Nordfrankens. — *Neues Jahrb. Geol. Palaontol. Monatshefte* (Stuttgart) Hf.2: 58–61.

KUKALOVA, J. 1968. Permian mayfly nymphs. — *Psyche* **75** (4): 310–327.

KUKALOVA-PECK, J. 1983. Origin of the insect wing and wing articulation from the arthropodan leg. — *Canad. Journ. Zool.* **61** (7): 1618–1669.

KUKALOVA-PECK, J. 1985. Ephemeroid wing venation based upon new gigantic Carboniferous mayflies and basic morphology, phylogeny, and metamorphosis of pterygote insects (Insecta, Ephemerida). — *Canad. Journ. Zool.* **63** (4): 933–955.

KUKALOVA-PECK, J. 1991. Fossil history and the evoluition of Hexapod structures. — In: I.D. Naumann (ed.). *The insects of Australia*, CSIRO, Melbourne Univ.: 141–179.

KUKALOVA-PECK, J., PECK, S.B. 1976. Adult and immature Calvertiellidae (Insecta: Palaeodictyoptera) from the Upper Palaeozoic of New Mexico and Czechoslovakia. — *Psyche* **83** (1): 79–83.

[KUSTAREVA, L.A.] КУСТАРЕВА Л.А. 1976. Поденки (Ephemeroptera, Ephemerellidae, Heptageniidae) рек Иссык-Кульской котловины [Mayflies (Ephemeroptera, Ephemerellidae, Heptageniidae) in rivers of Issyck-Kul Hollow]. — *Энтомол. Обозрение* [*Rev. Entomol. URSS*] **60** (1): 58–68.

LAFON, J. 1953. Note sur *Prosopistoma foliaceum* Fourcr. (Ephemeroptera). *Bull. Soc. Zool. France* **77** (5–6): 425–436.

LAMEERE, A. 1917. Étude sur l'evolution des Ephéméres. — *Bull. Soc. Zool. France.* **42**: 41–81.

LANDA, V. 1948. Contribution to the anatomy of Ephemerids larvae. I. Topography and anatomy of tracheal system. — *Vest. Cz. spol. zool.* **12**: 25–82.

LANDA, V. 1969a. Jepice – Ephemeroptera. — *Fauna ČSSR* **18**: 1–350.

LANDA, V. 1969b. Comparative anatomy of mayfly larvae (Ephemeroptera). — *Acta Entomol. Bohemosl.* **66**: 289–316.

LANDA, V. 1973. A contribution to the evolution of the order Ephemeroptera based on comparative anatomy. — In: W.L. Peters & J.G. Peters (eds.). *Proc. 1st Int. Conf. Ephemeroptera*, Florida A&M Univ. 17–20 August 1970. Leiden, Brill: 155–159.

LANDA, V. & SOLDÁN, T. 1985. Phylogeny and higher classification of the order Ephemeroptera: a discussion from the comparative anatomical point of view. Academia – Studie CSAV (Czechoslovak Acad. Sci.), No.4: 1–121.

LANDOLT, P. & SARTORI, M. & ELPERS, CH. & TOMKA, I. 1995. Biological studies of *Palingenia longicauda* (Olivier) (Ephemeroptera: Palingeniidae) in one of its last European refuges – Feeding habits, ethological observations and egg structure. — In: L.D. Corkum & J.H. Ciborowski (eds.). *Current directions in research on Ephemeroptera* (Proc. 7th Int. Conf. Ephemeroptera, 3–6 August 1992, Orono, Maine, USA). Canadian Scholar's Press Inc., Toronto.: 273–281.

LATREILLE, P.A. 1802–1805. *Histoire naturelle, générale et particulière des crustacés et des insectes.* Paris. T.1–14.

LATREILLE, P.A. 1810. *Considérations générales sur l'ordre naturel des animaux composant les classes des crustacés, des arachnides, et des Insectes.* Paris, F. Schoell: 1–444.

LATREILLE, P.A. 1833. Description d'un nouveau génre de crustacés. — *Ann. Mus. Hist. Nat.*, Paris, 3e Sér., **2**: 23–34.

LEACH, W.E. 1815. Entomology. — *Brewster's Edinburg Encyclopedia*, Ed. 1, **9** (1): 57–172.

LEHMKUHL, D.M. 1976. Additions to the taxonomy, zoogeography and biology of *Analetris eximia* (Acanthametrepodinae, Siphlonuridae, Ephemeroptera). — *Canad. Entomol.* **108**: 199–207.

LEHMKUHL, D.M. 1979a. The North American species of *Cinygma* (Ephemeroptera: Heptageniidae). — *Canad. Entomol.* **111** (6): 675–680.

LEHMKUHL, D.M. 1979b. A new genus and species of Heptageniidae (Ephemeroptera) from western Canada. — *Canad. Entomol.* **111** (7): 859–862.

LESTAGE, J.-A. 1917. Contribution á l'étude des larves des Ephéméres paléarctiques. — *Ann. Biol. lacustre* **8**: 213–459.

LESTAGE, J.-A. 1918. Les Ephéméres d'Afrique (Notes critiques sur les espèces connues). — *Rev. Zool. Africaine* **6**: 65–114.

LESTAGE, J.-A. 1921. Les Ephéméres indo-chinoises. — *Ann. Soc. Entomol. Belg.* **61**: 211–222.

LESTAGE, J.-A. 1922. Notes sur le génre *Nirvius* Navas (= *Ephemera* L.) [Ephemeroptera]. — *Bull. Soc. Entomol. France* № 16: 253–254.

LESTAGE, J.-A. 1923a. Étude sur les Palingeniidae (Ephéméres) et description de deux génres nouveaux et d'une espèce nouvelle de la Nouvelle Guinée. — *Ann. Soc. Entomol. Belg.* **63**: 95–112.

LESTAGE, J.-A. 1923b. L'Imbroglio Campsurien. Notes critiques les Campsurus (Ephemeroptera). — *Ann. Soc. Entomol. Belg.* **63**: 113–124.

LESTAGE, J.-A. 1924a. Les Ephéméres de l'Afrique du Sud. Catalogue critique & systematique des espèces connues et description de trois genera nouveaux et de sept espèces nouvelles. — *Revue Zool. Africaine* **12** (3): 316–352.

LESTAGE, J.-A. 1924b. Notes sur les Ephéméres de le Monographical Revision de Eaton. — *Ann. Soc. Entomol. Belg.* **64**: 33–59.

LESTAGE J.-A. 1924c. A propos du génre *Coenis* Steph. = *Brachycercus* Curt. (Ephemeroptera). — *Ann. Soc. Entomol. Belg.* **64**: 61–62.

LESTAGE, J.-A. 1924d. Les Ephéméres finnoises se M. le Docter J.E. Aro. — *Bull. Soc. ent. Belg.* **6**: 33-36.

LESTAGE, J.-A. 1930. Contribution à l'étude des larves des Ephéméroptères. V. - Les larves à trachéo-branchies ventrales. — *Bull. & Ann. Soc. Entomol. Belg.* **69**: 433–440.

LESTAGE, J.-A. 1931a. Notes sur le premier Brachycercidien decouvert dans le faune australienne *Tasmanocoenis Tonnoire* sp. nov. (Ephemeroptera) et remarques sur la famille des Brachycercidae Lest. — *Mem. Soc. Entomol. Belg.* 1930, **23**: 49–60.

LESTAGE, J.-A. 1931b. Contribution à l'étude des larves des Ephéméroptères. VII. — Le groupe Potamanthidien. — *Mem. Soc. Entomol. Belg.* 1930, **23**: 73–146.

LESTAGE, J.-A. 1931c. Contribution á l'tude des Ephéméroptères. VIII. Les Ephéméroptères du Chili. — *Bull. & Ann. Soc. Entomol. Belg.* **71**: 41–60.

LESTAGE, J.-A. 1935a. Contribution à l'étude des Ephéméroptères. IX. Le groupe Siphlonuridien. — *Bull. & Ann. Soc. Entomol. Belg.* **75**. 77–139.

LESTAGE, J. A. 1935b. Contribution à l'étude des Ephéméroptères. XII. Les composantes australiennes et néo-zélandaises du Groupe Siphlonuridien. — *Bull. & Ann. Soc. Entomol. Belg.* **75**: 346–358.

LESTAGE, J.-A. 1938a. Contribution à l'étude des Ephéméroptères. XVI. Recherches critiques sur le complexe amétrométrétopodiden. — *Bull. & Ann. Soc. Entomol. Belg.* **78**: 171–181.

LESTAGE, J.-A. 1938b. Contribution à l'étude des Ephéméroptères. XIX. L'inclusion des Behningiidae parmi les Oligoneuriidae. — *Bull. & Ann. Soc. Entomol. Belg.* **78**: 315–319.

LESTAGE, J.-A. 1939. Contribution à l'étude des Ephéméroptères. XXIII. Les Polymitarcidae de la faune africaine et description d'un gnre nouveua du Natal. — *Bull. & Ann. Soc. Entomol. Belg.* **79**: 135–138.

LESTAGE, J.-A. 1940. Contribution à l'étude des Ephéméroptères. XXIV. Un cas de non-agnathisme chez l'adult de *Paleoameletus primitivus* Trav. de l'Himalaya. — *Bull. & Ann. Soc. Entomol. Belg.* **80**: 118–124.

LESTAGE, J.-A. 1942. Contribution à l'étude des Ephéméroptères. XXV. Notes critiques sur les anciens Caenidiens d'Afrique et sur l'independance de l'evolution tricorythido-caenidienne. — *Bull. Mus. Roy. Hist. Nat. Belg.* **18** (48): 1–20.

LESTAGE, J.-A. 1945. Contribution à l'étude des Ephéméroptères. XXVI. Étude critique de quelques génres de la faune ethiopienne. — *Bull. & Ann. Soc. Entomol. Belg.* **81**: 81–89.

LEWIS, P.A. 1974. Taxonomy and ecology of *Stenonema* mayflies (Heptageniidae: Ephemeroptera). — *U.S.E.P.A. Environ. Monitoring Ser.* Rept. EPA-670/4–74–006: 1–81.

LEWIS, P.A. 1978a. On the use of pectinate maxillary spines to separate *Stenonema* and *Stenacron* (Ephemeroptera, Heptageniidae). — *Proc. Entomol. Soc. Washington* **80** (4): 655.

LEWIS, P.A. 1978b. An immature fossil Ephemeroptera (Ephemeridae) from the Ruby River Bassin (Oligocene) of Southwestern Montana. — *Ann. Entomol. Soc. Amer.* **71** (4): 479–480.

LIN, QI-BIN 1986. [Early Mesozoic fossil insects from South China] —*Palaeontologia Sinica* **170** (N.S. B21): i–iii, 1–121, 20 pl.

LIN, GI-BIN & HUANG, DI-YING 2001. Description of *Caenoephemera shangyuanensis*, gen. nov., sp. nov. (Ephemeroptera), from the Yixian Formation. — *Canad. Entomol.* **133** (6): 747–754.

LINNAEUS, C. 1758. *Systema Naturae per regna tria Naturae, secundum classes, ordines, genera, species, cum Characteribus, Differentiis, Synonymis, Locis.* T.I. [A photographic facsimile of the first volume of the tenth edition. London, Brit. Mus. (N. H.) 1956: 824 p.].

LUGO-ORTIZ, C.R. & MCCAFFERTY, W.P. 1995. Contribution to the taxonomy of the Leptohyphidae (Insecta: Ephemeroptera) of Central America. — *Stud. Neotrop. Fauna and Enveronment* **30**: 165–176.

LUGO-ORTIZ, C.R. & MCCAFFERTY, W.P. 1996b. Central American *Tortopus* (Ephemeroptera: Polymitarcyidae): A unique new species and new country records. — *Entomol. News* **107** (1): 23–27.

LUGO-ORTIZ, C.R. & MCCAFFERTY, W.P. 1998. First report of the genus *Siphlaenigma* Penniket and the family Siphlaenigmatidae (Ephemeroptera) from Australia. — *Proc. Entomol. Soc. Washington* **100** (2): 209–213.

MACLEAY, W.S. 1830. Explanation of the comparative anatomy of the thorax in winged insects, with a review of the present state of the nomenclature of its parts. — *Zool. Journ.* 1830: 145–179, Pl. V–VI.

MALZACHER, P. 1984. Die europäischen Arten der Gattung *Caenis* Stephens (Insecta: Ephemeroptera). — *Stuttg. Beitr. Naturk.*, 373: 1–48.

MALZACHER, P. 1986a. Diagnostik, Verbreitung und Biologie der europäischen Caenis-Arten (Ephemeroptera: Caenidae). — *Stuttgart. Beitr. Naturk.*, A (387): 1–41.

MALZACHER, P. 1986b. Caenidae aus dem Amazonasgebiet (Insecta, Ephemeroptera). — *Spixiana* **9** (1): 83–103.

MALZACHER, P. 1987. Eine neue Caeniden-Gattung *Afrocaenis* gen. nov. und Bemerkungen zu *Tasmanocaenis tillyardi* (Insecta: Ephemeroptera). — *Stuttg. Beitr. Naturk.*, Ser A, **407**: 1–10.

MALZACHER, P. 1991. Genital-Morphological features in the Caenidae. — In: J. Alba-Tercedor & A. Sanchez-Ertega (eds.). *Overview and strategies of Ephemeroptera and Plecoptera* (Proc. 6th Int. Congf. Ephemeroptera & 10th Int. Symp. Plecoptera, 24–30 July 1989, Granada, Spain), Sandhill Crane Press: 73–85.

MALZACHER, P. 1993. Caenidae der äthiopischen Region (Insecta: Ephemeroptera). Teil 2. Systematische Zusammenstellung aller bisher bekannten Arten. — *Mitt. Schweiz. Entomol. Ges.* **66**: 379–416.

MALZACHER, P. 1995. Caenidae from Madagascar (Insecta: Ephemeroptera). — *Stuttgart. Beitr. Naturk.* (Ser. A) **530**: 1–12.

MALZACHER, P. 1997. Relationships in the Caenidae (Insecta: Ephemeroptera). — In: P. Landolt & M. Sartori (eds.). *Ephemeroptera & Plecoptera. Biology-Ecology-Systematics.* (Proc. 8th Int. Conf. Ephemeroptera & 12th Int. Symp. Plecoptera 14–20 August 1995, Losanne, Switzerland). Mauron+ +Tinguely & Lacht SA, Fribourg / Switzerland: 550–553.

MARTINS-NETO, R.G. 1996. New mayflies (Insecta, Ephemeroptera) from the Santana Formation (Lower Cretaceous), Araripe Basin, northeast Brazil. — *Revista Espanola de Paleontologia* **1** (2): 177–192.

[MARTYNOV, A.V.] МАРТЫНОВ А.В. 1924. К пониманию жилкования и трахеации крыльев стрекоз и поденок [Sur l'interpretation de la nervuration et de la tracheation des ailes des Odonates et des Agnathes. — *Русск. Энтомол. Обозрение* [*Rev. Russe Entomol.*] 18: 145–174.

MARTYNOV, A.V. 1928. Permian fossil insects of North-East Europe. *Тр. Геол. Музея АН СССР* [*Trav. Mus. Geol. Acad. Sci. URSS*] **4**: 1–118, pl. I–XIX.

MARTYNOV, A.V. 1932. New Permian Palaeoptera with the discussion of some problems of their evolution. — Тр. Палеозоол. ин-та АН СССР [*"Trudy Paleozool. Inst. Acad. Nauk SSSR"*] **1**: 1–44, 2 pl.

[MARTYNOV, A.V.] МАРТЫНОВ А.В. 1938. Очерки геологической истории и филогении отрядов насекомых (Pterygota). Ч. I. Palaeoptera и Neoptera-Polyneoptera [Études sur l'histoire geologique et de phylogenie des ordes des insectes (Pterygota). I. Palaeoptera et Neoptera-Polyneoptera]. — Тр. Палеонтол. ин-та [*"Trudy Paleontol. Inst. Akad. Nauk SSSR"*] 7 (4): 5–149.

MASSELOT, G. & NEL, A. 1999. *Pseudokageronia thomasi* gen. nov., sp.nov., from Upper Miocene of Murat (France) [Ephemeroptera: Heptageniidae]. — *Ephemera* **1** (1): 90–94.

[MATSUMURA, S.] 1931. [*6000 illustrated Insects of Japan-Empire*] (in Japanese). Ephemerida: 1465–1480.

MAYO, V.K. 1968. Some new mayflies of the subfamily Leptohyphinae (Ephemeroptera, Tricorythidae). — *Pan-Pacif. Entomol.* **44** (4): 301–308.

MCCAFFERTY, W.P. 1968. A new genus and species of Ephemeridae (Ephemeroptera) from Madagascar. — *Entomol. Record and Journ. of Variation* **80**: 293.

MCCAFFERTY, W.P. 1970. Neotropical nymphs of the genus *Hexagenia*. — *Journ. Georgia Entomol. Soc.* **5**: 224–228.

MCCAFFERTY, W.P. 1971a. New genus of mayflies from eastern North America (Ephemeroptera: Ephemeridae). — *Journ. New York Entomol. Soc.* **79** (4): 45–51.

MCCAFFERTY, W.P. 1971b. New burrowing mayflies from Africa (Ephemeroptera: Ephemeridae). — *Journ. Entomol. Soc. South Afr.* **34**: 57–62.

MCCAFFERTY, W.P. 1972. Pentageniidae: a new family of Ephemeroidea (Ephemeroptera). — *Journ. Georgia Entomol. Soc.* **7**: 51–56.

MCCAFFERTY, W.P. 1973. Systematic and zoogeographic aspects of Asiatic Ephemeridae (Ephemeroptera). — *Oriental Insects* 7 (1): 49–67.

MCCAFFERTY, W.P. 1975. The burrowing mayflies of the United States. — *Trans. Amer. Entomol. Soc.* **101**: 447–504.

MCCAFFERTY, W.P. 1978. A natural subgeneric classification of *Ephemerella bartoni* and related species (Ephemeroptera: Ephemerellidae). — *Great Lakes Entomol.* **11** (3): 137–138.

MCCAFFERTY, W.P. 1979a. Evolutionary trends among the families of Ephemeroidea. — In: K. Pasternak & R. Sowa (eds.). *Proc. 2nd Int. Conf. Ephemeroptera,* Polish Acad. Sci., Lab. Water Biol., 23–26 August 1975. Panstwowe Wydawnictwo Naukowe, Warszawa – Krakow: 45–50.

MCCAFFERTY, W.P. 1979b. Variants of Hexagenia s.s. species affecting subgeneric diagnosis (Ephemeroptera: Ephemeridae). — *Proc. Entomol. Soc. Washington* **81** (3): 443–446.

MCCAFFERTY, W.P. 1987. New fossil mayfly in amber and its relationships among extant Ephemeridae (Ephemeroptera). — *Ann. Entomol. Soc. Amer.* **80**: 472–474.

MCCAFFERTY, W.P. 1989. Characterization and relationships of the subgenera of *Isonychia* (Ephemeroptera: Oligoneuriidae). — *Entomol. News* **100**: 72–78.

MCCAFFERTY, W.P. 1990. Insects from Santana Formation Lower Cretaceous, of Brazil. Ephemeroptera. — *Bull. Amer. Mus. Nat. Hist.* **195**: 20–50.

MCCAFFERTY, W.P. 1991a. Toward a Phylogenetic Classification of the Ephemeroptera (Insecta): A Comentary on Systematics. — *Ann. Entomol. Soc. Amer.* **84**: 343–360

MCCAFFERTY, W.P. 1991b. The cladistics, classification, and evolution of the Heptagenioidea (Ephemeroptera). — In: J. Alba-Tercedor & A. Sanchez-Ertega (eds.). *Overview and strategies of Ephemeroptera and Plecoptera* (Proc. 6th Int. Congf. Ephemeroptera & 10th Int. Symp. Plecoptera, 24–30 July 1989, Granada, Spain), Sandhill Crane Press: 87–102.

MCCAFFERTY, W.P. 1991c. Comparison of Old and New World *Acanthametropus* (Ephemeroptera: Acanthametropodidae) and other psammophilous mayflies. — *Entomol. News.* **102** (5): 205–214

MCCAFFERTY, W.P. 1991d. Synopsis of the Oriental mayfly genus *Eatonigenia* (Ephemeroptera: Ephemeridae). — *Oriental Insects*, **25**: 179–181

MCCAFFERTY, W.P. 1997a. Discovery and analysis of the oldest mayflies (Insecta, Ephemeroptera) known from amber. — *Bull. Soc. Hist. Nat. Toulouse* **133**: 77–82.

MCCAFFERTY, W.P. 1997b. Ephemeroptera. — In: *Nomina Insecta Nearctica*, a checklist of the insects of North America. Vol. 4: Non-holometabolous orders. Entomological Informational Service, Rockville, Maryland: 89–117 (Cit ex: McCafferty & Wang 2000).

MCCAFFERTY, W.P. & BAE, Y.J. 1990. *Anthopotamus*, a new genus for North American species previously known as *Potamanthus* (Ephemeroptera: Potamanthidae). — *Entomol. News* **101** (4): 200–202.

MCCAFFERTY, W.P. & BENSTEAD, J.P 2002. Cladistic resolution and ecology of the Madagascar genus *Manophyphella* Allen (Ephemeroptera: Teloganodidae). — *Ann. Limnol.* **38** (1): 41-51.

MCCAFFERTY, W.P. & BLOODGOOD, D.W. 1989. The female and male coupling apparatus in *Tortopus* mayflies. — *Aquatic Insects* **11** (3): 141–146.

MCCAFFERTY, W.P. & EDMUNDS, G.F., JR. 1973. Subgeneric classification of *Ephemera* (Ephemeroptera: Ephemeridae). — *Pan-Pacif. Entomol.* **49**: 300–307.

MCCAFFERTY, W.P. & EDMUNDS, G.F., JR. 1976a. Redefinition of the family Palingeniidae and its implication for higher classification of Ephemeroptera. — *Ann. Entomol. Soc. Amer.* **69** (3): 486–490.

MCCAFFERTY, W.P. & EDMUNDS, G.F., JR. 1976b. The larva of the Madagascar genus *Cheirogenesia* Demoulin (Ephemeroptera: Palingeniidae). — *Systematic Entomology* **1**: 189–194.

MCCAFFERTY, W.P. & EDMUNDS, G.F., JR. 1979. The higher classification of the Ephemeroptera and its evolutionary basis. — *Ann. Entomol. Soc. Amer.* **72**: 5–12.

MCCAFFERTY, W.P. & GILLIES, M.T. 1979. The African Ephemeridae. — *Aquatic Insects* **1** (3): 169–178.

MCCAFFERTY, W.P. & PROVONSHA, A.V. 1984. The first adult of *Spinadis* (Ephemeroptera: Heptageniidae). — *Entomol. News.* **95** (5): 173–179.

MCCAFFERTY, W.P. & PROVONSHA, A.V. 1985. Systematics of

Anepeorus (Ephemeroptera: Heptageniidae). — *Great Lakes Entomol.* **18**: 1–6.

MCCAFFERTY, W.P. & PROVONSHA, A.V. 1986. Comparative mouthpart morphology and evolution of the carnivorous Heptageiidae (Ephemeroptera). — *Aquatic Insects* **8** (2): 83–89.

MCCAFFERTY, W.P. & WANG, T.-Q. 1994a. Phylogenetics and the classification of the *Timpanoga* complex (Ephemeroptera: Ephemerellidae). — *Journ. North Amer. Benthol. Soc.* **13** (4): 569–579.

MCCAFFERTY, W.P. & WANG, T.-Q. 1994b. Relationships of the genera *Acanthametropus, Analetris*, and *Siphluriscus*, and re-evaluation of their higher classification. — *Great Lake Entomol.* **27** (4): 209–215.

MCCAFFERTY, W.P. & WANG, T.-Q. 1995. A mew genus and species of Tricorythidae (Ephemeroptera: Pannota) from Madagascar. — *Annls. Limnol.* **31** (3): 179–183.

MCCAFFERTY, W.P. & WANG, T.-Q. 1997. Phylogenetic systematics of the family Teloganodinae (Ephemeroptera: Pannota). — *Ann. Cape Prov. Mus. Nat. Hist.* **19**: 387–437.

MCCAFFERTY, W.P. & WANG, T.-Q. 1998. New name for a generic homonym in Teloganodinae (Ephemeroptera). — *Entomol. News* **109** (5): 344.

MCCAFFERTY, W.P. & WANG, T.-Q. 2000. Phylogenetic systematics of the major lineages of pannote mayflies (Ephemeroptera: Pannota). — *Trans. Amer. Entomol. Soc.* **126** (1): 9–101.

MCDUNNOUGH, J. 1923. New Canadian Ephemeridae with notes. — *Canad. Entomol.* **55**: 39–50.

MCDUNNOUGH, J. 1925. New Canadian Ephemeridae with notes. III. — *Canad. Entomol.* **57**: 168–176 185–192.

MCDUNNOUGH, J. 1931a. New species of North American Ephemeroptera. — *Canad. Entomol.* **63**: 82–93.

MCDUNNOUGH, J. 1931b. The eastern North American species of the genus *Ephemerella* and their nymphs (Ephemeroptera). — *Canad. Entomol.* **63**: 187–197, 201–216.

MCDUNNOUGH, J. 1932. Further notes on the Ephemeroptera of the north shore of the Gulf of St. Lawrence. — *Canad. Entomol.* **64**: 78–81.

MCDUNNOUGH, J. 1933. The nymph of *Cinygma integrum* and description of a new heptagenine genus. — *Canad. Entomol.* **65**: 73–76.

MCLACHLAN, R. 1873. *Oniscigaster wakefieldi*, a new genus and species of Ephemeridae from New Zealand. — *Entomol. Month. Mag.* **10**: 108–110.

MCLACHLAN, R. 1875. A sketch of our present knowledge of the neuropterous fauna of Japan (excluding Odonata and Trichoptera). — *Trans. Entomol. Soc. London*: 167–190.

[MESHKOVA, N.P.] МЕШКОВА Н.П. 1961. О личинках *Ephemeropsis trisetalis* Eichwald (Insecta, Ephemeroptera). — *Палеонтол. Журнал* [*Paleontol. Journal*], № 4: 164–168.

MOLINERI, C. 1999a. Descripción de una nueva especie de *Haplohyphes* Allen (Ephemeroptera: Leptohyphidae) y clave para separar los imagos machos del género. — *Acta Zool. Lilloana* **45** (1): 29–32.

MOLINERI, C. 1999b. Revision of the genus *Tricorythopsis* (Ephemeroptera: Leptohyphidae) with the description of four new species. — *Aquatic Insects* **21** (4): 285–300.

MOLINERI, C. 2001a. A new genus of Leptohyphidae. — In: E. Dominguez (ed.). *Trends in research in Ephemeroptera and Plecoptera* (Proc. IX Int. Conf. on Ephemeroptera & XIII Int. Symp. on Plecoptera 16–23 Aug. Tafi de Valle, Tucuman, Argentina). Kluwer Academic/Plenum Press, N.Y.: 337–345.

MOLINERI, C. 2001b. *Traverhyphes*: A new genus of Leptohyphidae for *Leptohyphes indicator* and related species (Insecta, Ephemeroptera). — *Spixiana* **24** (2): 129–140.

MOLINERI, C. 2001c. El género *Tricorythopsis* (Ephemeroptera: Leptohyphidae): nuevas combinaciones y descripción de nuevas especies y estadios. — *Rev. Soc. Entomol. Argentina* **60** (1–4): 217–238.

MOLINERI, C. 2002. Cladistic analysis of the South American species of *Tricorythodes* (Ephemeroptera: Leptohyphidae) with the description of new species and stages. — *Aquatic Insects* 24 (4): 273–308.

MOLINERI, C. 2003a. Revision of the South American species of *Leptohyphes* Eaton (Ephemeroptera: Leptohyphidae) with key to the nymphs. — *Studies on Neotropical Fauna and Environment* **38** (1): 47–70.

MOLINERI, C. 2003b. *Haplohyphes* (Ephemeroptera: Leptohyphidae), new species and stage descriptions with a key to separate the species of the genus. — *Zootaxa* **263**: 1–11.

MOLINERI, C. & DOMINGUEZ, E. 2003. Nymph and egg of *Melanemerella brasiliana* (Ephemeroptera: Ephemerelloidea: Melanemerellidae), with comments on its systematic position and the higher classification of Ephemerelloidea. — *Journ. N. Amer. Benthol. Soc.* **22** (2): 263–275.

MOLINERI, C. & PETERS, J.G. & ZUÑIGA DE CARDOSO, M.C. 2001. A new family, Coryphoridae (Ephemeroptera: Ephemerelloidea), and description of the winged and egg stages of *Coryphorus*. — *Insecta Mundi* **15** (2): 117–122.

MORTON, K.J. 1921. A new species of mayfly, *Palingenia* (sensu lato), from Mesopotamia. — *The Entomologist*, 56: 177–180, Pl. II.

MOTAS, C. & BĂCESCO, M. 1937. La decouverte en Roumanie d'une nymphe d'Ephemere appartenant au gnre *Behningia* J.A. Lestage 1929. — *Ann. Sci. Univ. Jassy* **24** (2): 25–29.

MÜLLER, A.H. 1965. Insectenreste aus der Trias von Thüringen (Buntsandstein, Keuper). — *Geologie* 14 (7): 865–877.

NAVÁS, L. 1912a. Insectos neuropteros nuevos. — *Verh. VIII Int. Zool. Kongr.*, Graz 1910: 746–751.

NAVÁS, L. 1912b. Quelques Névroptères de la Siberie méridionale-orientale. — *Русск. Энтомол. Обозрение* [*Rev. Russe Entomol.*] **12** (3): 414–422.

NAVÁS, L. 1912c. Notes sur quelques Névroptères d'Afrique. II. — *Rev. Zool. Africaine* **1**: 401–410.

NAVÁS, L. 1913. Algunos órganos de la alas de los insectos. — *2nd Int. Congr. Entomol.*, Oxford, August 1912, vol.**2**: 178–186.

NAVÁS, L. 1915. Neue Neuropteren. Erste Serie. — *Entomol. Mitt.*, 4: 146–153.

NAVÁS, L. 1920. Insectos Sudamericanos. 3a Serie. — *Anales de la Sociedad Cientifica Argentina* **90**: 52–72 (Cit. ex: Hubbard 1990).

NAVÁS, L. 1922a. Efemeropteros nuevos o poco conocidos. — *Bol. Soc. Entomol. Espana* **5**: 54–63.

NAVÁS, L. 1922b. Insectos de Fernando Poo. — *Treballs del Museu de Ciències Naturals de Barcelona*, 4: 107–116. (Cit. ex: Hubbard 1990).

NAVÁS, L. 1924. Insectos de la America Central. — *Broteria* **21**: 55–86. (Cit. ex: Hubbard 1990).

NAVÁS, L. 1932. Insecta orientalia. X Series. — *Mem. Pontifica Acad. Sci. Nuovi Lincei* (2) **16**: 921–949.

NAVÁS, L. 1936. Insectes du Congo Belge. Serie IX. — *Rev. Zool. & Bot. Africaine* **28**: 333–368 (Cit. ex: Hubbard 1990).]

NEEDHAM, J.G. 1905. Ephemeridae. — *Bull. New York State Mus.* **86**: 17–62, pl. 4–12 (Cit. ex: Hubbard 1990).

NEEDHAM, J.G. 1909. Studies of aquatic insects: a peculiar new May fly from Scandaga Park. — *New York State Mus. Bull.* **134**: 71–75.

NEEDHAM ,J.G. 1920. African stone-flies and May-flies collected by the American Museum Congo Expedition. — *Bull. Amer. Mus. Nat. Hist.* **43**: 35–40.

NEEDHAM, J.G. 1927. The Rocky Mountain species of the mayfly genus Ephemerella. — *Ann. Entomol. Soc. Amer.* **20**: 107–117.

NEEDHAM, J.G. 1932. Three new American Mayflies (Ephemeropt.). — *Canad. Entomol.* **64**: 273–276.

NEEDHAM, J.G. & BETTEN C. 1901. Aquatic insects of the Adirondacks. — *Bull. New York State Mus.* **47**: 384–612, 37 pl.

NEEDHAM, J.G. & MURPHY, H.E. 1924. Neotropical mayflies. — *Bull. Lloyd Libr.* **24**: 1–79.

NEEDHAM, J.G. & TRAVER, J.R. & HSU, YIN-CHI 1935. *The biology of mayflies, with a systematic account Noth American species.* New York, Comstock Publ. 16, 759p.

NEWMAN, E. 1853. Proposed division of Neuroptera into two classes. — *Zoologist* **11** (appendix): 181–204.

[NOVIKOVA, E.A. & KLUGE, N.J.] НОВИКОВА Е.А., КЛЮГЕ Н.Ю. 1987. Систематика рода *Baetis* (Ephemeroptera, Baetidae) с описанием нового вида из Средней Азии [Systematics of the genus Baetis (Ephemeroptera, Baetidae), with description of new species from Middle Asia]. — *Вестник Зоол.* [*"Vestnik Zool."*] № 4: 8–19.

[NOVIKOVA, E.A. & KLUGE, N.J.] НОВИКОВА Е.А., КЛЮГЕ Н.Ю. 1994. Поденки подрода *Nigrobaetis* (Ephemeroptera, Baetidae, *Baetis*). [Mayflies of the subgenus *Nigrobaetis* (Ephemeroptera, Baetidae, *Baetis*)]. — *Энтомол. Обозрение* [*Rev. Entomol. URSS*] **73** (3): 623–644.

OLIARINONY, R. & ELOUARD, J.-M. 1997. Biodiversité aquatique de Madagascar: 7 - *Ranorythus*, un nouveau génre de Tricorythidae définissant la nouvelle sous-famille der Ranorythinae (Ephemeroptera, Pannota). — *Bull. Soc. Entomol. France* **102** (5): 439–447.

OLIARINONY, R. & ELOUARD, J.-M. & RABERIAKA, N. 1998a. Biodiversité aquatigue de Madagascar: 8 - *Spinirythus*, un nouveau génre de Tricorythidae (Ephemeroptera Pannota). — *Bull. Soc. entomol. France* **103** (3): 237–244.

OLIARINONY, R. & ELOUARD, J.-M. & RABERIAKA, N. 1998b. Biodiversité aquatigue de Madagascar: 19 - neuf nouvelles espèces de *Tricorythus* Eaton [Ephemeroptera, Pannota, Tricorythidae]. — *Rev. Fr. Entomol.* (N.S.) **20** (3): 73–90.

OLIARINONY R., SARTORI M., ELOUARD J.-M. 2000. Premiere description des larves et des oeufs du malgache *Madecassorythus* (Ephemeroptera, Tricorythidae). — *Mitt. Schwiez. Entomol. Ges.*, 73: 369–378.

PACKARD, A.S. 1883. The systematic position of the Orthoptera in relation to other orders of Insects. — *Third Report of U.S. Entomol. Commission*: 286–345.

PACKARD, A.S. 1886. A new arrangement of the orders of insects (Summary of "Zoology for high schools and colleges", 5th ed.) — *Amer. Naturalist* **20** (9): 808.

PACLT, J. 1994. *Ephacerella*, a replaced name for *Acerella* Allen 1971 (Ephemeropterta), nec Berlese 1909 (Protura). — *Entomol. News* **105** (5): 283–284.

PATTERSON, C. & ROSEN, D.E. 1977. Review of ichthyodectiform and other Mesozoic teleost fishes and the theory and practice of classifying fossils. — *Bull. Amer. Mus. Nat. Hist.* **158** (2): 85–172.

PENNIKET, J.G. 1962. Notes on New Zealand Ephemeroptera. III. A new family, genus and species. — *Rec. Canterbery Mus.*, 7 (5): 389–398.

PENNIKET, J.G. 1966. Notes on New Zealand Ephemeroptera. IV. A new siphlonurid subfamily: Rallidentinae. — *Rec. Canterbury Mus.* **8**: 163–175.

PEREIRA, S.M. & DA SILVA, E.R. 1990. Nova especie de *Campylocia* Needham & Murphy 1924 com notas biologicas (Ephemeroptera, Euthyplociidae). — *Bol. Mus. Nac.* (Rio de Janeiro), nov.ser., No.336: 1–12.

PESCADOR, M.L. 1985. Systematics of the Nearctic genus *Pseudiron* (Ephemeroptera: Heptageniidae: Pseudironinae). — *Florida Entomol.* **68** (3): 432–444.

PESCADOR, M.L. & BERNER, L. 1981. The mayfly family Baetiscidae (Ephemeroptera). Part II. Biosystematics of the genus *Baetisca*. — *Trans. Amer. Entomol. Soc.* **107**: 163–228.

PESCADOR, M.L. & EDMUNDS, G.F., JR. 1994. New genus of Oligoneuriidae (Ephemeroptera) from South America. — *Ann. Entomol. Soc. Amer.* **87** (3): 263–269.

PESCADOR, M.L. & ELOUARD, J.M. 2001. New generic status of *Madeconeuria* from Madagascar (Ephemeroptera: Oligoneuriidae). — In: *2001 International Joint Meeting*. X Int. Conf. on Ephemeroptera & XIV Int. Symp. on Plecoptera: Program and Abstract Book. Univ. Studi Perugia: 37.

PESCADOR, M.L. & PETERS, W.L. 1974. The life history and ecology of *Baetisca rogersi* Berner (Ephemeroptera: Baetiscidae). — *Bull. Florida State Mus. Biol. Sci.* **17** (3): 151–209.

PESCADOR, M.L. & PETERS, W.L. 1980. A revision of the genus *Homoeoneuria* (Ephemeroptera: Oligoneuriidae). — *Trans. Amer. Entomol. Soc.* **106**: 357–393.

PETERS, J.G. & PETERS, W.L. 1986. Leg abscission and adult *Dolania* (Ephemeroptera: Behningiidae). — *Florida Entomol.* 69 (1): 246–250.

PETERS, W.L. 1980. Phylogeny of the Leptophlebiidae (Ephemeroptera): an introduction. — In: J.F. Flannagan & K.E. Marshall (eds.). *Advances in Ephemeroptera Biology* (Proc. 3rd Int. Conf. Ephemeroptera, 4–10 July 1979, Winnipeg, Canada). Plenum Press, New York: 33–41.

PETERS, W.L. 1981. *Coryphorus aquilis*, a new genus and species of Tricorythidae from the Amazon basin (Ephemeroptera). — *Aquatic Insects*, 3 (4): 209–217.

PETERS, W.L. & CAMPBELL, I.C. 1991. Ephemeroptera (mayflies). — In: I.D. Naumann & P.B. Cane (eds.). *The insects of Australia*. Melbourne Univ. Press & UCL Press Ltd.: 279–293.

PETERS, W.L. & GILLIES, M.T. 1991. The male imago of *Protobehningia* Tshernova from Thailand (Ephemeroptera, Behningiidae). — In: J. Alba-Tercedor & A. Sanchez-Ertega (eds.). *Overview and strategies of Ephemeroptera and Plecoptera* (Proc. 6th Int. Congf. Ephemeroptera & 10th Int. Symp. Plecoptera, 24–30 July 1989, Granada, Spain), Sandhill Crane Press: 207–216.

PETERS, W.L. & GILLIES, M.T. 1995. Square facets in a hexagonal world. Current directions in research on Ephemeroptera. — In: L.D. Corkum & J.H. Ciborowski (eds.). *Current directions in research on Ephemeroptera* (Proc. 7th Int. Conf. Ephemeroptera, 3–6 August 1992, Orono, Maine, USA). Canadian Scholar's Press Inc., Toronto.: 371–375.

PETERS, J.G. & HUBBARD, M.D. 1977. Authorship of the families Siphlonuridae, Palingeniidae and Potamanthidae. — *Eatonia* No.23: 1–2.

PETERS, W.L. & HUBBARD, M.D. 1989. Names and authorship of two family-groups in the Ephemeroptera. — *Journ. New York Entomol. Soc.* **97** (1): 115.

PETERS, W.L. & JONES, J. 1973. Historical and biological aspects of the Blackwater River in nortwestern Florida. — In: W.L. Peters & J.G. Peters (eds.). *Proc. 1st Int. Conf. Ephemeroptera*, Florida A&M Univ. 17–20 August 1970. Leiden, Brill: 242–253.

PETERS, W.L. & PETERS, J.G. 2000. Discovery of a new genus of Leptophlebiidae: Leopthophlebinae (Ephemeroptera) in Cretaceous amber from New Jersey. — In: D. Grimadi (ed.). *Studies on fossils in amber, with particular reference to the Cretaceous of New Jersey.* Backhuys Publ., Leiden, The Netherlands: 127–131.

PHILLIPS, J.S. 1930. A revision of New Zealand Ephemeroptera.

— *Trans. New Zealand Inst.* **61**: 271–390.

PHILLIPS, J.S. 1931. Studies of New Zealand mayfly nymphs. — *Trans. Entomol. Soc. London* **79**: 399–422.

PICTET F.J. 1843–1845. *Histoire naturelle générale et particulaire des Insectes Névroptères – Famille des Ephemerines.* — Genève/Paris: 319 pp., 47 tabs.

PIERCE, F.J. 1945. Two new fossils from the upper Miocene of the Puente Hills. — *Bull. South. California Acad. Sci.* **44**: 3–6.

PROVONSHA A.V. 1990. A revision of the genus *Caenis* in North America (Ephemeroptera: Caenidae). — *Trans. Amer. Entomol. Soc.* **116** (4): 801–884.

PROVONSHA, A.V. & MCCAFFERTY, W.P. 1985. *Amercaenis*: new Nearctic genus of Caenidae (Ephemeroptera). — *Int. Quart. Entomol.* **1** (1): 1–7.

PROVONSHA, A.V. & MCCAFFERTY, W.P. 1995. New brushlegged caenid mayfly from South Africa (Ephemeroptera: Caenidae). — *Aquatic Insects* **17** (4): 241–251.

PUTHZ, V. 1973. Eintagsfliegen (Ephemeroptera) aus Südamerika. — *Opus. Zool. (Magy)* **12** (1–2): 91–97 (Cit. ex: Реферативный Журнал 1977:8E89.).

PUTHZ, V. 1975. Eine neue Caenidengattung aus dem Amazonasgebiet (Insecta: Ephemeroptera: Caenidae). — *Amazoniana* **5**: 411–415.

[RASNITSYN, A.P.] Расницын А.П. 1980. Подкласс Lepismatona (=Apterygota s.str.). — In: Историческое развитие класса насекомых. — *Тр. Палеонтол. ин-та АН СССР ["Trudy Paleontol. Inst. Akad. Nauk SSSR"]* **175**: 22–24.

RASNITSYN, A.P. 2002. In: A.P. Rasnitsyn & D.L.J. Quicke (eds.). *History of insects.* Kluwer Acad. Publ., Dordrecht, Boston, London: 517 pp.

RIEK, E.F. 1955. Revision of the Australian mayflies (Ephemeroptera). I. Subfamily Siphlonurinae. — *Austral. Journ. Zool.* **3**: 266–280.

RIEK, E.F. 1963. An Australian mayfly of the family Ephemerellidae (Ephemeroptera). — *Jour. Entomol. Soc. Queensland*, 2: 48–50.

RIEK, E.F. 1970. Ephemeroptera. — In: Mackerras I.M. (ed.). *The Insects of Australia.* CSIRO and Melbourne Univ. Press: 224–240.

RIEK, E.F. 1973. The classification of the Ephemeroptera. — In: W.L. Peters & J.G. Peters (eds.). *Proc. 1st Int. Conf. Ephemeroptera,* Florida A&M Univ. 17–20 August 1970. Leiden, Brill: 160–179.

RIEK, E.F. 1976. An unusual mayfly (Insecta: Ephemeroptera) from the Triassic of South Africa. — *Palaeontol. Africana* **19**: 149–151.ALLEN, R.K. & EDMUNDS, G.F., JR.

ROBACK, S.S. 1966. The Catherwood Foundation Peruvia-Amazon expedition. VI – Ephemeroptera nymphs. — *Monogr. Acad. Nat. Sci. Philad.* **14**: 129–199.

[ROHDENDORF, B.B.] РОДЕНДОРФ Б.Б. 1957. Палеонтологические исследования в СССР. — *Тр. Палеонтол. ин-та АН СССР ["Trudy Paleontol. Inst. Akad. Nauk SSSR"]* **66**: 100 pp.

[ROHDENDORF, B.B.] РОДЕНДОРФ Б.Б. 1968. Направления филогенетического развити крылатых насекомых (Insecta Pterygota) [Trends of phylogenetical development in pterygote insects]. — *Журнал Общей Биологии [Journal of General Biology]* **29** (1): 57–66.

[ROHDENDORF, B.B.] РОДЕНДОРФ Б.Б. 1977. О рационализации названий таксонов высокого ранга в зоологии [On the codification of names of higher taxa in zoology]. — *Палеонтол. Журнал [Paleontol. Journal]* № 2: 14–22.

SARTORI, M. & ELOUARD, J.-M. 1999. Biodiversité aquatique de Madagascar 30: le génre *Cheirogenesia* Demoulin 1952 (Ephemeroptera, Palingeniidae). — *Rev. Suisse Zool.* **106** (2): 325–337.

SARTORI, M. & ELOUARD, J.-M. & RUFFIEUX, L. & L'EPLATTENIER, G. 1999. Description des femelles et morphologie des oeufs de quelques espèces de *Proboscidoplocia* (Ephemeroptera, Ephemeroidea). — *Mitt. Schweiz. Entomol. Ges.* **72**: 55–63.

SARTORI, M. & SOWA, R. 1992. New data on some *Rhithrogena* species from the Near- and Middle East (Ephemeroptera; Heptageniidae). — *Aquatic Insects* **14** (1): 31–40.

SCHOENEMUND, E. 1929. *Habroleptoides* eine neue Ephemeropteren-Gattung. — *Zool. Anz.* **80**: 222–232.

SCHOENEMUND, E. 1930a. Die Unterscheidung der Ephemeropteren-Gattungen *Heptagenia* und *Ecdyonurus*. — *Zool. Anz.* **90**: 45–48.

SCHOENEMUND, E. 1930b. Eintagsfliegen oder Ephemeroptera. — In: Dahl. *Die Tierwelt Deutschlands und der angrenzenden Meeresgebiete.* **19**. Teil. Gustav Fischer Verlag. Jena: 107 S.

SCOTT, D.C. & BERNER, L. & HIRSCH A. 1959. The nymph of the mayfly genus *Tortopus* (Ephemeroptera: Polymitarcidae). — *Ann. Entomol. Soc. Amer.* **52**: 205–213

SCUDDER, S.H. 1880. The Devonian Insects of New Brunswick. — *Anniversary Mem. Boston Soc. Nat. Hist.*: 1–41 1 pl.

SCUDDER S.H. 1890. *The fossil insects of North America, with notes on some european species. II. The Tertiary insects.* N.Y., McMillan et C°.

SELLARDS, E.H. 1907. Types of Permian Insects. Part II. Plectoptera. — *Amer. Journ. Sci.* (4) **23**: 345–355.

SELLARDS, E.H. 1909. Types of Permian Insects. Part III. Megasecoptera, Oryctoblattinidae and Protorthoptera. — *Amer. Journ. Sci.* (4) **27**: 151–173.

SELYS-LONCHAMPS, E., DE 1888. Catalogue raisonné des orthoptères et des Névroptères de Belgique. — *Ann. Soc. Entomol. Belg.* **32**: 103–203.

[SHAROV, A.G.] ШАРОВ А.Г. 1957. Типы метаморфоза насекомых и их взаимоотношения (по сравнительно-онтогенетическим и палеонтологическим данным) [Types of insect metamorphosis and their relationship]. — *Энтомол. Обозрение [Rev. Entomol. URSS]* **36** (3): 569–576.

[SINITSHENKOVA, N.D.] СИНИЧЕНКОВА Н.Д. 1973. Личинки поденок палеарктических видов рода *Rhithrogena* Eaton (Ephemeroptera, Heptageniidae) [Mayfly larvae of the Palaearctic species of the genus *Rhithrogena* Eaton (Ephemeroptera, Heptageniidae)]. — *Вестн. Моск. ун-та ["Vestnik Moscow. Univ."]* № 5: 9–17.

[SINITSHENKOVA, N.D.] СИНИЧЕНКОВА Н.Д. 1976a. Новые раннемеловые поденки (Insecta, Ephemeroptera) из Восточного Забайкалья. — *Палеонтол. Журнал [Paleontol. Journal]* № 2: 85–93.

[SINITSHENKOVA, N.D.] СИНИЧЕНКОВА Н.Д.. 1976b. Поденки рода *Iron* Eaton (Ephemeroptera, Heptageniidae) фауны Кавказа [Mayflies of the genus *Iron* Eaton (Ephemeroptera, Heptageniidae) in the fauna of the Caucasus]. — *Энтомол. Обозрение [Rev. Entomol. URSS]* **60** (4): 853–862.

[SINITSHENKOVA, N.D.] СИНИЧЕНКОВА Н.Д. 1979. Новое семейство палеодиктиоптер из карбона Сибири [A new family of Paleodictioptera from the Carboniferous of Siberia]. — *Палеонтол. Журнал [Paleontol. Journal]* № 2: 74–89.

[SINITSHENKOVA, N.D.] СИНИЧЕНКОВА Н.Д. 1981. К познанию личинок поденок рода *Epeorus* Eaton (Ephemeroptera, Heptageniidae) [To the knowledge of nymphs of the genus *Epeorus* Eaton (Ephemeroptera, Heptageniidae)]. — *Энтомол. Обозрение [Rev. Entomol. URSS]* **60** (4): 813–823.

[SINITSHENKOVA, N.D.] СИНИЧЕНКОВА Н.Д. 1982. Новый вид поденок *Epeorus rautiani* sp. n. из Забайкалья (Ephemeroptera, Heptageniidae) [A new species of mayflies *Epeorus rautiani* sp.n. from Transbaikal (Ephemeroptera, Heptageniidae). — *Бюлл. Моск. о-ва испыт. природы*, отд. биол. [*Bull. Moskow Soc. of Naturalists*, Biol.Ser.] **87** (5): 52–58.

[SINITSHENKOVA, N.D.] СИНИЧЕНКОВА Н.Д. 1985. Юрские поденки (Ephemerida - Ephemeroptera) Южной Сибири и Западной Монголии. — *Тр. Палеонт. ин-та АН СССР ["Trudy Paleontol. Inst. Akad. Nauk SSSR"]* **211**: 11–23.

[SINITSHENKOVA, N.D.] СИНИЧЕНКОВА Н.Д. 1986. Поденки. Ephemerida (= Ephemeroptera). — In: Насекомые в раннемеловых экосистемах западной Монголии [Insects in the Early Cretaceous ecosystems of West Mongolia]. — *Тр. Совместной Советско-Монгольской палеонтол. экспед. [Trans. of Joint Soviet-Mongolian Paleontol. Exped.]* **28**: 45–46.

[SINITSHENKOVA, N.D.] СИНИЧЕНКОВА Н.Д. 1989. Новые мезозойские поденки (Ephemeroptera) из Монголии. — *Палеонтол. Журнал [Paleontol. Journal]* № 3: 39–41.

[SINITSHENKOVA, N.D.] СИНИЧЕНКОВА Н.Д. 1990. Поденки. Ephemeroidea. — In: Поздне-мезозойские насекомые Восточного Забайкалья [Late Mesozoic insects of Eastern Transbaikalia]. — *Тр. Палеонтол. ин-та АН СССР ["Trudy Paleontol. Inst. Akad. Nauk SSSR"]* **239**: 13–20.

[SINITSHENKOVA, N.D.] СИНИЧЕНКОВА Н.Д. 1991. Новые мезозойские поденки из Забайкалья и Монголии [New Mesozoic mayflies from Transbaikalia and Mongolia]. — *Палеонтол. Журнал [Paleontol. Journal]* № 1: 120–123.

[SINITSHENKOVA, N.D.] СИНИЧЕНКОВА Н.Д. 2000a. Новые поденки из верхнемезозойского забайкальского местонахождения Черновские копи (Insecta: Ephemerida = Ephemeroptera) [New mayflies from Upper Mesozoic Transbaikalian locality Chernovskie Kopi (Insecta: Ephemerida = Ephemeroptera). — *Палеонтол. Журнал [Paleontol. Journal]* № 1: 63–69.

SINITSHENKOVA, N.D. 2000b. The first fossil prosopistomatid mayfly from Burmese amber (Ephemeroptera; Prosopistomatidae). — *Bull. nat. Hist. Mus. Lond. (Geol.)* **56** (1): 25–28.

SINITSHENKOVA, N.D. 2000c. New Jersey amber mayflies: the first North American Mesozoic members of the order (Insecta; Ephemeroptera). — In: D.A. Grimaldi (ed.). Studies on fossils in amber, with particular reference to the Cretaceous of New Jersey. Backhuys Publ., Leiden: 111–125.

SINITSHENKOVA, N.D. 2000d. A review of Triassic mayflies, with description of new species from Western Siberia and Ukraine (Ephemerida = Ephemeroptera). — *Paleontol. Journ.* **34**, Suppl. 34: S275–S283.

SNODGRASS, R.E. 1935. *Principles of insect morphology*. New York: 667 pp.

SÖDERSTRÖM, O. & NILSSON, J. 1986. Redescription of *Parameletus chelifer* Bengtsson and *P. minor* (Bengtsson), with keys to nymphal and adult stages of the Fennoscandian species of Siphlonuridae (Ephemeroptera). — *Entomol. Scand.* **17** (1): 107–117.

SOLDÁN, T. 1978a. New genera and species of Caenidae (Ephemeroptera) from Iran, India and Australia. — *Acta Entomol. Bohemoslavaca* **75** (2): 119–129.

SOLDÁN, T. 1978b. Revision of the genus *Palingenia* in Europe (Ephemeroptera, Palingeniidae). — *Acta. Entomol. Bohemoslovaca* **75**: 272–284.

SOLDÁN, T. 1979a. Struktur und Funktion der Maxillarpalpen von *Arthroplea congener* (Ephemeroptera, Heptageniidae). — *Acta Entomol. Bohemoslovaca* **76**: 300–307.

SOLDÁN, T. 1979b. Internal anatomy of *Dolania americana* (Ephemeroptera: Behningiidae). — *Ann. Entomol. Soc. Amer.* **72**: 636–641.

SOLDÁN, T. 1979c. The structure and development of female internal reproductive system in six European species of Ephemeroptera. — *Acta Entomol. Bohemoslov.* **76**: 353–365.

SOLDÁN, T. 1981. Secondary sexual characters in mayfly larvae and their evolutionary significance (Ephemeroptera). — *Acta Entomol. Bohemoslov.* **78**: 140–142.

SOLDÁN, T. 1983. Two new species of *Clypocaenis* (Ephemeroptera, Caenidae) with a description of adult stage and biology of the genus. — *Acta Entomol. Bohemoslovaca* **80** (3): 195–206.

SOLDÁN, T. 1986. A revision of the Caenidae with ocellar tubercles in the nymphal stage (Ephemeroptera). — *Acta Univer. Carolinae - Biologica 1982–1984*: 289–362.

SOLDÁN, T. & LANDA, V. 1991. Two new species of Caenidae (Ephemeroptera) from Sri Lanka. — In: J. Alba-Tercedor & A. Sanchez-Ertega (eds.). *Overview and strategies of Ephemeroptera and Plecoptera* (Proc. 6th Int. Congf. Ephemeroptera & 10th Int. Symp. Plecoptera, 24–30 July 1989, Granada, Spain), Sandhill Crane Press: 235–246.

SOLDÁN, T. & PUTZ, M. 2000. Karyotypes of some Central European mayflies (Ephemeroptera) and their contribution to phylogeny of the order. — *Acta Soc. Zool. Bohem.* **64**: 437–445.

SOWA, R. 1970. Sur la taxonomie de *Rhithrogena semicolorata* (Curtis) et de quelques espèces voisines d'Europe continentale (Ephemeroptera: Heptageniidae). — *Rev. Suisse Zool.* **77** (4), No.56: 895–920.

SPIETH, H.T. 1933. The phylogeny of some mayfly genera. — *Journ. New York Entomol. Soc.* **16**: 237–391.

SPIETH, H.T. 1940. The genus *Ephoron*. — *Canad. Entomol.*, **72**: 109–111.

SPIETH, H.T. 1941. Taxonomic studies on the Ephemeroptera. II. The genus *Hexagenia*. — *Amer. Midland Naturalist* **26**: 233–280.

SPIETH, H.T. 1943. Taxonomic studies on the Ephemeroptera III. Some interesting ephemerids from Surinam and other Neotropical localities. — *Amer. Mus. Novit.* **1244**: 1–13.

SPIETH, H.T. 1947. Taxonomic studies on the Ephemeroptera IV. The genus *Stenonema*. — *Ann. Entomol. Soc. Amer.* **40**: 87–122.

STANICZEK, A.H. 1997. The morphology of *Siphleanigma janae* Penniket (Ephemeroptera, Siphlaenigmatidae), and its significance for the ground plan of the Baetoidea. — In: P. Landolt & M. Sartori (eds.). *Ephemeroptera & Plecoptera. Biology-Ecology-Systematics* (Proc. 8th Int. Conf. Ephemeroptera & 12th Int. Symp. Plecoptera 14–20 August 1995, Losanne, Switzerland). Mauron+Tinguely & Lacht SA, Fribourg / Switzerland: 536–549.

STANICZEK, A.H. 2000. The mandible of silverfish (Insecta: Zygentoma) and mayflies (Ephemeroptera): its morphology and phylogenetic significance. — *Zool. Anz.* **239**: 147–178.

STANICZEK, A.H. & BECHLY, G. 2002. First fossil record of the mayfly family Baetiscidae from Baltic amber (Insecta: Ephemeroptera). — *Stuttgarter Beiträge zur Naturkunde* (B) Nr.322: 1–11.

STEPHENS, J.F. 1835. *Illustrations of British Entomology. Mandibulata*, **6**: 53–70, pl.29.

STRENGER, A. 1953. Zur Kopfmorphologie der Ephemeridenlarven. Erster Teil. *Ecdyonurus* und *Rhithrogena*. — *Oster. Zool. Z.* **4**: 191–228.

STUDEMANN, D. & LANDOLT, P. 1997a. Eggs of Ephemerellidae (Ephemeroptera). — In: P. Landolt & M. Sartori (eds.). *Ephemeroptera & Plecoptera. Biology-Ecology-Systematics* (Proc. 8th Int. Conf. Ephemeroptera & 12th Int. Symp. Plecoptera 14–20 August 1995, Losanne, Switzerland). Mauron+ +Tinguely & Lacht SA, Fribourg / Switzerland: 362–381.

STUDEMANN, D. & LANDOLT, P. 1997b. A phylogenetic system for the European species of *Siphlonurus* (Ephemeroptera, Siphlonuridae). — In: P. Landolt & M. Sartori (eds.). *Ephemeroptera & Plecoptera. Biology-Ecology-Systematics* (Proc. 8th Int. Conf. Ephemeroptera & 12th Int. Symp. Plecoptera 14–20 August 1995, Losanne, Switzerland). Mauron+ +Tinguely & Lacht SA, Fribourg / Switzerland: 554–560.

STUDEMANN, D. & LANDOLT, P. & TOMKA, I. 1987. Complément à la description de *Arthroplea congener* Bengtsson 1908 (Ephemeroptera) et à so statut sytématique. — *Bull. Soc. Friburg. Sci. Nat.* **76** (1–2): 144–167.

STUDEMANN, D. & LANDOLT, P. & TOMKA, I. 1988. Morphology and taxonomy of imagines and eggs of Central and Northern European Siphlonuridae (Ephemeroptera). — *Mitt. Schw. Entomol. Ges.* **61**: 303–328.

SUTER, P.J. 1984. A redescription of the genus *Tasmanocoenis* Lestage (Ephemeroptera: Caenidae) from Australia. — *Trans. R. Soc. Austr.* **108**: 105–111.

SUTER, P.J. 1993. *Wundacaenis,* a new genus of Caenidae (Insecta: Ephemeroptera) from Australia. — *Invertebrate Taxonomy* **7**: 787–803.

SUTER, P.J. 1999. *Irpacaenis*, a new genus of Caenidae (Ephemeroptera) from Australia. — *Australian Journ. Entomol.* **38**: 159–167.

THENIUS, E. 1979. Lebensspuren von Ephemeropteren-Larven aus dem Jung-Tertiär des Wiener Beckens. — *Ann. Naturhist. Mus. Wien* **82**: 177–188.

THEW, T.B. 1960. Revision of the genera of the family Caenidae (Ephemeroptera). — *Trans. Amer. Entomol. Soc.* **86**: 187–205.

THOMAS, A.G.B. & DIA, A. 1982. *Ecdyonurus* (?) *znojkoi* Tshernova 1938: redescription et appartenance générique réelle (Ephemeroptera, Heptageniidae). — *Bull. Soc. Hist. Nat. Tolouse* **118**: 297–303.

TIENSUU, L. 1935. On the Ephemeroptera fauna of Laatokan Karjala (Karelia Ladogensis). — *Ann. Entomol. Fenn.* **1**: 3–23.

TILLYARD, R.J. 1921. A new genus and species of May-fly (Order Plectoptera) from Tasmania, belonging to the family Siphlonuridae. — *Proc. Linn. Soc. New South Wales* **46**: 409–412, pl.34.

TILLYARD, R.J. 1923. Descriptions of two new species of mayflies (order Plectoptera) from New Zealand. — *Trans. New Zealand Inst.* **54**: 226–230.

TILLYARD, R.J. 1932. Kansas Permian Insects. Part 15. The order Plectoptera. — *Amer. Journ. Sci.* (5) **23**: 97–135, 237–272.

TILLYARD, R.J. 1933. The mayflies of Mount Kosciusko region. I. (Plectoptera). Introduction and family Siphlonuridae. — *Proc. Linn. Soc. New South Wales* **58**: 1–32, pl. 1.

TILLYARD, R.J. 1936a. The Trout-Food Insects of Tasmania. Part II. - A monograph of the Mayflies of Tasmania. — *Pap. & Proc. Roy. Soc. Tasmania* 1935: 23–59.

TILLYARD, R.J. 1936b. Kansas Permian Insects. Part 16. The order Plectoptera (contd.): The family Doteridae, with a note on the affinities of the order Protohymenoptera. — *Amer. Journ. Sci.* (5) **32**: 435–453.

[TIUNOVA, T.M.] ТИУНОВА Т.М. 1991. Описание имаго *Potamanthellus rarus* (Ephemeroptera, Neoephemeridae) из Приморья [Description of imago *Potamanthellus rarus* from Primorye]. — *Зоол. Журнал* [*Zool. Zhurnal*] **70** (7): 136–138.

TIUNOVA, T.M. & KLUGE, N.J. & ISHIWATA, SHIN-ICHI. Revision of the East Palaearctic *Isonychia* (Ephemeroptera). — *Canad. Entomol.* (in press).

TIUNOVA, T.M. & LEVANIDOVA, I.M. 1989. Description of new mayfly species (Ephemeroptera) from the Soviet Far East. — *Aquatic Insects* **11** (4): 241–245.

TOMKA, I. & ELPERS, C. 1991. Problems in the phylogeny of the Ephemeroptera. — In: J. Alba-Tercedor & A. Sanchez-Ertega (eds.). *Overview and strategies of Ephemeroptera and Plecoptera* (Proc. 6th Int. Congf. Ephemeroptera & 10th Int. Symp. Plecoptera, 24–30 July 1989, Granada, Spain), Sandhill Crane Press: 115–134.

TOMKA, I. & HASLER, W. 1978. Einsatz des Rasterelektronenmikroskops bei taxonomischen Studien an Eintafsfliegen (Ephemeroptera). — *Bull. Soc. Fribourg Sci. Nat.* **67** (2): 144–151.

TRAVER, J.R. 1933. Heptagenine mayflies of North America. — *Journ. New York Entomol. Soc.* **41**: 105–125.

TRAVER J.R. 1935. Two new genera of North American Heptageniidae (Ephemerida). — *Canad. Entomol.* **67**: 31–38.

TRAVER, J.R. 1938. Mayflies of Puerto Rico. — *Journ. Agric. Univ. Puerto Rico* **22**: 5–42, pl. 1–3.

TRAVER, J.R. 1939. Himalayan mayflies (Ephemeroptera). — *Ann. Mag. Nat. Hist.* (11) **4**: 32–56.

TRAVER, J.R. 1944. Notes on Brazilian mayflies. — *Bol. Mus. Nac. Zool. Rio de Janeiro*, N.S., 22: 2–53.

TRAVER, J.R. 1950. Notes on Neotropical mayflies. Part IV. Family Ephemeridae (continued). — *Rev. Entomol., Rio de Janeiro* **21** (3): 593–614.

TRAVER, J.R. 1958. The subfamily Leptohyphinae (Ephemeroptera: Tricorythidae). Part I. — *Ann. Entomol. Soc. Amer.* **51**: 491–503.

[TSHERNOVA, O.A.] ЧЕРНОВА О.А. 1928. Материалы к познанию поденок бассейна р. Оки [Beitrag zur Kenntnis der Ephemeropteren des Oka-Bassin]. — *Работы Окской биол. ст.*, Муром [*Arbeiten der Biologischen Oka-Station*, Murom] **5** (2–3): 113–115.

[TSHERNOVA, O.A.] ЧЕРНОВА О.А. 1934. Новый вид и род широко распространенной поденки северо-восточной половины СССР [Eine neue weiterverbreitete Ephemeropterengattung aus nordlichen Gebieten der UdSSR]. — *Докл. АН СССР* [*Comptes Rendus Acad. Sci. URSS*] **4** (4): 240–243.

[TSHERNOVA, O.A.] ЧЕРНОВА О.А. 1937. Поденки Днiпра [Die Eintagsfliegen des Dnjepr-Flusses]. — *Тр. гiдробiол. ст. АН УССР* [*Trav. Stat. Hydrobiol. Acad. Sci. RSS Ukraine*] № 15: 3–23.

[TSHERNOVA, O.A.] ЧЕРНОВА О.А. 1938. Новое семейство поденок (Ephemeroptera) / Sur une nouvelle familie Ephemeroptera. — *Изв. АН СССР*, сер. биол. [*Bull. Acad. Sci. URSS*, ser.biol.] № 1: 129–137.

[TSHERNOVA, O.A.] ЧЕРНОВА О.А. 1948. О новом роде и виде поденок из бассейна Амура (Ephemeroptera, Ametropodidae). — *Докл. АН СССР* [*"Doklady Akad. Nauk SSSR"*] **60** (8): 1453–1455.

[TSHERNOVA, O.A.] ЧЕРНОВА О.А. 1949. Нимфы поденок притоков Телецкого озера и р. Бии. — *Труды Зоологического инститyта АН СССР* [*"Trudy Zool. Inst. Akad. Nauk SSSR"*] 7 (4): 139–158.

[TSHERNOVA, O.A.] ЧЕРНОВА О.А. 1952. Поденки (Ephemeroptera) басс. р. Амура и прилежащих вод и их роль в питании амурских рыб. — Труды Амурской ихтиологической экспедиции 1945–1949, т.3. ["Trudy Amur. ichtiol. exped.", Vol.3]. — *Материалы к познанию флоры и фауны СССР, издаваемые Московским обществом испытателей природы* [*"Materialy k poznaniyu flory i fauny SSSR, izdavaemye Moskovskim obschestvom ispytateley priridy"*] nov.ser. **32** (47): 229–360.

[TSHERNOVA, O.A.] ЧЕРНОВА О.А. 1960. О положении семейства Neoephemeridae в системе отряда поденок (Ephemeroptera). — *Научн. докл. высшей школы*. Биол. [*"Nauchn. doklady vysshey shkoly"*. Biol.] № 24: 52–58.

[TSHERNOVA, O.A.] ЧЕРНОВА О.А. 1961. О систематическоми положении и геологическом возрасте поденок рода *Ephemeropsis* Eichwald (Ephemeroptera, Hexagenitidae) [On taxonomical position and geological age of the genus *Ephemeropsis* Eichwald (Ephemeroptera, Hexagenitidae)]. —

Энтомол. Обозрение [*Rev. Entomol. URSS*] **40** (4): 858–869.

[TSHERNOVA, O.A.] ЧЕРНОВА O.A. 1962a. Личинка поденки из Неогена Западной Сибири (Ephemeroptera, Heptageniidae) [Ephemeroptera larvae from the Neogene of West Siberia (Ephemeroptera, Heptageniidae)]. — *Зоол. Журнал* [*Zool. Zhurnal*] **41** (6): 943–945.

[TSHERNOVA, O.A.] ЧЕРНОВА O.A. 1962b. Отряд Ephemeroptera. Поденки. — *Основы палеонтологии*. Членистоногие трахейные и хелицеровые. (ред. Б.Б. Родендорф). М., изд. АН СССР [*"Osnovy Paleontologii"*. Arthropoda Tracheata & Chelicerata. Publ. Acad. Sci. USSR]: 55–64.

[TSHERNOVA, O.A.] ЧЕРНОВА O.A. 1967. Поденка современного семейства в юре Забайкалья [May-fly of the recent family in Jurassic deposits of Transbaikalia (Ephemeroptera, Siphlonuridae)]. — *Энтомол. Обозрение* [*Rev. Entomol. URSS*] **46** (2): 322–326.

[TSHERNOVA, O.A.] ЧЕРНОВА O.A. 1968. Новая поденка из Каратау. — *Юрские насекомые Каратау*, М., изд. АН СССР [*"Jurskie naskomye Karatau"*, Moskow, Publ. Acad. Sci. USSR]: 23–25.

[TSHERNOVA, O.A.] ЧЕРНОВА O.A. 1969. Новые раннеюрсмкие поденки (Ephemeroptera, Epeoromimidae и Mesonetidae) [New Early Jurassic may-flies (Ephemeroptera, Epeoromimidae и Mesonetidae)]. — *Энтомол. Обозрение* [*Rev. Entomol. URSS*] **48** (1): 153–161.

[TSHERNOVA, O.A.] ЧЕРНОВА O.A. 1970. О системе отряда поденок (Ephemeroptera), ископаемых и современных [On the classification of fossil and recent Ephemeroptera]. — *Энтомол. Обозрение* [*Rev. Entomol. URSS*] **49** (1): 124–145.

[TSHERNOVA, O.A.] ЧЕРНОВА O.A. 1971. Поденка из ископаемой смолы меловых отложений полярной Сибири [Mayfly from fossil pitch of Polar Siberia]. — *Энтомол. Обозрение* [*Rev. Entomol. URSS*] **50** (3): 612–618.

[TSHERNOVA, O.A.] ЧЕРНОВА O.A. 1972. Некоторые новые азиатские виды поденок (Ephemeroptera, Heptageniidae, Ephemerellidae) [Some new species of may-flies from Asia (Ephemeroptera, Heptageniidae, Ephemerellidae)]. — *Энтомол. Обозрение* [*Rev. Entomol. URSS*] **51** (3): 604–614.

[TSHERNOVA, O.A.] ЧЕРНОВА O.A. 1974. Родовой состав поденок сем. Heptageniidae в Голорктике и Ориентальной области [The generic composition of may-flies of the family Heptageniidae (Ephemeroptera) in the Holarctic and Oriental Region]. — *Энтомол. Обозрение* [*Rev. Entomol. URSS*] **53** (4): 801–814.

[TSHERNOVA, O.A.] ЧЕРНОВА O.A. 1976. Определительная таблица родов поденок сем. Heptageniidae Голарктики и Ориентальной области по личинкам [A key to the larvae of genera of Heptageniidae (Ephemeroptera) from Holarctic and Oriental Region]. — *Энтомол. Обозрение* [*Rev. Entomol. URSS*] **55** (2): 332–346.

[TSHERNOVA, O.A.] ЧЕРНОВА O.A. 1977. Своеобразные новые личинки поденок (Ephemeroptera: Palingeniidae, Behningiidae) из юры Забайкалья . — *Палеонтол. Журнал* [*Paleontol. Journal*] № 2: 91–96.

[TSHERNOVA, O.A.] ЧЕРНОВА O.A. 1980. Отряд Ephemerida Latreille, 1810. — In: Историческое развитие класса насекомых. — *Тр. Палеонтол. ин-та АН СССР* [*"Trudy Paleontol. Inst. Akad. Nauk SSSR"*] **175**: 31–36.

[TSHERNOVA, O.A.] ЧЕРНОВА O.A. 1981. К систематике имаго поденок рода *Epeorus* Eaton 1881 (Ephemeroptera, Heptageniidae) [On the systematics of adult mayflies of the genus *Epeorus* Eaton 1881 (Ephemeroptera, Heptageniidae)]. — *Энтомол. Обозрение* [*Rev. Entomol. URSS*] **60** (2): 323–336.

[TSHERNOVA, O.A. & BAJKOVA, O.JA.] ЧЕРНОВА O.A., БАЙКОВА О.Я. 1960. Об одном новом роде поденок (Ephemeroptera, Behningiidae) [On a new genus of may-flies (Ephemeroptera, Behningiidae)]. — *Энтомол. Обозрение* [*Rev. Entomol. URSS*] **39** (2): 410–416.

[TSHERNOVA, O.A. & KLUGE N.J. & SINITSHENKOVA, N.D. & BELOV, V.V.] ЧЕРНОВА O.A., КЛЮГЕ Н.Ю., СИНИЧЕНКОВА Н.Д., БЕЛОВ В.В. 1986. Отряд Ephemeroptera – Поденки. — *Определитель насекомых Дальнего Востока СССР*, Л., Наука [*"Opredelitel' nasekomyh Dalnego Vostoka SSSR"*, Leningrad, Nauka] Vol.1: 99–142.

[TSHERNOVA, O.A. & SINITSHENKOVA, N.D.] ЧЕРНОВА O.A., СИНИЧЕНКОВА Н.Д. 1974. Новый род и вид ископаемых поденок семейства Hexagenitidae (Ephemeroptera) с юга европейской части СССР и связь их с современными поденками [A new genus and species of the family Hexagenitidae (Ephemeroptera) from the south of the European part of the USSR and their relations recent ones]. — *Энтомол. Обозрение* [*Rev. Entomol. URSS*] **53** (1): 130–136.

TSUI, P.T.P. & HUBBARD, M.D. 1979. Feeding habits of the predaceous nymphs of *Dolania americana* in northwestern Florida (Ephemeroptera: Behningiidae). — *Hydrobiologia* **67**: 119–123.

TULLOCH, G.S. 1929. The proper use of the terms parapsides and parapsidal furrows. — *Psyche* **36**: 376–382.

UENO, M. 1928. Some Japanese mayfly nymphs. — *Mem. Coll. Sci. Kyoto Imp. Univ.* (B) **4** (1): 19–63.

UENO, M. 1931. Contribution to the knowledge of Japanese Ephemeroptera. — *Annot. Zool. Japan.* **13**: 189–226, pl. 12–13.

UENO, M. 1969. Mayflies (Ephemeroptera) from various regions of Southeast Asia. — *Oriental Insects* **3**: 221–238.

UJHELYI, S. 1960. *Metreletus hungaricus* sp.n., eine neue Eintagsfliege (Ephemeroptera) aus Ungarn. — *Acta Zool. Acad. Sci. Hungar.* **6** (1–2): 199–209

ULMER, G. 1904. Ephemeriden. — In: *Ergebnisse der Hamburger Magalhaensische Sammelreise* **7**: 1–8.

ULMER, G. 1914. Ordn. Ephemeroptera (Agnatha), Eintagsfliegen. — In: P. Brohmer. Fauna von Deutschland: 95–99. (Cit. ex: Hubbard 1990).

ULMER, G. 1916. Ephemeropteren von Äquatorial-Afrika. — *Arch. Naturgesch.* (A) **81**: 1–19.

ULMER, G. 1920a. Neue Ephemeropteren. — *Arch. Naturgesch.* (A) 1919, **85** (11): 1–80.

ULMER, G. 1920b. Übersicht über die Gattungen der Ephemeropteren nebst Bemerkungen über einzelne Arten. — *Stett. Entomol. Zeit.* **81** (1–2): 97–144.

ULMER, G. 1920c. Über die Nymphen einiger exotischer Ephemeropteren. — *Festschrift fur Zschokke* **25**: 1–25.

ULMER, G. 1921. Über einige Ephemeropteren-Typen älterer Autoren. *Arch. Naturgesch.* (A) **87** (6): 229–267.

ULMER, G. 1924a. Einige alte und neue Ephemeropteren. — *Konowia* **3** (1/2): 23–37.

ULMER, G. 1924b. Trichopteren und Ephemeropteren. — In. Wiss. Ergebn. von F. Werner unternomm. zool. Exped. anglo. — *Denkschr. Mathem.-Naturw. Klasse Akad. Wiss. Wien* **99**: 1–9 (Cit. ex: Hubbard 1990).

ULMER, G. 1924c. Ephemeriden von den Sunda-Inseln und den Philipinen. — *Treubia* **6** (1): 28–91.

ULMER, G. 1924d. Eine merkwurdige Ephemeriden-Nymphe aus der Wolga. — *Работы Волжской Биол. Ст.*, Саратов [*"Raboty Volzhskoy Biol. St."*, Saratov] **7** (3): 1–8.

ULMER, G. 1925. Beiträge zur Fauna sinica. 3. Trichopteren und Ephemeropteren. — *Arch. Naturgesch.* (A) **91** (5): 20–110.

ULMER, G. 1936. Neue chinesische Ephemeropteren, nebst Übersicht über die bisher aus China bekannten Arten. — *Peking Nat. Hist. Bull.* **10**: 201–215.

ULMER, G. 1938. Chilenische Ephemeropteren, hauptsachlich aus dem Deutschen Entomologischen Institut, Berlin-Dahlem. —

Arb. Morphol. Taxon. Entomol. Berlin-Dahlem **5** (2): 85–108.

ULMER G. 1939. Eintagsfliegen (Ephemeropteren) von den Sunda-Inseln. — *Arch. Hydrobiol.* Suppl.16: 443–692, fig. 1–469, 4 tab.

ULMER, G. 1942. Alte und neue Eintagsfliegen (Ephemeroptera) aus Süd- und Mittelamerika (Teil I). — *Stettiner Entomol. Zeit.* **103**: 98–128.

ULMER, G. 1943. Alte und neue Eintagsfliegen (Ephemeropteren) aus Sud- und Mittelamerika (Teil II). — *Stettiner Entomol. Zeit.* **104**: 14–46.

VAYSSIÈRE, A. 1881. Étude sur l'etat parfait du *Prosopistoma punctifrons*. — *Ann. Sci. Nat. Zool.* Ser.6, **11**: 1–15, Pl.I.

VAYSSIÈRE, A. 1890. Monographie zoologique et anatomique du génre *Prosopistoma* Latr.. — *Ann. Sci. Nat. Zool.* Ser.7, **9**: 19–87, Pl. 2–5.

VIA, L. & CALSADA, S. 1987. Arthropods fossiles de triasicos de Alcover-Montral I. Insectos. — *Cuadernos Geol. Iberica* **11**: 273–280.

WALSH, B.D. 1862. List of the Pseudoneuroptera of Illinois contained in the cabinet of the writer, with descriptions of over forty new species, and notes on their structural affinities. — *Proc. Acad. Nat. Sci. Philadelphia*: 361–402.

WALSH, B.D. 1863. Observations on certain N. A. Neuroptera, by H. Hagen, M. D., of Koenigsberg, Prussia; translated from the original French ms., and published by permission of the author, with notes and descriptions of about twenty new N. A. sp. of Pseudoneuroptera. — *Proc. Entomol. Soc. Philadelphia* **2**: 167–272.

WALTZ, R.D. & MCCAFFERTY, W.P. 1987d. New genera of Baetidae for Nearctic species previously included in *Baetis* Leach (Ephemeroptera). — Ann. Entomol. Soc. Amer. **80** (5): 667–670.

WANG, T.-Q. & MCCAFFERTY, W.P. 1995a. Specific assignment in *Ephemerellina* and *Vietnamella* (Ephemeroptera, Ephemerellidae). — *Entomol. News* **106** (4): 193–194.

WANG, T.-Q. & MCCAFFERTY, W.P. 1995b. Relationships of the Athropleidae, Heptageniidae, and Pseudironidae (Ephemeroptera: Heptagenioidea). — *Entomol. News* **106** (5): 251–256.

WANG, T.-Q. & MCCAFFERTY, W.P. 1995c. First larval descriptions, new species and evaluation of the southeast Asian genus *Atopopus* (Ephemeroptera, Heptageniidae). — *Bull. Soc. Hist. Nat. Toulouse* **131**: 19–25.

WANG, T.-Q. & MCCAFFERTY, W.P. 1996a. New genus of Teloganodinae (Ephemeroptera: Pannota: Ephemerellidae) from Sri Lanka. — *Bull. Soc. Hist. Nat. Toulouse* **132**. 15–18.

WANG, T.-Q. & MCCAFFERTY, W.P. 1996b. New diagnostic characters for the family Baetidae (Ephemeroptera). — *Entomol. News* **107** (4): 207–212.

WANG, T.-Q. & MCCAFFERTY, W.P. 1996c. Redescription and reclassification of the South American mayfly *Melanemerella brasiliana* (Ephemeroptera: Leptophlebiidae). — *Entomol. News* **107** (2): 99–103.

WANG, T.-Q. & MCCAFFERTY, W.P. & BAE, Y.J. 1997. Sister relationships of the Neoephemeridae and Caenidae (Ephemeroptera: Pannota). — *Entomol. News* **108** (1): 52–56.

WANG, T.-Q. & MCCAFFERTY, W.P. & EDMUNDS, G.F., JR. 1995. Larva and adult of *Teloganella* (Ephemeroptera: Pannota) and assessment of familial classification. — *Ann. Entomol. Soc. Amer.* **88** (3): 324–327.

WESTWOOD, J.O. 1840. *An introduction to the modern classification of insects.* — Longman, Orme, Brown, Green, and Longmans, London. Vol.1–2.

WHALLEY, P.E.S. & JARZEMBOWSKI, E.A. 1985. Fossil insects from the lithographic limestone of Montsech (Late Jurassic – Early Cretaceous), Lerida Province, Spain. — *Bull. Brit. Mus. Nat. Hist. (Geol.)* **38** (5): 381–412.

WHITING, E.R. & LEHMKUHL, D.M. 1987a. *Raptoheptagenia cruentata*, gen. nov. (Ephemeroptera: Hepatgeniidae), new association of the larvae previously thought to be Anepeorus with the adult of Heptagenia cruentata Walsh. — *Canad. Entomol.* **119**: 405–407.

WHITING, E.R. & LEHMKUHL, D.M. 1987b. *Acanthomola pubescens*, a new genus and species of Heptageniidae (Ephemeroptera) from western Canada. — *Canad. Entomol.* **119**: 409–417.

WIERSEMA, N.A. & MCCAFFERTY, W.P. 2000. Generic revision of the North and Central American Leptohyphidae. — *Trans. Amer. Entomol. Soc.* **126** (3+4): 337–371.

WILLIAMSON, H. 1802. On the *Ephoron leukon*, usually called the white fly of the Passaick River. — *Trans. Amer. Philos. Soc.* **5**: 71–73.

WU, W.-Y. & YOU, D.-S. 1986. A new genus and species of Potamanthidae from China (Ephemeroptera). — *Acta Zootaxonomica Sinica* **11**: 401–405.

YOON, I.B. & BAE, Y.J. 1984. The classification of Heptageniidae (Ephemeroptera) in Korea. — *Entomol. Res. Bull., Korean Entomol. Inst., Korea Univ.* **10**: 1–34.

YOON, I.B. & BAE, Y.J. 1985. The classification of Ephemeroidea (Ephemeroptera) in Korea. — *Entomol. Res. Bull., Korean Entomol. Inst., Korea Univ.* **11**: 93–109.

YOON, I.B. & BAE, Y.J. 1988. The classification of the Ephemerellidae (Ephemeroptera) in Korea. — *Entomol. Res. Bull., Korean Entomol. Inst., Korea Univ.* **14**: 23–44.

YOU, DA-SHOU 1984. A revison of genus *Potamanthodes* with a description of two new species (Ephemeroptera, Potamanthidae). — In: V. Landa, T. Soldan & M. Tonner (eds.). *Proc. 4th Int. Conf. Ephemeroptera,* Bechyne, 4–10 Sptember 1983. Czechoslovak Acad. Sci., Cseske Budejovice: 101–107.

[YOY, DA-SHOU & SU, CUI-RONG 1987. A new species of *Vietnamella* from China]. — *Acta Zootaxonomica Sinica* **12** (2): 176–180 (in Chinese).

[ZHANG, JUN. 1988. On the discovery of *Eatonigenia* (Ephemeroptera: Ephemeridae) in China]. — *Journ. Nanjing Normal Univ., Nat. Sci.* **3**: 67–68 (In Chinese).

ZHOU, CHANG-FA & PETERS, J. 2003. The nymph of *Siphluriscus chinensis* and additional imaginal description: a living mayfly with Jurassic origin (Siphluriscidae new family. Ephemeroptera). *Florida Entomologist* **86**: 345 352.

ZHOU, CHANG-FA & ZHENG, LE-YI 2000. A new species of the genus Neoephemera McDunnough from China (Ephemeroptera: Neoephemeridae) — Aquatic Insects **23** (4): 327–332.

ZIMMER, C. 1897. Die Facettenaugen der ephemeriden. — *Zeitschr. Wiss. Zool.* **63**: 236–262, Taf.XII–XIII.

ZLOTY, J. 1996. A revision of the Nearctic *Ameletus* mayflies based on adult males, with description of seven new species (Ephemeroptera: Ameletidae). — *Canad. Entomol.* **128**: 293–346.

ZLOTY, J. 1997. Larvae and adults of *Ameletus* mayflies (Ephemeroptera, Ameletidae) from Alberta. — *Canad. Entomol.* **129**: 251–289.

ZURWERRA, A. & TOMKA, I. 1985. *Electrogena* gen. nova, eine neue Gattung der Heptageniidae (Ephemeroptera). — *Entomol. Ber. Luzern* 1985 (13): 99–104.

ALPHABETIC INDEXES

Index of species names

Species names belonging to Turbanoculata and Leptophlebia/fg1 apart of those mentioned in the text, are not listed here. As in the whole book, species names are given in their original combinations and with original endings. For each species name, after a dash ("–"), its place in the book is indicated: For each type species of a genus-group name (valid or invalid) is given a corresponding genus-group name after the words "**typus nominis**" (i.e. type species of the name ...). In a few cases instead of "typus nominis" is written "**typ.nom.**". For each non-type species listed as belonging to a certain taxon is given a name of this taxon after the preposition "**in**". For a junior synonym is given the oldest synonym after the abbreviation "**syn.**". Abbreviation "INC.SED." means "INCERTAE SEDIS".

autumnalis [*Ecdyonurus*] – in Ecdyonurus/fg2
autumnalis [*Ephemerella*] – syn. *spinifera* [*Ephemerella*]
autumnalis [*Siphlonurus*] – in Siphlonurus/fg4
axillaris [*Ephemera*] – in Ephemera/fg10 INC.SED.
axillata [*Caenis*] – in Furcatergaliae INC.SED.
ayadi [*Rhithrogena*] – in Rhithrogena/fg4
azerbajdshanicus [*Ecdyonurus*] – in Electrogena/g(1)

B

bachofeni [*Heptagenia*] – in Radulapalpata INC.SED.
baekdu [*Ecdyonurus*] – in Notacanthurus/g(1)
bajaensis [*Caenis*] – in Caenis/f6=g4
bajkovae [*Ecdyonurus*] – syn. *zhiltzovae* [*Paracinygmula*]
bajkovae [*Rhithrogena*] – in Rhithrogena/fg4
bajkovi [*Baetisca*] – syn. *lacustris* [*Baetisca*]
balcanicus [*Metretopus*] – in Metreletus/g(1)
baltica [*Cinygma*] – in Radulapalpata INC.SED.
barbaroides [*Siphlonurus*] – in Siphlonurus/fg4
barbarus [*Siphlonurus*] – in Siphlonurus/fg4
barbus [*Tricorythodes*] – in Tricorythodes/fg1
baritu [*Haplohyphes*] – in Haplohyphes/g(1)
barnardi [*Afronurus*] – in Afronurus/g1
barnardi [*Ephemerellina*] – typus nominis *Ephemerellina*
bartoni [*Ephemerella*] – typus nominis *Dentatella*
basalis [*Baetis*] – in Siphloplecton/fg(1)
basalis [*Ephemerella*] – in Drunella/g1
basiri [*Rhithrogena*] – in Rhithrogena/fg3 INC.SED.
baskale [*Oligoneuriella*] – syn. *tskhomelidzei* [*O.*]
basuto [*Caenis*] – in Caenis/f6=g4
bauernfeindi [*Serratella*] – in Ephemerella/fg3 INC.SED.
baumanni [*Leptohyphes*] – syn. *mirus* [*Leptohyphes*]
becki [*Baetisca*] – in Baetisca/f3=g2
bednariki [*Stenonema*] – in Maccaffertium/g(1)
beipiaoensis [*Mesoneta*] – in Euplectoptera INC.SED.
belfiorei [*Caenis*] – syn. *pseudorivulorum* [*Caenis*]
belfiorei [*Ecdyonurus*] – in Ecdyonurus/fg2
belgica [*Torleya*]
— typus nominis *Torleya*
— syn. *major* [*Ephemerella*]
bella [*Caenis*] – in Caenis/f6=g4
bella [*Heptagenia*] – in Afghanurus/g2
bellieri [*Baetis*] – in Ecdyonurus/fg2
bellum [*Stenonema*] – syn. *mexicana* [*Heptagenia*]
bellus [*Ameletus*] – in Ameletus/fg2
bellus [*Tortopus*] – in Tortopus/g(1)
benedicta [*Hexagenia*] – syn. *albivitta* [*Baetis*]
bengalensis [*Ecdyonurus*] – in Radulapalpata INC.SED.
benkerti [*Geisfeldiella*] – typus nominis *Geisfeldiella*
bequerti [*Adenophlebia*] – in Notonurus/g
berenice [*Siphlonurus*] – syn. *typicus* [*Siphlurus*]
bernardezi [*Epeorus*] – syn. *torrentium* [*Epeorus*]
berneri [*Baetisca*] – in Baetisca/f3=g2
berneri [*Brachycercus*] – in Brachycercus/f1=g3
berneri [*Caenis*] – in Caenis/f6=g4
berneri [*Ephemerella*] – in Ephemerella/fg3 INC.SED.
berneri [*Leptohyphes*] – in Leptohyphes/fg1 INC.SED.
berneri [*Isonychia*] – in Isonychia/fg2
berolinensis [*Ephemera*] – syn. *venosa* [*Ephemera*]
beskidensis [*Caenis*] – syn. *pseudorivulorum* [*Caenis*]
beskidensis [*Rhithrogena*] – in Rhithrogena/fg4
betteni [*Potamanthus*] – syn. *verticis* [*Baetis*]
beybienkoi [*Epeoromimus*] – in Euplectoptera INC.SED.
bibisica [*Tritogenesia*] – typus nominis *Tritogenesia*
bicaudata [*Oligoneuriella*] – in Oligoneuriella/g2
bicaudatum [*Afroptilum*] – typus nominis *Micksiops*
bicolor [*Ephemerella*] – in Eurylophella/fg2
bicolor [*Ichthybotus*] – in Ichthybotus/fg(1)
bicolor [*Neoephemera*] – typus nominis *Neoephemera*
bicolor [*Palingenia*] – in Isonychia/fg2
bicoloroides [*Ephemerella*] – in Eurylophella/fg2
bicornis [*Caenis*] – in Caenis/f6=g4
bicornis [*Ephemerella*] – syn. *cryptomeria* [*Ephemerella*]
bifasciatum [*Centroptilum*] – typus nominis *Centroptiloides*
bifasciatus [*Ecdyonurus*] – in Radulapalpata INC.SED.
bifurcata [*Ephemerella*] – syn. *sachalinensis* [*E.*]
bifurcatus [*Epeorus*] – in Epeorus/fg4
biguttatus [*Misthodotes*] – in Permoplectoptera
bilineata [*Baetis*] – in Hexagenia/fg3
bilineatus [*Ecdyonurus*] – in Ecdyonurus INC.SED.
billi [*Proboscidoplocia*] – in Proboscidoplocia/g(1)
binerve [*Rhithrogena*] – in Epeiron/g(1)
binotata [*Rhithrogena*] – syn. *lepnevae* [*Rhithrogena*]
binotatus [*Siphlurus*] – in Siphlonurus/fg4
bipunctata [*Cinygma*] – in Arthroplea/fg(1)
bipunctata [*Heptagenia*] – syn. *flava* [*Heptagenia*]
bipunctatus [*Ecdyonurus*] – syn. *terminatum* [*Palingenia*]
birdi [*Stenonema*] – syn. *femorata* [*Baetis*]
birmanica [*Anagenesia*] – in Anagenesia/g2
birmanus [*Polymitarcys*] – in Polymitarcys/f3=Ephoron/g4
bisetosa [*Clypeocaenis*] – typus nominis *Clypeocaenis*
bishopi [*Caenoculis*] – typus nominis *Caenoculis*
bishopi [*Thalerosphyrus*] – in Ecdyonurus/fg1 INC.SED.
bispina [*Ephemerella*] – syn. *walkeri* [*Ephemerella*]
bispinosus [*Epeorus*] – in Epeorus/fg4
blanda [*Ephemera*] – in Ephemera/fg11
boanovae [*Lachlania*] – in Lachlania/g1
bocainensis [*Campylocia*] – in Campylocia/g(1)
bogoescui [*Rhithrogena*] – in Rhithrogena/fg4
boja [*Ephemerella*] – in Rhionella/g(1)
bollenganus [*Ecdyonurus*] – syn. *helveticus* [*Ecdyurus*]
boluensis [*Electrogena*] – in Electrogena/g(1)
borakensis [*Ephemerella*] – syn. *submontana* [*E.*]
borealis [*Eatonia*] – typus nominis *Eatonia*
borealis [*Heptagenia*] – in Metretopus/fg2
boreus [*Prosopistoma*] – in Prosopistoma/f1=g2
borneonia [*Epeorella*] – typus nominis *Epeorella*
borysthenica [*Oligoneuriella*] – typ.nom. *Oligoneurisca*
bothmeri [*Ecdyonurus*] – in Electrogena/g(1)
braaschi [*Ecdyonurus*] – in Electrogena/g(1)
braaschi [*Rhithrogena*] – in Rhithrogena/fg4
brasiliana [*Melanemerella*] – typus nominis *Melanemerella*
brasilianus [*Campsurus*] – in Campsurus/fg3
braueri [*Phacelobranchus*]
— typus nominis *Phacelobranchus*
— syn. *trisetalis* [*Ephemeropsis*]
brenneriana [*Rhithrogena*] – syn. *alpestris* [*Rhithrogena*]
brevicauda [*Ephemera*] – in NOMINA DUBIA
brevipes [*Caenis*] – in Caenis/f6=g4
brevissimus [*Leptohyphes*] – in Leptohyphes/fg1 INC.SED.
brincki [*Ephemerellina*] – in Melanemerella/fg1
britoi [*Palaeobaetodes*] – in Hexagenites/fg1
brocha [*Eburella*] – typus nominis *Eburella*
brodskyi [*Rhithrogena*] – in Rhithrogena/fg4

elongata [*Bantisca*] – typus nominis *Bantisca*
elongata [*Jarmila*] – typus nominis *Jarmila*
elongatula [*Leptophlebia*] – in Ephemerella/fg3 INC.SED.
elongensis [*Afronurus*] – in Afronurus/g1
elouardi [*Caenis*] – in Caenis/f6=g4
emmavillensis [*Cloeon*] – in Euplectoptera INC.SED.
emersoni [*Campsurus*] – in Campsurus/fg3
endensis [*Rhithrogena*] – in Rhithrogena/fg4
enoensis [*Eurylophella*] – in Eurylophella/fg2
eocaenicus [*Parabaetis*] – typus nominis *Parabaetis*
eophilum [*Ephoron*] – in Polymitarcys/f3=Ephoron/g4
epeorides [*Ecdyonurus*] – in Ecdyonurus/fg2
ermolenkoi [*Epeorus*] – in Belovius/g2
erratus [*Epeorus*] – in Radulapalpata INC.SED.
erytrophthalma [*Ephemera*] – syn. *ignita* [*Ephemera*]
escambiensis [*Baetisca*] – typus nominis *Fascioculus*
escomeli [*Spaniophlebia*] – in Spaniophlebia/g(1)
essequibo [*Campsurus*] – in Campsurus/fg3
ethiopicus [*Thalerosphyrus*] – in Ecdyonurus/fg1 INC.SED.
etowah [*Cercobrachys*] – typus nominis *Cercobrachys*
eugeniae [*Rhithrogena*] – in Epeiron/g(1)
eugenii [*Ameletus*] – syn. *inopinatus* [*Ameletus*]
euphratica [*Drunella*] – in Ephemerella/fg3 INC.SED.
europaeus [*Brachycercus*] – in Brachycercus/f1=g3
euterpe [*Ephemerella*] – syn. *maculata* [*Ephemerella*]
evae [*Afromera*] – in Afromera/g1
evanidus [*Campsurus*] – in Campsurus/fg3
excelsus [*Ecdyonurus*] – in Radulapalpata INC.SED.
excisa [*Rhithrogena*] – in Rhithrogena/fg4
excrucians [*Ephemerella*] – typus nominis *Ephemerella*
exiguum [*Stenonema*] – in Maccaffertium/g(1)
exilis [*Rhithrogena*] – in Rhithrogena/fg3 INC.SED.
eximia [*Analetris*] – typus nominis *Analetris*
eximius [*Leptohyphes*] – typus nominis *Leptohyphes*
expectata [*Rhithrogena*] – in Rhithrogena/fg4
expectans [*Potamanthus*] – in Ephemera/fg10 INC.SED.
explanatus [*Siphlurites*] – typus nominis *Siphlurites*
explicatus [*Tricorythus*] – typus nominis *Tricorythodes*
exquisitus [*Ameletus*] – in Ameletus/fg2
ezoensis [*Ephemerella*]–syn. *aurivillii* [*Chitonophora*]
ezoensis [*Ephemerella trispina*] nom.praeocc.
— syn. *triacantha* [*Ephemerella*]

F

facilis [*Ameletus*] – syn. *vancouvernensis* [*Ameletus*]
fallacina [*Tricorythodes*] – in Tricorythodes/fg1
fallax [*Baetis*] – in Electrogena/g(1)
fallax [*Tricorythodes*] – syn. *minutus* [*Tricorythodes*]
falsus [*Ameletus*] – in Ameletus/fg2
famulans [*Iron*] – in Radulapalpata INC.SED.
farsi [*Ecdyonurus*] (in litt.) – in Ecdyonurus/fg2
fasciata [*Caenis*] – in Furcatergaliae INC.SED.
fasciata [*Rhithrogena*] – in Rhithrogena/fg4
fasciatus [*Bleptus*] – typus nominis *Bleptus*
fasciatus [*Waltzoyphius*] – typus nominis *Waltzoyphius*
fascioculatus [*Ecdyonurus*] – syn. *affinis* [*Ecdyonurus*]
fattigi [*Isonychia*] – syn. *bicolor* [*Palingenia*]
fazi [*Siphlonurus*] – syn. *ventilans* [*Siphlonella*]
felsinea [*Caenis*] – syn. *luctuosa* [*Oxycypha*]
femina [*Caenis*] – in Caenis/f4=g2 INC.SED.
feminina [*Ephemerella*] – syn. *rotunda* [*Ephemerella*]
femorale [*Dicercomyzon*] – typus nominis *Dicercomyzon*
femorata [*Baetis*] – in Stenonema/fg3
femorata [*Asiatella*] – in Rhionella/g(1)
femorisetosa [*Clypeocaenis*] – in Clypeocaenis/g3
fennica [*Palmenia*]
— typus nominis *Palmenia*
— syn. *chelifer* [*Parameletus*]
ferreri [*Potamanthus*] – syn. *lutea* [*Ephemera*]
ferruginea [*Ephemera*] – syn. *sulphurea* [*Ephemera*]
ferruginea [*Isonychia*] – syn. *ignota* [*Baetis*]
ferruginea [*Rhithrogena*] – in Rhithrogena/fg4
ferrugineus [*Rhoenanthus*] – syn. *magnificus* [*Rh.*]
ferruginus [*Leptohyphes*] – syn. *zalope* [*Leptohyphes*]
fictus [*Tricorythodes*] – in Tricorythodes/fg1
fictus [*Tricorythopsis*] – in Tricorythopsis/g(1)
fiorii [*Rhithrogena*] – in Rhithrogena/fg4
fittkaui [*Caenis*] – in Caenis/f6=g4
fittkaui [*Homoeoneuria*] – typus nominis *Notochora*
flabellum [*Wundacaenis*] – in Caenis/g4=g2 INC.SED.
flava [*Heptagenia*] – in Heptagenia/f8=g7
flavata [*Heptagenia*] – in Heptagenia/f8=g7
flaveola [*Ephemera*] – syn. *verticis* [*Baetis*]
flavescens [*Palingenia*] – typus nominis *Heptagenia*
flavianula [*Heptagenia*] – in Rhithrogena/fg3 INC.SED.
flavicans [*Ephemera*] – syn. *lutea* [*Ephemera*]
flavida [*Baetis*] – typus nominis *Siphlonurus*
flavilinea [*Ephemerella*] – in Drunella/g1
flavimanus [*Ecdyonurus*] – in Radulapalpata INC.SED.
flavipennis [*Ephemera*] – syn. *longicauda* [*Baetis*]
flavipennis [*Ironodes*] – in Ironodes/g(1)
flavitincta [*Ephemerella*] – syn. *grandis* [*Ephemerella*]
flavitinctus [*Ameletus*] – in Nesameletus/f2=Metamonius/g2
flavomaculatus [*Ecdyonurus*] – syn. *joernensis* [*Ecdyurus*]
flavus [*Brachycercus*] – in Brachycercus/f1=g3
flavus [*Ecdyonurus*] – in Ecdyonurus/fg1 INC.SED.
flexus [*Siphlurus*]
— typus nominis *Siphloplecton*
— syn. *basale* [*Baetis*]
flinti [*Leptohyphes*] – typus nominis *Allenhyphes*
floreus [*Afronurus*] – in Afronurus/g1
floridense [*Stenonema*] – in Stenacron/g(1)
floridicola [*Brachycercus*] – in Brachycercus/f1=g3
floripara [*Ephemerella*] – in Ephemerella/fg3 INC.SED.
flosaquae [*Ephemera*] – syn. *longicauda* [*Ephemera*]
flowersi [*Compsoneuria*] – in Afghanurus/g1 INC.SED.
flowersi [*Nixe*] – in Afghanurus/g2
flowersi [*Thalerosphyrus*] – in Ecdyonuroides/g(1)
fluminum [*Baetis*] – syn. *dispar* [*Baetis*]
foliaceum [*Binoculus*] – in Prosopistoma/f1=g2
fonticola [*Rhithrogena*] – in Rhithrogena/fg4
forcipata [*Caenis*] – syn. *latipennis* [*Caenis*]
forcipatus [*Afrocercus*] – typus nominis *Afrocercus*
forcipula [*Baetis*] – in Ecdyonurus/fg2
formosana [*Ephemera*] – in Ephemera/fg10 INC.SED.
formosanus [*Chirotonetes*] – in Isonychia/fg2
formosicola [*Rhithrogena*] – syn. *parvus* [*Ecdyonurus*]
formosus [*Ameletus*] – in Ameletus/fg2
formosus [*Potamanthus*] – typus nominis *Potamanthodes*
fracta [*Electrogena*] – in Ecdyonurus/fg1 INC.SED.
fradleyi [*Rhithrogena*] – syn. *germanica* [*Rhithrogena*]
fragilis [*Ametropus*] – typus nominis *Ametropus*
fragilis [*Iron*] – in Iron/g2

keiseri [*Manohyphella*] – typus nominis *Manohyphella*
kennedyi [*Heptagenia*] – in Afghanurus/g1 INC.SED.
keralensis [*Cinygmina*] – in Afronurus/g1
kerklotsi [*Ecdyonurus*] – in Radulapalpata INC.SED.
khyberensis [*Eatonia*] – typ.nom.*Eatonia* (nom.praeocc.)
kiangsuensis [*Ecdyonurus*] – in Radulapalpata INC.SED.
kiangsuensis [*Isonychia*] – in Isonychia/fg2
kiboensis [*Ephemerythus*] – in Ephemerythus/fg1
kibunensis [*Ecdyonurus*] – in Ecdyogymnurus/g(1)
kidahi [*Elassoneuria*] – in Elassoneuria/g2
kihada [*Heptagenia*] – in Kageronia/g(1)
kimminsi [*Rhithrogena*] – in Rhithrogena/fg4
kimminsis [*Caenis*] – in Caenis/f4=g2 INC.SED.
kirgisicus [*Iron*] – in Iron/g2
kirinensis [*Ephemera*] – in Ephemera/fg11
kivuensis [*Caenis*] – typus nominis *Caenomedea*
klausnitzeriana [*Rhithrogena*] – in Rhithrogena/fg4
klugei [*Ecdyonurus*] – in Afghanurus/g1 INC.SED.
knowlesi [*Caenis*] – in Caenis/f6=g4
kogistana [*Ephemerella*] (in litt.) – in Torleya/g2
kohnoi [*Ephemerella*] – in Drunella/g1
kootenai [*Cinygmula*] – in Cinygmula/g1
kopetdagi [*Caenis*] – in Caenis/f6=g4
koreanicus [*Iron*] – syn. *aesculus* [*Epeorus*]
koshunensis [*Ephemera*] – in Ephemera/fg10 INC.SED.
kozhovi [*Ephemerella*] – in Ephemerella/fg5
krieghoffi [*Chitonophora*] – syn. *mucronata* [*Chitonophora*]
krueperi [*Potamanthus*] – in Ecdyonurus/fg2
kuccharonis [*Ephemera*] – syn. *orientalis* [*Ephemera*].
kugleri [*Afronurus*] – in Electrogena/g(1)
kungu [*Caenis*] – in Caenis/f6=g4
kuraensis [*Ecdyonurus*] – in Electrogena/g(1)
kurenzovi [*Cinygma*] – in Cinygmula/g1
kurobensis [*Cinygma*] – in Radulapalpata INC.SED.
kuwamayai [*Ephemera*] – syn. *strigata* [*Ephemera*]
kwangsuensis [*Potamanthodes*] – in Rhoenanthus/g1
kyotoensis [*Ameletus*] – in Ameletus/fg2
kyotoensis [*Heptagenia*] – in Heptagenia/f8=g7

L

L-nigrum [*Epeorus*] – in Epeorus/fg3 INC.SED.
labiatus [*Ameletus*] – in Ameletus/fg2
laciniosa [*Rhithrogena*] – in Rhithrogena/fg4
lactata [*Ephemerella*] – syn. *ignita* [*Ephemera*]
lactea [*Oxycypha*] – in Caenis/f6=g4
lactella [*Caenis*] – syn. *horaria* [*Ephemera*]
lacusalbinae [*Ameletoides*] – typus nominis *Ameletoides*
lacuscoerulei [*Tasmanophlebia*] – in Tasmanophlebia/g1
lacustris [*Baetisca*] – in Baetisca/f3=g2
lacustris [*Caenis*] – in Brachycercus/f1=g3
lacustris [*Ephemerella*] – in Ephemerella/fg3 INC.SED.
lacustris [*Siphlonurus*] – in Siphlonurus/fg4
lacustris [*Tasmanophlebia*] – typ.nom. *Tasmanophlebia*
ladakhensis [*Notacanthurus*] – in Notacanthurus/g(1)
ladogensis [*Polymitarcys*]
— syn. *nigridorsum* [*Eopolymitarcys*]
lagueatus [*Mongologenites*] – typ.nom. *Mongologenites*
lagunitas [*Iron*] – in Epeorus/fg3 INC.SED.
lahaulensis [*Epeorus*] – in Radulapalpata INC.SED.
lameerei [*Lithoneura*] – typus nominis *Lithoneura*
lanceolatus [*Tricorythus*] – in Tricorythus/fg5
lanchi [*Neopotamanthodes*]
— typus nominis *Neopotamanthodes*
— syn. *magnificus* [*Rhoenanthus*]
landai [*Cinygmina*] – in Afronurus/g1
landai [*Rhithrogena*] – in Rhithrogena/fg4
lankensis [*Ephemera*] – in Ephemera/fg10 INC.SED.
lapidula [*Ephemerella*] – syn. *flavilinea* [*Ephemerella*]
larssoni [*Succinogenia*] – typus nominis *Succinogenia*
lata [*Ephemerella*] – in Drunella/g1
lata [*Mesoneta*] – typus nominis *Furvoneta*
lata [*Palingenia*] – typus nominis *Anagenesia*
lateralis [*Baetis*] – typus nominis *Electrogena*
latericius [*Iron*] – syn. *maculatus* [*Iron*]
latifolium [*Epeorus*] – typus nominis *Belovius*
latifrons [*Cinygmula*] – syn. *hirasana* [*Cinygma*]
latipennis [*Caenis*] – in Caenis/f6=g4
latipennis [*Palingenia*] – typus nominis *Campsurus*
latipes [*Ephemerella*] – in Drunella/g1
latum [*Protereisma*] – in Permoplectoptera
latus [*Siphlurus*] – syn. *armatus* [*Siphlonurus*]
latus [*Tricorythus*] – typus nominis *Tricorythurus*
laurencae [*Cheirogenesia*] – in Cheirogenesia/g(1)
laurentiana [*Baetisca*] – in Baetisca/f3=g2
lawrencei [*Oligoneuriopsis*] – typ.nom. *Oligoneuriopsis*
lenati [*Stenonema*] – in Maccaffertium/g(1)
lenoki [*Ephemerella*] – typus nominis *Uracanthella*
leonardi [*Cratoligoneuriella*] – typ.nom. *Cratoligoneuriella*
lepidus [*Ironodes*] – in Ironodes/g(1)
leplattenierae [*Proboscidoplocia*] – in Proboscidoplocia/g1
lepnevae [*Ephemerella*] – in Drunella/g1
lepnevae [*Rhithrogena*] – in Rhithrogena/fg4
lepton [*Stenonema*] – syn. *terminatum* [*Palingenia*]
lerida [*Mesopalingea*] – typus nominis *Mesopalingea*
lestagei [*Behningia*] – in Behningia/fg3
lestes [*Leptohyphes*] – in Leptohyphes/fg1 INC.SED.
leucon [*Ephoron*] – typus nominis *Ephoron*
leucophthalma [*Ephemera*] – syn. *sulphurea* [*Ephemera*]
leucoptera [*Anagenesia*] – in Anagenesia/g2
levanidovae [*Ephemerella*] – in Cincticostella/g3
levanidovae [*Iron*] – syn. *aesculus* [*Epeorus*]
levanidovi [*Cinygmula*] – in Cinygmula/g1
levis [*Epeorus*] – in Afronurus/g1
levis [*Ephemerella*] – in Ephemerella/fg3 INC.SED.
lichyi [*Tricorythodes*] – in Tricorythodes/fg1
liebenauae [*Caenis*] – in Caenis/f6=g4
lieftincki [*Heptagenia*] – in Radulapalpata INC.SED.
lieftincki [*Prosopistoma*] – in Prosopistoma/f1=g2
ligata [*Heptagenia*] – in Radulapalpata INC.SED.
limai [*Protoligoneuria*] – typus nominis *Protoligoneuria*
limbata [*Ephemera*] – typus nominis *Hexagenia*
limbata [*Heptagenia*] – Radulapalpata INC.SED.
limnobium [*Ephoron*] – in Polymitarcys/f3=Ephoron/g4
linae [*Madecassorythus*] – in Madecassorythus/fg2
lineata [*Ephemera*] – in Ephemera/fg11
lineata [*Ephemerella*] – syn. *temporalis* [*Ephemerella*]
lineatus [*Ameletus*] – in Ameletus/fg2
liniti [*Leptohyphes*] – in Leptohyphes/fg2
linnaeanus [*Siphlurus*] – syn. *alternata* [*Baetis*]
lita [*Ephemerella*] – in Danella/g(1)
litaninensis [*Campsurus*] – in Campsurus/fg3
lithophilus [*Tineites*] – in Permoplectoptera
lithuanica [*Ephemerella*] – syn. *karelica* [*Eurylophella*]

martinus [*Iron*] – in Epeorus/fg3 INC.SED.
martynovae [*Ephemeropsis*] – in Hexagenites/fg1
matilda [*Isonychia*] – syn. *bicolor* [*Palingenia*]
matitensis [*Afronurus*] – in Ecdyonurus/fg1 INC.SED.
maxillaris [*Trichogenia*] – typus nominis *Trichogenia*
maxima [*Caenis*] – typus nominis *Caenomera*
maxima [*Ephemerella*] – syn. *aurivillii* [*Chitonophora*]
mccaffertyi [*Proboscidoplocia*] – in Proboscidoplocia/g(1)
mcfarlanei [*Rallidens*] – typus nominis *Rallidens*
media [*Ephemera*] – in Ephemera/fg10 INC.SED.
media [*Teloganopsis*] – typus nominis *Teloganopsis*
mediopunctatum [*Ecdyonurus*] – in Maccaffertium/g(1)
medius [*Potamanthus*] – syn. *myops* [*Ephemera*]
medius [*Siberiogenites*] – in Siberiogenites/g
megalopteria [*Mirawara*] – in Mirawara/g(1)
melanobranchus [*Leptohyphes*]
— in Leptohyphes/fg1 INC.SED.
melanocephalus [*Campsurus*] – in Campsurus/fg3
melanurus [*Ephemeropsis*] – syn. *trisetalis* [*Ephemeropsis*]
mellea [*Baetis*] – syn. *lutea* [*Ephemera*]
melli [*Thalerosphyrus*] – in Radulapalpata INC.SED.
mendesi [*Brasiliocaenis*] – in Brasiliocaenis/g(1)
merga [*Protobehningia*] – in Protobehningia/g(1)
meridensis [*Tintorina*] – typus nominis *Tintorina*
meridies [*Caenopsella*] – typus nominis *Caenopsella*
meridionalis [*Pseudiron*] – syn. *centralis* [*Pseudiron*]
meririvulanum [*Stenonema*] – in Maccaffertium/g(1)
mesoleucus [*Potamanthus*] – in Ephemerella/fg3 INC.SED.
mesopotamica [*Palingenia*] – typus nominis *Mortogenesia*
metlacensis [*Iron*] – in Epeorus/fg3 INC.SED.
metriotes [*Stenonema*] – syn. *mexicana* [*Heptagenia*]
mexicana [*Heptagenia*] – in Maccaffertium/g(1)
mexicana [*Hexagenia*] – typus nominis *Pseudeatonica*
meyeri [*Campsurus*] – in Campsurus/fg3
meyi [*Ecdyonurus*] – in Electrogena/g(1)
michaeli [*Leptohyphes*] – in Leptohyphes/fg1 INC.SED.
micheneri [*Ephemerella*] – in Ephemerella/fg3 INC.SED.
micus [*Ameletus*] – in Ameletus/fg2
midas [*Siphlonuroides*] – in Parameletus/fg2
middendorfi [*Ephemeropsis*] – syn. *trisetalis* [*E.*]
mikhaili [*Torleya*] – in Torleya/g2
mikulskii [*Oligoneuriella*] – syn. *pallida* [*Oligoneuria*]
miliaria [*Ordella*] – in Caenis/f6=g4
mimus [*Cinygma*] – in Cinygmula/g1
minazuki [*Rhithrogena*] – in Radulapalpata INC.SED.
minerva [*Heptagenia*] – in Leucrocuta/g2
mingo [*Hexagenia*] – syn. *limbata* [*Ephemera*]
minima [*Ephemera*] – in NOMINA DUBIA
minima [*Euthyplocia*] – typus nominis *Exeuthyplocia*
minima [*Rhithrogena*] – in Rhithrogena/fg3 INC.SED.
minimella [*Ephemerella*] – in Eurylophella/fg2
minimus [*Ameletus*] – in Ameletus/fg2
minimus [*Leptohyphes*] – in Tricorythopsis/g(1)
minnetonka [*Stenonema*] – in Stenacron/g(1)
minnoi [*Siphlonurus*] – in Siphlonurus/fg4
minor [*Heptagenia*] – in Radulapalpata INC.SED.
minor [*Iron*] – in Radulapalpata INC.SED.
minor [*Palingenia*] – in Anagenesia/g2
minor [*Potameis*] – in Parameletus/fg2
minus [*Protereisma*] – in Permoplectoptera
minuta [*Cinygmula*] – in Cinygmula/g1
minuta [*Iron*] – in Iron/g2
minuta [*Mesoneta*] – in Euephemeroptera INC.SED.
minutus [*Brachycercus*] – in Brachycercus/f1=g3
minutus [*Palaeoanthus*] – in Palaeoanthus/fg(1)
minutus [*Tricorythodes*] – in Tricorythodes/fg1
mirabilis [*Cinygma*] – syn. *congener* [*Arthroplea*]
mirca [*Tricorythodes*] – in Tricorythodes/fg1
mirifica [*Lithoneura*] – in Syntonoptera/fg1
mirus [*Leptohyphes*] – in Tricorythodes/fg1
mirus [*Siphlurus*] – in Siphlonurus/fg4
mithras [*Leptohyphes*] – in Haplohyphes/g(1)
mitificus [*Nixe*] – in Afghanurus/g2
modesta [*Ephemera*] – syn. *orientalis* [*Ephemera*]
modestus [*Heptagenia*] – in Maccaffertium/g(1)
moe [*Caenis*] – in Caenis/f4=g2 INC.SED.
moesta [*Caenis*] – syn. *luctuosa* [*Oxycypha*]
moffatae [*Ephemerella*] – in Ephemerella/fg3 INC.SED.
mogutshevae [*Archaeobehningia*]
— in Euarthropoda INC.SED.
molequs [*Tricorythodes*] – in Tricorythodes/fg1
molinai [*Heptagenia*]
— typus nominis *Dictyosiphlon*
— syn. *needhami* [*Murphyella*]
molita [*Ephemerella*] – in Torleya/g2
mollipes [*Leptohyphes*] – in Leptohyphes/fg2.
mollitia [*Ephemerella*] – in Ephemerella/fg3 INC.SED.
mongolica [*Heptagenia*] – syn. *joernensis* [*Ecdyurus*]
mongolica [*Mesoneta*] – in Euplectoptera INC.SED.
mongolica [*Oligoneuriella*] – syn. *pallida* [*Oligoneuria*]
monserrati [*Rhithrogena*] – in Rhithrogena/fg4
monta [*Ameletus*] – syn. *similor* [*Ameletus*]
montana [*Acerella*] – in Notacanthella/g(1)
montana [*Baetis*] – in NOMINA DUBIA
montana [*Caenis*] – in Caenis/f4=g2 INC.SED.
montanus [*Ameletus*] – in Ameletus/fg2
montanus [*Iron*] – in Iron/g2
montanus [*Siphlonurus*] – in Siphlonurus/fg4
montanus [*Tricorythodes*] – in Tricorythodes/fg1
monticolis [*Ecdyonurus*] – in Electrogena/g(1)
montivagus [*Ameletus*] – in Ameletus/fg2
mooiana [*Ephemera*] – in Ephemera/fg11
moreae [*Ecdyonurus*] – in Ecdyonurus/fg2
moravica [*Kukalova*] – typus nominis *Oboriphlebia*
morrisoni [*Epeorus*] – in Sibirigena/g(1)
mortua [*Ephemera*] – syn. *cellulosa* [*Ephemera*]
mota [*Yaurina*] – in Yaurina/g(1)
muchei [*Siphlonurus*] – in Siphlonurus/fg4
mucronata [*Chitonophora*] – in Ephemerella/fg5
muehlenbergi [*Afronurus*] – in Afronurus/g1
muelleri [*Ecdyonurus*] – in Afghanurus/g1 INC.SED.
muelleri [*Montralia*] – typus nominis *Montralia*
multinervosa [*Ephemera*] – typus nominis *Paedephemera*
multisetosa [*Clypeocaenis*] – in Clypeocaenis/g3
mulaiki [*Tricorythodes*] – in Tricorythodes/fg1
multinervosa [*Ephemera*] – syn. *cellulosa* [*Ephemera*]
munda [*Hexagenia*] – syn. *limbata* [*Ephemera*]
munionga [*Coloburiscoides*] – in Coloburiscoides/g(1)
mutilus [*Campsurus*] – in Campsurus/fg3
murdocki [*Leptohyphes*] – in Leptohyphes/fg1 INC.SED.
murihiku [*Nesameletus*] – in Nesameletus/f2=Metamonius/g2
musseri [*Leptohyphes*] – in Leptohyphes/fg1 INC.SED.

rarus [*Neoephemeropsis*] – syn. *chinensis* [*Potamanthellus*]
rautiani [*Epeorus*] – syn. *anatolii* [*Epeorus*]
reatina [*Rhithrogena*] – in Rhithrogena/fg4
reaumuri [*Eucharidis*]
— typus nominis *Eucharidis*
— syn. *lutea* [*Ephemera*]
recticostalis [*Siberiogenites*] – in Siberiogenites/g
rectus [*Prodromus*] – typus nominis *Prodromites*
recurvata [*Hexagenia*] – typus nominis *Litobrancha*
regia [*Caenis*] – syn. *longus* [*Tricorythus*]
reisi [*Ephemeropsis*] – syn. *trisetalis* [*Ephemeropsis*]
reissi [*Caenis*] – in Caenis/f6=g4
relevata [*Australiphemera*] – typ.nom. *Australiphemera*
remensa [*Ephemera*] – in Ephemera/fg11
remota [*Baetis*] – in Coloburiscus/fg3
renata [*Brasiliocaenis*] – in Brasiliocaenis/g(1)
ressli [*Ecdyonurus*] – in Electrogena/g(1)
reticulata [*Cinygmula*] – in Cinygmula/g1
reticulata [*Ephemera*] – syn. *lutea* [*Ephemera*]
reticulata [*Hexagenia*] – syn. *schoutedeni* [*Ephemera*]
reticulatus [*Tricorythus*] – in Tricorythus/fg5
reversalis [*Heptagenia*] – syn. *cruentata* [*Heptagenia*]
rhenana [*Oligoneuria*] – typus nominis *Oligoneuriella*
rhenanus [*Ecdyonurus*] – syn. *insignis* [*Heptagenia*]
rhenicola [*Caenis*] – syn. *pusilla* [*Caenis*]
rheophilus [*Iron*] – in Ironopsis/g2
rhithralis [*Epeorus*] – in Epeorus/fg3 INC.SED.
ridens [*Caenis*] – typus nominis *Amercaenis*
rieki [*Pseudocaenis*] – in Caenis/g4=g2 INC.SED.
rigida [*Hexagenia*] – in Hexagenia/fg3
rivulicolus [*Ecdyonurus*] – syn. *vicaria* [*Baetis*]
rivulorum [*Caenis*] – in Caenis/f6=g4
rivulorum [*Ecdyonurus*] – syn. *lateralis* [*Baetis*]
rivuscellana [*Electrogena*] – syn. *ujhelyii* [*Ecdyonurus*]
robacki [*Leptohyphes*] – in Leptohyphes/fg1 INC.SED.
robusta [*Caenis*] – in Caenis/f6=g4
robusta [*Palingenia*] – in Anagenesia/g2
robusta [*Pentagenia*] – syn. *vittigera* [*Palingenia*]
robusta [*Rhithrogena*] – in Rhithrogena/fg4
rodocki [*Heptagenia*] – in Afghanurus/g1 INC.SED.
roeselii [*Palingenia*]
— typus nominis *Jolia*
— syn. *ignota* [*Baetis*]
rogersi [*Baetisca*] – in Baetisca/f3=g2
rohdendorfi [*Rhoenanthus*] – syn. *coreanus* [*Potamanthus*]
rolandi [*Rhithrogena*] – in Rhithrogena/fg4
rolandi [*Tricorythus*] – in Tricorythus/fg5
rolstoni [*Leptohyphes*] – in Leptohyphes/fg1 INC.SED.
romantzovi [*Ephemera*] – in Ephemera/fg11
rosacea [*Hexagenia*] – syn. *limbata* [*Ephemera*]
rosae [*Spinirythus*] – in Spinirythus/g(1)
rosea [*Ephemera*] – syn. *ignita* [*Ephemera*]
rosea [*Heptagenia*] – in Afghanurus/g1 INC.SED.
rossenrayensis [*Protereisma*] – in Permoplectoptera
rossi [*Ephemerella*] – in Ephemerella/fg3 INC.SED.
rossi [*Myanmarella*] – typus nominis *Myanmarella*
rossica [*Lethaea*] – syn. *trisetalis* [*Ephemeropsis*]
rossicus [*Ecdyonurus*] – syn. *fuscogrisea* [*Ephemera*]
rossicus [*Phtharthus*]
— typus nominis *Phtharthus*
— syn. *netshaevi* [*Phtharthus*]
rothschildi [*Ecdyonurus*] – in Ecdyonurus/fg2
rotunda [*Ephemerella*] – in Ephemerella/fg3 INC.SED.
rotundatus [*Siberiogenites*] – in Siberiogenites/g
rotundus [*Siphangarus*] – typus nominis *Siphangarus*
rougemonti [*Cinygmula*] – in Cinygmula/g1
rubescens [*Cinygma*] – in Anteritorna INC.SED.
rubescens [*Cloe*] – in Baetisca/f3=g2
rubeus [*Epeorus*] – in Belovius/g2
ruber [*Ecdyonurus*] – syn. *modestus* [*Heptagenia*]
rubicunda [*Rhithrogena*] – in Rhithrogena/fg4
rubidus [*Iron*] – in Epeorus/fg3 INC.SED.
rubiens [*Potamanthellus*] – in Euplectoptera INC.SED.
rubrofasciatus [*Ecdyonurus*] – in Afghanurus/g2
rubromaculata [*Cinygmina*] – in Afronurus/g1
rubromaculata [*Heptagenia*] – syn. *modestus* [*Heptagenia*]
rubroventris [*Heptagenia*] – in Afghanurus/fg1 INC.SED.
rufa [*Ephemera*] – in NOMINA DUBIA
rufa [*Ephemerella*] – syn. *punctisetae* [*Drunella*]
rufa [*Isonychia*] – in Isonychia/fg2
rufescens [*Ephemera*] – syn. *ignita* [*Ephemera*]
ruffieuxae [*Proboscidoplocia*] – in Proboscidoplocia/g(1)
ruffii [*Ecdyonurus*] – in Ecdyonurus/fg2
rufous [*Potamanthus*] – syn. *myops* [*Ephemera*]
rugosa [*Caenis*] – in Caenis/f6=g4
rupestris [*Pristiplocia*] – typus nominis *Pristiplocia*
russevi [*Ecdyonurus*] – in Ecdyonurus/fg2
rusticalis [*Heptagenia*] – in Afghanurus/g2
rusticus [*Anepeorus*] – typus nominis *Anepeorus*
rutila [*Caenis*] – in Caenis/f6=g4
ryszardi [*Rhithrogena*] – in Rhithrogena/fg4

S

sabahensis [*Stygifloris*] – typus nominis *Stygifloris*
sabinas [*Leptohyphes*] – in Leptohyphes/fg1 INC.SED.
sacculobranchis [*Tricorythodes*] – in Tricorythodes/fg1
sachalinensis [*Ephemera*] – in Ephemera/fg11
sachalinensis [*Ephemerella*] – in Drunella/g1
sadleri [*Isonychia*] – syn. *bicolor* [*Palingenia*]
salvini [*Heptagenia*] – in Afghanurus/g1 INC.SED.
salviniae [*Homoeoneuria*] – typus nominis *Homoeoneuria*
samalorum [*Ecdyonurus*] – syn. *ujhelyii* [*Ecdyonurus*]
samochai [*Sigmoneuria*] – in Heptagenia/f8=g7
sampsoni [*Exeuthyplocia*] – typus nominis *Afroplocia*
sanctagabriel [*Iron*] – in Epeorus/fg3 INC.SED.
sangangensis [*Afronurus*] – in Radulapalpata INC.SED.
sangangensis [*Potamanthodes*] – in Rhoenanthus/g1
sangangensis [*Rhithrogena*] – in Rhithrogena/fg4
sanguinea [*Rhithrogena*] – syn. *impersonata* [*Heptagenia*]
sanjianfangensis [*Mesobaetis*] – in Euplectoptera INC.SED.
santarita [*Tricorythodes*] – in Tricorythodes/fg1
sanukensis [*Siphlurus*] – in Siphlonurus/fg4
sapporensis [*Siphlurus*] – in Cinygmula/g1
sarae [*Tortopus*] – in Tortopus/g(1)
sardoa [*Baetis*] – syn. *fallax* [*Baetis*]
saskatchewanensis [*Lachlania*] – in Lachlania/g1
satsuki [*Rhithrogena*] – syn. *tetrapunctigera* [*Rhithrogena*]
sauteri [*Ephemera*] – in Ephemera/fg10 INC.SED.
savignyi [*Palingenia*] – in Polymitarcys/f3=Ephoron/g4
savoiensis [*Rhithrogena*] – in Rhithrogena/fg4
sayi [*Isonychia*] – in Prionoides/g(1)
scalaris [*Ecdyonurus*] – in Ecdyogymnurus/g(1)
schoutedeni [*Ephemera*] – typus nominis *Eatonica*

sp.O4 – in Oligoneuria/f5=g6
sp.O5 – in Oligoneuria/f5=g6
sp.S1 – in Siphlonurus/fg4
sp.S2 – in Siphlonurus/fg4
sp.S3 – in Parameletus/fg2
sp.T1 – in Teloganodes/fg1
sp.T2 – in Teloganodes/fg1
sp.T3 – in Teloganodes/fg1
sp.T4 – in Teloganodes/fg1
sp.U1 – in Uracanthella/g1
sp.U2 – in Uracanthella/g1
sp.U3 – in Uracanthella/g1
sp.V1 – in Tasmanophlebia/g1
sparsatus [*Ameletus*] – in Ameletus/fg2
speciosa [*Ephemera*] – in NOMINA DUBIA
speciosa [*Ephemera*] nom.praeocc. – in Permoplectoptera
speciosum [*Siphloplecton*] – in Siphloplecton/fg(1)
speciosus [*Rhoenanthus*] – typus nominis *Rhoenanthus*
spectabilis [*Compsoneuria*] – typus nominis *Compsoneuria*
spectabilis [*Siphlonurus*] – in Siphlonurus/fg4
sphyxia [*Manohyphella*] – in Melanemerella/fg1
spiculosa [*Ephemerella*] – in Ephemerella/fg3 INC.SED.
spilosa [*Ephemera*] – in Ephemera/fg10 INC.SED.
spinifera [*Ephemerella*] – in Drunella/g1
spinosa [*Caenis*] – in Caenis/f6=g4
spinosa [*Cinygmula*] – syn. *levanidovi* [*Cinygmula*]
spinosa [*Ephemerella*] – in Ephemerella/fg3 INC.SED.
spinosa [*Ephemerella*] nom.praeocc. – syn. *ikonomovi* [*E.*]
spinosa [*Heptagenia*] – in Afghanurus/g1 INC.SED.
spinosus [*Leptohyphes*] – in Leptohyphes/fg1 INC.SED.
spinosus [*Siphlurus*] – syn. *aestivalis* [*Siphlurus*]
spodiocephala [*Anagenesia*] – in Anagenesia/g2
squamatus [*Ecdyonurus*] – in Electrogena/g(1)
srinagari [*Caenis*] – in Caenis/f6=g4
stackelbergi [*Rhithrogena*] – in Himalogena/g(1)
stapfi [*Misthodotes*] – in Permoplectoptera
starmachi [*Ecdyonurus*] – in Ecdyonurus/fg2
steinmanni [*Iron*] – syn. *alpicola* [*Heptagenia*]
stigma [*Ephemera*] – in NOMINA DUBIA
stocki [*Aphelophlebodes*] – typus nominis *Aphelophlebodes*
straminea [*Baetis*] – syn. *sulphurea* [*Ephemera*]
straeleni [*Ephemerythus*] – typus nominis *Tricomerella*
strenua [*Rhithrogena*] – in Rhithrogena/fg4
striatum [*Ephemera*] nom. praeocc. – syn. *brunnea* [*E.*]
striatus [*Campsurus*] – in Campsurus/fg3
strigata [*Ephemera*] – typus nominis *Sinephemera*
strugaensis [*Caenis*] – in Caenis/f6=g4
stubbei [*Ecdyonurus*] – syn. *joernensis* [*Ecdyurus*]
stygiatus [*Tricorythodes*] – in Tricorythodes/fg1
subaequalis [*Heptagenia*] – in Cinygmula/g1
subalpinus [*Ameletus*] – in Ameletus/fg2
subalpinus [*Ecdyonurus*] – in Ecdyonurus/fg2
subangulata [*Rhithrogena*] – in Rhithrogena/fg3 INC.SED.
subcostalis [*Potamanthus*] – in Rhoenanthus/g1
subflavus [*Afronurus*] – in Afronurus/g1
subfusca [*Baetis*] – syn. *venosa* [*Ephemera*]
subfuscus [*Ecdyonurus*] – in Ecdyonurus/fg1 INC.SED.
sublineata [*Rhithrogena*] – in Rhithrogena/fg4
sublongicauda [*Palingenia*] – in Palingenia/f4=g3
submontana [*Ephemerella*] – in Drunella/g1
submontanus [*Ecdyonurus*] – in Ecdyonurus/fg2
subnotatus [*Ameletus*] – typus nominis *Ameletus*
subpallidus [*Iron*] – in Epeorus/fg3 INC.SED.
subsolana [*Ephemerella*] – in Ephemerella/fg3 INC.SED.
subspinosa [*Nixe*] – syn. *zhiltzovae* [*Paracinygmula*]
subvaria [*Ephemerella*] – in Ephemerella/fg3 INC.SED.
succinus [*Leptohyphes*] – syn. *zalope* [*Leptohyphes*]
sudafricanum [*Centroptilum*] – typus nominis *Afroptilum*
suffusus [*Ameletus*] – in Ameletus/fg2
suffusus [*Iron*] – in Epeorus/fg3 INC.SED.
sukashii [*Siphlonisca*] – syn. *shima* [*Rhoenanthus*]
sulphurea [*Ephemera*] – in Heptagenia/f8=g7
sumatranus [*Chirotonetes*] – in Isonychia/fg2
sumatranus [*Ecdyonurus*] – typus nominis *Ecdyonuroides*
supposita [*Ephemera*] – in Ephemera/fg10 INC.SED.
suzukianum [*Cinygma*] – syn. *tobiironis* [*Ecdyonurus*]
suzukiella [*Kageronia*]
— typus nominis *Kageronia*
— syn. *kihada* [*Heptagenia*]
suzukii [*Dipteromimodes*]
— typus nominis *Dipteromimodes*
— syn. *tipuliformis* [*Dipteromimus*]
svenhedini [*Ephemerella*] – in Ephemerella/fg3 INC.SED.
swammerdamiana [*Ephemera*] – syn. *longicauda* [*Ephemera*]
swammerdiana [*Ephemera*] – syn. *longicauda* [*Ephemera*]
swatensis [*Ephemerella*] – in Ephemerella/fg3 INC.SED.
sylvestris [*Tricorythus*] – in Tricorythus/fg5
sylvicola [*Baetis*] – in Epeorus/fg4

T

tacajalo [*Leptohyphes*] – in Leptohyphes/fg2
tadzhikorum [*Heptagenia*] – syn. *perflava* [*Heptagenia*]
taeniata [*Ephemerella*] – syn. *aurivillii* [*Chitonophora*]
taimyricum [*Palaeocloeon*] – typus nominis *Palaeocloeon*
talea [*Lachlania*] – in Lachlania/g1
taprobanes [*Povilla*] – in Asthenopus/fg1
tarda [*Cinygma*] – in Cinygmula/g1
tardata [*Caenis*] – in Caenis/f6=g4
tarsalis [*Atopopus*] – typus nominis *Atopopus*
tarsos [*Leptohyphes*] – in Leptohyphes/fg1 INC.SED.
tarteri [*Ameletus*] – in Ameletus/fg2
tateyamana [*Rhithrogena*] – in Radulapalpata INC.SED.
tatrica [*Rhithrogena*] – syn. *loyolaea* [*Rhithrogena*]
taurisca [*Rhithrogena*] – in Rhithrogena/fg4
tauroides [*Madecocercus*] – typus nominis *Madecocercus*
teberdensis [*Rhithrogena*] – in Rhithrogena/fg4
temerata [*Polymitarcys*] – syn. *savignyi* [*Palingenia*]
temporalis [*Ephemerella*] – in Eurylophella/fg2
tenax [*Ephemerella*] – syn. *triacantha* [*Ephemerella*]
tenella [*Eurycaenis*] – in Furcatergaliae INC.SED.
tenera [*Palingenia*] – in Anagenesia/g2
tenuis [*Iron*] – syn. *fragilis* [*Iron*]
teresa [*Ephemerella*] – in Ephemerella/fg3 INC.SED.
terminata [*Palingenia*] – in Maccaffertium/g(1)
tertia [*Oboriphlebia*] – in Permoplectoptera
tertius [*Ameletus*] – in Ameletus/fg2
tertius [*Epeoromimus*] – typus nominis *Foliomimus*
tertius [*Metretopus*] – in Metretopus/fg2
tessellata [*Baetis*] – syn. *vicaria* [*Baetis*]
tetrapunctigera [*Rhithrogena*] – in Rhithrogena/fg3 INC.SED.
texanus [*Tricorythodes*] – in Tricorythodes/fg1
thailandensis [*Serratella*] – syn. *gosei* [*Ephemerella*]
thalia [*Isonychia*]) – syn. *georgiae* [*Isonychia*]

validus [*Ameletus*] – in Ameletus/fg2
validus [*Chirotonetes*] – in Anteritorna INC.SED.
vancouvernensis [*Ameletus*] – in Ameletus/fg2
varia [*Ephemera*] – in Ephemera/fg11
variabilis [*Hexagenia*] – syn. *limbata* [*Ephemera*]
variabilis [*Tricorythus*] – in Tricorythus/fg5
varicauda [*Caenis*] – typus nominis *Tricorythus*
variegatum [*Afroptilum*] – in Afroptiloides/g(1)
variegatum [*Prosopistoma*] – typus nominis *Prosopistoma*
varium [*Stenonema*] – syn. *modestus* [*Heptagenia*]
vayssiere [*Proboscidoplocia*] – in Proboscidoplocia/g(1)
vedana [*Ephemera*] – syn. *expectans* [*Potamanthus*]
velma [*Isonychia*] – in Isonychia/fg2
velmae [*Ephemerella*] – in Ephemerella/fg3 INC.SED.
velox [*Ameletus*] – in Ameletus/fg2
velteni [*Balticobaetisca*] – typus nominis *Balticobaetisca*
venosa [*Ephemera*] – typus nominis *Ecdyonurus*
venusta [*Hexagenia*] – syn. *limbata* [*Ephemera*]
ventilans [*Siphlonella*] – typus nominis *Siphlonella*
verisimilis [*Ephemerella*] – in Eurylophella/fg2
vernalis [*Ameletus*] – in Ameletus/fg2
vernalis [*Cinygma*] – in Cinygmula/g1
vernalis [*Ephemerella*] – syn. *rotunda* [*Ephemerella*]
vernalis [*Siphlonurus*] – syn. *aestivalis* [*Siphlurus*]
verrierae [*Dicercomyzon*] – in Dicercomyzon/fg(1)
verrucosa [*Ephemerela*] – syn. *nuda* [*Ephemerella*]
verticis [*Baetis*] – typus nominis *Anthopotamus*
vescus [*Leptohyphes*] – in Allenhyphes/g1
vespertina [*Ephemera*] – typus nominis *Leptophlebia*
vicaria [*Baetis*] – in Maccaffertium/g(1)
vicinus [*Afghanurus*] – typus nominis *Afghanurus*
vietnamensis [*Ecdyonuroides*] – in Ecdyonuroides/fg(1)
violacea [*Sparrea*] – syn. *japonicus* [*Chirotonetes*]
violaceus [*Campsurus*] – in Campsurus/fg3
violettae [*Ranorythus*] – typus nominis *Ranorythus*
vipavensis [*Electrogena*] – in Electrogena/g(1)
virgo [*Ephemera*] – typus nominis *Polymitarcys*
viridescens [*Palingenia*] – syn. *limbata* [*Ephemera*]
viridis [*Heptagenia*] – in Afronurus/g1
virilis [*Rhithrogena*] – in Rhithrogena/fg3 INC.SED.
viriosus [*Leptohyphes*] – syn. *minimus* [*Leptohyphes*]
viscata [*Ephemerella*] – Ephemerella/fg1 INC.SED.
vitalisi [*Polyplocia*] – typus nominis *Polyplocia*
vitalisi [*Rhoenanthus*] – syn. *magnificus* [*Rhoenanthus*]
vitrea [*Ephemera*] – in NOMINA DUBIA
vitrea [*Palingenia*] – in Proepeorus/g(1)
vitta [*Rhithrogena*] – in Rhithrogena/fg3 INC.SED.
vittigera [*Palingenia*] – typus nominis *Pentagenia*
volitans [*Heptagenia*] – syn. *fuscogrisea* [*Ephemera*]
volsellus [*Tricorythopsis*] – in Tricorythopsis/g(1)
vschivkovae [*Isonychia*] (in litt.) – in Isonychia/fg2
vulcanus [*Nesameletus*] – in Nesameletus/f2=Metamonius/g2
vulgaris [*Tricorythus*] – in Tricorythus/fg5
vulgata [*Ephemera*] – typus nominis *Ephemera*
vulpecula [*Rhithrogena*] – syn. *semicolorata* [*Baetis*]
vulturnus [*Leptohyphes*] – in Leptohyphes/fg1 INC.SED.

W

wabasha [*Stenonema*] – syn. *mexicana* [*Heptagenia*]
wahensis [*Ephemerella*] – in Torleya/g1 INC.SED.
wakefieldi [*Oniscigaster*] – typus nominis *Oniscigaster*
walkeri [*Ephemerella*] – in Drunella/g1
walkeri [*Potamanthus*] – syn. *verticis* [*Baetis*]
wallacei [*Spinadis*]
— typus nominis *Spinadis*
— syn. *simplex* [*Heptagenia*]
walleyi [*Ameletus*] – in Ameletus/fg2
wanquanensis [*Ephemera*] – in Ephemera/fg11
walshi [*Heptagenia*] – in Leucrocuta/g2
wappaei [*Ephemera*] – in Campsurus/fg3
wautieri [*Ecdyonurus*] – syn. *ruffii* [*Ecdyonurus*]
wayah [*Ephemerella*] – syn. *tuberculata* [*Ephemerella*]
weewa [*Hexagenia*] – syn. *limbata* [*Ephemera*]
werestschagini [*Heptagenia*] – syn. *simplicioides* [*H.*]
weyenberghii [*Hexagenites*]
— typus nominis *Hexagenites*
— syn. *cellulosa* [*Ephemera*]
weyenberghi [*Mesephemera*] – in Permoplectoptera
wilsoni [*Ephemerella*] – syn. *coloradensis* [*Ephemerella*]
winkleri [*Isonychia*] – in Isonychia/fg2]
wittmeri [*Ecdyonurus*] – in Ecdyonurus/fg1 INC.SED.
wolosatkae [*Rhithrogena*] – in Rhithrogena/fg4
wouterae [*Prosopistoma*] – in Prosopistoma/f1=g2
wuchowensis [*Ephemera*] – in Ephemera/fg11

X

xiasimaensis [*Cinygma*] – in Radulapalpata INC.SED.
xiaosimaensis [*Ephemerellina*] – in Ephemerella/fg3 INC.SED.
xylobiontum [*Asthenopodichnium*]
— typus nominis *Asthenopodichnium*

Y

yadongxiasimaensis [*Afronurus*] – in Radulapalpata INC.SED.
yangi [*Anagenesia*] – in Chankagenesia/g(1)
yangi [*Caenis*] – in Caenis/f4=g2 INC.SED.
yaoshani [*Ephemera*] – in Ephemera/fg11
yixingensis [*Cinygmina*] – in Afronurus/g1
yixingensis [*Ephemerella*] – in Uracanthella/g1
yacutinga [*Tricorythopsis*] – in Tricorythopsis/g(1)
yadohglinensis [*Cinygmula*] – in Radulapalpata INC.SED.
yanahuicsa [*Haplohyphes*] – in Haplohyphes/g(1)
yangi [*Caenis*] – in Caenis/f6=g4
yapa [*Yaurina*] – in Yaurina/g(1)
yooni [*Potamanthus*] – in Rhoenanthus/g1
yosemite [*Ephemerella*] – syn. *grandis* [*Ephemerella*]
yoshidae [*Ecdyonurus*] – in Afronurus/g1
yoshinoensis [*Ephemerella*] – in Drunella/g1
yoshinoensis [*Siphlonurus*] – in Siphlonurus/fg4
yougoslavicus [*Iron*] – in Ironopsis/g2
youi [*Neopotamanthus*] – typus nominis *Neopotamanthus*
youngi [*Caenis*] – in Caenis/f6=g4
youngi [*Iron*] – syn. *albertae* [*Iron*]
youngi [*Neoephemera*] – in Neoephemera/fg2
yunnanensis [*Potamanthodes*] – in Rhoenanthus/g
yura [*Tricorythodes*] – in Tricorythodes/fg1
yuta [*Yaurina*] – typus nominis *Yaurina*

Z

zachvatkini [*Cinygmula*] – syn. *levis* [*Epeorus*]
zaitzevi [*Epeorus*] – in Epeorus/fg4
zalesskyi [*Misthodotes*] – in Permoplectoptera
zalope [*Leptohyphes*] – in Leptohyphes/fg2
zanga [*Oligoneuriella*] – syn. *tskhomelidzei* [*Oligoneuriella*]
zapekinae [*Ephemerella*] – in Torleya/g2
zebrata [*Baetis*] – in Electrogena/g(1)
zelinkai [*Rhithrogena*] – in Rhithrogena/fg4

Index of supraspecies taxa names

This index includes hierarchical and circumscriptional supraspecies taxa names mentioned in this volume. Ranking family-group names are not included (letter "f" after the genus-group name indicates the whole complex of family-group names formed from this genus-group name). For each name after a dash ("–"), its place in the book is indicated: for a valid name – taxon number in brackets; for invalid name – higher taxon name after the preposition "**in**" or the oldest synonym after the abbreviation "**syn.**"; abbreviation "**INC.SED.**" means "INCERTAE SEDIS".

Index of terms

Page number directly after the term refers to the Chapter II – Morphology of mayflies; number in square brackets and page number after it refer to the Index of characters.

LIST OF FIGURES AND TABLES

LIST OF ABBREVIATIONS

design. – subsequently designated by ...
design. orig. – original designation (DESIGNATIO ORIGINALIS)
INC. SED. – INCERTAE SEDIS
incl. – including (INCLUSO) – oldest genus-group names of directly subordinated taxa ecept for nominative one
NOM. NUD. – NOMEN NUDUM
nom. praeocc. – preoccupied name (NOMEN PRAEOCCUPATUM)
sine – without, or excluding – oldest genus-group names of sister taxa
syn. obj. – objective synonym (SYNONYMUM OBJECTIVUM)
synn. obj. – objective synonyms (SYNONYMA OBJECTIVA)
syn. subj. – subjective synonym (SYNONYMUM SUBJECTIVUM)
synn. subj. – subjective synonyms (SYNONYMA SUBJECTIVA)

In "References" and "Material examined":
⊙ – egg(s) (OVUM)
Ծ – larva(e) (the symbol proceeds from baby's dummy)
Ծ♂ – male larva(e)
Ծ♀ – female larva(e)
◎ – subimago(es) of both sexes
◎♂ – male subimago(es)
◎♀ – female subimago(es)
⊕ – imago(es) of both sexes
⊕♂ – male imago(es)
⊕♀ – female imago(es)
◉♀ – adult female(s) of species which female has no imaginal moult

In "References" only:
* – diagnosis of the taxon as a whole
' – description of some representatives of the taxon

In "Material examined" only:
Ծ–◎–⊕ – reared imago(es) with larval and subimaginal exuviae
Ծ–◎ – reared subimago(es) with larval exuviae
◎–⊕ – reared imago(es) with subimaginal exuviae
Ծ/◎/⊕ – mature larva(e) with developed subimaginal and imaginal parts
Ծ/◎ – mature larva(e) with developed subimaginal parts
◎/⊕ – subimago(es) with developed imaginal parts
⊕♀/⊙ – female imago(es) with eggs
◉♀/⊙ – female adult(s) of non-moulting species with eggs
Ծ♀/◎♀/⊙, or Ծ♀/◎♀/⊕♀/⊙ – mature larva(e) with eggs
Ծ♀–◎♀–⊕♀/⊙ – reared female imago(es) with eggs
Ծ–◎–⊕/⊙ – all stages of both sexes associated by rearing (larvae of various instars, imagoes with larval and subimaginal exuviae, and eggs extracted from females)

Batch number: 10800765

Printed by Printforce, the Netherlands